13th International Conference on Textile Composites (TEXCOMP-13)

IOP Conference Series: Materials Science and Engineering
Volume 406

Milan, Italy
17 – 19 September 2018

ISBN: 978-1-5108-7194-6
ISSN: 1757-8981

Printed from e-media with permission by:

Curran Associates, Inc.
57 Morehouse Lane
Red Hook, NY 12571

Some format issues inherent in the e-media version may also appear in this print version.

Printed by Curran Associates, Inc. (2018)

For permission requests, please contact the Institute of Physics at the address below.

Institute of Physics
Dirac House, Temple Back
Bristol BS1 6BE UK

Phone: 44 1 17 929 7481
Fax: 44 1 17 920 0979

techtracking@iop.org

Additional copies of this publication are available from:

Curran Associates, Inc.
57 Morehouse Lane
Red Hook, NY 12571 USA
Phone: 845-758-0400
Fax: 845-758-2633
Email: curran@proceedings.com
Web: www.proceedings.com

Table of contents

Volume 406

13th International Conference on Textile Composites (TEXCOMP-13)

17–19 September 2018, Milan, Italy

Accepted papers received: 1 August 2018
Published online: 21 September 2018

Mechanical behaviour of fiber-reinforced polymer composites in the presence of voids

Properties and Characterization

Modelling

Damage characterization

Infusion and Permeability

Manufacturing

IOP Conf. Series: Materials Science and Engineering **406** (2018) 011001 doi:10.1088/1757-899X/406/1/011001

Preface

Valter Carvelli

Chairman of TEXCOMP-13
Politecnico di Milano, Italy

TexComp is a unique biannual international conference series fully dedicated to textile composites.

The TexComp conference series started in 1992 at KU Leuven, Belgium, and since then TexComp toured around the world.

The **13th International Conference on Textile Composites (TEXCOMP-13)** was organized by 'Politenico di Milano', under the auspicious of several well-known leading societies and organizations in the expanding world of composite materials.

TEXCOMP-13 continued the fine scientific tradition of previous TexComp conferences fostering the fundamental understandings as well as application oriented developments of textile composites for several industrial contexts.

The tradition of TexComp in hosting world-leader experts of textile composites continued in Milan with an intensive program of presentations including distinguished plenary and invited speakers, scientists, engineers and designers, both of Academia and Industry from all over the world. We were particularly pleased to have contributions from some pioneers in the textile composites community, together with several young researchers. Almost one third of the presentations by young researchers demonstrated the attitude of TEXCOMP-13 in recognizing the importance of the young generation for the advancement of the research and the development of textile composites.

These proceedings offer a snapshot of the strong research presented in plenary, invited and regular presentations at TEXCOMP-13. The conference covered a broad range of topics related to the textile composites: from the textile reinforcements to the manufacturing processes, from the mechanical properties of the materials to the structural performance modelling; from the nano- to the macro-scale studies.

The conference contained special sessions, organized by leading researchers, devoted to important topics in the areas of: virtual modelling, forming of reinforcements, effect of voids on the mechanical behavior, and fatigue performance.

TEXCOMP-13 was also the proper stage to present the recent developments of the very active research in the field of natural textile composites and nano-engineered composites.

The Chairman is very grateful and expresses his sincere acknowledgments to the members and colleagues of the Local Scientific Committee and the TexComp International Scientific Committee for the continuous support.

IOP Conf. Series: Materials Science and Engineering **406** (2018) 011002 doi:10.1088/1757-899X/406/1/011002

Peer review statement

All papers published in this volume of *IOP Conference Series: Materials Science and Engineering* have been peer reviewed through processes administered by the proceedings Editors. Reviews were conducted by expert referees to the professional and scientific standards expected of a proceedings journal published by IOP Publishing.

IOP Conf. Series: Materials Science and Engineering **406** (2018) 012001 doi:10.1088/1757-899X/406/1/012001

An integrated methodology for fatigue life prediction of carbon fiber-reinforced polymer matrix composites under constant and variable *R*-ratio loadings

M Kawai

Department of Engineering Mechanics and Energy, University of Tsukuba, Tsukuba 305-8573, Japan

mkawai@kz.tsukuba.ac.jp

Abstract. The progress in the anisomorphic constant fatigue life (CFL) diagram approach for efficiently identifying the fatigue of carbon fiber-reinforced composites under constant waveform loading over the whole range of mean stress is reviewed. The asset of this approach is the concept of critical stress ratio. The anisomorphic formulation of CFL diagram is flexible enough not only to increase the number of segments to accommodate the transient change in sensitivity to mean stress, but also to take a generalized form that considers the effect of temperature as well as the effect of variability in fatigue life. The anisomorphic approach is shown to succeed in describing the asymmetric and nonlinear CFL diagrams for carbon fiber composites of different kinds. It is suggested that the classic procedure for prediction of spectrum loading fatigue life of composites using the Miner rule with the rain flow cycle counting method is powered by integrating with the anisomorphic CFL diagram approach.

1. Introduction

Prediction of variable loading fatigue life of a given material raises three important issues to be taken into account [1-6]. A spectrum load that a material is required to withstand during service is a series of waveforms of changing stress level and stress ratio. It is complicated. To simplify the problem of fatigue life prediction, the spectrum loading is replaced with a set of constant waveform loading. The values of fatigue life for the reduced constant waveform loadings not only of different levels of peak/valley stress but also of different levels of mean stress are required to estimate their damaging effects. The ultimate state of damage accumulation has to be defined to judge fatigue failure under the spectrum loading. Typically, the rain flow cycle counting method [7, 8], a constant fatigue life diagram [9, 10] and the Miner rule [11, 12] are employed in the classic engineering scenario.

The understanding of the effects of constant stress level and stress ratio loading on fatigue life of composites has successfully been developed [13, 14]. The comprehensive investigation into the effect of stress ratio on constant waveform loading fatigue life of CFRPs over the whole range of mean stress has lead to significant progress in modeling of a CFL diagram that is suitable for composites.

In regard to the effect of variable loading on fatigue life of composites, however, we are still limited to an incomplete understanding the phenomenon and thus demand to establish an appropriate methodology for it [1-11]. To develop our understanding of the effect of load sequence on variable waveform loading fatigue life of CFRPs and to establish a method for predicting it as efficiently as possible, we need to study not only the effect of variation in alternating stress amplitude, but also the effect of variation in mean stress, i.e. the variation in *R*-ratio, on fatigue failure in CFRPs.

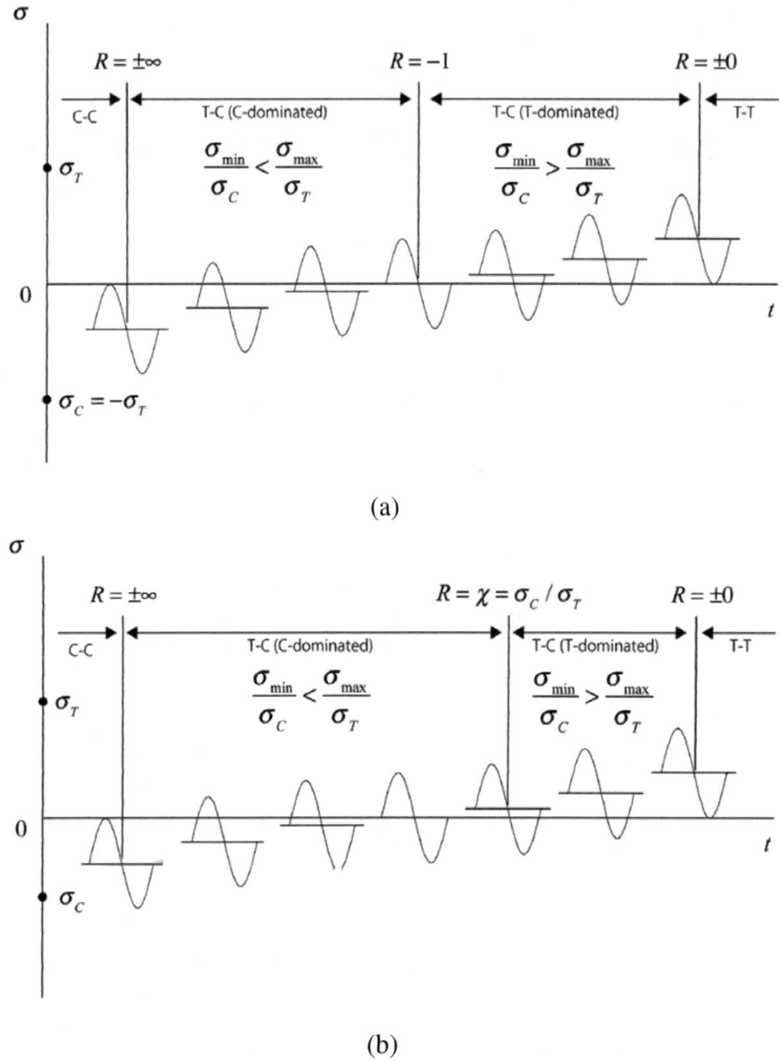

Figure 1. Classification of loading waveforms; (a) Traditional (b) Non-traditional

This article reviews the progress of the anisomorphic CFL diagram approach for carbon fiber-reinforced composites that has been made in the last decade [15-18], and demonstrates not only that it allows fully identifying the fatigue of composites under constant waveform loading, but also that it is flexible enough to be customized to take into account the effect of temperature [19, 20] as well as the effect of variability in fatigue life [21, 22]. It is further demonstrated that the classic fatigue life prediction methodology inviting the anisomorphic CFL diagram as its member scenario is a promising tool that allows consistently and moderately conservative prediction of fatigue life of CFRPs under variable waveform loading.

2. Anisomorphic CFL diagram approach

For metals of equal strength in tension and compression, as shown in figure 1(a), the particular value of stress ratio $R = -1$ is usually chosen to identify their fundamental S-N curves that are used to construct the Goodman diagrams for them. In contrast to metals, most fiber-reinforced composites exhibit different strengths in tension and compression. For those composites, therefore, it is suitable to replace figure 1(a) with figure 1(b). If the tensile strength is larger than the compressive strength of a composite, the margin to static failure is larger in tension. Then, the value of mean stress of equal distance from the tensile and compressive strengths deviates from zero to a positive number as illustrated in figure 1(b). This suggests dividing T-C loading into tension-dominated T-C loading and compression-dominated T-C loading by means of the particular waveform whose mean stress takes a value of equal distance from the tensile and compressive strengths. The particular waveform corresponds to the critical stress ratio that is given by the ratio of compressive strength to tensile strength, and the use of it as the basis for construction of a CFL diagram is the distinguishing feature of the anisomorphic formulation [15-22].

2.1. Two-segment anisomorphic CFL diagram

In the two-segment anisomorphic formulation [15, 16], the CFL envelope for a given constant value of life is composed of two smooth curves associated with tension-dominated fatigue failure and compression-dominated fatigue failure, respectively. They are smoothly connected with each other at a point on the radial straight line with a constant amplitude ratio

$$\frac{\sigma_a}{\sigma_m} = \frac{1-\chi}{1+\chi} \tag{1}$$

where χ is the critical stress ratio equal to the ratio of compressive strength σ_C (< 0) to tensile strength σ_T (> 0); i.e., $\chi = \sigma_C / \sigma_T \in (-\infty, 0)$. The two member curves of each anisomorphic CFL envelope on the two segments can be described, respectively, by means of the following piecewise defined functions:

[I] Tension-dominated segment ($\sigma_m^{(\chi)} \leq \sigma_m \leq \sigma_T$):

$$\frac{\sigma_a - \sigma_a^{(\chi)}}{\sigma_a^{(\chi)}} = \left(\frac{\sigma_m - \sigma_m^{(\chi)}}{\sigma_T - \sigma_m^{(\chi)}} \right)^{2-(\psi_\chi)^{k_T}} \tag{2}$$

[II] Compression-dominated segment ($\sigma_C \leq \sigma_m < \sigma_m^{(\chi)}$):

$$\frac{\sigma_a - \sigma_a^{(\chi)}}{\sigma_a^{(\chi)}} = \left(\frac{\sigma_m - \sigma_m^{(\chi)}}{\sigma_C - \sigma_m^{(\chi)}} \right)^{2-(\psi_\chi)^{k_C}} \tag{3}$$

where $\sigma_a^{(\chi)}$ and $\sigma_m^{(\chi)}$ represent the alternating stress and mean stress components of the maximum fatigue stress $\sigma_{max}^{(\chi)}$ for fatigue loading at the critical stress ratio χ. The exponents k_T and k_C play the role to adjust the rate of change in the shape of a CFL curve from a straight line to a parabola on the tension-dominated and compression-dominated segments, respectively. The exponent ψ_χ represents the fatigue strength ratio associated with loading at the critical stress ratio χ, and it is defined as

$$\psi_\chi = \frac{\sigma_{max}^{(\chi)}}{\sigma_T} \tag{4}$$

The exponent ψ_χ is treated as a continuous function of the number of cycles to failure N_f; i.e.,

$$2N_f = f_\chi(\psi_\chi) \tag{5}$$

Figure 2 shows the two-segment anisomorphic CFL diagram identified for a woven CFRP laminate at RT [19]; dashed lines and symbols indicate the predicted and observed results, respectively. It was assumed that $k_T = k_C = 1$. From figure 2, it is seen that the two-segment anisomorphic CFL diagram approach succeeds in adequately describing the experimental CFL diagram over the whole range of mean stress in the tested range of fatigue life. The experimental results show that the CFL diagram for the woven CFRP laminate at RT is asymmetric about the alternating stress axis, and the peaks of the CFL envelopes for different constant values of life lie almost on the radial line of the constant

IOP Conf. Series: Materials Science and Engineering **406** (2018) 012001 doi:10.1088/1757-899X/406/1/012001

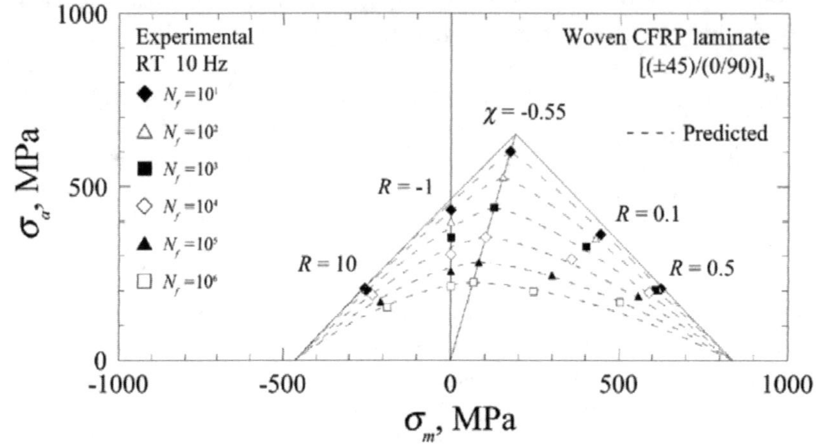

Figure 2. The two-segment anisomorphic CFL diagram for a woven carbon fabric composite laminate at RT [Kawai and Matsuda (2012)].

amplitude ratio associated with fatigue loading at the critical stress ratio. The latter observation justifies the concept of critical stress ratio in the anisomorphic formulation of a CFL diagram. The two-segment anisomorphic CFL diagram approach has successfully been applied to non-woven CFRP laminates as well [15, 16].

2.2. Three-segment anisomorphic CFL diagram

A class of composite laminates exhibits a significant transition in mean stress sensitivity in fatigue between T-T and C-C loadings [17, 18]. In that case, it may be an oversimplification to rely on only the S-N curve for the critical stress ratio to construct the CFL diagram over the whole range of mean stress.

To cope with a significant distortion of a CFL diagram due to a large change in mean stress sensitivity, a transitional segment may be inserted between the two segments that are associated with T-T and C-C dominated fatigue failures, respectively [17]. It has a role of connecting the two neighboring segments, for simplicity, with the aid of linear interpolation. To define the transitional segment, another reference stress ratio is introduced which is called the sub-critical stress ratio and designated by χ_s. The critical and sub-critical stress ratios divide the CFL diagram into three segments. The CFL curves in the three segments are described by means of the following piecewise-defined function over the domain of mean stress:

[I] Tension dominated zone ($\sigma_m^{(\chi)} \le \sigma_m \le \sigma_T$):

$$\frac{\sigma_a - \sigma_a^{(\chi)}}{\sigma_a^{(\chi)}} = \left(\frac{\sigma_m - \sigma_m^{(\chi)}}{\sigma_T - \sigma_m^{(\chi)}} \right)^{2-(\psi_\chi)^{k_T}} \tag{6}$$

[II] Tension-compression transitional zone ($\sigma_m^{(\chi_s)} \le \sigma_m < \sigma_m^{(\chi)}$):

$$\frac{\sigma_a - \sigma_a^{(\chi)}}{\sigma_a^{(\chi)} - \sigma_a^{(\chi_s)}} = \frac{\sigma_m - \sigma_m^{(\chi)}}{\sigma_m^{(\chi_s)} - \sigma_m^{(\chi)}} \tag{7}$$

[III] Compression dominated zone ($\sigma_C \le \sigma_m < \sigma_m^{(\chi_s)}$):

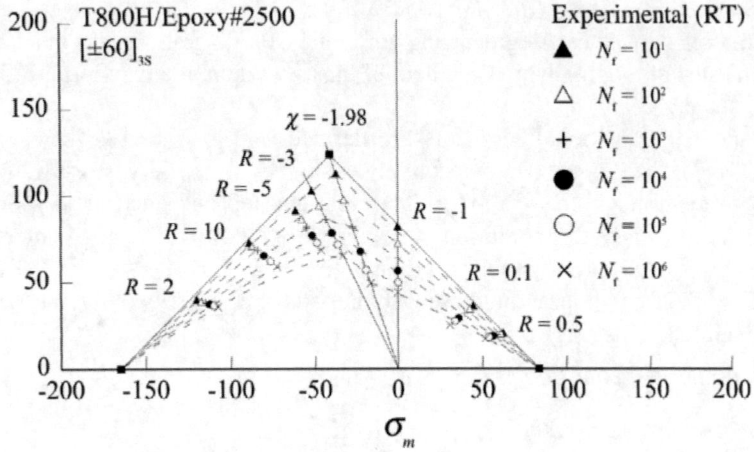

Figure 3. The three-segment anisomorphic CFL diagram for a carbon/epoxy [±60]3S laminate [Kawai and Murata (2010)].

$$\frac{\sigma_a-\sigma_a^{(\chi_s)}}{\sigma_a^{(\chi_s)}}=\left(\frac{\sigma_m-\sigma_m^{(\chi_s)}}{\sigma_C-\sigma_m^{(\chi_s)}}\right)^{2-(\psi_{\chi_s})^{k_C}} \tag{8}$$

where $\sigma_a^{(\chi)}$, $\sigma_m^{(\chi)}$, $\sigma_a^{(\chi_s)}$ and $\sigma_m^{(\chi_s)}$ represent the alternating and mean stress components of the maximum fatigue stresses $\sigma_{max}^{(\chi)}$ and $\sigma_{max}^{(\chi_s)}$ that are associated with fatigue loadings at the critical and sub-critical stress ratios χ and χ_s, respectively.

The two variable exponents ψ_χ and ψ_{χ_s} involved by the piecewise-defined function for the three-segment anisomorphic CFL diagram approach play the roles which are similar to the role of ψ_χ in the two-segment anisomorphic CFL diagram approach. The critical fatigue strength ratio ψ_χ for the three-segment anisomorphic CFL diagram is defined by equation (4). The sub-critical stress ratio is defined as

$$\psi_{\chi_s}=\begin{cases}\dfrac{\sigma_{max}^{(\chi_s)}}{\sigma_T}, & \chi\le\chi_s\le 0\\[2ex] \dfrac{\sigma_{min}^{(\chi_s)}}{\sigma_C}, & -\infty\le\chi_s<\chi\end{cases} \tag{9}$$

As in the two-segment anisomorphic CFL diagram approach mentioned above, the critical and sub-critical fatigue strength ratios ψ_χ and ψ_{χ_s} are treated as continuous monotonic functions of N_f; i.e., $\psi_\chi = f_\chi^{-1}(2N_f)$ and $\psi_{\chi_s} = f_s^{-1}(2N_f)$.

Figure 3 shows the three-segment anisomorphic CFL diagram for a non-woven carbon/epoxy [±60]3S laminate, along with the experimental CFL data [17]. A transitional segment was inserted between the T-T and C-C dominated segments. The modification is found effective to improve the accuracy of prediction of the distorted CFL curves for the [±60]3S laminate. The value of the shape transition exponent $k_T = 0.2$ retards the change in shape from a straight line to a parabola in the T-T dominated range $\chi < R < 1$. The adjustment of the rate of shape change successfully improves the accuracy of prediction of the CFL curves for T-T loading.

2.3. Four-segment anisomorphic CFL diagram

Two-side modification to the two-segment anisomorphic CFL diagram results in a four-segment version that removes the step of judging which segment of the two-segment anisomorphic CFL diagram should be modified [18].

The right and left transitional segments are defined as $[\chi, \chi_R] = \{R: -\infty < \chi \le R \le \chi_R \le 0\}$ and $[\chi_L, \chi] = \{R: -\infty \le \chi_L \le R \le \chi < 0\}$, respectively, using two auxiliary stress ratios χ_L and χ_R in conjunction with the critical stress ratio χ. The right and left auxiliary stress ratios χ_R and χ_L are determined so as to satisfy the condition $-\infty \le \chi_L \le \chi \le \chi_R \le 0$. The total interval of mean stress $[\sigma_C, \sigma_T]$ is thus separated into four subintervals: I. $[\sigma_m^{(\chi_R)}, \sigma_T]$; II. $[\sigma_m^{(\chi)}, \sigma_m^{(\chi_R)}]$; III. $[\sigma_m^{(\chi_L)}, \sigma_m^{(\chi)}]$; IV. $[\sigma_C, \sigma_m^{(\chi_L)}]$. The CFL envelopes on these subintervals are described by means of the following piecewise-defined functions:

[I] Tension-dominated zone ($\sigma_m^{(\chi_R)} \le \sigma_m \le \sigma_T$):

$$\frac{\sigma_a - \sigma_a^{(\chi_R)}}{\sigma_a^{(\chi_R)}} = \left(\frac{\sigma_m - \sigma_m^{(\chi_R)}}{\sigma_T - \sigma_m^{(\chi_R)}}\right)^{2-\psi_{\chi_R}^{k_T}} \tag{10}$$

[II] Right transitional zone ($\sigma_m^{(\chi)} \le \sigma_m < \sigma_m^{(\chi_R)}$):

$$\frac{\sigma_a - \sigma_a^{(\chi_R)}}{\sigma_a^{(\chi_R)} - \sigma_a^{(\chi)}} = \frac{\sigma_m - \sigma_m^{(\chi_R)}}{\sigma_m^{(\chi)} - \sigma_m^{(\chi_R)}} \tag{11}$$

[III] Left transitional zone ($\sigma_m^{(\chi_L)} \le \sigma_m < \sigma_m^{(\chi)}$):

$$\frac{\sigma_a - \sigma_a^{(\chi)}}{\sigma_a^{(\chi)} - \sigma_a^{(\chi_L)}} = \frac{\sigma_m - \sigma_m^{(\chi)}}{\sigma_m^{(\chi_L)} - \sigma_m^{(\chi)}} \tag{12}$$

[IV] Compression-dominated zone ($\sigma_C \le \sigma_m < \sigma_m^{(\chi_L)}$):

$$\frac{\sigma_a - \sigma_a^{(\chi_L)}}{\sigma_a^{(\chi_L)}} = \left(\frac{\sigma_m - \sigma_m^{(\chi_L)}}{\sigma_C - \sigma_m^{(\chi_L)}}\right)^{2-\psi_{\chi_L}^{k_C}} \tag{13}$$

where ψ_{χ_R} and ψ_{χ_L} designate the fatigue strength ratios associated with the right and left auxiliary stress ratios, χ_R and χ_L, respectively, and they are defined as

$$\psi_{\chi_R} = \frac{\sigma_{max}^{(\chi_R)}}{\sigma_T} \tag{14}$$

$$\psi_{\chi_L} = \frac{\sigma_{min}^{(\chi_L)}}{\sigma_C} \tag{15}$$

The variables $\sigma_a^{(\chi)}$, $\sigma_m^{(\chi)}$, $\sigma_a^{(\chi_R)}$, $\sigma_m^{(\chi_R)}$, $\sigma_a^{(\chi_L)}$, and $\sigma_m^{(\chi_L)}$ represent the alternating stress and mean stress components of the maximum fatigue stresses for fatigue loadings at the characteristic stress ratios.

Figures 4(a) and 4(b) show the four-segment anisomorphic CFL diagrams for on-axis and off-axis loadings of a unidirectional carbon/epoxy laminate, respectively [18]. For construction of these four-segment anisomorphic CFL diagrams, the standard auxiliary stress ratios $\chi_R = 0$ and $\chi_L = -\infty$ were used. As expected, the right and left transitional segments accommodate the anisomorphic CFL diagram to a large change in mean stress sensitivity from a higher level in tension-dominated fatigue loading to a lower level in compression-dominated fatigue loading, and thus the addition of these transitional segments to the anisomorphic CFL diagram successfully improves the accuracy of description.

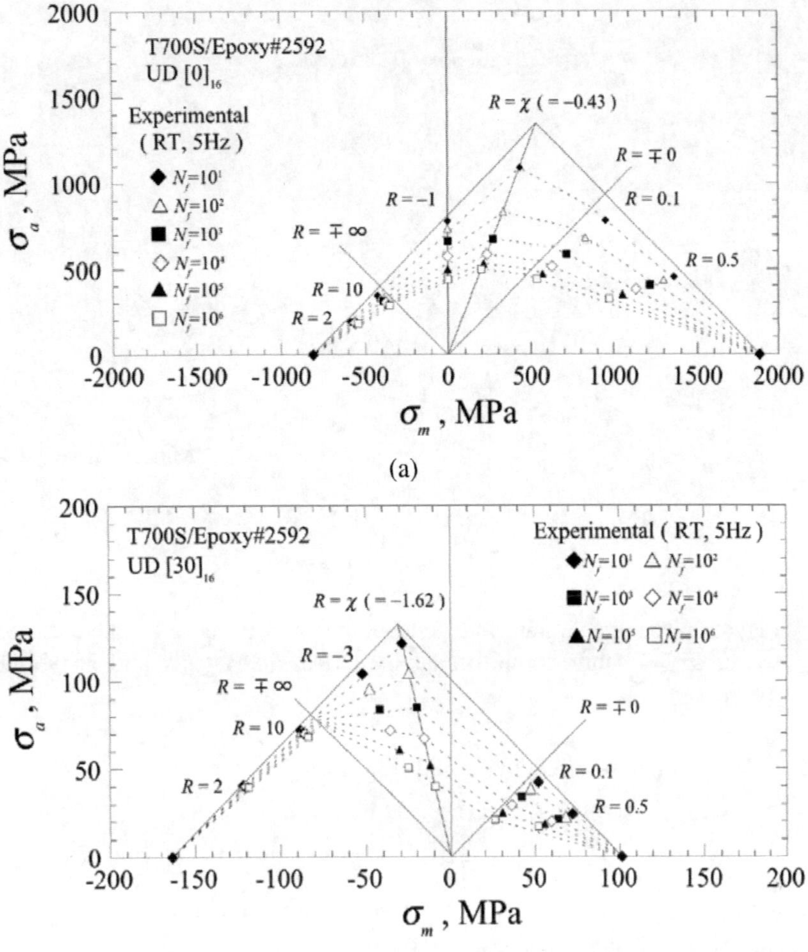

(a)

(b)

Figure 4. The four-segment anisomorphic CFL diagrams identified for unidirectional carbon/epoxy laminates; (a) $[0]_{16}$, (b) $[30]_{16}$ [Kawai and Ito (2014)].

3. Temperature-dependent anisomorphic CFL diagram approach

Consideration of the temperature dependence of the static strengths in tension and compression and of the reference S-N relationship for the critical stress ratio allows generalization of the anisomorphic CFL diagram approach for a given composite into a temperature-dependent version [19, 20]. In this attempt, the temperature dependences of tensile and compressive strengths are first modeled on the basis of the static test data at different temperatures. They determine the temperature dependence of the critical stress ratio as

$$\chi(T) = \frac{\sigma_C(T)}{\sigma_T(T)} \tag{16}$$

To predict the reference S-N relationship for the critical stress ratio at any temperature in a given range, we assume a grand master S-N curve for the critical stress that is described by means of a function

$$P_{LM}^{(F)} = g(\Psi_\chi) \tag{17}$$

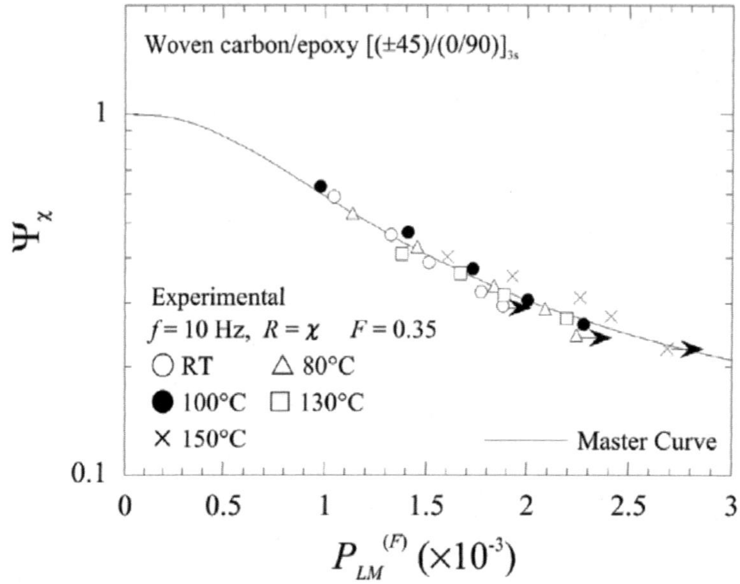

Figure 5. The temperature-independent master S-N curve for the critical stress of a woven carbon fabric composite laminate [Kawai, Matsuda and Yoshimura (2012)].

where Ψ_χ is the modified fatigue strength ratio [23] for the critical stress ratio

$$\Psi_\chi = \frac{\sigma_a^{(\chi)}}{\sigma_T - \sigma_m^{(\chi)}} \tag{18}$$

and $P_{LM}^{(F)}$ is the life-temperature parameter of the Larson-Miller type [24]

$$P_{LM}^{(F)} = T(F + \log_{10} N_f) \tag{19}$$

Figure 5 shows the grand master S-N curve for the critical stress for a woven CFRP laminate [20]. The fatigue data for the critical stress ratios at different temperatures are doubly scaled in terms of stress and life using the modified fatigue strength ratio $\Psi_\chi(T)$ and the LM parameter for fatigue $P_{LM}^{(F)}$. The normalized grand master S-N relationship $\Psi_\chi(T) - P_{LM}^{(F)}$ can successfully be identified.

Figures 6(a) and 6(b) show the predicted anisomorphic CFL diagrams for the woven CFRP laminate at RT and 100°C, respectively [20]. It is seen that the predicted CFL diagram agrees with the experimental CFL diagram, regardless of temperature. These comparisons demonstrate that the proposed temperature-dependent anisomorphic CFL diagram approach is useful for predicting the CFL diagrams for the woven CFRP laminate at different temperatures with reasonable accuracy over the whole range of mean stress.

4. Probabilistic anisomorphic CFL diagram approach

Using the statistical distributions for the static strength data in tension and compression and for the fatigue life data at the critical stress ratio, we can construct the anisomorphic CFL diagram for any constant value of probability of failure [21, 22]. The probabilistic anisomorphic CFL diagram allows predicting the P-S-N curves for any values of stress ratio.

The percentiles of the static strengths in tension and compression can be calculated using the distribution functions fitted to the observed static strength data. By contrast, however, it is not straightforward to deal with the P-S-N curve for the critical stress ratio since the critical stress ratio is

IOP Conf. Series: Materials Science and Engineering **406** (2018) 012001 doi:10.1088/1757-899X/406/1/012001

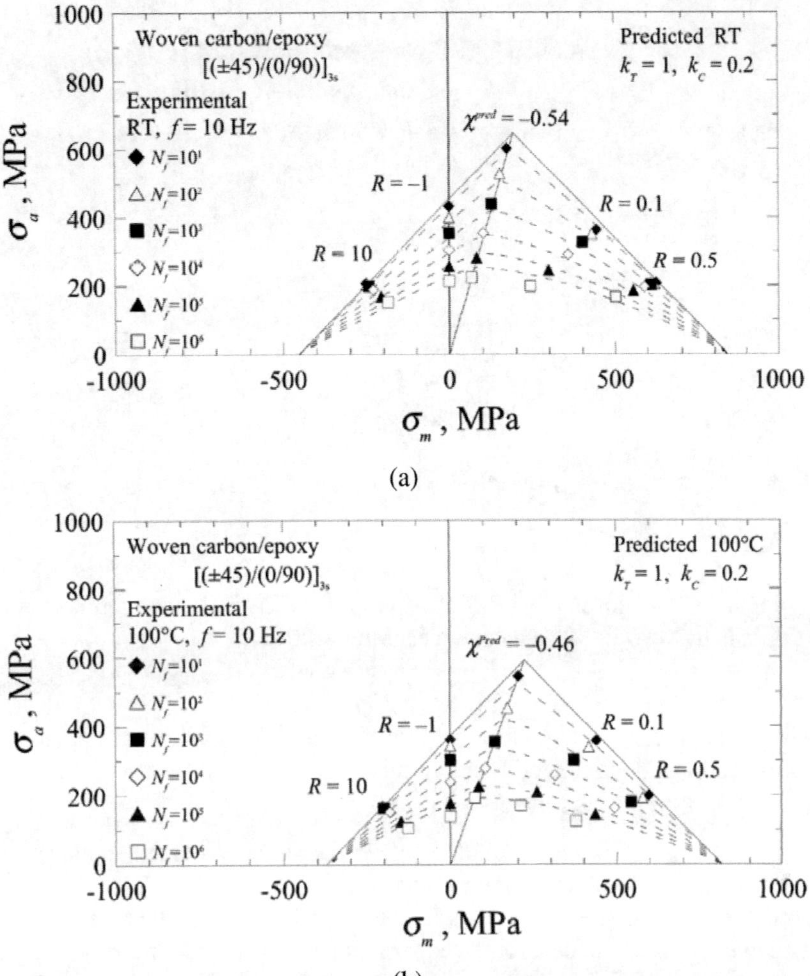

Figure 6. The predicted anisomorphic CFL diagrams for a woven carbon fabric composite laminate; (a) RT, (b) 100°C [Kawai, Matsuda and Yoshimura (2012)].

not a constant but a random variable. The modified fatigue strength ratio and fatigue life are random variables as well. To cope with the stress-ratio and probability dependence of the P-S-N curve for the critical stress ratio, we take advantage of the modified fatigue strength ratio that is useful for horizontal scaling to define a normalized master S-N curve. The modified fatigue strength ratio for any percentile of the critical stress ratio $\chi(P)$ is defined as follows:

$$\Psi_P^{(\chi(P))} = \begin{cases} \dfrac{\sigma_a^{(\chi(P))}}{\sigma_T(P) - \sigma_m^{(\chi(P))}}, & -1 \le \chi(P) < 0 \\ \dfrac{\sigma_a^{(\chi(P))}}{|\sigma_C(P)| + \sigma_m^{(\chi(P))}}, & -\infty < \chi(P) \le -1 \end{cases} \tag{20}$$

The non-dimensional S-N curve for the critical stress ratio for any probability of failure is described by means of a function

$$2N_f^{(\chi(P))}(P) = g_P(\Psi_P^{(\chi(P))}) \tag{21}$$

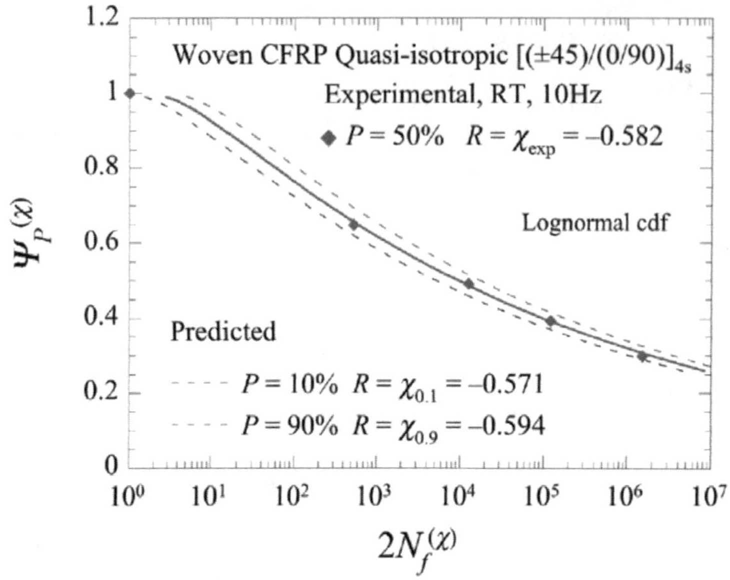

Figure 7. The normalized P-S-N curves for the critical stress ratio of a woven CFRP laminate [Kawai, Matsuda and Yoshimura (2012)].

Using the median S-N curve as the basis for transformation and assuming a small change in $\chi(P)$ with P, we can reduce equation (21) to the following form:

$$2N_f^{(\chi(P))}(P) = g_{0.5}(\Psi_P^{(\chi(P))}) \circ \Lambda_f(P) \tag{22}$$

where the function $\Lambda_f(P)$ represents the ratio of Pth percentile to median that can uniquely be determined using the cumulative distribution of fatigue life.

Figure 7 shows the normalized P-S-N curves for the critical stress ratio that are based on a lognormal distribution fitted to the fatigue life data for the experimental critical stress ratio χ_{exp} [21]. The symbols indicate the median of the experimental fatigue life data, and the solid curve indicates the fitting to them. Figures 8(a) and 8(b) show the anisomorphic CFL diagrams for the different constant values of failure probability P = 10% and 90%, respectively. It is seen that the anisomorphic CFL envelopes predicted for different constant values of failure probability agree well with the experimental CFL data for R = 0.1 and 10, regardless of the probability of failure. Note that the experimental results for R = 0.1 and 10 were not used for the verification observed above.

5. Integration into a classic spectrum loading fatigue life prediction methodology

Kawai and Ishizuka [26] have performed variable R-ratio loading tests and evaluated the classical fatigue life prediction methodology. One of the variable R-ratio loading tests they performed is illustrated in figure 9. Two waveforms of different R-ratios that are equal in life are alternately repeated; i.e., $R_1 \neq R_2$, $N_f^{(R_1)} = N_f^{(R_2)} = N_f^*$, $\sigma_{\max}^{(R_1)} \neq \sigma_{\max}^{(R_2)}$ and $\sigma_{\min}^{(R_1)} \neq \sigma_{\min}^{(R_2)}$. Since the alternation is along a vertical line as shown in figure 9(b), it was called a vertical alternating R-ratio (ARR) loading. It is a simple block loading. Nevertheless, this sort of variable loading tests allows us to examine the effect of simultaneous alternation in stress level and R-ratio on fatigue life.

Figure 10 shows the correlation between the predicted and observed fatigue life values for different vertical ARR loadings [26]. Note that the fatigue life predicted for vertical ARR loading that is shown in figure 10 was made by applying the Miner rule to the reduced waveforms counted by the rain flow method. The values of fatigue life for the counted constant waveforms were predicted using the anisomorphic CFL diagram that was constructed for the composite used in the ARR loading tests.

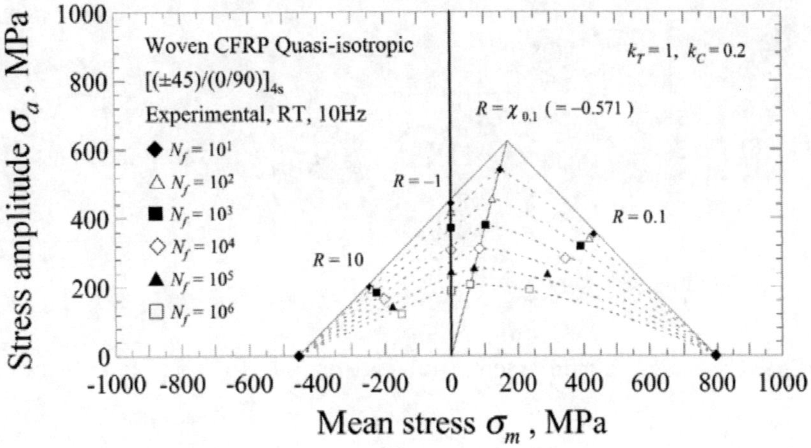

(a)

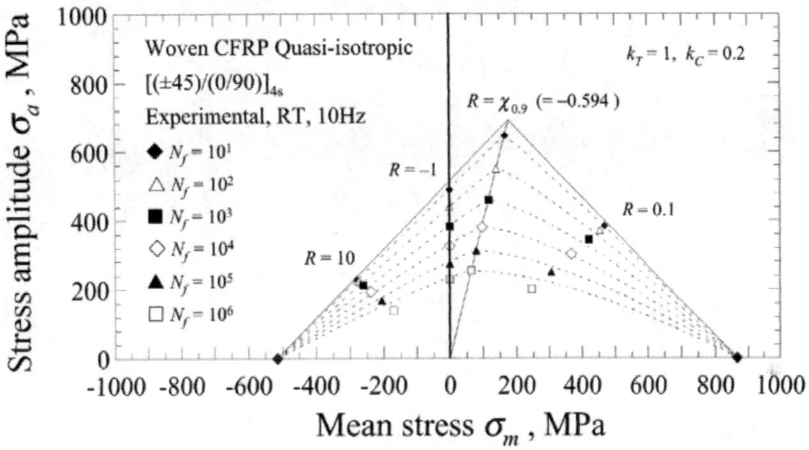

(b)

Figure 8. The probabilistic anisomorphic CFL diagrams for a woven CFRP laminate; (a) P = 10%, (b) P = 90% [Kawai and Yano (2015)].

From figure 10, it is seen that all the data points are distributed in the left border of the range of a factor of two. The horizontal axis indicates the predicted fatigue life. From figure 10, therefore, we can see that the variable loading fatigue life prediction methodology allows predicting moderately conservative values of fatigue life for the vertical ARR loadings, regardless of the pair of R-ratios. The same feature was observed in the range of probability of failure lower than 90%. A moderately conservative prediction of fatigue life is favorable from an engineering point of view. The correlation results in figure 10 thus prove the potential usefulness of the integration of the Miner rule, the rain flow cycle counting method and the anisomorphic CFL diagram approach in the context of the variable waveform loading fatigue life prediction of composites.

6. Conclusions

A series of attempts at systematic modeling of a full shape of CFL diagram suitable for carbon fiber-reinforced composites that have been made by the author in the last decade was reviewed with validation at each stage of sophistication. The asset of the anisomorphic modeling of CFL diagram for composites, which distinguishes it from the attempts by other researchers, is to take into account the

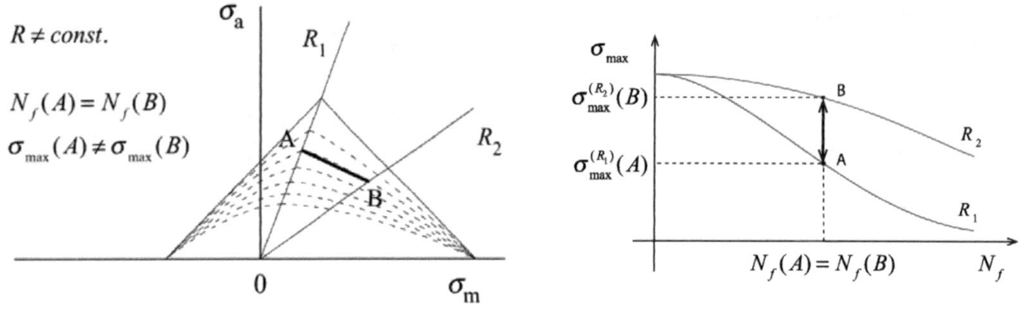

(a) (b)

Figure 9. Vertical alternating R-ratio loading; (a) σ_m-σ_a plane, (b) S-N plane [Kawai and Ishizuka (2018)].

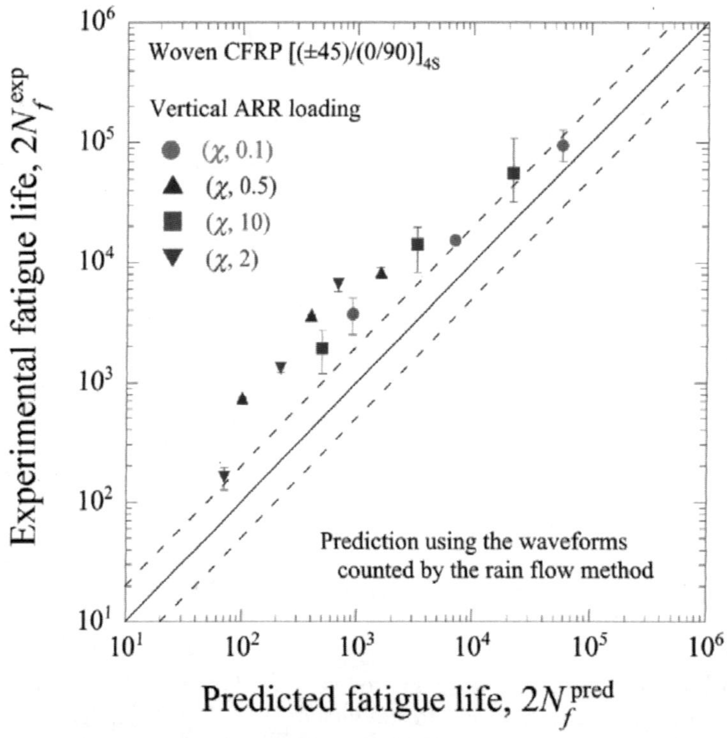

Figure 10. Comparison between the predicted and observed fatigue lives for different vertical ARR loadings [Kawai and Ishizuka (2018)].

fatigue loading at the critical stress ratio under which damage accumulation occurs most significantly. The two-, three- and four-segment anisomorphic CFL diagrams have been formulated, and they can be used properly taking into account the significances of the asymmetry and nonlinearity in CFL curves for the composites to consider. The more the number of segments increases, the more the fatigue data are required for identification. Nevertheless, it is unique in its general formulation not only that allows identification by means of a limited amount of test data but also that reduces to a linear model of the Goodman type on one extreme and to a parabolic model of the Gerber type on the other, and thus allows of any in between. Another feature of the anisomorphic formulation is the ease of generalization that makes it possible for us to consider not only the effect of temperature but also the probability of failure.

A classical fatigue life prediction methodology consists of the three elements; typically 1) the rain flow cycle counting method to reduce a given loading spectrum into the effective constant *R*-ratio waveforms, 2) a CFL diagram approach to estimate the values of life for the counted effective waveforms of different constant values of *R*-ratio, and 3) the Miner rule to judge fatigue failure in composites under the given spectrum loading. This methodology is favored, because of its simplicity, in engineering fatigue analysis of composites. The join of the anisomorphic CFL diagram approach in the classic three-element methodology is expected to establish a powerful tool for engineering fatigue and reliability analyses of composites that undergo complicated cyclic loading.

Acknowledgments
The research on which this review article is based has been carried out with my students at the University of Tsukuba, mainly with the financial support of the University of Tsukuba and the Ministry of Education, Culture, Sports, Science and Technology of Japan. The author is grateful to all the members in my lab who have contributed to the work.

References

[1] Bond I P 1999 *Composites Part A* **30** 961
[2] Filis P A, Farrow I R and Bond I P 2004 *International Journal of Fatigue* **26** 565
[3] Philippidis T P and Vassilopoulos A P 2004 *Composites Part A* **35** 657
[4] Sutherland H J and Mandell J F 2005 *Wind Energy* **8** 93
[5] Passipoularidis V A and Philippidis T P 2009 *International Journal of Fatigue* **31** 408
[6] Wesphal T and Nijssen R P L 2014 *Journal of Physics: Conference Series* **555** 012107
[7] Endo T, Mitsunaga K and Nakagawa H 1967 *Preliminary Proceedings of the Chugoku-Shikoku District Meeting*, p 41
[8] Endo T, Matsuishi M, Mitsunaga K, Kobayashi K and Takahashi K 1974 *Kyushu Institute of Technology Academic Repository* **28** 33
[9] Goodman J 1899 *Mechanics Applied to Engineering* (Longman Green, Harlow, UK)
[10] Kawai M 2010 *Fatigue Life Prediction of Composites and Composite Structures* Chapter 6, ed A P Vassilopoulos (Woodhead Publishing Limited, Abington Hall, Abington, Cambridge, CB21 6AH, UK) p 177
[11] Palmgren A Z 1924 *Zeitschrift des Verein Deutcher Ingenieure* **68** 339
[12] Miner M A 1945 *ASME J. Appl. Mech.* **12** A159
[13] Harris B 2003 *Fatigue in Composites* (Cambridge, Woodhead Publishing Limited, UK)
[14] Vassilopoulos A P 2010 *Fatigue Life Prediction of Composites and Composite Structures* (Woodhead Publishing Limited, UK)
[15] Kawai M 2006 *Proceedings of the Fifth Asian-Australasian Conference on Composite Materials (ACCM-5)*, Hong Kong
[16] Kawai M and Koizumi K 2007 *Composites Part A* **38** 2342
[17] Kawai M and Murata T 2010 *Composites Part A* **41** 1498
[18] Kawai M and Itoh N 2014 *Journal of Composite Materials* **48** (5) 571
[19] Kawai M and Matsuda Y 2012 *Composites Part A* 43 647
[20] Kawai M, Matsuda Y and Yoshimura R 2012 *Composites Part A* **43** 915
[21] Kawai M and Yano K 2016 Composites Part A **80** 244
[22] Kawai M and Yano K 2016 *International Journal of Fatigue* **83** 323
[23] Kawai M 2004 *Composites Part A* **35**(7-8) 955
[24] Larson F R and Miller A 1952 *Transactions of the ASME* **74** 765
[25] Kawai M and Yang K and Oh S 2015 *Journal of Composite Materials* **49** (27) 3387
[26] Kawai M and Ishizuka Y 2018 *International Journal of Fatigue* **112** 36

IOP Conf. Series: Materials Science and Engineering **406** (2018) 012002 doi:10.1088/1757-899X/406/1/012002

Hierarchical design of structural composite materials down to the nanoscale via experimentation and modelling

L. Gorbatikh[1*], Q. Liu[1], V. Romanov[1], M. Mehdikhani[1], A. Matveeva[1], O. Shishkina[1], A. Aravand[1], B. Wardle[2], I. Verpoest[1], S. V. Lomov[1]

[1]Department of Materials Engineering, KU Leuven, Kasteelpark Arenberg 44, B-3001 Leuven, Belgium

[2]Department of Aeronautics and Astronautics, Massachusetts Institute of Technology, 77 Massachusetts Avenue, Cambridge, MA 02139, USA

* Larissa.Gorbatikh@kuleuven.be

Abstract. We envision the next generation composite materials as hierarchically structured down to the nanoscale. The importance of nanoscale features has been long recognized from studies of naturally occurring composites. Thanks to their hierarchical organization spanning through multiple scales they show exemplary resilience to failure combined with a plethora of different functions. In man-made composites, nanostructure can be introduced through modifications of the matrix, interfaces and fibers. Here we review our research efforts in the field of nano-engineered composites, focusing on the mechanical performance of fiber-reinforced plastics modified with carbon nanotubes. Our research is a combination of experimental and computational studies.

1. Introduction

There has been immense interest in the use of carbon nanomaterials for reinforcement of plastics and their composites. Figure 1 shows evolution of the number of articles referring to carbon nanotubes (CNTs), graphene and composites in the last three decades. The number of articles published in 2017 on a combined topic of "carbon nanotubes" and "composites" is close to 10 000, and the one on "graphene" and "composites" is 14 000 (source: Web of Science). There has been an accelerated growth in graphene research related to composites triggered in the year of 2010. We like to believe that this remarkable productivity in research results is linked to the prior knowledge acquired on CNTs that is now being transferred to studies of graphene based composites.

The focus of this brief review is a special class of composite plastics that combine reinforcing components of two scales (microscopic fibers and nanotubes) and are known as hierarchical composite materials [1]. These materials have a complex structure with several levels of organization and distinguishable macroscopic, mesoscopic, microscopic, and nanoscopic features. Figure 2 illustrates these structural levels on an example of a nano-engineered alumina fiber/epoxy composite. One can distinguish layers of a woven fabric on the meso-level, individual fibers inside yarns of the fabric and CNT bundles grown on the fibers on the micro-level. By zooming in to the nanoscale one could examine internal structure of the CNT bundles (not shown on the figure) and even structure of individual nanotubes.

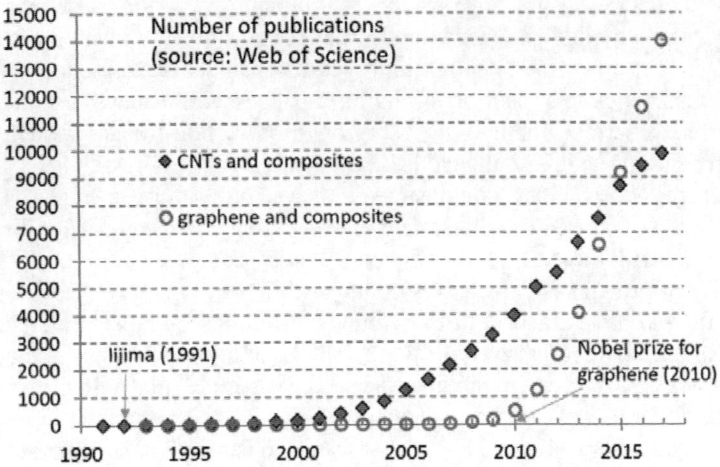

Figure 1. The number of journal articles per year in composites research with reference to carbon nanotubes and graphene.

2. Nano-engineering of fiber-reinforced polymer composites

Different routes can be followed for introduction of CNTs in fiber-reinforced plastics. For example, CNTs can be dispersed in the matrix, added in fiber sizing/coatings, directly grown on fibers/fabrics (as shown in Figure 2) or integrated in interleaves between plies. Benefits for the mechanical performance and challenges in processing for these strategies have been a subject of many reviews [1, 2]. Direct growth of CNTs on fibers is a promising route considering that it allows to bypass typical problems of CNT agglomeration and filtration in approaches based on CNT dispersion.

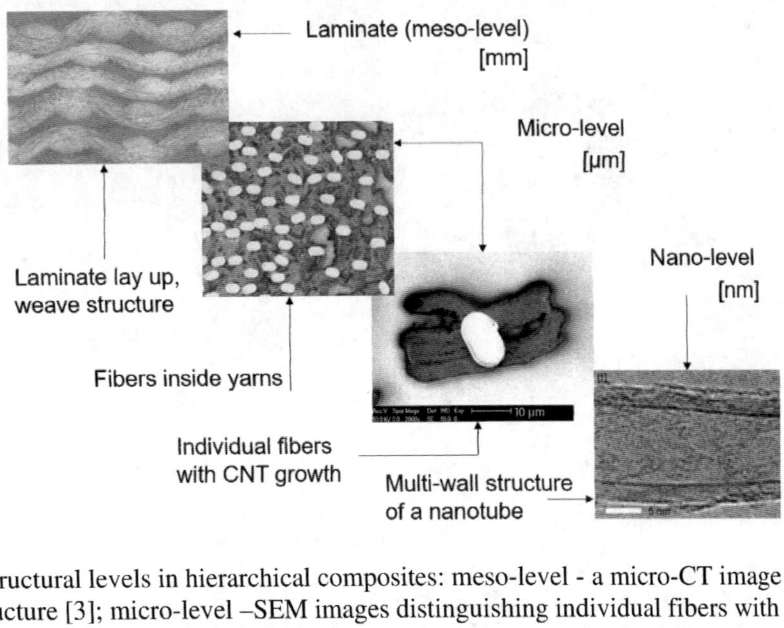

Figure 2. Structural levels in hierarchical composites: meso-level - a micro-CT image resolving textile structure [3]; micro-level –SEM images distinguishing individual fibers with aligned nanotube growth [4]; nano-level - a TEM image of a multi-wall nanotube grown on the fiber [5].

IOP Conf. Series: Materials Science and Engineering **406** (2018) 012002 doi:10.1088/1757-899X/406/1/012002

Moreover, the route of CNT grafting provides additional benefits such as control of CNT alignment and possibility to achieve a high CNT content in the composite. Composites with these CNT grafted fibers (also known as 'fuzzy' fibers) show improved mechanical performance (fiber–matrix interface strength, interlaminar toughness, bearing strength, [5, 6]). In earlier days researchers working in this direction had to face a serious challenge: CNT growth on carbon fibers was detrimental to the fiber strength, but in the meantime this challenge has been successfully addressed [7].

Processing of the fuzzy fiber composites with vacuum infusion, light RTM and autoclave manufacturing needs, in some cases, adjustment of processing parameters. The fibre volume fraction in composites produced with these techniques is controlled by compressibility of the fibre reinforcement. Since the presence of grown CNT decreases compressibility of dry fibres/fabrics [8, 9], it is likely to lead to a lower fiber volume fraction in the composite. Internal geometry of fuzzy fabrics and their composites is also different: yarns are typically swelled due to the CNT grafting leading to a laminate thickness increase [3]. Changes in the material microstructure may alter the process of damage development and final failure of the composite. Thus it is a combined effect of the nanoscale reinforcement and microstructural changes that is responsible for the changes in the mechanical performance.

3. Modelling of fiber-reinforced composites with carbon nanotubes

Modelling approaches have been indispensable in helping to understand the mechanical behavior of such complex materials as nano-engineered composites. The number of design parameters is so large that it is not feasible to investigate them all experimentally. In [10-12] we proposed a computational approach to predict the effect of CNTs on stresses in hierarchical composites (Figure 3). The approach is based on the mesh superposition technique in the framework of the finite element analysis.

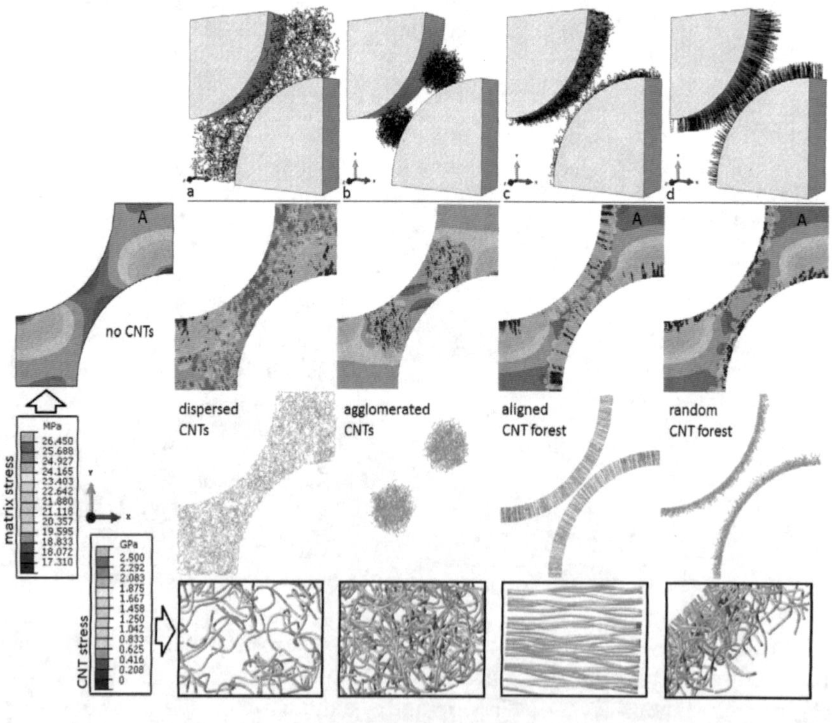

Figure 3. Geometrical models of the composites with different CNT assemblies and corresponding stress fields in the matrix and CNTs [10].

IOP Conf. Series: Materials Science and Engineering **406** (2018) 012002 doi:10.1088/1757-899X/406/1/012002

It allows modelling of composites with a large variety of CNT assemblies: CNTs that are dispersed in the matrix with different degrees of agglomeration, CNTs that are grown on fibres with different degrees of alignment and CNTs that are concentrated in the interphase region mimicking fibre sizings or coatings (Figure 3). The models have shown that CNTs introduce strong heterogeneity in stress and strain fields at the microscale. This heterogeneity is closely linked to the morphology of the CNT distribution [10-13].

Strains predicted by the models are consistent with the strain maps obtained experimentally using digital image correlation in combination with SEM [4] (Figure 4). Both experimental and modelling studies confirmed that aligned CNT forests constrain tensile deformation in their longitudinal direction (Figure 4). The strain in the transverse direction, on the other hand, is not constrained. Thus, an impregnated CNT bundle behaves as a quasi-unidirectional nanocomposite, transverse properties of which are dominated by the matrix behavior.

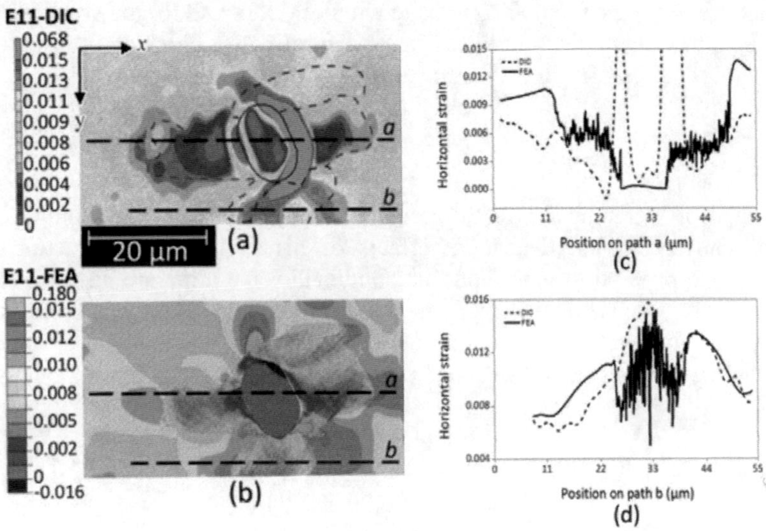

Figure 4. Comparison of the strain maps measured using digital image correlation (a) and maps predicted by the two scale FEM analysis (b); and comparison of the measured and predicted strain profiles along paths *a* and *b* (c,d) [4].

Recently we extended capabilities of our models towards predictions of the strength and toughness [14], by accounting for three damage mechanisms: 1) fibre/matrix debonding, 2) CNT/matrix debonding and 3) damage in the matrix. Computational experiments showed that damage diffusion (crack branching) is the major toughening mechanism in composites with CNT grafted fibres, while debonding of CNTs and that of the fibers play secondary roles. Figure 5 shows microscopic damage in the CNT-modified composite and the reference case (pristine composite without CNTs) at the time of composite failure. When CNTs are absent, damage starts at the fibre/matrix interface and then kinks into the matrix, resulting in a very localized fracture. In the case of fuzzy fibers we see significant damage diffusion in resin rich zones between CNT forests, which is accompanied by CNT debonding. The latter also consumes energy and further toughens the material. All this leads to a dramatic improvement of the transverse strength and toughness (Figure 5d).

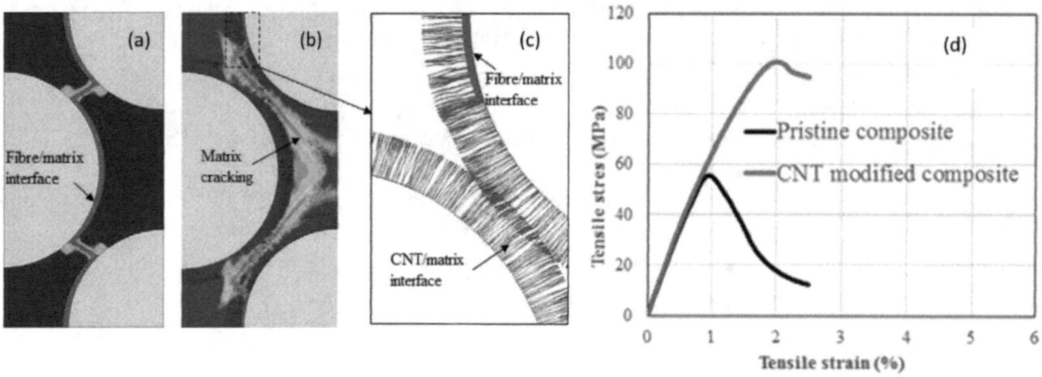

Figure 5. Damage in (a) the pristine composite without CNTs and (b) the composite with fuzzy fibers; (c) zoom in view on the interfacial damage in the composite with CNTs; (d) predicted stress–strain curves for the two composite types [14].

4. Outlook

Research in the field of hierarchical structural composites is in its very beginning. More research is needed to better understand interdependent effects of nano- and micro-structures on composite properties as well as to proceed to more material combinations and hierarchical levels as we see it in nature. Although not discussed here, the emergence of carbon nanomaterials is presenting us with unique opportunities not only to develop light, strong and tough composites but also smart composites that could self-manufacture, self-sense, self-heal and perform many other practical functions ([15-17]).

Acknowledgments
The authors acknowledge the support from C24/16/021 project, funded by KU Leuven Research Council. S.V. Lomov holds the Toray Chair for Composite Materials at KU Leuven.

References

[1] Gorbatikh L, Wardle BL, Lomov SV, Hierarchical lightweight composite materials for structural applications. 2016 *MRS Bulletin* **41**(9) 672

[2] Gorbatikh L and Lomov SV 2018 2.15 Damage in Architectured Composites. In: Beaumont, P.W.R. and Zweben, C.H. (eds.) *Comprehensive Composite Materials* II. vol. 2, 291–306. Oxford: Academic Press.

[3] Aravand MA, Shishkina O, Straumit I, Liotta AH, Wicks SS, Wardle B, Lomov SV, Gorbatikh L 2016 Internal geometry of woven composite laminates with ''fuzzy'' carbon nanotube grafted fibers. *Composites Part A* **88** 295–304

[4] Mehdikhani, M, Matveeva A., Aravand A, Wardle B., Lomov SV, Gorbatikh L 2016 Strain mapping at the micro-scale in hierarchical polymer composites with aligned carbon nanotube grafted fibres. *Composites Science and Technology* **137** 24–34

[5] Garcia EJ, Wardle BL, Hart AJ, Yamamoto N 2008 Fabrication and multifunctional properties of a hybrid laminate with aligned carbon nanotubes grown in situ. *Composites Science and Technology* **68** 2034–2041

[6] Wicks SS, Wang W, Williams MR., Wardle BL 2014 Multi-scale interlaminar fracture mechanisms in woven composite laminates reinforced with aligned carbon nanotubes. *Composites Science and Technology* **100** 128–135

[7] Li R, Lachman N, Florin P, Wagner HD, Wardle BL 2015 Hierarchical carbon nanotube carbon fibre unidirectional composites with preserved tensile and interfacial properties. *Composites Science and Technology* **117** 139–145

IOP Conf. Series: Materials Science and Engineering **406** (2018) 012002 doi:10.1088/1757-899X/406/1/012002

[8] Lomov SV, Wicks S, Gorbatikh L, Verpoest I, Wardle BL 2014 Compressibility of nanofibre grafted alumina fabric and yarns: aligned carbon nanotube forests. *Composites Science and Technology* **90** 57–66
[9] Lomov SV, Gorbatikh L, Verpoest I 2011 A model for the compression of a random assembly of carbon nanotubes, *Carbon* **49** 2079 –2091
[10] Romanov VS, Lomov SV, Verpoest I, Gorbatikh L 2015 Modelling evidence of stress concentration mitigation at the micro-scale in polymer composites by the addition of carbon nanotubes. *Carbon* **82** 184
[11] Romanov VS, Lomov SV, Verpoest I, Gorbatikh L 2015 Inter-fiber stresses in composites with carbon nanotube grafted and coated fibers. *Composites Science and Technology* **114** 79–86
[12] Romanov VS, Lomov SV, Verpoest I, Gorbatikh L 2014 Can carbon nanotubes grown on fibers fundamentally change stress distribution in a composite? *Composites: Part A* **63** 32–34
[13] Romanov VS, Lomov SV, Verpoest I, Gorbatikh L 2015 Stress magnification due to carbon nanotube agglomeration in composites. *Composite Structures* **133** 246–256
[14] Liu Q, Lomov S.V., Gorbatikh L, Carbon nanotubes grafted on fibers diffuse damage in composites leading to their improved toughness at the micro-scale. *Proceeding of ECCM18 - 18th European Conference on Composite Materials*, Athens, Greece, 24-28th June 2018
[15] Buschhorn ST, Kessler SS, Lachmann N, Gavin J, Thomas G, Wardle BL, Electrothermal icing protection of aerosurfaces using conductive polymer nanocomposites. *54th AIAA/ASME/ASCE/AHS/ASC Structures, Structural Dynamics, and Materials Conference*, Structures, Structural Dynamics, and Materials and Co-located Conferences (AIAA 2013-1729)
[16] Lee J, Ni X, Daso F, Xiao X, King D, Sánchez Gómez J, Varela TB, Kessler SS, Wardle BL 2018 Advanced carbon fiber composite out-of-autoclave laminate manufacture via nanostructured out-of-oven conductive curing. *Composites Science and Technology* (in print)
[17] Lee J, Stein IY, Kessler SS and Wardle BL 2015 Aligned Carbon Nanotube Film Enables Thermally Induced State Transformations in Layered Polymeric Materials. *ACS Appl. Mater. Interfaces* **7** (16) 8900–8905

History and future of composites forming analysis

R Akkerman[1,2], S P Haanappel[3] and U Sachs[1]

[1]TPRC, PO Box 770, 7500 AT Enschede, The Netherlands
[2]University of Twente, PO Box 217, 7500AE Enschede, The Netherlands
[3]Aniform, Palatijn 15, 7521PN Enschede, The Netherlands

Abstract. The developments of composites forming analyses are summarized, from a historic perspective and towards the expected future. While the recent methods are already accepted in the market, there is a number of challenges ahead in simulation methods and experimental characterisation, to reach even better predictions of defects and part distortions.

1. Introduction

Rapid manufacturing processes such as hot press forming of thermoplastic composites are key technologies for large scale composite applications to emerge. Typically, tool manufacturing is time consuming and the tooling costs are high, having a large impact on the development time and cost of new press formed parts. For products to be successful on the market, trial-and-error development should be prevented. Composites forming simulations provide a means to prevent or reduce these costly iterations to a bare minimum, and are in this sense crucial for a mature position of composites on this market for high volume manufacturing. Overseeing the past and the present, it can be observed that the applications of composite materials go hand-in-hand with the developments in simulation tools. It is to be expected that this also holds for the upcoming stage of coming to maturity, requiring predictive design tools and quality control which can be supported by further improved and extended simulation methods.

2. History

Fibre composites provide the potential for structural optimisation by means of orienting the fibres in their most favourable direction. The positioning of the fibres themselves, however, depends strongly on the manufacturing process. Fibre placement, for instance, provides great flexibility in fibre positioning, primarily constrained by the continuity of the fibres which should not change orientation in a stepwise manner from one location to another. A higher speed of fibre deposition can be achieved by using multiple tows simultaneously as in braiding, by using broader tape or by using textiles such as woven fabrics. Each approach generates its own constraints on the achievable fibre orientation distribution over the part to be manufactured.

Woven fabrics further provide great ease of handling and, depending on the fabric architecture, good drapeability. The same goes for random mats of continuous or discontinuous fibres, in which case there is little use of preferred fibre orientations, however. Early applications of fibre composites relied on manual lay-up of such mats and fabrics, on which a thermoset resin was applied to impregnate the fibres. By means of manual lay-up, the laminator can prevent the occurrence of defects such as wrinkling to some extent and, in case of fabrics, can steer the fibres in favourable directions. Early work on the analysis of drapeability of fabric plies dates back to the 1950s [1], more or less simultaneous to the development of the materials and processes involved. Ever since, the composites research and development have kept a close link between materials, processes and design aspects, including the underlying analysis methods. In this case, mapping schemes were developed to project a pin-jointed net on the tool geometry, providing a means to predict the fibre orientations and fabric shear as a result of draping for specific geometries in closed form. Since the pioneering analytical work by Mack and Taylor, many authors used the 'fishnet approach' for numerical simulations of the draping process, e.g. [[2]-[7]]. Typically, these drape models start from an arbitrary initial point and two initial fibre directions. Further points are then generated at a fixed equal distance from the previous points, creating a mesh of quadrilateral cells.

There is no unique solution for this geometrical drape method, as the solution depends on the chosen sequence of cells to be generated. This problem is generally solved by defining two fibre paths on the drape surface, from which the cells will be generated. Bergsma [6] introduced 'strategies' in order to find solutions for the drape algorithm, without pre-defining such fibre paths, and also included a mechanism to incorporate the locking phenomenon in his drape simulations. Iterative approaches to account better for the actual shear characteristics of the fabrics have been proposed [8]-[9], using local energy minimisation for each new set of cells to be added to those generated previously. In this way, the material behaviour can be represented to a certain extent, while retaining the very fast solver times of the geometric approach.

These essentially geometry-based methods cannot easily represent the actual material behaviour in great detail. Not accounting for the through-thickness interactions between the layers, are they not very well suited for simulations of forming multi-layered laminates at once, in which the inter-ply friction plays an important role. Also the interactions with blank holders and tool surfaces cannot easily be included in these mapping schemes. Such interactions can be implemented more easily in Finite Element representations, at the cost of significantly higher computation time.

Alternatively, a fabric can be represented as a collection of particles [10] with e.g. elastic or plastic interactions, representing energy contributions due to thread repelling, thread stretching, thread bending, thread trellising and gravity. Constraints can be used to prevent intersection of the particles with rigid bodies such as the surface on which the fabric is to be draped. Fast algorithms have been developed to simulate cloth behaviour on dressed virtual humans in real time, see e.g. [11]. Intended for virtual impressions of apparel fabric rather than detailed studies of the local deformations, these methods have not seen implementation in the technical industry to the authors' knowledge.

Finite Element representations for composites forming processes have been developed from the 1990s. The early formulations adopted orthotropic elastic material behaviour in continuum shell elements [12]-[14] to model e.g. the draping of a cloth on a table. An early multi scale approach was presented in [15], using a thermomechanically coupled scheme for an anisotropic fluid with anisotropic thermal properties, where these properties were solved using micromechanical modelling of RVEs with the appropriate fibre orientations. Alternatively, truss based schemes have been presented, with the trusses representing the stiff fibres (albeit in tension) and an underlying membrane representing the inplane shear behaviour [16],[17]. Also mesoscopic analyses have been performed, to analyse the interaction of fibres, possibly impregnated and possibly with the resin rich areas. In these, the material behaviour of the tows is non-trivial and not easily characterised experimentally. Nevertheless, the simulations provide useful insight in the origins of macroscopically observed behaviour.

Fairly rapidly, the technology of forming simulation software and associated material characterisation has found its way into commercial software and services, supporting composite manufacturing of press formed fibre reinforced thermoplastic parts. Early approaches adopted the geometric mapping schemes (in e.g. Fibersim and similar packages), later also specific tools became available using explicit FE solvers (PAM-Form) and implicit FE solvers (Aniform). The large deformations and the extremely anisotropic behaviour of composite materials in forming conditions give rise to significant challenges with respect to stable convergence and accuracy [18]. These FE models are primarily used on the macroscopic scale, treating a ply or a group of plies as an anisotropic continuum, represented in plate or shell elements.

Long [19] presented an overview of the deformation mechanisms in composite laminates during forming. Figure 1 provides an illustration of such a classification. Of these mechanisms, the current FE

IOP Conf. Series: Materials Science and Engineering **406** (2018) 012003 doi:10.1088/1757-899X/406/1/012003

descriptions typically include trellis shear (described as a viscous, elastic or viscoelastic phenomenon), inter-ply and tool/ply friction (described with a coefficient of friction or a viscous slip coefficient, which may depend on rate, pressure and temperature) and bending (usually described as an elastic phenomenon with an appropriate bending modulus). All require specific experimental techniques and procedures to determine the relevant material property data.

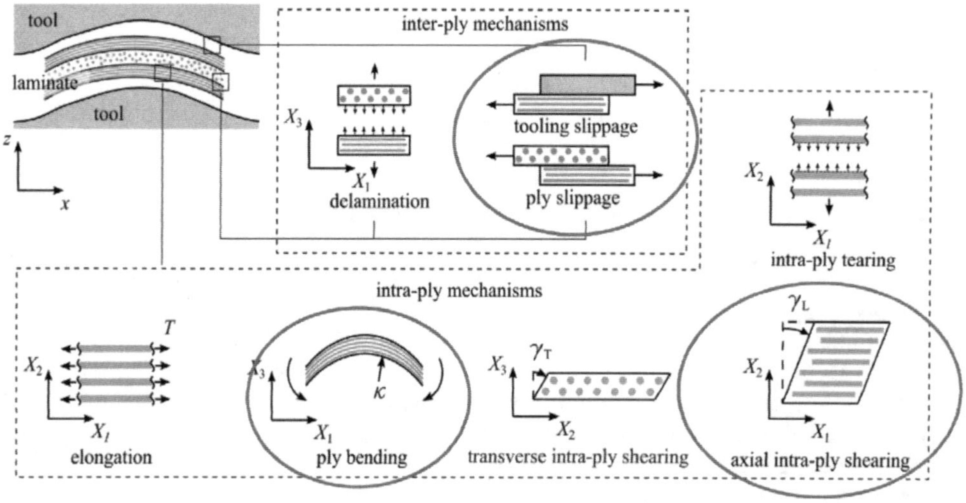

Figure 1. Deformation mechanisms in (unidirectional) composite laminates.

Early publications by Boisse and co-workers focused on the response of fabrics due to bi-axial tension in the fibre directions [14], for which an experimental set-up was developed and the nonlinear response was measured and included in the forming simulation model. Subsequently, the focus moved towards the inplane trellis shear response, in particular on normalising the results for different specimen sizes and on developing standards for obtaining operator and equipment independent test results. The modelling efforts were in particular devoted to rate independent behaviour of dry fabrics, typically described in terms of the normalised shear force as a polynomial function of the shear angle [20]. Benchmarking activities were performed not only on inplane shear [21] but also on friction [22] and forming simulations and experiments on a benchmark 'double dome' geometry [23]-[27].

3. Current trends in material modelling

While the process simulations have found their way into industrial application, this does not imply that the composites forming software is infallible. Not all possible deformation mechanisms are included in the current software, and there is ample room for improvement of the models of the mechanisms that are included already.

Thermoplastic laminates are conventionally made by consolidation of plies at high temperature and pressure, in order to minimise porosity in the laminate. It may come as no surprise that this leads to residual stresses that will attempt to de-consolidate the laminate when returning to melt temperature in the forming process. Early publications [28] addressed this phenomenon, but the effects of this changing state on the other deformation mechanisms have been hardly investigated. Slange [29] shows in a recent publication that the preconditioning of the consolidated laminates (drying but also annealing at high temperatures) has a strong effect on the amount of deconsolidation. Accurate quantitative description and predictive modelling of these phenomena requires significant further research efforts.

Inter-tow slippage is another mechanism that is observed to play a role [30] during laminate tensioning, but rarely modelled. An early continuum FE model by Ten Thije capturing this kind of fibre slip in non-crimp fabrics [31] has found little follow up, and also mesoscopic descriptions are found primarily in the research domain, but not in generically available design for composite manufacturing software.

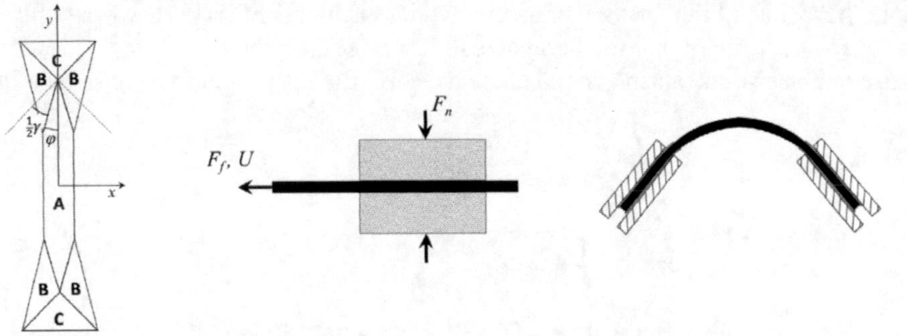

Figure 2. State of the art material characterisation methods (schematically) for uniaxial bias extension, tool/ply (and ply/ply) friction, bending (torsional rheometry).

The usually implemented mechanisms include trellis shear (described as a viscous, elastic or viscoelastic phenomenon), friction (described with a coefficient of friction or a viscous slip coefficient, which may depend on rate, pressure and temperature) and bending (usually described as an elastic phenomenon with an appropriate bending modulus). Figure 2 schematically illustrates some typical experiments for the associated material characterisation. Although good progress has been made over the years, the constitutive models are simplified representations of the actual material behaviour. This is, however, closely linked to the availability of accepted material characterisation methods and the associated material property databases for the growing range of materials available on the market. Further, the models employed so far are primarily isothermal, which is a fair approximation in many cases. However, the need for non-isothermal models is apparent when trying to predict shape distortions due to process induced residual stresses. The same holds for defect prediction. Forming induced large wrinkles and folds can be predicted well by the current approach, but wrinkles smaller than the element size require further advanced criteria, simulations and related material property data.

Inplane shear testing can be considered the most researched subject in fabric composite characterisation. Picture Frame (PF) and Uniaxial Bias Extension (UBE) tests are most often applied in this respect. The former generally suffers from lack of control of the fibre tension, which is likely to affect the shear response, whereas the latter involves non-uniform deformations which complicates the analysis of the results. In addition, UBE inherently invokes fibre tension together with shear, such that the 'zero-tension' situation cannot be tested with this method. In its simplest form, the effects of fibre tension on the shear behaviour are neglected and the resistance to shear is described as a path independent function of the local shear angle only. The accepted approach uses a power equilibrium method to derive the material property data from a bias extension test. Adding simple force equilibrium considerations to this approach [32] leads to tight constraints for the force-displacement and stress-strain relations of which the consequences are not fully understood to date. Even this very basic case has not been fully resolved yet.

Harrison et al. [33] presented a Biaxial Bias Extension (BBE) test by which the effect of tension on the shear response can be quantified under constant transverse loads. PF experiments with tensioners were presented by Nosrat-Nezami et al [34], showing significant non-linear effects of both uniaxial and biaxial tension on the shear response. The effects of these interactions on the forming behaviour was shown in [35] for rate independent materials by means of FE simulations using an Abaqus user material model.

Considering the contact area between successive plies and between the laminate and the tool, it comes as no surprise that friction has a large effect on the formability of composite laminates. Where friction of dry fabrics may still be described with a rate independent Amonton/Coulomb coefficient of friction, this is not applicable to impregnated fabric composites. In this case, the steady state tool/ply friction can be modelled well by means of hydrodynamic lubrication theories on the meso scale, predicting the macroscopic sliding resistance with the textile geometry, the matrix viscosity and the exerted normal pressure as the input [36]. Ply/ply friction measurements of textile plies in general depict a similar hydrodynamic behaviour. Friction measurements were benchmarked in [37], where only few set-ups were able to characterise thermoplastic laminates in forming conditions.

Sliding friction measurements in general show a clearly time dependent response (see figure 3).

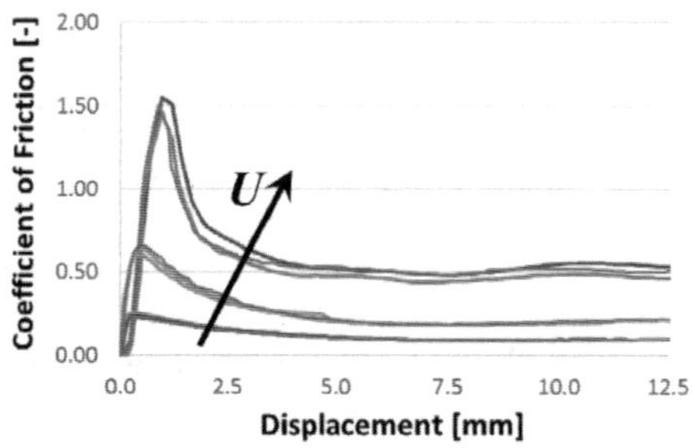

Figure 3. Sliding friction test results for tool/ply contact of 5HS C/PPS at 310°C, 50kPa and displacement rates of 20, 100, 500 mm/min.

The frictional force increases gradually to a steady state value for very low sliding velocities, but shows an (increasing) overshoot as the sliding velocity increases. Both the peak friction and the steady state friction are in the hydrodynamic range for both ply/ply and tool/ply friction, as demonstrated in e.g. [38]. However, a single coefficient of friction is clearly not equal to the transient response as actually measured. Further, the assumption of hydrodynamic lubrication due to flow of the matrix material is obviously not applicable for temperatures below T_m (assuming semi-crystalline matrices). In this case another mechanism applies, that of a solid fibre reinforced polymer sliding on a metallic substrate. Also this situation is temperature dependent (although less pronounced than above T_m), certainly when the polymer crosses the glass transition during the process. Finally, the condition of the tool surface will affect the sliding response below T_m. Careful characterisation is required for accurate process modelling, which needs to be able to account for all relevant details!

Bending of laminates is the third deformation mechanism usually considered. Standard shell elements would describe the behaviour of a laminate as being elastic, i.e. rate independent. As interply slip dominates the laminate bending behaviour, it is obvious that the resin viscosity plays an important role in this mechanism, which hence can be expected to be rate dependent. The common quasi static flexural tests, monitoring the bending of a laminate under its own weight, cannot easily be interpreted to determine such rate dependency. Rheometer bending tests such as proposed in [39] do offer this possibility. The characteristic responses can be captured in a viscoelastic model, essentially leading to a relation between the bending moment, the curvature and the transverse shear rate. Alternatively, a multi-mode nonlinear viscoelastic model can be formulated and fitted on bending experiments [40] to describe the relation between the bending moment and the evolution of the bending angle. As indicated in [41], the obvious approach requires a time dependent Mindlin-type formulation, specific for fibre composite laminates in forming conditions.

As a final remark on the continuously advancing field of constitutive modelling and material characterisation, also the translation from experiment to material property data can be non-trivial. Not every combination of characterisation experiment and constitutive model has a closed form relation between the data measured, with a unique solution ('best fit') for the material property data to be determined. This often calls for nonlinear iterative procedures, either using approximate analytical expressions or more elaborate numerical analyses simulating the experiment in more or less detail. These parameter identification methods (fitting material property data such that simulations of experiments agree with measured force-displacement data) provide a way to determine these data for arbitrary

IOP Conf. Series: Materials Science and Engineering **406** (2018) 012003 doi:10.1088/1757-899X/406/1/012003

constitutive equations, also those including nonlinear temperature and rate dependency. The results of such analyses may, however, depend on the initial guess, as there is often no guaranteed unique solution for the nonlinear equations involved. In general, as in any parameter fitting procedure, interpolation within the measured range is usually acceptable, but extrapolation beyond this range is debatable.

4. Towards defect prediction methods

Considerable progress has been made over the past two decades to enable predictive simulations of thermoplastic composite stamp forming processes with largely reliable outcome, provided the material property data has been measured and modelled accurately and provided that the boundary conditions imposed during the process are properly modelled. This serves primarily the purpose of predicting large processing defects such as wrinkles and folds, spanning multiple elements. Defects of a smaller size are not easily predicted yet, which attracts the attention of various research groups.

In a broad sense, processing induced defects include ply splitting/transverse cracks, matrix degradation, voids, dry spots, delamination, surface defects (due to severe traction or resin squeeze out), contamination, fibre distortions (folds, wrinkles and waviness) and fibre breakage. All defects affected by the local pressure during forming require an accurate pressure prediction, which in turn is strongly dependent on the thickness variation due to shear deformation of the laminate and the local cavity thickness (in particular when using two sided rigid tooling). This is well within range of current modelling capabilities. Second order effects such as tool deflections are outside the range of phenomena than can be predicted with conventional analyses.

Accurate descriptions of some of these defects require full thermomechanical modelling, including the prediction of residual stresses at the meso or micro level as well as the criteria for crack initiation and growth at all temperatures involved in the case of transverse cracks and delaminations. The latter criteria may require simple temperature dependent transverse strength values and/or critical energy release rates (possibly as a function of the consolidation state, fibre angles, …) over the same range from room temperature up to the forming temperature. This is beyond current modelling capabilities. Nevertheless, the current simulation results can already be utilised to check for indications for certain defects. Large transverse strains indicate ply splitting in unidirectional plies, high fibre stresses on rounded edges indicate resin squeeze out (figure 4), the possibility of fibre breakage is clearly indicated by (extremely) high fibre stresses and low pressure is an indicator for poor consolidation quality (including porosity and surface defects, see figure 5).

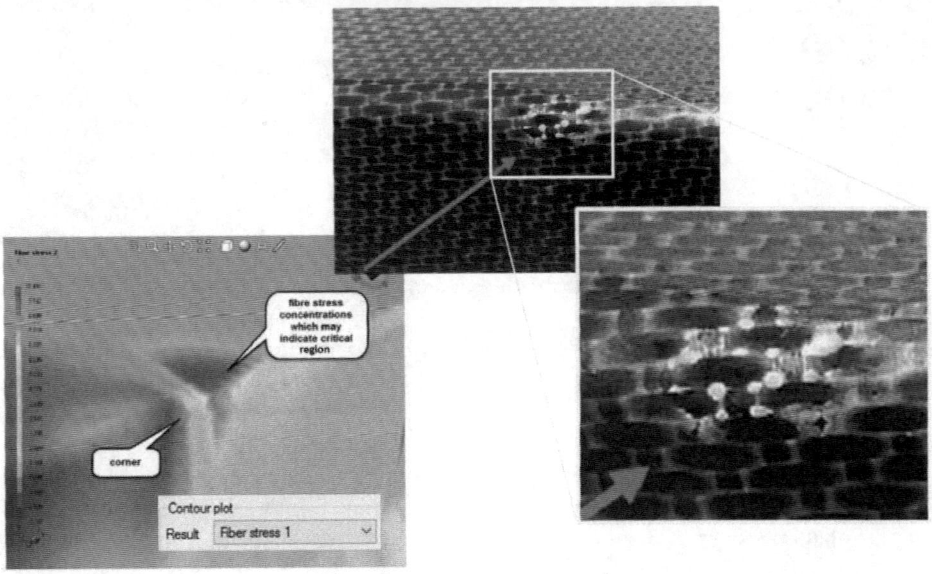

Figure 4. Predicted fibre stresses (left) are an indicator for resin squeeze out (right).

IOP Conf. Series: Materials Science and Engineering **406** (2018) 012003 doi:10.1088/1757-899X/406/1/012003

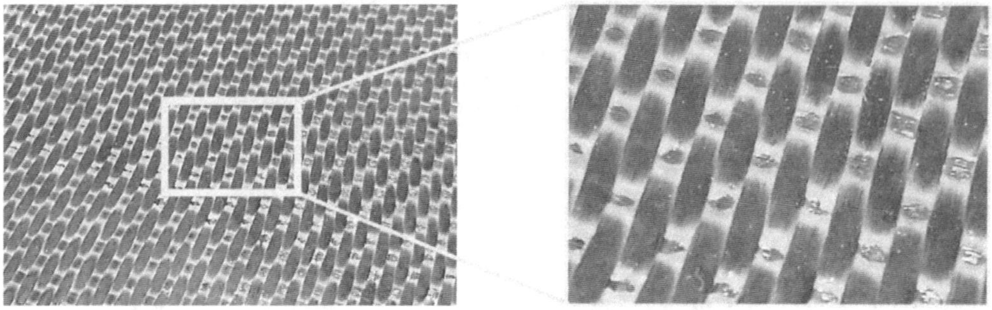

Figure 5. Poor consolidation quality due to insufficient laminate pressure.

The defects having received most attention, nevertheless, are wrinkles of different scales: from fibre waviness in- and out-of-plane to small wrinkles in some plies or in the laminate as a whole up to large folds in the final product. Already in an early stage, Gutowski et al [42] addressed laminate wrinkling during double diaphragm forming as caused by compression when deforming a laminate on a doubly curved tool, and using a mapping based scheme to quantify the amount of compression. Nonlinear FE simulations are well able to predict this kind of buckling phenomena, provided there are sufficient elements to describe the typical shape (see figure 6).

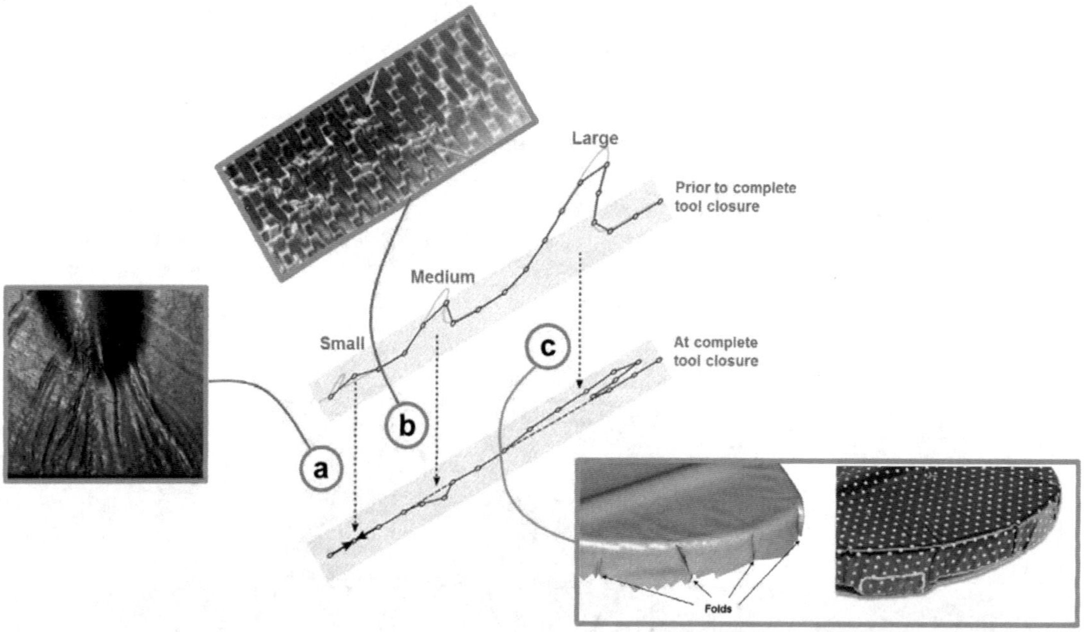

Figure 6. Development of fibre distortion defects of different sizes during tool closure.

Wrinkles smaller than an element can obviously not be described in an ordinary element displacement field ('a' in figure 6). However, the associated compressive stresses may well show up in the final result, which can be used as an indicator showing the risk for such small wrinkles/waviness. To date, to the authors' knowledge, there is not yet a reliable criterion and associated material characterisation technique to predict the waviness of this kind using current FE simulations of the composites forming process. A first attempt towards a Forming Limit Diagram was presented in [46], illustrating the challenges yet to overcome. Slightly larger wrinkles ('b' in figure 6) may disappear in the time step when the tools are totally closed, but which still show up in the actual stamp formed part as inplane waviness. These are easily observed in unidirectional materials, but can also be distinguished fabric composites with some more effort (see the 'Medium' example in figure 6). Simulation results typically show checkerboard patterns of the inplane shear angle when this phenomenon occurs. The larger wrinkles ('c' in figure 6) can be captured fairly well in current FE tools, despite all limitations of the constitutive models employed (see also [43]), and although a sensitivity study in [44] showed that the predicted deformations and defects are 'the result of a delicate balance between the shear, bending, and friction mechanisms considered'. Characterising the material in the right range of deformations and deformation rates may conceal underlying imperfections, such as using an elastic model for the essentially visco-elastic material behaviour during bending.

5. Conclusion

As shown in this paper, composites forming simulations have well advanced over the past decades, leading to accessible 'design for composite manufacturing' software tools which are maturing rapidly, enabling cost-effective composite part development. Ongoing efforts on material property databases will further support this advancement. The next stages in defect prediction methods are challenging, with respect to constitutive modelling, Finite Element formulations, material characterisation techniques and reliable defect indicators. The current research efforts worldwide, however, stimulated by the growth of thermoplastic composite applications, lead to good prospects for continued progress in the years to come.

Acknowledgements

The authors gratefully acknowledges the financial as well as technical support from the industrial and academic members of the ThermoPlastic composites Research Center (TPRC) as well as the support funding from the Province of Overijssel for improving the regional knowledge position within the Technology Base Twente initiative.

References

[1] Mack C and Taylor H (1956), 'The fitting of woven cloth to surfaces', *Journal of Textile Institute*, 47, 477-487.

[2] Robertson RE, Hsiue ES and Yeh GSY (1984), 'Fibre rearrangements during the moulding of continuous fibre composites II', *Polymer Composites*, 5, 191-197.

[3] Smiley AJ and Pipes RB (1988) 'Analysis of the diaphragm forming of continuous fiber reinforced thermoplastics', *Journal of Thermoplastic Composite Materials*, 1, 298-321.

[4] Heisey FL and Haller KD, 'Fitting woven fabric to surfaces in three dimensions', *Journal of the Textile Institute*, 1988, 2, 250-263.

[5] Long AC and Rudd CD (1994) 'A simulation of reinforcement deformation during the production of preforms for liquid moulding processes', *Journal of Engineering Manufacture*, 208, 269-278.

[6] Bergsma O (1995), Three-dimensional Simulation of Fabric Draping, PhD thesis Delft University of Technology.

[7] Trochu F, Hammami A and Benoit Y (1996), 'Prediction of fibre orientation and net shape definition of complex composite parts', *Composites Part A*, 27, 319-328.

[8] Long AC, Clifford MJ, Harrison P and Rudd, CD (2001), 'Modelling of draping and deformation for textile composites', *ICMAC–international conference for manufacturing of advanced composites*, Belfast, IOM Communications, 66-76.

[9] Sharma SB and Sutcliffe MPF (2003), 'Draping of woven fabrics: Progressive drape model', *Plastics, rubber and composites*, *32*(2), 57-64.

[10] Breen D, House D and Wozny M (1994), 'A particle-based model for simulating the draping behavior of woven cloth', *Textile Research Journal*, 64, 663-685.

[11] Cordier F and Magnenat-Thalmann N (2002), 'Real-time animation of dressed virtual humans' *Computer Graphics Forum (Eurographics 2002)*, 21, 327-336.

[12] Chen B, and Govindaraj M (1995), 'A physically based model of fabric drape using flexible shell

theory', *Textile Research Journal*, 65, 324-330.
[13] Kang T and Yu W (1995), 'Drape simulation of woven fabric by using the finite element method', *Journal of the Textile Institute*, 86, 635-648.
[14] Boisse P, Borr M, Buet K and Cheroaut A (1997), 'Finite element simulation of textile composite forming including the biaxial fabric behaviour', *Composites Part B*, 28, 453-464.
[15] Hsiao SW and Kikuchi N (1999), 'Numerical analysis and optimal design of composite thermoforming process', *Computer methods in applied mechanics and engineering*, *177*(1-2), 1-34.
[16] Lin H, Wang J, Long AC, Clifford MJ and Harrison P (2007), 'Predictive modelling for optimization of textile composite forming', Composites Science and Technology, 67(15-16), 3242-3252.
[17] Skordos AA and Sutcliffe MP (2008), 'Stochastic simulation of woven composites forming', *Composites Science and Technology*, *68*(1), 283-296.
[18] Ten Thije RHW, Akkerman R and Huétink J (2007), 'Large deformation simulation of anisotropic material using an updated Lagrangian finite element method', *Computer methods in applied mechanics and engineering*, *196*(33-34), 3141-3150.
[19] Long, AC (Ed.) (2014), Composites forming technologies. Elsevier.
[20] Harrison P, Fernandez Alvarez M, Anderson D (2017), 'Towards comprehensive characterisation and modelling of the forming and wrinkling mechanics of engineering fabrics', *International Journal of Solids and Structures*, accepted for publication.
[21] Cao J, Akkerman R, Boisse P, Chen J, Cheng HS, de Graaf EF, Gorczyca JL, Harrison P, Hivet G, Launay J, Lee W, Liu L, Lomov SV, Long A, de Luycker E, Morestin F, Padvoiskis J, Peng XQ, Sherwood J, Stoilova Tz, Tao XM, Verpoest I, Willems A, Wiggers J, Yu TX and Zhu B (2008), 'Characterization of mechanical behavior of woven fabrics: experimental methods and benchmark results', *Composites Part A: Applied Science and Manufacturing*, *39*(6), 1037-1053.
[22] Sachs U, Akkerman R, Fetfatsidis K, Vidal-Sallé E, Schumacher J, Ziegmann G, Allaoui S, Hivet, G, Maron B, Vanclooster K and Lomov SV (2014) 'Characterization of the dynamic friction of woven fabrics: experimental methods and benchmark results', *Composites Part A: Applied Science and Manufacturing*, *67*, 289-298.
[23] Willems A, Lomov SV, Verpoest I, Vandepitte D, Harrison P and Yu WR (2008), 'Forming simulation of a thermoplastic commingled woven textile on a double dome', *International Journal of Material Forming*, *1*(1), 965-968.
[24] Khan MA, Mabrouki T, Vidal-Sallé E and Boisse P (2010), 'Numerical and experimental analyses of woven composite reinforcement forming using a hypoelastic behaviour. Application to the double dome benchmark', *Journal of materials processing technology*, *210*(2), 378-388.
[25] Peng X and Rehman ZU (2011), 'Textile composite double dome stamping simulation using a non-orthogonal constitutive model', *Composites Science and Technology*, *71*(8), 1075-1081.
[26] Rietman B, Haanappel SP, ten Thije RHW and Akkerman R. (2012), 'Forming simulation sensitivity study of the double-dome benchmark geometry', In *Key engineering materials* (Vol. 504, pp. 301-306), Trans Tech Publications.
[27] Harrison P, Gomes R and Curado-Correia N (2013), 'Press forming a 0/90 cross-ply advanced thermoplastic composite using the double-dome benchmark geometry', *Composites Part A: Applied Science and Manufacturing*, *54*, 56-69.
[28] Soulat D, Cheruet A and Boisse P (2006), 'Simulation of continuous fibre reinforced thermoplastic forming using a shell finite element with transverse stress', *Computers & structures*, *84*(13-14), 888-903.
[29] Slange TK, Warnet LL, Grouve WJB and Akkerman R (2018), 'Deconsolidation mechanisms during stamp forming of C/PEEK blanks, *Composites Part A: Applied science and manufacturing*, accepted for publication.
[30] Boisse P, Hamila N, Guzman-Maldonado E, Madeo A, Hivet G and Dell'Isola F (2017), 'The

bias-extension test for the analysis of in-plane shear properties of textile composite reinforcements and prepregs: a review', *International Journal of Material Forming*, *10*(4), 473-492.

[31] Ten Thije RHW, Loendersloot R and Akkerman R (2005), Drape simulation of non-crimp fabrics, in *ESAFORM 2005: 8th ESAFORM Conference on Material Forming*, The Publishing House of the Romanian Academy.

[32] Akkerman R (2018) A continuum mechanics analysis of shear characterisation methods, *Composites Part A: Applied Science and Manufacturing*, *109*, 131-140.

[33] Harrison P, Abdiwi F, Guo Z, Potluri P and Yu, WR (2012), 'Characterising the shear–tension coupling and wrinkling behaviour of woven engineering fabrics', *Composites Part A: Applied science and manufacturing*, *43*(6), 903-914.

[34] Nosrat-Nezami F, Gereke T, Eberdt C and Cherif C (2014), 'Characterisation of the shear–tension coupling of carbon-fibre fabric under controlled membrane tensions for precise simulative predictions of industrial preforming processes', *Composites Part A: Applied Science and Manufacturing*, *67*, 131-139.

[35] Komeili M and Milani AS (2016), 'On effect of shear-tension coupling in forming simulation of woven fabric reinforcements', *Composites Part B: Engineering*, *99*, 17-29.

[36] Ten Thije RHW, Akkerman R, Ubbink M and Van der Meer L (2011), 'A lubrication approach to friction in thermoplastic composites forming processes', *Composites Part A: Applied Science and Manufacturing*, *42*(8), 950-960.

[37] Sachs U, Akkerman R, Fetfatsidis K, Vidal-Sallé E, Schumacher J, Ziegmann G, Allaoui S, Hivet G, Maron B, Vanclooster K and Lomov, SV (2014), 'Characterization of the dynamic friction of woven fabrics: experimental methods and benchmark results', *Composites Part A: Applied Science and Manufacturing*, *67*, 289-298.

[38] Akkerman R, Ten Thije R, Sachs U and De Rooij M (2010), 'Friction in Textile Thermoplastic Composites Forming', In *Proceedings of the 10th International Conference on Textile Composites-TEXCOMP* (Vol. 10, pp. 271-279).

[39] Sachs U and Akkerman R (2017), 'Viscoelastic bending model for continuous fiber-reinforced thermoplastic composites in melt', *Composites Part A: Applied Science and Manufacturing*, *100*, 333-341.

[40] Ropers S, Kardos M and Osswald TA (2016), 'A thermo-viscoelastic approach for the characterization and modeling of the bending behavior of thermoplastic composites', *Composites Part A: Applied Science and Manufacturing*, *90*, 22-32.

[41] Boisse P, Colmars J, Hamila N, Naouar N and Steer Q (2018), 'Bending and wrinkling of composite fiber preforms and prepregs. A review and new developments in the draping simulations', *Composites Part B: Engineering*, *141*, 234-249.

[42] Wolthuizen DJ, Schuurman J and Akkerman R (2014), 'Forming limits of thermoplastic composites', In Key Engineering Materials (Vol. 611, pp. 407-414), Trans Tech Publications.

[43] Sjölander J, Hallander P and Åkermo M. (2016), 'Forming induced wrinkling of composite laminates: A numerical study on wrinkling mechanisms', *Composites Part A: applied science and manufacturing*, *81*, 41-51.

[44] Gutowski TG, Dillon G, Chey S and Li H (1995), 'Laminate wrinkling scaling laws for ideal composites', *Composites Manufacturing*, *6*(3-4), 123-134.

[45] Boisse P, Colmars J, Hamila N, Naouar N and Steer Q (2018), 'Bending and wrinkling of composite fiber preforms and prepregs. A review and new developments in the draping simulations', *Composites Part B: Engineering*.

[46] Haanappel SP (2013), *Forming of UD fibre reinforced thermoplastics*, PhD thesis University of Twente.

[47] Haanappel SP, Ten Thije RHW, Sachs U, Rietman B and Akkerman R (2014), 'Formability analyses of uni-directional and textile reinforced thermoplastics', *Composites Part A: Applied science and manufacturing*, *56*, 80-92.

IOP Conf. Series: Materials Science and Engineering **406** (2018) 012004 doi:10.1088/1757-899X/406/1/012004

Stress analysis of 3D textile composites using high performance computing: new insights and challenges

M Keith Ballard and John D Whitcomb

Aerospace Department, Texas A&M University, College Station, TX 77840, USA

Abstract. A finite element analysis (FEA) framework that leverages high-performance computing (HPC) was used to analyze a non-idealized 2x2 orthogonally woven textile model, which was created using VTMS. Since the geometry was not idealized, simple boundary conditions were applied to a large analysis region and an interior subregion was used for investigating the stresses. The variation of the cross-sectional area was investigated, showing that the area varied by 26% within the tows near the top and bottom of the textile model. However, tows near the middle of the textile experienced very little variation in the cross-sectional area. The locations of severe stresses were investigated for two configurations: uniaxial tension along the binder direction and uniaxial tension transverse to the binder direction. It was shown that transverse tension in the tows perpendicular to the load was the most severe stress for both configurations, but the second most severe stress differed for the two configurations. When the load was aligned with the binders' paths, the second most severe type of stress was transverse tension and longitudinal shear within the binders near the region where they traversed through the thickness of the textile. However, when the load was perpendicular to the binders' paths, a severe longitudinal shear stress formed in the wefts.

1. Introduction

Two-dimensional textile composites have been routinely used within the aerospace industry, but three-dimensional textiles have received a much slower adoption due to their challenging characterization and manufacturing. Three-dimensional (3D) textiles use through thickness tows that tie the layers together to offer increased toughness, out-of-plane properties, and impact resistance. Three-dimensional woven composites have been used for some applications, such as blade containment systems and body armor, and they offer the potential for multifunctional applications, such as integrating active cooling, self-healing, or a tunable antenna into the material. [1] [2] [3] [4] However, the complex geometry makes it difficult to create realistic textile models and a valid mesh for finite element analysis (FEA). Additionally, the resulting system of equations can be relatively expensive to solve due to the large number of degrees of freedom required. Consequently, there are relatively few works in the literature on the topic, though some researchers have successfully created models for a unit-cell of a 3D woven composite. [5] [6] [7] In 2012, Drach et. al. used the tool DFMA [8], which is very similar to the one used in this study, to create a non-idealized unit-cell for a 3D textile [5], and that model was later used for a few other studies. [9] In 2017, Ballard et. al. studied a small unit cell of a 3D textile model, characterizing the simulated tow geometry, quantifying the distribution of stress, and investigating the locations of stress concentrations. [10] [11] However, the textile geometry was thinner than the textiles often used in industry, the tow paths were assumed to be 2D when assigning the material properties in the tows, and the unit-cell was so small that the boundary conditions probably introduced artifacts even in the interior of the model. Some groups have used non-standard FEA models, such as the voxel approach used by Green et. al. [6]

Learning from the shortcomings in Ref. [10] and [11], this paper will consider a more realistic, thicker textile. Orthogonally woven binders cross two wefts before traveling through the thickness, sometimes referred to as a 2x2 orthogonal weave. Tow paths will be 3D instead of 2D, and a local coordinate system will be assigned to each node instead of one coordinate system per element. To avoid boundary condition effects, a much larger analysis region will be considered, and a subset of the region will be used for post-processing to ensure artifacts due to the in-plane boundary conditions are avoided. This paper provides a characterization of the variation of the cross-sectional area of the tows and an investigation of stress concentrations for two load configurations, including an explanation of how load is transferred to cause the concentrations.

This paper is separated into two sections. The first section describes the geometry, mesh, boundary conditions, clipped region used for post-processing, and material properties used for this study. This is followed by the results section, which quantifies the variation of the cross-sectional area in the tows and explores the locations and causes of stress concentrations.

2. Finite Element Model

This section begins with a description of the method used to generate the textile geometry. Next, an overview of the steps used to create a conforming mesh and a description of the resulting mesh is provided. The boundary conditions for the two configurations are given. To avoid boundary effects, the boundary conditions are applied to a larger analysis region, and the analysis region is clipped down to a subregion, which will be used for all discussion in the results section. Finally, the material properties used for each constituent are presented.

2.1. Textile Geometry

A software tool called VTMS was used to create the geometry for a 3D textile. The Air Force Research Lab (AFRL) developed VTMS for modeling the effects of processing on the tow geometry, allowing the generation of realistic textile models. It models the effects of processing through the use of "digital chains", which consist of rigid spheres along an elastic wire. Ideally, a digital chain should be used for each fiber within the tow, but this is currently computationally prohibitive. Initially, the digital chains have a lot of space between them. The displacements at the ends of the digital chains are specified to be zero, and a contact analysis is used to solve for the final configuration. During the contact analysis, a tension is specified within the digital chains and a pair of rigid planes is used to compress the top and bottom of the model. The contact analysis prevents overlap between the rigid spheres of the digital chains, and tension in the chains encourages them to move towards the midplane. It is useful to begin with one digital chain per a tow until the unrealistic space between the tows is removed, and afterwards, the single digital chains are replaced with more digital chains. Each single digital chain was replaced with ten digital chains for this model. With multiple digital chains per a tow, the cross-section of the tows can deform during the contact analyses. After compaction and the final configuration is reached, cross-sectional profiles are created along each tow path by fitting a closed spline around the bundle of digital chains that define the tow cross-section for each sample point along the tow path. Typically, between two and five hundred cross-sectional profiles are sampled for each tow. Once a solid geometry has been created for each tow, interpretations between the tows are removed by adjusting the geometry of each tow. The algorithm used in VTMS to remove interpenetrations is not precise enough to allow the creation of a conforming, non-intersecting volume mesh for each tow. Consequently, the cross-sections for each tow are shrunk towards the centroid by a small amount, resulting in some matrix material between each tow. For the final steps in VTMS, the surface geometry of each tow is discretized and clipped to a smaller region to avoid the spurious effects near the edges of the initial model.

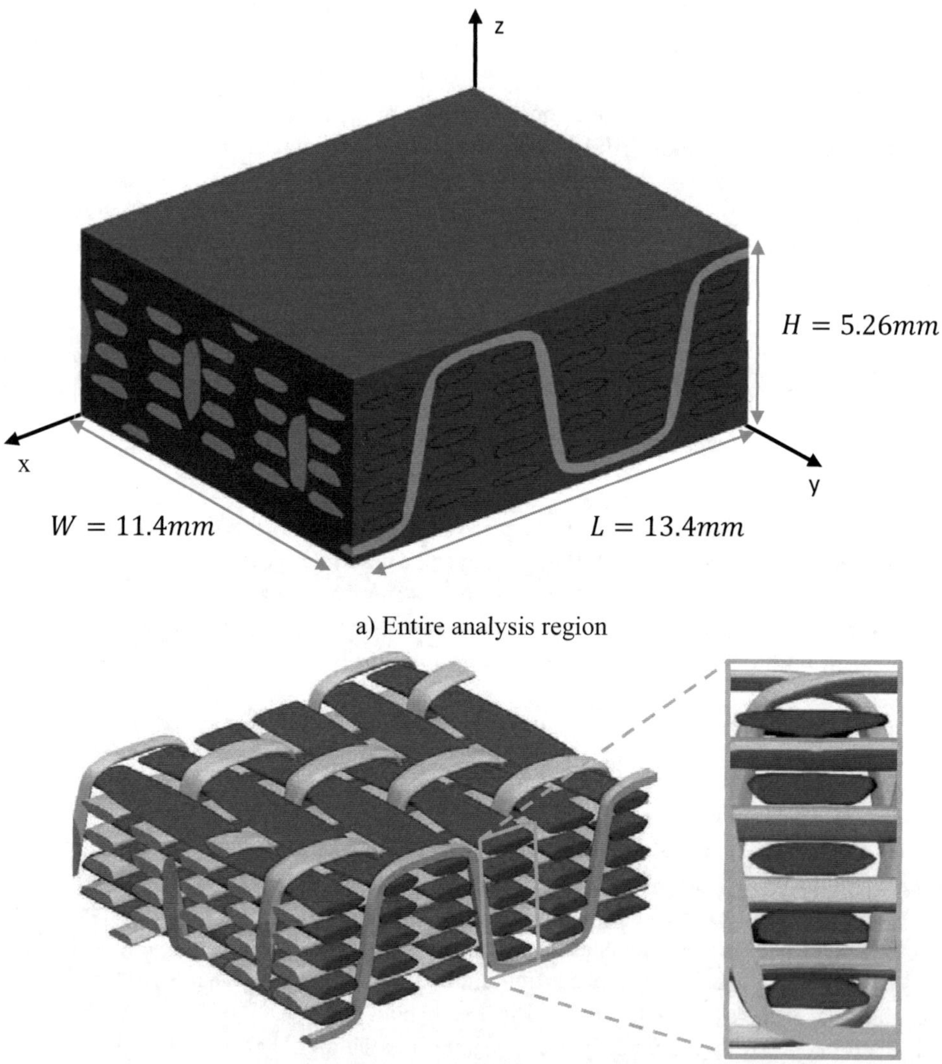

a) Entire analysis region

b) Analysis region with the matrix removed to show tows

Figure 1. Dimensions of 3D textile model (units shown in meters)

Figure 1a shows the resulting model with the matrix added, along with the dimensions of the model. The model consists of four layers of warps and five layers of wefts through the thickness, and the binders are in a 2x2 pattern where they cross two wefts at the top or bottom of the weave before traversing through the thickness. Figure 1b shows the tow architecture more clearly by removing the matrix and highlights the

IOP Conf. Series: Materials Science and Engineering **406** (2018) 012004 doi:10.1088/1757-899X/406/1/012004

variation of cross-section shapes within the tows. Since VTMS uses a rigid mold to compact the textile geometry, the outermost tows experience more deformation than the tows nearest the midplane, which is shown in Figure 1b. The extreme deformation of the outermost tows is reduced by compacting one layer at a time, but the outermost tows will still experience more deformation than the interior tows when the binders are compacted. Creating realistic 3D textile models remains an important challenge for the research community. Tools like VTMS and a very similar software called DFMA provide a method for producing non-idealized textile geometries, which is a step in the right direction, but significant work remains to ensure that the geometries closely resemble physical specimens.

2.2. Mesh

This work uses a standard FEA formulation, which requires a conforming mesh. The VTMS software includes meshing for the independent mesh method [12], but the resulting mesh is not usable herein because it is nonconforming. Instead, an in-house tool was used to create refined surface meshes of the tow geometries. The surface meshes were then used to create volume meshes of the tows and matrix using quadratic tetrahedrals, leveraging a general tetrahedral meshing library called TetGen [13]. Figure 2 illustrates the typical mesh refinement. Tows generally have 5 to 10 quadratic tetrahedrals through the thickness of the tows. A higher mesh refinement exists near binder tows. The final mesh used for this paper consists of 20 million nodes and 15 million quadratic tetrahedrals. It should be noted that the surface geometry that is taken from VTMS is faceted, resulting in unrealistic sharp corners between facets. Ideally, the faceted geometry should be smoothed to avoid unrealistic stress concentrations near sharp corners, but this is not done in this study. At the level of mesh refinement used in this study, it is not clear whether the faceted geometry has a significant effect.

The in-house meshing tool uses a cubic spline fit of the tow centroids to define each tow path. The material coordinate system is determined at each node within a tow mesh by finding the closest point on the spline of the tow path and using the tangent vector of the spline as the fiber direction, which defines the local x-axis. The local y-axis is assigned to be parallel to the cross product between the global z-axis and the tangent vector. The final local z-axis is defined by the cross product of the local x and y-axes, which defines a right-handed coordinate system.

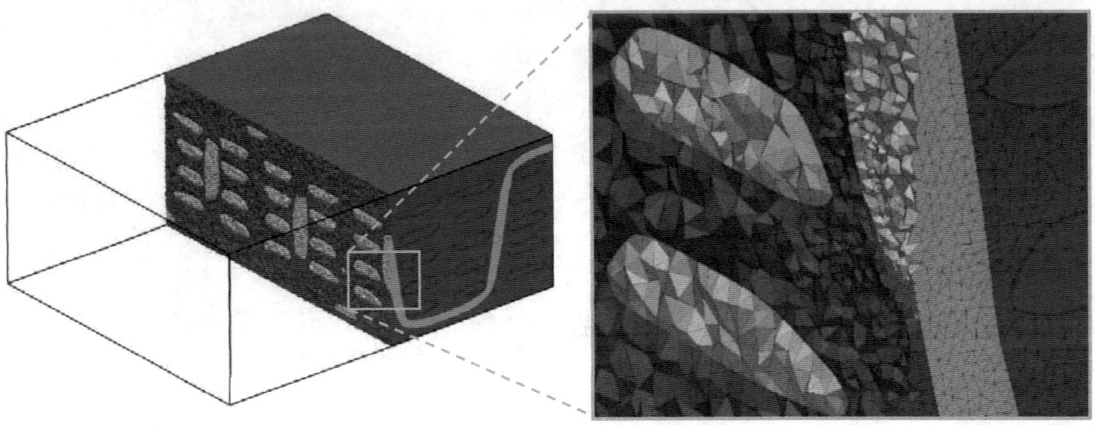

Figure 2. Illustration of mesh refinement

IOP Conf. Series: Materials Science and Engineering **406** (2018) 012004 doi:10.1088/1757-899X/406/1/012004

2.3. Boundary Conditions and Clipped Analysis Region

Ideally, periodic boundary conditions would be applied to a textile unit-cell, but the geometry from VTMS is not periodic since it creates a model by virtually simulating textile processing. Consequently, a larger section of the textile is modeled, simple boundary conditions are applied to the larger analysis region, and a subregion is used for post-processing to reduce artifacts due to the boundary conditions.

Three faces of the full analysis region (the $x = 0, y = 0$, and $z = 0$ planes) are assumed to be planes of symmetry. Refer to the coordinate system in Figure 1a. This results in the following boundary conditions: $u(0, y, z) = 0$, $v(x, 0, z) = 0$, and $w(x, y, 0) = 0$. In addition to these boundary conditions, two loading configurations are considered: uniaxial tension along the global x-axis and uniaxial tension along the global y-axis. The loading consists of specified displacements to result in a 1% volume average strain. Specifically, this requires $u(13.4\, mm, y, z) = 0.134\, mm$ for tension along the x-axis and $v(x, y = 11.4\, mm, z) = 0.114\, mm$ for tension along the y-axis. All other boundaries are traction free.

As mentioned before, the boundary conditions are applied to the full analysis region, and the analysis region is clipped for post-processing to reduce any artifacts near the boundaries that emerge from non-periodic boundary conditions. Figure 3 shows the full analysis region and the subset of the region used for post-processing. The clipped region used for post-processing is only approximately a full unit-cell, since the textile geometry is non-periodic. The in-plane boundaries the of the clipped analysis region are not planar since the tetrahedral elements are not cut by the clipping process. Any element with a node within the clipped analysis region is kept, while all others are removed, resulting in a rough surface on each clip plane, which can be observed in Figure 3b or Figure 3c.

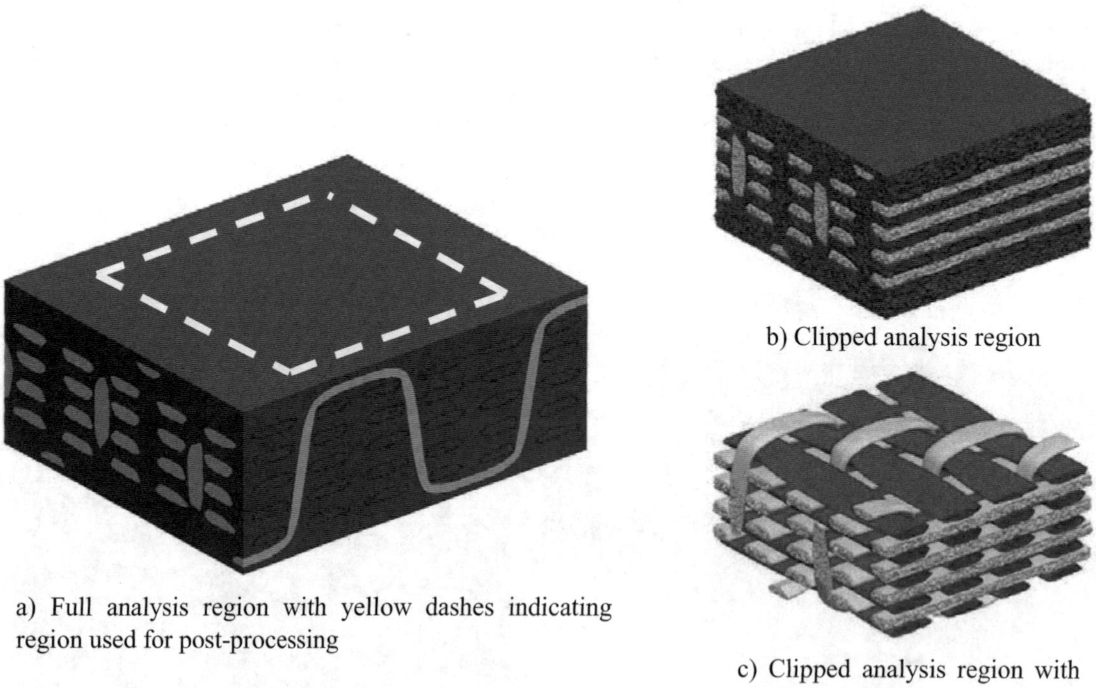

a) Full analysis region with yellow dashes indicating region used for post-processing

b) Clipped analysis region

c) Clipped analysis region with matrix removed

Figure 3. Illustration of the full analysis region and clipped region used for post-processing

2.4. Material Properties

For this study, the tows are assumed to consist of 60% IM7 graphite fiber and 40% 5220-4 epoxy by volume. An ensemble of random fiber/matrix models was used to predict the effective tow properties, similar to the method described in Ref. [14]. The material properties for the tows and neat matrix are shown in Table 1. Since this is an early study into the stress distributions of 3D textile models, linear elasticity was assumed.

Table 1. Material properties

IM7/5220-4 Tows	
E_1 (GPa)	167
E_2 (GPa)	10.3
G_{12} (GPa)	5.42
G_{23} (GPa)	3.32
ν_{12}	0.31
ν_{23}	0.54

5220-4 Epoxy Matrix	
E (GPa)	3.45
G (GPa)	1.28
ν	0.35

3. Results

The results of this paper are organized into two sections. The first section characterizes the tow architecture, and the second section investigates the stress distributions within the tows.

3.1. Characterization of Tow Architecture

Since the textile geometry is the result of simulated processing, the cross-section of the tows can vary along the tow paths due to a variety of reasons. During compaction, the cross-sectional area is likely to be reduced where tows contact other tows. Additionally, the shape of the tows can vary significantly when it changes direction, such as when the binder begins or ends traveling through the thickness of the textile. Characterizing the variation of the cross-sections of the tows is a first step toward understanding how closely the model resembles actual specimens. This section quantifies the variation of cross-sectional area and explains how aspects of VTMS's method for creating textile models led to the variations.

For the model considered in this paper, adjacent binders have a different phase of undulation through the textile, but every fourth binder has the same phase. Every other binder will traverse through the thickness of the textile at the same gap between weft tows, and so a pair of such binders is more likely to have similar cross-sectional areas. Figure 4 shows the variation of the cross-sectional area of the first and third binder, which are shaded in the figure as green and purple and labeled 1 and 3. The cross-sectional area is maximum where the binders are midway through the thickness of the textile, since these regions are not affected significantly by compaction, resulting in the area at the midpoint through the thickness remaining almost unchanged throughout the compaction process. The minimum occurs where the binders transition to and from traveling through the thickness. The low cross-sectional area at these points are due to several factors, such as the lower radius of curvature and the tension in the binders causing the binders to strongly contact neighboring wefts. Additionally, it was observed that digital chains tend to pass through each other in these regions, resulting in significant tow interpenetrations. It is unclear which factors are most significant, but the cross-sectional area decreases by 19% to 24% as the binders transition into traveling through the thickness, such as at point A in Figure 4. Where the binders are on the top or bottom of the textile, the cross-sectional area varies by at most 18%. It is unclear if this much variation is realistic or merely an artifact of the method used by VTMS to create the textile geometry.

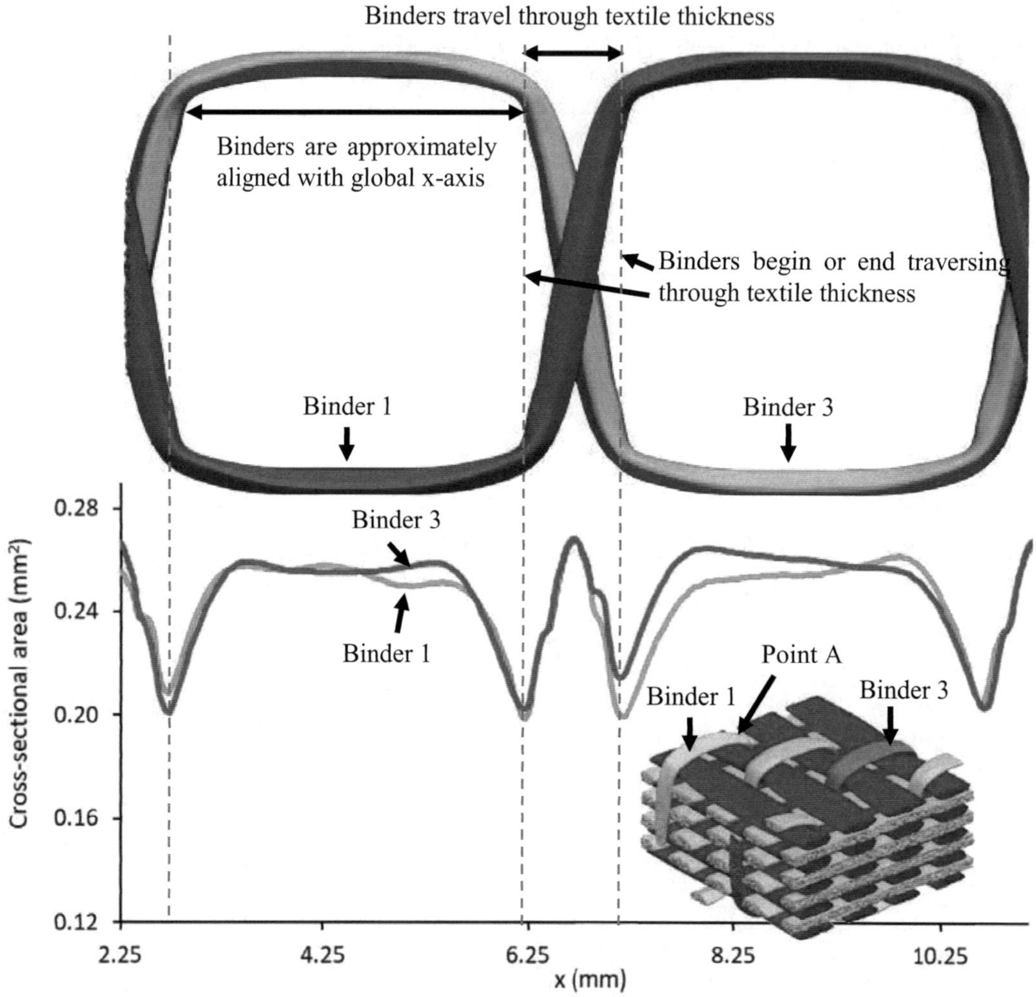

Figure 4. Variation of the cross-sectional area in the first and third binders

Figure 5 shows the variation of the cross-sectional area in the warp tows. Figure 5a shows that the cross-sectional area for the inner two layers of warps (green) remains relatively constant, varying at most by 4.5%. However, Figure 5b shows that the warps on the top and bottom of the textile experience significant changes in cross-sectional area. There are two types of variation of interest. One type of variation is that the cross-sectional area of a single tow changes along the tow path, which for the outer layer of warps is at most 11%. Another type of variation is that the cross-sectional area varies between tows in the same layer, which is at most 15% for the model presented in this paper. Both types of variations will exist in actual specimens, but comparisons to MicroCT scans would be needed to quantify the accuracy of the model geometry. However, these results do show that the cross-sectional area of tows in a model created by VTMS largely depend on the distance from the boundaries where the compaction planes contact the model. The deformation due to the rigid planes at the top and bottom boundaries does not cause significant deformation well into the interior of the model. Tows near the boundaries will be highly deformed, while tows in the interior will experience little deformation. The variation of the cross-sectional area in the wefts were not shown for conciseness, but they follow the same trend. The outermost layers varied by up to 26%, while the innermost layer varied by less than 2%.

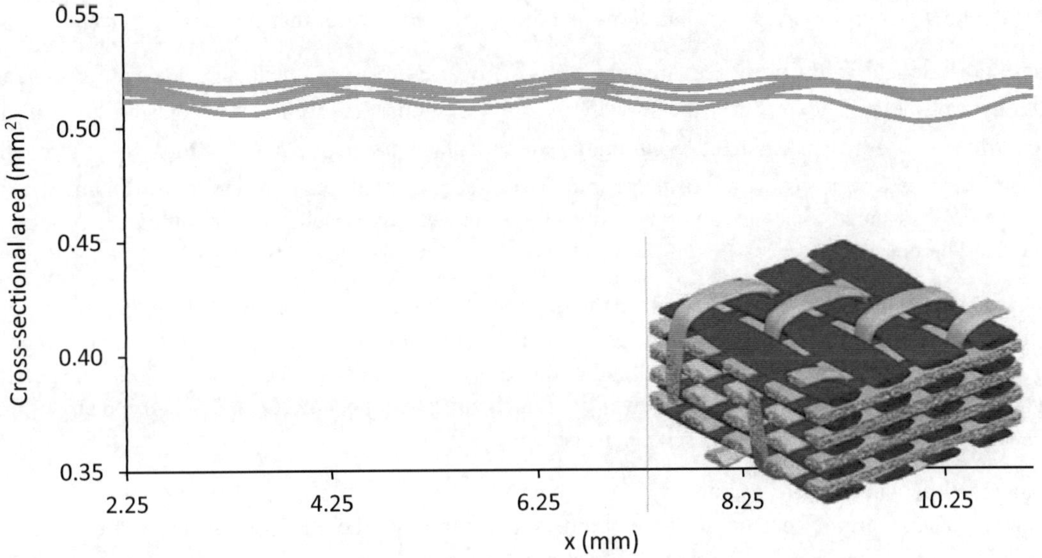

a) Interior warp tows

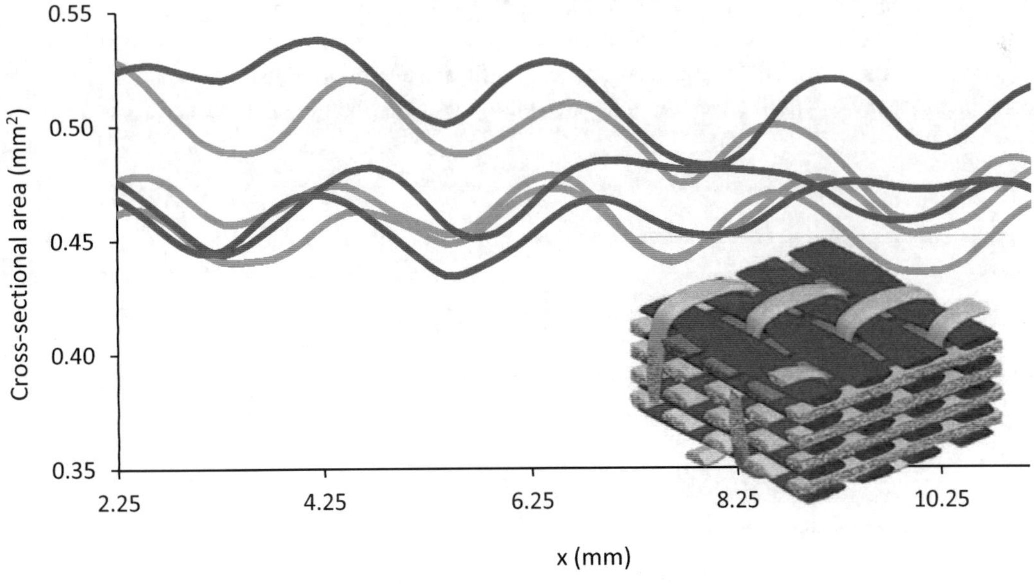

b) Top and bottom layers of warps

Figure 5. Variation of the cross-sectional area in the warps

IOP Conf. Series: Materials Science and Engineering **406** (2018) 012004 doi:10.1088/1757-899X/406/1/012004

3.2. Analysis of a 3D Textile Subjected to Uniaxial Tension

The 3D textile model was subjected to two types of load: uniaxial tension along the global x-axis and uniaxial tension along the global y-axis. This paper assumes linear elasticity, so the behavior of progressive damage cannot be understood from these results, but for each load configuration, this section aims to investigate the locations of stress concentrations and the local load transfer that caused the concentrations.

Stresses in the global coordinate system are denoted as σ_{ij}, while stresses in the local coordinate system are denoted by a prime, σ'_{ij}. To quantify the severity of stresses within the tows, components of stress in the local coordinate system are normalized by a nominal strength and denoted by a hat, $\hat{\sigma}'_{ij}$. Ideally, the strengths would be based on a series of experiments or microscale damage analyses. However, for this paper, the strengths will be assumed to be the same as the tows tested in Ref. [6], which have similar elastic properties to those used herein.

3.2.1 Tension Along X-Axis. Figure 6 shows the three most severe normalized stresses within the clipped analysis region for an applied volume average strain along the global x-axis, $\langle \varepsilon_{xx} \rangle$, of 1%. The stresses shown are in the local coordinate system, and each stress component is normalized by the respective strength, which provides a measure of the severity of each stress component. Overall, Figure 6 shows that wefts and binders experience the most severe stresses.

The binders experience severe σ'_{zz} and σ'_{xz} components of stress, as shown in Figure 6b and Figure 6c respectively. The severe σ'_{zz} occurs where the binders travel through the thickness of the textile model, as shown in Figure 6b. To show the locations of stress concentrations more clearly, Figure 7 shows $\hat{\sigma}'_{zz}$ within an x-z slice of one of the binders. In this region, the local z-axis almost aligns with the global x-axis, which is the direction of the load, refer to Figure 7b. Additionally, there are no wefts in the region where the binders travel through the thickness of the textile. Consequently, the binders take on much of the load via tension along the local z-axis. The peak stresses occur where wefts are nearby to transfer the load to the binders. This component of stress will likely cause transverse matrix cracking within the binders.

IOP Conf. Series: Materials Science and Engineering **406** (2018) 012004 doi:10.1088/1757-899X/406/1/012004

a) $\hat{\sigma}'_{yy}$

b) $\hat{\sigma}'_{zz}$

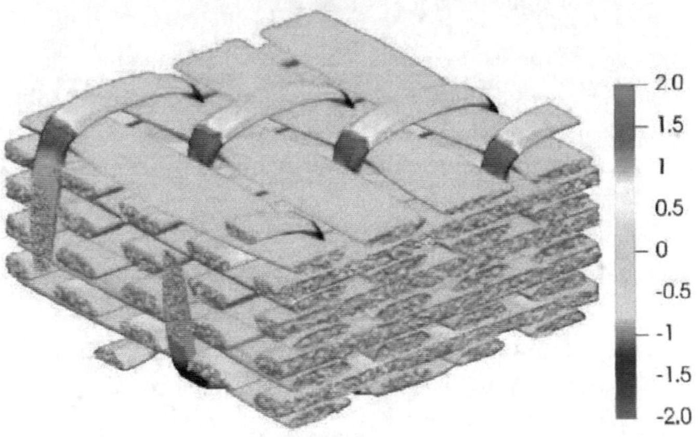

c) $\hat{\sigma}'_{xz}$

Figure 6. Contours of three most severe stress components in the local coordinate system normalized by the respective strength for uniaxial tension along the global x-axis

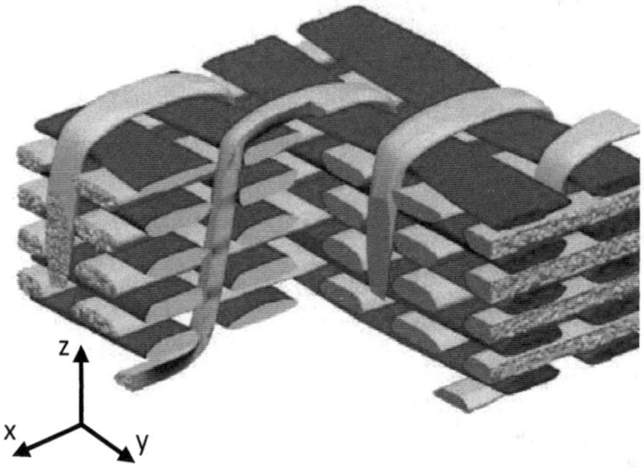

a) Location of x-z cross-section

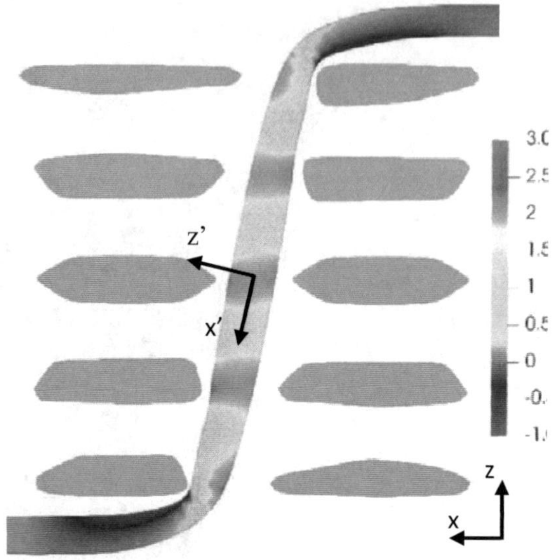

b) Normalized transverse normal stress $\hat{\sigma}'_{zz}$

Figure 7. $\hat{\sigma}'_{zz}$ contours for an x-z cross-section of a selected binder illustrating locations of severe σ'_{zz} in binders

In addition to the severe transverse tension, the binders experience a severe shear stress, σ'_{xz}, where they begin and finish traveling through the thickness of the textile, as shown in Figure 6c. The shear stress develops to maintain equilibrium as the binders shift between carrying a significant amount of load at the top and bottom of the textile, where the fibers of the binders are aligned with the load direction, to carrying much less load as they travel through the thickness of the textile.

Though the binders experience severe stresses, the wefts experience the most severe stress, namely $\hat{\sigma}'_{yy}$, within the textile, as shown in Figure 6a. For the wefts, the local y-axis is closely aligned with the global x-axis, which is the direction of the applied load for this configuration, so it is expected that the transverse tension in the wefts will be severe. However, the severity of the transverse tension varies dramatically depending on the location within the wefts. Figure 8 shows the local and global stress components of interest for an x-z cross-section centered on a binder tow, which is illustrated in Figure 7a. Figure 8a shows the $\hat{\sigma}'_{yy}$ contours in the wefts and binder in the x-z cross-section. In most locations, $\hat{\sigma}'_{yy}$ remains between 2 and 3. At some points when the wefts come close to a binder, $\hat{\sigma}'_{yy}$ reaches values near 4.5, such as point A in Figure 8a, but at other points near a weft tow, no stress concentration occurs, such as point B in Figure 8a. The common factor for areas of elevated stress is a sharp cross-sectional shape of the weft. When the cross-section is similar to a rectangle with rounded corners, there is no significant stress concentration, but when the cross-section comes to sharper point, a significant stress concentration occurs. The sensitivity of stress concentrations to the tow cross-sectional shape emphasizes the importance of creating textile models with realistic tow geometries.

IOP Conf. Series: Materials Science and Engineering **406** (2018) 012004 doi:10.1088/1757-899X/406/1/012004

There is one situation where the tow shape is less prone to cause a stress concentration for this configuration: where a binder crosses over a weft tow. In this region, $\hat{\sigma}'_{yy}$ in the weft drops to about 0, such as point C in Figure 8a. The drop in transverse tension is due to a compressive σ_{xx} (global) in the region where a binder crosses the wefts at the top or bottom of the textile, such as region D highlighted in Figure 8b. The compressive stress forms as load is transferred to or from the binder via shear, since the binder is orders of magnitude stiffer in the global x-direction than the surrounding material at the top and bottom.

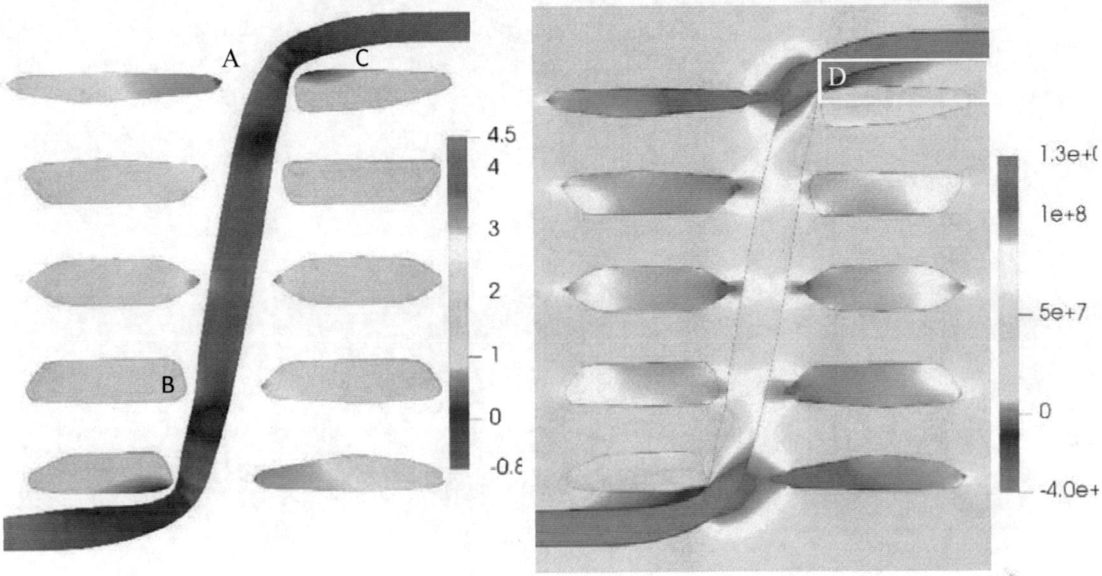

a) Local normalized transverse normal stress $\hat{\sigma}'_{yy}$ b) Global σ_{xx}

Figure 8. An x-z cross-section illustrating locations of severe local transverse tension in wefts

3.2.2 Tension Along Y-Axis Figure 9 shows three most severe normalized stresses within the clipped analysis region for an applied volume average strain along the global y-axis, $\langle \varepsilon_{yy} \rangle$, of 1%. Again, the normalized stresses shown are in the local coordinate system. Overall, Figure 9 shows that the warps and binders experience severe transverse tension ($\hat{\sigma}'_{yy}$), while the wefts experience severe longitudinal shear ($\hat{\sigma}'_{xy}$ and $\hat{\sigma}'_{xz}$).

The most severe stress in the textile is $\hat{\sigma}'_{yy}$ in the binders and warps. For both types of tows, the local y-axis is closely aligned with the global y-axis, which is the direction of the load for this configuration. To illustrate where the peak stresses occur in the textile, Figure 10 shows $\hat{\sigma}'_{yy}$ for one binder and row of warp tows, along with the local coordinate system for a point in the binder. The figure shows that the stress concentrations form where warps and binders come closest, such as point A in Figure 10. Additionally, stress concentrations form where the binder begins or ends traveling through the thickness of the textile, such as points B in Figure 10.

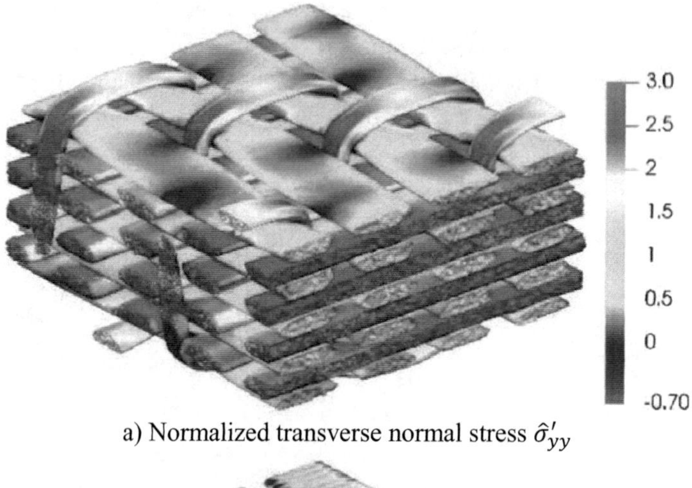

a) Normalized transverse normal stress $\hat{\sigma}'_{yy}$

b) Normalized longitudinal shear stress $\hat{\sigma}'_{xy}$

c) Normalized longitudinal shear stress $\hat{\sigma}'_{xz}$

Figure 9. Contours of each stress component in the local coordinate system normalized by the respective strength for the case of uniaxial tension along the global y-axis

In summary, the transverse normal stress $\hat{\sigma}'_{yy}$ was the most severe component for both configurations. In both cases, the peak $\hat{\sigma}'_{yy}$ occurred within the tows that are perpendicular to the load. However, when the load is along the global x-axis, the stress concentrations only form within the wefts, but when the load is along the y-axis, the stress concentrations form within the binders and warps. Additionally, the stress concentrations are more severe for the case of tension along the x-axis.

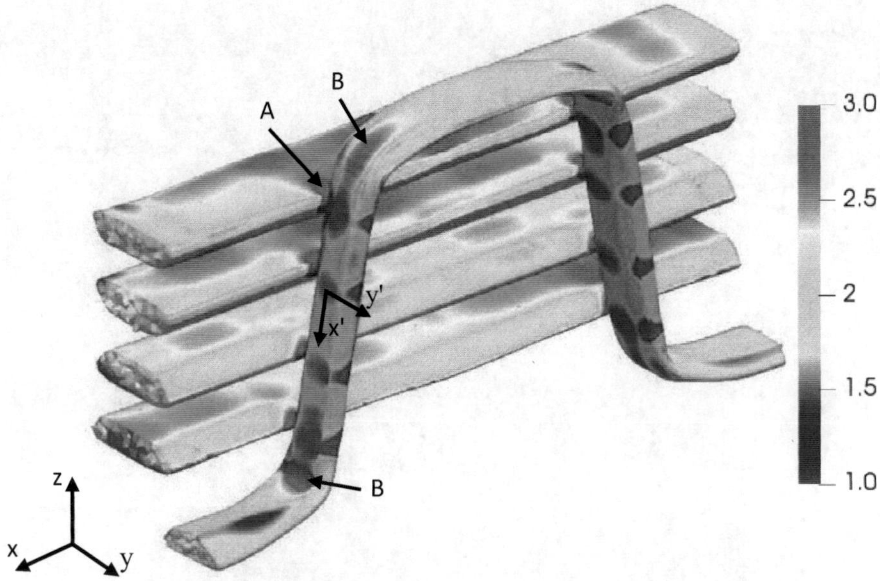

Figure 10. $\hat{\sigma}'_{yy}$ for a selected binder and row of warp tows

The second most severe type of stress in the textile is the longitudinal shear stresses, $\hat{\sigma}'_{xy}$ and $\hat{\sigma}'_{xz}$, within the wefts. To more clearly show the locations of severe shear stress, Figure 11 shows the magnitude of the normalized longitudinal shear stress, $\hat{\sigma}'_{s} = \left(\hat{\sigma}'^{2}_{xy} + \hat{\sigma}'^{2}_{xz}\right)^{1/2}$, for the volume of wefts with a value of $\hat{\sigma}'_{s} >$ 1.2. The entire weft tows are shown semi-transparently for context of where stress concentrations occur. Figure 11 shows that the severe shear stresses in the wefts occur in the top and bottom layers of tows and near locations where binders cross the wefts. It should be noted that the stress concentrations do not occur uniformly throughout the model due to variations in the textile geometry and mesh refinement.

In summary, the second most severe stress for both configurations was the longitudinal shear stress, but the locations of severe shear stress differed for the two types of loads. When the load was along the global x-axis, the severe shear stresses developed in the binders, but when the load was along the global y-axis, the severe shear stresses developed in the wefts. This indicates that the progression of damage might depend on the direction of load relative to the binders' paths.

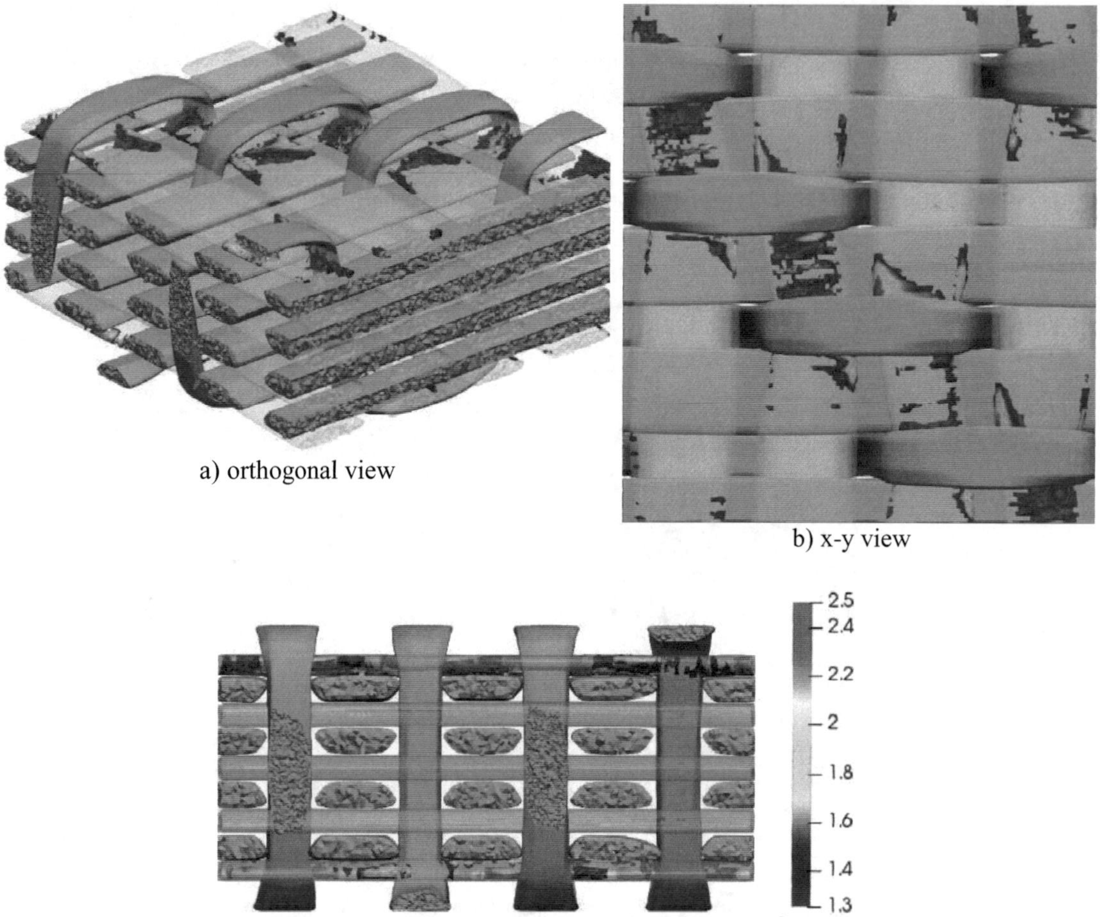

a) orthogonal view

b) x-y view

c) y-z view

Figure 11. Magnitude of normalized longitudinal shear stress, $\hat{\sigma}'_s = \left(\hat{\sigma}'^2_{xy} + \hat{\sigma}'^2_{xz}\right)^{1/2}$, for the volume of wefts with a value of $\hat{\sigma}'_s > 1.2$ (entire weft surfaces are shown as semi-transparent)

4. Conclusions

An in-house FEA code that leverages high-performance computing was used to analyze a large non-idealized 3D textile model with a 2x2 orthogonal weave created using VTMS and an in-house meshing tool. Due to the complexity of 3D textiles, the geometry of the model was characterized by quantifying the variation in the cross-sectional area along the tow paths. It was shown that the cross-sectional area in tows near the top and bottom boundary of the model exhibited large amounts of variation up to 26%. On the other hand, tows near the middle of the weave exhibit relatively little variation. In reality, tows experience some amount of variation in the cross-sections, but about the same number of fibers exist for any slice, which causes the fiber volume fraction to vary along the tow path. The range of variation of cross-sectional area in the model indicates that the variation of fiber volume fraction should be accounted for and could significantly affect the locations of stress concentrations. It will be important for future works to quantify the variation of the cross-sectional shape and compare the variation of tow architecture to actual specimens.

The locations of severe stresses were investigated for two configurations: uniaxial tension along the global x-axis and uniaxial tension along the global y-axis. Under tension in the x-direction, the wefts experienced the most severe stress in the textile, exhibiting a high transverse tension along the local y-axis, σ'_{yy}. The severe σ'_{yy} will likely lead to initial matrix cracking in the wefts. It is important to note that the largest σ'_{yy} in the wefts developed when wefts came close to a binder and the cross-sectional shape of the weft came to a sharp corner. The sensitivity of the stress concentrations to the tow shape highlight the need to create textile models with realistic tow architectures if the progression of damage is to be accurately predicted. Additionally, the binders exhibited a severe transverse tension and longitudinal shear, namely σ'_{zz} and σ'_{xz}, near where they traverse through the thickness of the textile. Unless matrix cracking in the wefts relieves the stress concentrations in the binders, the binders are likely to experience matrix cracking and longitudinal shear failure early in the progression of damage within the textile.

For the case of uniaxial tension along the global y-axis, transverse tension, σ'_{yy} in the warps and binders was the most severe type of stress in the textile. This is very similar to the previous configuration, though the severe stress is in different types of tows. Similarly, transverse matrix cracking is likely to occur in the binders and warps initially. However, the second most severe type of stress is very different from the previous configuration. When the load is transverse to the binders, a severe longitudinal shear stress develops in the wefts. Though the initial type of damage expected matches for the tension along the x- and y-directions, namely transverse matrix cracking, the next most severe types of stress are fundamentally different between the two cases due to the presence of binders. Without the binders, the severe shear stress is not expected to develop, showcasing the need for understanding what role binders play in distributing load within complex 3D textile geometries.

Creating realistic 3D textile models remains a challenge for the community. This paper took a step forward by using a large non-idealized model, a refined mesh that required the use of high-performance computing, clipping the analysis region to avoid boundary effects, and investigating the locations of stress concentrations with more detail. However, it was also shown that the shape of tows can be important, the variation of the fiber volume fraction should be accounted for, and understanding the role of the binders requires further study.

Acknowledgements

This work was supported by the Department of Defense (DoD) through the National Defense Science & Engineering Graduate Fellowship (NDSEG) Program. In addition, the authors acknowledge the Texas A&M Supercomputing Facility (http://sc.tamu.edu/) for providing computing resources used in conducting the research reported in this paper.

References

[1] A. Esser-Kahn, P. Thakre, H. Dong, J. Patrick, V. Vlasko-Vlasov, N. Sottos, J. Moore and S. White, "Three-Dimensional Microvascular Fiber-Reinforced Composites," *Advanced Materials,* vol. 23, p. 3654–3658, 2011.

[2] R. Gibson, "A review of recent research on mechanics of multifunctional composite materials and structures," *Composite Structures,* vol. 92, pp. 2793-2810, 2010.

[3] D. Hartl, G. Huff, H. Pan, L. Smith, R. Bradford, G. Frank and J. Baur, "Analysis and Characterization of Structurally Embedded Vascular Antennas Using Liquid Metals," in *SPIE Sensors and Smart Structures Technologies for Civil, Mechanical, and Aerospace Systems*, Las Vegas, NV, 2016.

[4] S. Rawal, D. Barnett and D. Martin, "Thermal Management for Multifunctional Structures," *IEEE Transactions on Advanced Packaging,* vol. 22, no. 3, pp. 379-383, 1999.

[5] A. Drach, B. Drach and I. Tsukrov, "Processing of fiber architecture data for finite element modeling of 3D woven composites," *Advances in Engineering Software,* vol. 72, p. 18–27, 2014.

[6] S. D. Green, M. Y. Matveev, A. C. Long, D. Ivanov and S. R. Hallett, "Mechanical modelling of 3D woven composites considering realistic unit cell geometry," *Composite Structures,* vol. 118, pp. 284-293, 2014.

[7] S. Soghrati, A. R. Najafi, J. H. Lin, K. M. Hughes, S. R. White, N. R. Sottos and P. H. Geubelle, "Computational analysis of actively-cooled 3D woven microvascular composites using a stabilized interface-enriched generalized finite element method," *International Journal of Heat and Mass Transfer,* vol. 65, pp. 153-164, 2013.

[8] G. Zhou, X. Sun and Y. Wang, "Multi-chain digital analysis in textile," *Journal of Composites Science and Technology,* vol. 64, pp. 239-244, 2003.

[9] I. Tsukrov, B. Drach, H. Bayraktar and J. Goering, "Modeling of Cure-Induced Residual Stresses in 3D Woven Composites of Different Reinforcement Architectures," *Key Engineering Materials,* Vols. 577-578, pp. 253-256, 2014.

[10] M. K. Ballard, J. S. McQuien and J. D. Whitcomb, "Analysis of three-dimensional woven composites," in *21st International Conference on Composite Materials*, Xi'an, 2017.

[11] M. K. Ballard and J. D. Whitcomb, "Prediction of Tow Architecture and Stress Distributions for a 3D Woven Composite," in *American Society of Composites*, West Lafayette, Indiana, 2017.

[12] E. V. Iarve, D. H. Mollenhauer, E. G. Zhou, T. Breitzman and T. J. Whitney, "Independent mesh method-based prediction of local and volume average fields in textile composites," *Composites Part A: Applied Science and Manufacturing,* vol. 40, no. 12, pp. 1880-1890, 2009.

[13] H. Si, "TetGen, a Delaunay-Based Quality Tetrahedral Mesh Generator," *ACM Transactions on Mathematical Software,* vol. 41, no. 2, 2015.

[14] M. K. Ballard, W. R. McLendon and J. D. Whitcomb, "The influence of microstructure randomness on prediction of fiber properties in composites," *Journal of Composite Materials,* vol. 48, no. 29, pp. 3605-3620, 2014.

IOP Conf. Series: Materials Science and Engineering **406** (2018) 012005 doi:10.1088/1757-899X/406/1/012005

How to evaluate the fatigue life of textile composite structures by the numerical analysis and the experimental data with scatter

Masaru Zako[1,2]
[1] *Graduate School of Engineering, Osaka University, Osaka, Japan*
[2] *Expert adviser of ITOCHU Techno-Solutions Corporation, Japan*

zako@mit.eng.osaka-u.ac.jp

Abstract. The fatigue test data and Goodman diagram have employed in order to evaluate the safety for a long period of time. Since the discrepancies exist in the fatigue test data, a modified Goodman diagram with a failure probability and a confidence level has been proposed in this paper. The confidence level is a function of total number of test data. The proposed procedure has an advantage which can be applied to a small number of specimens. As a numerical example, a hydrogen gas tank for automobile made of CFRP has been picked up. FEM has been applied to obtain the stress under a pressure in the tank. The in and out of gas to tank is considered as a fatigue phenomenon. The safety of tank for repeated use has been evaluated by the comparison of maximum stress generated in the tank and the modified Goodman diagram with a probability of failure. It is recognized that the proposed procedure is a useful and reasonable method for safety evaluation.

1. Introduction

There are many kinds of fabrics, matrices and textiles, and lots of combinations of them. As they progress rapidly, many researchers and designers on them have the demand which needs to estimate their properties in a short time. It is not easily to prepare a large number of specimens and to conduct the experiments within a fixed period of time. The designers for textile and fibrous composites have the requirements which obtain not only the mechanical properties but also the characteristic on fatigue within a given period and with a small number of samples. However, it is difficult to obtain the characteristics on fatigue within a short time and to estimate the variation in data with a small number of specimens. On the other hand, ASME (Sec.III) [1] has proposed the code which shifts S-N curve to safety side. ASME Sec.III code consists of two processes. At first, S-N curve is got by the experimental data and the least squares method. A new S-N curve is obtained by shifting 1/2 to the stress amplitude or strain amplitude. And the other S-N curve is also obtained by shifting 1/20 to the number of cycle to failure. Using these two S-N curves, the conservative S-N curve is finally adopted as the design curve. This procedure is very simple and is widely used. It has a problem, however, that the shifting value does not change even if the variation in an experimental data is reduced by a modification or an improvement of production. The scatter of data can evaluate quantitatively with the theory of reliability. The probability of failure is equivalent to the safety factor, and the confidence level is concerned with the number of specimens. If it is possible to apply them to the treatment of fatigue test data, it will be a very useful method for designer to determine S-N curve with probability of failure and confidence level. The object of this study is to propose a procedure of determination of the shifting value based on reliability in order to satisfy designer's requirements.

IOP Conf. Series: Materials Science and Engineering **406** (2018) 012005 doi:10.1088/1757-899X/406/1/012005

2. Probability with confidence level

2.1 Number of samples based on reliability

Figure 1 shows the relation between the number of specimens and the accumulated probability with a confidence level based on Median rank [2]. It is recognized that MIL specs requires 298 specimens and 28 specimens for A and B values, respectively. The confidence level has also a great influence on the number of specimens. Though it is difficult for a small number of specimens to obtain the sufficient probability, the probability may be revised by the employment of Figure 1. Equation (1) shows the relation between the number of specimens and the probability with a confidence level.

$$n = \frac{\ln \gamma}{\ln(1-F)} \tag{1}$$

Where n is the total number of specimens, γ is the significance level and F is the probability of failure.

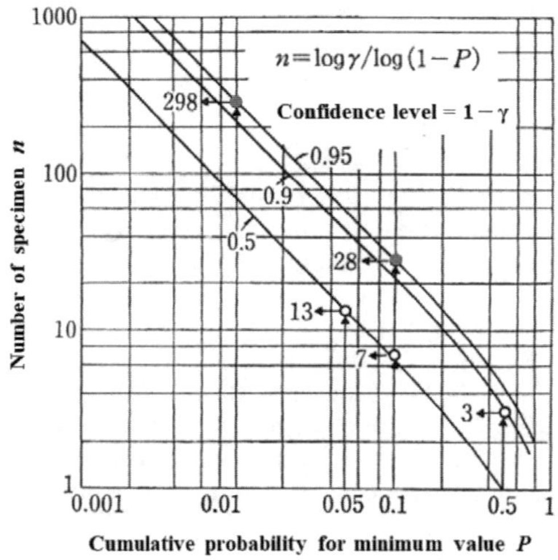

Figure 1. Relation between the number of test specimens and the probability with a confidence level

2.2 Equivalent stress (strain) amplitude

Figure 2 shows the typical example of S-N curves for S25C [3] and GFRP [4] obtained by bending fatigue test. GFRP consists of a glass roving cloth and polyester resin, and the volume fraction of fiber (V_f) is 35%. It is recognized that the test data on stress amplitude have scatter for number of cycles to failure, and that the distribution curves are not the same ones as shown in Figure 2(a). This tendency for S25C was made clear by Nishijima and he published as P-S-N diagram [3]. However, the distribution curves for the stress amplitude on a number of cycles to failure are same. GFRP has the same tendency as shown in Figure 2(b). This fact is very important to the arrangement of data.

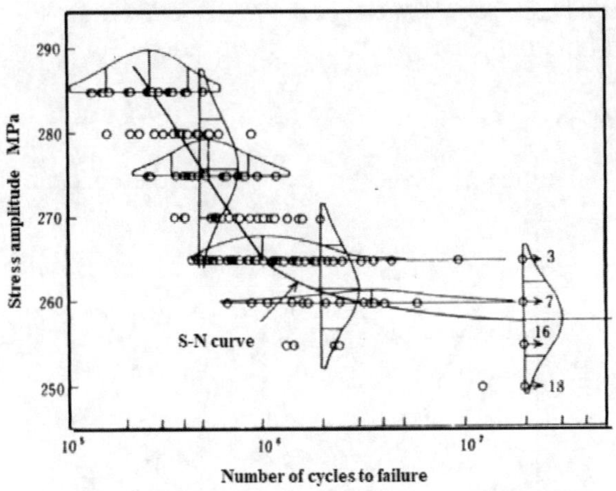

(a) Bending fatigue test result for S25C [3]

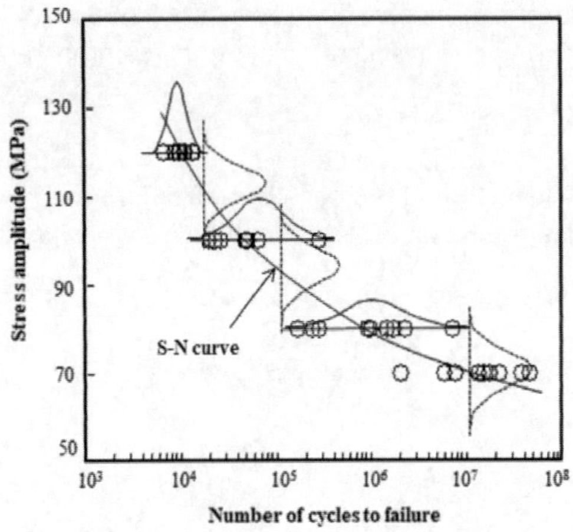

(b) Bending fatigue test result for GFRP [4]

Figure 2. Typical example of bending fatigue test data for S25C and GFRP

Figure 3 shows a procedure how to arrange the experimental data and how to get the distribution curve for the stress amplitude (or the strain amplitude) [5]. The oblique solid lone is an analytical result which obtains from the least squares method. Figure 3(a) shows how to shift the stress amplitude at an arbitrary number of cycles (*Nc*). The stress amplitude (S_1) at a number of cycles (N_1) can convert into the stress amplitude (*Sc*) at an arbitrary number of cycles (*Nc*) by Equation (2).

$$Sc = S1\left(\frac{Nc}{N_1}\right)^{m_0} \tag{2}$$

Where *Sc* is equivalent stress amplitude for the convenience and m_0 is the gradient of line.
Figure 3(a) indicates a logical meaning that the specimen will be broken at *Nc*, if the stress amplitude is changed from *S1* to *Sc*. Figure 3(b) shows how to gather the experimental data in a number of cycles (*Nc*) in order to analyze the distribution curve of equivalent stress amplitude.

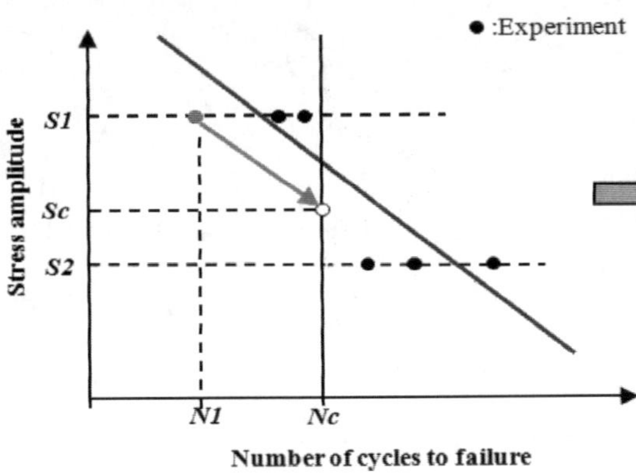

(a) Shifting procedure of data

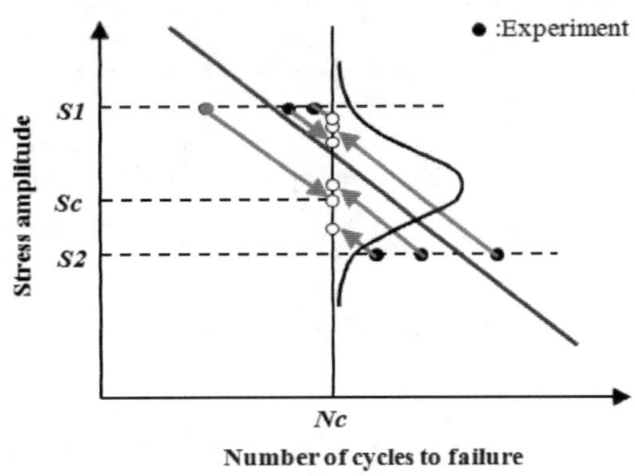

(b) Distribution analysis after shifting data

Figure 3. Procedure how to arrange the experimental data and how to get the distribution curve

2.3 S-N Curve for design with a probability of failure and a confidence level

The fatigue test data has been arranged for the analysis on probability of failure as shown in Figure 3. This procedure has an advantage which is able to get the probability of failure even if numer of specimens is few. The procedure has also the other advantage which can improve S-N curve in case of getting new test data. The estimated probability of failure $\hat{P}_f$ can be obtained by Equation (3).

$$\Pr[\hat{P}_f < P_f] = 1 - \gamma \tag{3}$$

Where P_f is the probability of failure and γ is the significance level. Figure 4 shows the concept of determination of stress amplitude for design.

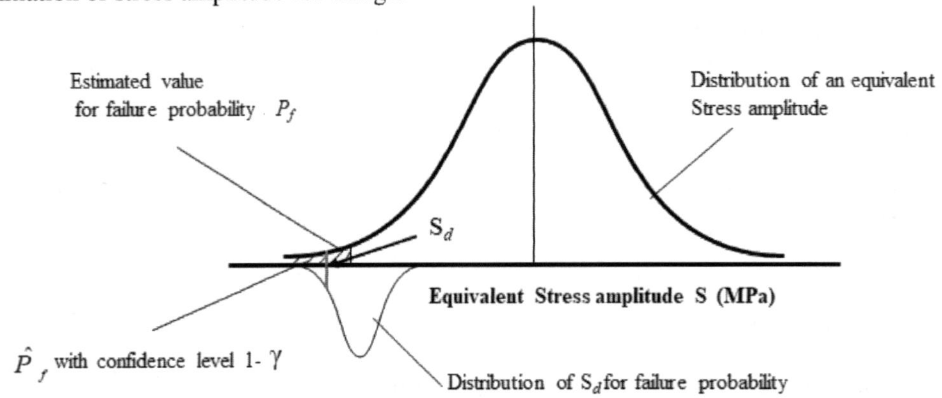

Figure 4. The concept of determination of stress amplitude for design based on estimated probability of failure and confidence level

The designer supposes some serious considerations and will determine an allowable stress amplitude (S_d). Of course, S_d will be determined under the consideration which the estimated failure probability is lower than P_f with confidence level (1-γ). It means that the designer deterrmines a probability of failure with confidence level instead of safety factor. The estimated probability of failure is calculated by one-side interval estimation as shown in Figure 4. The allowable stress amplitudes can be obtained by Equation (4). When the distribution of the equivalent stress amplitude follows the normal distribution, the allowable stress amplitude S_d is calcuted by Equation (4-a). In cases of log-normal and Weibull distributions, the allowable stress amplitudes (S_d) for each distribution are obtained by Equations (4-b) and (4-c), respectively.

$$S_d = \mu - \left\{\Phi^{-1}(1-P_f) + \frac{\Phi^{-1}(1-\gamma)}{n}\sigma\right\} \tag{4-a}$$

$$S_d = \exp\left[\mu - \left\{\Phi^{-1}(1-P_f) + \frac{\Phi^{-1}(1-\gamma)}{n}\sigma\right\}\right] \tag{4-b}$$

$$S_d = \frac{\mu}{\Gamma(1+1/\alpha)}\left\{\frac{\chi^2_\gamma(2n)}{2n\ln(1-P_f)^{-1}}\right\}^{-1/\alpha} \tag{4-c}$$

(4)

Where, μ is the mean value, σ is the standard variation, α is the shape parameter of Weibull distribution, χ^2 is chi-square, γ is the significance level and n is the total number of specimens, respectively. S_d for the other distributions can get by the statistics procedure as same as Equation 4. The type of distribution is judged by the coefficient of correlation. Figure 5 shows the concept of S-N curve for design which is obtained from the above mentioned procedure.

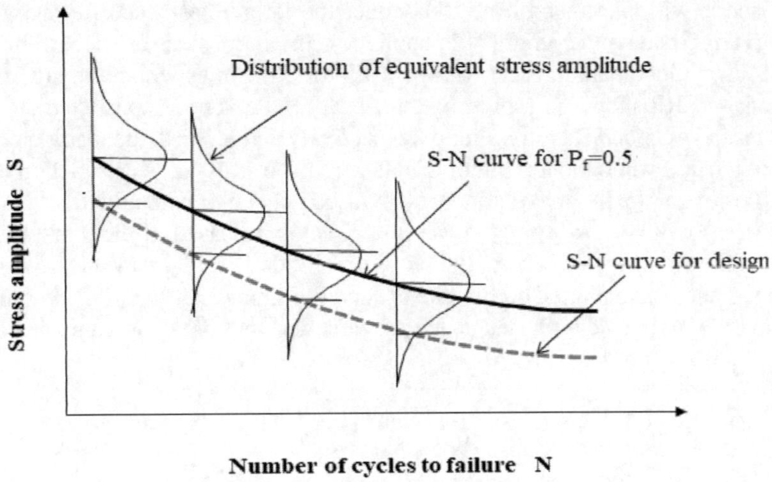

Figure 5. S-N curve for design obtained by a probability of failure and a confidence level [6]

Figure 6 shows the fatigue test data under repeated tensile loading condition for unidirectional CFRP[7] and S-N curve obtained by the proposed procedure. The specimens are made of carbon fiber (T-700SH) and epoxy, and V_f is 60%. The computer system is useful for the caluculation of probability of failure. The numerical analysis of probability of failure has been intoroduced in a computer system. S-N curves with each probability of failure as shown in Figure 6 are obtained by computer syatem. Using the system, S-N cueve with probability of failure can be easily obtaiened. It is also recognized that the proposed procedure is useful for designer of structure because the statistical analysis reflects the total number of specimens as the confidence level.

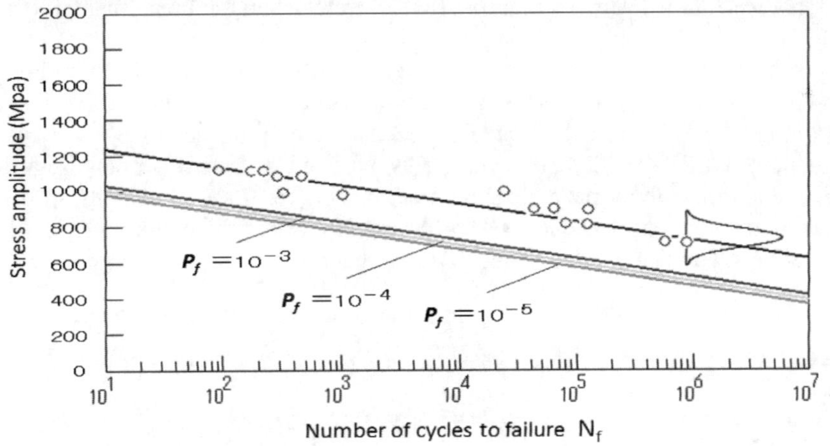

Figure 6. An example of S-N curve for each probability of failure when the confidence level is 0.95 for CFRP (T700SH/Epoxy and 60% V_f)

3. A proposal of safety evaluation for CFRP hydrogen gas tank

3.1 Goodman diagram

Type 4 tank applies to a hydrogen gas tank to reduce the weight of automobile. Type 4 is the composites tank which is wound CFRP around a plastic liner tank. The putting in and taking out hydrogen gas looks like fatigue phenonemon. The high pressure gas safety institute of Japan (KHK) has established a standard for safety evaluation. As an example, the number of repeat cycles for automobile is required

30,000 cycles. Since it will take much time to conduct the fatigue test, the effective method for the evaluation is required. Goodman diagram[8] is applied to the judgement of occurrence of failure by fatigue. Figure 7 shows Goodman diagram with $P_f = 0.5$ for 10^5 cycles which obtains from the fatigue test of unidirectional CFRP shown in Figure 6. In Figure 7, the vertical axis is stress amplitude, the horizontal axis is mean stress and R is stress ratio, respectively. When R=0, the loading condition means one side loading of tension when the minimum load is zero. The stress condition act on tank is one side repeated tension. For example, the maximum stress will be generated when the tank is filled with gas, and the minimun stress will be zero when the gas is completely released. If the maximum stress act on the tank is *S* and the minimum stress is zero, the stress amplitude *Sa* and the mean stress *Sm* will be *S*/2, respectively. The value has just plotted on the line of R=0 in Figure 7. If the plotted point is lower than Goodman diagram of $N_f=10^5$ cycles, it means that the tank does not break down up to 10^5 cycles.

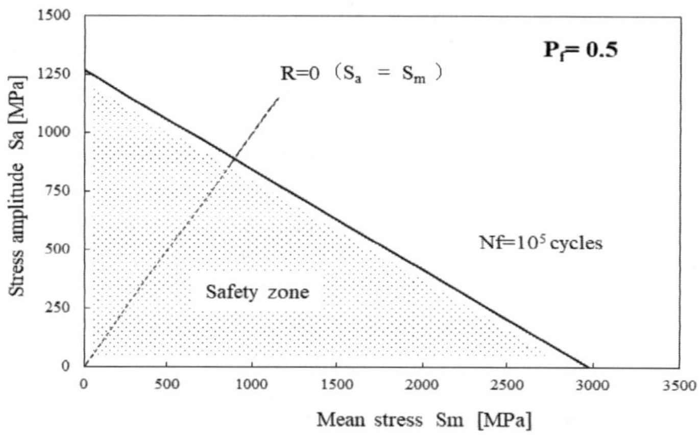

Figure 7. Goodman diagram for $N_f=10^5$ cycles of an unidirectional CFRP

3.2 Evaluation a hydrogen gas tank by Goodman diagram with probability of failure Figure 8 shows a hydrogen gas tank of type 4. The tank is 600 mm length with a plastics inner tank 240mm diameter, the capacity is 15 L and the pressure is 80Mpa. The tank is made up of the composite structure which is wound unidirectional CFRP (T700S/epoxy and 60%V_f) around a plastic liner tank. Figure 8 displays a quarter of the section of tank to make clear of stracture. Figure 9 shows Goodman diagram for the probability of failure with confidence level 0.95 for 30,000 cycles which obtains from Figure 6. Three probability of failures, which are 10^{-3},10^{-4} and 10^{-5} have been picked up as the example.

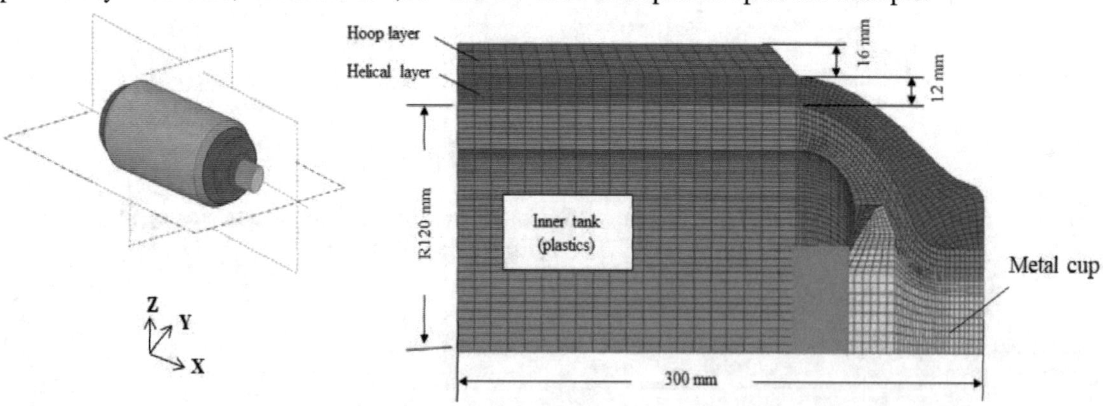

Figure 8. A quarter section of a hydrogen gas tank for an automobile and mesh size for analysis

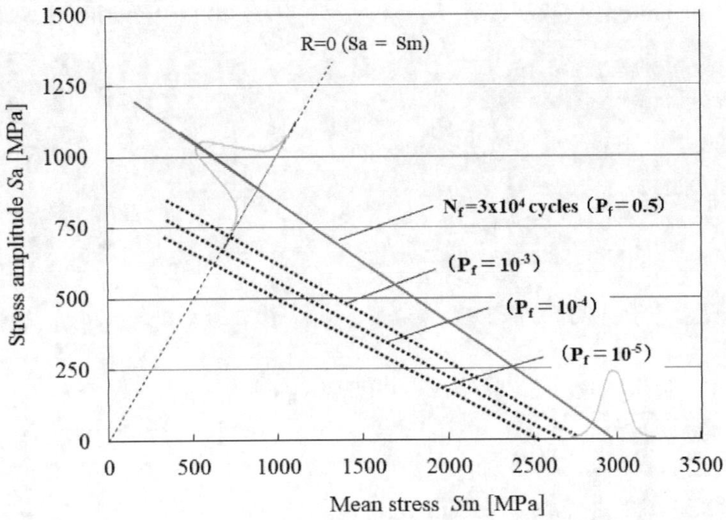

Figure 9. Goodman diagram for 30,000 cycles with probability of failure (P_f) (confidence level is 0.95)

According to the regulation of KHK, the margin factor is set to 1.25. When 80MPa is assumed to tank pressure, 100MPa should be applied as the internal pressure. The numerical analysis has been carried out for full scale model and SACOM (FEM program code) has been employed. Since the outer layer has been wound unidirectional CFRP, the strength of unidirectional fiber reinforced plastics dominates the burst pressure of tank. Figure 10 shows the distributions of S_L,S_T and S_{LT} under 100MPa. From the numerical results, it is recognized that the maximum stress S_L has been generated in hoop layer and the maximum stress is about 760MPa. When the gas is completely released, the stress ratio R becomes zero. The stress amplitude equals to the mean stress and they are a half of the generated maximum stress. In this case, they are 380MPa and is plotted on Goodman diagram as point A in Figure 11(a).
It is shown that the hydrogen tank will be safe up to 30,000cycle with the confidence level 0.95, becouse the plotted point has a distance from the line of P_f=10^{-5}. As the other example, we consider the situation which make repeated use of a half consumption. In this case, the maximum stress is 760MPa and the minimum sterss is 380MPa. The mean stress is 570MPa and the stress amplitude is 190MPa, respectively. The point B in Figure 11(a) indicates the situation of a half consumption. Point B has a considerable distance from the line of P_f =10^{-5} than point A. It is estimated that the operation B is a durable type than the operation A. Figure 11(b) shows the result of evaluation for 10^6 cycles. In this case, the operation A contacts with the line of P_f = 10^{-5}. It means that the tank will break at 10^6 cycles with the probability of failure P_f =10^{-5} and the operation B is durable even 10^6 cycles. It is also recognized that the proposed procedure is simple and reasonable.

(a) Distribution of S_L (b) Distribution of S_T

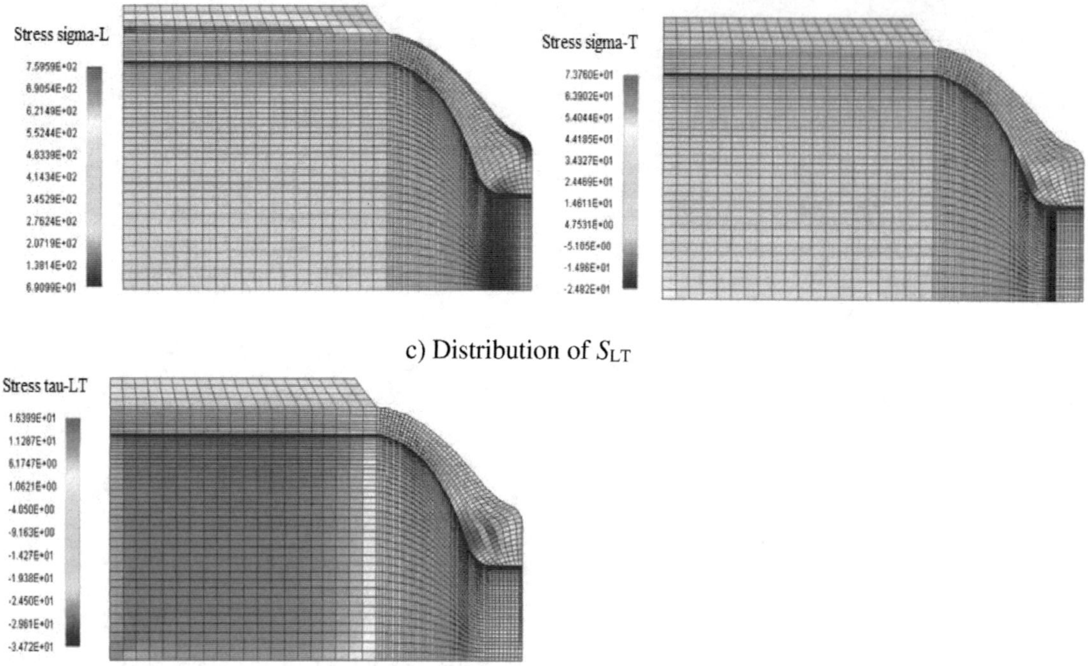

Figure 10. Numerical results of the distribution of S_L, S_T and S_{LT} under 100 MPa

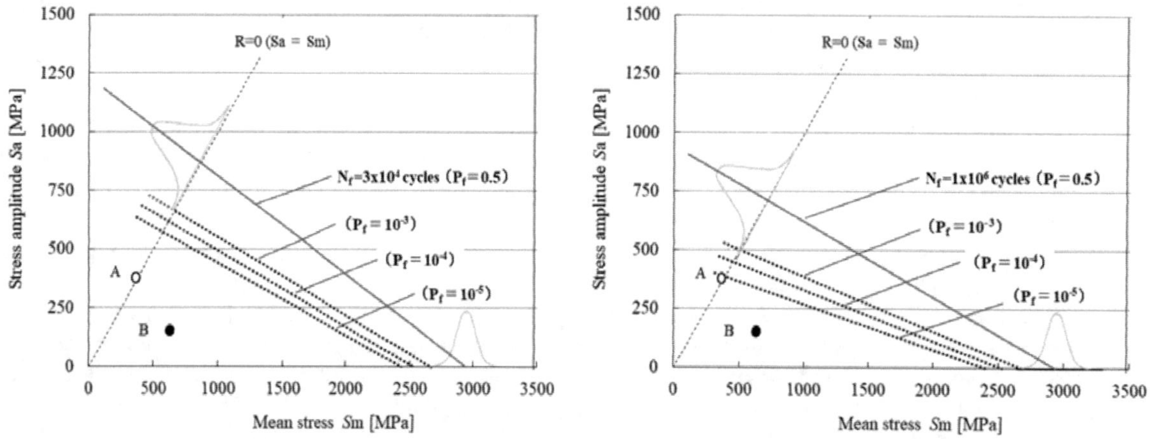

(a) Evaluation for 3x10^4 cycles (b) Evaluation for 10^6 cycles

Figure 11. Results of the safety evaluation by Goodman diagram with probanility of failure

4. Conclusion

An effective procedure to evaluate the safety for structures has been proposed. The proposed procedure consists of two parts. One is to decide the allowable stress with confidence level which is the function of the total number of specimens. The other is to evaluate the fatigue life by Goodman diagram with an estimated failure of probability. It is recognized that the proposed procedure is logical and simple. A hydrogen gas tank made of CFRP has been picked up as an example. The following points are clarified.

(1) S-N curve with probability of failure can be regulated by the equivalent stress amplitude. Goodman diagram with the probability of failure is very useful for judgment of the safety of structure.

IOP Conf. Series: Materials Science and Engineering **406** (2018) 012005 doi:10.1088/1757-899X/406/1/012005

(2) The confidence level is useful index for composites because it is applicable even if the number of specimens is small due to progress rapidly.
(3) The probability of failure will be useful and reasonable index instead of a safety factor.

In the future, the verification of Goodman diagram for textile composites is necessary. The fracture in textile composites consists of some damages like matrix crack, debonding, delamination and fibre breaking etc. Damage mechanics can be applied to the judgment of occurrence and distinguish of complex damage modes. If the occurrence of these damages during fatigue make clear experimentally, the explicit estimation for life will be possible by the combination of damage mechanics and the data. In addition, the computer system is efficient, because the complicated formulation of statistics and FEM are included in the analyses. ITOCHU Techno-Solutions Corporation will publish the system as "Composites dream ver.4" in the near future.

Acknowledgments
A part of this research was sponsored by the project of NEDO in Japan. The author wishes to express their appreciations to Mr. M.Sugawara and Dr. M.Kawahara for the contribution of production of hydrogen gas tank. The author also wishes to extend my thanks to Dr.T.Kurashiki, Dr. K.Hanaki, Dr. X.Li and Dr.Mukouyama for their works on the analysis of fatigue data.

References
[1] ASME 1980 *Boiler and Pressure Vessel code. Sec.III Div.1 AppendixII*
[2] Ichikawa M. et al 1987 *Practical Reliability of Engineering for mechanical and Civil Engineer* (Yokenndo) p 60
[3] Nishijima S. 1981 *ASTM STP* p75
[4] Tanimoto T. et al 1981 *Proceeding of US-Japan conference on composite materials* p145
[5] Zako M. Kurashiki T. and Hanaki S. 2001 *A Statistical decision making procedure of S-N curve with fatigue limit* JSTM Vol50, No.3 p279
[6] Zako M. Takano N. Ichikawa M. and Hanaki S. 1995 An evaluation System for Advanced materials based on Reliability Engineering *JSME* p1909
[7] NEDO report in Japan 2014 *Fatigue test data of CFRP* p52
[8] Goodman J. 1899 *Mechanics Applied to Engineering* Longman (London Green & Company)

IOP Conf. Series: Materials Science and Engineering **406** (2018) 012006 doi:10.1088/1757-899X/406/1/012006

Influence of voids presence on mechanical properties of 3D textile composites

Y Chen [1,2], D Vasiukov [1,2]*, C-H Park [1,2]

[1] IMT Lille Douai, Institut Mines-Télécom, France
[2] Université de Lille, France

[*] dmytro.vasiukov@imt-lille-douai.fr

Abstract. Presence of the voids in final part made of textile reinforced materials is one of the main process-induced type of defects, which is governed by a set of manufacturing parameters. This work focuses on investigating the effect of voids on mechanical properties of the 3D textile reinforced composites. This study includes the following steps: manufacturing of composite plates with RTM process, mechanical analysis, and multi-scale material modeling. The internal structure has been characterized from μCT scans and scanning electron microscopy observations. The influences of micro- (intra-yarn) and meso- (inter-yarn) porosities are discussed. Anisotropic damage model has been implemented into Fast Fourier Transform solver for simulation of non-linear response of the 3D interlock composite. Analysis includes the averaged stresses in different material phases (warp, weft, and matrix) and local concentration of stress/damage field near voids.

1. Introduction

Resin Transfer Molding (RTM) is used in producing geometrically complex shapes made of textile composites in a cost-effective way. However, one of the possible drawbacks could be a presence of the voids induced into the final part during manufacturing. In general, voids are not expectable as they can reduce functionality of the composite structure affecting its elastic properties, strength, electrical resistivity etc. Commonly voids are considered as the origin of local stress concentrations and hence of crack initiation in the textile composites. Both elastic and damage behaviors of the composites can be affected by presence of voids. In this work, the focus is placed on the 3D interlock composites with voids induced by RTM process.

Voids can be generally separated into two groups: intra-yarn voids and inter-yarn voids. The former exist inside each yarn and usually has line-like geometries elongated in the direction of adjacent fibers; while the latter are located between yarns whose geometries are mostly spherical or ellipsoidal.

Due to the complex 3D feature of both, the textile architecture and the void defects, a reliable characterization requires a 3D observation. X-ray micro-computed tomography is one of the most powerful tools for this purpose [1,2]. Voids and fibrous architectures can be identified using some specific image processing algorithms. A mesoscopic method has been proposed by [3] to separate the yarns from the matrix. The identified mesoscopic geometry then feeds further numerical modeling of mechanical behavior of composites (see e.g. [4]).

Once the different phases are identified in a unit cell, different numerical methods can be used to simulate the local response of the composite. Finite element method (FEM) is the most commonly used, but it requires a relatively tedious preprocessing: meshing. For highly heterogeneous microstructures, such as the textile composites herein, conformal FE meshes are difficult to be achieved and have already been the subject of many researches [5–7]. Regular meshing based on the initial image discretization is an alternative way to simplify this processing, yet it requires a large number of elements to reduce the unexpected effect due to the non-conformal discretization of the phase interfaces (see e.g. [4,8] for an illustration).

As an alternative to FEM, Fast Fourier Transform (FFT) based methods [9–14] may be used due to their simplicity of meshing and efficiency of parallel computation. The FFT methods intrinsically use regular grids to describe the microstructural geometry. In addition, through a reference material and the related Green operator, the local problem can be separately solved at each discretization point for each iteration, so the FFT models can be efficiently parallelized. Therefore, the FFT methods are very convenient for image-based modeling.

In the present work, the mesoscopic image segmentation technique [3] will be employed. Yarns with different orientations (warps and wefts) are separated. The anisotropy inside yarns will also be considered using the local orientation identified by the structure-tensor method. The microstructures will be described by regular grids generated from the initial image. The FFT based method will be coupled with a continuum damage model [4] to simulate the damage evolution in the composites.

2. Materials and experiments

2.1. Interlock textile and RTM process

The interlocks studied in this work are usually classified as 3D orthogonal interlock with layer-to-layer binding (Figure 1) in which binding warp yarns connect by two adjacent layers. These textiles were manufactured at ENSAIT (Roubaix, France). The areal density (ρ_A) is 0.266 g/cm2. Yarns are made of E-glass fibers with density ρ_f = 2.61 g/cm3, E = 72 GPa and v = 0.2. Textiles were impregnated with thermoset resin (Prime 27) which has the following properties: ρ_m = 1.11 g/cm3, E = 3 GPa and v = 0.3.

Figure 1. Repetitive Unit Cells of 3D interlock textile architecture.

The RTM process described in this paper is assisted by constant pressure injection. Table 1 consists the manufacturing parameters for two textiles described in this paper. Figure 2 describes the adopted manufacturing scheme.

Table 1. Parameters for the manufacturing of the composite plate.

	Twill – 3D
Injection pressure	1.5 Bar
Length of injection	26 cm
Time of injection	585 s
Resin velocity	0.5 mm/s

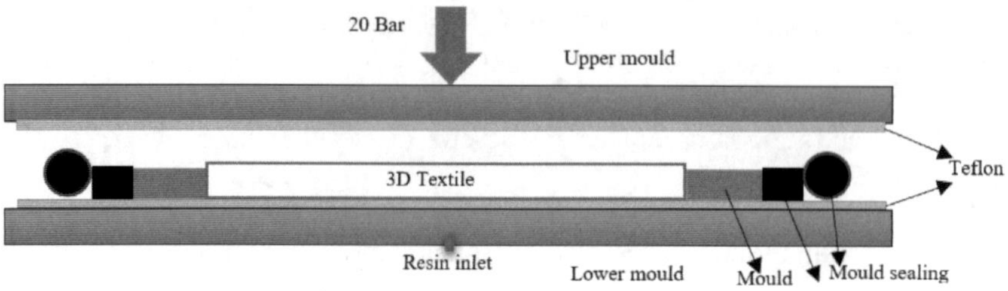

Figure 2. Protocol for RTM.

2.2. *μCT and image processing*

The tomographic images were acquired on the ISIS4D platform (LML/LaMcube, France). The scanning parameters are given in Table 2. The sample was scanned with 1440 projection images over a 360° rotation. A projection image was the average of 6 frames to reduce the random noises. The spot size of about 2 μm was chosen to achieve the required voxel sizes.

Table 2. Imaging parameters used during the CT scan

Sample	Voxel size (μm)	Tension (kV)	Current (μA)	Frame rate (f/s)	Scan time (min)	Dimension of reconstructed image (mm^3)
RVEU	8.79	80	68	10	20	13.71x3.14x9.99

Three-dimensional images were reconstructed using the software of RXSolution©. Due to the high density contrast between the glass fibers and epoxy matrix, beam hardening was observed within the reconstructed images (unpublished). Therefore, a beam-hardening reduction algorithm was applied during the reconstruction procedure. Physical filter (e.g. aluminum) was not used during the scans but would be another option to reduce the beam hardening effect. The finale reconstructed images (Figure 3.a) cover one RVE of the fibrous architecture.

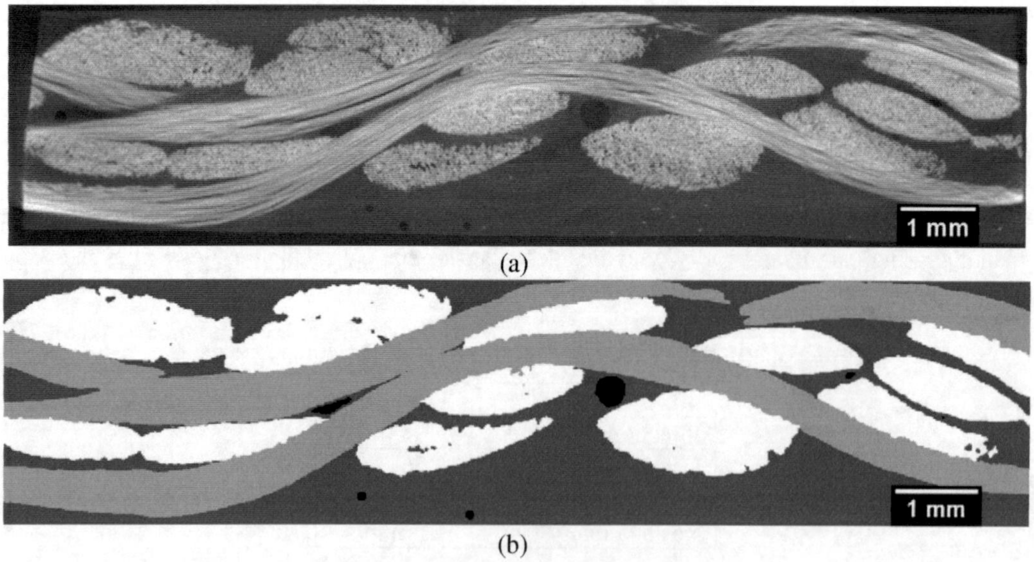

Figure 3. (a) cross-sectional slice of the reconstructed image; (b) segmented image using the structure tensor based method.

2.3. μCT and image processing

The post-processing of the obtained images aims at generating unit cells for numerical simulations. A mesoscopic modeling strategy for textile composites using an image processing technique based on structure tensor of image gray levels [3] has proved its efficiency (for example [4]). In the present work, the same approach is implemented in MATLAB routine, which allows us to identify inter-yarn voids, matrix, warps, and wefts. In our implementation, the structure tensor is calculated at each voxel position within a small local window (7 voxels per side in the present case). Therefore, the resulted description of the microstructure (image segmentation result) has the same resolution of the initial image (Figure 3.b). Three parameters are used for the image segmentation: average gray level (AVG), anisotropy degree (β) and characteristic angle (γ). The characteristic angle is chosen as the angle between the local orientation vector and the X-axis (warp direction), so that the warps and wefts can be distinguished.

The resulted unit cell contains about 450 million elements (1407x323x987 voxels). Even though the FFT based methods have been proven to provide a great advantage for parallel computation [15], such simulations are quite expensive that require a huge number of processors. To economize the computational cost, we choose to decrease the unit cell resolution (i.e. number of elements). Unit cells with four different resolutions have been generated (Table 3).

Table 3. The resolutions and element sizes of the unit cells generated from the same CT image of the 3D interlock composite.

Name	w21	w11	w5	w3
Resolution (voxels)	15x69x49	29x131x93	64x291x207	107x483x343
Element size (μm)	185	97	44	26

3. Numerical models

3.1. FFT based methods

Since the first conference paper of [9] (called "basic scheme", see Table 4), the FFT based methods have been improved, extended to many nonlinear behavior (viscoelasticity, plasticity, *etc*) and have become more and more attractive for image-driven modeling of materials properties, thanks to their simplicity and efficiency. The variables to be solved in FFT based methods are strains or stresses, instead of displacement. Note that no tangent calculation is necessary, which is different from the FEM scheme. In fact, by introducing a polarization term ((3) of Table 4), the local problem is transformed into Lippmann-Schwinger equation, leading to a convolution product with the Green operator: $\boldsymbol{\varepsilon}(\boldsymbol{x}) = \mathbf{E} - (\boldsymbol{\Gamma}_0 * \boldsymbol{\tau})(\boldsymbol{x})$. With periodic boundary conditions, the problem can be transformed into Fourier space, turning the convolution product to a simple multiplication ((5) of Table 4). Such an algorithm is very suitable for parallel computation, because: (i) the computations of (2,3,5) in Table 4 are local and can be solved separately for each material point; (ii) the FFT and inverse FFT are not local, but various packages (*e.g.* parallel FFTW) are available to efficiently complete these two steps in a parallel way.

A code (AMITEX, [16]) developed in CEA France is employed in the present work. This code provides a possibility to massively parallelize the computation in a large number of processors. AMITEX is based on the basic scheme [10]. In addition, it incorporates a convergence acceleration procedure, inspired from the FEM code CAST3M [17], as well as a modified Green operator, same as that proposed by [18]. As a result, the convergence in AMITEX is faster than the original basic scheme and much less sensitive to the choice of reference material. For instance, the computation with the highest unit-cell resolution (~18 million elements for w3 in Table 4) has been completed within 60 hours using 12 processors.

Table 4. The basic scheme of the FFT method proposed by [10].

Initialization		$\boldsymbol{\varepsilon}^0(\boldsymbol{x}) = \boldsymbol{E}$	(1)
Loop			
$t = k$	Compute the stress through a constitutive law in real space (UMAT)	$\boldsymbol{\sigma}^k = \boldsymbol{\sigma}(\boldsymbol{\varepsilon}^k(\boldsymbol{x}))$	(2)
	Convergence test		
	Compute the polarization tensor	$\boldsymbol{\tau}^k(\boldsymbol{x}) = \boldsymbol{\sigma}^k(\boldsymbol{x}) - \boldsymbol{C}_0 : \boldsymbol{\varepsilon}^k(\boldsymbol{x})$	(3)
	FFT transform	$\boldsymbol{\tau}^k(\boldsymbol{x}) \rightarrow \hat{\boldsymbol{\tau}}^k(\boldsymbol{\xi})$	(4)
$t = k + 1$	Solve the problem via the Green operator in Fourier space	$\hat{\boldsymbol{\varepsilon}}^{k+1}(\boldsymbol{\xi}) = -(\hat{\boldsymbol{\Gamma}}_0 \cdot \hat{\boldsymbol{\tau}}^k)(\boldsymbol{\xi})$	(5)
	Inverse FFT transform	$\hat{\boldsymbol{\varepsilon}}^{k+1}(\boldsymbol{\xi}) \rightarrow \boldsymbol{\varepsilon}^{k+1}(\boldsymbol{x})$	(6)

3.2. Local constitutive laws

The composite microstructure is discretized into three types of elements, whose behaviors follow different local constitutive laws: inter-yarn voids with no stiffness, matrix as isotropic damageable, and yarns (warps and wefts) as anisotropic damageable. As a first attempt, the influence of intra-yarn is neglected. The yarns are considered as homogeneous media, whose anisotropic properties are estimated by: simplified micromechanical models [20,21] for elastic and longitudinal strength in tension (X_{LT}), experimental for longitudinal strength in compression (X_{LC}), and numerical simulation based on micro-mechanical analysis for transversal and in-plane shear (X_{TT}, X_{TC}, X_S) as shown in [24]. This model has been readily used by several authors [4,15,21]. It requires the volume fraction of fibers in the yarn to be evaluated, which is approximately estimated to be 60% from SEM observations in the present study. The results are given in Table 5. The fracture toughness are chosen to be 11 kJ/m^2 and 0.2 kJ/m^2 for the tensile failure in longitudinal and transverse directions respectively.

Table 5. Properties of the homogenized yarns (unit in MPa):

E_{11}	$E_{22} = E_{33}$	$\nu_{12} = \nu_{13}$	ν_{23}	$G_{12} = G_{13}$	X_{LT}	X_{LC}	X_{TT}	X_{TC}	X_S
44400	11640	0.26	0.33	4480	1500	1000	60	150	50

The CDM model used by [4] is employed to simulate the initiation and propagation of damage in the composite. This model is implemented in a user subroutine UMAT supported by AMITEX. Recently, a similar FFT model was also proposed by [21] to simulate the damage evolution in virtual/idealized composites. Yet, a different iterative scheme was used in their work, which is based on a variational formulation of the FFT method [13].

3.3. Loading and boundary conditions

The boundary conditions in FFT methods are intrinsically imposed as periodic. However, the unit cells of real microstructures are never perfectly periodic, which will inevitably induce artifacts on the simulated stress/strain field. Particularly in the case of damage modeling, this edge effect must be treated with more attention, because the geometrical discontinuity of microstructure at unit cell boundaries lead to unrealistic stress concentrations hence damage initiation. In order to reduce the edge effect, we add marginal layers around the unit cells, as shown in Figure 4. These marginal layers are assigned with different elastic (non-damageable) properties according to the loading direction: for loading in X-axis (weft direction), the elastic modulus of margin Y will be set to zero, while that of margin X larger than zero; inversely for loading in Y-axis (warp direction).

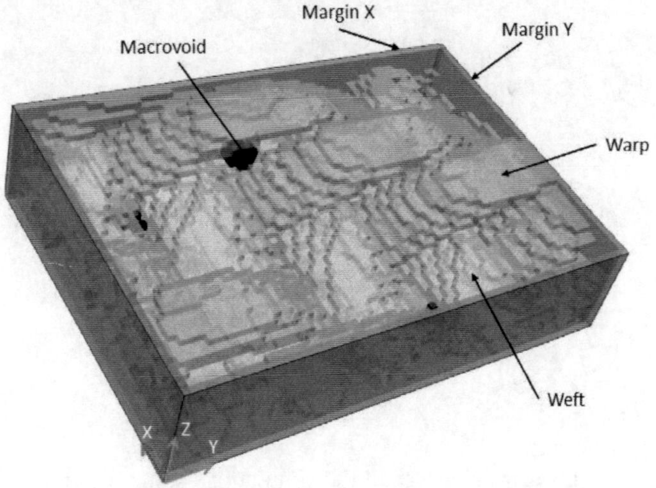

Figure 4. 3D rendering of the unit cell with the resolution of 15x69x49 voxels (w21), the matrix is not shown in the figure.

The marginal layers with non-zero elastic properties can be understood as intermedia to uniformly transmit the load (stress in the present case) over the boundary surfaces. Therefore, their stiffness should be as high as possible. However, a too high stiffness of the marginal layers induces difficulties to the convergence of the calculation (reason unclear for now). The effect of this stiffness of the marginal layers will be discussed in the following section.

The unit cells will be subjected to incremental strains with 100 time steps up to 1.5% in either X-axis or Y-axis.

4. Results and discussions

4.1. Effect of the stiffness of marginal layers

We use the unit cell of the lowest resolution (w21 from Table 3) to analyze the effect of the stiffness of marginal layers. Six simulations are carried out with different stiffness of marginal layers. The elastic coefficient of matrix ($E = 3$ GPa, $\nu = 0.3$) is chosen as a reference for this analysis. The macroscopic stress-strain curves of the six simulations are presented in Figure 5. The strength of the material increases with the stiffness of marginal layers and trends to be stable after 60 GPa (C0x20). On the other hand, a higher marginal stiffness (> 150 GPa, C0x50) makes the computation not convergent after the damage is activated. As a compromise, we choose the elastic modulus of the marginal layers as 60 GPa for further simulations.

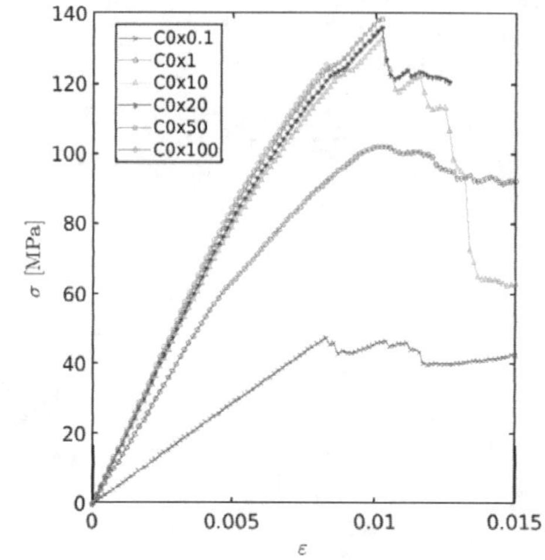

Figure 5. Macroscopic stress-strain curves of the simulations using different stiffness of marginal layers for tension in warp direction.

4.2. Effect of unit-cell resolution

The study of the voxel size on the macroscopic behavior is shown here. For this purpose four unit cells of different resolutions generated from the same CT image (see Table 2 in section 2). Note that the used damage model incorporates a regularization scheme based on Bazant's crack band model [22] so that the mesh dependency on dissipated fracture energy is minimized. The macroscopic stress-strain curves are shown in Figure 6. One interesting finding is that the elastic behavior is not dependent on the image resolution, i.e. the stress-strain curves are superimposed. Both the damage initiation (the beginning of the nonlinearity of the curves) and the material strength increase with the image resolution. This effect becomes very small when the image resolution is high enough (e.g. element size of 44 μm for w5 from Table 3).

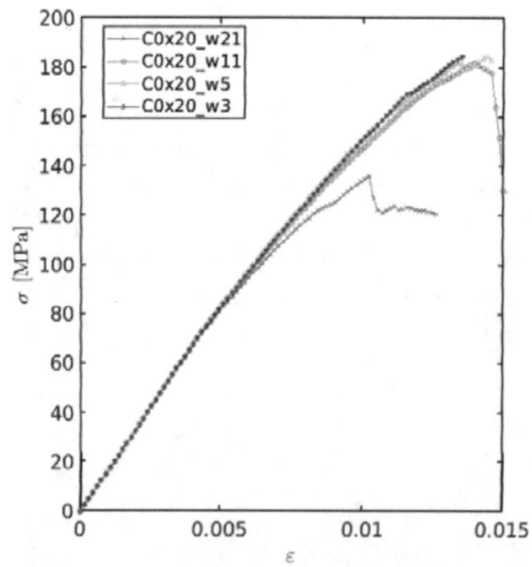

Figure 6. Macroscopic stress-strain curves of the simulations using different image resolutions under tension in weft direction.

IOP Conf. Series: Materials Science and Engineering **406** (2018) 012006 doi:10.1088/1757-899X/406/1/012006

4.3. Stress evolution and damage field

Figure 7 shows the fractured zones in matrix and transversal yarns, i.e. warps for tension in weft direction and verse-versa. Under both loads, the damage zones are less observed in matrix than in transversal yarns. The damage localized near the composite borders is linked to the geometrical discontinuity. Inside the composite volume, the transversal damage is well reproduced under both loads. For tension in warp direction, damage seems more severe between yarns (inter-yarn damage) than inside yarns (intra-yarn damage). The inter-yarn damage is mostly observed in the connecting regions between warps and wefts. On the contrary, the tension in weft direction seems to produce more intra-yarn damage than inter-yarn damage. This difference may be explained by the difference in waviness of warps and wefts. In fact, the out-of-plane undulation of warps can induce much stronger shear stress concentrations at connecting region under tension in their longitudinal direction than in their transversal direction.

Regarding inter-yarn void presence effect on the mechanical response, the following conclusions can be drown. There is no significant effect on the macroscopic elastic properties was observed under presented loading cases. The stress concentration in the adjacent yarns has been increased locally, however, the final fractured zones seems to be strongly influenced by the transverse properties and internal architecture of textile composites. On the other hand, the transverse properties of yarns are strongly affected by intra-yarn voids. To confirm this preliminary conclusion, further detailed investigations are necessary.

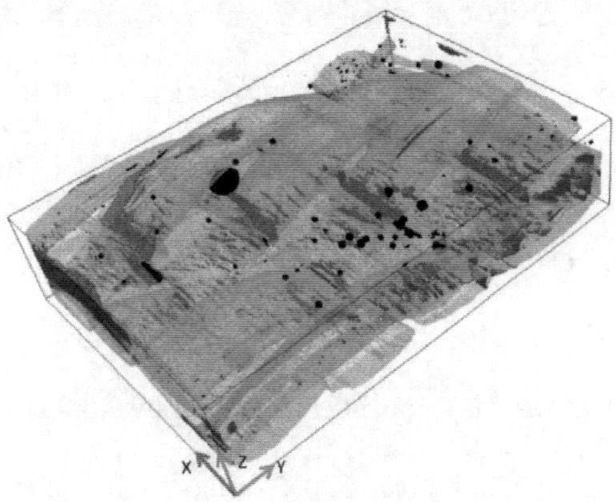

(a)

(b)

Figure 7. Fractured zones, whose damage parameter is larger than 0.8, in matrix (blue) and transversal yarns (red) at tension level of $\varepsilon = 1.35\%$: (a) under tension in warp direction, (b) under tension in weft direction.

5. Conclusions

An FFT based method has been employed with coupled CDM model to predict the damage evolution in 3D interlock composite including voids induced by RTM manufacturing process.

Geometrical model has been reconstructed from micro CT data. In order to reduce the edge effect due to the geometrical discontinuity at the image borders, marginal layers have been added, and the effect of the stiffness on the simulation result has been studied suggesting optimal values.

Unit-cells with different resolutions have been generated from the same CT data in order to show the sensitivity of numerical solution to unit-cell resolution. It was observed that elastic response is not sensitive to unit-cell resolution, while the nonlinear (damage) behavior can be strongly affected. The efficiency of the FFT based methods is demonstrated for using high unit-cell resolution in the context of damage prediction.

A preliminary conclusion has been drawn on the influence of voids on damage evolution in the composite. The presence of the inter-yarn voids has induced stress concentrations in adjacent yarns affecting local stress field and thus the damage initiation. However, the transverse properties and out-of-plane undulation of yarns are more critical than the inter-yarn voids for the elementary properties used in the simulations. More sophisticated models at the microscopic length scale are required to study the relative importance of intra- and inter-yarn voids on damage behavior of the composite.

Acknowledgment
The Nord-Pas-de-Calais Region and the European Community (FEDER funds) partly funds the X-ray tomography equipment ISIS4D platform (LML/LaMcube, France). The authors want to thank Lionel GELEBART from CEA Saclay for fruitful discussions and helps on AMITEX. Prof. S.V. Lomov who coordinates the Composite Materials Group and holds Toray Chair for Composite Materials @ KU Leuven is gratefully acknowledged.

References

[1] Stock SR. Recent advances in X-ray microtomography applied to materials. Int Mater Rev 2008;53:129–81. doi:10.1179/174328008X277803.

[2] Maire E, Withers PJ. Quantitative X-ray tomography. Int Mater Rev 2014;59:1–43. doi:10.1179/1743280413Y.0000000023.

[3] Straumit I, Lomov S V., Wevers M. Quantification of the internal structure and automatic generation of voxel models of textile composites from X-ray computed tomography data. Compos Part A Appl Sci Manuf 2015;69:150–8. doi:10.1016/j.compositesa.2014.11.016.

[4] Liu Y, Straumit I, Vasiukov D, Lomov S V., Panier S. Prediction of linear and non-linear behavior of 3D woven composite using mesoscopic voxel models reconstructed from X-ray micro-tomography. Compos Struct 2017;179:568–79. doi:10.1016/j.compstruct.2017.07.066.

[5] Schöberl J. An advancing front 2D/3D-mesh generator based on abstract rules. Comput Vis Sci 1997;1:41–52. doi:10.1007/s007910050004.

[6] Coffman VR, Reid ACE, Langer SA, Dogan G. OOF3D: An image-based finite element solver for materials science. Math Comput Simul 2012;82:2951–61. doi:10.1016/j.matcom.2012.03.003.

[7] Legrain G, Cartraud P, Perreard I, Moes N. An X-FEM and level set computational approach for image-based modelling: Application to homogenization. Int J Numer Methods Eng 2011;86:915–34.

[8] Doitrand A, Fagiano C, Irisarri FX, Hirsekorn M. Comparison between voxel and consistent meso-scale models of woven composites. Compos Part A Appl Sci Manuf 2015;73:143–54. doi:10.1016/j.compositesa.2015.02.022.

[9] Moulinec H, Suquet P. A FFT-based numerical method for computing mechanical properties of composite materials from images of their microstructure. R Pyrz (Ed), Microstruct Interact Compos Mater Kluwer Acad Publ Dordrecht, 1995:235–46.
[10] Moulinec H, Suquet P. A numerical method for computing the overall response of nonlinear composites with complex microstructure. Comput Methods Appl Mech Eng 1998;157:69–94. doi:10.1016/S0045-7825(97)00218-1.
[11] Brisard S, Dormieux L. FFT-based methods for the mechanics of composites: A general variational framework. Comput Mater Sci 2010;49:663–71. doi:10.1016/j.commatsci.2010.06.009.
[12] Gélébart L, Mondon-Cancel R. Non-linear extension of FFT-based methods accelerated by conjugate gradients to evaluate the mechanical behavior of composite materials. Comput Mater Sci 2013;77:430–9. doi:10.1016/j.commatsci.2013.04.046.
[13] Zeman J, de Geus TWJ, Vondřejc J, Peerlings RHJ, Geers MGD. A finite element perspective on nonlinear FFT-based micromechanical simulations. Int J Numer Methods Eng 2017;111:903–26. doi:10.1002/nme.5481.
[14] Schneider M, Merkert D, Kabel M. FFT-based homogenization for microstructures discretized by linear hexahedral elements. Int J Numer Methods Eng 2017;109:1461–89.
[15] Chen Y. Damage mechanisms in SiC/SiC composite tubes : three-dimensional analysis coupling tomography imaging and numerical simulation, PhD Thesis. Université Paris-Est, 2017.
[16] Home — AMITEX_FFTP 2.3 documentation http://www.maisondelasimulation.fr/projects/amitex/html/.
[17] Cast3M http://www-cast3m.cea.fr/.
[18] Willot F. Fourier-based schemes for computing the mechanical response of composites with accurate local fields. Comptes Rendus - Mec 2015;343:232–45. doi:10.1016/j.crme.2014.12.005.
[19] Chamis C. Mechanics of Composite Materials : Past , Present , and Future. J Compos Technol Res 1989;11:3–14. doi:10.1520/CTR10143J.
[20] Rosen B. Mechanics of composite strengthening. In: Fiber composite materials. Metals Park (OH): ACM; 1964.
[21] Wang B, Fang G, Liu S, Fu M, Liang J. Progressive damage analysis of 3D braided composites using FFT-based method. Compos Struct 2018;192:255–63. doi:10.1016/j.compstruct.2018.02.040.
[22] Bazant ZP, Oh BH. Crack band theory for fracture of concrete. Matériaux Constr 1983;16:155--177.
[23] Fang G, El Said B, Ivanov D, Hallett SR. Smoothing artificial stress concentrations in voxel-based models of textile composites. Compos Part A Appl Sci Manuf 2016;80:270–84. doi:10.1016/j.compositesa.2015.10.025.
[24] D. A. Vajari. A micromechanical study of porous composites under longitudinal shear and transverse normal loading, Composite Structures, 2015; 125: 266-276.

IOP Conf. Series: Materials Science and Engineering **406** (2018) 012007 doi:10.1088/1757-899X/406/1/012007

Predicting the effect of voids on mechanical properties of woven composites

R. S. Choudhry[1*], Tahir Sharif[1], Kamran A. Khan[2], Sohaib Z Khan[3], Abid Hassan[4] and Muhammad A Khan[5]

[1] Department of Mechanical Engineering & Built Environment, College of Engineering, University of Derby, Derby, England, UK
[2] Department of Aerospace Engineering, Khalifa University of Science Technology and Research (KUSTAR), UAE
[3] Department of Mechanical Engineering, Faculty of Engineering, Islamic University of Madinah, Madinah PO Box 170, Kingdom of Saudi Arabia
[4] Pakistan Petroleum Limited, Pakistan
[5] School of Aerospace Transport and Manufacturing, Cranfield University, Cranfield, UK

*R.Choudhry@derby.ac.uk

Abstract. An accurate yet easy to use methodology for determining the effective mechanical properties of woven fabric reinforced composites is presented. The approach involves generating a representative unit cell geometry based on randomly selected 2D orthogonal slices from a 3D X-ray micro-tomographic scan. Thereafter, the finite element mesh is generated from this geometry. Analytical and statistical micromechanics equations are then used to calculate effective input material properties for the yarn and resin regions within the FE mesh. These analytical expressions account for the effect of resin volume fraction within the yarn (due to infiltration during curing) as well as the presence of voids within the composite. The unit cell model is then used to evaluate the effective properties of the composite.

1. Introduction

Woven fabric reinforced composites at micro level are heterogeneous materials however, the modelling of structural response using FEA or analytical techniques is generally carried out at ply level and requires the user to assume an effectively homogenized material with a set of effective material properties for this homogenized material. The number of constants required depends on the assumed behaviour. The most common approximation used is orthotropic elastic and requires input of nine material constants. Besides the resin and fibre types and their relative volume fraction, these constants are also sensitive to other factors such as weave architecture, extent of compaction, void content and extent of resin infiltration within yarn. In case of existing materials, the constants can be evaluated using standard ASTM tests. Performing these tests however, is often, time consuming, expensive, requires specialized test fixtures, and in some case suffer from poor repeatability. In particular, the reliance on testing is not practical for development of new materials or optimization of manufacturing (e.g. curing) parameters (such as temperature and pressure etc.). In these cases, ideally one needs a capability to theoretically

IOP Conf. Series: Materials Science and Engineering **406** (2018) 012007 doi:10.1088/1757-899X/406/1/012007

predict the properties of resultant composite even before the material is made, with only few tests required at the end to validate the model.

Methods of determination of effective engineering constants of composites using micro-mechanical modelling are now sufficiently well developed for both unidirectional as well as fabric reinforced composites. Micro-mechanics based strength prediction of composites on other hand is more challenging. Failure initiation is often well estimated but final failure seldom is; this is because final failure in composites is progressive and heterogenous in nature and not necessarily periodic. Strength prediction is even more difficult for composites with a significant voids content or with those having matrix toughening inclusions.

Several analytical as well as FEA based modelling methodologies have been presented by various authors for micro-mechanics based estimation of composite properties e.g. [1–9]. The analytical methods vary widely in terms of accuracy and input data requirements. There always seems to be an inevitable trade-off between simplicity and correctness. Techniques such as Periodic Microstructure Modelling (PMM) [6] provide quite an accurate estimate for unidirectional fibre reinforced composites but are difficult to implement and need a lot of input parameters, which at times may be difficult to obtain. Analytical methods for woven composites are even more difficult to implement however some of them are now available in the form of free to use codes. Barbero for example has made his algorithm [10] available to wider audience in the form of online software CADEC. More recently, analytical methods have also been presented and validated for orthogonal 3D woven composites [11,12]. Analytical models are quicker for evaluating property trends and thus better suited to inform development of new materials. Their accuracy is often limited because of the assumptions made and all properties are not predicted equally well with a single model, thus requiring use of multiple models for various problems. Most of the analytical models are less generic and change in parameters such as weave architecture for example may require a significant development effort rather than just specification of some new parameters.

FEA methods although more time consuming to implement and solve can be used to reliably and accurately predict the effective properties of any proposed weave architecture. The accuracy of FEA model in principle depends on a) accuracy of the geometric model and mesh used to construct the RVE, b) correct representation and implementation of periodic boundary conditions for the unit cell. While the latter is a mathematical constraint and with due diligence one can insure sufficient correctness, the former depends highly on available resources. Firstly, the resources for capturing and representing the internal architecture correctly and secondly the computational resources for the solving a highly accurate mesh. Given the fact that resources are always finite, reasonable assumptions must be made to simplify the representation of internal architecture. The geometric model and subsequent FEA mesh for FE micro-mechanics is often produced employing major simplifications which include,

a) modelling the yarn as homogenous solid (as opposed to modelling individual fibres within a yarn),
b) ignoring the weave compaction that takes place during manufacturing and curing process, i.e. pre-cure weave architecture as opposed to post cure,
c) assuming no infiltration of resin within the yarn and assuming resin and fibre to be perfectly bonded and dispersed.
d) ignoring voids as well as any inclusions within matrix.

All these simplifications can severely reduce the accuracy of the model unless these are accounted for in some way. Thus, in this paper we will present a methodology to account for these simplifications in a consistent manner using the example of a satin weave composite. The aim is to give recommendations so that the material designers and modelling community can make informed choices about the assumptions that they take while developing and using an FE based micro-mechanics model of textile composites.

IOP Conf. Series: Materials Science and Engineering **406** (2018) 012007 doi:10.1088/1757-899X/406/1/012007

2. Methodology

The modelling methodology proposed in this study can be outlined as a five-step process. These steps are demonstrated by applying on satin weave reinforced phenolic pre-preg (ACG SL246/40) composite. Full material specifications and details of manufacturing and layup are available in our earlier publication [13]. The test panels used for the study were vacuum bagged and cured using QuickStep ™. The volume fraction of fibers (V_{fe}) achieved was 39.9%, volume fraction of matrix (V_m) was 54.8% and the average void content was 5.3%.

2.1. Defining the Representative Volume Element (RVE)

Woven composites are characterized by a weave pattern that repeats in-plane periodically for each lamina. In this study the notation used by Barbero [10] is employed to define the RUC in terms of Harness/Shift/Interlace numbers. The representative volume element (RVE) used for FE micromechanics for a plain weave is 1/16th of the RUC using symmetry and anti-symmetry considerations. In case of other weave patterns however, the RUC may not be reduced any further and in such cases the RUC and RVE are the same. Since in this study a satin weave is chosen as an example thus the term RVE and RUC are used interchangeably.

The RUC of the fabric used was 8/5/1 (Harness/Shift/Interlace). In real composites this RUC weave pattern repeats exactly, however other parameters such as the yarn thickness along fill and warp or the thickness of matrix region between layers, the cured shape of the yarn, the gaps between tows and any voids or air pockets will not repeat exactly. Small variations are always expected for these parameters. Therefore as suggested in [13], an approximate representation of the RVE, which is correct in an average sense is better than using an RVE directly generated from a scan of one very small region of the specimen. This is also important from new materials development point of view where the emphasis is not on getting an exact property value, rather, it is on generating realistic property bounds or ranges. In practice the RVE may be generated using the post-cure architecture of the desired woven composite by averaging the RVE parameters taken from a series of internal cross-sectional views generated from a technique such as X-ray microtomography or cross-sectional microscopy. In case of new material development study, the post-cure geometry used should be approximated from post cure geometry of a similar type of fabric resin combination. Alternately a procedure such as a one outlined in [8] can be used to model the compaction and consolidation of the proposed weave during curing. In this study the RVE was identified using average RVE parameter values from X-ray micro tomographic scans of the example composite. The main RVE parameters are summarized in table 1.

2.2. Creating a geometric model of the RVE

The RVE parameters such as the ones defined in table 2, can be used to generate the solid model of an RVE using any commercial solid modelling package and then this can be imported in FEA software such as ABAQUS ™. In this study however, a dedicated textile composite solid modeller and pre-processor called TexGen was used for ease of model and mesh generation. TexGen not only makes it simple to generate the geometric model, it also makes it easy to assign materials properties, apply orientations and create boundary conditions. Thus, the entire pre-processing is done in this environment. As is customary, while generating the RVE the cured yarn as opposed to individual fibres within the yarn are modelled and the yarn is approximated as an effectively homogenous albeit orthotropic solid with a lenticular shape as defined by average parameters defined in table 1. Modelling the individual fibres in yarn with a tow count of 3K or more in each yarn is firstly highly complicated with respect to the process required for geometric model generation and secondly even if one were to generate such a model the resulting mesh will be computationally very expensive. Thus, in this study a secondary micro-mechanical model is proposed to generate the desired input mechanical properties of the yarn. This is explained in following section. It should also be noted that individual voids are not modelled discretely using the geometric model in this study. Instead as will be explained later an approach to account for these voids in an average sense is introduced and discussed later.

Table 1. Key RVE properties

Parameter	Mean value
Width of warp yarn '*Wp*'	0.41 (mm)
Width of weft yarn 'Wt'	0.45 (mm)
Resin interface between consecutive yarns 'hint'	0.01 (mm)
Height of yarns 'hy'	0.12 (mm)
Thickness of resin layer between two consecutive lamina 'hr'	0.05 (mm)
Volume of fibers in the UC (V_f)	1.472 mm3
Volume of yarns in the UC (V_y)	1.895 mm3
Fibre volume fraction of the yarn (V_{fy})	0.7768
Resin Modulus of Elasticity (E_m), GPa)	3.60
Resin Poisson Ratio (υ_m)	0.35
Fibre Modulus of Elasticity (E_f), GPa)	73
Fibre Poisson ratio (υ_f)	0.23

2.3. *Calculating Material property inputs for the RVE*

Once the geometric model of the RVE is created correct material properties need to be assigned to each zone. As mentioned above the yarn as well as matrix have been idealized as homogenous materials. In reality, yarn is a composite of individual fibres, infiltrated resin within yarn as well as voids. Matrix on the other hand is a composite of resin, voids as well as resin inclusions such as toughening micro or nano particles in some cases. This results in the yarn and matrix properties to be different from those of dry fibre and pure cured resin. In this study we suggest using secondary micro-mechanics model to generate these properties.

2.3.1. *Generating effective properties for yarn*

The yarn can be approximated as a unidirectional fibre reinforced composite, as long as the material properties of yarn are assigned in a local coordinate system that rotates with the yarn along each material integration point within the FE mesh. Thus, existing analytical micro-mechanics based relations for effective properties of UD composites may be used for estimating these properties. Alternatively, an FE based micro mechanics model of UD composites may be used.

In this study a transverse isotropic behaviour was assumed for the yarn and analytical relations that were used are summarized in [13, Table 6]. In these equations the elastic modulus and Poisson ratio of neat cured resin (E_m, v_m) and dry fibre (E_f, v_f) are used as input. These equations also require the input of fibre volume fraction within yarn (V_{fy}). It must be noted that V_{fy} is different from the overall fibre volume fraction of composite. This is due to the fact that yarn is approximated as lenticular in shape (see e.g. figure 1) and due to this the total volume of fibres in the Unit Cell is different from the total volume of yarn in the unit cell. For, details of the calculation procedure the reader is referred to [13].

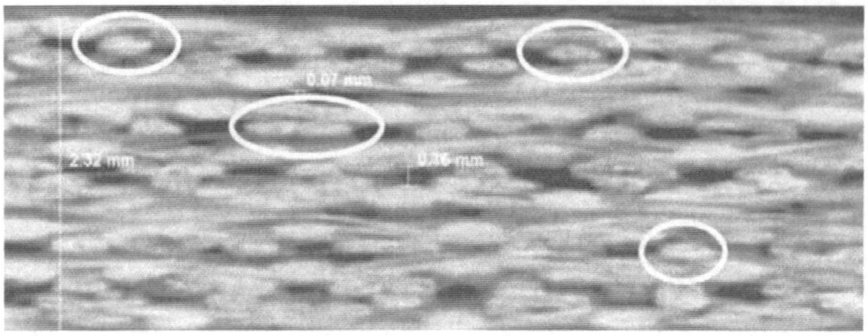

Figure 1. X-ray tomography image (right) [14,15] showing the lenticular shape of yarn in a cured composites (example woven composites satin weave ACG SL246/40)

2.3.2. Void compensation

The presence of voids (as well as any possible particulate inclusions) are expected to have a direct influence on resin properties (E_m, v_m), no effect on fibre properties (E_f, v_f) and an indirect effect on yarn properties (due to the change in input matrix properties in the relations outlined earlier). The dispersion of these voids or possible particulate inclusions are generally random and therefore, the response of matrix is expected to remain isotropic. In this study the resin used does not have any inclusions and thus the properties should only be modified to reflect the effect of voids.

A simple semi-empirical strategy was used to compensate for voids. The properties of resin in the presence of void may be calculated by assuming that the matrix properties can simply be reduced in proportion to the fraction of the void content present in the matrix. Thus, the void compensation factor may be calculated as an empirical factor, $V_{cf} = \frac{V_m}{(V_m+V_o)}$. The matrix input properties are then simply calculated by multiplying this factor with the properties of neat resin. The same reduced properties are also used in the analytical relations for calculating the yarn properties.

Table 2. Constituents input properties for RVE using corrected fiber volume fraction in yarn and with void compensation for matrix properties

	Material Parameter	Value
Matrix	Modulus of Elasticity (E_m, GPa)	3.285
	Poisson Ratio (υ_m)	0.319
	Shear Modulus (G_m, GPa)	1.245
Yarn	Longitudinal Modulus of Elasticity ($E_{x(y)}$, GPa)	57.438
	In-plane Transverse Modulus of Elasticity ($E_{y(y)}$,GPa)	24.284
	Out-of-plane Transverse Modulus of Elasticity ($E_{z(y)}$,GPa)	24.284
	In-plane Poisson Ratio ($\upsilon_{xy(y)}$)	0.250
	Out-of-plane Poisson Ratio ($\upsilon_{xz(y)}$)	0.250
	Out-of-plane Poisson Ratio ($\upsilon_{yz(y)}$)	0.876
	In-plane Shear Modulus ($G_{xy(y)}$, GPa)	7.468
	Out-of-plane Shear Modulus ($G_{xz(y)}$, GPa)	7.468
	Out-of-plane Shear Modulus ($G_{yz(y)}$, GPa)	6.472

2.4. Finite Element Analysis (FEA)

The geometric model after assignment of material properties and orientations is meshed using solid elements (see figure 2). Periodic boundary conditions are also applied to the RVE at this stage and an input deck is generated for solution in commercial FEA software like ABAQUS. The details of FEA are covered in [13] and thus not repeated here.

Figure 2. Finite element mesh of the RVE using 8 node cubic 3D solid elements with reduced integration (C3D8R) in ABAQUS

3. Results and Conclusions

In table 3, A comparison of FEA properties predicted using the above approach after void compensation and correction for fibre volume fraction in yarn is made with the analytical predictions obtained from the CADEC software as well as with experimental values for two data points.

Table 3. Effective orthotropic properties predicted in Case 3

Woven composite effective property	FEA micromechanics	Analytical Micro-mechanics (CADEC)	Experimental
Modulus of Elasticity in Warp Direction, E_x^0	20.33 GPa	20.62 GPa	20.512 GPa
Modulus of Elasticity in Weft Direction, E_y^0	20.03 GPa	20.98 GPa	-
Out-of-plane Elastic Modulus, E_z^0	8.312 GPa	*	-
In-plane Shear Modulus, G_{xy}^0	3.843 GPa	4.828 GPa	3.855 GPa
Out-of-plane Shear Modulus, G_{zx}^0	2.504 GPa	*	-
Out-of-plane Shear Modulus, G_{yz}^0	2.510 GPa	9.073 GPa	-
In-plane Poisson Ratio, v_{xy}^0	0.16	0.156	-
In-plane Poisson Ratio, v_{yx}^0	0.15	*	-
Out-of-plane Poisson Ratio, v_{xz}^0	0.47	*	-
Out-of-plane Poisson Ratio, v_{zx}^0	0.19	*	-
Out-of-plane Poisson Ratio, v_{yz}^0	0.48	0.156	-
Out-of-plane Poisson Ratio, v_{zy}^0	0.20	*	-

IOP Conf. Series: Materials Science and Engineering **406** (2018) 012007 doi:10.1088/1757-899X/406/1/012007

This comparison shows that the proposed methodology is able to produce highly accurate results. The advantage of using this methodology over the analytical methodologies are multiple. Firstly, the out of plane properties and shear properties can be predicted with same degree of accuracy using this approach. Secondly, the approach can easily be extended to account for hybrid weaves. This is a big advantage and can prove very useful in evaluating new and novel material combinations.

References

[1] Adumitroaie A and Barbero E J 2011 Beyond plain weave fabrics - I. geometrical model *Compos. Struct.*

[2] Adumitroaie A and Barbero E J 2011 Beyond plain weave fabrics - II. Mechanical properties *Compos. Struct.*

[3] Barbero E J, Trovillion J, Mayugo J A and Sikkil K K 2006 Finite element modeling of plain weave fabrics from photomicrograph measurements *Compos. Struct.*

[4] Barbero E J, Lonetti P and Sikkil K K 2005 Finite element continuum damage modeling of plain weave reinforced composites *Compos. Part B Eng.*

[5] Adumitroaie A and Barbero E J 2012 Stiffness and strength prediction for plain weave textile reinforced composites *Mech. Adv. Mater. Struct.*

[6] Barbero E J, Damiani T M and Trovillion J 2005 Micromechanics of fabric reinforced composites with periodic microstructure *Int. J. Solids Struct.*

[7] Barbero E J, Lonetti P and Sikkil K K 2005 Finite element continuum damage modeling of plain weave reinforced composites *Compos. Part B Eng.*

[8] Farooqi J K and Sheikh M A 2006 *Finite element modelling of thermal transport in ceramic matrix composites* (University of Manchester, Manchester, UK)

[9] Sherburn M 2007 *Geometric and Mechanical Modelling of Textiles* (University of Nottingham)

[10] Barbero E J 2011 Introduction to Composite Materials Design *CRC Press*

[11] Mahmood A, Wang X and Zhou C 2013 Generic stiffness model for 3D woven orthogonal hybrid composites *Aerosp. Sci. Technol.*

[12] Shah S Z H, Choudhry R S and Khan L A 2017 Challenges in Compression Testing of 3D Angle-Interlocked Woven-Glass Fabric-Reinforced Polymeric Composites *J. Test. Eval.*

[13] Choudhry R S, Khan K A, Khan S Z, Khan M A and Hassan A 2016 Micromechanical modeling of 8-harness satin weave glass fiber-reinforced composites *J. Compos. Mater.* **51** 705–20

[14] Choudhry R S 2009 *Characterisation and modelling of impact damage in bonded joints of woven fibre polymeric composites* (University of Manchester, Manchester, UK)

[15] Choudhry R S, Hassan S F, Li S and Day R 2015 Damage in single lap joints of woven fabric reinforced polymeric composites subjected to transverse impact loading *Int. J. Impact Eng.*

IOP Conf. Series: Materials Science and Engineering **406** (2018) 012008 doi:10.1088/1757-899X/406/1/012008

X-ray μCT based assessment of thermal cycling induced cracks in non-crimp 3D orthogonal woven composite materials with porosity

M. Gigliotti[1*], Y. Pannier[1], Y. Sinchuk[1], R. Antoranz-Gonzalez[1], M.C. Lafarie-Frenot[1], S.V. Lomov[2]

[1]Institut Pprime, University of Poitiers, ISAE-ENSMA, CNRS 3346 - Department of Physics and Mechanics of Materials - 1, Avénue Clément Ader F-86962 Futuroscope Chasseneuil, France

[2]Department of Materials Engineering, KU Leuven Kasteelpark Arenberg 44 bus 2450, 3001 Leuven, Belgium

* marco.gigliotti@ensma.fr

Abstract. This work focuses on X-ray micro-Computed Tomography (μCT) assessment of thermal cycling behavior of a carbon fiber/epoxy matrix composite material reinforced with a non-crimp 3D orthogonal woven preform (3DNCOW) with porosity. Porosity levels and damage mechanisms – i.e. matrix cracking - induced by thermal cycling are characterized by X-ray micro-computed-tomography (μCT). Qualitative and quantitative descriptions of the morphology and the evolution of cracks with thermal cycling are carried out through the analysis of μCT scans of samples at different cycle numbers, with emphasis on the influence of porosity on the damage mechanisms induced by thermal cycling. X-ray μCT image-based Finite Element models are employed to provide rational interpretation to the observed phenomenological scenario.

1. Introduction

Due to their low through-thickness mechanical properties, 2D laminated composites have somehow limited application associated to quite poor mechanical behavior along the transversal direction ([1]).

To overcome such shortcomings various types of composite materials with three-dimensional (3D) fiber structures incorporating fibers in the direction normal to the plane of the material, commonly known as 3D fiber textile composites, have been developed. Among others, the 3D non-crimp orthogonal woven (3DNCOW) developed at North Carolina State University [2-4] stands out for allowing industrial-scale production of unitized thick preforms by using 3D weaving machines which are fully automated. To be fully exploitable, the mechanical behavior of such material needs to be deeply characterized. Extensive characterization of the quasi-static tensile and fatigue tension behavior of such material has been carried out and reported in references [5-7]. Not much work has been carried out concerning the response of such materials to environmental solicitations and thermal cycling.

The aim of this work is to provide X-ray micro-Computed Tomography (μCT) assessment of thermal cycling behavior of 3DNCOW material with porosity. Following an experimental work by the same authors [8], the present paper reports first some experimental evidence ([8]) in Sections 2 and 3, concerning:

- porosity levels and matrix cracking induced by thermal cycling and characterized by X-ray micro-computed-tomography (μCT),
- qualitative and quantitative descriptions of the morphology and the evolution of cracks with thermal cycling carried out through the analysis of μCT scans of samples at different cycle numbers.

Emphasis is put on the influence of porosity on the damage mechanisms induced by thermal cycling due to constrained thermal deformation of polymer matrix between carbon yarns.

X-ray μCT image-based Finite Element models are then employed in Section 4 for simulations: the aim of this activity is not to show the capability to perform detailed simulation of the complete thermal cycling behavior but the emergence of the employment of realistic materials models with defects necessary to provide rational interpretation to the observed phenomenological scenario and for the development of proper Virtual Testing Tools.

It should be emphasized that there exist extensive literature on thermal cycling of composite laminates ([9]) made by pre-pregs with continuous long-fibres. In this case, thermal cycling solicitations induce regular matrix cracking patterns in the composite plies whose density increases with increasing number of cycles: for laminates with simple stacking sequence, the specificity of the cracking pattern allows employing semi-analytical methods based on the concept of Energy Release Rate for composite laminate configurations [10].

With respect to these literature studies, the issue of thermal loading of 3D textile materials involves higher complexity, intrinsic to the complexity of the fibrous texture ([11-14]): the employment of dedicated detailed FE models is of paramount importance to understand the complex damage phenomenological scenario.

2. Material, samples, experimental setup, image-based FE model construction

This section illustrates the material, the samples, the experimental setup and the basic steps leading to the image-based FE model construction. More details about the description of the experimental setup and the employed material samples are given in reference [8].

2.1. Material, samples, experimental setup

The studied material is a non-crimp 3D orthogonal woven carbon/epoxy composite (3DNCOW, commercial name is 3WEAVE®). It was manufactured in 2011 by 3TEX, Inc by using a Vacuum Assisted Resin Transfer Molding (VARTM) process [2-4].

The epoxy resin used in this material is the West System 105 with 209 Extra Slow Hardener, with a glass transition temperature (T_g) equal to around 54°C.

Figure 1 illustrates a material Unit Cell (UC): the preform is composed of three warp layers (red) and four fill (yellow) layers, separated by two Z yarns (green), the presence of which leads to some spaces, respectively between the warp and weft yarns.

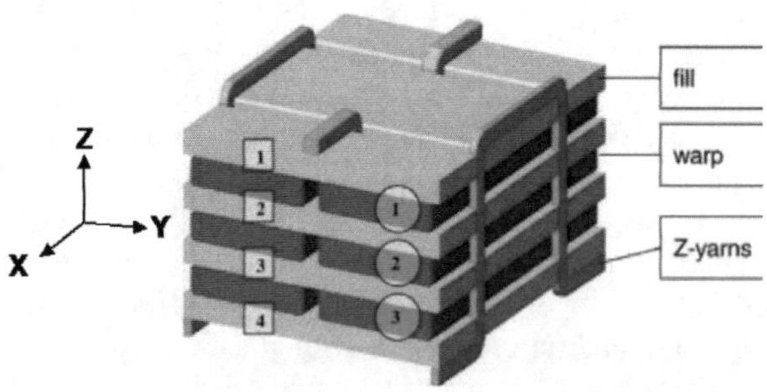

Figure 1. Schematic representation of the Unit Cell architecture of the 3DNCOW fabric (see also references [5-8]).

For the experimental activity, two different series of 8 samples were therefore cut in these regions, whose dimensions were determined based on the repetition of one (1 UC) or two (2x2 UC) geometric unit cells along the width and the length of the specimens. Table 1 reports experimental measurements of the specimens' dimensions.

Table 1. Average specimens' dimensions (8 samples per series, ± means standard deviation).

	1 UC	2×2 UC
X (mm)	5.7 ± 0.3	10.8 ± 0.3
Y (mm)	5.0 ± 0.1	10.8 ± 0.1
Z (mm)	2.7 ± 0.01	2.7 ± 0.01

Specimens were subjected to triangular thermal cycles (+50°C/-50°C, the maximum temperature slightly lower than T_g), with constant heating and cooling rates of 4°C/min, up to 1400 cycles, corresponding to laboratory conditions.

The X-ray micro-tomography imaging technique employed to visualize the 3D architecture of the composite material as well as the different defects present in the virgin sample or induced by thermal cycling employs UltraTom CT scanner, manufactured by RX-Solutions, allowing achieving a minimal resolution of around 0.25 µm. A sealed type micro-focus X-ray source with Tungsten target was used. Filament voltage and current were 60kV and 72µA leading to a focal spot around 5µm. For each specimen, 1440 radiographies were recorded with a Varian flat panel of 1920×1536 pixels. Reconstruction was achieved by FBK algorithm with XAct RX-Solutions software.

All the details of the image segmentation protocol including segmentation of matrix, yarns, porosity and cracks are reported in reference [8]: for the purpose of illustration, Fig. 2 shows a 2D slice of a 1 UC sample with a detail of the different phases to be segmented (Fig. 2a) and the corresponding gray values histogram (Fig. 2b).

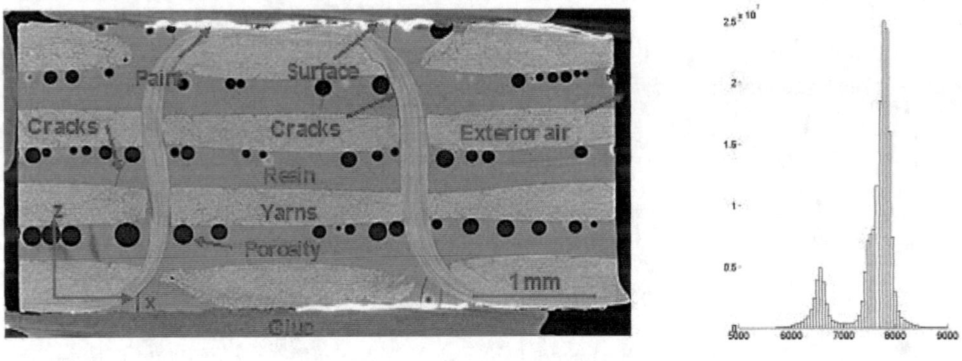

a) b)

Figure 2. Example of a 1 UC 2D slice obtained from the 3D tomographic reconstruction a) detail of the different phases to be segmented b) corresponding histogram of gray values (see also reference [8]).

The range between 5000 and 10000 gray levels is relatively narrow in comparison with the full image dynamic range (0 to 65536), showing that the contrast between the different phases is quite low. The darkest gray values correspond to the lowest intensity values (more to the left in the histogram) while the lightest ones correspond to highest intensities (more to the right in the histogram). Fig. 3b shows that there are two well-differentiated peaks; the first one corresponding to porosity and exterior air, the second one including matrix and yarns. However, since the matrix is slightly darker than the yarns, a left dissymmetry of the second peak can be noted. The cracks gray levels result between these two main peaks.

All segmentations are performed with the aid of Avizo® segmentation editor (a commercial editor that facilitates Boolean operations on labeled materials) and by employing a semi-manual procedure (see also reference [8] for more details).

One of the main interests of 3D scan of composite samples with complex texture is to get the real geometry of the composite. Figure 3 illustrates a 3D reconstruction of a 1 UC sample obtained by segmenting a XRay μCT scan of a virgin specimen. The shapes of the specimen and the yarns are obtained here by contouring them manually in different orthogonal slices. The final 3D shapes are then obtained by interpolations between successive contours.

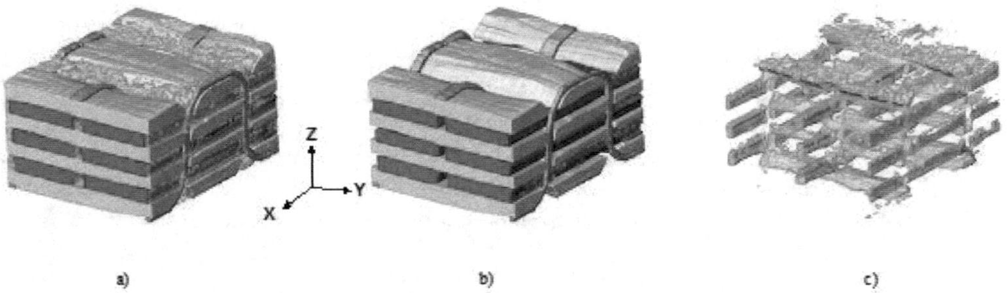

Figure 3. 3D representation of the different constituents of a real 3DNCOW Unit Cell after segmentation: a) full ample, b) sample without matrix, c) matrix with porosity ([8]).

Figure 3 illustrates the different constituents of the sample: red and yellow entities illustrate, respectively, warp and fill yarns, green entities illustrate the yarns in the through-the-thickness direction while blue entities illustrate the polymer matrix with porosity.

Table 2 resumes average and standard deviation global porosity volume fraction values (% V_{fp}) for 1 UC and 2×2 UC samples.

Table 2. Average and standard deviation global porosity volume fractions values for 1 UC and 2×2 UC samples.

Size	% V_{fp}	Resolution
1 UC	0.99 ± 0.06	8.34 μm
2×2 UC	0.85 ± 0.03	8.34 μm

2.2. μCT image-based FE model construction

The basic principles of the μCT image-based FE model construction are presented in detail – for a different textile material – in references [13, 14]. Generally, the technique can be referred as a "global-local" approach, since it allows simulating hierarchically - at different scales – the behavior of a tested sample. The approach involves:

- the employment of a voxel mesh FE model at the sample scale; this is referred as global model,
- the employment of smooth interface reconstruction followed by tetrahedral volumetric meshing at the yarn/matrix interface scale; this is referred as local model.

The FE analysis carried out by the global model at the sample scale allows identifying critical zones that are then analyzed in more detail through the use of the local model, which is refined both in terms of resolution and in quality of the mesh.

All models are developed within the framework of the ABAQUS® FE commercial software. Global and local models are linked by using the ABAQUS® submodeling technique, since the submodels have relatively small influence on the higher level models behavior. Simulation at the higher scale is performed independently from the lower scale model. By employing the node-based submodeling technique, the submodel boundary conditions are taken from the output file of the upper level simulation. Boundary condition setting needs to select all external nodes as set of driven nodes: then, ABAQUS® automatically interpolates the boundary conditions at driven nodes based on nodal values of a global (upper level) simulation.

2.2.1. Global model. X-ray µ-CT data of a 1 UC sample (8.34 µm resolution) are used for the construction of the global model at the sample scale. The 3D image has dimension 702×343×630 voxel (in X, Y, Z axis directions, respectively), with voxel size 0.00834 mm. The first necessary step of model creation is segmentation of the image. For the purpose of model construction, the ImageJ® software is employed for segmentation, by a semi-manual procedure including the selection of yarn cross-section regions and manual improvement by filling of tiny artificial gaps between the yarns and yarns interpenetrations. The pore segmentation was performed by separating the pores from the initial image using brightness/contrast adjustment, then by smoothing the pores by median filter application and by removing very tiny pores; finally, the pores were combined with the result of tow segmentation.

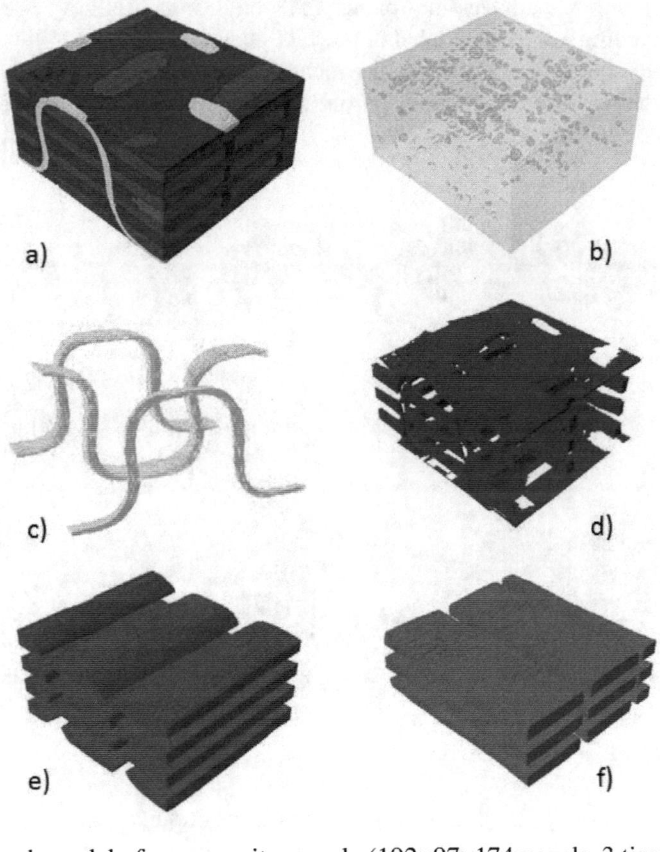

Figure 4. Voxel mesh model of a composite sample (192×97×174 voxels, 3 times decreased initial resolution): a) full sample b) porosity c) through-the-thickness yarns (with corresponding orientations) d) matrix phase e) warp and f) fill yarns.

The polymer matrix is considered as an isotropic material, while yarns are considered as orthotropic, whose behaviour is calculated by analytical and numerical homogenisation tools [13, 14].

For yarn orientation recognizing and assignment, it is assumed that the orientations of warp and fill yarns are along the X and Y directions, while concerning through-the-thickness yarns orientation, the orientation of the yarn central line is calculated along a discrete set of axial points: the yarn axes obtained by skeletonizing 2D cross-sections along the XZ plane. After that, manual selection of some marker points along the yarn axis is carried out and the coordinates of such points are stored in text format file using ImageJ® scripting. The orientation at each yarn segment between two marker points is calculated as the orientation of the straight line going through two marker points. The obtained values are used in Abaqus ORIENT subroutine for getting material orientation at any point of the vertical tows.

A MATLAB® script is implemented for generating FE model as Abaqus .INP files from 3D image segmentation results: the resulting FE mesh has regular grid structure where sets of elements correspond to different voxels colours. Decreasing image resolution with the aim of reducing the model size is acceptable as long as models with decreased resolution give comparable results with respect to models with full resolution. In the present case, it is found that the initial resolution can be reduced by 3 without degrading the quality of the output result. Fig. 4 illustrates in detail the resulting voxel mesh FE model: this model employs 192×97×174 voxels with a 3 times reduced initial resolution.

A non-porous sample can be obtained by numerically filling all pores with material matrix elements or by idealisation of the texture geometry (Fig. 5): in this last case, starting from 2D μCT manual tracing by splines and lines along real images in Abaqus CAE environment allows building a CAD sample draw. By doing this drawing it is assumed that the surfaces between warp and fill yarns are planar, which is close to the real microstructure. The employment of semi-idealised model allows adopting tet-meshing strategies. The resolution of the non-porous semi-idealised model in Fig. 5 is very close to the voxel mesh based model in Fig. 4.

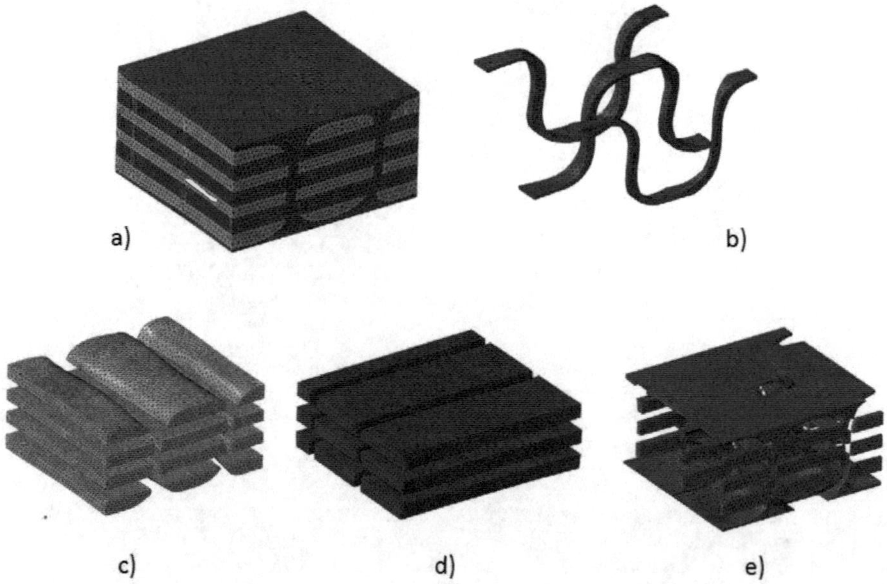

Figure 5. Semi-idealised tet-mesh model of a composite sample: a) full sample b) through-the-thickness yarns c) fill yarns d) warp yarns e) non-porous matrix phase.

2.2.2. Local model. At the local level (yarn/matrix interface), the model image is taken from the initial image of the whole composite sample (Fig. 6a), tetrahedral meshing with smooth material interface is used, with refined representation of the material interfaces and void contours (Figs. 6b and 6c).

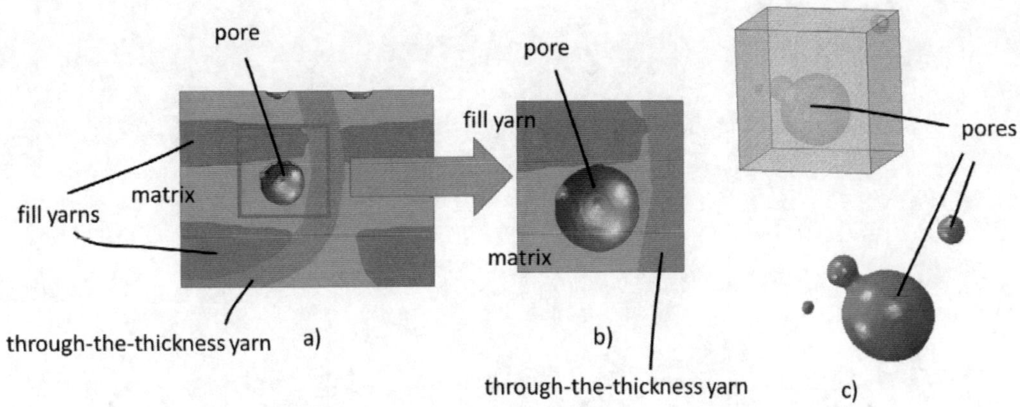

Figure 6. Local model: a) initial image taken from the image of the whole sample b) 2D model section c) 3D representation of the void included in the local model

The model with pores in Fig. 6 contains 1592120 elements, the equivalent model without pores (obtained by pore filling) contains 1462717 elements (more elements due to pore filling, less mesh refinement at the voids contour).

3. Thermal cycling induced cracks: experimental evidence

The experimental scenario of thermal cycling induced cracks is exhaustively reported in reference [8], here the main results of the experimental activity are recalled.

Fig. 7 illustrates the 3D representation of 1 UC and 2×2 UC samples and the development of matrix cracks (in blue) after 200, 800 and 1400 thermal cycles. After 200 thermal cycles (Figs. 7a and 7b), few small cracks can be observed in each sample. These first cracks are observed in an inter-warp plane, close to the surface of the sample. These critical regions correspond to zones with severe yarn curvature. Crack onset at specific critical locations is also affected by the presence of random defects (texture defects, porosity …).

After 800 thermal cycles (Figs. 7c and 7b) the crack system is much more developed: for both samples, almost all the critical zones close to the curved Z yarns are cracked. The cracks are observed along the Z-yarns in the inter-warp planes and also on the upper and lower surfaces of the specimen in the inter-fill matrix-rich zones.

After 1400 thermal cycles (Figs. 7e and 7f), significant crack propagation can be noted: almost all the vertical interfaces of the Z-yarns are fully cracked, and the cracks present in the upper and lower inter-fill matrix layers tend to coalesce.

The crack patterns and their evolution with the number of cycles appear very similar in 1 UC and the 2×2 UC samples: the cracking scenario (highly schematically) includes:

- crack onset at the most critical sites (interfaces between curved z yarns and matrix), multiplication of cracks up to saturation of all the "available" critical sites,
- crack propagation mainly along the thickness direction.

This scenario is quite schematic but is observed almost systematically. However, no information is gained about the effect of pores.

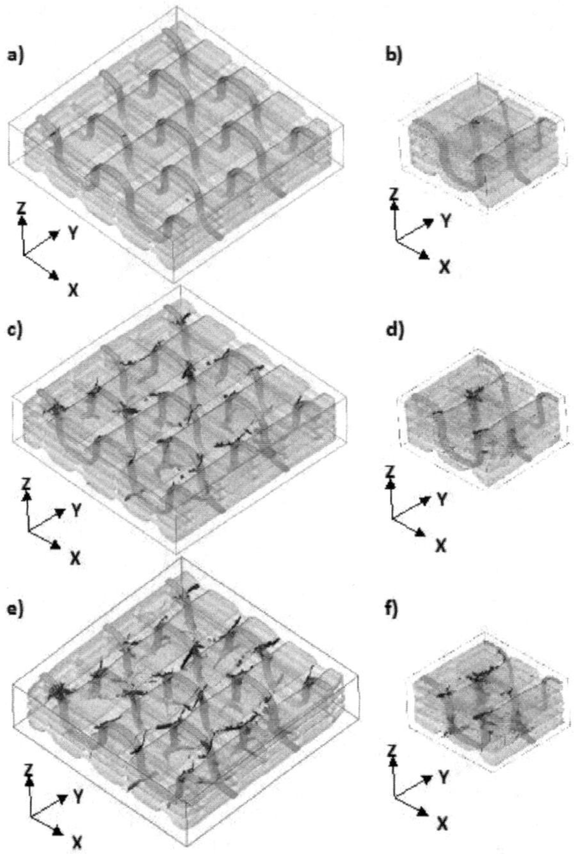

Figure 7. Cracking system for 2×2 UC (left), and 1 UC (right) samples; a) and b) 200 cycles, c) and d) 800 cycles, e) and f) 1400 thermal cycles. The resolution of the tomography scans is 8.34 μm (see also reference [8]).

Figure 8 shows a 2D tomographic section of the material after 1400 thermal cycles, visualizing a z-yarn in an inter-warp plane (section AA), and two orthogonal sections BB and CC. The comparison between BB and CC sections points out the deformations of the fill yarns and the z-yarns induced by the weaving near the exterior surfaces of the plate. The cracks present in this specimen after 1400 cycles are displayed in these images in blue, green and red colors: blue cracks initiate and propagate inside the specimen at the interface of the z-yarns, red cracks develop in the surface inter-fill matrix rich regions. The green colour allows differentiating the intra z-yarn cracks that connect the red and the blue cracks.

Figures 8a and 8b show clearly the presence of a pore is correlated to a highly cracked zone: moreover, when a pore is close to this zone, the crack is slightly deviated from the interface or passes through the porosity. The same observations can be made for the cracks in the resin pockets within the inter-fill planes, an example of which is given in Fig. 8c. The tortuous trajectory of the cracks is deviated locally by the presence of a nearby void.

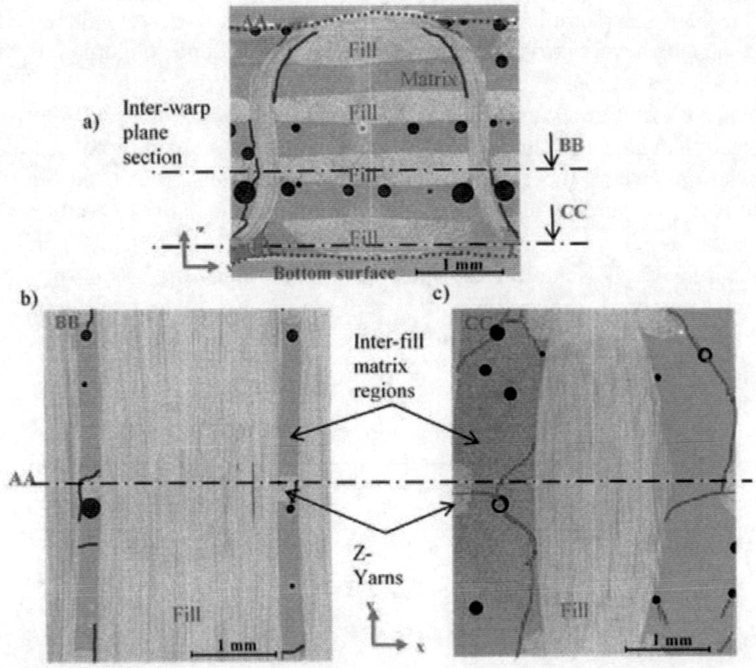

Figure 8. Crack morphology in a) through-the-thickness and b), c) in-plane slices after 1400 thermal cycles along through-thickness z yarns in concave zones (blue), in surface resin rich region (red), and inside z yarn (green) (see also reference [8]).

4. Discussion based on FE model simulations

As shown in Section 3, μCT experimental measures allow identifying zones which are particularly sensible to thermal cycling, that is, prone to develop damage due to thermal cycling: Fig. 9 illustrates the results of μCT scan carried out on the 1 UC sample after 1400 thermal cycles.

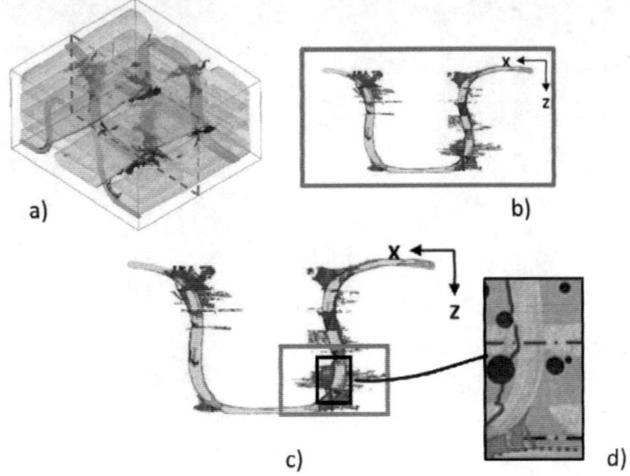

Figure 9. μCT images of 1 UC sample after 1400 thermal cycling: a) μCT scan of the whole sample b) severely damaged yarn c) concentration of damage in a zone with severe curvature d) in the presence of porosity (1 void close to the curved yarn).

The scan of the complete sample (Fig. 9a) allows capturing a severely damaged yarn (Fig. 9b), particularly in a zone with severe curvature (Fig. 9c) and in the presence of some porosity (Fig. 9d): one relatively big void (diameter around 0.25 mm) is close to the curved yarn (Fig. 9d). This zone has been the location of damage onset at a early number of thermal cycles (around 200 thermal cycles) and is much more damaged than adjacent curved yarns without voids. Moreover, voids act as crack attractors.

In order to better understand these issues, numerical simulations are carried out by employing the µCT image-based FE (global and local) models described in Section 2. In this analysis, the entire sample is subjected to a static step involving a temperature difference, ΔT (ΔT = 100°C), between 50°C (close to the material T_g, at which the sample is assumed to be stress free) and -50°C (minimum temperature value during thermal cycling); the analysis is purely elastic, the material properties in Table 3 are employed: the matrix is isotropic, the yarn is orthotropic; the yarn properties are calculated by numerical homogenization [13, 14].

Table 3. Material properties for FE simulations.

	Matrix	Yarn (longitudinal)	Yarn (transverse)
Young's Modulus [MPa]	3600	147500	17600
Poisson's ratio [–]	0.3	0.23	0.26
Coefficient of Thermal Expansion [K^{-1}]	5×10^{-5}	0.45×10^{-6}	17.3×10^{-6}
Density [g/cm^3]	1.23	1.56 (rule of mixtures)	

The global analysis carried out by the voxel-model containing porosity (Fig. 10a and 10b) allows identifying the zone of critical stress concentration (round circle in Fig. 10a, Fig. 10b) which is the same captured experimentally by µCT scans; in this zone the average Von Mises stress is around 70 MPa.

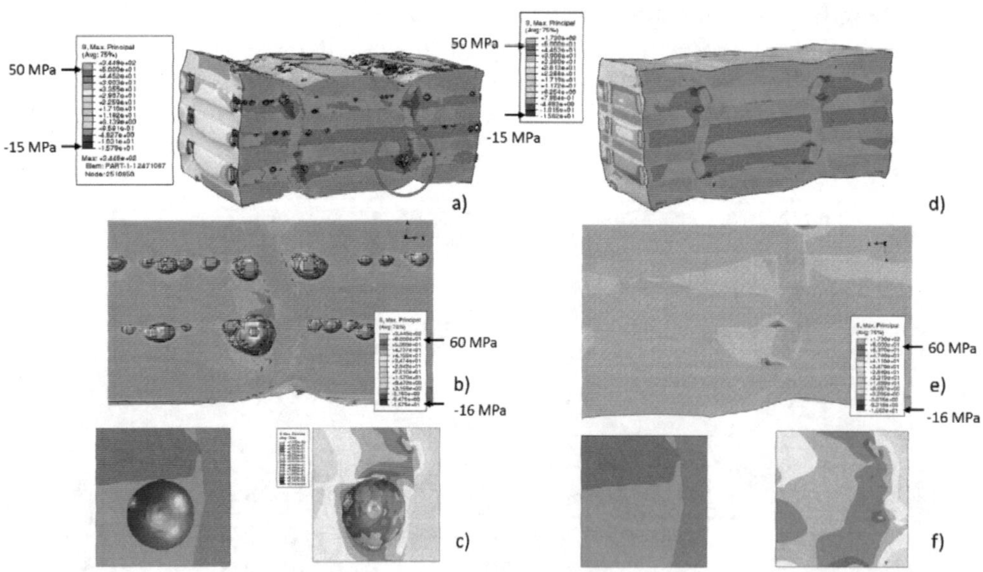

Figure 10. FE simulations (static elastic analysis (ΔT = 100°C): a) global voxel-model with porosity b) detail of a severely stressed zone (max Von Mises stress) c) local tet-mesh model of the critical zone d) global semi-idealised tet-mesh model without porosity e) detail of a severely stressed zone f) local tet-mesh model of a critical zone.

A more detailed analysis carried out by the refined local model (Fig. 10c) allows understanding that the stress concentration is particularly high in the vicinity of the void and on the void external surface (with Von Mises stress values between 57 MPa and 88 MPa at the void external surface): stress concentrations tend to "deviate" from the yarn/matrix surface to the void external surface. The observed behaviour is related to the consistent difference in the coefficients of thermal expansion of matrix and tows, the tow regions represent a true constraint to the free thermal expansion (or contraction of the matrix); this constraint effect is then amplified by the presence of pores.

The same analysis carried out on the semi-idealised tet-mesh model without porosity (Figs. 10d and 10e) does not allow discriminating between different yarns and – within a severely curved yarn – does not allow discriminating between curved zones with close values of curvatures: the average Von Mises stress in these zones is around 60 MPa that is 15% lower than that captured by the voxel-mesh model in the most critical zone with porosity (comparison between Figs. 10b and 10e). An analysis carried out by a local tet-mesh model of a critical zone (Fig. 10f) confirms this trend; moreover – differently from the model with porosity (Fig. 10c) - stress concentrations lay entirely on the yarn/matrix interface (with values around 65 MPa, that is 30% lower than those calculated by the model with porosity on the external pore surface). This analysis is purely comparative between different configurations since the critical values of stress are not available for these materials.

The analysis illustrated in Fig. 11 is of different type: this analysis is carried out by employing the local tet-mesh model in a material with porosity and consists in calculating the elastic energy release rate when passing from a configuration where a crack initially located very close to the fiber/matrix interface (Fig. 11a, configuration O) either propagates along the fiber/matrix interface (Fig. 11b, configuration A) either deviates to a close pore (Fig. 11c, configuration B): the analysis is comparative (no critical energy release rate values are available), the three cracks configurations have the same surface (equal to around 0.004 mm^2).

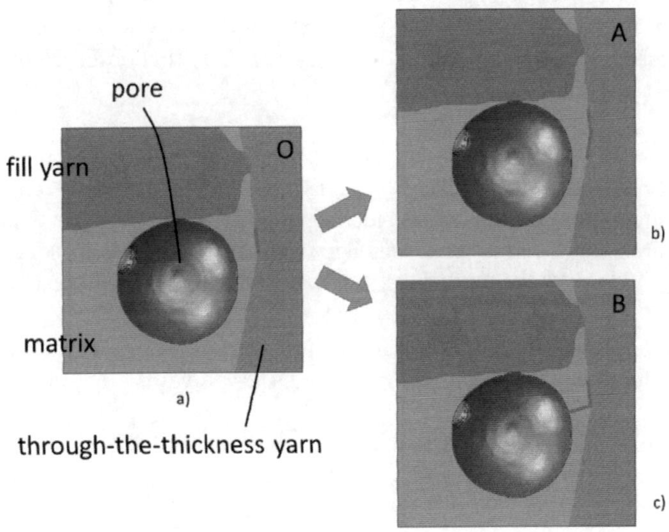

Figure 11. Schematic illustration of the calculation of the energy release rate a) configuration O, crack emanating close to the matrix/yarn interface b) configuration A, crack propagating along the matrix/yarn interface c) configuration B, crack deviating towards a void.

It is calculated that the energy release rate for passing from O to B is around 50% lower than passing from O to A which means that the transition O to B is preferred from an energetic point of view (based on a purely elastic analysis).

Fig. 12 illustrates an analysis qualitatively similar to that illustrated in Fig. 10: in this case the local tet-mesh model with porosity is used to calculate the elastic energy release rate when passing from a configuration where a crack initially located (originating from) on the pore surface (Fig. 12a, configuration O') propagates along the fiber/matrix interface (Fig. 12b, configuration C).

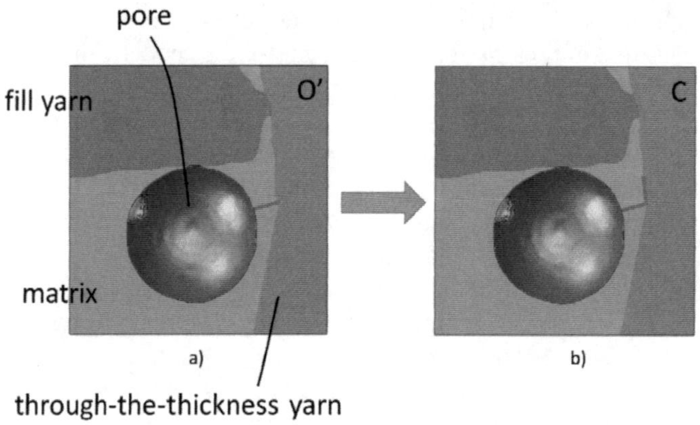

Figure 12. Schematic illustration of the calculation of the energy release rate a) configuration O', crack emanating from a void b) configuration C, crack propagating along the matrix/yarn interface.

It is calculated that the energy release rate for passing from O' to C is around 70% higher than passing from O to A and 40% higher than passing from O to B: this means that the transition O' to C is very unlikely to occur based from an energetic point of view (purely elastic analysis). This has a plausible physical explanation: in the material/structure illustrated in Fig. 12, a crack emanating from a void is "arrested" by the presence of a nearby yarn (effect very similar to that observed in structures with cracks propagating towards rigid crack arresters).

The conclusion of this numerical study is that:

- based on a simple elastic analysis, the global voxel-mesh model built on μCT images, including all defects captured by the μCT resolution (e.g. voids, texture defects), is able to catch the location of stress concentrations, thus identifying the locations where damage is likely to onset, when compared to μCT images of damaged samples. The critical zones include pores in the vicinity of a yarn/matrix interface (stress concentrations are calculated on the pore external surface),
- by the same analysis, the global semi-idealised tet-mesh model without defects is not able to discriminate between similar critical zones (e.g. curved yarns are almost all critical at the same extent),
- the global voxel-mesh model based on μCT images and the related local tet-mesh model (extracted from the global model after identification of criticalities) predicts average Von Mises stresses that are around 15% higher than those calculated in the same zones by the semi-idealised tet-mesh model. The defect plays a decisive role in the determination of stress concentrations that promote then damage onset and propagation,
- based on simple energetic considerations and on the explicit calculation of energy release rate by means of the local tet-mesh model with porosity, the simulation predicts that a crack emanating from a yarn/matrix interface is more likely to deviate towards a void (when the void is close to the interface) than propagating along the yarn/matrix interface. This analysis depends obviously on the distance of the pore from the yarn/matrix interface, but this study has not been carried out, only a comparative calculation has been performed,
- based on simple energetic considerations and on the explicit calculation of energy release rate by means of the local tet-mesh model with porosity, the simulation predicts that a crack emanating from a pore is not likely to propagate along the yarn/matrix interface, since the yarn

acts as a "crack arrester". Therefore either a crack is not likely to onset at a pore surface, either a crack that onsets at a pore surface are arrested by the presence of a yarn.

5. Conclusions and perspectives

The experimental results of thermal cycling (up to 1400 thermal cycles) between 50°C and -50°C show that the crack patterns and their evolution with the number of cycles appear very similar in 1 UC and the 2×2 UC samples: the cracking scenario (highly schematically) includes:

- crack onset at the most critical sites (interfaces between curved z yarns and matrix), multiplication of cracks up to saturation of all the "available" critical sites,
- crack propagation mainly along the thickness direction.

In this scenario, the presence of a pore is correlated to a highly cracked zone: moreover, when a pore is close to this zone, the crack is slightly deviated from the interface or passes through the porosity. The tortuous trajectory of the cracks is deviated locally by the presence of a nearby void.

The numerical analysis carried out by employing μCT image-based FE models at the sample (global) scale (voxel-mesh and semi-idealised tet-mesh models) and at the matrix/yarn interface (local) scale (tet-mesh model) allows concluding that:

- based on a simple elastic analysis, the global voxel-mesh model based on μCT images, including all defects captured by the μCT resolution (e.g. voids, texture defects), is able to catch the location of stress concentrations, thus identifying the locations where damage is likely to onset, when compared to μCT images of damaged samples. The critical zones include pores in the vicinity of a yarn/matrix interface (stress concentrations are calculated on the pore external surface): by the same analysis, the global semi-idealised tet-mesh model without defects is not able to discriminate between similar critical zones (e.g. curved yarns are almost all critical at the same extent),
- based on simple energetic considerations and on the explicit calculation of energy release rate by means of the local tet-mesh model with porosity, the simulation predicts that a crack emanating from a yarn/matrix interface is more likely to deviate towards a void (when the void is close to the interface) than propagating along the yarn/matrix interface: the simulation predicts also that a crack emanating from a pore is not likely to propagate along the yarn/matrix interface, since the yarn acts as a "crack arrester". Therefore, either a crack is not likely to onset at a pore surface, either a crack that onsets at a pore surface is arrested by the presence of a yarn.

Work is in progress to enhance both the experimental capability to catch the damage scenario through the realization of *in-situ* tests capable to capture the details of damage onset and propagation and the numerical capability to perform multi-physical simulations taking into account realistic material model behavior. These efforts should allow gaining a better understanding of the phenomenological scenario and pave the way to the development of Virtual Testing tools.

Acknowledgements

This work was partially funded by the French Government program "Investissements d'Avenir" (LABEX INTERACTIFS, reference ANR-11-LABX-0017-01; EQUIPEX GAP, reference ANR-11-EQPX-0018). S.V. Lomov holds the Toray Chair for Composite Material at KU Leuven, the support of which is gratefully acknowledged. We acknowledge Dr. A. E. Bogdanovich and Dr. D. Mungalov for providing (in 2011, at the time affiliated with 3TEX, NC, USA) the material used in this work.

References

[1] Miravete A 1999 in *3-D Textile Reinforcements in Composite Materials*, Woodhead Publishing, Ed. Miravete A, ISBN: 9781855733763
[2] Mohamed MH, Bogdanovich AE, Dickinson LC, Singletary JN and Lienhart RB 2001 *SAMPE Journal* **37** 8-17
[3] Dickinson L and Bogdanovich AE 2002 In: *IMECE'02, ASME international mechanical engineering congress & exposition*, New Orleans, 2002

[4] Bogdanovich AE and Mohamed MH 2009 *SAMPE Journal 45* 8
[5] Bogdanovich AE, Karahan M, Lomov SV and Verpoest I 2013 *Mechanics of Materials* **62** 14
[6] Karahan M, Lomov SV, Bogdanovich AE, Mungalov M and Verpoest I 2010 *Composites Part A: Applied Science and Manufacturing* **41** 1301
[7] Karahan M, Lomov SV, Bogdanovich AE and Verpoest I 2011 *Composites Science and Technology* **71** 1961
[8] Gigliotti M, Pannier Y, Antoranz Gonzalez R, Lafarie-Frenot MC and Lomov SV 2018 *Composites Part A: Applied Science and Manufacturing* **112** 100
[9] Rouquie S, Lafarie-Frenot MC, Cinquin J and Colombaro AM 2005 *Composites Science and Technology* **65** 403
[10] Lafarie-Frenot MC, Hénaff-Gardin C and Gamby D 2001 *Composites Science and Technology* **61** 2327
[11] Zhang C, Binienda WK, Morscher GN, Martin RE and Kohlman LW 2013 *Composites Part A: Applied Science and Manufacturing* **46** 34
[12] Tsukrov I, Bayraktar H, Giovinazzo M, Goering J, Gross T, Fruscello F and Martinsson L 2011 *International Journal of Fracture* **172** 209
[13] Sinchuk Y, Pannier Y, Gueguen M, Tandiang D and Gigliotti M 2017 *International Journal of Solids and Structures* **Accepted In Press**
[14] Sinchuk Y, Pannier Y, Gueguen M and Gigliotti M 2017 *Proc IMechE Part C: Journal of Mechanical Engineering Science* **Accepted In Press**

IOP Conf. Series: Materials Science and Engineering **406** (2018) 012009 doi:10.1088/1757-899X/406/1/012009

Effect of intra-ply voids on the homogenized behavior of a ply in multidirectional laminates

A Matveeva[1,*], D Garoz[2,3], R D B Sevenois[2,3], M Zhu[3,4], L Pyl[5], W Van Paepegem[2] and L Farkas[1]

[1] Siemens Industry Software NV, Interleuvenlaan 68, 3001, Leuven, Belgium
[2] Ghent University, Department of Materials, Textiles and Chemical Engineering – MaTCh, Tech Lane Ghent Science Park – Campus A, Technologiepark-Zwijnaarde 903, 9052 Zwijnaarde (Gent), Belgium
[3] SIM vzw, Technologiepark 935, 9052 Zwijnaarde (Gent), Belgium
[4] KU Leuven, Department of Materials Engineering, Kasteelpark Arenberg 44, 3001 Leuven, Belgium
[5] Department of Mechanics of Materials and Constructions (MeMC), Faculty of Engineering Sciences, Vrije Universiteit Brussel (VUB), Pleinlaan 2, 1050 Brussels, Belgium

* anna.matveeva@siemens.com

Abstract. This work focuses on the effect of intra-ply voids on the homogenized nonlinear behavior of a ply in multidirectional composites. Voids were modeled explicitly on the fiber scale and linked to the ply-scale by the recently developed two-scale framework which couples Classical Laminate Theory on the macro-scale with Finite Element analysis on the micro-scale. Laminates $[\pm45]_{2s}$ and $[\pm67.5]_{2s}$ were used as validation cases. The computed homogenized behavior of plies with and without voids for each laminate were compared against existing experimental data on manufactured plates. The nonlinearity of the homogenized stress-strain curves of all models is in a good agreement with experiments up to 1% of applied deformation for a laminate $[\pm45]_{2s}$ and up to 0.4% for a laminate $[\pm67.5]_{2s}$. The effect of voids was assessed only virtually and it is shown that 4% of void content decreases the ply strength by 30%, transversal Young's and shear moduli by around 10% and 8% respectively, whereas longitudinal stiffness is only slightly affected by the presence of voids. This work is the first step towards automatization of the virtual identification of the complete set of damage-plasticity parameters for the LMT-Cachan damage model accounting for the presence of intra-ply voids.

1. Introduction

The effect of voids on the mechanical behavior of the fiber reinforced composites, including multidirectional laminates and textile based composites attracts attention of the composite community since many years. A recent review of the effect of voids on the mechanical properties of composites can be found in [1]. One of the main conclusions is that, despite the extensive amount of available research, further assessment of the effect of voids is needed. This especially in automotive and aerospace industry, where voids are the key defects in manufacturing techniques such as out-of-autoclave (OoA) and automated tape laying (ATL).

In [2] a multiscale approach to account for effect of voids is presented. The local material properties including transverse strength and Mode I fracture toughness are computed in the presence of voids at the scale of fibers (micro-scale) and then subsequently homogenized to the ply-scale (macro-scale) and

IOP Conf. Series: Materials Science and Engineering **406** (2018) 012009 doi:10.1088/1757-899X/406/1/012009

assigned to "weak volumes", representing the presence of voids. Extended Finite Element (XFEM) is employed at the ply-scale to analyze the crack density evolution in cross-ply laminates with and without the presence of voids. This approach allows to track initiation and propagation of multiple cracks in a ply, as well as to link them to inter-laminar cracks/delamination [3]. The alternative to the time-consuming simulations with XFEM, is to use Continuum Damage Models (CDM) at the ply-scale, where the properties of the ply/or the region of interest are degraded based on a damage evolution law, which can be modified according to the void content at the lower scale.

A ply damage model developed at LMT-Cachan [4–6] and implemented in the FE code Siemens PLM Software LMS Samtech SAMCEF [7] couples the damage evolution in the transverse and in-plane shear directions taking into account the plastic behavior of the reinforced polymer. It is also possible to address the damage evolution in the fiber direction and through-the-thickness. The model has been validated for intra-laminar behavior at the coupon level for UD plies [8]. The model requires several input parameters characterizing elastic behavior, damage, plasticity and final ply failure. The parameter identification requires experimental tests on different laminates, such as quasi-static tensile and compression tests on $[0/90]_{4s}$ laminates and cyclic tensile tests on $[\pm45]_{2s}$ and $[\pm67.5]_{2s}$ laminates [4, 8, 9]. The effect of voids on the composite mechanical properties and on the set of parameters for LMT-Cachan model has been investigated in [10]. A clear influence of voids on the damage initiation and propagation and hence on the parameters of the LMT-Cachan damage model was established. Such experimental assessments of the effect of voids is very challenging, especially if to compare the effect against the reference material without voids. By changing the manufacturing parameters, e.g. by lowering the autoclave pressure and the cure temperature, and ensuring the presence of voids in a composite, the behavior of its constituents, mainly the matrix and the fiber/matrix interface might be affected [11]. To decouple the influence of voids from the influence of manufacturing parameters is not a straightforward task. Instead, a virtual parameter identification can offer a promising alternative, however the validation of the methodology is challenging for the same reasons. In [12] Garoz et al. provides a methodology to identify parameters of the LMT-Cachan damage model by means of dedicated micro-mechanical virtual tests with detailed constituent material models, representative geometry and specific load conditions. In order to verify the approach, Garoz et al. compared the ply model with the identified parameters (based on physical tests) against the micro-mechanical virtual tests on the $[\pm45]_{2s}$ and $[\pm67.5]_{2s}$ laminates. The two approaches gave the same results, except for some small discrepancies in the transverse stress-strain curve of the virtual test on the $[\pm67.5]_{2s}$ laminate. In the current study, the proposed workflow in [12] was further modified by considering the presence of voids at the micro-scale level.
To assess the methodology, firstly, micro-mechanical virtual tests on a void-free material were employed with the objective to compare the homogenized behavior of plies in $[\pm45]_{2s}$ and $[\pm67.5]_{2s}$ laminates under the static tensile loading against existing experimental data on the laminates. Secondly, voids were embedded at the micro-scale and their effect on the homogenized ply behavior was virtually assessed.

2. Material and Experiments
The composite laminates used in this work are manufactured from unidirectional (UD) prepregs PYROFIL™ TR 360E250S supplied by Honda R&D Co., Ltd. and Mitsubishi Chemical Corporation [13]. The prepregs were made of polyacrylonitrile (PAN) based carbon fibers embedded in PYROFIL #360 resin with a volume fraction of 60%. Several plates were made of UD prepregs with a stacking sequence $[0]_4$, $[90]_8$, $[\pm45]_{2s}$, $[\pm67.5]_{2s}$. The plates were tested under static tension and measured stress-strain curves served as reference data for modelling results. Several tests on constituents were performed to obtain the input data for the models.

2.1. Input data for the micro-scale model
- Fiber properties

The elastic properties of the carbon fiber are reversed engineered from the 3D homogenized elastic tensor of the UD ply [14]. Full 3D orthotropic elastic stiffness tensor of the ply is computed by measuring the propagation velocity of both the longitudinally and transversally polarized bulk waves at various symmetry planes of a UD Carbon/Epoxy laminate using ultrasonic insonification. Determined Young's moduli, shear moduli and Poisson's ratios can be found in table 1. For the in-depth discussion on the effect of high frequencies produced by the ultrasound on the elastic properties of the material under investigation, the reader is referred to [14].

- Matrix behavior in tension

To test the matrix behavior in tension, static tensile tests on the matrix were performed according to the ASTM D638 using a testing machine Instron 5567 with mechanical wedge-type clamps and the test speed 2 mm/min. The tests were accompanied by a mechanical extensometer and a Digital Image Correlation (DIC) Limess VIC 3D system, in 2D mode with one camera. 5 samples were tested, all failing in the gauge section with a brittle failure. The moduli of elasticity were calculated as a slope of the stress-strain curves between 0.1% and 0.3% strain. Table 1 contains average values and standard deviations of Young's modulus and maximum strain at failure in tension, measured by DIC.

- Matrix behavior in shear

To test the matrix behavior in shear, static tensile tests on the matrix were performed according to the ASTM D5379 using a testing machine Instron 5800R with Iosipescu test setup and the test speed 1 mm/min. The tests were accompanied by a DIC system: VIC 3D system, 3D mode with two cameras. The shear modulus was calculated as a maximum progressive slope of the longitudinal stress-strain curve using DIC. Table 1 contains average values and standard deviation of shear modulus measured by DIC.

- Interface

The properties of the interface were not measured for this set of fibers and the matrix. Interfacial strength and toughness are taken from [15] for a similar carbon/epoxy material.

Material properties of the constituents are summarized in table 1.

Table 1. Material properties of constituents

Fibers [14]	
Property	**Value**
Young's moduli	
E_{11} (GPa)	23.3
E_{22} (GPa)	18.4
E_{33} (GPa)	18.4
Poisson's ratio	
ν_{12}(-)	0.02
ν_{13}(-)	0.02
ν_{23}(-)	0.27
Shear moduli	
G_{12} (GPa)	36.9
G_{13} (GPa)	36.9
G_{23} (GPa)	7.2
Fiber radius	
r (μm)	3.6

Matrix	
Property	**Average value (std)**
Young's modulus	
E (MPa)	2.84 (0.1)
Shear modulus	
G (MPa)	0.97 (0.013)
Poisson's ratio	
ν (-)	0.46[a]
Strain at failure in tension	
ε_f^t (%)	2.63 (0.01)
Strain at failure in shear	
γ_f^s (%)	11 (0.03)

Interface [15]	
Property	**Value**
Interface maximum strength	
τ_n (MPa)	50
τ_t (MPa)	75
Interface critical strain energy release rates	
G_{IC} (N/mm)	0.002
G_{IIC}, G_{IIIC} (N/mm)	0.006
Mixed-mode interaction parameter	
η (–)	1.45

[a] *matrix Poisson's ratio was calculated to fulfil the isotropic assumption of* $G=E/(2(1+\nu))$

2.2. Data used for validation of the macro-scale behavior

- UD laminate

Several plates made of 4-ply UD prepregs with a stacking sequence $[0]_4$ and $[90]_8$ were tested under tension to derive longitudinal and transverse Young's moduli of a ply. A testing machine Instron 4505 with mechanical wedge-type clamps and the test speed 2 mm/min was equipped with Instron extensometer and digital image correlation (DIC system): Limess VIC 3D system, 3D mode with two cameras. Transverse modulus in table 2 was calculated from results on $[90]_8$ laminate as a slope of the stress-strain curve using DIC between 0.05% and 0.1% strain. Longitudinal modulus in table 2 was calculated from results on $[0]_4$ laminate as a slope of the stress-strain curve using DIC between 0.1% and 0.3% strain. DIC Poisson's ration was calculated using DIC response in linear stress-strain curve from the results on $[0]_4$ laminate based on longitudinal and transverse strain available from DIC measurements.

- Laminates $[\pm45]_{2s}$ and $[\pm67.5]_{2s}$

Produced laminates with a stacking sequence $[\pm45]_{2s}$ and $[\pm67.5]_{2s}$ were tested according to ASTM D3039. Laminates $[\pm67.5]_{2s}$ were tested in different loading phases, instead of immediate test to failure. A testing machine Instron 4505 with mechanical wedge-type clamps and the test speed 2 mm/min was equipped with Instron extensometer and digital image correlation (DIC system): VIC 3D system, 3D mode with two cameras. Shear modulus in table 2 was calculated from results on $[\pm45]_{2s}$ laminate as a slope of the shear stress-strain curve using DIC between 0.2% and 0.6% strain. Measured stress-strain curves are presented in the result section (figure 6, figure 7, figure 8) in a local coordinate system (CS) of a ply, the transformation was done as proposed in [4].

Table 2. Material properties of a UD ply

UD (ASTM + DIC)	
Property	Average value (std)
Young's moduli	
E_{11} (GPa)	127.52 (1.24)
E_{22} (GPa)	8.49 (0.721)
Poisson's ratio	
ν_{12}(-)	0.34 (0.0143)
In-plane shear modulus	
G_{12} (GPa)	4.11 (0.23)
Maximum shear stress	
τ_{12}^{max} (MPa)	52.27 (0.95)
Maximum shear strain	
γ_{12}^{max} (%)	3.66 (0.96)

3. Virtual Material Characterization with the VMC ToolKit

The Simcenter™ VMC ToolKit of Siemens PLM Software is a set of software tools developed for virtual material characterization (VMC) to replace physical tests with virtual experiments. The VMC ToolKit allows detailed material modelling at different scales and covers wide range of supported composite types, such as UD plies, 2D/3D woven, non-crimp, weft-knitted, 2D/3D braided fabrics. More detailed concept of the tool can be found in [16]. Several validation cases to predict elastic properties of different composites are demonstrated in [17–19]. Current results are based on the recent extension of the VMC ToolKit towards prediction of non-linear and damage behavior accounting for the presence of voids at the fiber scale.

3.1. Micro-scale FE models with the presence of voids

To estimate the "as-manufactured" properties of specimens, *Random Packing* (RP) functionality of the VMC ToolKit was extended with a feature of *Void Generation* at the fiber scale. The matrix is modeled as a solid body with voids of different shapes (cylindrical and spheroidal) and different sizes. The percentage of voids in the model and their spatial distribution can vary according to user input. UD fibers remain to have constant diameters and random distribution in the matrix with predefined volume fraction. The algorithm [20] was employed to randomly distribute fibers in the matrix with a modification of randomizing the minimum distance between fibers [21]. Fibers and cylindrical voids are generated as an extrusion of corresponding circular cross sections through the unit cell thickness and spheroidal voids are revolved according to their equatorial radius (minor radius) and the distance from the center to pole along the symmetry axis (major radius). Radii of voids, as well as a through-the-thickness position of spheroidal voids can be defined either randomly or manually as a user input.

Three different test cases of periodic unit cells of randomly distributed fibers in a matrix are generated: matrix being modelled without voids, matrix containing spheroidal voids with void fraction of 2.3% and matrix containing spheroidal voids and cylindrical voids with void fractions of 2.2% and 1.9% respectively (see figure 1). The cylindrical void was placed in the center of the unit cell with a radius of 3 microns. The size of spheroidal voids was randomly chosen between 1 and 3 microns for both minor and major radii. The dimension of the voids and their volume fractions were chosen randomly from the range of values found in literature [1]. In all cases the size of the unit cell is 38.5 x 38.5 x 3.85 μm^3, the radius of the fibers is 3.6 μm and the fiber volume fraction is fixed to 60%.

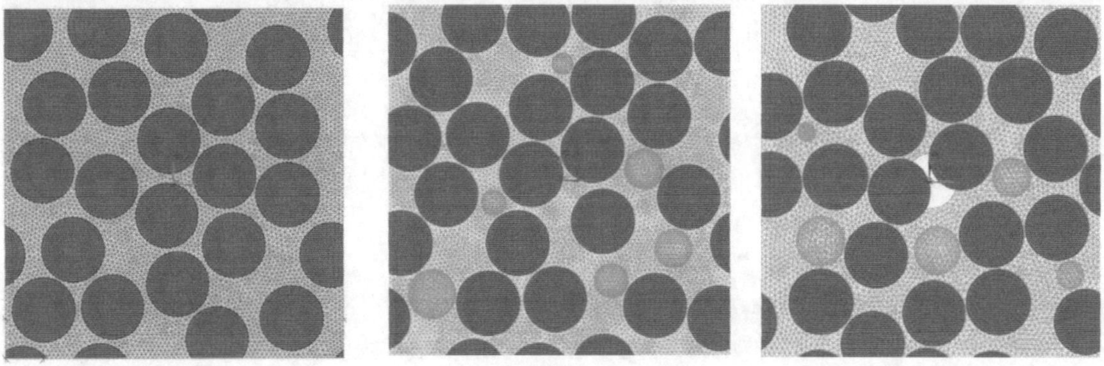

(a) (b) (c)

Figure 1. (a) RP model without voids; (b) RP model with 2.3% of spheroidal voids; (c) RP model with 2.2% of spheroidal voids and 1.9% of cylindrical voids.

In all three models, the matrix is discretized using linear 4-node tetrahedal solid elements and fibers using the 8-node hexagonal solid elements, both with an average element size of 0.5 micron. The interface between fibers and the matrix is modeled by cohesive elements: zero-thickness 8-node hexagonal cohesive elements automatically generated. Discretization of the unit cells with second order elements and further mesh-sensitivity analysis is under investigation.

3.2. The applied material behavior's

The fibers were modeled by transversely isotropic material without damage with elastic properties summarized in table 1. The behavior of the matrix is modeled with elasto-plastic constitutive model (properties listed in table 1). The plastic behavior of the matrix is described with Raghava yield criteria and tabulated engineering stress-strain curve for tension. The ratio between compression and traction limit for Raghava yield criteria is fixed to a default value of 1.3 due to absence of the test data in compression. The damage model for the matrix is an isotropic material model with damage. Once the equivalent plastic strain has reached the limit, a static damage, d^s, is put to 1 and the true damage is obtained by solving the following law, expressing the damage rate:

$\dot{d} = \frac{1}{\tau_c}(1 - \exp(-a_c\langle d^s - d\rangle_+))$, τ_c and a_c are parameters for the delay law, set to 0.001 and 1 without any physical relevance, used for solver stability. The elastic Hooke tensor is degraded by scaling with *(1-d)*. The equivalent plastic strain at failure was obtained by comparing the stress-strain curves on a single element patch test of a pure matrix against experimental data.Considering that the main deformation mechanism of the laminates under consideration is shear, equivalent plastic strain at failure was obtained by calibrating the material data in terms of shear strain to failure. Dotted curves (figure 2) are the updated laws of available test data in shear and in tension, obtained by loading a single element model of pure epoxy in tension and in shear with periodic boundary conditions. The extension of the stress-strain curves introduces artificial behavior in the unit cell, allowing the unit cell to deform much more in tension than in reality and therefore to reach higher stresses. To overcome this limitation, an application of a different plasticity-damage model for the matrix to be considered. In the Section 4 it will be shown, that the strain at failure of the matrix set to 11% led to a significant difference in strength of the ply against the experimental curve. One of the possible explanations is that at the micro-scale matrix should be capable to bare larger deformations before the failure. In [22], a matrix tested at the micro-scale shows a much-higher strain to failure than in conventional mechanical setups. Therefore an assumed and increased value of strain to failure up to 58% was also considered in this work. This value comes from a different epoxy matrix [19].

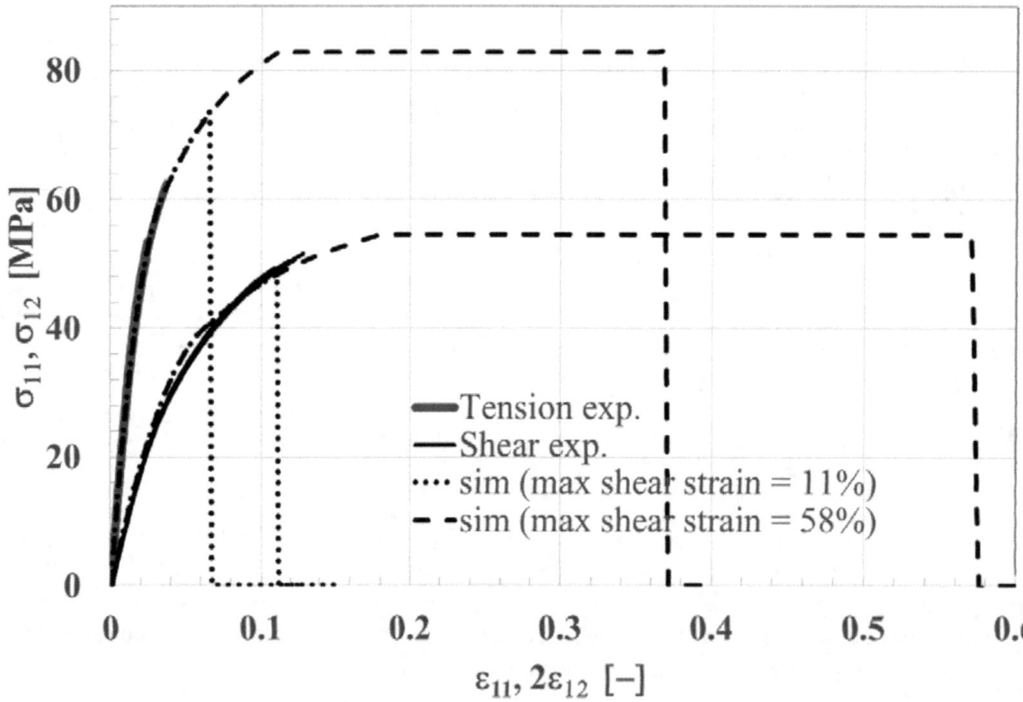

Figure 2. Strain-stress behavior of the epoxy matrix in experiments and simulations for tension and shear

The interface between fibers and the matrix is modelled with a damage interface material model called cohesive zone model. The cohesive behavior is governed by a damage initiation criterion and a damage evolution criterion, both are energy-based formulations. The bi-triangular cohesive parameters are listed in table 3, which are based on the experimental data presented in table 1. The elastic interfacial

IOP Conf. Series: Materials Science and Engineering **406** (2018) 012009 doi:10.1088/1757-899X/406/1/012009

characteristics are chosen by trial-and-error: they should be rather high to model continuity and not too high to avoid numerical divergence.

Table 3. Input data for the interface material model

Property	Units	Value
Transverse Stiffness (K03)	N/mm^3	2e+07
Shear Stiffness (K02 and K01)	N/mm^3	1.5e+07
Critical strain energy release rates (GIc, GIIc, GIIIc)	N/mm	0.002 0.006 0.006
Energy Threshold (Y0S)	N/mm	6.25e-05
Coupling Coefficient (DCOU)	-	1.45
Law Option		Bi-triangular
Time for Delay (TAU)	s.	0.001

3.3. Loading for linear and nonlinear homogenization

In order to compute linear-elastic properties of a ply, six load cases are applied on the generated micro-scale unit cells: three in tension and three in shear under periodic boundary conditions. The process works also on non-symmetrical meshes, which becomes valuable on complex models [23, 24]. The loading of a unit cell is done by imposing mean strains on a special *Homogenization* element implemented in SAMCEF and dedicated to perform homogenization of a unit cell. To assess material variability in terms of fiber and void distribution, ten configurations for each test case were generated. The average linear-elastic properties and standard deviations are computed and compared with available experimental data.

In [12], Garoz et al. developed a process for micro-mechanical tests capturing the damage mechanisms at the micro-scale and specific load conditions coupled with the Classical Laminate Theory (CLT) on the laminate scale. The concept is implemented to determine the initiation and propagation of the cracks in the ply and to mimic the virtual tests on multidirectional laminates. The current implementation is limited to laminates with $[\pm45]_{2s}$ and $[\pm67.5]_{2s}$ stacking sequences. This is linked to the identification of damage-plasticity material parameters for the LMT-Cachan damage model covering transverse and shear material behavior.

First the CLT is utilized to obtain a link between applied deformations of a laminate in longitudinal direction and local strain field in each ply in their local coordinate systems. Secondly, the ply is modeled as a unit cell with randomly distributed fibers in a matrix, as described in the section 3.1 and 3.2, and loaded with pre-calculated strains equivalent to the macro-strain loading. The load conditions are introduced using three dummy nodes per direction x; y; z. Each simulation provides stress-strain fields, from where homogenized stress-strain curves are derived. GUI for the stiffness and damage homogenization pre-processing steps are shown in Figure 3.

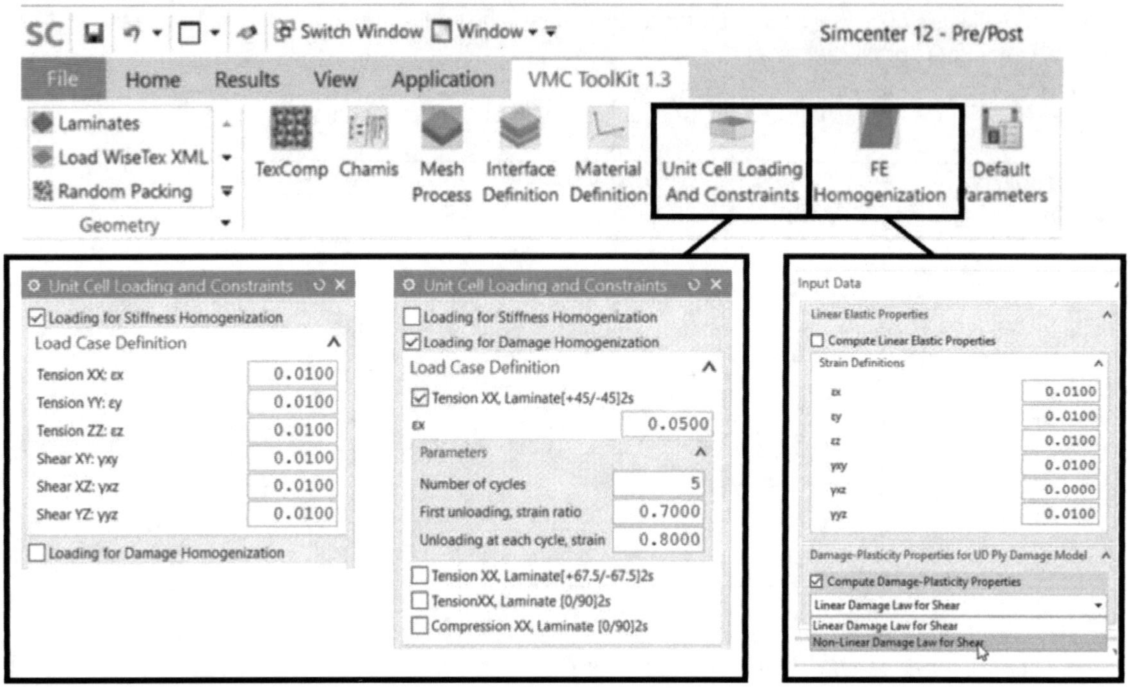

Figure 3 Graphical User Interface of the VMC ToolKit: *Unit Cell Loading and Constraints* and *FE Homogenization*

4. Results and discussion

4.1. Effect of voids on the linear elastic properties

The results on the linear elastic properties are summarized in figure 5. Longitudinal Young's modulus shows a difference of 6.5% between numerical data and experimental measurements. The effect of voids on the longitudinal Young's modulus is negligible. Longitudinal shear moduli display 9% difference between the 0.0%-void simulation and experimental results. The model with a void content of 4.1% shows a decrease in longitudinal in-plane shear modulus by 1.7 % and in longitudinal out-of-plane shear modulus by 3.0% in comparison with the 0.0%-void model. The difference between in-plane and out-of-plane longitudinal shear moduli can be eliminated by increasing number of fibers in the unit cells or by increasing number of tested configurations for each test case.

Values for transverse Young's moduli in the 0.0%-void model are higher than the experimental results by 14%. For transverse shear moduli, the experimental data is not available. Both, transverse Young's moduli and transverse shear moduli appear to decrease with void content by 10.0% and 7.86%, respectively.

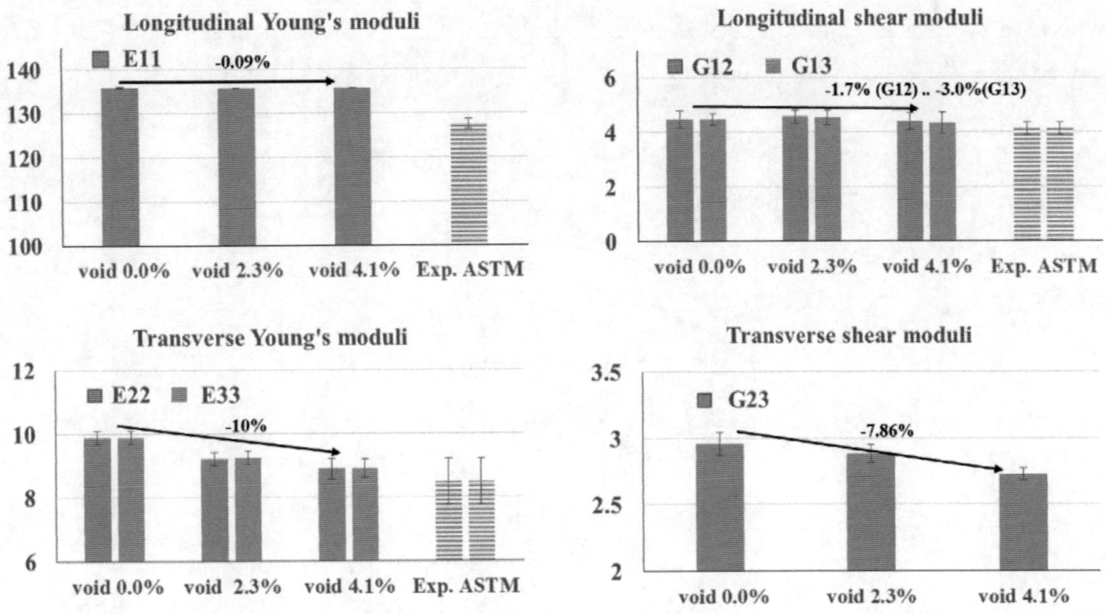

Figure 4. Effect of voids on linear elastic properties of a UD ply

4.2. Effect of voids on the nonlinear behavior of a ply

Figure 6 contains in-plane shear stress-strain curve of a [-45] ply in a laminate $[\pm 45]_{2s}$ when the shear strain at failure of the matrix was set to 11%. A significant difference in the maximum stress is observed in comparison with experimental results. The same figure demonstrates the damage development at different values of macro-strain of the model containing 4.1% of both cylindrical and spheroidal voids. Damage in the matrix started in few locations (point 1), joining in one crack (point 2) but not at the cylindrical void, where the highest stress concentrations are expected. The complete failure occurred at around 1.3% of applied shear deformations (point 3) driven only by the damage in the matrix and only a few debondings were observed. The maximum stress in the void-free model is 25% higher than in the model containing 4.1% voids.

IOP Conf. Series: Materials Science and Engineering **406** (2018) 012009 doi:10.1088/1757-899X/406/1/012009

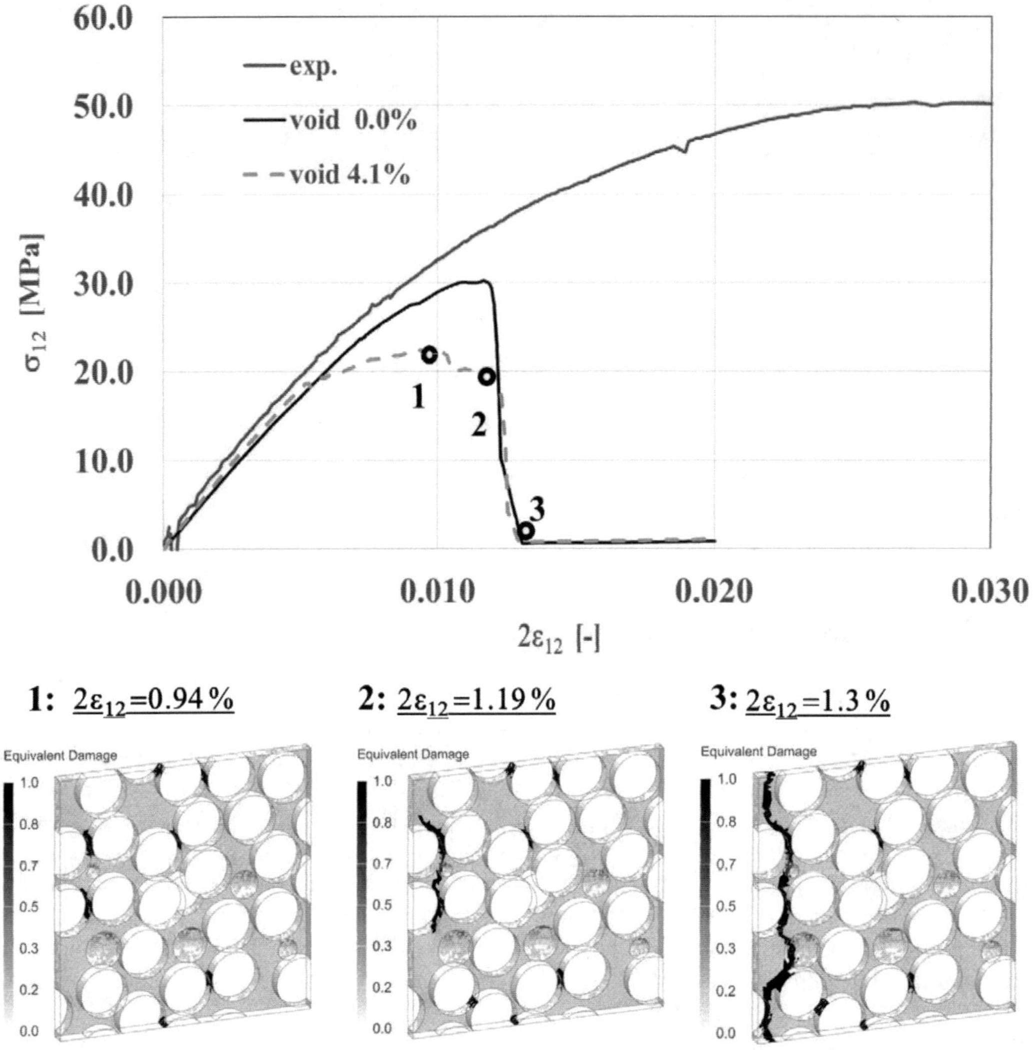

Figure 5. Homogenized stress-strain curves in shear direction in [-45] ply from the virtual test on the $[\pm45]_{2s}$ laminate (shear strain at failure of the matrix was set to 11%) and damage development in the model containing 4.1% voids

One of the possible reasons that the models with strain to failure of 11% fail so early is because the cracks are restricted by the neighbouring plies only within the linear CLT allowing cracks to progress freely through the structure and causing early failure. Further investigation of this phenomena is needed. Another reason for the discrepancy with the experimental results, as was mentioned earlier (Section 3.1), is that the deformations of the matrix at the micro-scale up to failure can be higher than 11% (scale-effect). To analyze the effect of strain to failure of the matrix on the homogenized behavior of a ply, the computations are repeated with a strain at failure in a matrix of 58% presented in figure 7. Point 1 corresponds to the formation of few fiber/matrix debondings at the locations of highest stress concentrations. With increased applied deformation, debonds are starting to merge into a single debond in the vicinity of the cylindrical void (point 2). Further increased applied deformation results in the continuation of the debond growth with a formation of cracks in the matrix, bridging debonds (point 3). The maximum stress value of the model with voids is 30% higher than in the model without voids. The difference in shear strength between models with 0%, 2.3% and 4.1% of void content and experimentally measured strength is 11.7 %, -9.2 % and -20.7 % respectively.

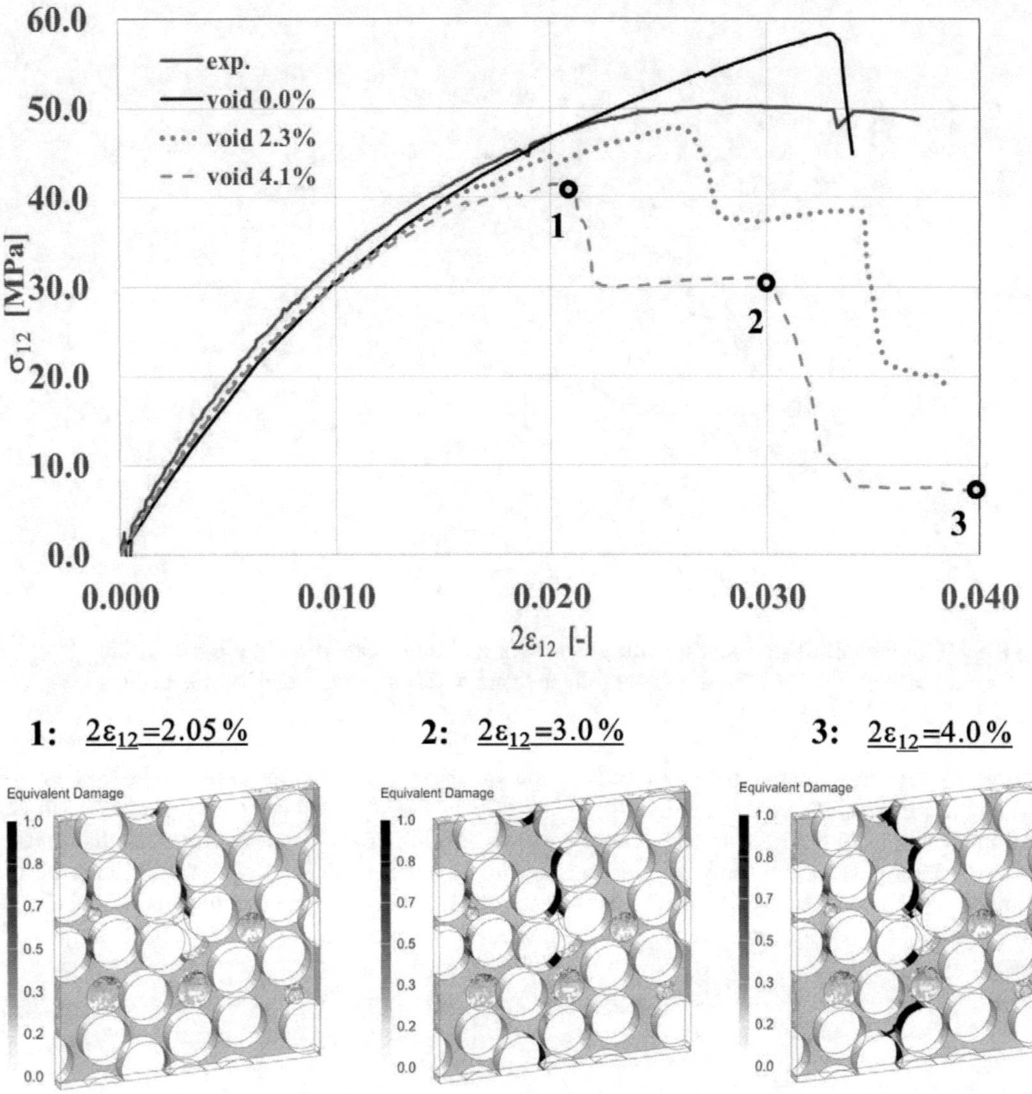

Figure 6. Homogenized stress-strain curves in shear direction in [-45] ply from the virtual test on the $[\pm 45]_{2s}$ laminate (shear strain at failure of the matrix was set to 58%) and damage development in the model containing 4.1% voids

Figure 8 presents results of virtual testing on laminates $[\pm 67.5]_{2s}$. Homogenized stress-strain curves in shear and in transverse direction are compared between all three cases and experimental data. The difference in shear strength between models with 0%, 2.3% and 4.1% of void content and experimentally measured strength is 2.6%, 7.6% and -5.4% respectively. The difference in transverse strength between models with 0%, 2.3% and 4.1% of void content and experimentally measured strength is 19%, 9.2% and -11.8% respectively.

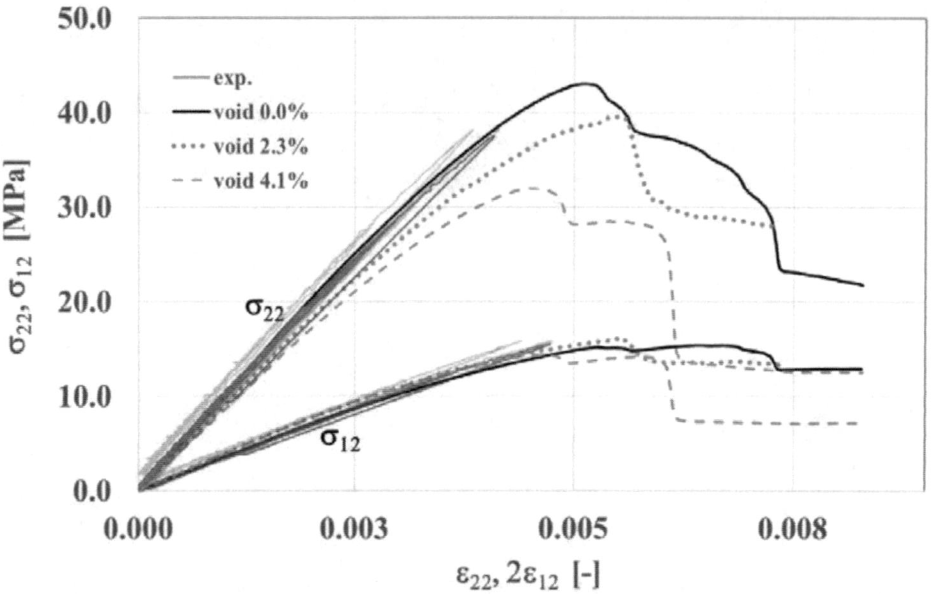

Figure 7. Homogenized stress-strain curves in shear and transverse direction in [-67.5] ply from the virtual test on the [±67.5]$_{2s}$ laminate (shear strain at failure of the matrix was set to 11%).

The summary of maximum ply stresses is presented in figure 9. The decrease of strength between the void-free model and model with 4.1 % voids is 30% in shear directions for laminate [±45]$_{2s}$ and 25% for laminate [±67.5]$_{2s.}$ in transverse direction. The effect of voids on the maximum stresses in shear direction for laminate [±67.5]$_{2s}$ is less pronounced and equals to 7%. Important to note that periodic boundary conditions are currently applied on the micro-scale models, which results in a periodic repetition of the damage across the ply. Therefore, strength values presented here should not be treated as absolute values, because the ply cracking considering crack evolution would affect the predicted material behavior.

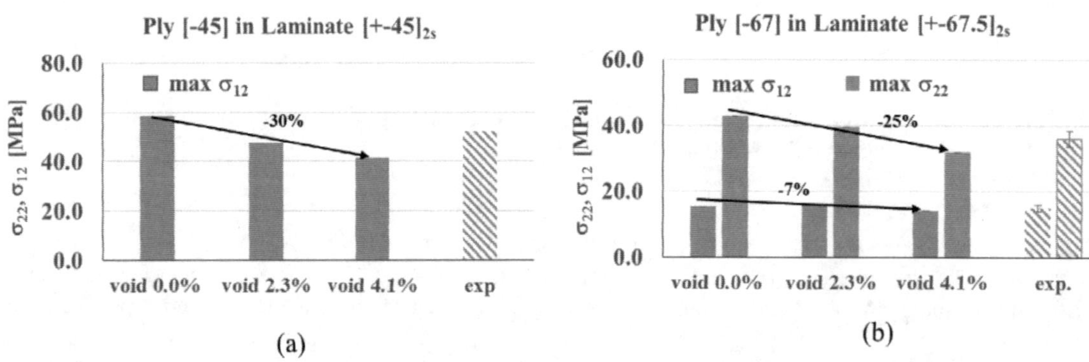

Figure 8. Effect of voids on maximum shear and transverse stress in a UD ply: (a) in ply [-45] from the virtual test on the [±45]$_{2s}$ laminate; (a) in [-67.5] ply from the virtual test on the [±67.5]$_{2s}$ laminate

5. Conclusions and future work

To estimate the "as-manufactured" properties of specimens, the Simcenter™ VMC ToolKit was extended with a feature of void generation at the fiber scale together with a special set of loadings and homogenization in order to perform virtual testing on multidirectional laminates. Simulations on the laminates $[\pm45]_{2s}$ and $[\pm67.5]_{2s}$ are conducted with and without voids, that are explicitly modeled at the fiber scale with different void type and content. In order to achieve this, several unit cells were generated and loaded with a complex equivalent macro-strain fields, which are computed from deformations of a ply using CLT.
First, the effect of voids was assessed on the linear elastic properties of the ply. For this, three tensile and three shear load cases were performed in order to build the full stiffness tensor and to compute linear elastic constants. It is shown that the presence of voids does not affect much the longitudinal elastic properties, and decreases the transversal Young's and shear moduli by around 10% and 8% respectively (for a void content of 4.1%). The effect of voids was as well analyzed on the strength/damage behavior of the ply. The nonlinearity of the homogenized stress-strain curves of all models is in a good agreement with experiments up to 2% of shear macro-strain for a laminate $[\pm45]_{2s}$ and up to 0.4% for a laminate $[\pm67.5]_{2s.}$ The introduction of 4.1% of voids led to a decrease of maximum shear stresses in a laminate $[\pm45]_{2s}$ of approximately 30%. The effect of voids in the laminate $[\pm67.5]_{2s}$ is showing the decrease of transverse strength in a ply by 25% and the decrease of shear strength in a ply by 8%. Due to a limitation of the methodology with respect to a crack evolution through the laminate, these values rather indicate a possible trends on the effect of voids on the material failure. This paper is a demonstrator of a virtual material characterization process based on an engineering workflow implemented in Simcenter™ VMC ToolKit that predicts the macro-scale effect of key micro-scale material features, facilitating efficient material performance assessment in a bottom-up approach or facilitating virtual material optimization in a top-down approach.

Acknowledgments
The authors gratefully acknowledge SIM (Strategic Initiative Materials in Flanders) and VLAIO (the Agency Flanders Innovation & Entrepreneurship) for their support of the IBO project M3Strength (Grant no.140158), which is part of the research program MacroModelMat (M3), coordinated by Siemens (Siemens PLM Software, Belgium). The authors express their gratitude towards Honda R&D Co., Ltd., and Mitsubishi Chemical Corporation for their generosity in supplying material for this research.

References
[1] Mehdikhani M, Gorbatikh L, Verpoest I and Lomov S V 2018 Voids in fiber-reinforced polymer composites: a review on their formation, characteristics, and effects on mechanical performance *J. Compos. Mater. In print*
[2] Mehdikhani M, Petrov N A, Straumit I, Melro A R, Lomov S V and Gorbatikh L 2018 The effect of voids on matrix cracking in composite laminates revealed through combined modeling at the micro- and meso-scales *In preparation*
[3] Petrov N A, Gorbatikh L and Lomov S V 2017 A parametric study assessing performance of eXtended Finite Element Method in application to the cracking process in cross-ply composite laminates, *Composite Structures.* **187**, 489-497
[4] Ladevèze L and LeDantec E 1992 Damage modelling of the elementary ply for laminated composites," *Composites science and technology*, **43**(3), 257–267
[5] Ladevèze P 1992 A damage computational method for composite structures, *Computers and Structures*, **44**, 79–87
[6] Ladevèze P, Allix O, Gornet L, Lévêque D and Perret L 1998 A computational damage mechanics approach for laminates: Identification and comparison with experimental results, *Damage Mechanics in Engineering Materials*, 481–500
[7] Delsemme J P, Bruyneel M, Jetteur P, Magneville B, Naito T and Urushiyama Y 2015 Progressive damage modeling in composites: from aerospace to automotive industry, *Proc* in

Third Int. conf. on Buckling and Post-buckling Behavior of Composite Laminated Shell Structures, p 22-26, Braunschweig, Germany

[8] Bruyneel M, Delsemme J, Goupil A, Jetteur P, Lequesne C, Naito T and Urushiyama Y 2014 Damage modeling of laminated composites: validation of the intra-laminar law of SAMCEF at the coupon level for UD plies," in *Proc of the 16th European conf. on composite materials*, Seville, Spain

[9] Malgioglio F, Carella-Payan D, Magneville B and Farkas L 2016 Parameter Identification for interlaminar damage modelling of composites, *In Proc of the 6th ECCOMAS Thematic Conf. on the Mech. Response of Composites,* Eindhoven, Netherlands

[10] Steensels E 2017 Mechanical damage in carbon fibre-reinforced composites: experimental observations and model validation, *Maser Thesis, K.U. Leuven*

[11] Mehdikhani M, Standaert A, Steensels E, Vallons K, Gorbatikh L and Lomov S. V 2018 Multi-scale digital image correlation for detection and quantification of matrix cracks in carbon fiber composite laminates in the absence and presence of voids, *Compos. Part B-eng. under review.*

[12] Garoz D, Gilabert F A, Sevenois R D B, Spronk S W F and Van Paepegem W 2017 Material parameter identification of the elementary ply damage mesomodel using virtual micro-mechanical tests of a carbon fiber epoxy system, *Composite Structures*, **181**, 391–404

[13] Mitsubishi Chemical Corporation, https://www.m-chemical.co.jp/en/index.html

[14] Sevenois R D B, Garoz D, Verboven E, Spronk S W F, Gilabert F A, Kersemans M, Pyl L and Van Paepegem W 2018 Multiscale Approach for Identification of Transverse Isotropic Carbon Fibre Properties and Prediction of Woven Elastic Properties using Ultrasonic Identification, *under review*

[15] Arteiro A, Catalanotti G, Melro A, Linde P and Camanho P 2014 Micro-mechanical analysis of the in situ effect in polymer composite laminates, *Comp. Struct,* **116**, 827–840

[16] Farkas L, Vanclooster K, Erdelyi H, Sevenois R D B, Lomov S V, Naito T, Urushiyama Y and and Van Paepegem W 2016 Virtual material characterization process for composite materials: an industrial solution," in *Proc. of the 17th European. Conf. on Composite Materials* ,26–30, *Munich*

[17] Catera P G, Gagliardi F, Mundo D, De Napoli L, Matveeva A and Farkas L 2017 Multi-scale modeling of triaxial braided composites for FE-based modal analysis of hybrid metal-composite gears," *Comp. Struct.*, **182**, 116–123

[18] Matveeva A, Tabatabaei A, Sevenois R D B, Farkas L, Van Paepegem W and Lomov S V 2017 Assessement of different meso-modelling techniques for 2D woven composites, *in Proc* of the *NAFEMS conf: Multiscale and Multiphysics Modeling & Simulation - Innovation Enabling Technologies, At Columbus, OH*, USA

[19] Shishkina O, Matveeva A, Wiedemann S., Hoehne K, Wevers M, Lomov S V, Farkas L 2018 X-Ray Computed Tomograhy-based FE-Homogenization of sheared organo sheets, in *Proc. of the 18th European Conf. on Composite Materials, Athens, Greece*

[20] Melro A, Camanho P and Pinho S 2008 Generation of random distribution of fibres in long-fibre reinforced composites, *Comp. Science and Technology, 68,* 2092–2102

[21] Romanov V, Lomov S V, Swolfs Y, Orlova S, Gorbatikh L and Verpoest I 2013 Statistical analysis of real and simulated fibre arrangements in unidirectional composites, *Composites Science and Technology*, **87**, 126–134

[22] X. Morelle 2015 Mechanical characterization and physics-based modeling of highly-crosslinked epoxy resin, PhD Thesis, UCL-Université Catholique de Louvain, Belgium

[23] Garoz D, Gilabert F A, Sevenois R D B, Spronk S. W. F and Van Paepegem W 2018 Consistent Application of Periodic Boundary Conditions in Implicit and Explicit Finite Element Simulations of Damage, t*o be submitted*

[24] Kassem G. A. 2009 Micromechanical material models for polymer composites through advanced numerical simulation techniques, *PhD Thesis*, RWTH Aachen University

IOP Conf. Series: Materials Science and Engineering **406** (2018) 012010 doi:10.1088/1757-899X/406/1/012010

Analysis of void morphology in composite laminates using micro-computed tomography

Mahoor Mehdikhani[1,2], Nghi Q Nguyen[1], Ilya Straumit[1], Larissa Gorbatikh[1,2], Stepan V Lomov[1,2*]

[1] KU Leuven, Department of Materials Engineering, Kasteelpark Arenberg 44, 3001, Leuven, Belgium
[2] SIM vzw, Technologiepark 935, B-9052 Zwijnaarde, Belgium

[*]stepan.lomov@kuleuven.be

Abstract. VoxTex software for quantification of micro-CT images provides useful tools for analysis of voids in fibre reinforced composites. These tools are applied in the current work to (1) carbon fibre laminate (-45/0/45/-20/0/20/90/0)so, made with automated dry fibre placement and vacuum assisted resin transfer moulding and to (2) carbon fibre laminates [0/90]ns, made by prepreg technology and cured in autoclave witha regime proliferating void formation. The materials contain very different porosity levels (below 1% for (1) and few percent in (2)). The segmentation and image processing algorithms allow identification of the overall void content and quantification of the void dimensions and their orientation in the form of distribution functions. These distributions are presented and their characteristic features are discussed. A typical peculiarity of segmented micro-CT images is presence of small (few pixels in size) voids, which persist even after the spurious "void pixels" are eliminated. Whether these micro-void-features represent real microscopic voids in the material is discussed based on the comparison with SEM images.

1. Introduction

Voids are common manufacturing defects in Fiber-Reinforced Composites (FRCs), forming inevitably in all processing techniques. They can cause degradation of mechanical properties, especially those controlled by matrix properties [1]. If their effect can be precisely quantified, limits allowing certain amount of voids can be determined, depending on the loading scenario and application. Evaluation of the voids' effect asks for their accurate characterization, meaning identifying their characteristics, i.e. shape, size, orientation, and spatial distribution. Once these characteristics are known, their effect on mechanical properties can be investigated experimentally. Moreover, when the void allowance limits are determined, deciding if a composite part is accepted/rejected needs accurate characterization of the voids in that part.

Quantification of the effect of voids is also possible, and maybe even less difficult, through simulations. However, the modelling approach also needs information on voids' characteristics to produce reliable results. Therefore, characterization of voids is essential for evaluation of their effects.

There has been some research conducted on the application of micro-computed tomography (micro-CT) to characterization of voids, see more in [1]. However, works that provide a complete set of data on void characteristics in carbon/epoxy laminates are still rare. In the present study, we perform a full 3D analysis of the voids in carbon/epoxy laminates with different stacking sequences using micro-CT. Micro-CT results are processed with *VoxTex*, an in-house software, to obtain details of voids' size, morphology, orientation, and spatial distribution.

2. Materials

There are two materials studied. One of them is an aerospace-grade carbon/epoxy composite made from unidirectional prepreg tapes. The lay-up is prepared with Automated Tape Laying (ATP) and the curing is performed in an autoclave, both at *SABCA Limburg NV, Belgium*. In order to produce laminates with voids for research purpose, the manufacturer's cure cycle was not followed, instead a low-pressure-temperature cure cycle was applied, with debulking prior to cure and vacuum application during cure. Two lay-ups were produced: $[0/90]_{2s}$, and $[0/90]_{4s}$. Micro-CT samples with the size of ~ 15 mm × 10 mm × 1.5 mm for the 8-ply lay-ups and ~ 15 mm × 10 mm × 3 mm for the 16-ply lay-up were cut from the produced plates.

The other material is produced at *NLR, The Netherlands*, with Automated Dry Fibre Placement (ADFP) followed by Vacuum Assisted Resin Transfer Moulding (VARTM). The composite reinforcement is Hexcel HiTape UD126 AS7/V800E/QD2/6.35mm, which is a unidirectional dry tow of AS7 carbon fibres. The areal density of the tow is 126 g/m2. The tow width and thickness is 6.35 mm (1/4 in.) and 0.135 mm, respectively. The dry preform is laid up as [-45/0/45/-20/0/20/90/0]so, where the subscript "so" stands for Symmetric Odd, indicating that there is only one 0° ply in the middle of the laminate. The preform is infused with HexFlow RTM 6 resin. The nominal fibre volume fraction of the final composite is 50%. This is the same material as studied in [2].

3. Micro-CT image acquisition and processing

The ATP samples were scanned with the *HECTOR* system at *Ghent University Centre for X-ray Tomography* (*UGCT*). The scan volume was 12 × 12 × 12 mm^3, and the voxel size was 6.55 µm. The scan time for each specimen was about 45 minutes. The reconstruction of the projection data to tomographic slice images was performed with the *Octopus* reconstruction software.

The ADFP material was scanned with a Skyscan 1172 X-ray computed tomography system in the Department of Materials Engineering of *KU Leuven*. 2D X-ray images were acquired for 180° of tomographic rotation with a rotation step of 0.2°. The nominal spot size and the exposure time were 11.27 mm and 1767 ms. The scan volume is 6.0 × 6.5 × 2.0 mm^3, and the voxel size was 2.0 µm.

The segmentation and data processing is carried out using *VoxTex* software [3] developed at *KU Leuven*. After segmentation, voids are fitted, in *VoxTex*, to equivalent ellipsoids. For each ellipsoid, the minor, medium, and major axes as well as the location of its centroid and orientation are calculated. The orientation of the voids is defined as the angle between their major axis and the scan direction.

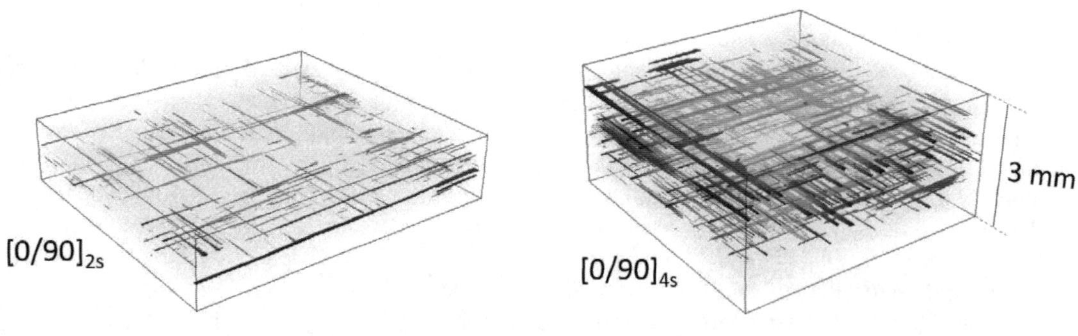

Figure 1. 3D representations of voids (in purple) in the two ATP laminates detected with micro-CT.

4. Voids characterization in ATP materials

The 3D representation of the segmented voids in the scanned laminates is visualized in figure 1. Voids, in purple, have elongated features orientated in the fiber directions in each ply of the laminates. There is a higher level of voidage in the $[0/90]_{4s}$ laminate (16 plies) than in the $[0/90]_{2s}$ laminate (8 plies).

For each void, fitted to an ellipsoid after segmentation, three size parameters are defined: semi-minor, semi-medium, and semi-major axis of the ellipsoid. The frequency distribution of these parameters is shown in figure 2.

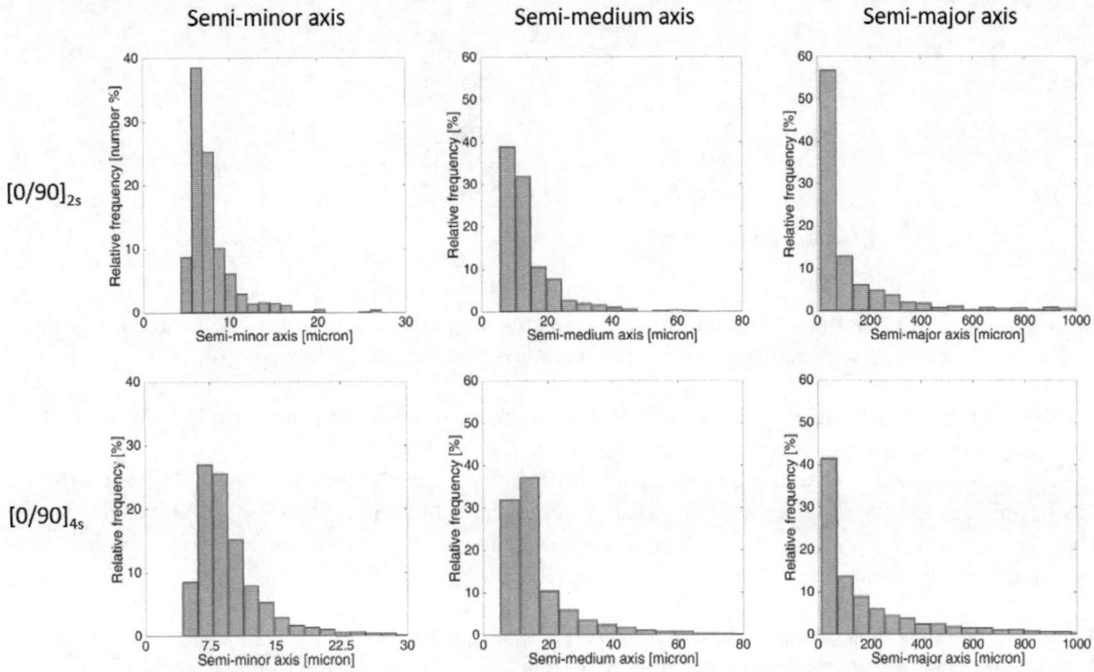

Figure 2. Frequency distribution of the size parameters of the detected voids.

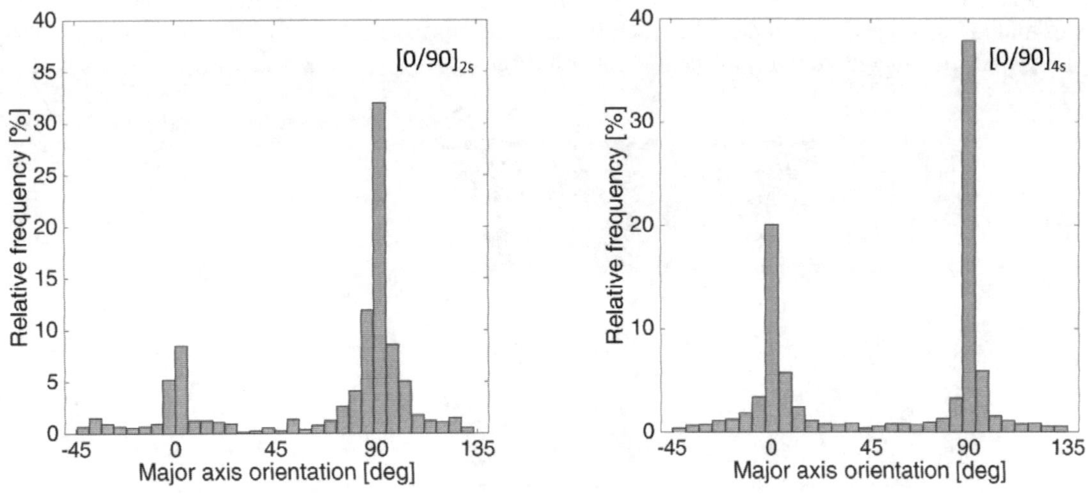

Figure 3. Frequency distribution of the orientation of the detected voids.

IOP Conf. Series: Materials Science and Engineering **406** (2018) 012010 doi:10.1088/1757-899X/406/1/012010

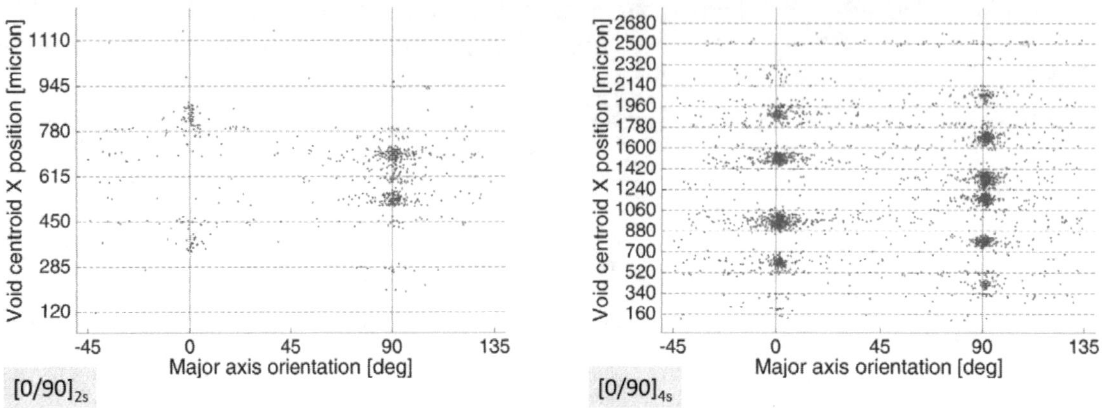

Figure 4. Voids' in-plane orientation, defined as the angle between the major axis and the scan orientation, through the thickness of different laminates.

The frequency distribution of voids' orientation is shown in figure. As can be noted, the majority of voids are oriented in the nominal fiber direction in the plies of each laminate.

The orientation of voids through the thickness of the laminate is shown in figure 4. The clusters in this figure indicate that the majority of voids in each ply have the same orientation, and it is close to the ply orientation. This figure also reveals that there exist more voids in the inner plies of the laminates than in the outer plies.

5. Voids characterization in ADFP material

The ADFP material has a very low void content, below 0.2%. Segmentation of the image for these small voids presents a challenge, as the size of some voids is on the level of the image noise. The image segmentation was enhanced (the image noise reduced) by applying Gaussian filter to the micro-CT image and by using a reduced greyscale dynamic range on original data set. The processed and thresholded images show agreement between the size and shape of clearly visible voids on the Micro-CT cross-section images and the segmented images (Figure 5).

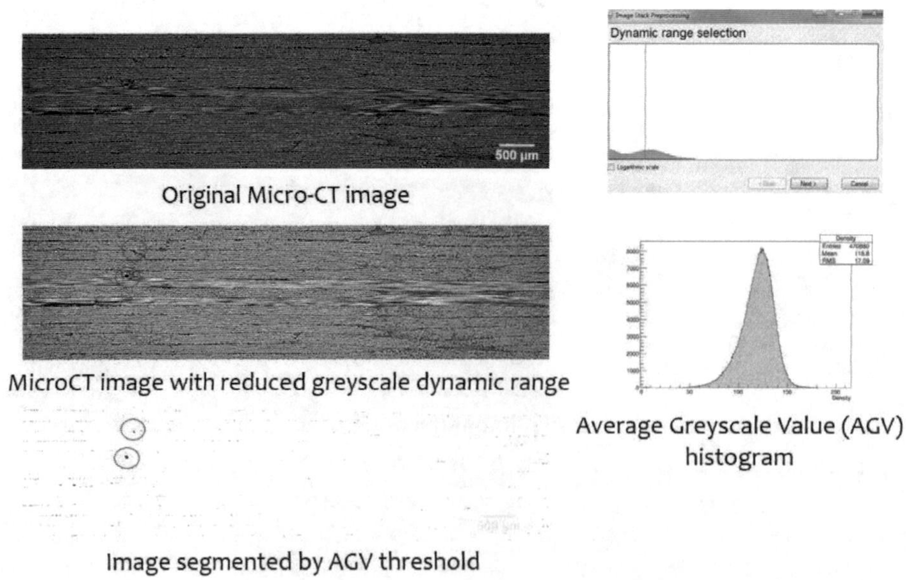

Figure 5. Image processing and thresholding for the ADFP material.

However, a large number of small voids that are seen in the segmented images and not clearly recognisable in the section are suspicious. The question is whether they are really present in the material, or simply a result of rudiments of the image noise? The former (they are real) seems to be supported by localization of these features: they clearly follow the tape layers boundaries. To verify or disprove their presence in the material, scanning electron microscopy (SEM) was performed (Figure 6). This additional characterization has confirmed that these micro-voids are indeed present in the laminate, and their dimensions are comparable with the dimensions seen in the segmented micro-CT images.

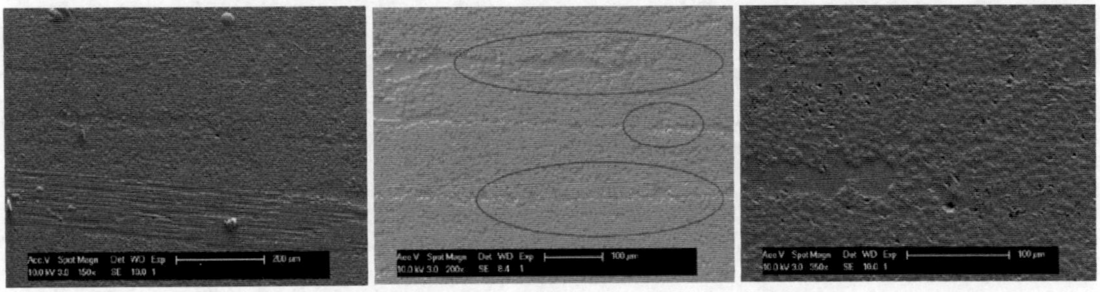

Figure 6. Micro-voids seen with SEM, ADFP material.

6. Conclusions

Micro-CT, and specifically VoxTex software, can be used not only to determine the void content in carbon fibre laminates composite but also other important information such as voids' spatial distribution and their size and shape. With this technique even very small voids (few pixels in size) in the composite with the overall void content below 0.2% could be captured.

Acknowledgments

The work leading to this publication has been funded by CANAL (CreAting Non-conventionAl Laminates) FP7 project, the Vietnamese Ministry of Education and Training, and by SIM (Strategic Initiative Materials in Flanders) and VLAIO (Flemish government agency for Innovation and Entrepreneurship), IBO project M3Strength, which is part of the research program MacroModelMat (M3) coordinated by Siemens (Siemens PLM Software, Belgium). The authors highly appreciate the collaboration of NLR (R. Klomp and Ch. Groenendijk) and SABCA Limburg (K. Verjans and S. Wathiong) in production of the carbon fiber laminates. The micro-CT images have been acquired partially by Inside Matters and partially on the X-ray computed tomography facilities at KU Leuven, financed by the Hercules Foundation and maintained under the supervision of Prof M. Wevers. Development of VoxTex software was supported by QUICOM project of FP7. Support of Toray Group of the Toray Chair Professorship in KU Leuven (S.V. Lomov) is gratefully acknowledged.

References

[1] Mehdikhani M, Gorbatikh L, Verpoest I and Lomov SV 2018 *J. Compos. Mater.* In print
[2] Nguyen N, Mehdikhani M, Straumit I, Gorbatikh L, Lessard L, Lomov SV 2018 *Composites Part A* **104** 14
[3] Straumit I, Lomov SV, Wevers M *Composites Part A* **69** 150

IOP Conf. Series: Materials Science and Engineering **406** (2018) 012011 doi:10.1088/1757-899X/406/1/012011

Multi-scale experimental and computational investigation of matrix cracking evolution in carbon fiber-reinforced composites in the absence and presence of voids

Mahoor Mehdikhani[1,2*], Nikolay Petrov[1], Larissa Gorbatikh[1,2], Stepan V Lomov[1,2]

[1] KU Leuven, Department of Materials Engineering, Kasteelpark Arenberg 44, 3001, Leuven, Belgium

[2] SIM vzw, Technologiepark 935, B-9052 Zwijnaarde, Belgium

* mahoor.mehdikhani@kuleuven.be

Abstract. Manufacturing defects such as voids can influence matrix cracking, which is one of the main forms of damage in fiber-reinforced laminates. Characterization and investigation of matrix cracks is not straightforward, neither experimentally nor computationally. In the current study, we develop a methodology for *in-situ* analysis of matrix cracking evolution in the absence and presence of voids. The cracks are detected using digital image correlation at three different scales: macro, meso, and micro. Additionally, a combined two-scale computational approach is established for prediction of the effect of voids on matrix cracking. Local properties of the material are calculated using computational micromechanics and then serve as input for the meso-scale model based on extended finite element method. The computational approach predicts the crack density evolution in function of the applied deformation. The results of both experimental and computational studies reveal that voids initiate matrix cracks, causing earlier damage in off-axis plies. They also result in higher crack density, for a given applied strain.

1. Introduction

Voids are manufacturing defects that are inevitably formed during production of Fiber-Reinforced Composites (FRCs). Although control and reduction of voids have been widely worked on and improved, recent manufacturing techniques, such as automated tape processing and out-of-autoclave curing still face void control as one of their main challenges. Voids can affect the mechanical properties of FRCs. They primarily degrade the matrix-dominated properties such as inter-laminar shear, compressive, and transverse tensile strength. The broad research on formation, characteristics, and mechanical effects of voids in FRCs is reviewed in [1].

Voids can influence the damage development in FRCs. In particular, matrix cracking that is one of the main damage modes in off-axis plies is sensitive to the presence of voids. It has been reported [2, 3] that voids can cause initiation of the matrix cracks, but with increase of loading, fewer voids remain that have not started any cracks and cracks start forming also in regions without voids. Hence, the effect of voids on matrix cracking fades. The initiation of matrix cracks from voids is attributed to the strain concentrations caused by them detected via Digital Image Correction (DIC) at the micro-scale [4].

In order to identify the significance of the effect of voids on matrix cracking evolution, matrix cracks need to be characterized during the mechanical loading. This is not straightforward especially for carbon FRCs because of their dark color. The available approaches for characterization of matrix cracks, including (edge replication) optical microscopy, electron microscopy, X-ray radiography or computed tomography, and ultrasonic C-scan, require stopping of the test and sometimes removing the sample before it can be characterized. Acoustic emission registration is an indirect measurement and requires interpretation of acoustic emission signals and identification of the damage type, which is challenging. Direct *in-situ* observation of the cracks has been shown feasible in transparent materials like glass FRCs. However, for carbon FRCs, this can be done, in the best case scenario, only on the specimen surfaces. It also requires manual counting of cracks that can be laborious and prone to errors, especially for unopened cracks or cracks with small opening. Therefore, a tool for detection and characterization of matrix cracks in *in-situ* captured data would be very helpful.

Due to difficulties in detection of matrix cracks as well as in controlling the parameters of voids, such as the void content, size, and distribution, a computational approach is a valuable tool for analysis of the effect of voids on matrix cracking. Though some research on prediction of the effect of voids on effective material properties of FRCs has been performed [5-7], their effect on statistically-controlled properties, such as matrix cracking has not been modeled, to the authors' knowledge. This is mainly because voids are microscopic features that interact with other heterogeneities in the material both at lower and higher scales (such microscopic fibers and cracks at the meso-scale). This issue can be resolved by a combined two-scale modeling methodology.

In the present study, we develop two methodologies to address the challenges in investigation of the effect of voids on matrix cracking in FRCs. The experimental methodology characterizes cracking in *in-situ* images captured at different scales during the test, using Digital Image Correlation (DIC). It is demonstrated on cross-ply aerospace-grade carbon/epoxy laminates and laminates with deliberately-produced voids. The computational methodology is developed in a combined two-scale framework. The degradation of local properties in the presence of voids is calculated, using computational micromechanics, and transferred to a laminate-level model that can simulate transverse cracking, using eXtended Finite Element Method (XFEM). Both methodologies allow investigation of the evolution of transverse cracking in function of applied strain, influenced by voids.

2. Materials

An aerospace-grade carbon/epoxy composite is studied. The unidirectional prepregs are made by *Cytec* from *Tenax® - E HTS40 F13 12K* carbon fibers and *CYCOM® 977-2* (toughened) epoxy resin. Cross-ply laminates are produced with automated tap laying and autoclave curing at *SABCA Limburg NV, Belgium*. The lay-ups are debulked at room temperature prior to cure. Two different cure cycles are defined: one high-pressure-temperature cycle to produce reference material and one low-pressure-temperature to produce imperfect material, including intra-laminar voids. $[90/0]_{2s}$ and $[90/0]_{4s}$ laminates with an average ply thickness of ~ 0.2 mm are produced. The fiber volume fraction is within 53-60% and the void content is around 1%.

3. Experimental and computational methods

3.1. Experimental methodology: in-situ characterization of cracking using DIC

In-situ characterization of transverse cracking is performed at three different length scales: macro, meso, and micro. The macro- and meso-scale images are acquired from $[90/0]_{2s}$ specimens under uniaxial tension. To this end, tensile specimens, from both reference and imperfect material, are prepared according to the standard ASTM D3039, one surface is paint sprayed as the required random speckle pattern for DIC, and one thickness is grinded and polished for edge crack monitoring. Macro-scale images are taken from the sprayed front surface of the specimen with a window size of ~ 50 mm × 40 mm and meso-scale images from the polished thickness edge with a window size of ~ 5 mm × 4 mm. Both sets of images are captured via *Limess* cameras with application of local LED lights.

The micro-scale analysis is similar to those performed in [8, 9]. The micro-scale images are acquired from a small $[90/0]_{4s}$ specimen loaded under 3-point bending inside a Scanning Electron Microscope (SEM). One of the thickness edges of the specimen is grinded and polished and the specimen is mounted in a Deben mini-tester stage, installed inside an SEM. A suitable study zone on the tensile edge of each reference and imperfect material is identified, from which SEM images are captured during loading. The micro-scale images have a window size of ~ 1500 μm × 1300 μm.

IOP Conf. Series: Materials Science and Engineering **406** (2018) 012011 doi:10.1088/1757-899X/406/1/012011

The transverse cracks are detected via DIC in the macro- and meso-scale images since they can hardly be observed with the naked eye. The transverse cracks appear as lines with high strains in the DIC strain maps and peaks in the DIC strain profiles along a horizontal line. The detection of these peaks is carried out semi-automatically using MATLAB. The number of peaks in each image divided by the length of the profile gives a crack density at that deformation step, which can be plotted versus applied strain, resulting in crack density evolution graphs. The details of the experimental methodology are explained in [10].

3.2. Computational methodology: two-scale modeling of voids' effect on cracking
Micro-scale finite element models of a composite containing randomly distributed fibers and a void are built. Carbon fibers have transversely isotropic elastic properties without damage and the epoxy matrix is assigned with an elasto-plastic damage model, developed in [11]. A reference model is also created. Effective elastic properties through homogenization and transverse strength through simulation of transverse tension are obtained for the models. These properties are then used in the meso-level model.

The meso-level model simulates a carbon/epoxy laminate with the size of 20 mm × 20 mm × 0.6 mm. The laminate consists of unidirectional plies with homogenized properties in the stacking sequence of [0°/90°/0°]. XFEM is assigned to the transverse ply, as described in [12]. Damage in XFEM is controlled by a bilinear cohesive law, which uses the transverse strength of the material at the micro-scale as the damage initiation threshold and Mode I fracture toughness for the damage progression. To model the presence of voids, local regions, called "weak volumes", are randomly distributed in the transverse ply. Different void contents can considered. A reference model is also created. For the meso-models with voids, the transverse stiffness of the transverse ply is reduced to that obtained via micro-modeling, based on the void content. The transverse strength and the Mode I fracture toughness of the weak volumes is assigned to those computed via micro-modeling, based on the void size of interest. Other properties of the transverse ply and all the properties of the longitudinal plies are set to those obtained from the reference micro-model. Displacement-based uniaxial tension is applied to the meso-level model. Both micro- and meso-level simulations are performed using *Abaqus FEA*. The computational methodology is schematically shown in figure 1. The details of the computational methodology are explained in [13].

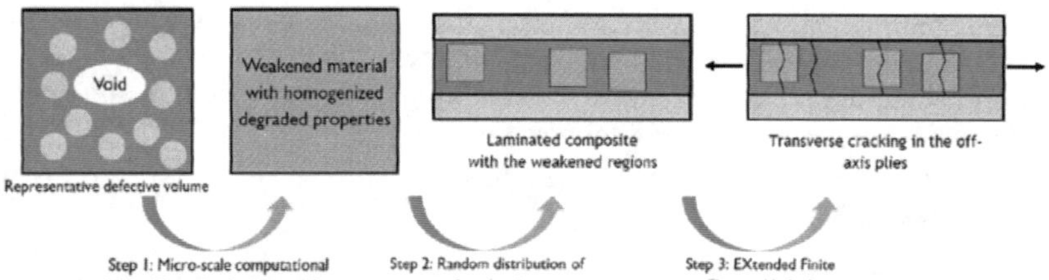

Figure 1. Schematics of the computational methodology for simulation of transverse cracking in the presence of voids using computational micromechanics and XFEM.

4. Results and discussions

4.1. Experimental results

4.1.1. Macro- and meso-scale analysis. Macro-scale horizontal strain (E_{xx}) maps are obtained via DIC on the images of the front surface. As can be observed in figure 2, the transverse cracks in the reference specimen tend to start from the edge and span, sometimes gradually, the width of the specimen. In the imperfect material, cracks initiate in early stages of loading at the specimen's edge, as visible in figure 3, but they slowly/rarely propagate toward the other edge. This may be because initiation of cracks is more favorable than their propagation due to the presence of the voids, as strain concentrators. The crack density along the mid-width line of the reference specimen is obtained via detection of peaks in the corresponding strain profiles (figure 3b) and plotted versus the applied strain in figure 3c.

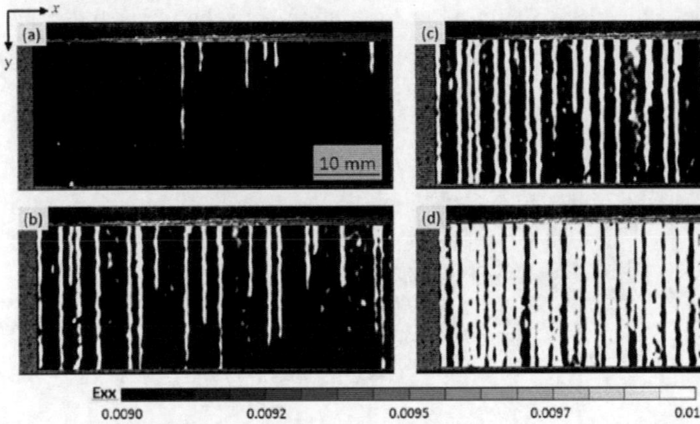

Figure 2. Cracking evolution in the outer ply of the [90,0]2s reference material under tension (applied in the horizontal direction), detected in the DIC horizontal strain maps at the applied strain of (a) 0.63%, (b) 0.77%, (c) 0.87%, and (d) 1.12%.

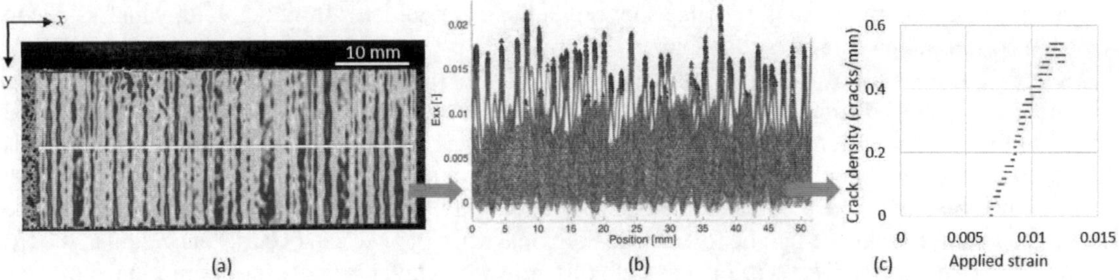

Figure 3. Macro-scale strain map at the last deformation step on the front surface of the $[90,0]_{2s}$ in the reference specimen; DIC strain profiles of all deformation steps along the mid-width line of the reference specimen; the processed crack density in function of the applied strain for the reference.

The meso-scale horizontal strain (E_{xx}) maps are obtained via DIC on the images of the thickness edge. These maps are used to extract the crack density at the edge of the specimen. This is performed for an outer ply of both materials and presented in figure 4. The figure shows that voids can cause earlier start of cracking and increase the cracking rate and saturation level in the outer ply.

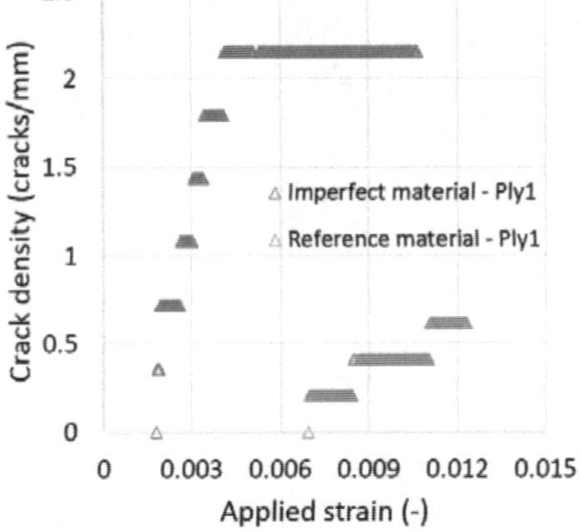

Figure 4. Crack density evolution in an outer ply (Ply1) of the reference and imperfect materials obtained via meso-scale DIC analysis.

IOP Conf. Series: Materials Science and Engineering **406** (2018) 012011 doi:10.1088/1757-899X/406/1/012011

4.1.2. Micro-scale analysis. Analysis at this scale is similar to that at the meso-scale, but allowing to see the fibers, voids, and cracks together. Therefore, a correlation between voids and crack formation can be made. For the reference specimen, no cracks are detected in the study zone on the second transverse ply from the bottom. In contrast, cracks do occur in the imperfect material. The first two cracks form from the two voids in the study zone and the second two cracks form in regions without voids. The cracks detected via DIC with their order of formation are shown in figure 5.

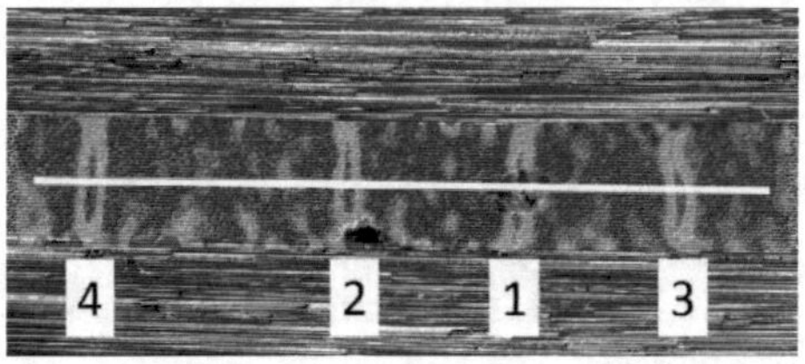

Figure 5. Transverse cracks detected via DIC at the micro-scale – the numbers show the sequence of crack formation, showing the first cracks initiating from voids.

4.2. Computational results

Micro-scale modeling reveals a reduction in transverse stiffness, transverse strength, and fracture toughness of the unidirectional ply. It also shows that the crack initiates from the void, which is due to the strain concentration caused by the void.

The total surface area of all cracks divided by the volume of the ply gives crack density for a given increment of the applied strain. This is calculated for all the meso-models and plotted versus the applied strain in figure 6. This figure shows that the initiation of cracking can be influenced by voids. This is particularly more significant for the model with the highest number of weak volumes, causing a high level of agglomeration. The predicted crack density close to the saturation level can be affected by the modeling parameters, e.g. the number of enriched regions, where cracks can occur. This region is shaded in figure 6. The detailed analysis of models with different void contents and sizes are presented in [13].

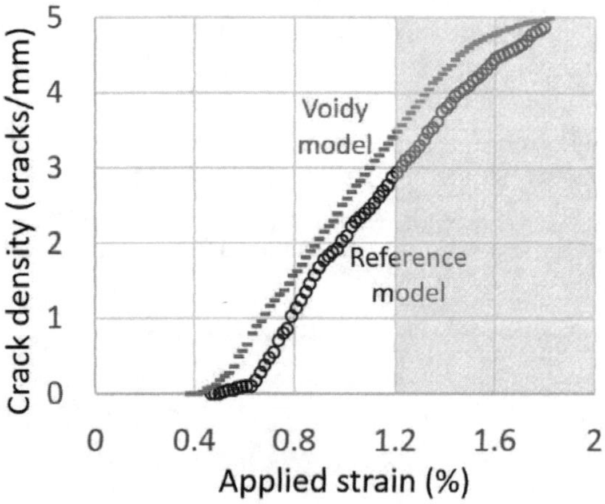

Figure 6. Predicted crack density evolution influenced by voids.

5. Conclusions

Two approaches, experimental and computational, for investigation of the effect of voids on matrix cracking are developed and examined. The experimental approach, using DIC, characterizes the cracks in the *in-situ* images acquired during mechanical tests at three scales: macro, meso, and micro. The detection and counting of cracks is performed semi-automatically using *MATLAB*. The computational approach is performed by linking micro- and meso-level finite element models. The degradation of the properties in the presence of voids is calculated via computational micromechanics. The meso-scale model takes the predictions of the micro-level model as input and simulates crack initiation and development using XFEM. Both approaches show influence of voids on transverse cracking, especially on its initiation.

Acknowledgments
The authors gratefully acknowledge *SIM* (Strategic Initiative Materials in Flanders) and *VLAIO* (Flemish government agency for Innovation and Entrepreneurship) for their support of the *IBO project M3Strength*, which is part of the research program *MacroModelMat* (*M3*), coordinated by Siemens (*Siemens PLM Software*, Belgium). The authors highly appreciate the collaboration of *SABCA* Limburg (K. Verjans and S. Wathiong) in production of the carbon fiber laminates. Stepan V. Lomov is a holder of *Toray* Chair for Composite Materials at KU Leuven, which provided partial funding of the work reported here.

References
[1] Mehdikhani M, Gorbatikh L, Verpoest I and Lomov SV 2018 *Voids in fiber-reinforced polymer composites: a review on their formation, characteristics, and effects on mechanical performance. J. Compos. Mater.* In print
[2] Huang Y, Varna J and Talreja R 2014 *Statistical methodology for assessing manufacturing quality related to transverse cracking in cross ply laminates. Compos. Sci. Technol.* **95** 100
[3] Sisodia S, Gamstedt EK, Edgren F and Varna J 2014 *Effects of voids on quasi-static and tension fatigue behaviour of carbon-fibre composite laminates*. J. Compos Mater. **49** 2137
[4] Aratama S, Hashizume R, Takenaka K, Koga K, Tsumura Y, Miyake T, Nishikawa M and Hojo M 2016 *Microscopic observation of voids and transverse crack initiation in CFRP laminates* Adv. Compos. Mater. **25** 115
[5] Ashouri Vajari D, González C, Llorca J and Legarth BN 2014 *A numerical study of the influence of microvoids in the transverse mechanical response of unidirectional composites. Compos Sci. Technol.* **97** 46
[6] Carraro PA, Maragoni L and Quaresimin 2015 M *Influence of manufacturing induced defects on damage initiation and propagation in carbon/epoxy NCF laminates. Adv. Manuf. Polym. Compos. Sci.* **1** 44
[7] Liebig WV, Viets C, Schulte K and Fiedler B 2015 *Influence of voids on the compressive failure behaviour of fibre-reinforced composites. Compos Sci Technol.* **117** 225
[8] Mehdikhani M, Aravand M, Sabuncuoglu B, Callens MG, Lomov SV and Gorbatikh L 2016 *Full-field strain measurements at the micro-scale in fiber-reinforced composites using digital image correlation. Compos. Struct.* **140** 192
[9] Mehdikhani M, Matveeva A, Aravand MA, Wardle BL, Lomov SV and Gorbatikh L 2016 *Strain mapping at the micro-scale in hierarchical polymer composites with aligned carbon nanotube grafted fibers. Compos. Sci. Technol.* **137** 24
[10] Mehdikhani M, Steensels E, Standaert A, Vallons K, Gorbatikh L and Lomov SV 2018 *Multi-scale digital image correlation for detection and quantification of matrix cracks in carbon fiber composite laminates in the absence and presence of voids controlled by the cure cycle. Compos. Part B-Eng.* under review
[11] Melro AR, Camanho PP, Andrade Pires FM and Pinho ST 2013 *Micromechanical analysis of polymer composites reinforced by unidirectional fibres: Part I – Constitutive modelling. Int. J. Solids. Struct.* **50** 1897
[12] Petrov NA, Gorbatikh L and Lomov SV 2018 *A parametric study assessing performance of eXtended Finite Element Method in application to the cracking process in cross-ply composite laminates. Compos. Struct.***187** 489
[13] Mehdikhani M, Petrov N, Melro AR, Lomov SV and Gorbatikh L 2018 *The effect of voids on matrix cracking in composite laminates revealed through combined modeling at the micro and meso-scales*. In preparation.

IOP Conf. Series: Materials Science and Engineering **406** (2018) 012012 doi:10.1088/1757-899X/406/1/012012

Examination of voids and geometry of bio-based braided composite structures

Garrett W. Melenka[1*], Brianna M. Bruni-Bossio[2], Cagri Ayranci[2], Jason P. Carey[2]

[1] Lassonde School of Engineering, York University, 4700 Keele St. North York, ON M3J 1P3

[2] Mechanical Engineering, University of Alberta, Donadeo Innovation Centre for Engineering, 9211-116 Street NW, Edmonton AB Canada T6G 1H9

* gmelenka@yorku.ca

Abstract. Braided composites are formed by interlacing continuous fibers into a textile pre-form and then impregnating the pre-form within a matrix material. Braid mechanical properties are manipulated through the selection of matrix, fiber and braid geometry. Braided composites are produced with conventional materials like carbon, aramid and glass fibers; however, they can also be produced using natural fibers such as jute, hemp, flax or regenerated cellulose. The mechanical properties of conventional and natural braided composites are highly affected by voids within the braided structure. The effect of voids on braided structures must be investigated to improve braided composite performance. A high-resolution micro-computed tomography (μCT) measurement method was utilized to quantify the size and distribution of voids within natural fiber- bio resin braided composite structures. Image processing techniques were employed to quantify void, matrix and fiber content within 35° and 45° braid samples. Accurate quantification of fiber, matrix and void volumes are crucial for evaluating the quality and repeatability of the braided composite manufacturing process. Reduction of voids and pores will improve braided composite mechanical properties and performance. Measurement of the constituents within braided composites is also necessary for the development of accurate models for predicting braid mechanical properties. Measurement of the internal microstructure of braided composites will allow for the development of improved analytical and numerical braided composite models.

1. Introduction

Braided composites are an advanced textile-based manufacturing process that create near-net-shape structures by interlacing continuous fiber yarns into a pre-form [1], [2]. Two dimensional (2D) tubular braids are commonly formed using a Maypole braiding machine while three dimensional (3D) braids are formed using Cartesian or rotatry braiding machines [3]. Braided composites are produced by impregnating (or wetting) the braided pre-form with a matrix material.

Presently, braided composites have been manufacturing using conventional composite fibers including carbon, aramid and glass fibers. The braid manufacturing technique can also be adapted to bio-based fibers such as hemp, jute, flax or cellulose [4]–[7]. Additionally, bio-based resins and polymers can also be used as the matrix material for braided structures [4], [8]. The ability to use a wide range of matrix and fiber materials demonstrates the versatility of the braided composite manufacturing process. Bio-based composites are an example of an emerging alternative to conventional composites that can be produced sustainably while yielding comparable mechanical properties [9]–[11].

Accurate measurement and analysis of braided composite structures is required so that these products can be used for structural applications and other strategically critical applications. Analytical and numerical models require data such as braid angle (θ), yarn width (W_y), yarn thickness (t), matrix volume fraction (v_m), fiber volume fraction (v_f) and void volume fraction (v_v) to accurately predict braid mechanical properties [12]–[15]. Geometric data of braided structures can be obtained using methods such as optical microscopy or scanning electron microscopy (SEM) however, these methods are limited to two dimensional (2D) cross-sections for analysis [16]. Optical microscopy and SEM measurements are also destructive measurement methods. The internal geometry of composite structures can be evaluated using an X-ray based micro-computed tomography (μCT) technique [17]–[19]. This approach allows for the non-destructive analysis of voids throughout the composite material micro-structure [20]. The μCT has advantages over conventional measurement methods since the three dimensional (3D) geometry of a sample can be analyzed. Braided and woven composite structures have complex 3D geometries and the μCT measurement method allows for the visualization and evaluation of these structures [16], [17], [21]. Additionally, the μCT measurement method allows for the quantification of voids and defects within composite structures [18], [20], [22]. The μCT method is advantageous over conventional resin burn off or acid digestion methods since the size and distribution of voids within braided structures can be obtained.

The study by Melenka *et al.* [22] examined the geometry and void content of a tubular braided composite manufactured using conventional aramid yarns within an epoxy matrix. A Diamond (1/1) braided geometry was examined and a single braiding angle was considered. The braid sample was examined using a voxel size of 18.2 μm^3 allowing for the entire braid cross-section to be examined however, details analysis of individual fiber tows was not possible. Additionally, the yarn size and shape of 3D braided composites was also examined by Ya *et al.* [18]. A resolution of 10 μm^3 was achieved in the study by Ya *et al.* The studies that have examined braided structures using the μCT method have focused on the overall braid geometry. A detailed investigation of a braid unit cell has not been performed using this measurement approach.

Voids and open pores that occur within braided composite structures will produce failure initiation sites within the braided composite structure and therefore have a negative impact on mechanical properties [20], [23], [24]. Using the μCT measurement method the size and distribution of voids within natural fiber- bio resin braided composite structures will be examined. The effect of braiding angle on the voids size and distribution will also be investigated. Resin, fiber and void volume fractions will be quantified using the μCT measurement method. Accurate quantification of fiber, matrix and void volumes are crucial for evaluating the quality and repeatability of the braided composite manufacturing process. Reduction of voids and pores will improve braided composite mechanical properties and performance. Measurement of the constituents within braided composites is also necessary for the development of accurate models for predicting braid mechanical properties.

2. Methods

2D tubular braided tubular composite samples were manufactured using a Maypole braiding machine (Steeger USA, K80–72, Steeger USA, Inman, SC). The samples were manufactured in a full-load configuration consisting of 36 yarn carriers resulting in a Regular (2/2) braid pattern [2]. Samples were produced with two different braiding angles (35° and 45°). Regenerated cellulose fibers (BioMid™, 1650 Denier, ENC International Inc, Burnaby, BC, Canada) formed the braided preform. After the braid performs were manufactured, the samples were placed onto 9.525mm (3/8") polytetrafluoroethylene (PTFE) mandrels and impregnated with a bio-based resin (SuperSap One, Entropy Resins, Bay City, MI). Each sample was cured at room temperature for seven days to reach full cure.

A high resolution micro-computed tomography technique was used to evaluate the geometry and void content of natural fiber/ bio-based resin braided composite structures. Current studies have examined conventional fibers and matrix materials using the μCT measurement method [18], [22]. The braid samples were evaluated using a SkyScan 1272 microtomograph (Bruker-MicroCT, Kontich, Belgium). Images of the braid samples were collected at a resolution of 4904x3280 pixels. The settings for the μCT scan of each braid sample are detailed in Table 1. The settings shown in Table 1 were selected to optimize contrast between the matrix and fiber materials. The scan time for each of the braid samples using the listed settings was approximately 60 minutes. The final image voxel size for each sample was set at 1.0 μm^3 to achieve sufficient resolution to observe voids and individual fiber filaments within the braided structures.

Table 1. Micro-computed tomography (μCT) scan parameters

μCT Parameter	Value
X-Ray Voltage	40 kV
Current	200 μA
Frame Averaging	2
Rotation Step	0.2°
Number of Projections	900
Projection Exposure Time	1316 ms
Detector Pixel Size	7.4 μm
Reconstructed Image Pixel Size	1.0 μm

A schematic of the tubular braid sample and sample mount are shown in Figure 2. The custom sample holder was designed using a computer aided design package (SolidWorks 2018, Dassault Systems, France). The sample mount was manufacturing using a 3D printer (Select Mini V2, MonoPrice, Brea, CA). As shown in Figure 2, the sample mount offset the centerline of the braided composite samples from the rotation axis of the μCT machine by 4.0mm. Offsetting the rotation axes of the braided composite sample allowed for a high resolution scan of a segment of each of the braid samples to be examined.

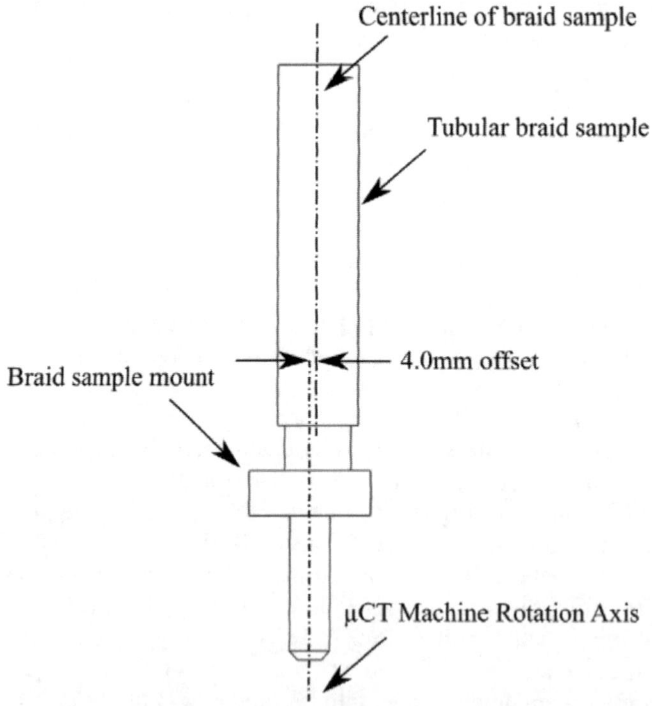

Figure 1. Tubular braided composite μCT sample mount

Radiograph images of the braided composite were collected for the two braid sample. Example radiographs of the 35° and 45° braid samples are shown in Figure 2. In this image, the braid samples can be seen off-set from the center of the radiograph image due to the sample mount shown in Figure 1. Large pores within the braided composite sample can be observed in the radiographs shown in Figure 2. As shown in Table 1, 900 radiographs were obtained for each braid sample. The collected radiographs allowed for a segment of each braid to be reconstructed for detailed analysis. The radiograph images were imported into a reconstruction software (NRECON 1.7.1.0, Bruker, Belgium) to create cross-sectional images of the braided composite as seen in Figure 3. In Figure 3, grayscale contrast indicating the matrix and fibers can be seen in the braid cross-sectional image as well as the varying braid geometry of the 35° and 45° braids. The grayscale variations in Figure 3 enable the quantification of the resin, matrix and void content within each braid sample. All reconstructed cross-sectional images were exported as 8-bit bitmap (BMP) images with an image size of 4904 x 4904 pixels for analysis.

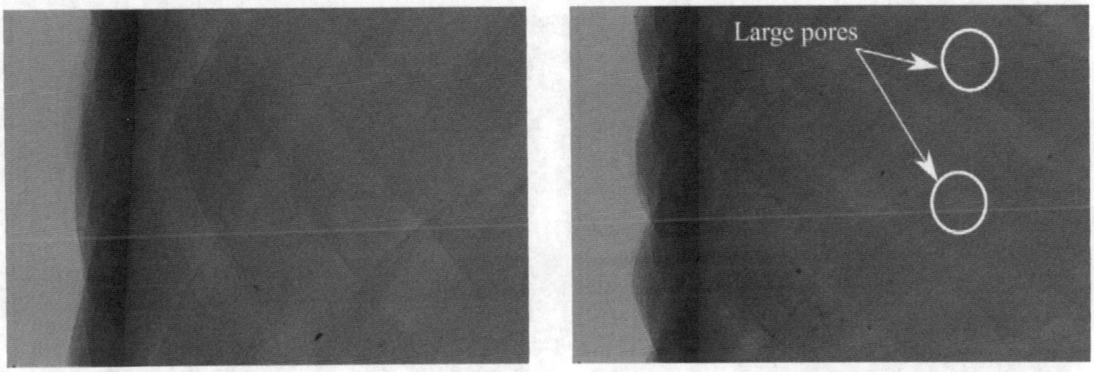

(a) (b)

Figure 2. Example radiographs of tubular braided composite samples (a) 35° braid (b) 45° braid. Voids, fibers and matrix material can be seen in each of the radiographs

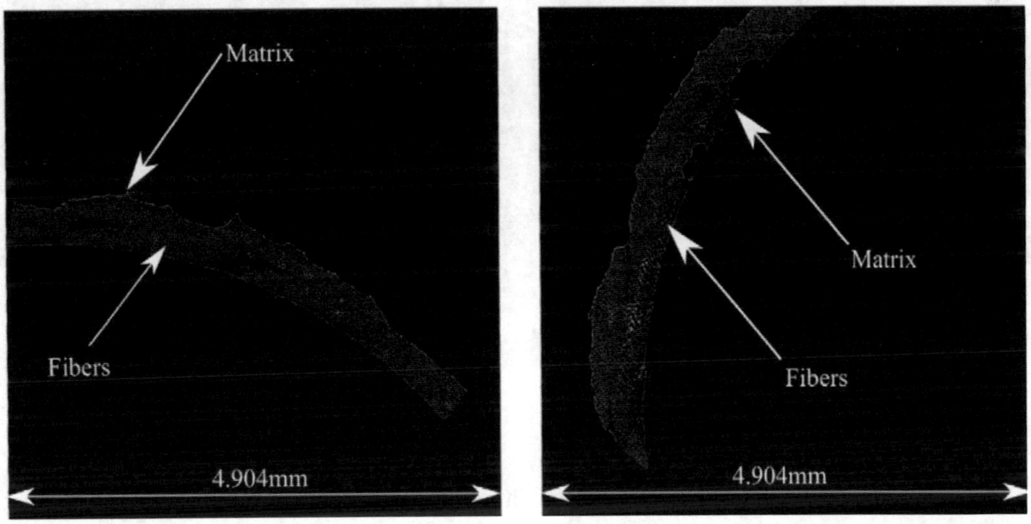

(a) (b)

Figure 3. Representative reconstructed cross-sectional images of tubular braid samples (a) 35° braid 1.0 μm scan resolution (b) 45° braid 1.0 μm scan resolution. Braid yarn geometry, voids and the reinforcing matrix can be seen with varying grayscale intensity.

2.1. Braid Morphology
Image segmentation was performed on the braid sample cross-sectional images to identify braid features [25]. Segmentation of the collected braid cross-sectional images was performed using an image analysis and visualization software package (CTan 1.16.90, Bruker microCT, Belgium). The image segmentation procedure identified matrix, fibers, and voids within the braid structure. The following image segmentation procedure was used to quantify the constituents of the braid sample. To remove noise from the braid cross-section images a Gaussian Blur operation was used with a circular structuring element of 1 pixel. The braid yarns were determined by thresholding the braid sample using a threshold range of 30-120. Similarly, the resin was identified by thresholding the braid sample using a threshold of 10-30. The segmented braid yarns and matrix for both 35° and 45° braid samples are shown in Figure 4. The identified fibers and matrix materials for both braid samples are shown in this figure.

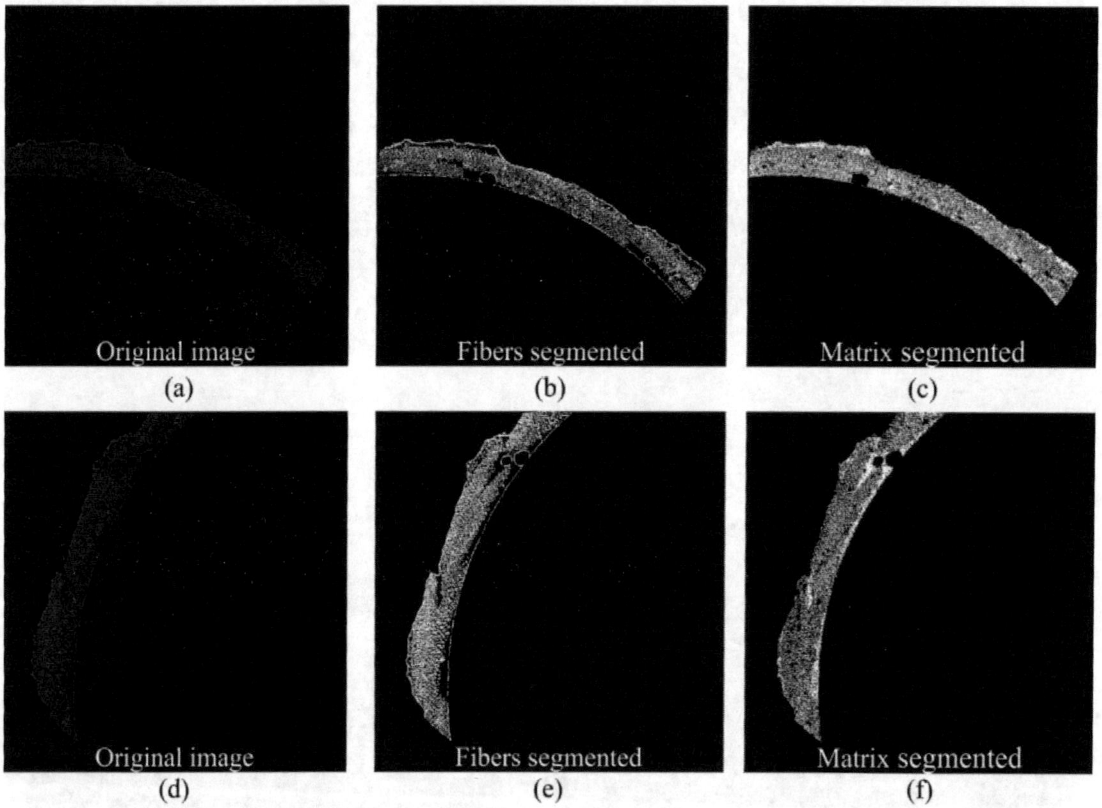

Figure 4. Matrix and fiber segmentation (a) 35° original braid cross section (b) 35° braid segmented yarns (c) 35° segmented matrix (d) 45° original cross-section (e) 45° segmented yarns (f) 45° segmented matrix

2.2. Yarn Morphology
Individual braid yarns were also examined for both the 35° and 45° braid angle samples. Individual yarns were identified using the region-of-interest "ROI" tool in CTan to identify a single fiber tow from within the braid geometry. An example of idealized braid yarns used in the prediction of the mechanical properties of Diamond and Regular braid are shown in Figure 5 [26]. The identified braid fiber tows can be seen in Figure 6. In Figure 6 (a) and (b) an individual fiber tow is identified for the 35° and 45° braid respectively. The braid tows were segmented to identify individual fibers as shown in Figure 6 (c) and (d). The identified individual fibers with the tow will be compared with the manufacture data to assess the number of fiber filaments within a single tow. Additionally, the geometry of an individual fiber tow

was also identified as shown in Figure 6 (e) and (f). The total cross-sectional area, major diameter, and minor diameter of the fiber tow are calculated from the segmented images shown in Figure 6 (e) and (f) to assess the effect of braiding angle on tow geometry.

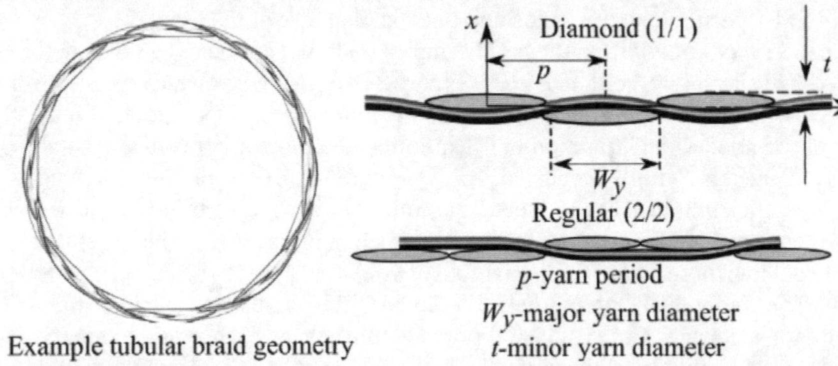

Figure 5. Example idealized braid geometries used for predicting mechanical properties

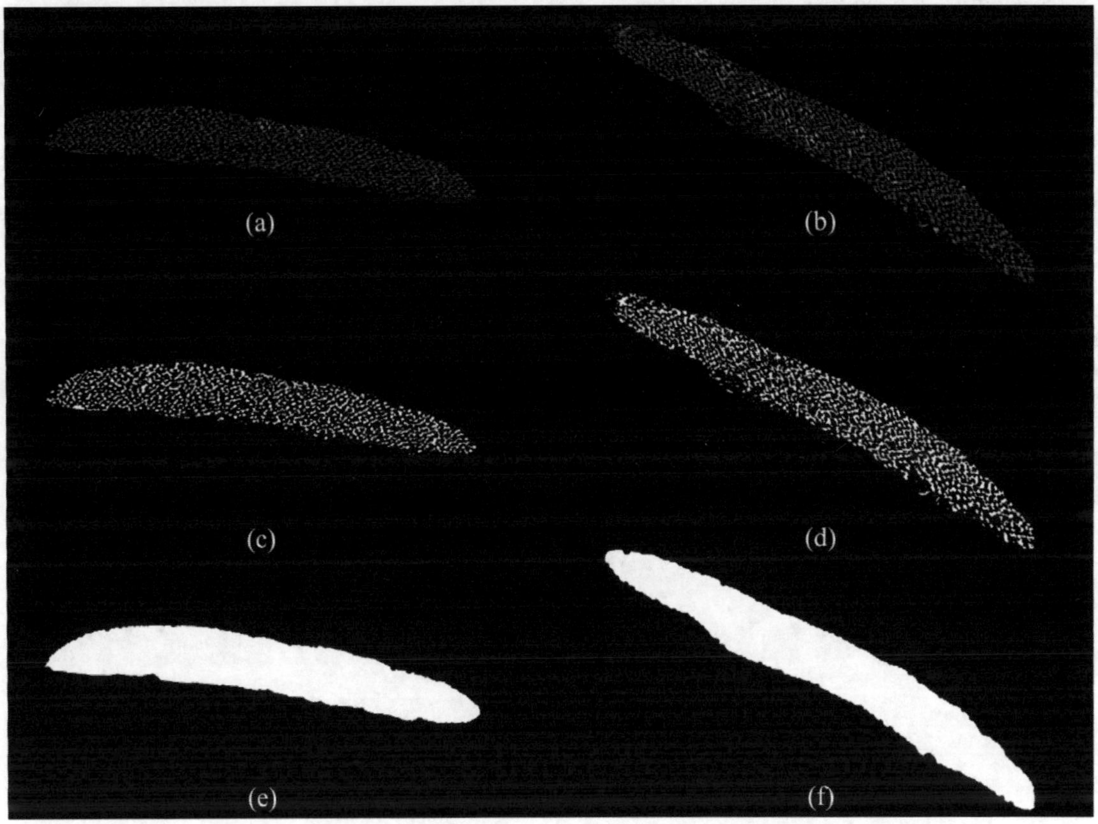

Figure 6. Identification of individual braiding yarns (a) 35° braid yarn (b) 45° braid yarn (c) Segmented 35° braid yarn where individual fibers are identified (d) Segmented 45° braiding yarn with identified fibers (e) Identified 35° fiber tow (f) Identified 45° fiber tow.

IOP Conf. Series: Materials Science and Engineering **406** (2018) 012012 doi:10.1088/1757-899X/406/1/012012

2.3. Braid Void Analysis

Voids and open pores were identified using an image segmentation algorithm. Examples of voids within the braided composite sample cross-section are shown in Figure 7. Enclosed voids are shown in Figure 7 (a) where the enclosed voids occur within the inner and outer diameters of the braid samples. Open pores, shown in Figure 7 (b), occur along the inside and outside diameter of the braid.

The identification process for voids within the braid samples is shown in Figure 8. Voids in the braid samples were identified by using a threshold of 10-120 (Figure 8 (b)). All voids were removed from the segmented image to allow for individual void identification (Figure 8 (c)). Open pores, Figure 8 (d), were identified using the "Shrink-wrap" function in CTan to locate all open pores that exist along the inner and outer braid diameters. Fully enclosed voids, Figure 8 (e), were identified using the "Despeckle-remove pores" function in CTan to identify all fully voids which are fully enclosed within the braid geometry. All enclosed voids and open pores can be seen in Figure 8 (f). The generated binary images shown in Figure 8 allow for the size and distribution of voids and pores to be analyzed throughout the 3D braid geometry. The "Individual Object Analysis" tool in CTan was used to calculate the volume, major diameter, and minor diameter of all voids and pores identified for each braid geometry. After each segmentation, a 3D volume was created using the "Double-Time Cubes" stereolithography (STL) algorithm in CTan. Finally, the 3D fiber, resin and void volumes were visualized using an open-source visualization software (ParaView 5.4.1, Kitware, Clifton Park, NY) to visualize the spatial distribution of the segmented braid features.

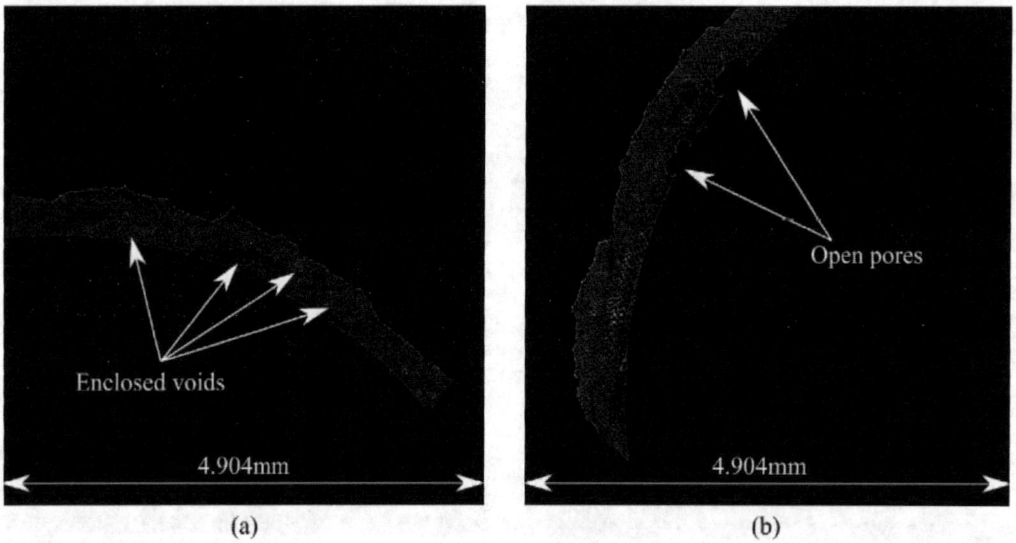

(a) (b)

Figure 7. Void Examples (a) Enclosed voids (b) Open pores

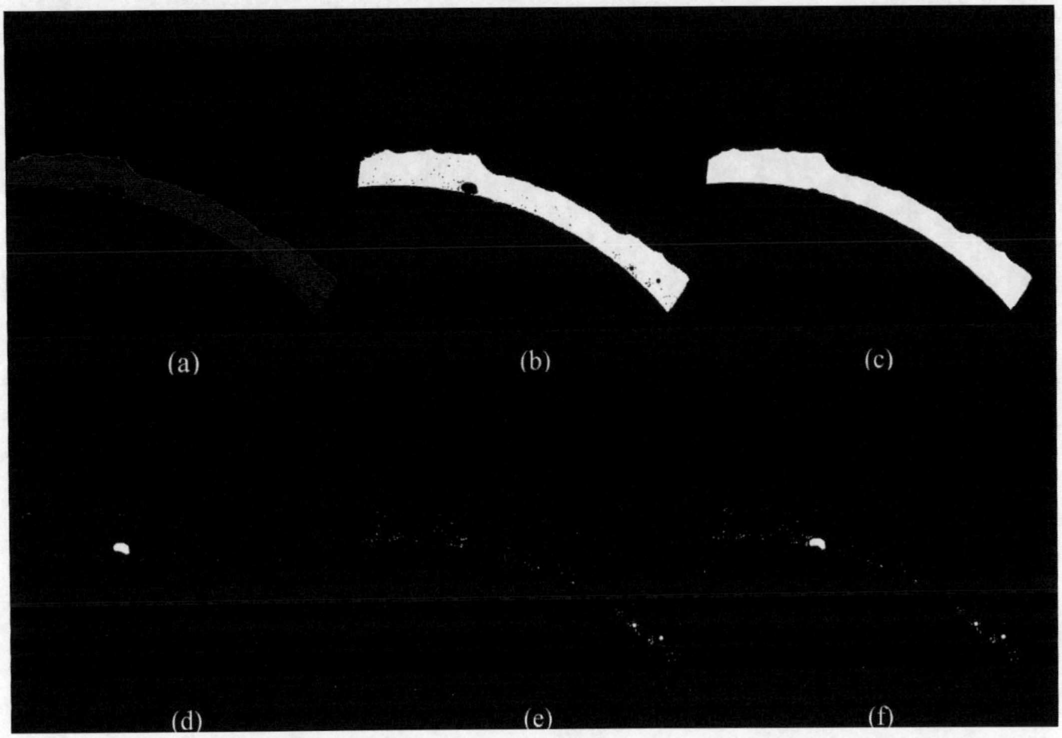

Figure 8. Braid void segmentation procedure (a) original braid cross-section, (b) thresholded cross-section showing enclosed voids and open pores, (c) braid geometry with open pores and enclosed voids filled, (d) identified open pores, (e) identified closed voids, (f) identified open pores and closed voids

3. Results and Discussion

3.1. Braid Morphology

The resulting 3D geometries for the 35° and 45° braid samples are shown in Figure 9. The open pore and enclosed void distribution is visualized for both braid geometries. In addition, the matrix and fiber distribution for both geometries are shown. The distribution of voids relative to the braid matrix and fiber volumes can also be seen. From Figure 9 it can be seen that enclosed voids commonly occur long the braid yarn paths. A significant portion of enclosed voids occur at the cross-over region between braiding yarns. Also seen in Figure 9 are several large pores which were formed along the inside and outside diameters of the braid. Large spherical pores occur along the inside diameter of the braid while elliptical or elongated pores occur along the outside diameter of each braid. The three dimensional geometries of the two braid samples can also be seen in Figure 9 (g) and (h).

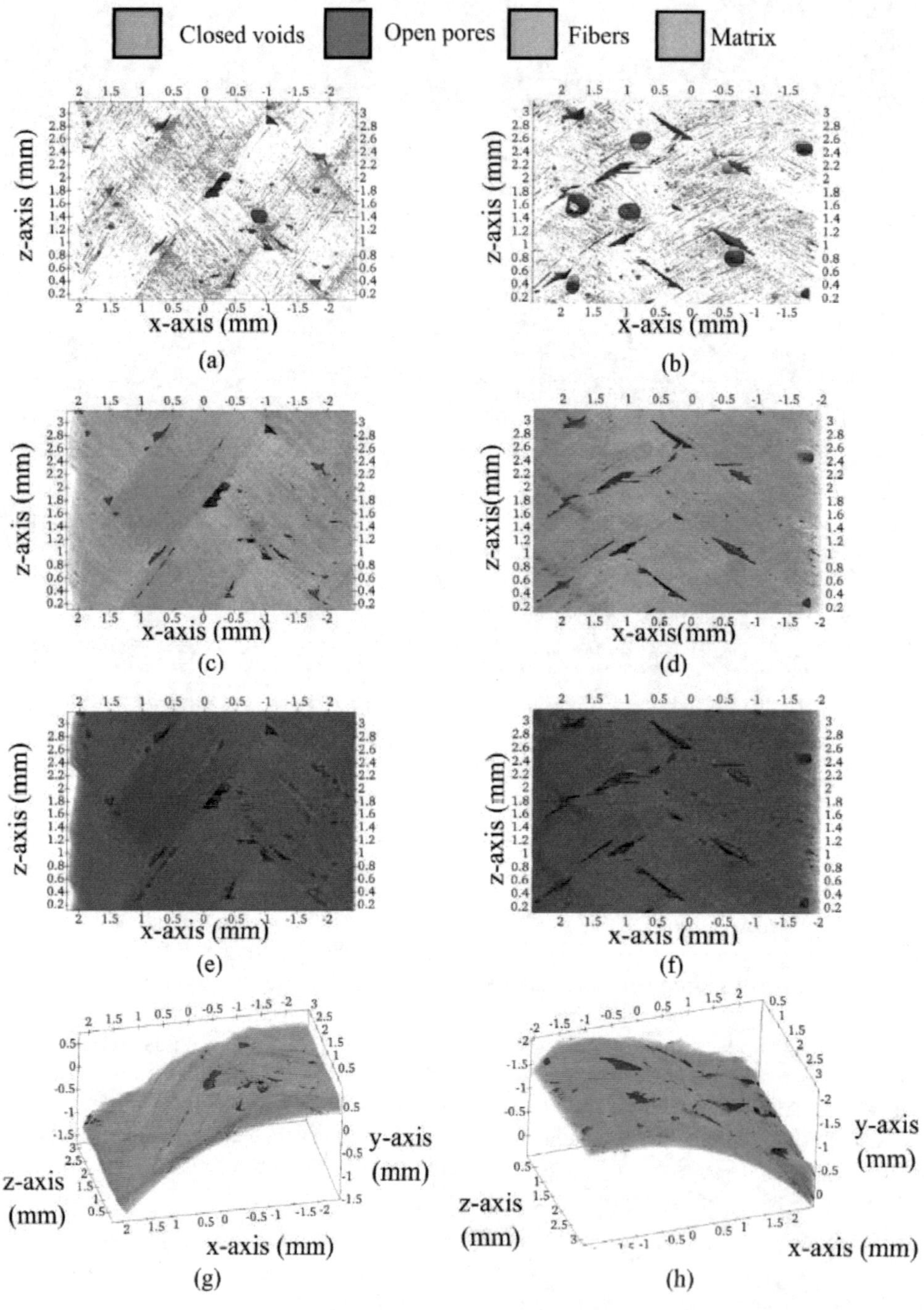

Figure 9. 3D visualization of braid geometry (a) 35° braid closed void and open pore distribution (b) 45° braid closed void and open pore distribution (c) 35° braid fiber distribution with voids shown (d) 45° braid fiber distribution with voids shown (e) 35° braid matrix distribution with voids shown (f) 45° braid with voids shown.

The total fiber volume V_f, matrix volume V_m, enclosed void volume V_{vc} and open pore volume V_{vo} and total void V_T volumes were calculated for each braid geometry. The calculated volumes for both braid samples are summarized in Table 2. The fiber volume fraction (v_f), matrix volume fraction (v_m)and closed void volume fraction (v_{vc}) and open pore volume fraction (v_{co}) were calculated using Equations (2) - (5). The total void volume fraction (v_T) is the sum of the open pore and closed void volume fractions as shown in Equation (1).

$$v_f = \frac{V_f}{V_f + V_m + V_{vc} + V_{vo}} \tag{2}$$

$$v_m = \frac{V_m}{V_f + V_m + V_{vc} + V_{vo}} \tag{3}$$

$$v_{vc} = \frac{V_{vc}}{V_f + V_m + V_{vc} + V_{vo}}$$

$$v_{vc} = \frac{V_{vc}}{V_f + V_m + V_{vc} + V_{vo}} \tag{4}$$

$$v_{vo} = \frac{V_{vo}}{V_f + V_m + V_{vc} + V_{vo}} \tag{5}$$

$$v_T = v_{vc} + v_{vo} \tag{6}$$

Table 2 shows that there is an increase in fiber volume and fiber volume fraction as braiding angle increases. Additionally, the 45° braid sample exhibited a greater number of open pores and total void volume. The increased number of open pores is a result of the increased braiding angle. Braid samples manufactured in this study were not cured under a vacuum and as a result air can become entrapped between the braid mandrel and fibers curing sample curing resulting in the large open pores observed. The matrix volume fraction was also observed to decrease with increasing braiding angle. The lower fiber volume fraction in the 35° results in an increase in the matrix volume fraction within the braid geometry. As fiber volume fraction increases, matrix volume fraction will decrease due to fiber compaction within the braid structure.

Table 2. Braid calculated fiber, matrix and void volumes

Sample	35° braid	45° braid
Fiber Volume V_f (mm^3)	2.0939	3.5266
Matrix Volume V_m (mm^3)	3.9081	3.3608
Closed Void Volume V_{vc} (mm^3)	0.0660	0.0663
Open Pore Volume V_{vo} (mm^3)	0.0153	0.0615
Total Void Volume (mm^3)	0.0812	0.1278
Total Volume (mm^3)	6.0832	7.0152
Fiber Volume Fraction v_f	34.42%	50.27%
Matrix Volume Fraction v_m	64.24%	47.91%
Closed Void Volume Fraction v_{vc}	1.08%	0.94%
Open Pore Volume Fraction v_{vo}	0.25%	0.88%
Total Void Volume Fraction v_T	1.34%	1.82%

3.2. Yarn Morphology

Individual fiber tows for the 35° and 45° braids were examined to determine the total fiber tow cross-sectional area and to determine the fiber tow major and minor diameters. The measured fiber tow area, major diameter and minor diameter for selected braid yarns are summarized in Table 3. The measured tow major diameter and minor diameter correspond to the yarn width (W_y) and thickness (t) measurements utilized by authors to describe braid yarn geometry for analytical and numerical modelling [26]–[28]. Accurate measurement of fiber tow dimensions is critical for accurate model development for tubular braided composites. Table 3 demonstrates that the fiber tow major, minor diameters and cross-sectional area vary with braiding angle. As seen in Table 3 the yarn major diameter increases within braiding angle. The increase in yarn major width or major diameter is a result on the increased fiber compaction that occurs with an increasing braid angle. The differences in yarn major and minor diameters can be observed in the yarn images shown in Figure 6. The geometry of braid yarns was also examined by Melenka *et al.* [22]. A single braid with a Diamond braiding pattern was examined. Accurate measurement of the effect of braiding angle on yarn dimensions is critical for the accurate prediction of braid mechanical properties.

Table 3. Braid fiber tow analysis

Sample	35° braid	45° braid
Fiber Tow Area (mm^2)	0.2928±0.0032	0.4043±0.0287
Fiber Tow Major Diameter (mm)	1.6946±0.0036	2.1189±0.0594
Fiber Tow Minor Diameter (mm)	0.2177±0.0038	0.2584±0.0217

Individual braid tows were also examined to identify the individual fiber filaments within a fiber tow. The identified filaments within a fiber tow can be seen in Figure 6 (c) and (d). The results from the braid tow analysis is shown in Table 4. The total number of identified filaments are summarized as well as the total filament area and average major and minor diameters of each of the individual filaments.

Table 4. Braid yarn filament analysis

Sample	35° braid	45° braid
Number of Filaments Identified	685±25	673±60
Filament Area (μm^2)	96.67±5.20	169.40±10.33
Filament Major Diameter (μm)	16.87±0.58	20.49±0.99
Filament Minor Diameter (μm)	7.64±0.27	9.52±0.49

The cellulose yarns that were used in this study have a fiber denier of 1650 and are reported to consist of 900 filaments with a filament diameter of 11μm. From Table 4 it can be seen that 685±25 and 673± 60 filaments were identified for the 35° and 45° fiber tows respectively. The measured number of fiber filaments was lower than the manufacturer specifications. The lower number of identified filaments within the fiber tow is most likely due to the image segmentation process. The proximity of the fiber filaments to one another can result in an underestimate of the total number of filaments within the fiber tow. Figure 10 shows a number of large objects that were identified from the image segmentation process. The presence of these large objects results in an underestimate of the true number of filaments within the braid tow. Improvement to the image segmentation algorithm and contrast within the fiber tow images will allow for better identification of individual filaments within the braid tow. The average major diameter of the tow filaments was found to be 16.86±0.57 and 20.48±0.98 μm for the 35° and 45° braids respectively. The measurements of the filament diameter also suggest that all individual yarns have not be identified as the reported diameter of the cellulose filaments is 11μm. The segmentation and analysis of individual filaments within the braid yarns could be improved by the addition of a contrast enhancement agents such as zinc iodide to aide in filament segmentation [16]. A higher resolution μCT scan with a resolution of less than 1.0μm may also be required to identify all tow filaments. Finally, the SkyScan 1272 μCT utilized in this study is capable of achieving a scan resolution of 0.25μm which will allow for improved resolution of individual fiber filaments.

IOP Conf. Series: Materials Science and Engineering **406** (2018) 012012 doi:10.1088/1757-899X/406/1/012012

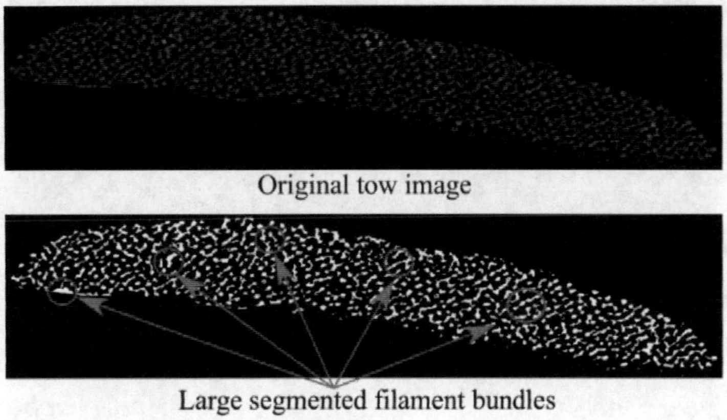

Figure 10. Fiber filament identification

3.3. Braid Void Analysis

The void distribution throughout both the 35° and 45° samples was examined to quantify the range of open pore and closed void size throughout both samples. The identified closed voids and open pores for both braiding samples are shown in Figure 11. Figure 11 demonstrates that there are significantly more closed voids within both the 35° and 45° braid samples than open pores. Measurement of the average closed void and open pore volumes are also summarized in Table 5. Figure 11 (a) and (b) show that a total of 2204 and 1862 closed voids were identified for the 35° and 45° braids respectively. Similarly, Figure 11 (c) and (d) shows that a total of 119 and 148 open pores were identified for the 35° and 45° braid samples. The void distribution is visualized for both braid geometries in Figure 9. This figure also demonstrates that there are a significantly more closed voids within both braid geometries than open pores.

The average open pore and closed pore volumes and maximum closed void and open pore volumes are summarized in Table 5. Table 5 demonstrates that the open pore volumes found within the two braid samples are an order of magnitude greater than the fully enclosed voids in both samples. The presence of open pores can be significantly minimized by utilizing a vacuum assisted resin transfer molding procedure (VARTM) to ensure improved permeation of the matrix material throughout the braided composite structures. A VARTM process would also assist in decreasing the quantity of closed voids that occur within the braided structure.

Table 5. Average closed void and open void volumes

Sample	35° braid	45° braid
Closed Void Average Volume (mm^3)	1.90E-05	2.50E-05
Maximum Closed Void Volume (mm^3)	7.40E-04	2.40E-03
Number of Identified Closed Voids	2204	1862
Open Pore Average Volume (mm^3)	1.23E-04	4.11E-04
Maximum Open Pore Volume (mm^3)	5.02E-03	9.00E-03
Number of Identified Open Pores	119	148

IOP Conf. Series: Materials Science and Engineering **406** (2018) 012012 doi:10.1088/1757-899X/406/1/012012

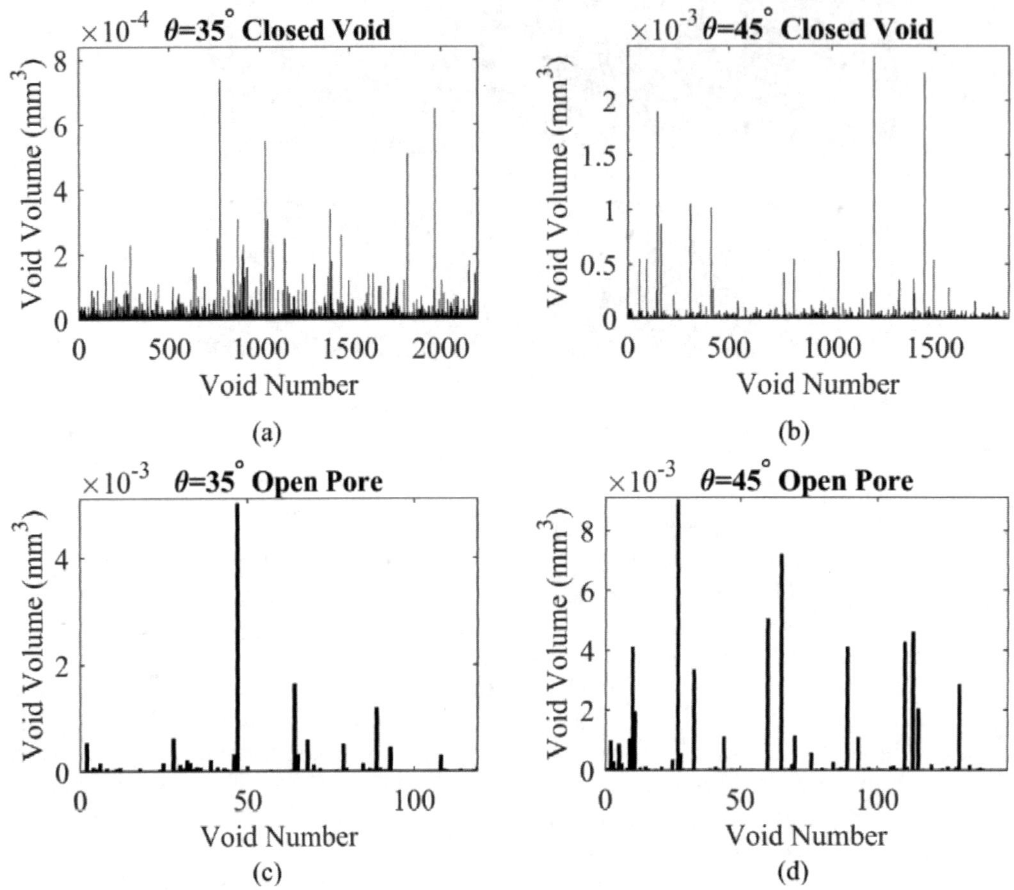

Figure 11. Identified closed void and open pores (a) 35° braid closed voids (b) 45° braid closed voids (c) 35° open pores (d) 45° open pores

4. Conclusions

Braided composite samples were examined using a high resolution μCT measurement technique to quantify the sample void, matrix and fiber contents. The braids used in this study were manufactured with braiding angles of 35° and 45° and were produced using natural cellulose fibers and a bio-based resin. The braid samples were scanned using a 1.0μm^3 voxel size to examine the microstructure of each braid sample. Two braiding angles were selected to allow for the effect of braid angle on geometry to be investigated. In addition, the void distribution throughout each sample was also examined. The total void volume fraction for the 35° and 45° braids was determined to be 1.34 and 1.82% respectively.

Individual fiber tows were also investigated for the manufactured 35° and 45° braid samples. The overall tow geometry was examined for the two manufactured braid geometries to demonstrate the effect of braid angle on fiber tow geometry. The major diameters for the 35° and 45° samples was measure to be 1.6946 and 2.1189 mm respectively. The results from the fiber tow measurement demonstrate the change in fiber tow geometry with braiding angle. Accurate measurement of fiber tow geometry is required to support analytical and numerical models for braided composite structures.

Finally, individual filaments were examined for fiber tows within each of the braid geometries. Individual fiber filaments were identified however the results from the filament analysis indicate that a higher resolution μCT scan or improve image segmentation process will be required for improved analysis of fiber filaments within braided structures.

This study demonstrates that the µCT measurement technique is a highly effective measurement tool for assessing the geometry of braided composite structures. Open pores and closed voids were identified for two braid samples and their distribution thought both samples was examined. The overall matrix and fiber content for both samples was also quantified. Individual fiber tows were also examined. The method demonstrated in this work will allow for assessing the manufacturing consistency of braided composite structures. Geometric measurements will also support analytical and numerical models used to predict braided composite mechanical properties.

Acknowledgments

The authors acknowledge the funding support from the Natural Sciences and Research Council (NSERC) Canada and for support from the Department of Mechanical Engineering of the Lassonde School of Engineering.

References

[1] D. Brunnschweiler, "Braids and Braiding," *J. Text. Inst. Proc.*, vol. 44, no. 9, pp. P666–P686, 1953.
[2] G. W. Melenka *et al.*, "Manufacturing processes for braided composite materials," in *Handbook of Advances in Braided Composite Materials: Theory, Production, Testing and Applications*, 2016, pp. 47–153.
[3] F. K. Ko, "Braiding," vol. 21, pp. 69–77, 2001.
[4] S. Kobayashi and K. Takada, "Processing of unidirectional hemp fiber reinforced composites with micro-braiding technique," *Compos. Part A Appl. Sci. Manuf.*, vol. 46, no. 1, pp. 173–179, 2013.
[5] S. Kobayashi, K. Takada, and D. Song, "Effect of Molding Condition on the Mechanical Properties of Bamboo- Rayon Continuous Fiber / Poly (Lactic Acid) Composites," *Adv. Compos. Mater.*, vol. 21, no. 1, pp. 79–90, 2012.
[6] A. Memon and A. Nakai, "The Processing Design of Jute Spun Yarn / PLA Braided Composite by Pultrusion Molding," *Adv. Mech. Eng.*, vol. 2013, pp. 1–8, 2013.
[7] I. I. Qamhia, S. S. Shams, and R. F. El-Hajjar, "Quasi-Isotropic Triaxially Braided Cellulose-Reinforced Composites," *Mech. Adv. Mater. Struct.*, vol. 22, no. 12, pp. 988–995, 2015.
[8] J. Datta and M. Włoch, "Selected biotrends in development of epoxy resins and their composites," *Polym. Bull.*, vol. 71, no. 11, pp. 3035–3049, 2014.
[9] P. Wambua, J. Ivens, and I. Verpoest, "Natural fibres: Can they replace glass in fibre reinforced plastics?," *Compos. Sci. Technol.*, vol. 63, no. 9, pp. 1259–1264, 2003.
[10] R. Dunne, D. Desai, R. Sadiku, and J. Jayaramudu, "A review of natural fibres, their sustainability and automotive applications," *J. Reinf. Plast. Compos.* , vol. 35, no. 13, pp. 1041–1050, 2016.
[11] F. P. La Mantia and M. Morreale, "Green composites: A brief review," *Compos. Part A Appl. Sci. Manuf.*, vol. 42, no. 6, pp. 579–588, 2011.
[12] J.-H. Byun, "The analytical characterization of 2-D braided textile composites," *Compos. Sci. Technol.*, vol. 60, no. 5, pp. 705–716, Apr. 2000.
[13] S. C. Quek, A. M. Waas, K. W. Shahwan, and V. Agaram, "Analysis of 2D triaxial flat braided textile composites," *Int. J. Mech. Sci.*, vol. 45, no. 6–7, pp. 1077–1096, 2003.
[14] M. M. Shokrieh and M. S. Mazloomi, "An analytical method for calculating stiffness of two-dimensional tri-axial braided composites," *Compos. Struct.*, vol. 92, no. 12, pp. 2901–2905, Nov. 2010.
[15] L. Xu, S. J. Kim, C. H. Ong, and S. K. Ha, "Prediction of material properties of biaxial and triaxial braided textile composites," *J. Compos. Mater.*, vol. 46, no. 18, pp. 2255–2270, 2012.
[16] L. P. Djukic, I. Herszberg, W. R. Walsh, G. a. Schoeppner, B. Gangadhara Prusty, and D. W. Kelly, "Contrast enhancement in visualisation of woven composite tow architecture using a MicroCT Scanner. Part 1: Fabric coating and resin additives," *Compos. Part A Appl. Sci.*

Manuf., vol. 40, no. 5, pp. 553–565, May 2009.

[17] R. G. Rinaldi, M. Blacklock, H. Bale, M. R. Begley, and B. N. Cox, "Generating virtual textile composite specimens using statistical data from micro-computed tomography: 3D tow representations," *J. Mech. Phys. Solids*, vol. 60, no. 8, pp. 1561–1581, 2012.

[18] J. Ya, Z. Liu, and Y. Wang, "Micro-CT Characterization on the Meso-Structure of Three-Dimensional Full Five-Directional Braided Composite," *Appl. Compos. Mater.*, pp. 1–18, 2016.

[19] P. Badel and E. Maire, "X-ray tomography analysis of the mechanical behaviour of reinforcements in composites," *Compos. Reinf. Optim. Perform.*, pp. 565–587, 2011.

[20] J. E. Little, X. Yuan, and M. I. Jones, "Characterisation of voids in fibre reinforced composite materials," *NDT E Int.*, vol. 46, pp. 122–127, Mar. 2012.

[21] F. Desplentere, S. V. Lomov, D. L. Woerdeman, I. Verpoest, M. Wevers, and a. Bogdanovich, "Micro-CT characterization of variability in 3D textile architecture," *Compos. Sci. Technol.*, vol. 65, no. 13, pp. 1920–1930, Oct. 2005.

[22] G. W. Melenka, E. Lepp, B. K. Cheung, and J. P. Carey, "Micro-computed tomography analysis of tubular braided composites," *Compos. Struct.*, vol. 131, pp. 384–396, 2015.

[23] N. Abdelal and S. L. Donaldson, "Comparison of methods for the characterization of voids in glass fiber composites," *J. Compos. Mater.*, p. 002199831771008, 2017.

[24] K. Xu and X. Qian, "An FEM analysis with consideration of random void defects for predicting the mechanical properties of 3d braided composites," *Adv. Mater. Sci. Eng.*, vol. 2014, 2014.

[25] R. C. Gonzalez, *Digital image processing*. Prentice Hall YR - 2002.

[26] G. W. Melenka and J. P. Carey, "Development of a generalized analytical model for tubular braided-architecture composites," *J. Compos. Mater.*, 2017.

[27] L. Xu, C. Z. Jin, and S. Kyu Ha, "Ultimate strength prediction of braided textile composites using a multi-scale approach," *J. Compos. Mater.*, no. January, p. 0021998314521062-, 2014.

[28] C. S. Lee, S. G. Kim, H. J. Kim, and S. H. Ahn, "Measurement of anisotropic compressive strength of rapid prototyping parts," *J. Mater. Process. Technol.*, vol. 187–188, pp. 627–630, 2007.

IOP Conf. Series: Materials Science and Engineering **406** (2018) 012013 doi:10.1088/1757-899X/406/1/012013

In-situ computer tomography for analyzing the effect of voids on the damage behavior of composite materials

R Protz*, I Koch and M Gude

TU Dresden, Institute of Lightweight Engineering and Polymer Technology (ILK), Holbeinstr. 3, 01307 Dresden, Germany

* richard.protz@tu-dresden.de

Abstract. Due to their high specific stiffness and strength composites are ideally suited for applications in machine building, vehicle constructions, wind power technology and aviation. In the manufacture of composite structures voids are an unavoidable fact. Several non-destructive techniques are potentially able of detecting defects, but only the exact knowledge of the effects of defects on the mechanical properties allows the definition of thresholds for the purpose of quality management. Here, an experimental program for characterizing the effect of voids on the damage behaviour is presented. Flat specimens with glass fibre non-crimp fabric reinforcement and epoxy matrix (GF-NCF/EP) are produced using vacuum assisted resin transfer moulding. By means of a specific adjustment of the process parameters, test specimens with three different void contents are produced and compared. For detailed analysis of damage evolution in the vicinity of voids, in-situ computed tomography (in-situ CT) is used, for the volumetric visualization of the material during tensile loading.

1. Introduction

Composites offer high levels of specific stiffness and strength as well as an adjustable energy absorption capacity. Thus, they are ideally preferred for applications in material and energy efficient lightweight structures as used in the wind energy utilization and in aviation [1, 2]. In the manufacture of composite structures voids of different size, geometry and orientation are an unavoidable fact [3, 6-10] and have been investigated by many authors. Investigations can be divided in studies on the mechanism of void formation [10-17] and in studies on the effect of voids on the mechanical behavior [3, 6-8].

The development of voids strongly depends on the manufacturing technology. The cause of voids in prepreg materials used in the autoclave process differs essentially from that of semi-finished products infiltrated in an infusion process [18]. Voids are formed in a prepreg laminate when the pressure applied to the liquid resin is lower than the opposing vapor pressure of moisture entrapped in the laminate [22, 23]. Investigations to void formation and void morphology in prepreg materials were published among others in Refs. [7, 8, 10, 19-24]. Prepreg materials are commonly used for investigating the influence of voids on the subsequent material behavior. During the infusion process (i.e. the various resin transfer moulding process, RTM), the resin is injected in a mould cavity containing the textile preform. The interaction of vacuum quality, pressure of resin-injection, permeability of the reinforcement material and viscosity of the resin all show major influences on content and phenomenology of the emerging voids. These parameters lead to variations of the flow rate in the fibre bundles and in the channels between two bundles. Different flow fronts of the resin support a partial infiltration leading to void encapsulation.

In the literature there are indications of a critical void contents from which an influence of the mechanical properties are identifiable in the range from 0.3 % to 6 % [20-22, 25-27]. This wide range results from the fact that apart from the pure void content, especially the size, the shape and the distribution of the voids as well as the type of load must be considered.

For the determination of void contents, void geometry and for the analysis of the influence of voids during mechanical loads, a large number of test methods such as microscopy, ultrasound technology, thermography, acoustic emission analysis and X-ray methods are already available. In order to avoid destruction of the non-transparent samples, void analysis under the microscope is limited to the cover layers and the longitudinal edges of the specimens [33, 34]. Under mechanical stress, voids can be detected by ultrasound or thermally excited lock-in thermography [35]. Also acoustic emission is often used for determining the effect of voids in composites on the damage behaviour [36]. With in-situ CT it is possible to efficiently analyse void content and morphology of voids and the influence of voids under static loading. As published in [32], the existing disadvantages of classical CT-based damage analysis, such as stress-strain-hysteresis effects, inaccuracies due to crack-closing phenomena after specimen relief and damage induction due to repeated expansion and contraction, are eliminated with this method. This also significantly reduces the experimental effort.

The focus in this paper is a detailed analysis of damage evolution in the vicinity of voids. The generated void volume levels in these studies were between 0.01 % and 2.9 % (as well as in the investigations of other authors) and thus typical for composites produced in resin transfer moulding process [6, 10-17, 25-29]. Praxis related glass fibre non crimp fabric reinforced epoxy composite (GF-NCF/EP) was considered. More detailed results of the author's studies on the influence of voids in the aforementioned material regarding its structural behaviour have been published in Refs. [5, 30, 31].

2. Specimen preparation

For the experimental and theoretical studies, composites with GF-NCF reinforcement of eight layers [90/-45/0/45/-45/0/45/90] are considered. The plies contain different amount of fibers (49 % in 0°, 23 % in -45°, 23 % in 45 and 5 % in 90°).

Using the vacuum assisted resin transfer moulding (VARTM) method, test plates were infiltrated in a flat steel mould with injection epoxy resin (EP) RIMR 135 with hardener RIMH 137 in a ratio of 100:30. This resin has a low viscosity and is cold curing within 24 h. For the purpose of reducing the process time, the infiltration as well as curing is carried out at 40 °C. A curing time of 10 h is followed by a tempering cycle of 6 h at 80 °C. After the heat treatment the resin shows a glass transition temperature of 85 °C. A volume fraction of about 33 % has been achieved according to ISO 1172.

To investigate the influence of production-related defects, voids with contents of 0.01 %, 1.43 % and 2.9 % have been systematically induced by adapted process parameters and modified VARTM-complex. The main influencing parameters for adjusting the void contents are the injection pressure on the sprue side and the vacuum pressure on the riser side, as this can vary the resin flow rate.

After infiltration and hardening test specimens with dimensions of 100 mm in length, 15 mm in width and 3 mm in thickness were cut from the test plate with water cooled diamond-tipped saw blade (Figure 1). In order to prevent failure caused by the fixing of the test machine aluminium-doublers on both ends of the flat specimen are used. This results in a free test length of 70 mm.

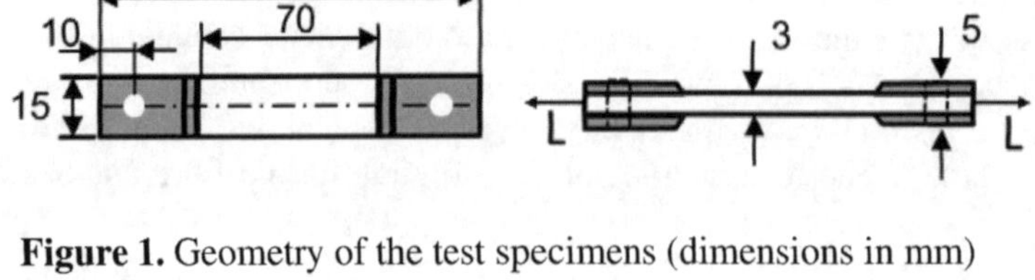

Figure 1. Geometry of the test specimens (dimensions in mm)

3. Test equipment and procedure

An in-situ computer tomograph (in-situ CT) is used for detailed analysis of the evolution of damage caused by voids under superimposed tensile loading at the micro level (Figure 2). This unique test complex consists of a tensile-compression torsion testing machine with a maximum force of 250 kN and a maximum torque of 2000 Nm and a computer tomograph FCTS 160-IS from Finetec, whose technical performance data are shown in Table 1. The right side of Figure 2 shows the alignment of the load train between the X-ray source and detector rotating around the loaded sample during the scan cycle. The measuring time is approx. 20 min at a resolution of 6.3 μm.

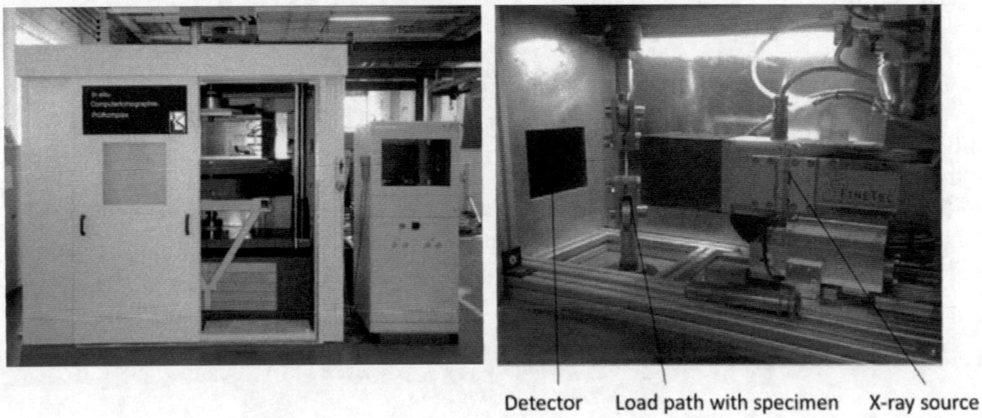

Figure 2. In-situ CT test rig (left) and test unit (right)

Table 1. Technical performance data of the in-situ CT test complex

Parameter	Characteristic
Acceleration voltage	30 - 160 kV
X-ray current	10 - 1000 μA
Target material	Tungsten
Min. focal spot	3 μm
Active detector area	405 mm x 290 mm
Detector resolution	3200 x 3200 pixels
Grayscale	14 bit
Min. measuring field diameter	85 mm
Max. Measuring field height	60 mm

Figure 3 shows the basic procedure of an in-situ CT examination on test specimens. Immediately after the specimen has been clamped, a first load-free scanning cycle takes place, which analyses the existing voids in the specimen. Then, analogous to quasi-static tensile tests, the force is measured continuously at a test speed of 2 mm/min until a force drop of 0.2 % of the current force is registered on the machine side. This indicates an increase in damage. The testing machine automatically relieves the load to 50 % of the force in the peak and remains force-controlled in the second scan cycle. After a waiting time of 90 seconds for strain retardation in the material, the CT scan starts. Within 20 minutes, the X-ray unit rotates around the sample and tomographs the damage. In the next step, the specimen is loaded in position control until another force drop occurs and the third scan cycle starts in force control at 50 % of the current force peak. These processes are repeated until the specimen fails.

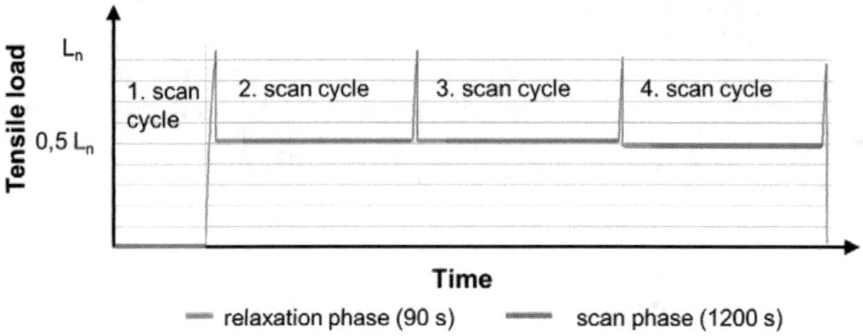

Figure 3. Basic procedure for in-situ CT examinations

4. Results

4.1. *Void analysis*

A void analysis was prepared in the first scan cycle without tensile loading. Figure 4 shows selected 3D tomograms specimens with low, middle and higher void content (V_V).

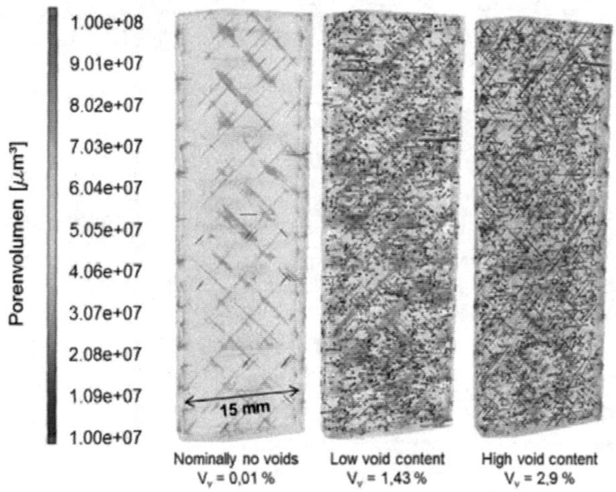

Figure 4. Typical 3D tomograms of nominal void-free (left), with few voids (centre) and many voids (right) of GF-NCF/EP specimens

In the nominal void-free test specimens (V_V = 0.01 %), isolated tube voids were detected between the fibre bundles of the 0°-layers and the 45°-layers. The determined average total void volume of 2.2×10^8 μm^3 corresponds to an average void volume content of 0.01 %.

Samples with low void volume contents of 1.43 % showed a homogeneous distribution of voids. With an average total volume of 3.7×10^{10} μm^3 voids, the majority of the voids have a void volume of less than 1×10^7 μm^3. In the fibre bundles of the outer layers, however, larger voids of 1.4×10^7 to 2.2×10^7 μm^3 were also detected in individual cases (Figure 5).

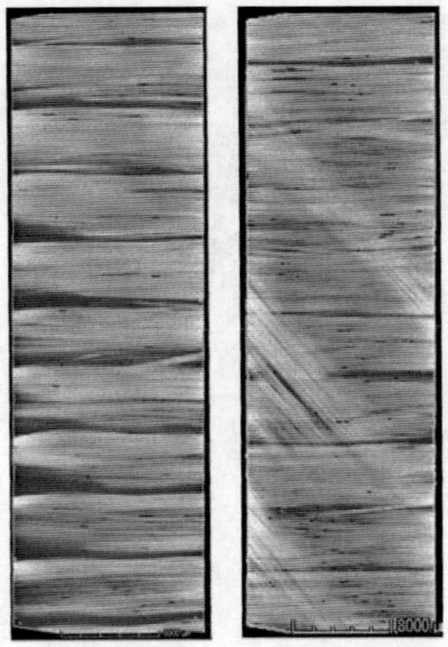

Figure 5. Highlighting of occasional larger voids in the 90°-layers

In the design of test specimens with a high void content, an average void volume content of 2.9 % with a total void volume of 2.0×10^{12} µm³. Here, too, most voids have a volume of less than 1.0×10^{7} µm³. Larger voids from 1.2×10^{7} to 1.5×10^{7} µm³ were detected in both the 0°-layers and the 45°-layers (Figure 6).

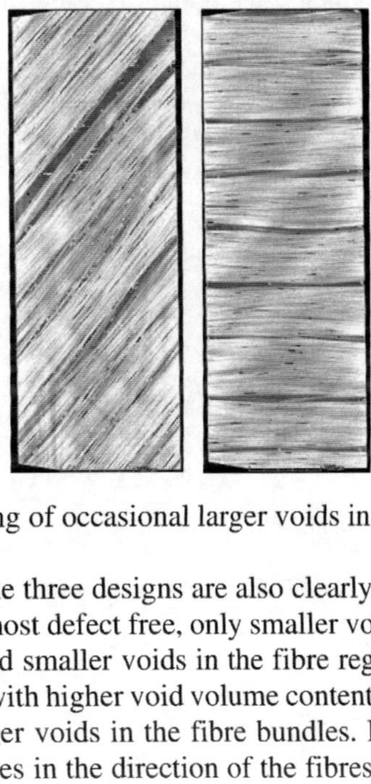

Figure 6. Highlighting of occasional larger voids in the 0°- and 45°-layers

The different void contents of the three designs are also clearly visible in the micrographs in Figure 7. The reference sample (left) is almost defect free, only smaller voids are visible. Large spherical voids in the intermediate fibre regions and smaller voids in the fibre regions are visible in samples with low void volume content. The samples with higher void volume content have significantly more voids across the cross-section but above all larger voids in the fibre bundles. In the upper 0°-layer, tube voids are often visible which orient themselves in the direction of the fibres. However, the majority of the voids can be found in the 45°-layers.

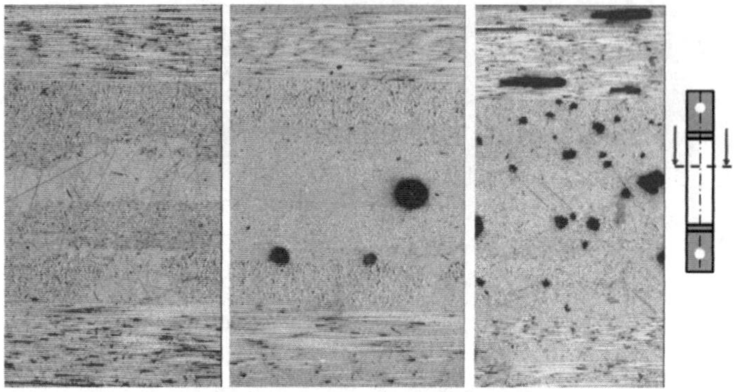

Figure 7. Selected micrographs for nominal void-free (left), with few voids (centre) and many voids (right) from GF-NCF/EP specimens

4.2. *In-situ CT*

The influence of voids on material mechanical behaviour was investigated with the aid of an in-situ CT. In favour of a more precise analysis of the fibres and voids, the known phenomena for the formation of inter-fibre breaks and delaminations under quasistatic tensile loading were not considered.

In the nominal defect free (low void content) reference samples, after the first force drop of 0.2 %, a first crack appears in a fibre bundle starting from the left edge, of the individual layer lieing in the direction of loading. If the crack is located at a point where the fibre strands of the individual layers touch, the crack propagates parallel to the fibre orientation of the adjacent layer. From scan cycle to scan cycle a zigzag growth of this crack can be observed. The fibre strand also tears from the opposite side, as individual filaments break. Further cracks become visible in the adjacent fibre bundle, which increase until the sample fails (Figure 8). The individual voids detected in the sample do not initiate any cracks themselves and influence the load-induced crack propagation in the reference test piece. Crack initiation always occurs from the outside to the inside until a large part of the rovings in the load direction is damaged to such an extent that the load can no longer be transferred. The location of the cracks and the location of the final specimen failure are spontaneous and unpredictable in the sense of this investigation.

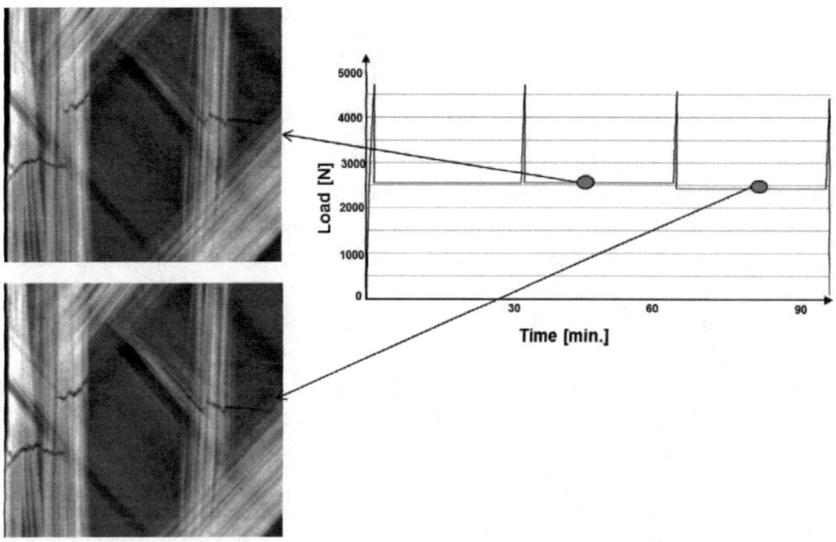

Figure 8. Typical crack progression in low-void reference specimen under tensile load

For the specimens examined here with void contents of 1.43 % and 2.90 %, the crack formation is far from the location of the largest void. Figure 9, for example, shows the three individual layers with the largest voids of a sample with a low void volume fraction detected by the CT software in the upper third of the test piece in pink. The two sectional images on the right show the crack formation locations removed from these voids.

Even with higher void contents, no correlation between the positions of the largest void and the location of crack formation can be seen (Figure 10).

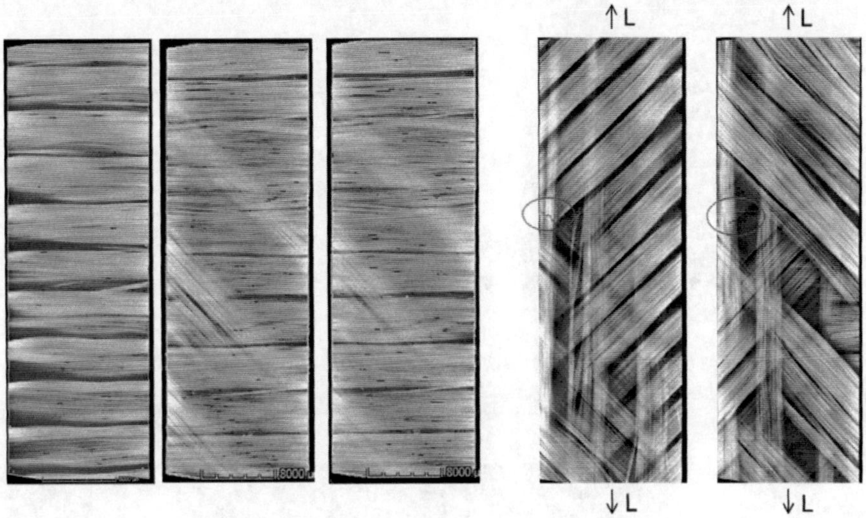

Figure 9. Location of the largest voids in GF-NCF/EP composites with low void content

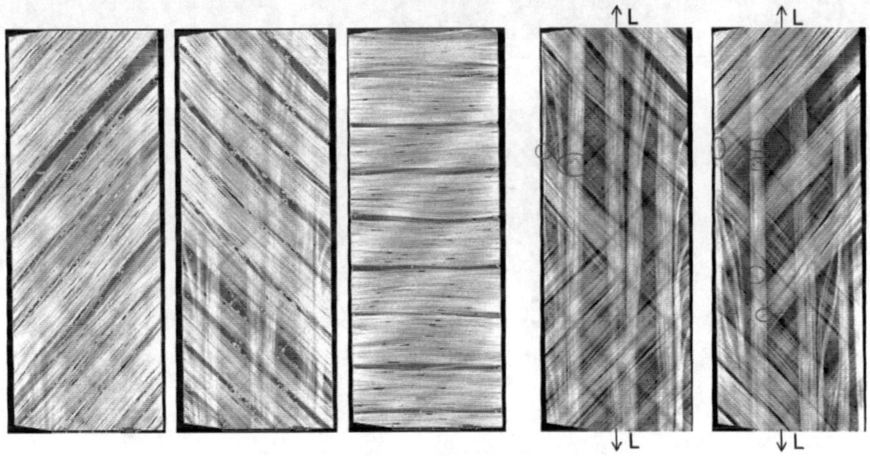

Figure 10. Location of the largest voids in GF-NCF/ EP composites with high void content

In the analysis of selected individual inter-fibre cracks, the influence of voids cannot be excluded beyond all doubt. Cracks can be observed in various cases which migrate unhindered through void-rich regions and are neither deflected to the nearest void nor stopped by voids (Figure 11). The crack path is furthermore influenced by the orientation of the fibre bundles. Only in resin rich areas, where the crack usually is oriented perpendicular to the loading direction, selected cracks are influenced by voids. However, there are examples where voids act as crack stoppers or crack formation could be observed in the immediate vicinity of voids (Figure 12).

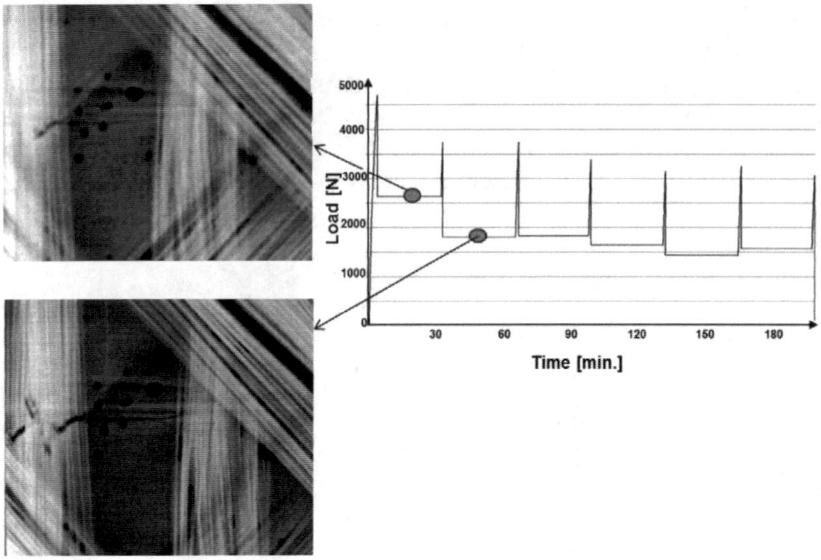

Figure 11. Example for voids not influencing the crack progression

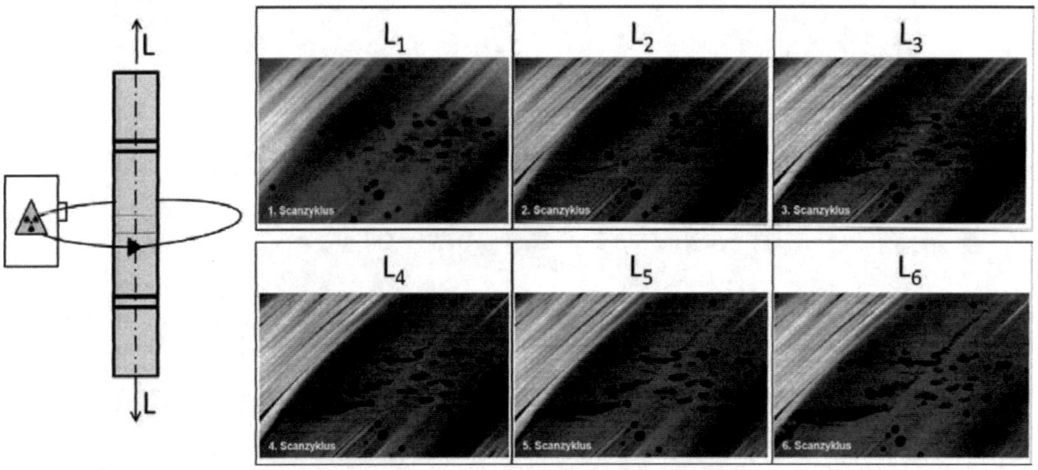

Figure 12. Example for voids acting as crack stoppers and source of cracks

5. Conclusion

An in-situ CT methodology was applied to determine the influence of voids on the damage behaviour of non-crimp glass fibre reinforced composites under tensile loading. It was found, that in samples with low void volume content cracks are initiated under tensile stress and their growth is not influenced by voids. As the void content increases, the cracks are increasingly deflected from existing voids. In the case of pointed tube voids, crack initiation indicates stress concentration in the vicinity of the void. However, also positive effects have been found as the growth of cracks may be stopped by spherical voids.

Acknowledgments

The authors gratefully acknowledge the financial support within the scope of research project PAK 267 "Effects of Defects" by the Deutsche Forschungsgemeinschaft (DFG).

IOP Conf. Series: Materials Science and Engineering **406** (2018) 012013 doi:10.1088/1757-899X/406/1/012013

References

[1] Hufenbach W, Gude M and Ebert C 2006 Tailored 3D-textile reinforced composites with load adapted property profiles for crash and impact applications *Composites* **6** (3) 8-13.

[2] Hufenbach W, 2007 editor. Textile Verbundbauweisen und Fertigungstechnologien für Leichtbaustrukturen des Maschinen- und Fahrzeugbaus Dresden: Progress media-Verlag, ISBN 978-3-00-022109-5.

[3] Kosmann N, Karsten JM, Schuett M, Schulte K and Fiedler B 2015 Determining the effect of voids in GFRP on the damage behaviour under compression loading using acoustic emission *Compos Part B* **70** 184-8.

[4] Rheinfurth M, Kosmann N, Sauer D, Busse G and Schulte K 2012 Lamb waves for noncontact fatigue state evaluation of composites under various mechanical loading conditions *Compos A Appl Sci Manuf* **43** (8) 1203-11.

[5] Hufenbach W, Gude M and Protz R 2012 Modeling of the strain rate dependent material behavior of 3D-textile composites with production and operational defects. *Adv Mat Res* **403-408** 651-5.

[6] Varna J, Joffe R, Berglund LA and Lundström TS 1995 Effect of voids on failure mechanisms in RTM laminates *Compos Sci Technol* **53** 241-9.

[7] Liu L, Guo ZS and Zhang B 2006 Experimental investigation of porosity and its effects on interlaminar shear strength in composite laminates In: *SAMPE 2006-Long Beach*, CA.

[8] Talreja R 2006 Effects on manufacturing induced defects on composite damage and failure. In: *The 27th Riso International Symposium*: Polymer Composite Materials for Wind Power Turbines 83-96.

[9] Kaleta J, Blazejewski W, Gasior P and Rybaczuk M 2010 Optimization of the IV generation tanks for hydrogen storage applied in vehicles. Modelling and experiment. In: Proceedings of the WHEC 25-31.

[10] Lundström TS and Gebart BR 1994 Influence from process parameters on void formation in resin transfer molding *Polym Compos* **15** (1) 25-33.

[11] Afendi Md, Banks WM and Kirkwood D 2005 Bubble free resin for infusion process *Compos Part A* **36** 739-46.

[12] Binetruy C, Hilaire B and Pabiot J 1998 Tow impregnation model and void formation mechanisms during RTM *J Compos Mater* **32** (3) 223-45.

[13] Patel N, Rothatgi V and Lee J 1995 Micro scale flow behavior and void formation mechanism during impregnation through a unidirectional stitched fiberglass mat *Polym Eng Sci* **35** (10) 837-51.

[14] Kang MK, Woo L and Hahn HT 2000 Formation of microvoids during resintransfer molding process *Compos Sci Technol* **60** 2427-34.

[15] Lee DH, Lee WI and Kang MK 2006 Analysis and minimization of void formation during resin transfer molding process *Compos Sci Technol* **66** 3281-9.

[16] Pujol S, Johnson MS, Warrior NA, Kendall KN and Hill DJ 2007. The effects of processing parameters on reactive epoxy adhesives. In: 16th International Conference on Composite Materials 1-9.

[17] Rius E, Achim V, Soukane S, Trochu F and Breard J 2006 Optimization of injection flow rate to minimize micro/macro-voids formation in resin transfer molded composites *Compos Sci Technol* **66** 475-86.

[18] Howe CA, Paton RJ and Goodwin AA 1197 A comparison between voids in RTM and prepreg carbon/Epoxy laminates. In: Proceedings of ICCM11, Gold Coast, Australia.

[19] Lambert J, Chambers AR, Sinclair I and Spearing SM 2012 3D damage characterisation and the role of voids in the fatigue of wind turbine blade materials *Compos Sci Technol* **72** 337-43.

[20] Olivier AP, Mascaro B and Margueres P 2007 CFPR with voids: ultrasonic characterization of localized porosity, acceptance criteria and mechanical characteristics *16. ICCM*, Japan.

[21] Chamber AR, Earl JS, Squires CA and Suhout MA 2006 The effect of voids on the flexural fatigue performance of unidirectional carbon fibre composites developed for wind turbine applications *Int J Fatigue* **28** 1389-98.

[22] De Almeida SFM and Neto ZSN 1994 Effect of void content on the strength of composite laminates *Compos Struct* **28** (2) 139-48.

[23] Browning CE 1986 Processing science of graphite/epoxy composites *Chem Eng Prog* **82** 41-9.

[24] Huang H and Talreja R 2005 Effects of void geometry on elastic properties of unidirectional fiber reinforced composites *Compos Sci Technol* **65** 1964-81.

[25] Schmidt F, Rheinfurth M, Horst P and Busse G 2012 Multiaxial fatigue behaviour of GFRP with evenly distributed or accumulated voids monitored by various NDT methodologies *Int J Fatigue* **43** 207-16.

[26] Guo Z, Liu L, Zhang B and Du S 2009 Critical void content for thermoset composite laminates. *J Compos Mater* **43** (17) 1775-90.

[27] Jeong H 1997 Effects of voids on the mechanical strength and ultrasonic attenuation of laminated composites **J** *Compos Mater* **31** 276-92.

[28] Bowles KJ and Frimpong S 1992 Void effects on the interlaminar shear strength of unidirectional graphite-fibre-reinforced composites. *J Compos Mater* **26** 1487-509.

[29] Rohatgi V, Patel N and Lee LJ 1996 Experimental investigation of flow-induced microvoids during impregnation of unidirectional stitched fiberglass mat *Polym Compos* **17** (2) 161-70.

[30] Protz R, Kosmann N, Gude M, Hufenbach W, Schulte K and Fiedler B 2015 Voids and their effects on the strain rate dependent and fatigue damage behaviour of non-crimp fabric composites. *Composites B* **83** 346-351.

[31] Protz R, Kosmann N, Fritsch D, Fey P, Adebahr W, Dietrich K, Gude M, Horst P, Kreutzbruck M, Schulte K, Busse G, Hufenbach W and Fiedler, B 2016 Influence of voids and impact damage on the fatigue behaviour of large scale composites *Mat.-wiss. u. Werkstofftech.* **47**, No. 11 1058-71.

[32] Böhm R, Stiller J, Behnisch T, Zscheyge M, Protz R, Radloff S, Gude M and Hufenbach W 2015 A quantitative comparison of the capabilities of in situ computed tomography and conventional computed tomography for damage analysis of composites *Composites Science and Technology*, **110**, 62-68.

[33] Maragoni L, Carraro PA, Peron M and Quaresimin M 2017 Fatigue behaviour of glass/epoxy laminates in the presence of voids *International Journal of Fatigue* **95** 18-28.

[34] Maragoni L, Carraro PA and Quaresimin M 2016 Effect of voids on the crack formation in a $[45/-45/0]_s$ laminate under cyclic axial tension *Composites A* **91** 493-00.

[35] Schmidt F, Rheinfurth M, Horst P and Busse G 2012 Multiaxial fatigue behaviour of GFRP with evenly distributed or accumulated voids monitored by various NDT methodologies *International Journal of Fatigue* **43** 207-16.

[36] Kosmann N, Karsten JM, Schuett M, Schulte K, and Fiedler B 2015 Determining the effect of voids in GFRP on the damage behaviour under compression loading using acoustic emission *Composites B* **70** 184-88.

IOP Conf. Series: Materials Science and Engineering **406** (2018) 012014 doi:10.1088/1757-899X/406/1/012014

Visual analysis of void and reinforcement characteristics in X-ray computed tomography dataset series of fiber-reinforced polymers

M. Schiwarth*, **J. Weissenböck, B. Plank, B. Fröhler, C. Heinzl and J. Kastner**

University of Applied Sciences Upper Austria - Wels Campus, Stelzhamerstraße 23, 4600 Wels, Austria

* Marcel.Schiwarth@fh-wels.at

Abstract. Fiber-reinforced polymers (FRPs) are of great importance in various industries because of their superior properties as compared to conventional materials, their versatile processing, and their wide application possibilities. To fulfil the high-quality standards in its respective applications, industrial 3D X-ray computed tomography (XCT) is increasingly used. It enables an accurate, non-destructive characterization of material features such as inclusions, voids, fibers, or other reinforcements, which is of core importance for material and component design. In this work we present FeatureAnalyzer, a generalization of the previously introduced PorosityAnalyzer tool, which allows to analyze dataset series as generated for exploring the parameter space of image processing workflows (including pre-filtering, segmentation, post-processing or quantification) applied to XCT datasets of fiber-reinforced polymers. With a scatter plot matrix (SPLOM), the characteristics of the features of interest may be examined in more detail regarding the used input and output parameters. Individual results may be selected in the SPLOM and analyzed using 2D slice views and 3D renderings. For this work, three different samples (*sample #1 - #3*) were scanned by means of XCT and were evaluated by using FeatureAnalyzer. The samples #1 and #2 have a porosity value of approximately 1.7 vol. %. By using the FeatureAnalyzer in combination with SPLOM, the threshold parameters could be analyzed before the over-segmentation of voids occurs. Additional evaluations by parallel coordinates clearly show, that *sample #2* has a higher number of spherical voids in the center of the specimen compared to *sample #1*. By evaluating the resin content of sample #3, the individual layer thickness could be measured. The source code of the tool is available on Github: https://github.com/3dct/open_iA/

1. Introduction and Motivation

Fiber-reinforced polymers (FRPs) are currently widely used because of properties like light weight and increased stiffness. With respect to these properties, FRPs have a broad range of application in common industry fields like aeronautics, sports, and increasingly in the automotive sector [1], [2]. Because of their superior properties, FRP composites can replace materials like aluminum and steel. For the design and analysis of new textile materials based on FRP composites, domain experts are interested in features like voids, inclusions, fibers, and epoxy resins to predict the behavior of materials. Especially the porosity is critical because the interlaminar shear strength of FRP composites is decreased by 7 % per 1 vol. % of porosity [2], [3]. The acceptance rate of porosity in aerospace is lower than 2 vol. % [2].

By analyzing identified epoxy rich areas in the FRP sample, domain experts can draw conclusions about the amount of layers of the surrounding fiber bundles, as well as about the properties of these fiber bundles. In order to get insights into the micro-structure, non-destructive testing (NDT) methods for materials characterization have been developed [4]. XCT is a NDT method, which provides high resolution 3D scans of the materials. Applied industrial 3D-X-ray computed tomography (XCT) gives insights into the micro-structure of FRP composites. XCT scans provide resolutions below 1 micron. The extraction or segmentation of interesting features such as voids and fibers in 3D volume datasets is performed using 3D image processing techniques. Typically, the following general characteristics of features like voids or fibers are of interest for domain experts: average feature volume, dimensions, feature count, and feature orientation. The main contribution of this work is to link the visual analysis of the 3D image segmentation process of extracted features to an interactive tool for visual analysis of feature characteristics. Moreover, FeatureAnalyzer supports the user in the visualization of specified feature regions. In the domain of FRP analysis, numerous tools for the extraction of e.g., fibers are available. For example, Salaberger et al. [5], [6], [7] developed an algorithm for extracting and quantifying fibers in FRPs. Plank et al. provide a review of porosity extraction algorithms in the field of carbon fiber-reinforced polymers [8]. However, to find a suitable feature extraction result, the domain experts need to compare segmentation algorithms with respect to the specified parameterization of the used algorithm. Aside other issues, the result of image segmentation strongly depends on the used pre-processing filters. Unfortunately, the evaluation of segmentation algorithms for feature extraction of XCT scans by analyzing parameters of applied filters with respect to the segmentation results is a time consuming process. The PorosityAnalyzer [9] supports domain experts in visual analysis of the 3D-image processing result in porosity analysis of XCT-data. Interactive parameter space analysis of segmented fibers and voids is provided by the FeatureScout [10].

Domain experts are also interested in features and reinforcements of FRP composites like inclusions, epoxy resins and fiber bundles. However, so far no software tool exists, which helps to visually analyze the parameter space of the image segmentation process for such features and their characteristics [11]. To address this research gap, we propose the FeatureAnalyzer tool as an enhancement and generalization to the PorosityAnalyzer. It assists domain experts in the evaluation of segmentation results for features. Mainly, the FeatureAnalyzer combines the aspects of the analysis of feature characteristics of interests to domain experts like volume, orientation, length, etc. with the interactive visualization of extracted feature regions in 2D and 3D by a link to the FeatureScout.

2. FeatureAnalyzer and FeatureScout

We present FeatureAnalyzer, a novel tool for the visual analysis of features from XCT scans of FRP samples. The aim of the FeatureAnalyzer is to provide support to domain experts in the evaluation of a 3D image segmentation workflow for the extraction of features like voids, inclusions epoxy rich areas as well as matrix reinforcements. As input data for the FeatureAnalyzer, a general dataset of originating from XCT scans of an FRP specimen, further a text file generated by other software tools can be applied: The FeatureAnalyzer is not constrained to a dataset containing voids only. This means that with the help of our tool, the domain experts can now analyze the 3D image segmentation process for any FRP feature. The workflow of the FeatureAnalyzer as illustrated in Figure 1 is described as follows: Building on the PorosityAnalyzer our tool consists of a computation module and an analysis module. The computation module allows to setup and run the offline computation of a user specified image processing pipeline. For each XCT dataset the image processing pipeline can be set up by combining various smoothing filters and segmentation filters. Furthermore, the computation module allows the user to sample the complete parameter space of all input parameters of the assembled image processing pipeline.

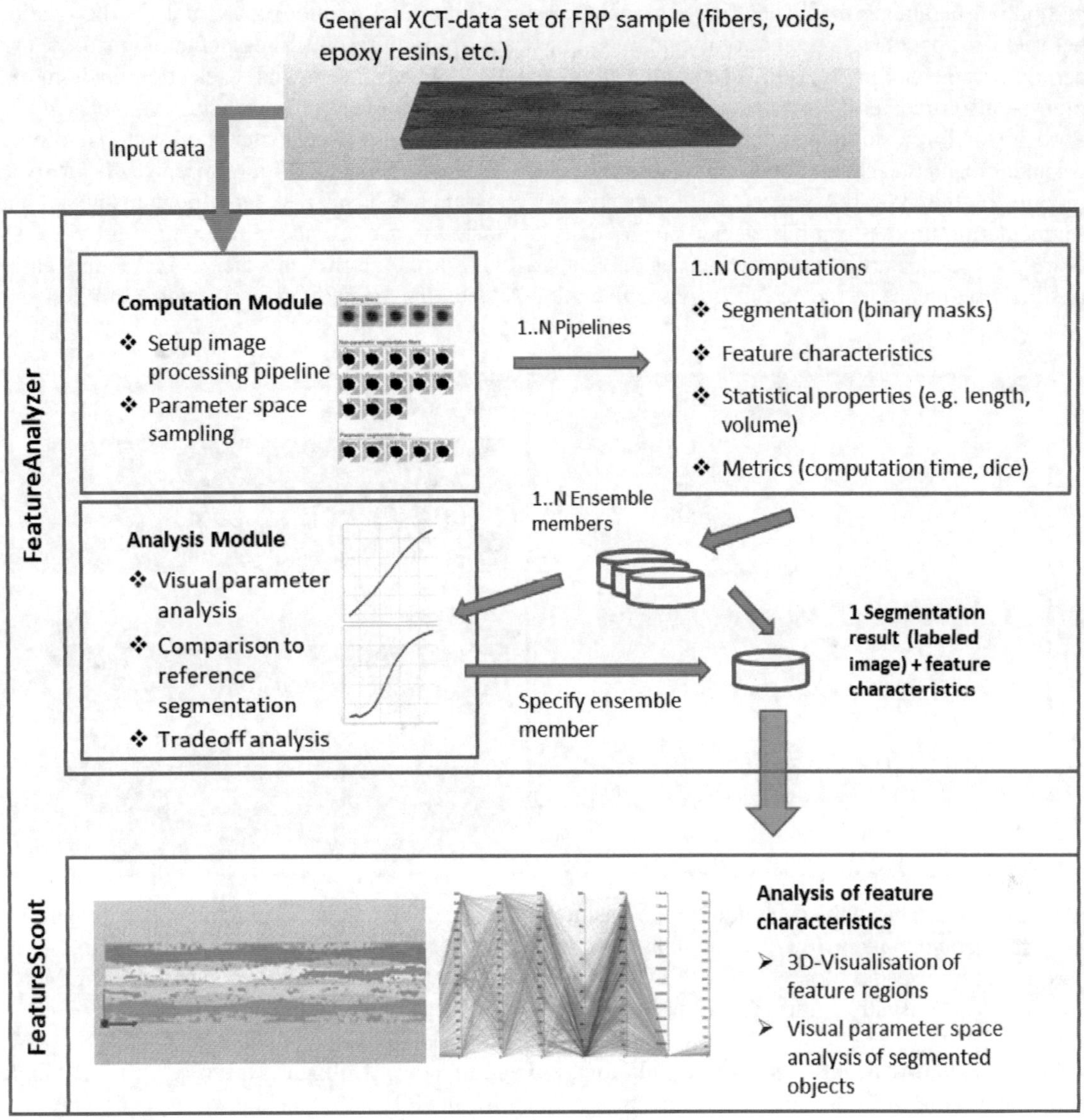

Figure 1. Analysis workflow of the FeatureAnalyzer. 3D volumetric data obtained by an XCT scan of an arbitrary FRP specimen serves as input for setup and analysis of the segmentation process in the FeatureAnalyzer. Detailed analysis of the computed feature characteristics for a specific segmentation is provided by the FeatureScout [10].

We call the segmentations from all computed runs ensemble members. Each ensemble members consists of a segmentation mask, a table of its feature characteristics, metrics like computation time and error metrics, as well as statistical properties like average volume and average length of the segmented features. The ensemble members serve as input for the analysis module. The analysis module provides visual support to domain experts in evaluating the 3D-image segmentation process with respect to the sampled input parameter range. Moreover, the analysis module allows domain experts a general analysis of the characteristics of a 3D-image segmentation result for features such as fibers, voids, etc., as mentioned before. The parameter space of related input parameters and computed output characteristics like feature count, volume and length as well as the metrics can be visually analyzed through the scatter plot matrix (SPLOM). The SPLOM consists of scatter plots visualizing the relation between every input and output parameter for a specified segmentation algorithm. Each scatter plot visualizes the relation between two parameters. Each point contained in a scatter plot represents a segmentation result with a certain parameter combination. The scatter plots are linked, which means that a selection made in one plot is also transferred to the other plots. In addition to the data previously available in the PorosityAnalyzer, domain experts can now analyze general statistical characteristics such as volume, orientation and the feature count of features to be extracted in the SPLOM. Figure 2 depicts the analysis module of the FeatureAnalyzer. As described by Weissenböck et al. [9] individual results of the segmentation process can be selected and visually compared to a reference segmentation provided by domain experts. Further metrics such as calculation time, error metrics and dice coefficients can be analyzed by means of the SPLOM. A segmentation selected in the SPLOM is visualized in 2D-slice views [9].

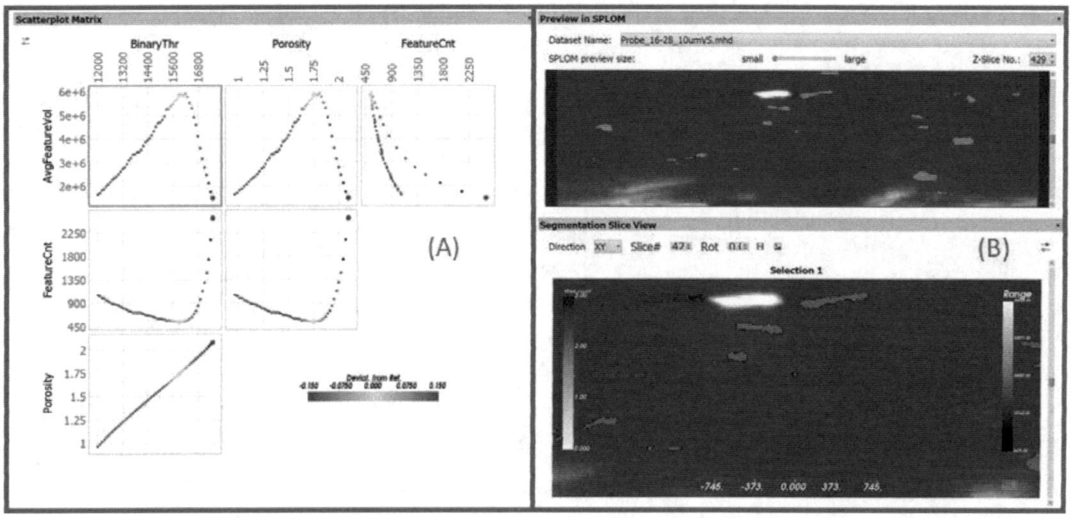

Figure 2. The analysis module of the FeatureAnalyzer for visual evaluation of the 3D-image segmentation workflow. Visual analysis of the parameter space of the segmentation results is presented by a scatter plot matrix (SPLOM) (A). The segmentation results selected in the SPLOM are visually compared to a user specified segmentation in 2D-slice views (B).

The FeatureAnalyzer provides a link to the FeatureScout: Domain experts specify a single segmentation result of a general FRP feature provided by an XCT scan, such as voids and epoxy rich areas or other FRP features. The specified segmentation result, which is based on a specific input and output parameter combination, is then passed to the FeatureScout. There, domain experts can visually analyze the parameter space, e.g., the characteristics such as volume, shape and position of extracted features by means of a scatter plot matrix (SPLOM) and parallel coordinates (PC). The SPLOM and the PC visualize relations between the feature characteristics. The SPLOM and PC are linked to the 3D viewer, which provides an interactive visualization of user specified feature regions. According to the requirements of the domain experts the FeatureScout also allows a color-coded classification of regions of interests as described by Weissenböck et al. [10].

3. Experimental setup and data parametrization

3.1. Sample description and XCT parameters

In this study, three different CFRP samples, manufactured by prepreg technology in plain weave style, were investigated. Sample type #1 and #2 are made of CYCOM® 977-2 [12] and contain porosity and some additional glass fiber rovings as shown in Figure 3 (A) and (B). The samples were cut out from a real world part. For this XCT investigation the sample size was 10 x 10 x 3.5 mm³. The sample of type #3 was made of out of 5 plies of PRG09229 - CC 120 ER450 43% as shown in Figure 3 (C). For this XCT investigations a sample size of 10 x 10 x 0.69 mm³ was cut out of a test plate. The ply thickness is about 0.138 mm.

XCT scans were performed on a Nanotom 180 NF device manufactured by GE phoenix|x-ray. The device uses a 180 keV nano-focus tube and a full digital 2304² pixel flat panel detector from Hamamatsu. As target material molybdenum was used. No pre- or post-filters were used for the scans. The exact measurement parameters are listed in Table 1:

Table 1. Description of XCT measurement parameters for Sample #1 - #3.

Sample Name	*Voxel size [µm]*	*Tube voltage [kV]*	*Tube current [µA]*	*Integration time [ms]*	*Projections*	*Average*	*Measurement time [min]*
Sample #1 & #2	10	60	290	500	2400	1 (continuous)	20
Sample #3	6	50	220	600	1700	5	105

The data specification and parametrization for the samples #1 - #3 are listed in Table. 2. The data encoding represents the range of the color intensities for gray values of a dataset.

Table 2. Description of the parameterization for Sample #1 - #3.

Sample Name	*Dimension [Voxels]*	*Components*	*Segmentation Threshold*	*Reference Porosity/ Volume [vol. %]*	*Voxel size [µm]*	*Data encoding (Color range)*
Sample #1	853 x 243 x 858	Voids	15846	1.74	10	16 Bits: 0..65355
Sample #2	915 x 202 x 846	Voids	15846	1.70	10	16 Bits: 0..65355
Sample #3	699 x 99 x 699	Epoxy resins	42300	9.57	6	16 Bits: 0..65355

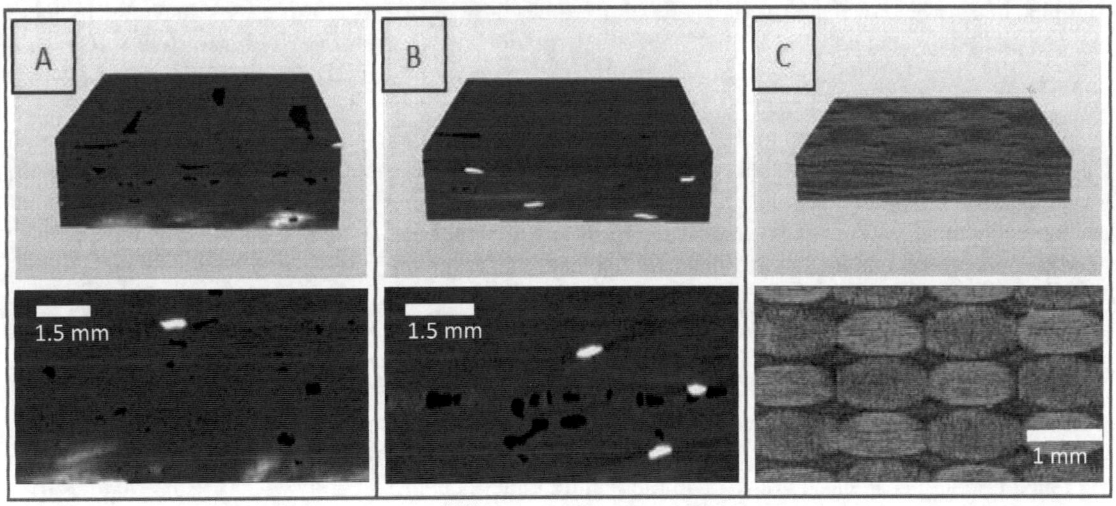

Figure 3: 3D visualization and 2D-slice views of volumetric datasets of FRP-samples obtained by XCT: (A) *sample #1*, (B) sample #2, (C) *sample #3*.

3.2. Parametrization of data and applied segmentation

For the void segmentation we analyzed the *samples #1 and #2*. To define the reference porosity value for further segmentation, we apply the segmentation of porosity through the ISO 50 method. As shown in Plank et al. [13], with high resolution scans the threshold for segmentation has quite a low impact on the evaluated porosity. For the used material system, a high resolution reference scan with 5 µm voxel size was performed and evaluated by applying the ISO50 threshold. These results were applied for the

further samples, which were scanned at 10 µm voxel size, as already introduced by Senck et al. [14]. The thresholds of the porosity segmentation of sample #1 and sample #2 were determined by ISO 50. The resulting porosity and the specified thresholds are also listed in Table 2 of section 3.1.
As a starting segmentation, a general binary threshold filter from the image segmentation and registration toolkit (ITK) was applied [15], [9]. For the comparison of the segmentation results with respect to the statistical output parameter results of both datasets, the upper gray value threshold range was sampled between 12000 and 17500 with a step size of 100 gray values.

To illustrate the analysis of the extraction of epoxy rich areas, our domain experts provide an XCT dataset consisting of plain weave fiber bundles. For the dataset *sample #3* a reference segmentation does not exist. Thus, the upper comparison threshold was specified to an estimated intensity value of 42300 with an obtained volume of epoxy rich areas of 9.57 vol. %. The image processing pipeline is set up as follows: For the preprocessing of the XCT data, the Gradient Anisotropic Diffusion filter is applied for edge-preserving noise removal [15]. To analyze the segmentation results, a general threshold filter was applied within a specified threshold range of gray value intensities. For the parametrization of the segmentation setup for epoxy rich areas, the lower threshold is set to 0 and the upper range of the threshold to be sampled is estimated between the range of 41000 and 44000 with a step size of 10.

4. Results and Discussion

In this section we discuss our findings on the application of the FeatureAnalyzer for porosity segmentation (see section 4.1) and the segmentation of epoxy rich areas, see section 4.2).

4.1. Analysis of void sample datasets

The relation of the gray value intensities to the parameters porosity as well as the feature count is shown in the scatter plots of Fig. 4. A clear over-segmentation of sample #1with respect to the analysis of the feature characteristics was considered. To analyze this aspect, the threshold value parameter was sampled in a range between 6000 and 34000 gray values in steps of 250, as this gray value range includes air and materials content. Furthermore, the segmentation masks are compared at different threshold values. The color bar indicates the deviation of the segmentation result to the reference porosity where a deviation greater than or equal to 0.2 % is specified as over-segmentation, color coded in dark red, and a deviation of smaller or equal to -0.2 % is specified as under-segmentation, color coded in dark blue. It can be seen that the porosity is constantly increasing with the binary threshold. From a threshold value of 18000 with a porosity of 2.24 vol. %, the feature count (FeatureCnt) is significantly increasing with an increasing binary threshold, which is also shown by the color bar. From that point, the feature volume is slightly decreasing. This shows that additionally a large number of small elements are classified, which explains the high feature count. The property feature count reaches its highest value at a threshold of 19750, the segmented porosity is of 7.72 vol. % for this threshold. From that point on, the feature count is strongly decreasing with increasing binary threshold, as the segmentation starts classifying the matrix as void, thus connecting objects of larger volume. Starting with a threshold of 22000, the average feature volume is significantly increasing with an increasing binary threshold. Further, the feature count is slightly decreasing, and approaching a nearly constant level at a comparatively small number. The obtained porosity is at 71.54 vol. %, and the feature count is approximately 11000. Thus we identify a single major region of a large, connected volume, which represents the FRP matrix, and further regions of several small elements are identified as seen in Figure 4 (2). At a threshold of 25250 the porosity reaches 90.33 vol. %. This means, that approximately 90 percent of the sample volume of the FRP dataset is classified. This is in good agreement with the illustration shown in Figure 4 (3), which shows that the entire FRP matrix is nearly classified as void at this threshold.

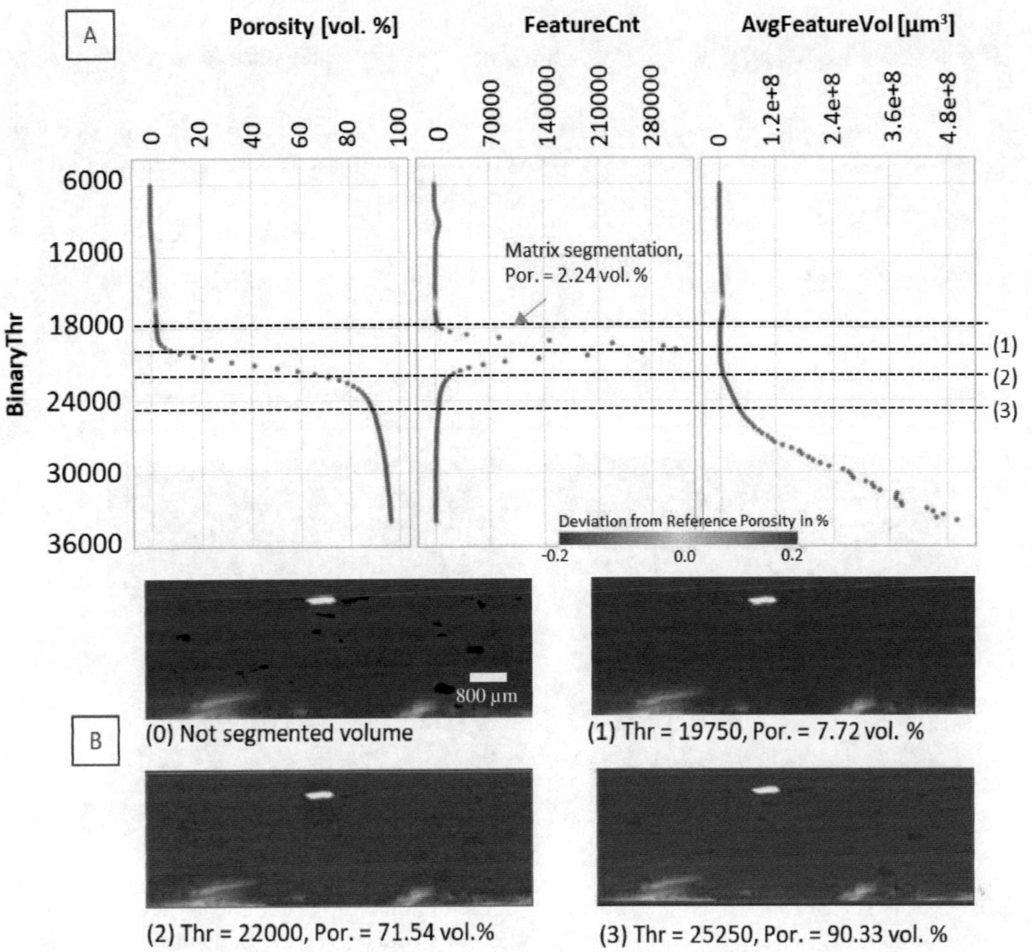

Figure 4: Figure: (A) scatter plots of *sample #1* visualizing the relation of the binary threshold (BinaryThr) to the feature count (FeatureCnt), porosity (Por.) and the average feature volume (AvgFeatureVol). (B) Comparison of the unsegmented data (0) to (1-3) the segmentation masks obtained at different thresholds (Thr) and their respective porosity (Por.).

For a detailed investigation of the segmentation results close to the reference porosity value in Fig. 5, the gray value intensities were sampled in the range between 12000 and 175000 in steps of 100. By means of the SPLOM, the relation of the binary threshold value *with* respect to properties such as the porosity, feature count (FeatureCnt) and the average feature volume (AvgFeatureVol) is visually analyzed. The SPLOM shows that the porosity is increasing along with the binary threshold. The feature count (FeatureCnt) is slightly decreasing when increasing the binary threshold to a certain point, and then starts increasing. At the same point the average feature volume reaches its maximum value. We can conclude that by increasing the binary threshold, an increasing number of voxels are classified as connected regions of voids, leading to an increase of the volume porosity, which we identified as over-segmentation. In Figure 5, we exemplarily show an over-segmentation of the porosity volume at a gray value of 17500. An under-segmentation of porosity is illustrated at the gray value intensity of 12000. The reference porosity is specified to the gray value intensity of 15900. The segmentation result of the reference porosity is shown by the yellow segmentation mask. The blue color indicates a decreased void size in comparison to the reference segmentation, whereas the red color indicates an increase of the segmented voids, as can be seen in Figure 5 (2) and Figure 5 (3), respectively.

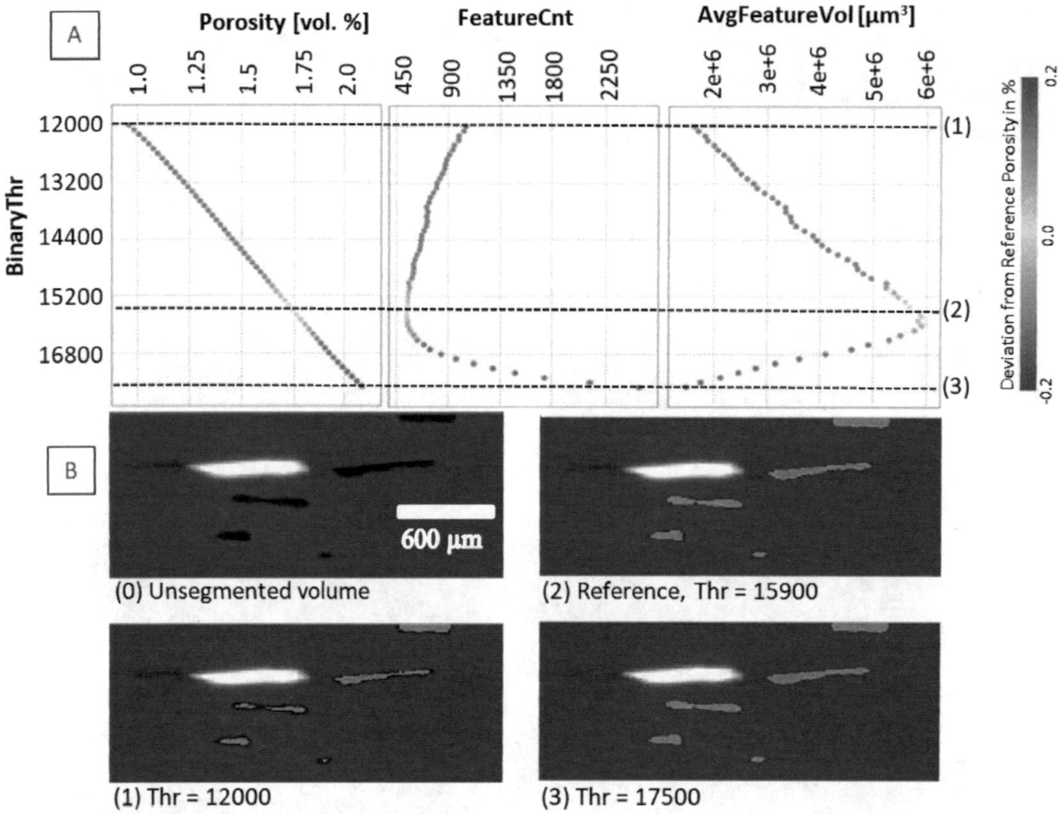

Figure 5: (A) The SPLOM illustrates obtained segmentations for the sample #1 at different binary thresholds (Thr) of 12000 (1), 15900 (2) and 17500 (3). (B) The 2D-slice previews of the segmentation masks of (1-3) are compared to the unsegmented XCTvolume.

For sample #2 the same reference threshold leads to a slightly different reference porosity, as discussed in section 4.2. The general trend of the threshold variations leads to similar results as depicted in Fig. 5. Thus, we scope our discussion to the analysis of the detailed feature characteristics of the resulting porosity extraction for both datasets. Therefore, the obtained segmentations at the reference gray value intensity of 15846 for sample #1 and sample #2 are sent to the FeatureScout. The feature characteristics for sample #1 and sample #2 were obtained by means of the Volume Graphics[1] defect detection tool in order to get additional calculated properties such as the sphericity. The segmented objects of a size smaller than 27 voxels are considered to be noise and are thus removed from further analysis. The feature characteristics of both datasets are compared by means of the parallel coordinates (PC), which are linked to the 3D view as illustrated in Figure 6. In the PC every axis represents a single characteristic of the segmentation result. Every segmented object is represented by a polyline connecting every property axis [10]. In our application case each connected polyline represents a single extracted void. The PC show equally distributed voids in X and Z directions. The position with respect to the thickness of the pore specimen of the located voids in the dataset is represented by the property Pos.Y. The sphericity represents the shape of the voids. A sphericity value of 1 indicates round voids with the shape of a sphere, whereas a sphericity value close to 0 indicates voids of flat shape. The selection in red represents

[1] Volume Graphics is a software for the analysis and visualization of industrial 3D computed tomography data. https://www.volumegraphics.com/

voids at the center of the specimen thickness. There, the voids are equally distributed with respect to the x- and z-directions. Analyzing the PC view in combination with the 3D viewer, the *sample #2* shows a higher number of defects with a higher sphericity between 0.5 and 0.7 located in the center of the sample compared to the *sample #1*. The 3D visualizations of the selected voids located in the center of the specimen confirm these findings.

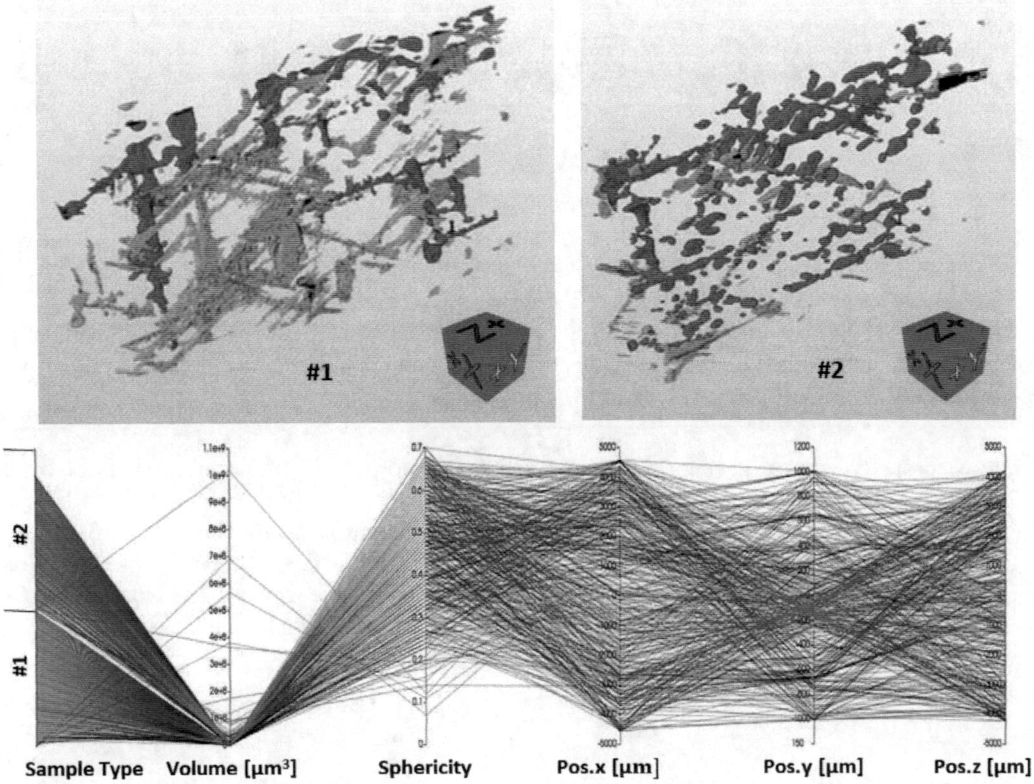

Figure 6: Comparison of the feature characteristics of extracted voids for sample #1 and sample #2 through Parallel Coordinates and 3D visualization at a reference threshold of 15846. The red lines represent a selection of voids located in the center of both datasets.

4.2. Analysis of the segmentation of epoxy rich areas

In this section, we demonstrate the analysis of the segmentation result of epoxy rich areas for the specified segmentation pipeline of the dataset *sample #3*. The result of the reference segmentation is determined empirically with the help of FeatureAnalyzer with respect to the specified parametrization as mentioned in section 3.2. The feature characteristics obtained for the reference segmentation at the threshold of 42300 are visually analyzed in the FeatureScout. The scatter plot illustrates the correlation between feature volume (Volume) and major length (MajorLength), where the major length represents the longest extent of a segmented object along a main axis. The feature volume with respect to the major length of the segmentation results are visually analyzed in the FeatureScout at the gray value intensities of 42300 as visualized in Figure 7. The segmentation results show a large count of unconnected regions of small volumes. These regions are probably resulting from voxels misclassified by the binary threshold algorithm. The most likely reason for those misclassifications of voxels are a high level of noise in the XCT dataset as well as additional measurement artifacts due to the low difference of absorption contrast between carbon fibers and epoxy resins. The reference segmentation depicts four connected regions of

IOP Conf. Series: Materials Science and Engineering **406** (2018) 012014 doi:10.1088/1757-899X/406/1/012014

large volumes, which are classified by different colors. The connected regions are visually identified as layers of epoxy rich areas.

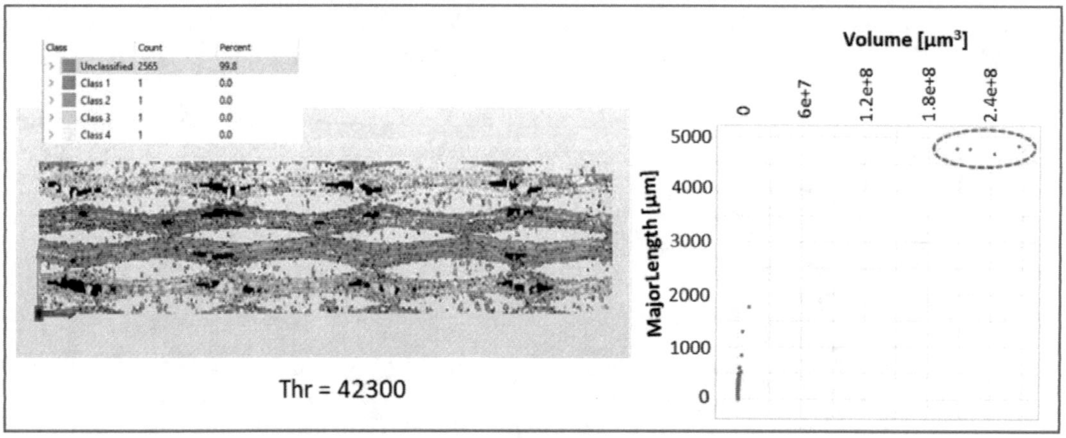

Figure 7: 3D-Visualisation of identified epoxy rich areas (marked with red) of sample #3 comparing the reference segmentation at a threshold (Thr) of 42300. The scatter plot visualizes the respective feature characteristics.

The volume property with respect to the sample center positions and sample dimensions of dataset *sample #3* is visualized in Figure 8. The figure illustrates four layers, equally spaced within the thickness of the sample, as can be seen through the property Ym. Here a mean layer thickness of 0.131 mm can be extracted, which fits well to the theoretical value of 0.138 mm as described in chapter 3.1. Small regions of misclassified voxels are equally spread in the center of the sample. The scatter plot (Xm versus Ym) in Figure 8 on the right, depicts that the layers range along the complete specimen length Xm from zero to four millimeters.

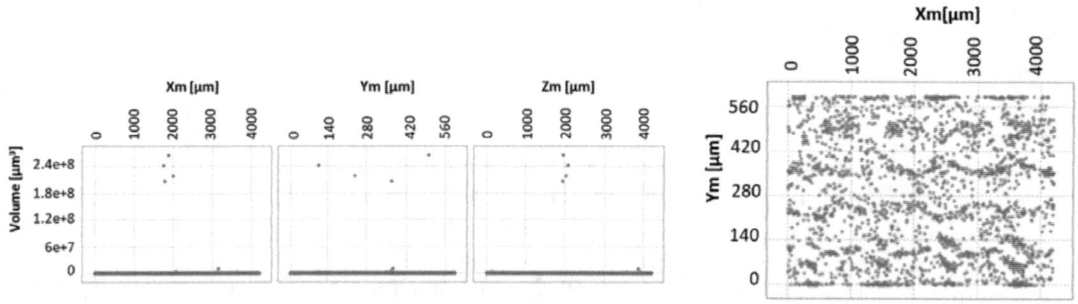

Figure 8: Scatter plots of the reference segmentation for the sample #3 obtained at a threshold of 42300.

5. Conclusion and future work

In this work we introduced FeatureAnalyzer, a novel tool to support domain experts in the field of 3D image processing of XCT data for identifying general features in FRPs like voids and other matrix components. FeatureAnalyzer allows the user to do a visual parameter analysis of the segmentation results through a scatter plot matrix. Further the segmentation results can be quantified by analyzing relation of input parameters with respect to calculated statistical output parameters such as average feature volume and feature counts of a segmentation result. Analyzing those parameters linked to the

2D slice view of a selection, helps domain experts to visually identify a certain gray value range of a suitable segmentation and to predict the behavior of the corresponding segmentations. The parameter studies, as shown in this work on sample #1, help to optimize the segmentation results, which can then be used as a basis for future feature extraction of internal microstructures such as pores. In a further step and for error estimations, results from FeatureAnalyzer in combination with FeatureScout should be considered for developing material models used for microstructure simulations in future tasks. We demonstrated the applicability of our tool on two application cases. The analysis of detailed feature characteristics of a single segmentation result from extracted features is done by means of the link of the FeatureAnalyzer to the FeatureScout. Visual parameter space analysis with FeatureScout by means of the SPLOM and the PC allows a color coded classification and 3D-visualisation of extracted feature regions with respect to characteristics such as volume, orientation and localization of each extracted feature. In the future we plan to added quantities like different shape factors to further characterize the segmented features. Additionally, we plan to modify our tool to be able to analyze extractions of fiber bundles.

Acknowledgements

This work was supported by the project "Interpretation and evaluation of defects in complex CFK structures based on 3D-CT data and structural simulation" (DigiCT-Sim; FFG proj. no. 862015) funded by the federal government of Upper Austria and FFG. Additional funding came from the FFG Bridge Early Stage project no. 851249: Advanced multimodal data analysis and visualization of composites based on grating interferometer micro-CT data (ADAM). Furthermore, we wish to thank the company's FACC AG and PEAK Technology for manufacturing the CFRP specimens.

References

[1] *Composites-Marktbericht 2017, retrieved from* https://www.carbon-composites.eu/media/2996/ccev-avk-marktbericht-2017.pdf.

[2] G. H. Michaeli W and L. Wolters, Technologie der Kunststoffe, München: Carl Hanser Verlag, 2003.

[3] E. A. Birt and R. A. Smith, "A review of NDE methods for porosity measurement in fibre-reinforced polymer composites", *Insight - Non-Destructive Testing and Condition Monitoring,* Vol. 46 (11) pp. 681-686, 2004.

[4] J. Kastner, B. Plank, D. Salaberger and J. Sekelja, "Defect and porosity determination of fibre reinforced polymers by x-ray Computed Tomography", 2nd Int. Symposium on NDT in Aerospace, Hamburg, Germany, 2010.

[5] D. Salaberger, K. A. Kannappan, J. Kastner, J. Reussner and T. Auinger, "Evaluation of Computed Tomography Data from Fibre Reinforced Polymers to Determine Fibre Length Distribution," *IPP,* Vol. 26(3), pp. 283-291, 2011.

[6] K. Robb, O. Wirjadi and K. Schladitz, "Fiber Orientation Estimation from 3D Image Data: Practical Algorithms, Visualization, and Interpretation," in *7th International Conference on Hybrid Intelligent Systems (HIS 2007)*, 2007, DOI: 10.1109/HIS.2007.26

[7] K. Schladitz, "Quantitative micro-CT", *Journal of Microscopy, Vol.* 243, pp. 111-117, 2011, DOI: https://doi.org/10.1111/j.1365-2818.2011.03513.x.

[8] A. Reh, B. Plank, J. Kastner, E. Gröller and C. Heinzl, "Porosity Maps - Interactive Exploration and Visual Analysis of Porosity in Carbon Fiber Reinforced Polymers," *Computer Graphics Forum,* Vol. 31(3), pp. 1185-1194, 2012.

[9] J. Weissenböck, A. Amirkhanov, E. Gröller, J. Kastner and C. Heinzl, "PorosityAnalyzer: Visual analysis and evaluation of segmentation pipelines to determine the porosity in fiber-reinforced polymers," in *2016 IEEE Conference on Visual Analytics Science and Technology (VAST)*, 2016, DOI: 10.1109/VAST.2016.7883516

[10] J. Weissenböck, A. Amirkhanov, W. Li, A. Reh, A. Amirkhanov, E. Gröller, J. Kastner und C. Heinzl, "FiberScout: An Interactive Tool for Exploring and Analyzing Fiber Reinforced Polymers“, in *2014 IEEE Pacific Visualization Symposium*, 2014, doi: 10.1109/PacificVis.2014.52.

[11] C. Heinzl and S. Stappen, "STAR: Visual Computing in Materials Science“, *Computer Graphics Forum,* Vol. 36(3), pp. 647-666, 2017.

[12] Cytec (2012, March 19): *CYCOM 977-2 EPOXY RESIN SYSTEM*, retrieved from https://www.cytec.com/sites/default/files/datasheets/CYCOM_977-2_031912.pdf

[13] B. Plank, G. Rao, J. Kastner - Evaluation of CFRP-Reference Samples for Porosity made by Drilling and Comparison with Industrial Porosity Samples by Means of Quantitative X-ray Computed Tomography - Proceedings 7th International Symposium for NDT in Aerospace, Bremen, Germany, pp. 10., 2015

[14] S. Senck, M. Scheerer, V. Revol, K. Dobes, B. Plank and J. Kastner, "Non-Destructive Evaluation of Defects in Polymer Matrix Composites for Aerospace Applications Using X-ray Talbot-Lau Interferometry and Micro CT,“ 58th AIAA/ASCE/AHS/ASC Structures, Structural Dynamics, and Materials Conference, AIAA SciTech Forum, Grapevine, USA, pp. 1-9, 2017.

[15] T. S. Yoo, M. J. Ackerman, W. E. Lorensen, W. Schroeder, V. Chalana, S. Aylward, D. Metaxas and R. Whitaker, „Engineering and Algorithm Design for an Image Processing API: A Technical Report on ITK - the Insight Toolkit,“ *Studies in health technology and informatics, Vol 2,* pp. 586-592, 2002.

IOP Conf. Series: Materials Science and Engineering **406** (2018) 012015 doi:10.1088/1757-899X/406/1/012015

Definition and quantification of drapeability through the measurement of constituent effects

M Christ[1*], A Herrmann[2]

[1]Faserinstitut Bremen e.V., Am Biologischen Garten 2, 28359 Bremen, Germany
[2]Universität Bremen, Postfach 33 04 40, 28334 Bremen, Germany

* christ@faserinsitut.de

Abstract. Although considerable work has been carried out in the field of drapeability research, both the definition of what drapeability comprises as well as the tools to measure and quantify it must be further developed. This paper presents a definition of drapeability that is meant to broaden its scope to include not only the important effects of shearing and wrinkling, but all constituent effects such as gap formation, undulations, loops (tow buckling) and fabric anisotropy. This definition is completed by an experimental test device to measure and quantify all of these effects in a single, standardized setup (available as the Textechno Drapetest).The Drapetest utilizes optic measurement technology coupled with digital image analysis software to detect and measure fibre positions and orientation, and fabric topology. For each of the defined drapeability effects characteristic values were generated from this optical data that can describe the intensity, distribution and extent of the effect in question. The device and methodology was empirically validated using a set of 72 specially manufactured glass fibre NCF. The interdependencies between textile design, process parameters and drapeability effects as well as a semi-analytical equation for them are shown. A method to use these semi-analytical equations in the composite design process is introduced [1].

1. Introduction

The usage of fibre reinforced plastics (FRP) is further increasing every year. It has by now become a standard material in all areas of aircraft design, in wind turbine blades and even serial production of automobiles. In parallel with this development the sophistication of the design and manufacturing processes has increased. Moving away from metal emulating designs with quasi-isotopic stacks of woven fabrics the manufacturing and design processes now increasingly incorporate fibre conscious technology and techniques to maximize the fibre based material's potential.

Including fabric specific material properties such as drapeability in the testing, design, manufacturing and quality control is one aspect of such fibre conscious design. The increasing sophistication of fabrics, e.g. usage of multiaxial non-crimp-fabrics instead of plain woven fabrics, creates the necessity to deepen the knowledge of the drapeability of the involved materials. This can be achieved through a re-evaluation of the definition of and how the choice of this definition effects testing and design methodologies.

This paper introduces a drapeability definition that includes all relevant effects. Furthermore a set of indicator numbers is shown that might be used to assess these drapeability effects quantitatively. These characteristic values can be measured by using the automated Textechno Drapetest. To validate the chosen drapeability definition empirical studies were conducted with said testing device to show usefulness of the indicator numbers. As a conclusion it is shown how these empirical results can be used to formulate a semi-analytical material model of all drapeability effects included in the definition.

IOP Conf. Series: Materials Science and Engineering **406** (2018) 012015 doi:10.1088/1757-899X/406/1/012015

2. Definition of Drapeability and Characteristic Values

Previous definitions of drapeability focussed on the shearing behaviour of fabrics and defined good drapeability as adherence to a double curved shape without the formation of wrinkles. The reason for this definition is historic. In tightly woven fabrics shearing is the only relevant deformation mechanism and fabric shear is that case the cause of wrinkles, so this narrow definition of drapeability can be applied correctly. However, in different fabrics, a large number of drapeability effects have been identified [2, 3], as shown in Figure 1 for NCF.

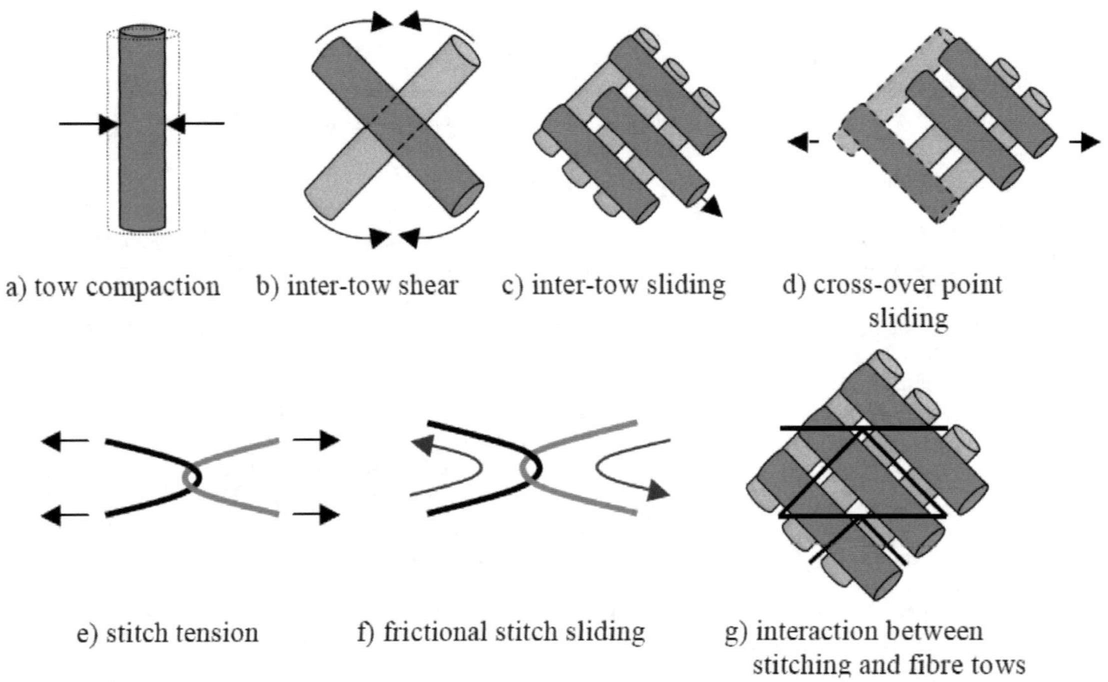

Figure 1. Deformation mechanisms in NCF [2]

Drapeability is in fact not the adherence of a fabric to a curved shape without a wrinkle. Drapeability must encompass all effects since even without a wrinkle changes in the fabric structure might be changing the preforms properties. Under the classical definition of drapeability, a fabric that shears considerably but avoids any wrinkles possesses “Good” drapeability. In reality though, a fabric that easily changes it fibre direction might lead to a preform with fibre orientations that differ greatly from the presupposed directions and thus also load directions. For sheared fabrics orthogonality can also not be assumed. A better approach to define drapeability would be to quantify effects and present a quantitative description of the observed effects.

Furthermore, looking at reinforcement fabrics with a wider scope, non-crimp-fabrics (NCF) must also be considered, and thus, other drapeability effects beyond shearing cannot be neglected anymore. In NCF the yarns can exhibit three modes of motion: rotational (by fabric shear), translational and axial.

To grasp the drapeability of a fabric it is necessary to define three terms: the draping parameters, the draping behaviour and the drapeability effects. The draping parameters are determined by the selection of fabric and its design, and by the requirements for the final component shape. They can usually be quantified as simple geometric parameters such as curvature. The interaction between the textile design parameters determine the fabric specific draping behaviour. The draping behaviour can be derived from empirical results (as long as not complete mechanical material law for textiles is available) in the form of functions that include the draping parameters as variables and the empirically determined material coefficients. The drapeability effects are the resulting effects overserved in the draped fabric. They can be characterised by a set of indicator numbers in intensity and extent, either as a result of empirical experiments or by solving the determined draping behaviour functions of the fabric used.

The importance of choosing a comprehensive drapeability definition is compounded by the fact that it influences the next steps of drapeability research. The design of a drapeability tester will be based in the chosen drapeability definition. In production, the quality testing of preforms will also only be able to find drapeability effects that have previously been incorporated in a definition. In fact, this can become a self-enforcing circular definition in which a testing device only detects those effects its underlying definition includes which in turn is validated by this testing. Instead the definition must be chosen as broadly as possible and a testing device must be able to test for all defined effects. Only after the fact of quantifying the effects in a fabric can it be concluded that in a particular case the definition of drapeability can be narrowed to a smaller subset of effects.

Table 1. Drapeability effects and their characterising indicator numbers

Drapeability effect	Indicator	Symbol	Unit
Gaps			
	Gap Percentage	G_P	%
	Gap Width	G_W	mm
	Gap Shape	G_S	-
Fibre Angle Change			
	Angle Difference	Δ	°
	Undulations	Δ_{loc}	°
Fabric Shape Change			
	Out-of-roundness	Θ	%
Out-of-plane effects			
	Waviness	Λ	mm
	Loops	Ω	mm²

3. Measurement technology

Fabric forming tests can be assorted according to their specificality: from the generic drapeability test setups to the more specific preforming test setups [4]. The generic test setups such as the trellis frame (also known as picture frame) test or the bias extension test attempt to elicit a singular drapeability effect in a fabric by creating a generic load condition e.g. a pure shear loading. This approach is similar to the test setups used to measure mechanical properties of stiff materials (laminates or metals), the bias extension test is close to the setup for measuring the shear modulus of biaxial laminates. The results of these generic drapeability tests is also similar to those for usual engineering materials as they essentially measure the shear modulus of a fabric. On the other end of the scale, preforming tests use a very specific geometrical condition to elicit all drapeability effects that will be present in a later application by using a test geometry that is as similar as possible to the final tool shape. The advantage of this approach is, that no question arises how to apply the laboratory experiments to the production environment because the laboratory geometry and production geometry are very similar or even the same. On the other hand, the results of one test can hardly be applied to another geometry because the observed effects are only applicable to the specific geometry used in this instance. Also preforming tests often rely on quantitative descriptions of the observed fabric state and do not lend themselves to gain insights into the material mechanics behind the formation of drapeability effects. In between these setups are those that use a generic reference body such as a hemisphere or tetrahedron to induce drapeability effects and then either record deformation forces or optically measure the fabric surface. They are more complex in their conditions and results as the generic test setups yet more universal than the preforming test.

IOP Conf. Series: Materials Science and Engineering **406** (2018) 012015 doi:10.1088/1757-899X/406/1/012015

The requirements for a drapeability test setup that allows for verification of the proposed drapeability definition can be summarised as such:

1. Measuring the fibre orientation as this is the most important attribute of reinforcement fabrics
2. Measuring the position of yarns to detect lateral yarn movement
3. Detection of out-of-plane yarns
4. Recording of necessary forming force
5. Using a generic test condition for universal application of test results
6. Measuring at different degrees of drape
7. Generating a defined clamping condition

The development of a drapeability tester to fulfil these requirements resulted in a test setup that takes the classical reference body setup as described above and equips it with measuring technology to quantify all resultant drapeability effects [5]. Going beyond the usual laboratory setups the technology is available in a guaranteed build and quality in the form of the Textechno Drapetest while the testing method is available as the DIN SPEC 8100 in an open access manner. This combination means that the drapeability test can be standardised to maximise interlaboratorial comparability.

The Drapetest relies heavily on optical image technology to detect and quantify the drapeability effects (see Figure 2). This has the advantage of offering the capabilities to detect all effects that are visible. To measure in-plane-effects a detail camera takes images of the fabric surface from which an image analysis software can discern small-scale effects. Out-of-plane effects are measured by scanning the fabric topology with a triangulation laser and generating a 3D-representation of it. Additionally, an overhead camera photographs the entire fabric sample to measure larger scale fabric deformations. The fabric sample is clamped by pneumatic tube, which means the clamping conditions are constant and user definable.

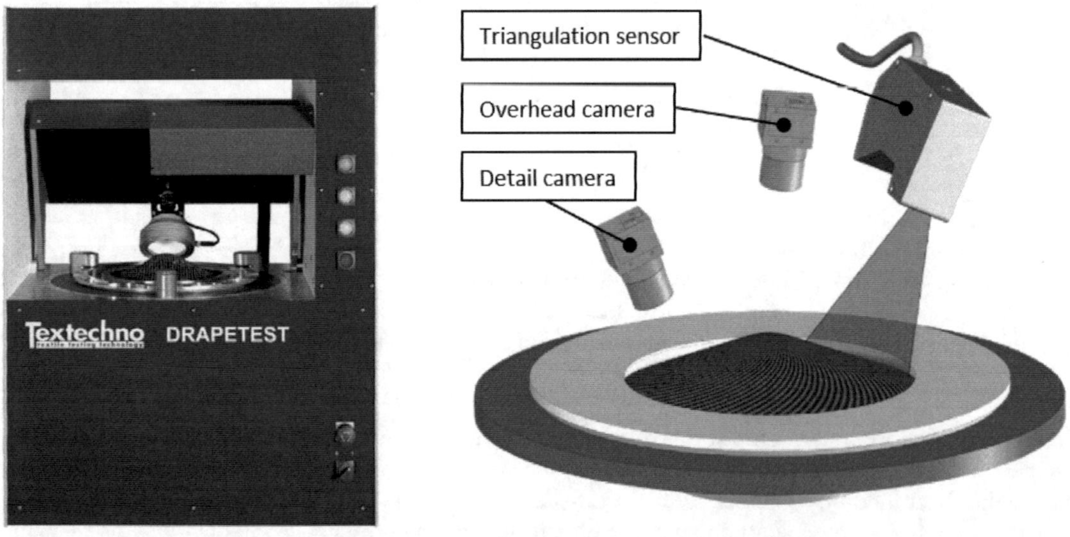

Figure 2. Drapeability tester Textechno Drapetest (left) and utilised optical measurement technology to quantify drapeability effects (right)

The images and height profiles recorded by the tester are analysed with an automated image analysis software. This is capable of detecting and measuring all the defined effects and generate the defined indicator numbers that are than communicated to the user. The indicator numbers are meant to give the user a clear and complete overview of the present drapeability effects and their intensity. By communicating these numbers this information can then be conveyed to other involved e.g. textile manufacturers or part designers.

IOP Conf. Series: Materials Science and Engineering **406** (2018) 012015 doi:10.1088/1757-899X/406/1/012015

4. Equations of constituent effects

The goal of the empirical testing was to validate the usefulness of the proposed drapeability definition. A useful definition can describe not only the drapeability effects but also the drapeability behaviour of a fabric. The drapeability behaviour describes how a specific fabric responds to the external constraints, (i.e. like the component geometry) due to its internal structure i.e. its textile design. To test this capability of the definition a set of NCF was manufactured whose design parameters were chosen to full out a fully parametric design space with the variables, stitch length (in machine direction MD), stitch point distance (in cross direction CD), weight per area, and stitching pattern. The fabrics were all of the two layer biaxial +45°/-45° type manufactured by Karl Mayer Malimo with a constant stitch yarn tension. These fabrics were tested with the Drapetest on a hemispherical reference body that had its height varied to investigate different degrees of drape. Due to the symmetrical nature of the reference body, the influence of fibre orientation to the curvature of the body could also be investigated. The clamping forces were kept constant.

To assess the drapeability behaviour each fabric is measured at ten degrees of drape with the reference body being moved outward of the textile plane in 10 mm steps. At each step the surface of the fabric sample is measured with the described optical technologies and quantitative data of the extent and intensity of each drapeability effect is gathered by the image analysis software. The formation of the drapeability effects can then be described as a function of the reference body displacement (i.e. the degree of drape) as shown in Figure 3 on the left side. Between a reference body displacement of 10 mm and 80 mm the results can be linearised and described by the linear slope $k_E = dE/dh$ for each effect E over the displacement of the reference body h. By varying a textile design parameter this slope is changed, so it can be used to describe the drapeability behaviour of a specific textile in regard to a certain drapeability effect. The k-coefficients may now be used as data points to model the drapeability of NCF as a function of a design parameter. As shown in Figure 3 on the right, for certain conditions a linear function can be chosen that models the influence of a design parameter onto the drapeability behaviour with a very high degree of confidence.

In the shown example the formation of gaps is dependent on the stitch length of the warp-knit stitching. A longer stitch length leads to a low amount of initial gaps, due to less stitch points, but shows a faster growth of gaps because of the lower stiffness of the reinforcing warp-knit structure.

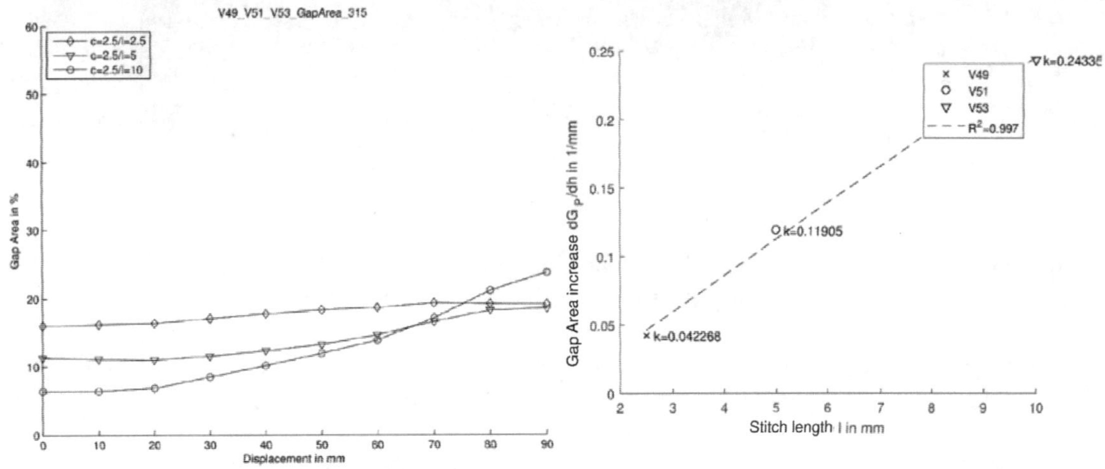

Figure 3. Graph of measured Gap Area over displacement for NCF with three different stitch length (left) and the linear function of the Gap Area Increase with the stitch length as a variable

In the next step these semi-analytic functions can be combined even further to model the entire multi-dimensional process space of draping NCF materials.

IOP Conf. Series: Materials Science and Engineering **406** (2018) 012015 doi:10.1088/1757-899X/406/1/012015

5. Drapeability Modelling

As described the slope coefficients *k* gathered during the empirical testing can be displayed as a function of a textile design parameter or even several design parameters. These functions include in effect a drapeability effect, one or more geometric parameters (the degree of drape and/or the fibre orientation) and one or more textile design parameters. This way, if all textile design and draping process parameters with their coefficients are combined, a model of the drapeability behaviour can be setup that is analytic in nature but based on empirically gained material laws. Due to its analytical nature such a model is simple and fast to calculate. No expertise of numerical methods such as FEM is required. The problem of measuring fabric properties as input parameters is also circumvented because all input parameters (the k-coefficients) are generated by one test, the standardised drapeability test with the Drapetest.

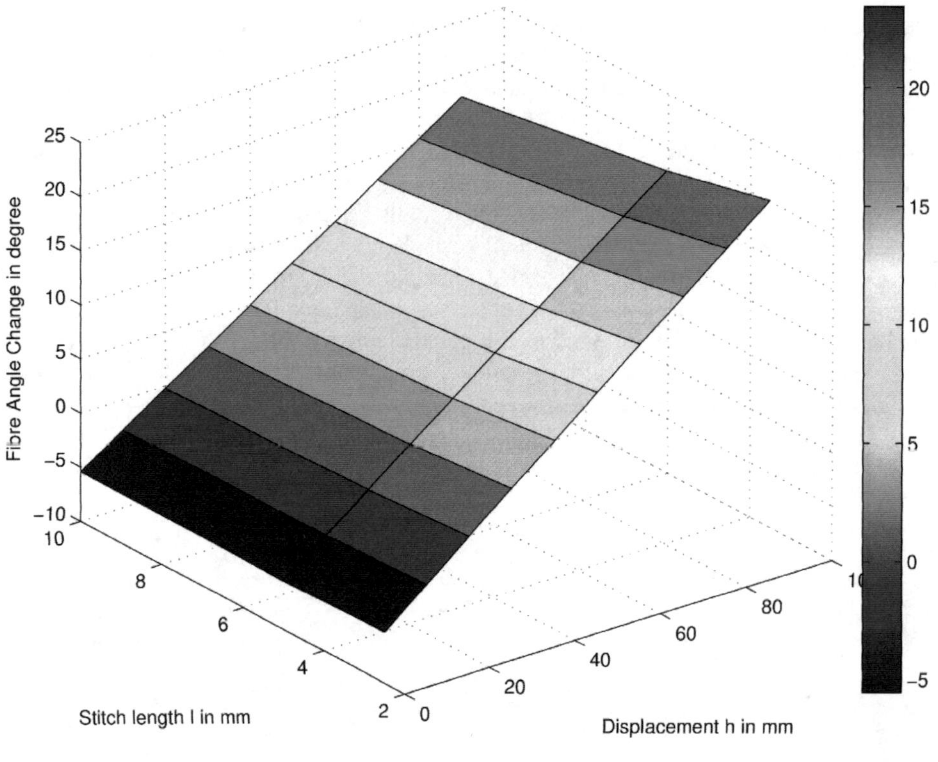

Figure 4. Graphical representation of the semi-analytical model of the fibre angle change in a NCF dependent on the stitch length l and the displacement h

The complexity of modelling the drapeability of NCF stems from its multi-dimensionality. The fabric's drapeability behaviour can be influenced by numerous interdependent design parameters. For the formation of the drapeability effects these parameters then interact further with the process parameters such as the tool curvature and the preform orientation. An example how a semi-analytical modelling of these interactions look is shown in Figure 4. For the sake of clarity only one process parameter, the reference body displacement and one design parameter, the stitch length. But for the semi-analytical model further combinations are possible but do not lend themselves to graphical representation (due to them existing in a space higher than the $\mathbb{R}^3$). Further refinement of the model and application to such conditions in which the drapeability behaviour in non-linear requires further material testing to populate the design space more thoroughly.

IOP Conf. Series: Materials Science and Engineering **406** (2018) 012015 doi:10.1088/1757-899X/406/1/012015

6. Conclusion

This paper concerned the challenge of defining, testing, and modelling the drapeability of non-crimp-fabrics (NCF). Classical definitions of drapeability can well be applied to woven fabrics but are insufficient for fabrics whose yarns not only possess rotational (through fabric shear) but also axial and lateral movability. In the classical definition the effects of shearing and subsequent wrinkling are considered to evaluate the drapeability of a fabric.

A novel definition of drapeability includes a number of further drapeability effects such as gaps and preform deformation due to axial yarn slippage. To evaluate the drapeability of a fabric all effects must be considered by using quantitative indicator numbers that are gained by empirical testing. A defined set of indicator numbers can be used to describe the state of the fabric and effects present in their extent and intensity.

A drapeability tester is introduced that is able to deliver these indicator numbers by measuring a fabric sample in different degrees of drape using optical measurement technology coupled with a digital image analysis software. The tester uses the classical reference body method but expands upon it by allowing for several degrees of drape through displacement of the reference body and a quantitative drapeability analysis by employing two cameras and a laser scanner.

Through experiments empirical relationships between various NCF design parameters and the defined drapeability effects could be found. These relationships can be expressed in the form of semi-analytical functions which can be combined into a model of the draping process of NCF. The model can used as a simple tool during material selection or, when more parameters are included, as the basis for a fast computer-based modelling of the complex problem of NCF drapeability.

References

[1] Christ M (2018) *Definition und Quantifizierung der Drapierbarkeit von multiaxialen Gelegen durch die Vermessung von Einzeleffekten* Bremen urn:nbn:de:gbv:46-00106559-11

[2] Creech G 2006 *Mesoscopic Finite Element modelling of non-crimp fabrics for drape and failure analyses* (Cranfield)

[3] Ermanni P 2004 *Textile Halbzeuge* in *Composites Technologien* (Zürich)

[4] Krieger H 2015 *Methode zur Auslegung von Gelegen mit lokal angepassten Fertigungsparametern für Hochleistungs-Faserverbundkunststoffe* (Aachen)

[5] Christ M, Miene A and Mörschel U 2017 Measurement and analysis of drapeability effects of warp-knit NCF with a standardised, automated testing device *J. App. Comp. Mat.* **24** 803

IOP Conf. Series: Materials Science and Engineering **406** (2018) 012016 doi:10.1088/1757-899X/406/1/012016

Multi-scale analysis of flax fibres woven fabrics for composite applications

A C Corbin[1]*, D Soulat[1], M Ferreira[1], A R Labanieh[1], X Gabrion[2], V Placet[2]

[1] ENSAIT, GEMTEX – Laboratoire de Génie et Matériaux Textiles, F-59000 Lille, France, 2 allée Louise et Victor Champier BP 30029 59056 Roubaix Cedex 1 France
[2] FEMTO-ST, CNRS/UFC/ENSMM/UTBM – Université de Bourgogne Franche-Comté 24 rue de l'Epitaphe 25000 Besançon France

* anne-clemence.corbi@ensait.fr

Abstract. Natural fibres are used as an alternative to man-made fibres in fibre reinforced composite materials in a variety of applications due to their good specific strength and stiffness properties and limited environmental impact. However, contrary to man-made fibres, used in the form of continuous filaments, natural fibres are extracted from plants in which they have a finite length. After their extraction form plants, they are paralyzed and twisted together to form roving or yarn. In this paper, the potential of manufacturing flax woven reinforcements optimized for composite applications is evaluated. A multi-scale analysis of the textile and mechanical properties is conducted to evaluate the impact of each transformation step and the respective contribution of fibre, yarn and fabric to the composite performance. The impact of a chemical treatment on the roving features is evaluated. Results show that this treatment doesn't impact significantly the fibre tensile strength and stiffness and improves its weavabilty. The difference between yarn and roving is also shown regarding the weavability and their impact on the composite tensile properties. The properties of these flax reinforced composites are finally compared with those available in the open literature for similar fibre composites. A promoting conclusion is then proposed.

1. Introduction

Nowadays, composite materials are more and more used to replace traditional materials, such as metals, ceramics or wood, due to their better specific properties. The organic matrix composite materials are made up of a reinforcement phase, and a matrix phase consisting of thermoplastic or thermosetting polymer. The carbon, glass and aramid fibres are the dominant fibres used as reinforcement. However, these fibres are expensive, difficult to recycle non-biodegradable and their handling and use can be dangerous for human health. A weak point of the glass fibre is also is quite high density. So, the plant fibre represents a promising eco-friendly alternative to the man-made fibres for textile reinforcement. However, they have with counterpart a lower resistance to temperature (which limits the level applied in several composite manufacturing processes), lower mechanical properties, an important moisture absorption capability and a year-to-year variability [1]. By contrast, the multi-scale analyses have shown that composite materials with natural fibre reinforcements have a variability in properties similar to those with carbon or glass reinforcements [2]. Thus, by using biobased reinforcement and matrix, the environmental impact, weight and cost of composite can be reduced and the thermal and acoustic properties improved [3]. Table 1 presents the specific gravity, specific strength, specific stiffness and strain at failure for flax and hemp fibre in comparison with glass and carbon fibre.

Table 1. Properties of several fibres commonly used in composite materials [4].

Fibre	Specific gravity ($g.cm^{-3}$)	Specific strength ($MPa.cm^3.g^{-1}$)	Specific stiffness ($GPa.cm^3.g^{-1}$)	Strain at failure (%)
Flax	1.45	550 – 1030	38 – 52	1.5 – 2
Hemp	1.48	370 - 600	27 – 44	1.6
Glass	2.55	780 – 940	27 – 29	3
Carbon	1.8	1900 - 2700	128	1.5 – 2.1

Currently, plant fibres are used for the manufacturing of reinforcements in form of chopped fibres nonwoven preforms and in the form of long fibres for UD preforms. But they are not commonly used in form of yarn or roving to manufacture woven fabric reinforcements that are optimized for composite applications. Only few studies report the behaviour of composite materials reinforced by woven fabrics [5,6]. In these works, the reinforcements were not necessarily well optimized for composite applications. With a multi-scale analysis, the parameters and properties of rovings, yarns, fabrics and composite can be optimized to obtain the better material which fits well with the requirements for this application. Plant fibres have a finite length. Therefore the spinning of the fibres is required after extraction from the plants, to paralyse and assembly the discontinuous fibres in a continuous reinforcement. A specific linear density is associated with the yarns. On the other side, the man-made fibres, such as glass or carbon, are produced under the form of filament (continuous fibre), so this spinning step is not needed. Assembling the fibres to form a roving or a yarn is carried out by twisting the fibres together to increase the inter-fibre friction or by gluing the fibres using a chemical agent. The twisting step has an impact on composite manufacturing and quality, because it can decrease the impregnation ability of yarns. Indeed, as the twist level increases, the yarn structure becomes more compact [7]. The twist level for roving is lower than that for yarn and a chemical treatment can be performed to impart the necessary resistance to the roving for weaving process. Different chemical treatments can be applied but they may have an impact on the internal structure of the fibres and on their mechanical properties [8]. The weaving process consists in warping yarn into a beam, installing the warp yarns under tension on the weaving loom then inserting a weft yarn successively though a shed created by raising and falling downward warp yarns alternately during the process according to weave pattern. Thus all these steps require sufficient strength and strain at failure for yarn and roving [9]. Therefore the potential of using roving to manufacture reinforcement by weaving process is examined in this paper and the impact of the use of chemical treatment is evaluated by conducting a multi-scale analysis and characterization up to the composite scale.

2. Materials and methods

Flax fibre in the form of two different half-products were used to manufacture different woven fabric reinforcements, namely yarn (Y) and roving without chemical treatment (R1) and with chemical treatment (R2) (Figure 1 (a), (b) and (c)). These yarns and rovings were provided by Linificio Canapificio Nazionale, an Italian Company.

Woven fabrics are produced on a manual dobby loom, a Leclerc Weavebird loom. Two plain weave fabrics are manufactured with 6 warps and 6 wefts per cm. For the first fabric (FLAX1), Figure 1 (d), yarn (Y) is used in warp and roving R1 in weft. While for the second fabric (FLAX2), Figure 1(e), yarn (Y) is also used in warp but roving R2 is used in weft. Weaving process requires applying high tension on warps (ends) during weaving that cannot be sustained by a roving. That is why yarn is used for the two fabrics in the warp direction. These produced fabrics are used to manufacture a composite with 4 plies using an epoxy resin, the GreenPoxy 56, provided by Sicomin. The composite is manufactured by thermocompression. It is cured at 130°C during 1 hour under a pressure of 3 bars. The fibre volume fraction of the obtained composite is approximately 45%. After manufacturing, the plates were stored in climatic chamber (T = 23°C, RH = 50%).

A multiscale analysis is conducted at each manufacturing stage in order to evaluate the impact of each stage on the reinforcement and composite properties, as well as the effect of the chemical treatment

IOP Conf. Series: Materials Science and Engineering **406** (2018) 012016 doi:10.1088/1757-899X/406/1/012016

of roving (R2) on composite behaviour. Therefore, the properties of the materials are characterized at each scale: fibre, yarn, fabric and composite.

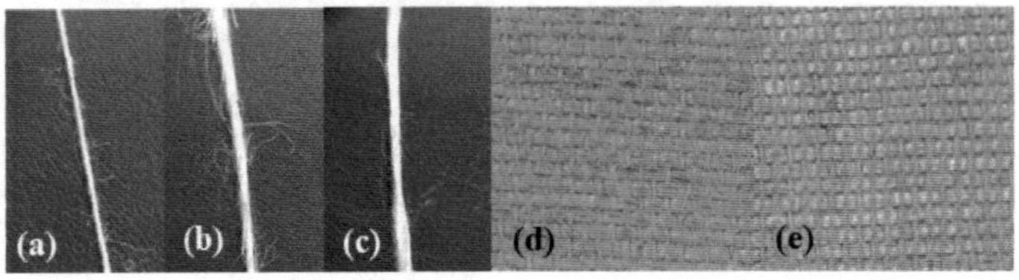

Figure 1. Pictures of yarns and fabrics: (a) Y yarn (b) R1 roving (c) R2 roving (d) F1 fabric (e) F2 fabric.

2.1. Fibre scale

The tensile properties of the flax fibres are identified by inverse method using the Impregnated Fibre Bundle Test (IFBT). Bensadoun and al. [10] have shown that IFBT method is a reliable and efficient method to determine the tensile properties of plant fibres, in particular when compared to the time-consuming and challenging single fibre tensile test (NF T25-501-2 standard [11]). To manufacture the IFBT specimens, bunches of roving were impregnated in an epoxy resin, the GreenPoxy 56, and cured at 60°C during 24 hours under a pressure of 2 bars. After manufacturing, the IFBT specimens were stored in a climatic chamber (T = 23°C, RH = 50%) during at least three weeks, to ensure that the composite material reaches the moisture equilibrium.

Tensile tests were further performed using a universal tensile electro mechanical machine (MTS Criterion 45) under a constant crosshead displacement rate of 1 mm/min and with a load cell of 5 kN full range. Fibre strength and stiffness were then determined by inverse method, using the rule of mixture and the recorded tensile stress-strain curves. The disorientation of the fibre, when compared to the longitudinal direction of the IFBT specimens, due to their twisting, is neglected, since twist level of the used rovings is very low.

2.2. Yarn scale

The textile and tensile properties of yarn and roving (R1 and R2) are also evaluated. For the textile properties, the linear density and twist level are measured according to NF G07-316 standard [12] and NF G07-079 standard [13] respectively. As well as the uniformity, hairiness index (H) and standard deviation of hairiness (sH) for the yarn and roving are measured via an Uster Tester machine. The uniformity is represented by the coefficient of variation of the weight or fibre number over the yarn length (CVm%). The hairiness index represents the total length of the protruding fibres with reference to the sensing length of 1cm of the yarn/roving.

The tensile test is also performed on a tensile machine, according to NF EN ISO 2062 standard [14] for a single yarn. In order to eliminate the effect of the sample length when comparing the tensile behaviour of a single yarn to the tensile behaviour of a yarn in the fabric structure, the same sample length of 200 mm is considered for both yarn and fabric samples in the tensile tests. The extension rate is set to 200 mm/min and a preload of 0.5 cN/Tex is applied for the yarn whereas no preload is applied for the roving. For each tensile test, the tenacity-strain curve is recorded. The tenacity of a yarn is defined as the force per linear density (Tex). These characterizations are conducted on yarns from the bobbins before weaving and on yarns extracted from the woven fabrics after weaving process.

2.3. Fabric scale

The textile and mechanical properties are also identified at the fabric scale. The thickness, areal density and air permeability are measured for both FLAX1 and FLAX2 according to NF EN ISO 5084 standard [15], NF EN 12127 standard [16] and NF EN ISO 9237 standard [17] respectively. The shrinkage percentage of warp and weft inside the fabric structure are also measured for both fabrics according to the NF ISO 7211-3 standard [18]. The bending and tensile behaviour are also characterized. A bending test is performed using a cantilever apparatus according to the ISO 4604 (05) standard [19], with a sample size of 300 length and 50 mm width. The tensile tests are performed according to the NF EN ISO 13934-1 standard [20] on a tensile machine, with an extension rate of 20 mm/min, gauge length of 200 mm, width of 50 mm and pre-tension of 5N. Sandpaper strips are glued to the jaws of the tensile machine clamps to improve gripping and avoid sample slippage. The force-displacement curve is recorded to calculate the breaking tenacity and strain at break.

2.4. Composite scale
The composite samples are manufactured using 4 plies of the flax fabrics and the GreenPoxy 56 epoxy resin, which is also used in IFBT tests. The impregnated reinforcements are cured during 1h at 130°C under a pressure of 3 bars. Tensile specimens were laser cut from the plates. The tensile test is performed on the composite samples according to NF EN ISO 527 standard [21], with a crosshead displacement-rate of 1 mm/min, a gauge length of 150 mm, a sample width of 15 mm and a sample thickness of 1.7 mm. The strength, tangent modulus (between 0% and 0.1% of strain) and strain at break are calculated from the recorded force-displacement curves. The identified properties of the manufactured composites are also compared with the properties of similar woven fabrics composites reported in the literature [5] and available on the market.

3. Results and discussion

3.1. Fibre scale
The specific strength and modulus of the fibres, back-calculated from IFBT, are listed in Table 2. The modulus 1 is calculated between 0% and 0.1% of tensile strain while modulus 2 is calculated between 0.3% and 0.5% of strain. Fibres without treatment come from R1 rovings and fibres with treatment comes from R2 rovings. Interestingly, Table 2 shows that the specific strength and modulus of the fibres, treated or not, are similar. A maximum difference of about 8 % on the mean values is observed. Thus, it can be concluded that the chemical treatment has no significant impact on the fibre tensile properties. As already observed for natural fibre [22], the first modulus is higher than the second one. In comparison with glass fibre, (properties in Table 1) flax fibre has a lower specific strength but it has a similar specific modulus. It confirms, once more, the interest of using the natural fibres like flax as a substitute to glass fibre in fibre reinforced plastic composite.

Table 2. Properties of flax fibres (back-calculated from IFBT method).

Fibre	Specific gravity ($g.cm^{-3}$)	Specific strength ($MPa.cm^3.g^{-1}$)	Specific modulus 1 ($GPa.cm^3.g^{-1}$)	Specific modulus 2 ($GPa.cm^3.g^{-1}$)
Flax fibres without chemical treatment (from R1 roving)	1.45 [4]	355 ± 21	35.2 ± 2.1	23.9 ± 0.7
Flax fibres with chemical treatment (from R2 roving)	1.45 [4]	377 ± 22	38.7 ± 2.8	21.4 ± 1.4

3.2. Yarn scale
The measured textile properties of the yarn (Y) and rovings (R1 and R2) are listed in Table 3. Their tensile behaviour is presented in Figure 2 through tenacity-strain curves. The yarn has a lower linear density than the roving, which is related to the lower number of fibres in its cross section. Therefore a higher twist level is generally inserted to the yarn. By twisting operation, the fibres take a spiral path in the yarn structure resulting in an increase of the inter-fibre compression and friction. That leads to a superior breaking tenacity and modulus when compared to the rovings (R1 and R2), as it can be seen on Figure 2. So, on the weaving loom, a high twist level is required for the warp yarn to decrease the

breakage and to sustain the necessary applied tension during weaving. However, the angle between the fibre and yarn main directions (i.e. the spiral angle) increases with increasing twist level, which can be detrimental for the longitudinal tensile strength and rigidity of UD composites. Furthermore, when the twist level increases the fibres are compressed and the yarn structure is more compacted. Thus, the interpenetration of the resin inside the yarn may be more difficult and result in higher void volume fraction inside the yarn. Conversely, twisting leads to a decrease of the hairiness index (H) of the yarn when compared to the roving (R1 and R2) which have a lower twist level. It can also be noticed that the yarn is more irregular (CVm%) than the rovings (R1 and R2). This is related to the spinning process conditions.

By comparing the two rovings (R1 and R2), it can be concluded that the chemical treatment induces a decrease in the linear density. This is attributed to the removal of impurities and of some of the fibres during the treatment. Even though this treatment leads to a slight decrease in the hairiness index, it nevertheless induces an increase in the standard deviation (sH).

Table 3. Textile properties of yarns.

	Flax yarns (Y)	Flax rovings without chemical treatment (R1)	Flax rovings with chemical treatment (R2)
Linear density (Tex)	173.4 ± 14.1	369.8 ± 48.8	304.4 ± 16.7
Twist level (turns/m)	231.1 ± 17.8	31.6 ± 2.5	35.7 ± 3.7
CVm% (%)	27.9	18.8	20.0
Hairiness H	7.4	21.9	18.8
sH(%)	3.4	7.8	8.5

Thus, even if, the chemical treatment doesn't have a significant impact on the textile properties of the roving, it induces, on the other hand, an obvious improvement in the breaking tenacity and stiffness. No significant change in the strain at break was observed (see Figure 2 (b)). Figure 2 (a) points out that the tangent modulus of the roving after treatment is close to the modulus of the yarn. So the treatment, even if inducing no significant effect at the fibre scale, induces an improvement of the tensile properties at the yarn scale (Figure 2. (b)). This is attributed to an increase in the cohesion at the interface between the fibres inside the yarn. Anyways, despite this hypothetical increase in cohesion, friction between fibres does still exist, since no impact is observed on the strain at failure of the yarns.

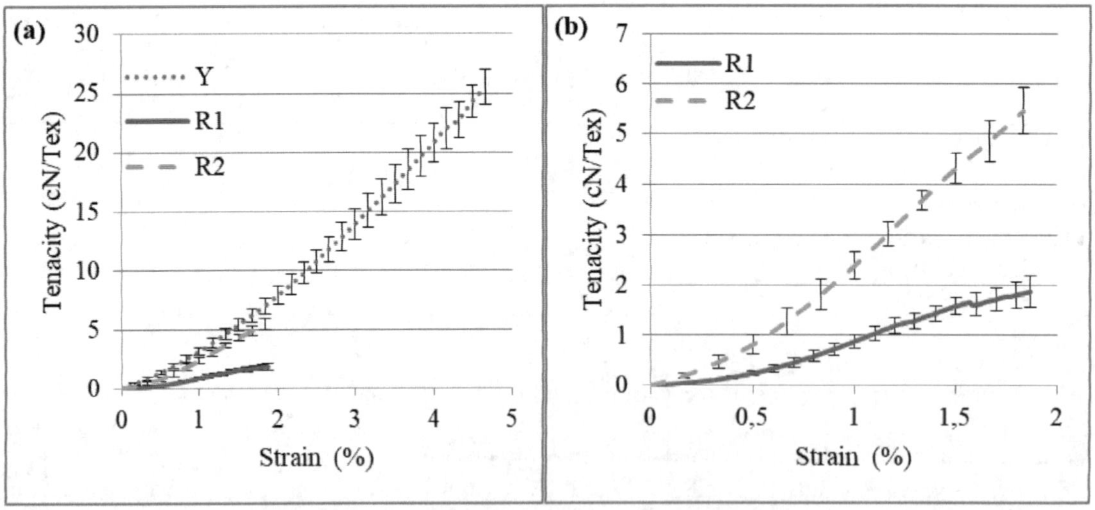

Figure 2. Tenacity vs. strain curves (a) Comparison of the 3 types of yarns and rovings (b) Zoom on rovings.

3.3. Fabric scale

The weavability of both rovings (R1 and R2) is evaluated for the two fabrics (FLAX1 (F1) and FLAX2 (F2)) by considering the number of breakage of the roving during the weaving process. 50% of roving R1 (without treatment) breaks by the weft insertion operation during weaving against less than 1% for R2 (with treatment). Thus, the chemical treatment of the roving permits to decrease the material waste: from 15% without treatment to only 3% with treatment. As a conclusion, the chemical treatment improves the roving behaviour during weaving process by sticking the fibres to the roving body accompanied with a lower hairiness index (Table 3). It finally results in a reduction of the rubbing damage caused by passing the roving over loom machineries and by yarn/roving contact.

The textile properties of the manufactured fabrics F1 and F2 are listed in Table 4. The bending rigidity values and tenacity-strain curves are presented in Figure 4 (a and b).

Table 4. Textile properties of fabrics.

	Thickness (mm)	Areal density (g/m^2)	Air permeability ($L/m^2/g$)	Warp shrinkage (%)	Weft shrinkage (%)
FLAX1 (F1)	1.00 ± 0.07	307.66 ± 9.20	1364.90 ± 253.59	2.72 ± 0.33	1.90 ± 0.18
FLAX2 (F2)	0.91 ± 0.03	288.43 ± 19.82	1934.60 ± 218.68	2.98 ± 0.25	2.56 ± 0.40

The thickness and the areal density of F1 are slightly higher than those of F2. This is related to the lower linear density of the roving R2, which is more compacted, as pointed out in Table 3. It also impacts the air permeability. A higher permeability index is measured for F2 when compared to F1, even if an identical warp and weft count per unit length and a yarn (Y) count are considered for both fabrics.

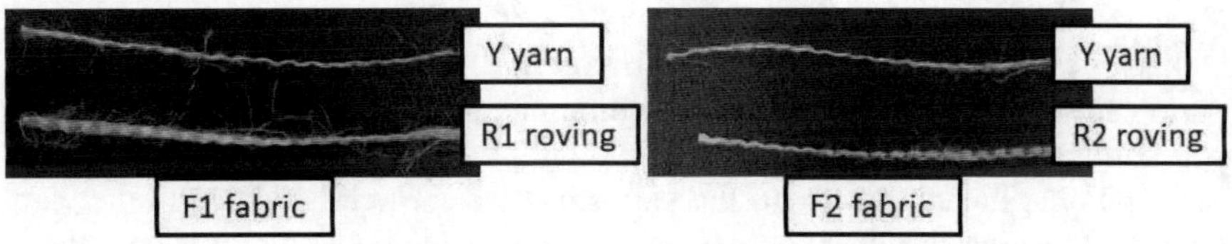

Figure 3. Visual aspect of yarns extracted from the fabrics

The weave diagram is the same for the two fabrics, it is a plain weave. In table 4, it can be observed that the warp shrinkage is more important than the weft shrinkage. This is due to the manufacturing process settings, in particular the yarn tension and the shed formation, in addition to a lower linear density for the yarn making it more flexible to bending when compared to roving. In F2 fabric, weft yarns (R2) are more compact and more rigid than R1 in F1 fabrics. Thus, warp and weft yarns in F2 are more crimped than in F1 and warp and weft shrinkages are higher in F2 than in F1 (Figure 3).

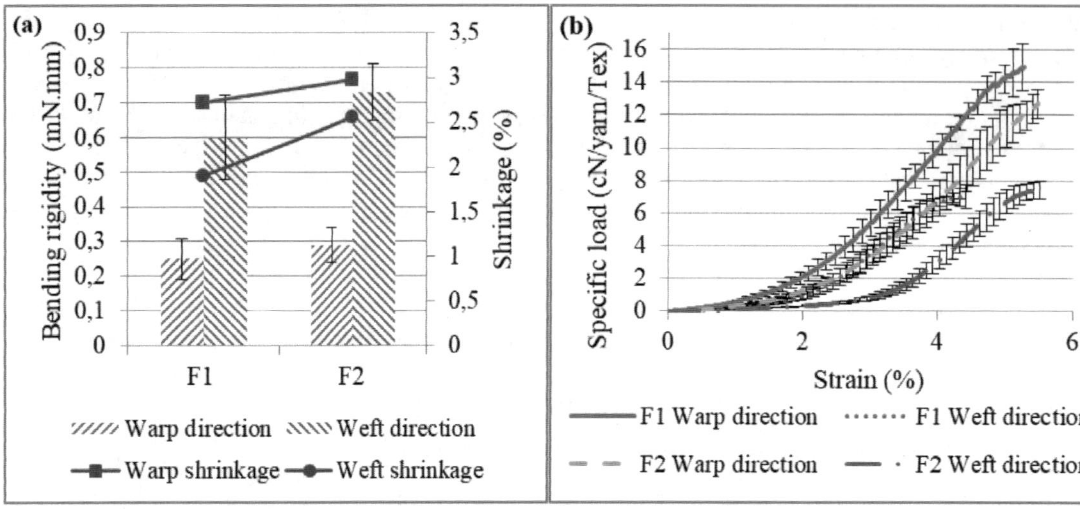

Figure 4. Mechanical properties of fabrics (a) bending rigidity properties and warp/weft shrinkages (b) tensile properties.

For both fabrics, the bending rigidity is more important in the weft direction than in the warp direction. However, the same material and yarn count per unit length is used in both warp and weft directions. Thus, this difference is related essentially to the higher linear density of the roving (R1 or R2) used in the weft direction in comparison with the yarn (Y) used in the warp direction for both fabrics. The higher linear density involves a higher number of fibres in the cross section and then a higher thread width and thickness resulting in a higher bending rigidity.

By comparing the two fabrics in term of bending rigidity, for the same materiel direction, and considering the shrinkage percentage in the given directions (Figure 4 (a)), it can be concluded that the fabric bending rigidity is linked directly to the yarn shrinkage, which is related to the yarn path inside the fabric structure. The high shrinkage percentage is accompanied with a higher bending rigidity.

The tensile behaviour of the two fabrics is shown on Figure 4(b) through a specific load-strain curve. The load is normalized (cN/yarn/Tex) by considering the number of yarns per cm and their counts (Tex) in both warp and weft directions. It permits to highlight the effect of yarn/roving structure and the effect of the roving treatment, in comparison to the yarn/roving tensile properties characterized before weaving.

The tensile load-strain curves can be divided in two parts: a non-linear part, referred to the fabric structural response which matches with the loss of crimp in the extension direction, followed by a linear part, where yarns become tight and respond directly to the applied load, referred to the yarn response. In warp direction, the shrinkage percentage is higher in F2 fabric than in F1. It results from a more extended non-linear part leading to a higher strain at break at the scale of the fabric. However, on the linear part, the tangent modulus in the warp direction is similar for the two fabrics. For both fabrics, properties are better in warp direction than in weft direction, which can be analysed with regard to the yarn scale (yarns used in warp direction for the two fabrics, are stronger with higher breaking specific load than rovings used in weft direction).

When comparing the fabric specific load in warp direction (Figure 5 (b)) to the yarn behaviour characterized at the yarn scale (Figure 2 (a)), a loss of breaking tenacity is notice. That is attributed to a decrease in twist and to damage caused by the high tension necessary applied on the warp yarn in the warping preparation process weaving processes. In weft direction, the difference in the shrinkage percentage between F1 and F2 is more important resulting in higher elongation due to the loss of the crimp in F2. Similarly, when comparing the fabric specific load in weft direction (Figure 4 (b)) with the roving behaviour characterized at the yarn scale, (Figure 2), higher breaking tenacity is imparted especially for R2. That is related to the impact of the other orthogonal yarns (warp) that compact and compress the roving leading to an increase in the inter-fibre cohesion in the roving structure giving better strength.

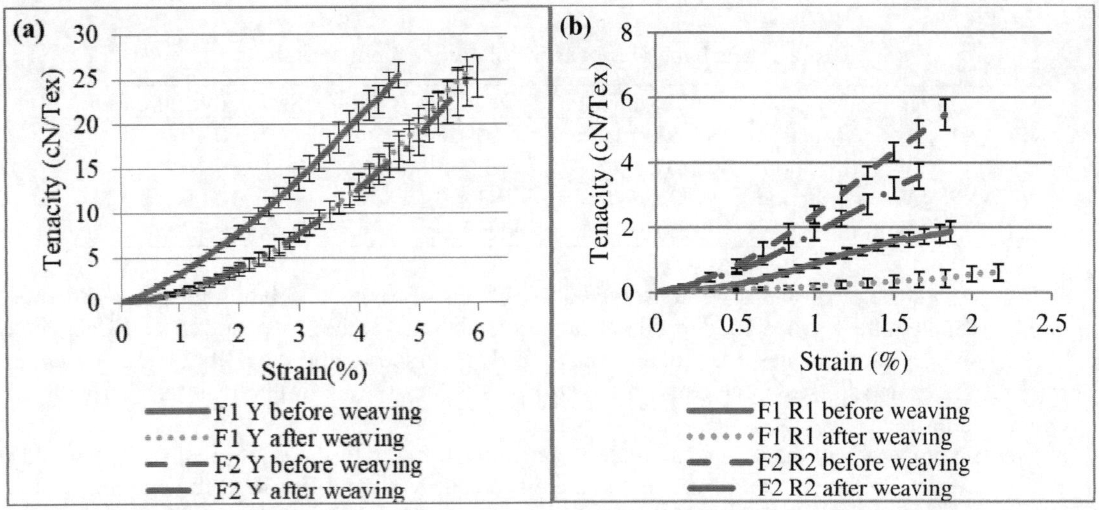

Figure 5. Influence of weaving process on yarns mechanical behaviour (a) warp yarns (b) weft yarns.

The impact of the weaving process on the yarn tensile properties is observable on Figure 5 (a) and Figure 5 (b) giving the tenacity-stain curves for the warp and weft yarns respectively. In this section, the tensile behaviour of a thread extracted from the manufactured fabric is compared to the behaviour of this thread before weaving. For warp yarns, the weaving process leads to an increase of the strain at break of the yarn and a decrease in the curve slopes in the first part. The maximal load for the warp yarn used for F1 and F2 fabrics are equivalent and their apparent rigidity in the linear part is similar. The breaking tenacity for the warp yarn extracted from the fabric is higher than the specific maximum load (cN/Tex) of the fabric in the warp direction. This is attributed to the supplementary loads induced by the micro bending in the crimping areas. The twist level of yarns was also measured on yarns extracted from the fabrics, and the properties are similar to those of yarns from the bobbins before weaving. Indeed, the twist level is not impacted by the weaving process.

For weft, the mechanical properties are affected by the weaving process and a different mechanical behaviour is observed between F1 and F2 fabrics. For F1 fabric, the strain at break is higher after weaving, but the maximal load and the two apparent initial rigidities are lower. In F2 fabric, the strain at break and maximal load in the weft direction also decrease after weaving but the apparent rigidity remains similar.

The weft rovings are more damaged than the warp yarns during weaving process due to the used technique for the weft insertion operation. The weft roving is winded into a small bobbin to be inserted inside a shuttle used to pass the weft thought the warp shed. This process induces a de-twisting of the weft reinforcement and a bending damage in view of winding a thick roving around the small and thin tube of the shuttle. The tenacity at break of the weft yarns extracted from the fabric is lower when compared to the strength of the fabric itself in the weft direction. So, warp yarns may increase the cohesion between the fibres inside the roving due to the applied transverse compression. It leads to a high strength at the scale of the preform.

3.4. Composite scale

The tensile properties of the composite samples made up of the two woven fabrics are presented in Table 5. C1 composite is reinforced by F1 fabric and C2 composite by F2 fabric. The obtained tensile properties in weft direction are higher than in warp direction due to higher fibre content in this direction.

This is attributed to the roving linear density which is higher in the weft direction than the one of yarn in warp direction. In the weft direction, the properties are similar for C1 and C2. Thus, it can be concluded that there is no significant impact of the chemical treatment on the composite behaviour.

Table 5. Mechanical properties of composite materials.

		Strength (MPa)	Modulus (GPa)	Strain at failure (%)
FLAX1	Warp direction	91.5 ± 1.0	9.0 ± 0.3	1.71 ± 0.10
(C1)	Weft direction	126.0 ± 6.0	12.0 ± 0.6	3.10 ± 0.10
FLAX2	Warp direction	90.0 ± 5.0	9.3 ± 0.1	1.82 ± 0.09
(C2)	Weft direction	118.0 ± 6.0	12.7 ± 0.3	2.44 ± 0.17

Even if a low tensile resistance was observed at the scales of roving and fabric in roving direction (Figure 2 (a) and Figure 4 (b)), a good tensile behaviour is obtained at the composite scale. Effectively, at the yarn and dry fabric scales, the tensile behaviour is governed by the inter-fibre friction. On the contrary, at the composite scale, the stress between fibres is transferred via the resin and through the fibre/matrix interface.

The tensile properties of the composite materials manufactured in this study were also compared to the tensile properties of other composite materials reinforced by natural fibres, such as flax and jute, listed in the open literature (Table 6), such as in Torres and al. [5] or in Bensadoun's Thesis [6]. These composites were also made up woven fabrics, like C1 and C2 in our study, or are stratification of unidirectional or quasi unidirectional fabrics, for Lineo reinforcement. The Lineo FlaxPly composite, from [5], is a crossply woven textile composite, reinforced by 4 plies of quasi-unidirectional fabric (95% of fibres in main direction and 5% of fibres in transverse direction) and the Lineo FlaxTape composite, from [6], is a crossply laminate, reinforced by 4 plies of unidirectional fabric (nonwoven with fibres totally aligned in the same direction). For these two composites, two plies are laminate in 0° direction and 2 plies in 90°, to obtain reinforcements in the twi main directions of the composite materials, like with woven fabrics. Currently, composites made with Lineo FlaxPly (with a crossply stacking sequence) are widely used for the manufacturing of plant fibres composites in industry. The areal density of F1 (310 g/m²) and F2 (290 g/m²) fabrics is positioned between the ones of the two Lineo reinforcements (180 and 200 g/m²) and Biotex preforms (from 400 to 550 g/m²) and similar to Libeco Flax Plain Weave (285 g/m²) used to manufacture the composites in the cited references. However, the fibre volume fraction is higher for C1 and C2 (about 45%) than the other composites (between 30 to 40%) [5]. C1 and C2 have higher strength and stiffness than the other composites except for the two composite reinforced by Lineo fabrics, where properties can be judged similar. These results are nevertheless promising for the future of the study. The use of FlaxTape reinforcements is currently the reference for natural fibres composites, due to the unidirectional structure: properties are not influenced by some out-of-plane orientations resulting from twisting of yarns or interlacement of warp or weft yarns, as it can occurs with quasi-unidirectional or woven fabrics. Indeed, properties of C1 and C2 composites, made up woven fabrics whose properties are not optimized, are in the same range than the best reinforcement currently available. Woven fabrics are advantageous for composite manufacturing because they are easier to handle and to drape.

IOP Conf. Series: Materials Science and Engineering **406** (2018) 012016 doi:10.1088/1757-899X/406/1/012016

Table 6. Comparison with other natural fibres composites materials from literature.

Reinforcement	Areal density (g/m^2)	Number of plies	Fibre volume fraction (%)	Strength (MPa)	Modulus (GPa)	Strain at failure (%)
Lineo FlaxPly [5]	180	4 $[0/90]_{2S}$	31	147.6 ± 8.3	8.5 ± 0.6	2.15 ± 0.10
Lineo FlaxTape [6]	200	4 $[0/90]_{2S}$	40	126 ± 7	14.5 ± 0.8	1.08 ± 0.16
Biotex Flax 7-1 Satin weave [5]	400	4	35	102.8 ± 3.2	8.2 ± 0.6	1.70 ± 0.11
Biotex Jute 2-2 Twill Weave [5]	550	4	36	63.1 ± 3.6	6.4 ± 0.6	1.21 ± 0.11
Biotex Jute Plain weave [5]	500	4	40	66.7 ± 3.5	6.4 ± 0.5	1.33 ± 0.11
Libeco Flax Plain Weave [6]	285	4	40	135 ± 18	12.6 ± 0.4	1.69 ± 0.32
Composite Evolution Flax 2-2 Twill [6]	400	6	40	120 ± 2	12.8 ± 0.4	1.65 ± 0.14
Hexcel/Safilin Flax 2-2 Twill [6]	200	4	40	126 ± 9	12.9 ± 1.1	1.52 ± 0.13

4. Conclusion

A multi-stage and multi-scale analysis was conducted to better understand the influence of the plant reinforcement features (including single fibre properties, roving, yarn and preform textile and mechanical properties) on the mechanical behaviour of the final composite.

So, the textile and mechanical properties at each manufacturing stage were determined. A chemical treatment is proposed and carried out on the roving to improve its weavability. It induced an increase in the tensile strength of the roving. An obvious decrease in the breakage and waste during the reinforcement processing is then noticed. It is hypothesized that this treatment increases the inter-fibre cohesion in the roving. Interestingly, this treatment doesn't affect neither the tensile properties of the fibres themselves nor those of the final composite.

The results obtained at the yarn scale suggest that the yarn/roving behaviour under tension is mainly governed by inter-fibres friction and the twist has an obvious impact to improve the breaking tenacity. Even if the chemical treatment significantly improves the roving tenacity, it cannot lead to the tenacity obtained with twisting and sometimes required for weaving, at least in one of two main material directions. That's why the roving (R1 and R2) were not used in the warp direction when manufacturing the fabrics.

At the fabric scale, several conclusions can be drawn. The chemical treatment performed on the roving brings more air permeability to the fabric, but has no significant impact on its mechanical behaviour. The fabric bending rigidity is highly influenced by the yarn linear density. A higher rigidity is obtained in the roving direction with higher yarn/roving linear density. During the weaving process, the yarns in warp direction are subjected to a cyclic extension that leads to a slight decrease in the yarn breaking tenacity but an increase in the strain at break. The roving is more affected by the weaving process due to the weft insertion operation using a shuttle. Finally, at the composite scale, quite high tensile stiffness and strength are obtained in both warp and weft directions. So, the results obtained at the scale of the composite using these flax preforms are very promising.

In the upcoming work more parameters will be considered in the multi-scale analysis, such as raw material type, yarn count, treatment type and weave.

Acknowledgements
"This project has received funding from the Bio Based Industries Joint Undertaking under the European Union's Horizon 2020 research and innovation program under grant agreement No 744349".

The authors sincerely thank the Linificio e Canapificio Nazionale Company from Italy for providing the flax yarns and rovings.

References

[1] Misnon M I, Islam M M, Epaarachchi J A and Lau K 2014 Potentiality of utilising natural textile materials for engineering composites applications *Mater. Des.* **59** 359–68

[2] Blanchard J M F A, Sobey A J and Blake J I R 2016 Multi-scale investigation into the mechanical behaviour of flax in yarn, cloth and laminate form *Compos. Part B Eng.* **84** 228–35

[3] Dittenber D B and GangaRao H V S 2012 Critical review of recent publications on use of natural composites in infrastructure *Compos. Part Appl. Sci. Manuf.* **43** 1419–29

[4] Pil L, Bensadoun F, Pariset J and Verpoest I 2016 Why are designers fascinated by flax and hemp fibre composites? *Compos. Part Appl. Sci. Manuf.* **83** 193–205

[5] Torres J P, Vandi L-J, Veidt M and Heitzmann M T 2017 The mechanical properties of natural fibre composite laminates: A statistical study *Compos. Part Appl. Sci. Manuf.* **98** 99–104

[6] Bensadoun F 2016 *In-service behaviour of flax fibre reinforced composites for high performance applications* (KU Leuven)

[7] Goutianos S and Peijs T 2003 The optimisation of flax fibre yarns for the development of high-performance natural fibre composites *Adv. Compos. Lett.* **12** 237–41

[8] Sepe R, Bollino F, Boccarusso L and Caputo F 2018 Influence of chemical treatments on mechanical properties of hemp fiber reinforced composites *Compos. Part B Eng.* **133** 210–7

[9] Rudov-Clark S, Mouritz A P, Lee L and Bannister M K 2003 Fibre damage in the manufacture of advanced three-dimensional woven composites *Compos. Part Appl. Sci. Manuf.* **34** 963–70

[10] Bensadoun F, Verpoest I, Baets J, Müssig J, Graupner N, Davies P, Gomina M, Kervoelen A and Baley C 2017 Impregnated fibre bundle test for natural fibres used in composites *J. Reinf. Plast. Compos.* **36** 942–57

[11] AFNOR 2015 *NF T25-501-2 - Reinforcement fibres - Flax fibres for plastics composites - Part 2 : determination of tensile properties of elementary fibres*

[12] AFNOR 1988 *NF G07-316 - Textiles - Tests of yarns - Determination of linear density.*

[13] AFNOR 2011 *NF G07-079 - Textiles - Testing threads - Determining the twisting of threads by untwisting/retwisting with a double re-test*

[14] AFNOR 2010 *NF EN ISO 2062 - Textiles - Yarns from packages - Determination of single-end breaking force and elongation at break using constant rate of extension (CRE) tester*

[15] AFNOR 1996 *NF EN ISO 5084 - Textiles - Determination of thickness of textiles and textile products*

[16] AFNOR 1998 *NF EN 12127 - Textiles - Fabrics - Determination of mass per unit area using small samples*

[17] AFNOR 1995 *NF EN ISO 9237 - Textiles - Determination of permeability of fabrics to air*

[18] AFNOR 2017 *NF ISO 7211-3 - Textiles - Woven fabrics - Construction - Methods of analysis - Part 3 : determination of crimp of yarn in fabric*

[19] AFNOR 2011 *ISO 4604 - ISO 4604 - Reinforcement fabrics - Determination of conventional flexural stiffness - Fixed-angle flexometer method*

[20] AFNOR 2013 *NF EN ISO 13934-1 - Tensile properties of fabrics - Part 1 : determination of maximum force and elongation at maximum force using the strip method*

[21] AFNOR 2009 *NF EN ISO 527 - Plastics - Determination of tensile properties*

[22] Placet V 2009 Characterization of the thermo-mechanical behaviour of Hemp fibres intended for the manufacturing of high performance composites *Compos. Part Appl. Sci. Manuf.* **40** 1111–8

IOP Conf. Series: Materials Science and Engineering **406** (2018) 012017 doi:10.1088/1757-899X/406/1/012017

A dissipative constitutive model for the hysteritical behaviour of a woven composite fabric under large strain

Y Denis[1], E Guzman-Maldonado[1], N Hamila[1], F Morestin[1]

[1] Université de Lyon, INSA-Lyon, LaMCoS UMR 5259, 20 avenue Albert Einstein, F69621 Villeurbanne Cedex, France.

Abstract. Draping composite reinforcement on non-developable shapes necessarily leads to deformations in the plane generating large shears between warp and weft. Energy dissipation appears since the sliding between yarns and between fibers creates friction effect. The theory presented here is a constitutive model describing the dissipative behaviour of a 2D composite woven fabric. The model is based on two innovative points which are the Green – Naghdi additive decomposition for anisotropic material and the theory of nested surfaces according with Mroz studies. Both points lead to define the strong non-linearity of the problem and the hysteretical behaviour during the unloading phase. The dissipation process driven by fibers friction is exclusively associated with the in-plane shear deformation mode. The model is calibrated using standard methods such as the Picture Frame test. This model is also discretised to be integrated in a finite element calculation software.

1. Introduction

The composite fields offer a wide range of material configuration. Depending on the applications, these kinds of material can be from different natures, being from different elements and get specific geometric and mechanical characteristics. Depending on the type of composite used, shaping can vary from simple and mastered processes to much more delicate processes which are still subjects to many stages of research and discussion. In many cases, the composite material consists of a reinforcement that gives most of the mechanical characteristics to the material and a matric that is injected onto the woven fabric to lock it. The object of this paper is then to study the dissipative behaviour of a woven fabric made by continuous fibre. One of the famous shaping process for composite reinforcement is the Resin Transfer Moulding [1], [2] which consist to shape the fabric and then inject the resin. The work in this paper will focus on the shaping phase where the fabric knows very large deformations and in some cases cyclical large deformations. Indeed, the material will undergo very high deformations to be superimposed on the shape of the matrix. Since the complexity of the piece's shape is in constant evolution, it is now, not unusual to get large variation of shear angle for example. However, to describe the shaping phase, only hyper-elastic or visco-elastic model already exist [3]–[8]. But these models do not consider dissipation. Then, for these reasons, the aim of this paper is to write a constitutive model which is considering the dissipation and the hysteritical behaviour of the material considering some assumptions. Indeed, during a cyclic loading, it is shown in this paper that the behaviour is following high non-linearities. Moreover,

since the fabric is composed of fibre, it will therefore deform according to different modes (elongations in the direction of yarns) and the in-plane shear mode [9]–[11]. As this type of material is usually made of fibres that have a very high stiffness (which prevents them from lying down) so most of the energy dissipation is due to the in-plane shear. Since the deformations are very high, the model presented here is written under large strains and considering the hypothesis that the fibres no not lengthen, the dissipation will be exclusively defined by a pure in-plane shear kinematics.
To write a model taking into this energy dissipation, the additive decomposition of Green-Naghdi applied for an anisotropic material under large strain will be used but no demonstrate here. The use of this decomposition gives a certain degree of freedom since the use of the intermediate configuration is possible and every fundamental tensor can be defined [12], [13]. On the other hand, this decomposition links all these tensors to the initial configuration, so that the formulation of this constitutive model is in total lagrangian formalism. Working on the initial configuration is easier since every quantity are known. Then, the first part of this paper is about the elaboration of an elasto-plastic model based on woks already done [14]–[16]. It is shown that it is working well only if the load is monotone, however it does not describe the reality. Secondly, the hysteritical effect of the material can be approach by several method such as the fractional derivative. However, every works from fractional derivative are made under the condition of small perturbation and seems to be not very stable when the loading is negative [17]. In addition, it works when the unloading would be linear but for a woven fabric under large strain, the behaviour during an unloading phase is asymptotic. Then, the method presented here is based on the Mroz [18], [19] works who propose to improve a model by adding internal surfaces on it. By changing the evolution law of each surfaces, the model can describe high non-linearity and the asymptotic and hysteritical behaviour. This is the strategy used here and it is shown that with only two nested surface, the result can be interesting.
Moreover, at the end of this article it is shown that all the constitutive model is only depending on only one variable. This is very interesting since a simple Newton-Raphson algorithm is enough to compute the model and find the solution. In addition, this model is only available for thick fabric. This strong hypothesis makes possible to simplify the study because of the in-plane stresses assumption.
Finally, at the end of this paper, a comparison between the experimental and the use of this model is made to show the capacity of this constitutive description of the material.

2. Model assumptions

The model presented in this paper can be only used under certain conditions:

- Traction-compression deformation modes (considered elastic) and in-plane shear (considered dissipative) are decoupled.
- Stress and strain tensors are assumed to be written with generalised quantities that have already been integrated into the thickness of the fabric.
- Since the carbon or glass fibres does not lengthen, the behaviour of the elongation/compression modes is purely elastic and does not dissipate energy. Therefore, after the first assumption, the dissipation process follows a kinematic of pure shear.
- The study presented here is made at the macroscopic scale. The fabric is then considered as an homogeneous material.

3. Kinematics of the deformations

In this section, the set of tensors necessary to write the model is defined. The coupling between the Green-Naghdi and the Kroner-Lee decomposition allows decoupling the total transformation and deformation tensors into an elastic part and a plastic part.

3.1. Decomposition and decoupling under large strain

The decomposition of Green-Naghdi is a relatively common formalism in the hypothesis of small perturbations and in the framework of crystalline materials. After a thermodynamical analysis inspired by works already done, the compatibility of this decomposition with the field of finite strain and anisotropic structures allows to write the formulation defined by equation 1.

$$\bar{\bar{E}} = \bar{\bar{E}}_e + \bar{\bar{E}}_p \tag{1}$$

Coupled with the multiplicative decomposition of Kröner-Lee formulated by equation 2, this couple permits de define tensors in the initial configuration (Lagrangian configuration).

$$\bar{\bar{F}} = \bar{\bar{F}}_e \cdot \bar{\bar{F}}_p \tag{2}$$

The use of such decomposition makes possible to use the method of the intermediate configuration involving a configuration where residual stresses and plastic strains remain figure 1.

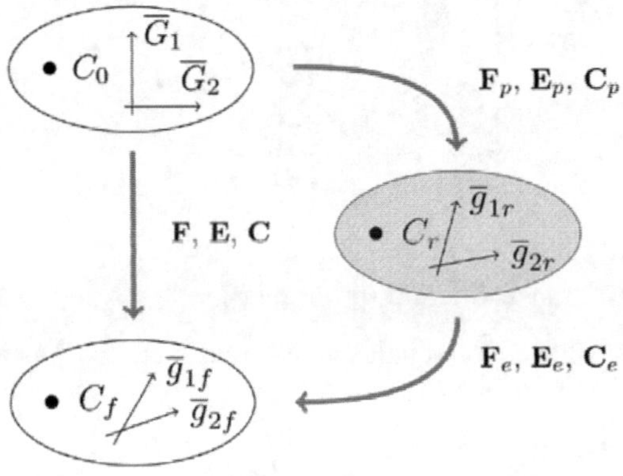

Figure 1. Intermediate configuration method and decoupling.

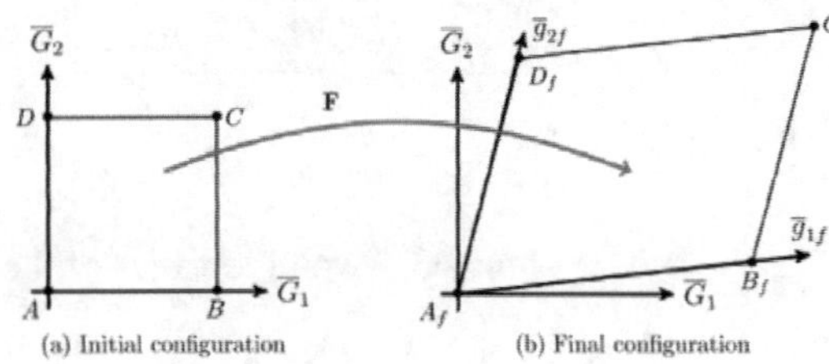

Figure 2. Kinematics of the total transformation.

The quantities defined on the figure 1 and figure 2 are:

- C_0: Initial material configuration
- C_r: Relaxed configuration of the material. It is the so called intermediate configuration.
- C_f: Final material configuration.
- $\bar{\bar{F}}, \bar{\bar{F}}_e, \bar{\bar{F}}_p$: Respectively, the gradient of the total transformation, the gradient of the elastic transformation and the gradient of the plastic transformation.
- $\bar{\bar{E}}, \bar{\bar{E}}_e, \bar{\bar{E}}_p$: Respectively, the total Green-Lagrange deformation tensor, its elastic component and its plastic component.
- $\bar{\bar{C}}, \bar{\bar{C}}_e, \bar{\bar{C}}_p$: Respectively, the total right Cauchy-Green tensor, its elastic component and its plastic component.

3.2. Description of the kinematics

Total transformation is imposed by a test or a load as it can be seen on the figure 2. A user imposes this tensor and it can be described as the equation 3.

$$\bar{\bar{F}} = F_{ij}(\vec{G}_i \otimes \vec{G}_j) \tag{3}$$

$\vec{G}_i$ and $\vec{G}_j$ are the initial warp and weft direction of the material.

IOP Conf. Series: Materials Science and Engineering **406** (2018) 012017 doi:10.1088/1757-899X/406/1/012017

Plastic transformation is imposed by the assumption written above. Considering the last condition, it follows a pure shear kinematics. Figure 3 shows two possibilities of pure shear kinematics. For a question of simplicity, the kinematics shown on figure 3b is chosen since it is symmetric. However, if the kinematics of figure 3a would have been chosen, the result would be the same since the difference is only a pure rotation and since the model must satisfy the principles of thermodynamics and must follow the principle of objectivity.

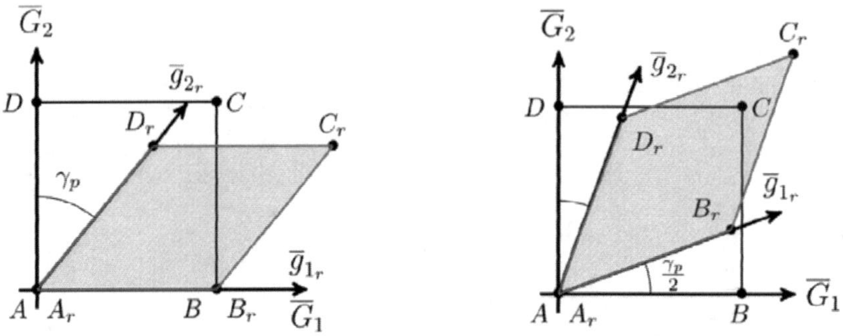

(a) Non symmetric kinematic (b) Symmetric kinematic

Figure 3. Two kinematics of the plastic transformation (pure shear).

The definition of this quantity is given by equation 4. The chosen kinematics gives a symmetrical tensor.

$$\overline{\overline{F}}_p = F_{pij}(\vec{G}_i \otimes \vec{G}_j) = \cos\left(\frac{\gamma_p}{2}\right)\left(\vec{G}_1 \otimes \vec{G}_1 + \vec{G}_2 \otimes \vec{G}_2\right) + \sin\left(\frac{\gamma_p}{2}\right)\left(\vec{G}_1 \otimes \vec{G}_2 + \vec{G}_2 \otimes \vec{G}_1\right) \quad (4)$$

Considering this tensor, it is now possible to calculate $\overline{\overline{C}}_p$ and $\overline{\overline{E}}_p$ given by equation 5 and equation 6.

$$\overline{\overline{C}}_p = \overline{\overline{F}}_p^{\,t} \cdot \overline{\overline{F}}_p = \overline{\overline{I}} + 2\cos\left(\frac{\gamma_p}{2}\right)\sin\left(\frac{\gamma_p}{2}\right)\left(\vec{G}_1 \otimes \vec{G}_2 + \vec{G}_2 \otimes \vec{G}_1\right) \quad (5)$$

$$\overline{\overline{E}}_p = \frac{1}{2}\left(\overline{\overline{C}}_p - \overline{\overline{I}}\right) = \frac{1}{2}\sin\left(\frac{\gamma_p}{2}\right)\left(\vec{G}_1 \otimes \vec{G}_2 + \vec{G}_2 \otimes \vec{G}_1\right) \quad (6)$$

Elastic deformation tensor can be deduced using equation 1 and equation 2. Using the Green-Naghdi additive decomposition for anisotropic field, if follows:

$$\overline{\overline{E}}_e = \frac{1}{2}\overline{\overline{F}}_p^{\,t} \cdot \left(\overline{\overline{F}}_e^{\,t} \cdot \overline{\overline{F}}_e - \overline{\overline{I}}\right) \cdot \overline{\overline{F}}_p \quad (7)$$

From an energetical point of view, the dual of the quantities presented above are used. Since the model is in total lagrangian, the tress tensor used for the elasto-plastic law described just after this section is Piola-Kirchhoff (PKII) written with the letter S in this paper. Its energetical strain dual is the tensor of Green-Lagrange E. The next section is about the elasto-plastic behaviour.

4. Elasto-plastic law

To write an elasto-plastic constitutive model, three quantities are necessary:

- A potential of free energy Ψ, which leads to define the stress tensor.
- A plasticity criterion f_S, directly connected to as dissipative potential D.

IOP Conf. Series: Materials Science and Engineering **406** (2018) 012017 doi:10.1088/1757-899X/406/1/012017

- A plastic flow law λ, defined from the derivation of the dissipative potential D by its stress space.

Moreover, some assumptions are added to write the inequality of Clausius-Duhem under large strain:

- Temperature is constant during the elastic transformation $T : \dot{T} = 0$.
- Temperature is homogeneous during the elastic transformation.
- A single internal variable is used here, the dissipative part of the Green-Lagrange tensor.

To determine the potential of free energy, it is necessary to start with the energy of the transformation Φ_0 which is the sum of an elastic potential Ψ and a dissipative potential D as following:

$$\Phi_0 = \Psi + D \tag{8}$$

This free energy potential leads to define the stress tensor considering the inequality of Clausius-Duhem under the assumption taken before:

$$\bar{\bar{S}} : \dot{\bar{\bar{E}}} - \mu \frac{d\Psi}{dt} \geq 0 \tag{9}$$

In this inequality it appears:

- The rate of total deformation $\dot{\bar{\bar{E}}}$.
- The surface density μ.
- The power per surface unit $\frac{d\Psi}{dt}$.

Using the principle of the local state and according to the additive decomposition of Green-Naghdi, it comes:

$$\left(\bar{\bar{S}} - \mu \frac{d\Psi}{\partial \bar{\bar{E}}_e} \right) : \dot{\bar{\bar{E}}}_e + \bar{\bar{S}} : \dot{\bar{\bar{E}}}_p \geq 0 \tag{10}$$

Given the formalism describes in the second section, the simplification of the inequality before leads to:

$$\bar{\bar{S}} = \mu \frac{d\Psi}{\partial \bar{\bar{E}}_e} \tag{11}$$

4.1. Calculation of the stress tensor: Piola-Kirchhoff II

To determine the stress tensor, it is necessary to define the potential of free energy firstly. As it was said in the introduction, there are two main in-plan deformation mode associated to this macroscopic model. One in elongation and one in shear. Since the assumption of decoupled mode is made, the behavior can be described using an elastic potential using these following invariants.

$$I_{11} = \bar{\bar{E}}_e : \left(\vec{G}_1 \otimes \vec{G}_1 \right) \tag{12}$$

$$I_{22} = \bar{\bar{E}}_e : \left(\vec{G}_2 \otimes \vec{G}_2 \right) \tag{13}$$

$$I_{12} = \bar{\bar{E}}_e : \left(\vec{G}_1 \otimes \vec{G}_2 \right) + \bar{\bar{E}}_e : \left(\vec{G}_2 \otimes \vec{G}_1 \right) \tag{14}$$

Then, it now possible to define the potential of free energy, and therefore the Piola-Kirchhoff tensor.

$$\Psi(I_{11},I_{22},I_{12})=\frac{1}{2}\cdot\left(K_1 I_{11}^2+K_2 I_{22}^2+K_{sh} I_{12}^2\right) \tag{15}$$

With K_1, K_2, K_{sh}, are material parameters concerning the rigidity of each transformation modes. By derivating equation 15 by the elastic part of the Green-Lagrange tensor, it comes equation 16.

$$\overline{\overline{S}}=\mu\begin{bmatrix} K_1\left(\overline{\overline{E}}_e:\left(\vec{G}_1\otimes\vec{G}_1\right)\right)\cdot\left(\vec{G}_1\otimes\vec{G}_1\right)+K_2\left(\overline{\overline{E}}_e:\left(\vec{G}_2\otimes\vec{G}_2\right)\right)\cdot\left(\vec{G}_2\otimes\vec{G}_2\right)+ \\ K_{sh'}\left(\overline{\overline{E}}_e:\left(\vec{G}_1\otimes\vec{G}_2+\vec{G}_2\otimes\vec{G}_1\right)\right)\cdot\left(\vec{G}_1\otimes\vec{G}_2+\vec{G}_2\otimes\vec{G}_1\right) \end{bmatrix} \tag{16}$$

It is important to notice that $K_{sh'} = 0.5 \cdot K_{sh}$. Indeed, because of the shear invariant definition and since the elastic part of Green-Lagrange tensor is symmetric, the component E_{e12} appears twice in each direction.

4.2. Plasticity criterion

As it was said before, the plasticity criterion can be written using the dissipative part D from equation 8. Given the macroscopic approach and being in total lagrangian (reasoning on the initial configuration), the definition of the dissipative potential is following:

$$D=\overline{\overline{S}}:\dot{\overline{\overline{E}}}_p \tag{17}$$

Using equation 6, the equation above can be rewritten:

$$D=\overline{\overline{S}}:\left[\frac{\dot{\gamma}_p}{2}\cos\left(\frac{\gamma_p}{2}\right)\left(\vec{G}_1\otimes\vec{G}_2+\vec{G}_2\otimes\vec{G}_1\right)\right] \tag{18}$$

As the variation of the dissipative part of Green-Lagrange is only composed by the components 12 and 21, then by duality, only the component 12 and 21 of the stress tensor will contribute to the energy. The plasticity criterion without hardening evolution is following:

$$f_s^0\left(\overline{\overline{S}}\right)=\left|\overline{\overline{S}}:\left(\vec{G}_1\otimes\vec{G}_2+\vec{G}_2\otimes\vec{G}_1\right)\right|-S_y \tag{19}$$

Since, f_s is depending on a yield S_y, there are different domains in which the value of the criterion can be located:

- The elastic domain such as $f_s < 0$
- The limit of the elastic domain such as $f_s = 0$
- The plastic domain such as $f_s > 0$

4.3. Hardening function

There is two hardening function for this kind of behaviour. The first one is the kinematic hardening which represents the displacement of the centre of the yield surface. This is commonly written trough a second-order tensor noted X and this tensor must be chosen to be able to define the reduce stress, commonly noted S – X inside the criterion:

$$f_s^X\left(\overline{\overline{S}},\overline{\overline{X}}\right)=\left|\left(\overline{\overline{S}}-\overline{\overline{X}}\right):\left(\vec{G}_1\otimes\vec{G}_2+\vec{G}_2\otimes\vec{G}_1\right)\right|-S_y \tag{20}$$

Such as:

$$\overline{\overline{X}}=\frac{1}{2}q(\gamma_p)\left(\vec{G}_1\otimes\vec{G}_2+\vec{G}_2\otimes\vec{G}_1\right) \tag{21}$$

Where $q(\gamma_p)$ is a function determinate experimentally which represent the evolution of the surface center. The second is the isotropic hardening which increase or decrease the plasticity yield S_y. The representative function is α and the formalism is following:

$$f_s^{\alpha}\left(\overline{\overline{S}}\right)=\left|\overline{\overline{S}}:\left(\vec{G}_1\otimes\vec{G}_2+\vec{G}_2\otimes\vec{G}_1\right)\right|-\left(S_y+\alpha\left(\gamma_p\right)\right) \tag{22}$$

Given that the projection is equal in both direction, it comes:

$$\overline{\overline{X}}:\left(\vec{G}_1\otimes\vec{G}_2+\vec{G}_2\otimes\vec{G}_1\right)=2\frac{1}{2}q(\gamma_p)=q(\gamma_p) \tag{23}$$

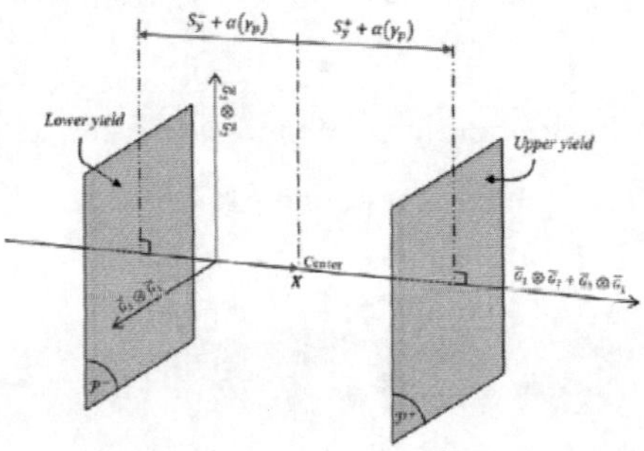

Figure 4. The yield criterion in the 3D stresses space (with γ_p supposed to be greater than zero).

Finally, the total yield criterion is only depending on a single variable which is the angle of plastic deformation γ_p.

$$f_s\left(\gamma_p\right)=\left|\overline{\overline{S}}:\left(\vec{G}_1\otimes\vec{G}_2+\vec{G}_2\otimes\vec{G}_1\right)-q\left(\gamma_p\right)\right|-\left(S_y+\alpha\left(\gamma_p\right)\right) \tag{24}$$

5. Calibration by a Picture Frame experimentation

The elasto-plastic model described above needs to be compared to experimental results to be calibrated. To do that, a Picture Frame experimentation is made imposing this following tensor of total transformation F:

$$\overline{\overline{F}}=\cos\left(\frac{\gamma_t}{2}\right)\left(\vec{G}_1\otimes\vec{G}_1+\vec{G}_2\otimes\vec{G}_2\right)+\sin\left(\frac{\gamma_t}{2}\right)\left(\vec{G}_1\otimes\vec{G}_2+\vec{G}_2\otimes\vec{G}_1\right) \tag{25}$$

Equation 25 coupled with equation 24 leads to:

$$f_s\left(\gamma_p\right)=\left|2\mu K_{sh}\left(\sin\left(\gamma_t\right)-\sin\left(\gamma_p\right)\right)-q\left(\gamma_p\right)\right|-\left(S_y+\alpha\left(\gamma_p\right)\right) \tag{26}$$

The experimental results are shown in figure 5:

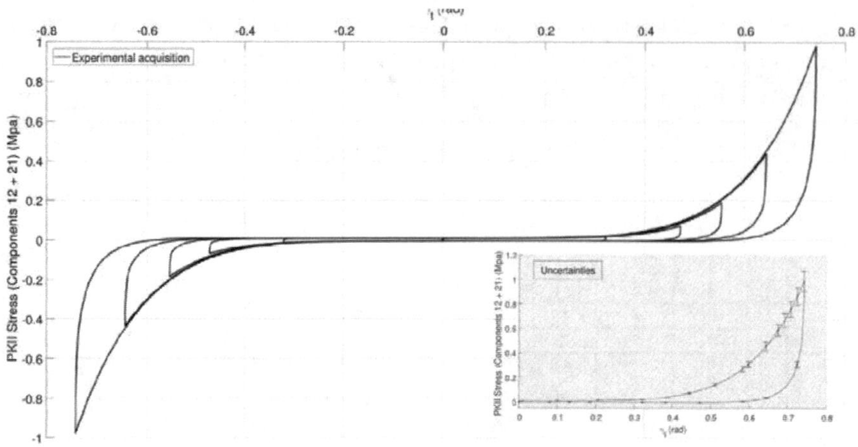

Figure 5. Experimental behaviour during cycling loading and its measurement uncertainties.

Through the behavior presented in figure 5, it is possible to deduce the equation describing the hardening functions (Kinematics and Isotropic).

$$q(\gamma_p) = \sum_{i=1}^{8} Q_i \gamma_p^i \tag{27}$$

$$\alpha(\gamma_p) = \sum_{i=1}^{8} A_i \gamma_p^i \tag{28}$$

Such as

Table 1. Kinematics and Isotropic hardening parameters.

i	Q_i	A_i	i	Q_i	A_i
1	0.01	0.0016	**5**	2.853	0.0013
2	-0.00008	-0.018	**6**	-0.004	3.962
3	0.0084	-0.0002	**7**	3.184	-0.00126
4	0.000982	0.926	**8**	1.810	1.907

Using the parameters presented in table 1 and $K_1 = K_2 = 10000$ and $K_{sh} = 142$ the simulation result is following:

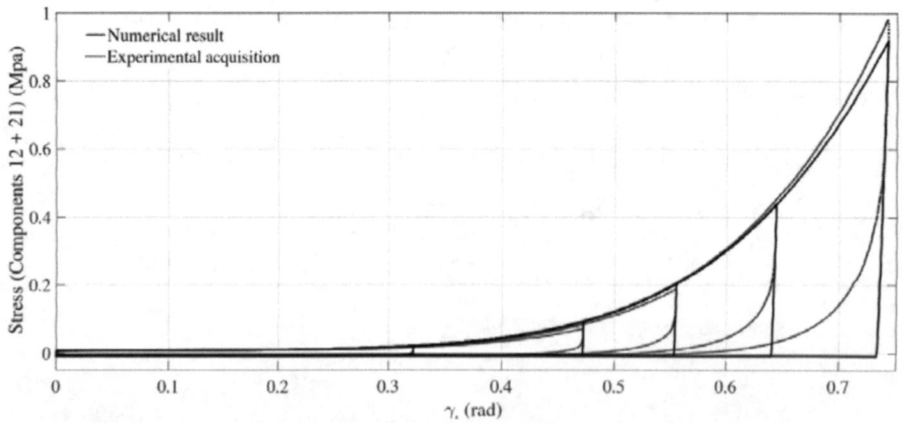

Figure 6. Experimental result and comparison with numerical result.

As it can be shown on the figure 6, the numerical result is satisfying during the loading phase but does not give efficient results during the unloading phase. To improve this model, the strategy adopted here is the use of nested surface as Mroz described it in 1967. Considering the figure 4, where only one surface was used and adding two other surfaces inside which are following specific evolution laws (linear) it would be possible to get the hysteretical behaviour of the material.

6. Improvement of the model using nested surfaces

As it was said before, the constitutive model is good enough for the loading phase but adapted for the unloading phase. In 1967 Mroz proposed a possibility to fit an elastoplastique behaviour with several surfaces. The idea in this section is to improve the general modelisation by adding two surfaces inside the surface presented before. The evolution laws of internal surfaces are supposed to be linear but are different following different case. Every laws and cases are explained below.

6.1. Formalism with three surfaces

The improved model is composed of three surfaces which are presented on figure 7.

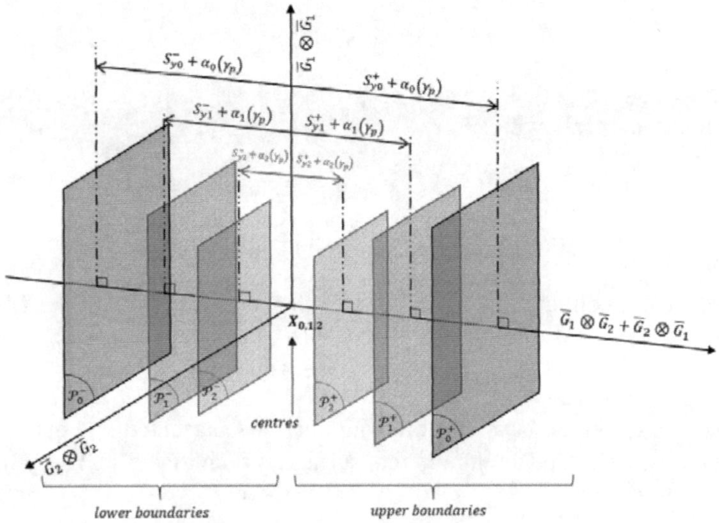

Figure 7. Improved constitutive model with two internal surfaces.

In figure 7, several surfaces are presented:

- The external yield surface (denoted 0) in black (or denoted 0) on the figure 7.
- An internal yield surface (denoted i with $i \in [1, n-1]$ with n the total number of internal surfaces in blue (or denoted 1) on the figure 7.
- The smallest yield surface (denoted n) in green (or denoted 2) on the figure 7.

The first possibility is when the smallest surface is active. This surface is active at the beginning of the simulation or at the transition between the loading and the unloading phase.

6.2. The smallest surface is active

In this case, only one surface has to be updated, and its yield function is described by this following equation:

$$f_s(\gamma_p) = \left|\bar{\bar{S}} : \left(\vec{G}_1 \otimes \vec{G}_2 + \vec{G}_2 \otimes \vec{G}_1\right) - q_n(\gamma_p)\right| - \left(S_{yn} + \alpha_n(\gamma_p)\right) \quad (29)$$

Thus, it is sufficient to update the active surface n by its proper linear evolution laws:

$$\begin{cases} \alpha_n(\gamma_p) = a_n \alpha_0(\gamma_p) \\ q_n(\gamma_p) = h_n(\Delta\gamma_p) \end{cases} \quad (30)$$

Then, the second possibility is when a surface i between the smallest and the external surfaces is active.

6.3. A middle surface is active

In this case, more surfaces have to be updated. Indeed, in figure 7 if the surface 1 is active, then the surfaces 1 and 2 must be updated. Concerning the active surface i, the specific laws are following:

$$f_s(\gamma_p) = \left| \bar{\bar{S}} : (\vec{G}_1 \otimes \vec{G}_2 + \vec{G}_2 \otimes \vec{G}_1) - q_i(\gamma_p) \right| - (S_{yi} + \alpha_i(\gamma_p)) \quad (31)$$

The coefficients are updated by these following equations:

$$\begin{cases} \alpha_i(\gamma_p) = a_i \alpha_0(\gamma_p) \\ q_i(\gamma_p) = h_i(\Delta\gamma_p) + q_i(\gamma_p - \Delta\gamma_p) \end{cases} \quad (32)$$

Concerning the surfaces j smaller than the active surface such as $j \in [i+1, n]$ their coefficients of the evolution laws are updated as following:

$$\begin{cases} \alpha_j(\gamma_p) = a_j \alpha_0(\gamma_p) \quad a_i > a_j > a_n \quad \forall j, a_j \in [0,1] \\ q_i(\gamma_p) = \bar{\bar{S}} : (\vec{G}_1 \otimes \vec{G}_2 + \vec{G}_2 \otimes \vec{G}_1) - \mathrm{sgn}(\Delta\gamma_p)(S_{yj} + \alpha_j(\gamma_p)) \end{cases} \quad (33)$$

As it can be seen, the evolution laws are made to keep the contact between the surface in order to be fully representative. The latest case is when the active surface is the external one.

6.4. The external surface is active

In this case, the external evolution laws are following the laws presented on equations 26, 27 and 28. However, it is still necessary to update the internal surfaces j such as $j = [1, n]$ by the same evolution laws presented in equations 33.

6.5. Final result of the final model

The result is presented on figure 8. As it can be shown, the model is compatible for positive and even negative loading and unloading. All parameters used on the constitutive equations and on the evolution laws are presented in table 2.

Table 2. Evolution laws coefficients.

$\|\gamma_t\|$ (Deg	a_0	a_1	a_2	h_0	h_1	h_2
[00, 19[	1	0.9	0.85	1	0.2	2
[19, 28[	1	0.9	0.75	1	1	2
[28, 32[	1	0.9	0.6	1	1.5	4
[32, 37[	1	0.9	0.5	1	2	8
[37, 40+[	1	0.99	0.45	1	4	10

The material coefficients are presented in table 3.

IOP Conf. Series: Materials Science and Engineering **406** (2018) 012017 doi:10.1088/1757-899X/406/1/012017

Table 3. Material coefficients.

	Value
K_1	10000
K_2	10000
K_{sh}	142
Sy_0	0.0081
Sy_1	0.006
Sy_2	0.0025

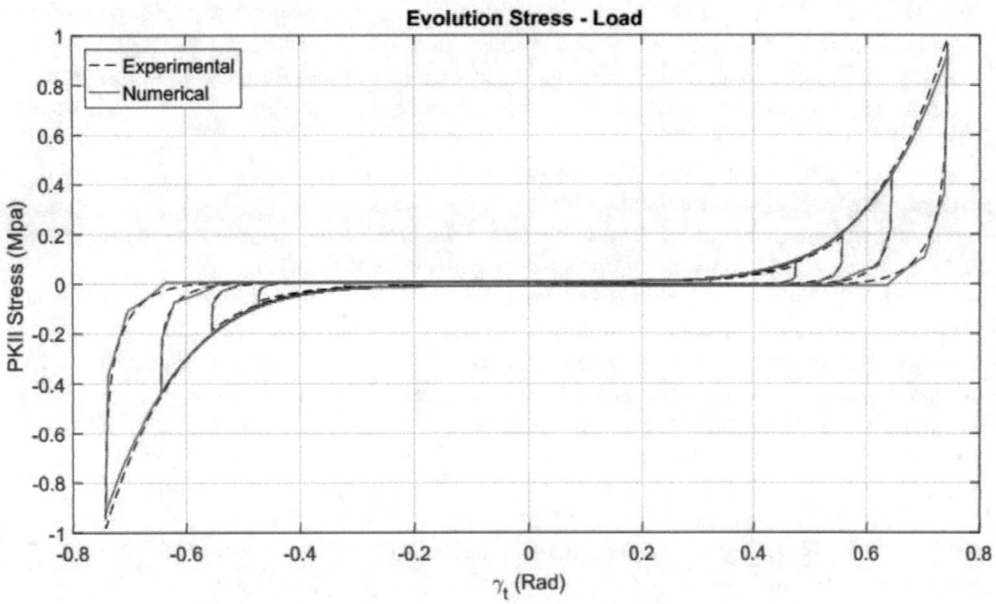

Figure 8. Final result of the improved constitutive model.

The model, using three nested surfaces makes possible to approach the experimental acquisition correctly. The result would be even better by adding other surfaces. However, it is necessary to find a compromise between the number of surfaces and the simulation time. Indeed, in this case, having very small and very close plasticity yields impose small increments at the beginning to avoid instabilities between the surfaces.

7. Conclusion

The principal objective of this paper is to propose an elasto-plastic constitutive model considering the hysteritical effect of the material during an unloading phase. To do that, two major innovations were written. Firstly, the additive decomposition of Green-Naghdi has been adapted for anisotropical material knowing high geometrical no-linearities (under large strains). On the other hand, the use of the Mroz theory about nested surfaces gives the possibility to approach the hysteritical behaviour of the material. As it is possible to see on the figure 8, this model is good enough, but it needs a large quantity of parameters which can be difficult to identify. However, it is very easy to compute this model since it depends by only one variable. Then, a simple Newton-Raphson algorithm permits to get a solution very quickly. Finally, this model is a first step to describe dissipative effect and high non-linearity for woven fabric during its shaping phase. It completes the models already written who are, for a majority of them, hyper-elastic or written in the context of small perturbation.

References

[1] Potter K, *The early history of the resin transfer moulding process for aerospace applications.* Composites Part A: Applied Science and Manufacturing, pp: 619-621, 1999.
[2] Parnas R, *Liquid composite molding*. Garner Publications, 2000.
[3] Komeili M, 'Mechanical behavior of woven fabric composites under meso-level uncertainties: modeling and sensitivity analysis', PhD Thesis, University of British Columbia, 2010.
[4] Komeili M and Milani A S, 'Shear response of woven fabric composites under meso-level uncertainties', *Journal of Composite Materials*, vol. 47, no. 19, pp. 2331–2341, 2013.
[5] Kawabata S, Niwa M and Kawai H, 'Finite-deformation theory of plain-weave fabrics - 3. The shear-deformation theory', *Journal of the textile institute*, vol. 64, no. 2, pp. 62–85, 1973.
[6] Peng X and Cao J, 'A continuum mechanics-based non-orthogonal constitutive model for woven composite fabrics', *Composites Part A: Applied Science and Manufacturing*, 2005.
[7] Xue P, Peng X and Cao J, 'A non-orthogonal constitutive model for characterizing woven composites', *Composites part A: Applied Science and manufacturing*, vol. 34, no. 2, pp. 183–193, 2003.
[8] Spencer A, Mulhern J and Rogers T, *A continuum theory of a plastic elastic fibre reinforced material*. Galway and Nottingham, 1969.
[9] Boisse P and Launay J, *Experimental analysis of the influence of tensions on in plane shear behaviour of woven composite reinforcements*. ScienceDirect, 2010.
[10] Kashani M, Rashidi A, Crawford B and Milani A S, 'Analysis of a two-way tension-shear coupling in woven fabrics under combined loading tests: Global to local transformation of non-orthogonal normalized forces and displacements', *Composites Part A: Applied Science and Manufacturing*, vol. 88, pp. 272–285, 2016.
[11] Komeili M and Milani M, 'On effect of shear-tension coupling in forming simulation of woven fabric reinforcements', *Composites Part B: Engineering*, vol. 99, pp. 17–29, 2016.
[12] Green A E and Naghdi P M, 'A general theory of an elastic-plastic continuum', *Archive for rational mechanics and analysis*, vol. 18, no. 4, pp. 251–281, 1965.
[13] Green A E and Naghdi P M, 'Some remarks on elastic-plastic deformation at finite strain', *International Journal of Engineering Science*, vol. 9, no. 12, pp. 1219–1229, 1971.
[14] Spencer A, 'Plasticity theory for fibre-reinforced composites', *Journal of engineering mathematics*, vol. 26, no. 1, pp. 107–118, 1992.
[15] Ziegler H, *A modification of Prager's hardening rule*. Brown University, pp: 55-65, 1959.
[16] Ziegler H, *The Theory of Plasticity: A Survey of Recent Achievements*. James Clayton Lecture, 1955.
[17] Krasnobrizha A, Rozycki P, Gornet L and Cosson P, 'Hysteresis behaviour modelling of woven composite using a collaborative elastoplastic damage model with fractional derivatives', *Composite Structures*, vol. 158, pp. 101–111, 2016.
[18] MRoz Z, *On the description of anisotropic workhardening*. Journal of Mechanic, Physics. Solids, 1967.
[19] MRoz Z, Dubey R and Shrivastava H, *A Non-Linear Hardening Model and its Application to Cyclic Loading*. Acta Mech, 1976.

IOP Conf. Series: Materials Science and Engineering **406** (2018) 012018 doi:10.1088/1757-899X/406/1/012018

Automated wet compression moulding of load-path optimised TFP preforms with low cycle times

J Fial*, M Harr, P Böhler and P Middendorf

Institute of Aircraft Design, University of Stuttgart, Pfaffenwaldring 31, 70569 Stuttgart, Germany

* fial@ifb.uni-stuttgart.de

Abstract. This paper describes the implementation of a wet compression moulding process using TFP preforms. The fibre paths of these preforms are derived from the actual load paths which are numerically analysed taking into account multiple loading conditions. Preforms are created using a Tajima embroidery machine. This approach allows an innovative way of textile forming in the following process step. This step is based on wet compression moulding using a flexible stamp made of silicone rubber. The wet compression moulding is implemented in the draping test rig of the Institute of Aircraft Design, University of Stuttgart. It is able to produce structures with varying thicknesses which therefore stand out for their high lightweight properties.

1. Introduction

This work presents a production process for structural carbon fibre reinforced parts. Based on tailored fibre placement (TFP) preforms, wet compression moulding is combined with a novel kind of forming process. The aim of this work consists out of three different key elements:

First of all, the goal is to create a light-weight composite structure. Thereby not only lightweight materials are used but the material layup is designed in a way that ideally suits the anisotropic material properties.

Another important point is controlling the forming behaviour of the textile during the process presented. The idea is to avoid complex draping properties and use easy to handle mechanisms instead.

The third important aspect is that the whole process of part manufacturing as well as the design shall reach a high degree of automation. The work presented is done in cooperation with industrial partners and the process is implemented using the example of an armrest designed for common aircrafts. This work is intended to serve as a pre study for a possible future series production.

1.1. Light-weight aspect of the envisaged structure

As indicated, high value is put on the light-weight character of the envisaged structure. Common industrial applications often seek to replace metal structures. Consequently, the target geometries do not always represent an ideal application for fibre reinforced materials whose anisotropy plays an important role. Therefore, the light-weight potential of the composites are rarely completely exploited [1]. Quasi isotropic textile lay-ups are one example for this subject.

The TFP process is used to meet this issue in the present work. Starting with a numerical analysis of the loading conditions of the structure, the load-paths are identified and serve as design pattern for the reinforcement yarns placed during the TFP process (see section 2.1). Variable thickness and stiffness structures are created and reinforcement material is only used where its mechanical properties are required. This has the positive side effect of reducing any unnecessary and expensive waste to a minimum.

1.2. Avoiding complex draping mechanisms and implementation of novel forming mechanisms
During every forming process of a reinforcement textile its architecture is changed in order to adapt to the newly imposed geometry. Complex mechanisms are common and defects such as wrinkles or gaps can occur [2]. The forming process presented is based on adapting the preform properties during the TFP process: The stitching substrate and the reinforcement yarns embroidered upon it are combined in a way that creates the simplest mechanisms possible. Those are folding for the base material and gliding in defined paths for the reinforcement yarns (cf. section 2.2.2).

1.3. Reproducible process and the test rig
The process described is built up to a large extend at the Institute of Aircraft Design (cf. section 3.) and thought to demonstrate the feasibility in an industrial environment. The creation of the preform takes place in a Tajima embroidery machine. The wet compression moulding process is integrated into the draping test rig using low-cost tools. This allows the investigation of the most critical process steps such as textile forming and impregnation. Other important aspects like the automated handling of the preforms and the cutting of the near-net-shape structure which is generated by the wet compression moulding are not implemented and considered trivial.

2. Methodology
This section aims to explain the basic methodologies which underlie the process presented, the two main aspects being:

- The deviation of stitching paths for the creation of the TFP preforms
- The optimal use of material properties regarding forming and mechanical behaviour

2.1. Deviation of stitching paths for the creation of TFP preforms
As indicated in section 1. , carbon fibre reinforcement yarns are placed upon a glass fibre fabric base material in dedicated paths. These paths are stress related and designed in a way that a minimum amount of material has to be used.

The starting point for the creation of these stitching paths is the numerical analysis of the target structure (left side of figure 1). This method (also described in [3]) takes into account multiple load cases and is able to derive the main load paths. These load paths are in a next step identified as the area where the reinforcement yarns have to be placed and interpreted as stitching paths for the reinforcement yarns during the TFP process. Subsequently a tool automatically builds up a mesoscopic model of the preform, including stitching yarns and physical boundary conditions like friction parameters [4] (cf. centre of figure 1). The following reverse mesoscopic finite element forming simulation finds the representation of the flat preform geometry (right side of figure 1). This serves as the pattern for the embroidering of the preforms during the TFP process.

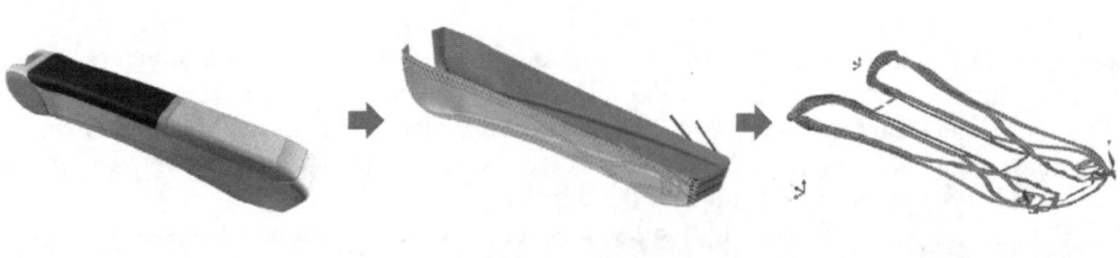

Figure 1. Derivation of the 2-dimensional stitching paths from the target structure (left). Based on a numerical analysis a mesoscopic textile model is created (middle). A reverse draping simulation creates the flat pattern for the creation of the 2-dimensional preform (right).

2.2. Optimal use of material properties regarding forming and mechanical behaviour

During the generation of the TFP based preforms, the application of different embroidering parameters can influence the preforms both in their mechanical as well as in their forming behaviour. In order to obtain optimal properties concerning these two aspects, they are analysed more detailed below.

2.2.1. Mechanical properties. The structure investigated consists out of a glass fibre fabric base material and carbon fibre reinforcement yarns. In the final structure the base material represents the geometry and is not designed to carry any loads. The reinforcement yarns on the other hand are designed to carry the stresses, incorporating different loading conditions. They are stitched according to the load-path distribution and a variable thickness structure is created. This means that there are areas of pure base material and areas with base material and one or more reinforcement yarns. The reinforcement yarns are stitched on the side of the base material which will later be on the inside of the structure. This ensures a constant quality of the visible part surface.

2.2.2. Forming properties. The cutting of the base material in areas of 2-dimensional curvatures is a way of avoiding all possible defects that can occur in a draping process (wrinkles, undulations, etc.). This is due to a local reduction of forming constraints for the textile which is formed. Regular reinforcement yarns stitched upon the base material do not influence this folding mechanisms and are formed in the exact same way.

Additional reinforcement yarns are deployed in a way that allows to directly influence the forming of the textile. These “manipulation yarns” are stitched upon the base material using two different sets of stitching parameters: Stitching with small stitching lengths and widths creates a strong bonding and prevents any relative movement. Stitching with high stitching lengths and widths and using low sewing yarn tension on the contrary means low friction which enables relative yarn movement through the stitching path. The corresponding forming mechanism is selectively manipulated yarn gliding [5].

2.2.3. Implementation. The implemented strategy is illustrated by figures 2 and 3 below. It consists of the following steps:

- Cutting of the base material (glass fibre fabric) in the edges and thereby implementation of folding instead of draping as forming mechanism.
- Stitching of the reinforcement yarns (carbon fibre tows) upon the base material along the previously deviated paths (see 2.1).
- Utilisation of some of the reinforcement yarns to span the cuts in the base material. Application of particular stitching parameters to allow local yarn gliding through the stitching for these manipulation yarns.
- Pulling of the manipulation yarns at the free ends and coeval folding of the base material. This creates the envisaged 3-dimensional geometry and closes the cuts in the final structure.

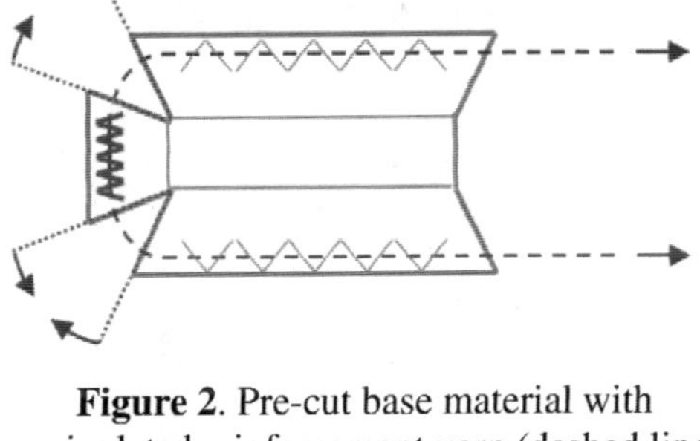

Figure 2. Pre-cut base material with manipulated reinforcement yarn (dashed line). Arrows indicate the yarn manipulation and the closing of the cuts in the base material.

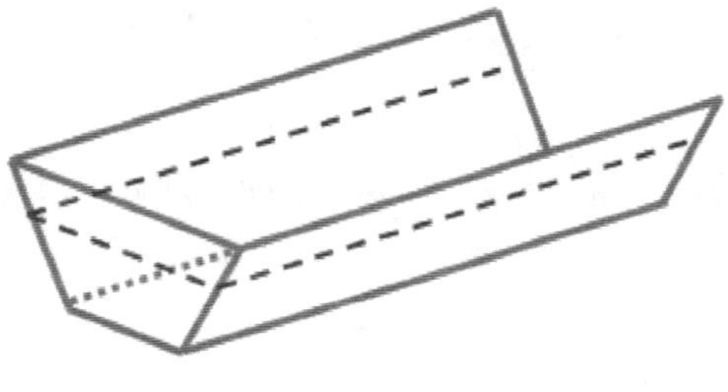

Figure 3. Final structure represented by the folded base material and one exemplary manipulated reinforcement yarn (dashed line).

3. Implementation of the production process

The whole process consists out of the preform production via TFP, handling steps and the wet compression moulding. One key element is an especially developed multi-purpose frame which is used throughout the whole process. This frame is presented in section 3.1 whereas the embroidering and the wet compression moulding are shown in sections 3.2 and 3.3.

3.1. The multi-purpose frame

The multi-purpose frame is essential for the three main steps described above. During each step it realises different tasks and has to fulfil different requirements. This is summarised in the following table:

Table 1. Summary of process steps, tasks and requirements of the frame throughout the process.

Process step	Task	Requirement
TFP preform generation	Supply of the base material	Low thickness due to TFP stitching machine geometry
Handling steps	Supply of the preform	Light-weight design
Wet compression moulding	Perform yarn manipulation	Compatibility to tooling geometry

The requirements shown in table 1 lead to the design of the frame which can be seen in figure 4. The frame consists of three 1 to 2 mm thick sheets of aluminium and is approximately 700 mm wide and long. In the middle part, magnetic clamping is realised which holds the base material in place (dashed area). In the outer area of the frame manipulation mechanisms are integrated.

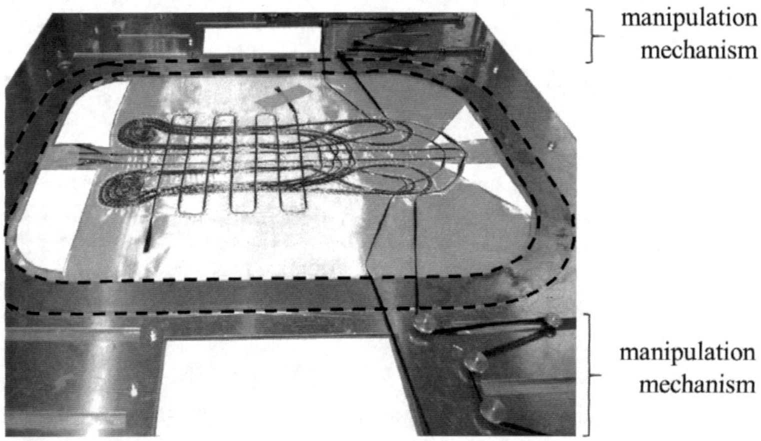

Figure 4. Multi-purpose frame with embroidered preform. Part of the yarns are connected with the manipulation mechanisms.

The manipulation mechanisms consist of slides which are inserted between the top and bottom aluminium sheets of the frame. At one end of a slide, the manipulation yarn is wound around and at the other side springs are attached (see figure 5). This applies a constant force upon the yarn and a passive and therefore failure tolerant yarn manipulation during the wet compression moulding process is guaranteed.

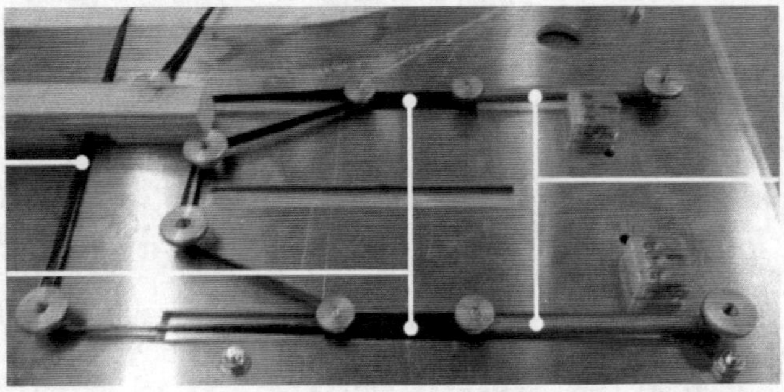

Figure 5. Manipulation mechanism consisting of slides applied with spring forces.

3.2. Embroidering process: Yarn deposition in manipulation mechanism

This section is dedicated to emphasise particular characteristics of the embroidering process of the manipulation yarns. Basic explanations on the TFP process itself are set aside.

As explained above, different stitching widths are used to create areas with high yarn fixation and areas where yarn gliding is enabled. Locally, the sewing yarn is cut, in order to create yarn paths that span the cuts in the base material (cf. figure 2). This is also done when depositing the manipulation yarn into the manipulation mechanism. In this region, the needle is in the topmost position and the fibre placement follows automatically the path indicated by the dashed line in figure 6. No further manual interference is necessary.

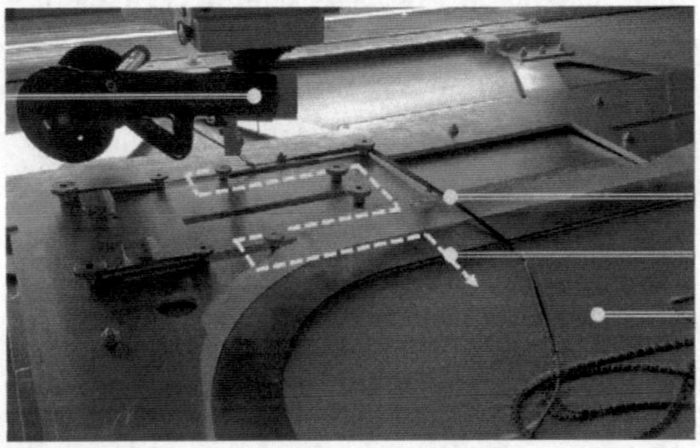

Figure 6. Embroidering process with yarn deposition in manipulation mechanism.

3.3. Wet compression moulding

The wet compression moulding takes place in the test setup shown in figure 7. This figure shows the draping test rig of the Institute of Aircraft Design that has been equipped with the necessary tooling and a material infeed system for the frame (including the preform). Subsequently, details on the process steps as well as on the stamp configuration are given in sections 3.3.1 and 3.3.2.

Figure 7. Test set-up for the wet compression moulding: On the left-hand side, the frame is placed upon the material infeed system which will transport the frame in and outside the tooling (right-hand side).

3.3.1. Process steps during wet compression moulding. Once the frame is put on the infeed system, epoxy resin is manually applied upon the preform. The centre part of the lower tooling (cf. figure 8) is heated and the frames slides into the tooling. Centring pins ensure a reproducible relative positioning of the tool and the frame.

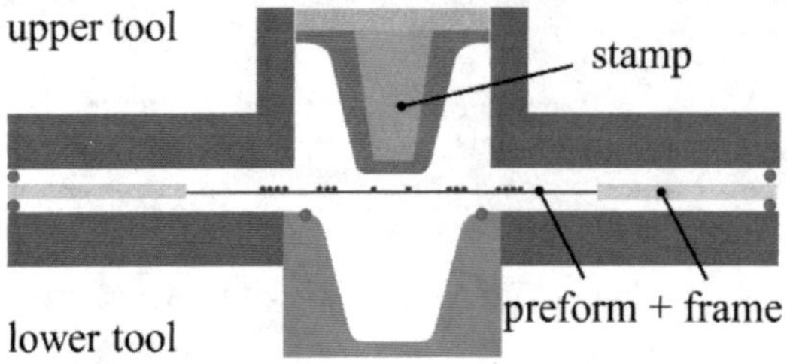

Figure 8. Schematic view of the wet compression moulding tooling, including the frame between the upper and the lower tool.

Proper sealing allows to depress the whole cavity. Inside the tooling a stamp is pushed down and promotes the folding of the preform. As described in section 2.2.3, the manipulation yarns are pulled simultaneously and the final part geometry with closed cuts is created. After curing, the cavity can be opened and the frame together with the near-net-shape structure slides out of the tooling.

3.3.2. Stamp configuration and preform compression. As indicated in figure 8, the stamp is built up of different materials. The centre material is stiff and allows the introduction of the necessary compression forces. The outside is covered with a flexible coating made of silicone rubber.

The flexible coating enables the stamp to adapt to the varying thickness of the preform. It pushes into regions of pure base material and also allows the compaction of areas with one or more layers of reinforcement yarns. Due to a hydrostatic-like pressure which builds up during compression, this effect can also be seen on the side walls of the preform. The goal is to thereby reach a constant fibre volume ratio throughout the whole structure.

4. Process evaluation

As stated in section 1, this process is thought to demonstrate the feasibility of the different techniques involved. Therefore, the means implemented, like e.g. the tooling, are low-cost and the wet compression moulding is not to be directly compared to a process used in industry. For these reasons, in this process evaluation the *qualitative* influence of different varied parameters is presented in this section.

There are a number of parameters which have an impact on the properties of the final structure. These parameters are set during different process steps which is shown in the following table:

Table 2. Overview of process parameters and corresponding process steps.

Process step	Process parameter
TFP preform generation	Stitching parameters like e.g. stitching length Number of reinforcement layers
Wet compression moulding preparation	Resin distribution on preform Temperature of the tooling
Wet compression moulding process	Stamp material Stamp pressure and blank holder forces Sealing and cavity depression

During the realisation of the process, the parameters shown in table 2 are studied. For some of them it is possible to quickly find an optimum which is shown in section 4.1. Others have to be investigated more carefully which is argued in 4.2.

IOP Conf. Series: Materials Science and Engineering **406** (2018) 012018 doi:10.1088/1757-899X/406/1/012018

4.1. Basic process parameters

The first basic parameters investigated are those set by the embroidery machine during the stitching in the region where the gliding of the manipulation yarn is necessary. The general idea is that the friction on the manipulation yarn has to be as low as possible. This is accomplished by a low sewing thread tension as well as high stitching lengths and widths. These parameters directly influence the capability of successfully closing the cuts during the forming of the TFP textile. This is shown in the following 2 figures:

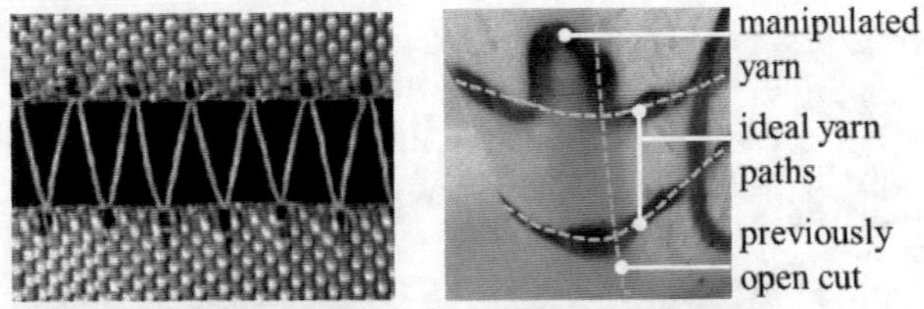

Figure 9. Effect of tight stitching pattern (left) on the gliding behaviour of manipulation yarns (right): Ideal yarn paths cannot be accomplished.

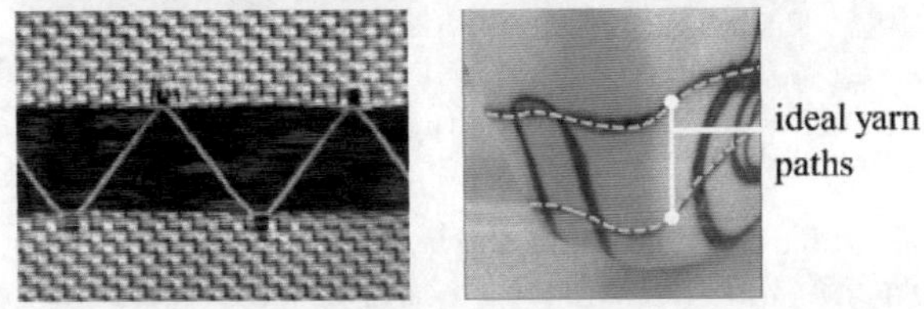

Figure 10. Effect of lose stitching pattern (left) on the gliding behaviour of manipulation yarns (right): Yarn paths correspond to the intended pattern.

Another basic parameter set concerns the implementation of blankholders. They are put in place on every side of the rectangular mould in order to give the textile a certain resistance when being formed into the cavity. The main thing to consider is having a constant blankholder force distribution on every side. This guarantees a symmetric forming result and moderate pressure is sufficient.

4.2. Process parameters with more complex influence on the result

The first parameters which are considered more complex are the type of resin and the corresponding forming temperature. These aspects directly influence the process cycle time and are important for the implementation of the process. Two types of epoxy resins are investigated and provided by the industrial partner Sika. As the target application is to be installed inside an aircraft cabin, also fire, smoke and toxicity related issues are taken into account. Various different parameter sets are studied and finally a forming temperature of 80°C in combination with the resin system Sika Epolam 2500 [6] shows a promising cycle time of 5min.

The application of the resin on the preform is equally important for the process quality. As indicated in section 3.3.1, this is done manually and directly influences the local impregnation level of the final structure. Epolam 2500 shows good results once the resin is distributed homogeneously amongst the whole preform surface. This is due to the characteristic short resin flow path of compression wet moulding which is mainly in thickness direction. The resin application as well as the corresponding local impregnation is shown in figure 11 below. In the centre picture there is also an example of an unsuccessful impregnation.

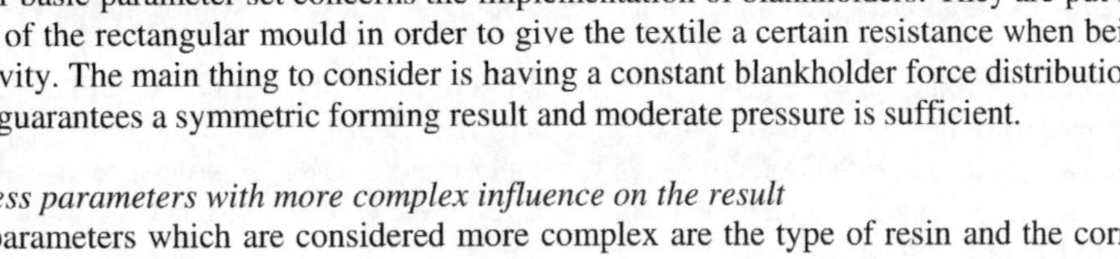

Figure 11. Impact of the resin distribution on local textile impregnation. An uneven distribution (left) leads to poor impregnation results (middle). A more homogeneous resin application leads to better results without local dry spots (right).

Another set of parameters investigated in this part of the work is the number of reinforcement yarn layers and the material of the flexible stamp coating. As the preform possesses areas without any reinforcement yarns, the number of embroidered layers defines the local change in thickness of the structure. In order to avoid areas of pure resin, the flexible stamp has to be soft enough to mould every detail of the preform. Accordingly, the bigger the difference in preform thickness, the softer the stamp has to be. At a maximum number of 3 layers, a silicone rubber shore hardness of 0 respectively 20 is tested for the coating of the stamp. The results are shown in the figure below:

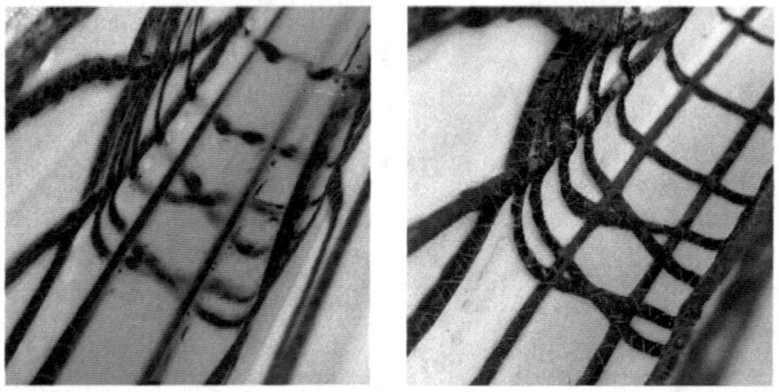

Figure 12. Resulting structure when wet compression moulding with high (left) and low (right) shore hardness stamp material. Details are moulded better with the soft material (right) which avoids pure resin areas (left).

The analysis of entirely embroidered areas of the final part also shows, that a fibre volume ratio of 58% is possible on the side wall of the structure. This indicates that the hydrostatic-like pressure created by the stamp material is successful.

Finally, the last parameters mentioned in table 2 - cavity pressure and sealing strategy - have to be taken into account. The sealing of the tooling needs special attention as during the process the cavity is depressed in order to reduce the void content in the part. The fact that the frame is put between the two moulds demands a double sealing towards the upper as well as the lower mould. Additionally, the stamp is controlled by actuators which are fed through the upper mould. In this region, special sealing is also essential. These precautions lead to a good result concerning the void content on the part surface which can be seen in figure 12 on the right.

However, the maximum part quality achievable with this process is limited as during these investigations a low-cost tooling is used. Additionally, it has to be mentioned that this analysis shall be used just for a first impression. A more differentiated examination, like a photomicrography analysis of the part, is necessary in order to get more reliable results.

5. Results and outlook

The positive result of this work is the successful implementation of a novel kind of production process for TFP based composite structures. It is shown, that direct yarn manipulation can be implemented in an automated forming process and thereby significantly reduce issues regarding draping defects. In connection with a silicone rubber based stamp, good part quality with varying thicknesses can be reached in a wet compression moulding process.

The process time for the wet compression moulding – including textile forming – is 5min. As other process steps are easy to perform in parallel, this is considered as the most important time factor.

For the resulting structure (see figure 13) the initial weight is reduced significantly. With two layers of embroidered reinforcement yarns, its weight is about 66g. Also, local fibre volume ratios in the side wall areas of the structure are verified to reach 58%. It has to be mentioned that the evaluation regarding mechanical performance is still ongoing. First testing results are promising but final conclusions cannot be made at this point.

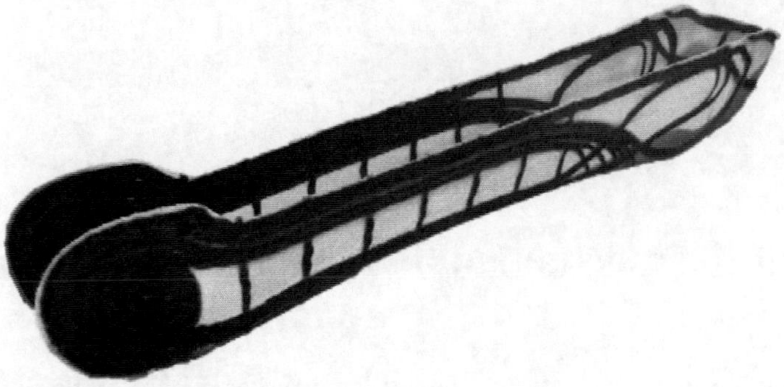

Figure 13. Armrest structure produced by the process presented.

If this feasibility study is to be further examined for industrial applications, several details have to be looked into in detail. Some of those details are e.g.:

- Proper metal tooling design for wet compression moulding. In this context, sealing techniques are of high importance. In [7] possible solutions are presented.
- Investigation of the fatigue behaviour of the silicone rubber material in context with high forming temperatures and contact with epoxy resin.
- Automation of manual tasks such as resin application.
- Inspection of current mechanical solutions (e.g. yarn manipulation mechanism) in regard to process robustness.

Acknowledgements
The research presented has been performed in the framework of the project "3D-TFP", funded by the Ministry of Science, Research and Arts Baden-Württemberg. We also thank our partners from Siemens AG, Recaro Aircraft Seating GmbH & Co. KG and Sika Deutschland GmbH who took part in this research project.

References

[1] R. Lässig and M. Eisenhut, *Serienproduktion von hochfesten Faserverbundbauteilen: Perspektiven für den deutschen Maschinen- und Anlagenbau*, 2012.
[2] F. Härtel, *Entwicklung eines Vorentwurfansatzes für die automatisierte Drapierung von Faserhalbzeugen,* Dissertation, Universität Stuttgart, 2015.
[3] D. Zink, C. Awe, and P. Middendorf, "Automated design approach and potential assessment of composite structures: fast analytical engineering tool for multiple load cases," ECCM 2017 - 17th European Conference on Composite Materials, 2017.
[4] P. Böhler, S. Carosella, C. Goetz et al., "Path Definition for Tailored Fiber Placement Structures Using Numerical Reverse Draping Approach," *Key Engineering Materials*, 651-653, pp. 446–451, 2015.
[5] J. Fial, P. Böhler, and P. Middendorf, "Advanced Forming of Tailored Textiles using locally modified Properties for optimised Lightweight Structures," *ECCM17 - 17th European Conference on Composite Materials*, 2016.
[6] Axson Technologies, "Epolam 2500-2501: Epoxid-Laminierharz für Warungs- und Reparaturarbeiten in Flugzeuginnenräumen," https://deu.sika.com/de/solutions_products/industrie/Tooling-und-Composites/tooling-composites.html.
[7] P. Bockelmann, *Process Control in Compression Molding of Composites,* Dissertation, Technische Universität München, 2017.

IOP Conf. Series: Materials Science and Engineering **406** (2018) 012019 doi:10.1088/1757-899X/406/1/012019

Evaluation of the mechanical behaviour of tufted preforms

Imen Gnaba*, Peng Wang, Xavier Legrand and Damien Soulat

University of Lille, ENSAIT, GEMTEX, F-59056, France

* imen.gnaba@ensait.fr

Abstract. 3-Dimensional fabrics have been developed in order to remedy the damage through-the-thickness of the laminate. An alternative to the multi-layered preforming is to use structures reinforced through-the-thickness in order to manufacture thicker and more complex pieces. Currently, the tufting technology is getting more interest due to its simplest and efficient process where it involves the insertion of binder threads via a single needle through the fabric. This technique of reinforcement through-the-thickness requires only one access to the preform which makes it suitable for three-dimensional shapes.
This study aims to improve the understanding of the mechanical behaviour of tufted preforms where an experimental investigation was performed in order to study the in-plane shear of tufted structures. The effect of the tufting process on the mechanical performance of dry fabrics is also highlighted in the present work.

Keywords: *Reinforcement through-the-thickness, Tufting, In-plane shear behaviour.*

1. Introduction

The large use of composite materials in several fields (aeronautics, space, defence…) is related not only to their excellent mechanical performance but also to their lightweight. The need for affordable composites that have significantly improved the through-the-thickness strength and resistance to delamination intensified the development of a wide variety of textile preforms [1].

The key of such growth consists on the overcome of the weak properties in the thickness of the laminate in terms of delamination and impact resistance [2-5]. The insertion of a binder through-the-thickness of the laminate provides connections between plies which improve the mechanical performance of the final structure [6-8].

The stitching technology is based on the sewing process which enquires a dual-threading system to link several fabric layers together by inserting stiff threads [9-12]. Although the tufting process represents the simplest one-sided stitching technique and it is specifically designed for the dry preform/liquid composite moulding process route. It involves the insertion of binder yarns, via a single needle, through-the-thickness of the laminate [13-15].

Tufted 3D-fibre composites and their mechanical properties have been deeply studied in literature [15-18]. Nevertheless, few studies [14, 19] focus on the impact of tufting on the mechanical behaviour of dry fabrics which is a key point for understanding the deformability of 3D fabrics during manufacturing [14].

This paper outlines the mechanical performance of dry tufted fabrics where an experimental investigation was performed in order to highlight the in-plane shear behaviour of tufted structures.

Therefore, a set of tufted and untufted specimens was conducted for the sake of studying the influence of the process parameters on the mechanical behaviour of dry preforms.

2. Materials and methods

2.1. Tufting process

Currently, tufting is becoming more widespread for through-the-thickness reinforcement. It is based on the sewing process which represents the simplest one-sided technique as illustrated in figure 1.

The insertion of tufts is carried out via a specific needle without generating any tension at the surface of the laminate which avoids thread crimping.

The tufting yarn remains in position due to the natural friction between the fabric and the thread. The threads can be fully inserted where the tufts exceed the depth of the laminate and the loops are formed on the underside of the structure or applied to a partial depth through the preform thickness.

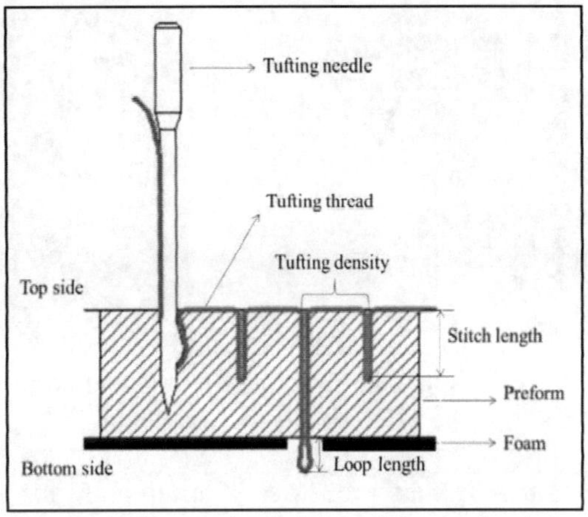

Figure 1. Illustration of the tufting technology.

Figure 2 presents the tufting machine developed in our laboratory in order to handle tufting structures. This bench is supplied by the following equipment:

- ***Tufting device***: made with a tufting needle, with a diameter of 2 mm, which is connected to a pneumatic jack to control the tufting deepness.
- ***Presser foot device***: controlled by another pneumatic jack allowing the application of a specific pressure during the tufting process.
- ***Feeding device***: delivers the tufting thread during manufacturing.

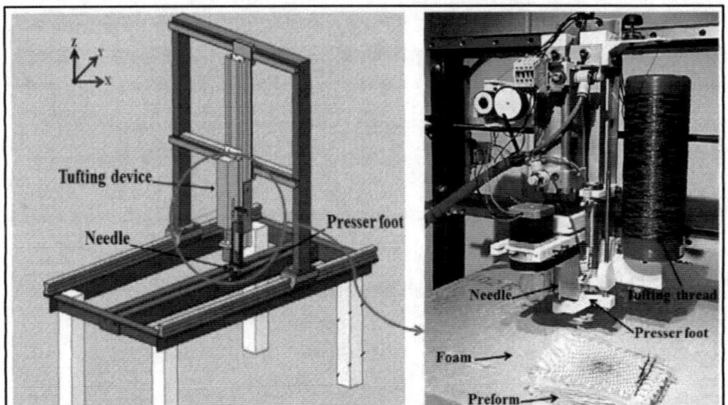

Figure 2. Tufting device (Gemtex-ENSAIT).

IOP Conf. Series: Materials Science and Engineering **406** (2018) 012019 doi:10.1088/1757-899X/406/1/012019

2.2. Preform manufacturing

The reinforcement fabric is an E-glass plain weave with an areal density 160±2 g/m². The tested preforms were laminated with a single layer oriented at ± 45°.

The tufting technique was handled by inserting carbon thread possessing specific features: 2 * 1K 15 S 67 Tex.

The scheme of the tufting pattern is shown in figure 3 (a) where d_x and d_y are respectively the tufting spacing in the x and y axes and an example of dry tufted preform is mentioned in figure 3 (b).

Table 1 illustrates the main properties of tested preforms in the present work.

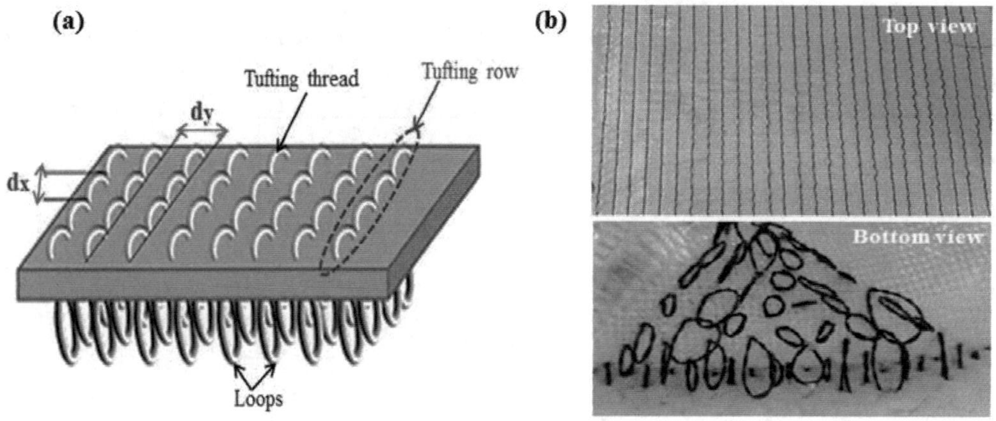

Figure 3. Illustrations of **(a):** Tufting pattern [20], **(b):** Top and bottom views of dry tufted preform.

Table 1. Main properties of tested preforms.

Characteristics	Untufted preform	Tufted preform
Areal density (g/m²)	622 ±2	722 ±2
Tufting spacing (mm)	-	$d_x = d_y = 10$

3. Mechanical characterisation

3.1. Experimental trials

An experimental investigation was performed in order to study the in-plane mechanical properties of tufted fabrics which are essential and complementary to understand the preforming behaviour.

In fact the tensile and bending characterisations of the same tufted fabrics were deeply described in [21, 22]. The results presented in [21, 22] show that the mechanical behaviours are highly influenced by the tufting process where the application of a specific tufting density improves the tensile properties as well as the bending characteristics which depends also on the areal density of the final structure.

The present study outlines particularly the in-plane shear behaviour of untufted and tufted fabrics. For this, a bias extension test was performed to characterise the shear behaviour of untufted and tufted preforms. Table 2 illustrates the main parameters of the bias extension test.

Table 2. The main parameters of the shear test.

Parameters	Value
Test speed (mm/min)	30
Preform size (mm²)	210 x 70
Number of tested specimen	4

In this experiment, a tensile test is conducted on a rectangular sample where the fabric is orientated at ± 45° relatively to the direction of the applied tensile load as mentioned in figure 4 (a). Great care should be taken to avoid any misalignment and fibre crimping.

During the bias extension test, three specific regions can be distinguished on the tested specimen as shown in figure 4 (b) [23-25]:

- ***Zone A***: remains undeformed where the specimen has a clamped end avoiding yarn slippage.
- ***Zone B***: one yarn direction is clamped at its end and the other direction is free.
- ***Zone C***: the yarns in this region are free at their ends. The non-sliding and the stretching of the specimen lead to a pure shear deformation characterised by a shear angle "γ" related to the fabric size "D" and the displacement "d". The theoretical shear angle can be calculated from equation (1), while the experimental shear angle is measured by optics measurements.

$$\gamma = \frac{\pi}{2} - 2 \times \cos^{-1}\left(\frac{D + d}{D \times \sqrt{2}}\right) \tag{1}$$

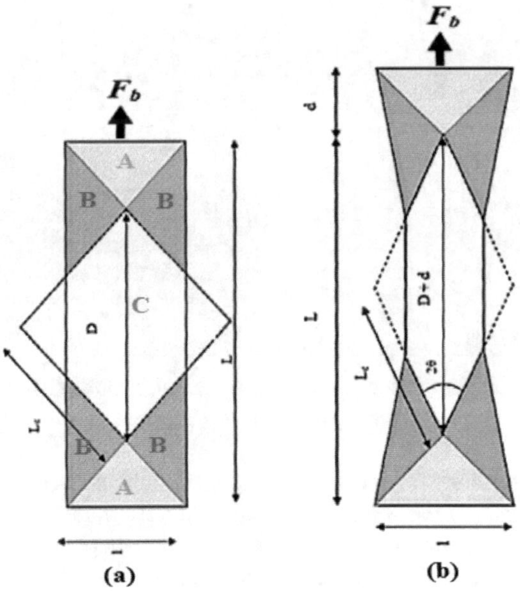

Figure 4. Bias extension test **(a):** Initial shape, **(b):** Deformed shape [23].

3.2. Results and discussions

From the bias extension test, it is possible to extract the main mechanical properties of the in-plane shear behaviour (load, displacement, shear angle…). Figure 5 (a) illustrates the typical load-displacement curves for the untufted and tufted preforms. It should be noted that each curve represents the average value of four tested specimens. The theoretical and measured in-plane shear angles are presented in figure 5 (b).

From the extracted data, it is obvious to mention that the evolution of the load is proportional to the variation of the strain where the rise of the displacement leads to an increase of the force as shown in figure 5 (a). This can be explained by the fact that the test direction is parallel to the direction of the tuft threads which creates tension deformation tufts and increases the shear force.

The experimental data show that the in-pane shear behaviour of untufted and tufted structures is quite similar where the maximum load and the stain at break are uniform for both tested preforms.

Hence, it is possible to say that the in-plane shear behaviour is not influenced by the addition of reinforcement through-the-thickness of the laminate.

Regarding the shear behaviour, it can be noted that the shear angle increases in accordance with the displacement. The shear angle-displacement curves reveal the same behaviour as the load-displacement curves where the curves are similarly coincident.

From a specific value of the shear angle, the measured angle is completely distinct to the theoretical angle (figure 5 (b)) which increases the shear rigidity. This high stiffness reflects the shear locking angle where there is no revolving movement within the fabric and the yarns are totally clamped. In this phase, the yarns are in lateral compression which allows reaching the shear load.

Thus, it is possible to conclude that tufting does not damage the in-plane mechanical performances of the final structure.

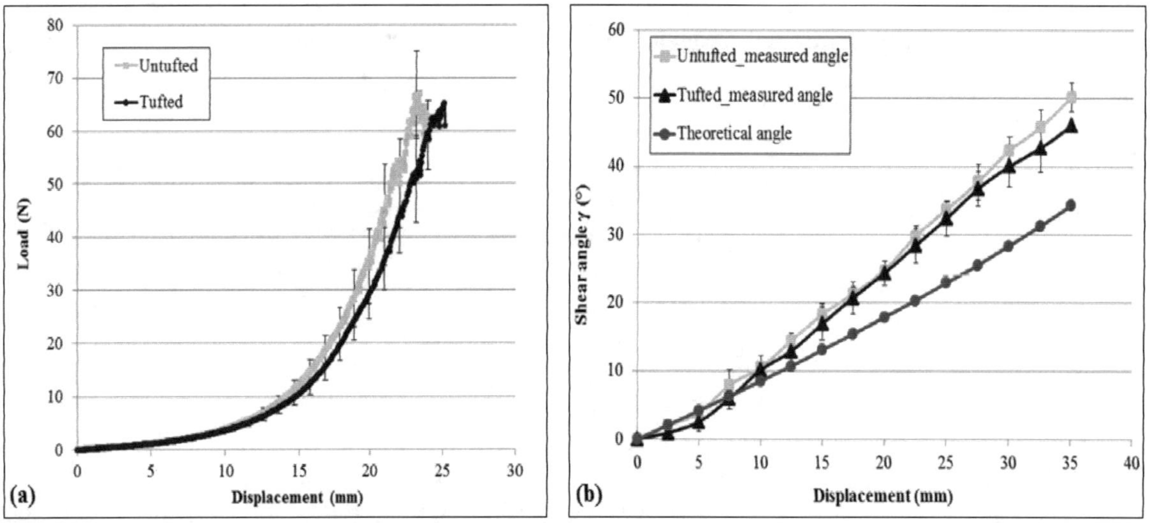

Figure 5. Illustrations of **(a):** In-plane shear load-displacement curves, **(b):** Theoretical and measured shear angles of tested preforms.

A macroscopic characterisation of the shape of the specimen, during the bias extension test, was conducted in order to understand the shear behaviour of untufted and tufted preforms. Figure 6 shows the shape deformation of tested samples during the shear trial where "d" represents the displacement.

The examination of the shape deformation reveals a significant deformation mechanism for both tufted and untufted specimens during the bias extension test. This deformation mechanism highlights the main shear properties where the different zones (A, B and C) are clearly identified with the occurrence of the wrinkling phenomenon at the end of the test

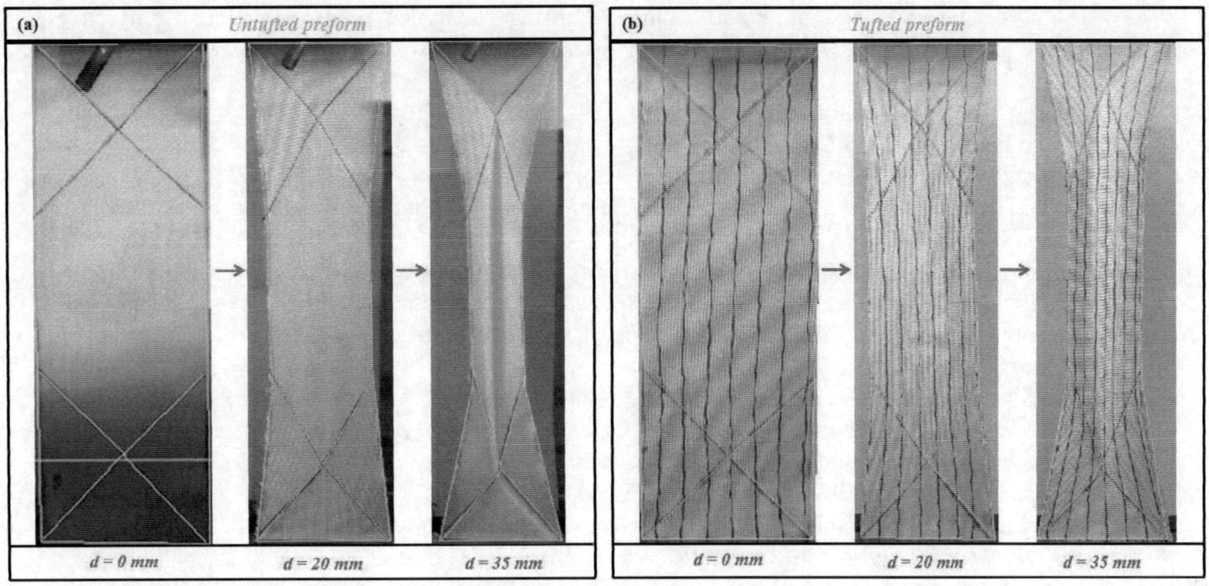

Figure 6. Shape deformation of **(a):** untufted preform, **(b):** Tufted preform.

4. Conclusion

The objective in this study was to better understand the impact of adding carbon tufts on the mechanical in-plane shear properties of glass weave reinforcement. Therefore, a bias extension test was performed for both untufted and tufted fabrics.

It was found that the increase of the displacement leads to an increase of the shear force which involves higher shear rigidity.

Regarding the shear angle which rises proportionally to the displacement up to a specific value where the angle becomes extremely large reflecting the shear locking angle.

From this research, it was proved that tufting does not necessarily weaken the in-plane properties of the final structure.

As future work, the studies about the in-plane shear using the picture frame is recommended in order to compare with results obtained by the bias extension test already carried out.

Subsequently, the preforming behaviour will be conducted by varying the shape of the punch (hemispherical, square box, tetrahedral…) to predict the feasible forming conditions.

Further efforts are ongoing in order to study the numerical in-plane shear behaviour of tufted structures.

Acknowledgments

This project is funded by the French National Research Agency (ANR) for the project “COMP3DRE”.

References

[1] Frank K Ko and Lynn Y W Textile Structural Composites: From 3-D to 1-D Fiber Architecture. Chapter 27 of the Structural Integrity of Carbon Fiber Composites, P.W.R. Beaumont, C. Soutis (eds.), DOI 10.1007/978-3-319-46120-5_27

[2] Potluri P, Hogg P, Arshad M, Jetavat D and Jamshidi P Influence of fiber architecture on impact damage tolerance in 3D woven composites 2012 Appl *Compos Mater.***19**(5):799–812

[3] Dau F, Dano ML and Duplessis-Kergomard Y Experimental investigations and variability considerations on 3D interlock textile composites used in low velocity soft impact loading 2016 *Compos. Struct.***153**:369–79

[4] Hallal A, Younes R, Fardoun F and Nehme S Improved analytical model to predict the effective elastic properties of 2.5D interlock woven fabrics composite 2012 *Compos. Struct.* **94**(10):3009–28

[5] Warren KC, Lopez-Anido RA and Goering J Experimental investigation of threedimensional woven composite 2015 *Compos. Part A: Appl Sci Manuf* **73**:242–59

[6] Mouritz AP Review of z-pinned composite laminates 2007 *Compos. A* **38** 12: 2383–2397

[7] Cartié DDR, Troulis M and Partridge IK Delamination of Z-pinned carbon fibre reinforced laminates 2006 *Compos. Sci. Technol.* **66** 6: 855-861

[8] Partridge IK and Cartié DDR Delamination resistant laminates by Z-Fiber® pinning: Part I manufacture and fracture performance 2005 *Compos. A* **36** 1:55–64

[9] Mouritz, AP and Cox BN Mechanistic approach to the properties of stitched laminates 2000 *Compos. A* **31(1)**: p. 1-27

[10] Tan KT, Watanabe N and Iwahori Y Stitch fiber comparison for improvement of interlaminar fracture toughness in stitched composites 2011 *J. Reinf. Plas. Compos.* **30** 2:99 109

[11] Aymerich F, Priolo P and Sun CT Static and Fatigue Behaviour of Stitched Graphite/Epoxy Composite Laminates 2003 *Compos. Sci. Technol.* **24**: 907–917

[12] Pelstring RM and Madan RC Stitching to improve damage tolerance of composites 1989 *In Proc. 34th Int. SAMPE Syrup.* pp 1519-1528

[13] Gnaba I, Legrand X, Wang P and Soulat D Through-the-thickness reinforcement for composite structures: A review 2018 *J. of Ind. Tex.*, https://doi.org/10.1177/1528083718772299

[14] Liu L S, Zhang T, Wang P, Legrand X and Soulat D Influence of the tufting yarns on formability of tufted 3-Dimensional composite reinforcement 2015 *Compos. Part A* **78**: 403 411

[15] Liu L S, Wang P, Legrand X and Soulat D Investigation of mechanical properties of tufted composites: Influence of tuft length through the thickness 2017 *Compos.Struc.***172**:221-228

[16] Henao A et al. Mechanical performance of through-thickness tufted sandwich structures 2010 *Compos. Struc.* Vol. **92**, pp. 2052-2059

[17] Dell'Anno G, Cartié DDR, Partridge IK and Rezai A Exploring mechanical property balance in tufted carbon fabric/epoxy composites 2007 *Compos. Part A*. **38**:2366–2373

[18] De Verdiere MC, Pickett AK, Skordos AA and Witzel V Evaluation of the mechanical and damage behaviour of tufted non crimped fabric composites using full field measurements 2009 *Compos. Sci. Technol.* **69**:131–138

[19] Saboktakin RA Integrity assessment of preforms and thick textile reinforced composites for aerospace applications. PhD Thesis 2013, École de technologie supérieure Université du Québec

[20] Ranz D, Cuartero J, Miravete A and Miralbes R Experimental research into interlaminar tensile strength of carbon/epoxy laminated curved beams 2017 *Compos.Struc.***164**:189-197, http://dx.doi.org/10.1016/j.compstruct.2016.12.010

[21] Gnaba I, Wang P, Legrand X and Soulat D 2018 *Proc. Int. Symp. on Aircraft Materials, ACMA 2018, (Compiégne)*

[22] Gnaba I, Wang P, Legrand X and Soulat D 2018 *Proc. Eur. Conf. on Composite Materials, ECCM18 (Greece)*

[23] Daghboudj S and Satha H Determination of the in-plane shear rigidity modulus of a carbon non crimp fabric from bias-extension data test 2014 *J.of Compos. Materials.* **48**(22) 2729 2736. DOI: 10.1177/0021998313502063

[24] Hosseini A, Kashani M H, Sassani F, Milani A S and Ko F K Identifying the distinct shear wrinkling behavior of woven composite preforms under bias extension and picture frame tests 2018 *Compos. Struc.***185**:764–773

[25] Launay J, Hivet G, Duong A V and Boisse P Experimental analysis of the influence of tensions on in plane shear behaviour of woven composite reinforcements 2007 *Compos. Sci. and Technol.*, 2007, **68** (2), pp.506. DOI:10.1016/j.compscitech.2007.06.021

Compaction behaviour of tow at meso-scale: experimental investigations of morphological evolution at dry, lubricated and saturated states

J Hemmer[1, 2*], A Sakkalatty Dharmalingam[1], A-S Lectez[1, 3], C Burtin[1], S Comas-Cardona[1], C Binetruy[1], T Savart[2]

[1]Institut de Recherche en Génie Civil et Mécanique (GeM) – Ecole Centrale de Nantes
[2]Plateforme CANOE
[3]Centre National d'Études Spatiales (CNES)

*julie.hemmer@ec-nantes.fr

Abstract. Tows, partially constrained (tows within textile architecture) or free (single tows) exhibit movements and deformations, due to the various mechanical solicitations applied during manufacturing. During weaving or stitching, single dry tows undergo local compaction due to interlacing or stitch, leading to the evolution of the tows cross-section while processing. Once the woven or non-crimp fabrics are produced, they are impregnated by a fluid: when considering the infusion process, the filling step leads to a global compressive unloading at macro-scale that modifies, in a saturated state, tows cross-section and tows spatial distribution within the fabric. These evolutions of the fibrous microstructure at mesoscopic scale impact both the mechanical properties and the permeability. Thus, this work proposes two novel experimental methodologies to quantify the morphological evolution of tows under compaction at dry, lubricated and saturated states.

1. Introduction

During textile fabrics and composites manufacturing, tows, partially constrained (tows within textile architecture) or free (single tows) exhibit movements and deformations, due to mechanical solicitations as compaction. Depending on the considered process, tows can either be in dry, lubricated or saturated states. For instance, during weaving or stitching, single dry tows undergo local compaction due to interlacing or stitch, leading to the evolution of the tows cross-section while processing. Several researchers have proposed different fabrics' mechanical modeling at various scales as presented and mentioned in, for example, the review articles [1, 2, 3]. However, very few authors measured and provided available data of the width and thickness evolution of tow under compression perpendicular to the fibers (as presented in [4]).

Woven or non-crimp fabrics are then impregnated by a fluid: when large composite parts (as wind blade turbine for instance) are considered, the infusion process is competitive on both mechanical and cost aspects. It is a closed mold composite process, where the bottom side of the mold is solid, and the top side is a vacuum bag. Because the vacuum bag is flexible, local compaction state of the porous fabric preform can easily be modified. In the saturated area of the part, far behind the flow front, a thickness rise of the fabrics is experimentally observed, due to the gradient of the resin pressure field [5, 6, 7]. The transient macroscopic thickness rise has been extensively studied but no microscopic evolution of the channels (meso-pores) and tows of a given sample has been proposed. However, the impact of the meso-pores distribution, size and location on the global in-plane permeability of fabric is well-known [8, 9]: the global fiber volume fraction seems not fine enough to represent the flow paths and velocity inside a dual-scale fabric.

This work proposes two novel experimental methodologies to quantify the morphological evolution of constrained and free tows under compaction at dry, lubricated and saturated states. On one side, the cross-sectional dimensions (thickness and width) of single tows are measured under compaction using a chromatic confocal scanner. This method allows defining a normalized bounding box cross-section (NCS) for tows, whose evolution with compaction differs depending on the experimental conditions (dry or lubricated state, free tow). On the other side, a set-up is developed to realize in situ downsized infusion inside an X-ray Computed Tomography (XCT) device. 3D images of the same stack are recorded under dry and saturated states, for a quasi-unidirectional non-crimp fabric (quasi-UD NCF). Due to the fluid pressure rise between these two states, the fibrous stack is submitted to an unloading compressive phenomenon. The image processing of the associated microstructures allows quantifying the influence of the compaction state on tows morphological evolution.

2. Characterization of single tow evolution at dry and lubricated states

The method and results presented in this section have been recently published in [4]. First, a method for measuring and comparing tows (or yarns) geometrical evolution during compaction perpendicular to the fibers is presented. Then, results about the compacity evolution are extracted for a wide range of compaction levels.

2.1. Materials of the study

Two E-glass fiber tows (1200 and 4800 tex), and two high strength (HS) carbon fiber tows (6k and 12k) have been tested. The fibers are continuous filaments. Their specifications and features are given in table 1. The initial cross sections, considered as rectangular bounding boxes (CS = w_0h_0 where w_0 is the initial width, and h_0 the initial thickness) are measured under a 2N compression force applied on the tow. Those 4 references have been chosen because they are constituted of 2 natures of fibers and they have different initial cross-section aspect ratios (AR) ranging from 3.1 to 12.2. The fibrous materials are tested in dry and lubricated states. For the lubricated states, the fiber tows are sprayed (below 5% in mass) with a lubricant constituted of aliphatic hydrocarbons and mineral oils (WD- 40 Company).

Table 1. Specifications of the materials of the study [4].

Material	Fiber nature	Tow features		
		Number of fibers per tow	Tow lineal weight (g/m)	Tow initial aspect ratio (w_0/h_0)
Glass 4800 tex	E-glass	8000	4.8	3.1
Glass 1200 tex	E-glass	2000	1.2	12.2
Carbon 12k	HS carbon	12000	0.90	5.0
Carbon 6k	HS carbon	6000	0.42	6.4

2.2. Methods

A compression device has been specially designed to apply uniaxial (z-direction) transverse compression on a sample tow under a chromatic confocal scanner (CCS Prima, STIL). This device is presented in figure 1. It is composed of a micrometric compact lab jack (LJ 750, Thorlabs), through which vertical displacements are applied, a 2 kN force cell (Type U3, HBM), a bottom polished steel plate on which lie the sample, and a top PMMA plate, bonded to a steel frame (two columns). Compaction forces from 2 to 150 N are consecutively applied on the samples in the z-direction. Each measurement is realized on the deformed sample few minutes after applying the force. The scan is

IOP Conf. Series: Materials Science and Engineering **406** (2018) 012020 doi:10.1088/1757-899X/406/1/012020

performed as shown in figure 2 at five different locations for each tow sample. Scanning the sample in the lateral direction (y-direction) provides its width as well as its thickness (figure 3). Three to five tow samples (5 cm long) are used for each tow reference. As the compaction force is increased on the free-edge (laterally unconstrained) tow, its thickness (z-direction, h) reduces and its width (y-direction, w) increases.

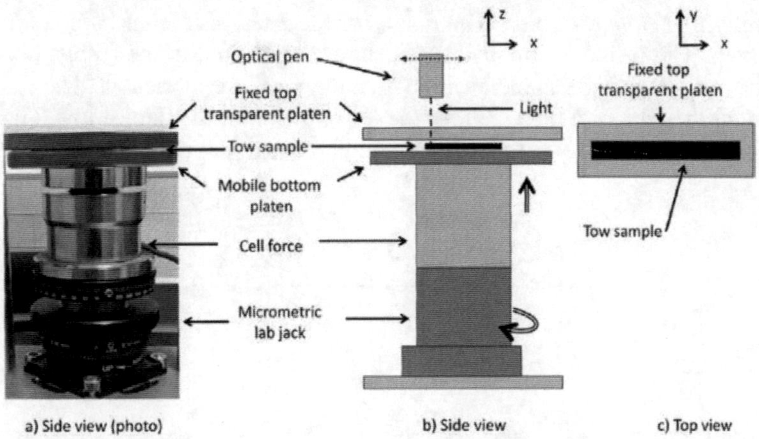

Figure 1. Compression set-up under chromatic confocal scanner [4].

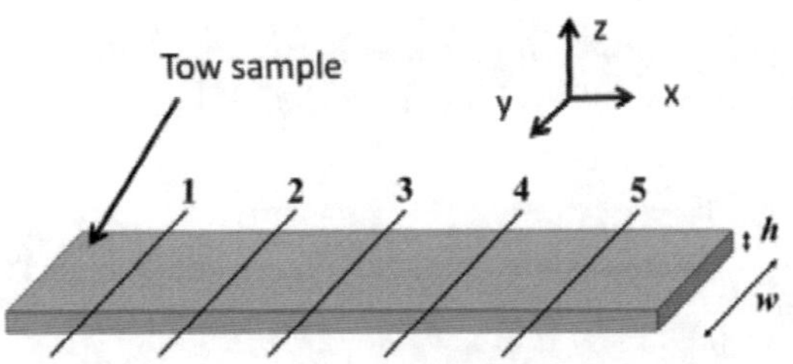

Figure 2. View of the tow sample and the 5 locations of the profiles [4].

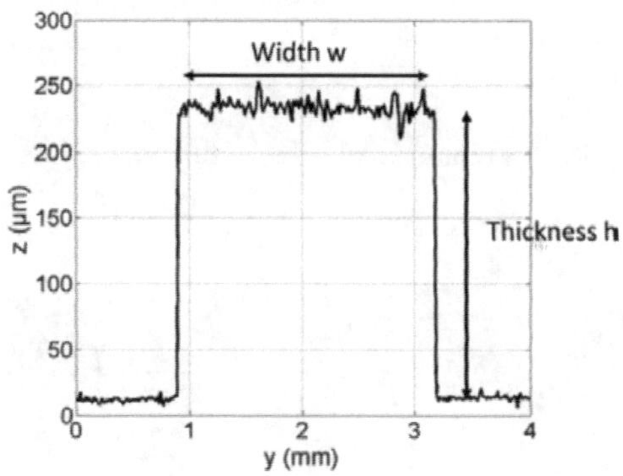

Figure 3. Example of a measured profile at a given location on the tow [4].

2.3. Results

Normalized bounding box cross-section (NCS) wh/w_0h_0 are measured and displayed. The thickness extension $\lambda_h = h/h_0$ is defined for sake of clarity. Fibrous (porous) tows exhibit some degree of compressible behavior as they are compacted. For instance, figure 4 illustrates the expected evolution of the NCS with respect to extension for various transverse (z-direction) compressibility behaviors of the tows. For highly compressible tows under uniaxial compaction, its width may not evolve, while the thickness will reduce and the material will densify (higher fiber volume fraction). For "incompressible" tows, as it is compacted in the thickness direction, the tow will flatten. The width will increase so as to maintain constant density (or constant fiber volume fraction). Some tows will behave in between those both extreme behaviors.

2.3.1. Dry state evolution. The results of the 4 tested references in a dry state are displayed in figure 5. The initially flattest tows (Carbon 6k and Glass 1200tex of aspect ratio 6.4 and 12.2 respectively) are experiencing a strong reduction of volume with limited lateral expansion as they are compacted. The flatter the tow initially, the less lateral expansion occurs as it is compacted. Those tows are densifying with compaction whereas the Glass 4800tex exhibits a flattening behavior.

2.3.2. Dry and saturated state evolution. The results of the 2 tested references in a lubricated state are displayed in figure 6. The results of the measurement are compared to the results obtained in the dry state. For both references the NCS is increasing when the tows are lubricated. The relative increase is more pronounced for the Glass 1200tex than for the Carbon 6k. For the flattest tow (Glass 1200tex), the lubrication allows for an enhanced flattening that could be due to the reduction of fiber/fiber friction.

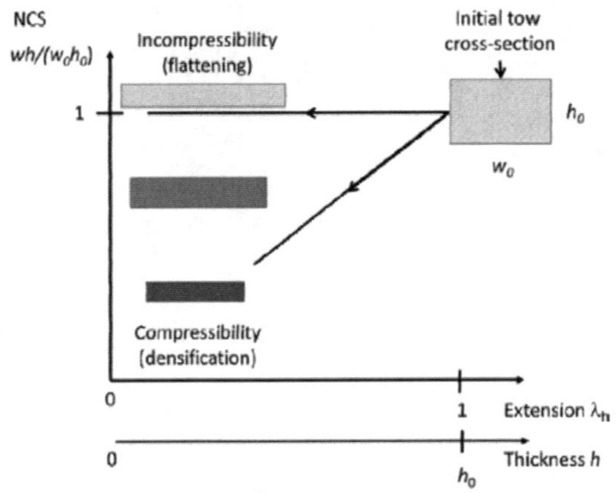

Figure 4. Evolution of the normalized bounding box cross-section (NCS) of the tows with respect to the extension or thickness for various compressibility behaviors [4].

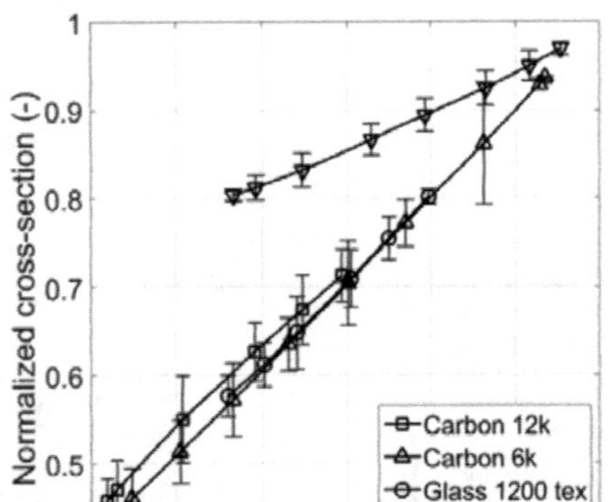

Figure 5. Normalized cross-section (NCS) with respect to extension for all tested dry tows [4].

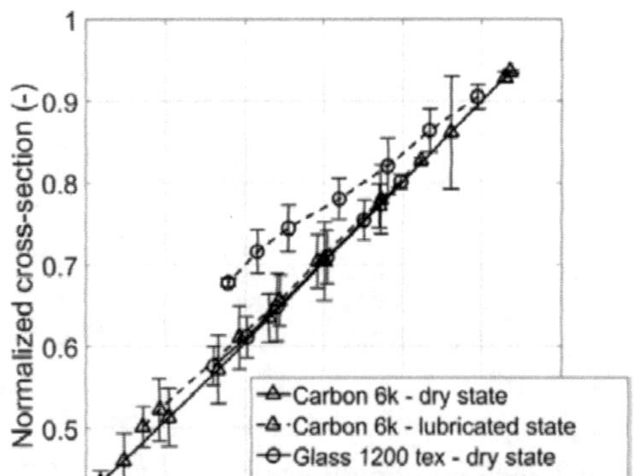

Figure 6. Normalized cross-section (NCS) with respect to z-extension for Carbon 6k and Glass 1200 tex in dry and lubricated states [4].

2.3.3. Conclusion. These methodology and analysis allow to follow free (not constrained in a fabric architecture) tows behavior for a wide range of compaction levels at dry and lubricated states. The four different tested tows showed various behaviors at dry state. The Glass 4800tex exhibits a flattening behavior, its tow initial aspect ratio being low (3.1) whereas the initially flattest tows (Carbon 6k and Glass 1200tex of aspect ratio 6.4 and 12.2 respectively) are experiencing a limited lateral expansion as they are compacted.

3. Characterization of tow evolution inside a fabric between dry and saturated states

The method and results presented in this section have been submitted in [10]. First, a method to realize and post-process an *in situ* downsized infusion inside a XCT device is presented and results about tow deformation between the dry and the saturated states (unloading compressive phenomenon) are detailed.

3.1. Material

The quasi-unidirectional non-crimp fabrics (quasi-UD NCF) selected for the study are manufactured from batches of Advantex® glass roving (Owens Corning). A single ply is made of two layers: one warp layer of UD glass tows (4800 tex, presented in the table 1) and one weft backing glass layer oriented at 80° (68 tex). Both layers are stitched (warp-knitting) with a very thin polyester yarn, symmetrical and spaced of 5 mm that constrains locally the 4800 tex tows in width direction.

To impregnate the fibrous reinforcement, technical glycerol from VWR is used in dilution with 5% of distilled water (viscosity of 0.3 Pa.s at 25°C).

3.2. Methods

XCT experiments are conducted to record a given dual-scale fibrous microstructure evolution under several compaction states: one a dry state, and another one at saturated state. The adapted setup to realize downsized infusion process is described in [10]. The experimental methodology to record representative 3D images is detailed below and the post-processing methods used to analyse them are also presented.

3.2.1. Experimental procedure. A controlled level of vacuum (60 mbar abs) is applied in a cavity (hermetically closed with a vacuum bag) where plies of quasi-UD NCF are previously laid down (figure 7a), fiber direction being along the X-axis. It has been established that one hour of relaxation is necessary to stabilize the dry fabric. After this relaxation time, two 3D images are recorded with a XCT device: the first one near the fluid inlet (in red in figure 7b) and the second one near the vacuum vent (in orange in figure 7b). The microstructure M obtained near the inlet at dry state is illustrated in figure 7b.

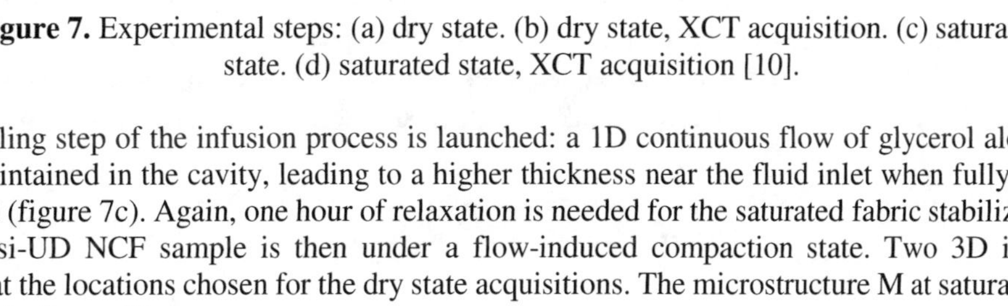

Figure 7. Experimental steps: (a) dry state. (b) dry state, XCT acquisition. (c) saturated state. (d) saturated state, XCT acquisition [10].

The filling step of the infusion process is launched: a 1D continuous flow of glycerol along the X-axis is maintained in the cavity, leading to a higher thickness near the fluid inlet when fully saturation is attained (figure 7c). Again, one hour of relaxation is needed for the saturated fabric stabilization. The same quasi-UD NCF sample is then under a flow-induced compaction state. Two 3D images are recorded at the locations chosen for the dry state acquisitions. The microstructure M at saturated state is illustrated in figure 7d.

IOP Conf. Series: Materials Science and Engineering **406** (2018) 012020 doi:10.1088/1757-899X/406/1/012020

3.2.2. XCT parameters. A XCT device (Micro-XCT400, Xradia company) with the objective Macro70 records and reconstructs the 3D images of the fibrous microstructure. The set of parameters (detailed in table 2) have been chosen to obtain the best ratio between a large enough region-of-interest (ROI) and sufficient resolution to determine a refined tow evolution. To ease the post treatment, the set of parameters is kept equal between the dry and the saturated states.

Table 2. Set of parameters for XCT acquisition [10].

	Value
Input parameters	
Number of images (-)	1947
Exposition time	20
Output parameters	
ROI (mm x mm x mm)	19 x 19 x 19
Resolution (μm/pixel)	10.6
Acquisition duration (h)	12

3.2.3. Image processing. Once the desired dry or saturated compaction state is reached in the XCT, image processing is used to detect the tow deformation. When reconstructed, the 3D image is a matrix Im(i,j,k), where i ∈ [1,m], j ∈ [1,n], and k ∈ [1,N]. A slice k is then an (m x n) image in the YZ-plane, and the 3D image is composed of N slices. Each value of the matrix, between 0 and 255, is a voxel gray-level value. Each k value corresponds to an X-location (a slice) in the fibrous reinforcement: X being the dimensionless position in the X-direction, X=0 is near the fluid inlet and X=1 is near the vacuum vent.

At a defined X-location (a slice YZ-plane), existing edge detection algorithms (for instance Canny or Sobel, as review in [11]) are not efficient to find the warp tows boundaries (figure 8c). This is mainly due to the weft backing glass layer of the fabrics, whose out-of-interest weft tows gray-levels are close to the warp tows gray-levels. Because the weft tows orientation is 80°, the horizontal channel in between two warp tows is easy to identify when moving toward the X-direction of a 3D image: an horizontal channel not detectable at a defined position X is visible at X+δX (figures 8a and 8b). A new image, considered as the concatenation of 100 slices (δx=100mm) around the slice k is then defined to get rid of the weft tows (figure 8d):

$$GLc_k(p) = \min_{k'=k-50\rightarrow k+50} GL_{k'}(p) \quad (1)$$

where GLc_k is the gray-level concatened image around the centered slice k (k-50 to k+50), p is the considered pixel, $GL_{k'}$ is the gray-level slice k'. Applying an edge detection to these concatened images enhance drastically the detection of tows boundaries, as illustrated in figure 8.

At a given X-location, each tow area is computed at dry ($A_{dry}(X,tow)$) and saturated ($A_{saturated}(X,tow)$) states and the swelling of particular tow ($\alpha(X,tow)$) is defined as its area evolution between these two states:

$$\alpha(X,tow) = \frac{A_{saturated}(X,tow) - A_{dry}(X,tow)}{A_{dry}(X,tow)} \quad (2)$$

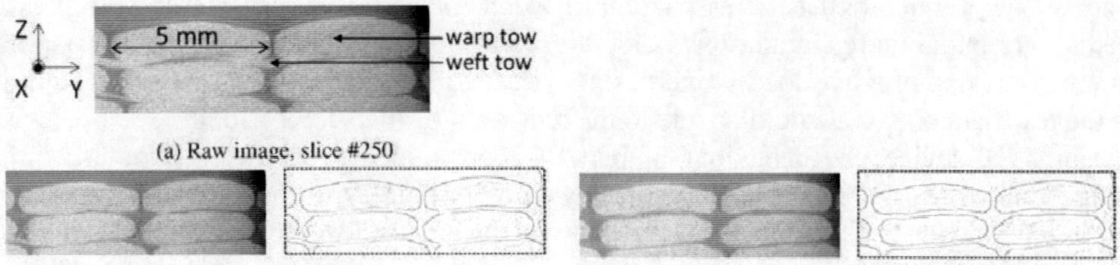

(a) Raw image, slice #250

(b) Raw image, slice #300 (c) After edge detection, slice #250 (d) Concatened image, slice #250 to #350 (e) After edge detection, slice #250 to #350

Figure 8. Illustration of image concatenation process and of its impact on tow detection enhancement [10].

3.3. Results
The results below are presented at a specific X-location (X=0.04), for a same fibrous microstructure recorded at dry and saturated states. Near the fluid inlet (X=0.04), the compressive unloading phenomenon induces the larger deformation along the stack thickness. Further analyses are detailed in [10].

3.3.1. Tow deformation. To enhance the visibility of each tow evolution, the centroids of dry tow and its corresponding saturated tow are superposed (figure 9c). Besides, the initial dry tow is represented in green, the final saturated tow is in pink, and the intersection between both is in white.

The tows area evolution ($\alpha(X, tow)$ *in %*) is not depending on the tow location inside the stack: tows near the vacuum bag and tows near the PVC plate evolve equally (figure 9c). The tow deformation occurs in the thickness direction (Z-axis); lateral tow deformation (Y-axis) is negligible. It could be explained by the stitch yarn (not visible in the recorded XCT images), whose limiting the spreading of tows during the compressive loading.

(a) X=0.04, dry, 60 *mbar* (b) X=0.04, saturated, 60 *mbar* (c) X=0.04, tows

Figure 9. Visualization of tows evolution between the dry and the saturated states. Percentages correspond to the values of α (X, tow) [10].

3.3.2. Conclusion. The proposed methodology and analysis allow following the behaviour of E-glass tows (4800 tex) inside a quasi-UD NCF, between the dry and the saturated states. Due to the fluid pressure rise between these two states, the fibrous stack is submitted to an unloading compressive phenomenon. Inside the stack, tows are undergoing large deformation: the area of tow can increase by 8%. Because the tows are partially constrained by the stitch yarn of the fabric, tow deformation occurs in the thickness direction (Z-axis) and lateral tow deformation (Y-axis) is negligible.

4. Conclusions and outlook
This study proposed two experimental setups to follow the tow behaviors under several boundaries conditions.

Using a chromatic confocal scanner, the behaviour of free tows has been characterized at both dry and lubricated states, for a wide range of compaction, perpendicular to fibers. The results show that for initial flatness tows (high cross-section aspect ratio), a densification is observed as the compressive load is increasing. For initial more circular tows (4800 tex, with low cross-section aspect ratio), a flattening is observed as the compressive load is increasing. Also, for the flattest tow, the lubrication tends to enhance the tow flattening that could be due to the reduction of fiber/fiber friction.

Using an XCT device, the behaviour of partially constrained tow (tow inside a specific fabric architecture, referred as 4800 tex) has been followed from the dry to the saturated states. The tows located near the fluid inlet of the infusion setup undergo the highest level of compressive unloading, due to the fluid pressure rise. The results shows that tows are undergoing large deformation: the tow area can increase by 8%. This tow deformation occurs only in the thickness direction (Z-axis) because the stitch yarn is partially constraining the tow in the width direction (Y-axis).

These results illustrate that the tows behaviours under loading or unloading compression perpendicular to fibers seem to depend on several parameters: the initial aspect ratio of the tow, the lubrication state (dry, lubricated or saturated), and the spatial constrains (single tow or tow in a fabric).

Ongoing work is focused on the homogenization of the results presented in this study. The confocal chromatic setup is being modified to able dry, lubricated and saturated experiments, where wide range of loading and unloading compaction will be applied to free and constrained tows.

Acknowledgments
The authors would like to thank CANOE for their financial support so as the funding provided by the French ANRT by grants under CIFRE funding program. Jean-Michel Lebrun is gratefully acknowledged for its help to design setups.

References

[1] Boisse P, Colmars J, Hamila N, Naouar N and Steer Q 2018 *Compos B Eng* **141** 234-49
[2] Gereke T, Döbrich O, Hübner M, Cherif C 2013 *Compos Appl Sci Manuf* **46** 1–10
[3] Syerko E, Comas-Cardona S, Binetruy C 2012 *Compos Appl Sci Manuf* **43** 1365–88
[4] Dharmalingam A S, Hemmer J, Lectez A-S, Binetruy C, Comas-Cardona S 2018 *Composites Part B* **148** 235-42
[5] Tackitt K D, Walsh S M 2005 *Materials and Manufacturing Processes* **20** 607–627
[6] Yenilmez B, Senan M, Sozer E M 2009 *Composites Science and Technology* **69**
[7] Yang J, Xiao J, Zeng J, Jiang D, Peng C 2012 *Applied Composite Materials* **19** 443–58
[8] Salvatori D, Caglar B, Teixido H, Michaud V 2018 *Composites Part A* **108** 41–52
[9] Caglar B, Orgéas L, Rolland du Roscoat S, Sozer E M, Michaud V 2017 *Composites Part A* **99** 1–14
[10] Hemmer J, Burtin C, Comas-Cardona S, Binetruy C, Savart T, Babeau A 2018 *submitted in Composites Part A*
[11] Pal N R 1993 *Pattern recognition* **26** 1277–94

IOP Conf. Series: Materials Science and Engineering **406** (2018) 012021 doi:10.1088/1757-899X/406/1/012021

Forming low-cost, high quality carbon tows for automotive application

T Sharif [1,2]*, P Potluri[2], R S Choudhry[1], A Dodworth[3]

[1] Department of Mechanical Engineering & Built Environment, College of Engineering and Technology University of Derby, England, UK
[2] School of Materials, The University of Manchester, England, UK
[3] Bright Lite Structures (BLS), UK

*t.sharif@derby.ac.uk

Abstract. Carbon fiber reinforced composites are widely used in many industries due to their high performance. Its application in the aerospace industry has increased significantly, however, in mass produced automobile sector it is still limited. The current production of carbon fiber tow is slow and capital intensive. Thus, carbon manufactures produce higher tow counts to increase production rate to reduce its cost. In order to offset the higher cost of carbon fiber composite, an innovative and unique approach has been developed. The higher tow count carbon spools are split into smaller tow counts. Due to the delicate nature of carbon fiber, it is important to control the filamentation during that process. Different splitting process line strategies have been developed in this research work for understanding the process limitations and challenges involved. The process was made feasible for production by developing a fully automated process line with a laser feedback system. The system splits a 12K spool into two 6K tows. The quality of the 6K split tows has been determined statistically by recording real time data from the laser during the splitting process. It was demonstrated that the proposed process effectively controls filamentation and produces consistent tow quality.

1. Introduction

The superior mechanical performance of carbon composites has significantly increased its application in the aerospace industry. However, its usage remains limited within high volume automotive sector and other clean energy applications due to its costs. Currently, the carbon precursor is almost 50% of overall cost while other 50% is conversion and surface treatment [1] . The carbon manufacturers are developing new methods to produce low-cost precursor and investigating economical production methods. However, it has not reached to the threshold where it can be mass applied.

The production of carbon fiber is slow and capital intensive. Therefore, tows with a higher count (number of filaments) are produced to reduce its cost. The 3k carbon spool is almost three times expensive than a 12k spool. There are several industrial applications where finer tows are desirable due to better drapability and mechanical performance such as discontinuous robotic preforming. The tow count has a profound influence on the performance of discontinuous fiber composites. Rondeau [2] has compared the stiffness and strength of 12k, 6k and 3k discontinuous fiber composites to unidirectional continuous prepreg counterparts. It has been observed that the finer tow count showed better stiffness and strength retention compared to the 12k and 6k composites. As the fiber bundle size decreases, it splits into smaller sub-units during the spraying process and provides the homogenous distribution of filaments on the tool surface. A robotic based preforming system, Programmable Powder Preform

Process (P4), has been developed by the automobile sector for low-cost, low waste and high production [3][4] . Initially, this process was conceived for glass fiber reinforcement, however due to the ongoing development in this process, it is now also suitable for carbon fiber reinforcement.

One of the proposed approach to the reduce the cost of low-count fiber is to possibly split higher tows into finer tows. This research is focused on developing an innovative splitting method to be applied on the P4 preforming system. There is limited research available on the carbon fiber splitting process. Peter A. Kiss et al in his patent [5] has designed a splitting process machine which splits 48K tow into 7-10 strands using various combination of smooth grooved guide and Splitter bars having crown radii. and groove depths. He documented the challenges faced such as fiber crossovers, fiber breakage, and creation of fiber fuzz. His patent design could not achieve complete success. Another research conducted on 'Low-cost carbon fiber production' attempted several techniques to split large tows[1]. However specific details of the process and production success were not made public.

In this research a unique approach has been adopted to split 12k carbon tow into two separate 6k tows. The split 6k tows were wound back onto separate spools by the rewinding machine. Tow splitting is a complex process in which the optimisation of the process parameters plays an important role to facilitate the splitting process. Different designs of the splitting process line have been developed in this research work in order to understand the process limitations. This includes the use of high frequency pneumatic muscles to oscillate the splitting blade, and a static blade splitting system. A fully automated process line with a laser feedback system has been developed to achieve optimisation. Finally, the quality of the split tows has been determined statistically by recording the real- time data during the splitting process.

2. Material and Methods

2.1. Material

The 12K carbon fiber spools (T700sc) was procured from Toray torayca. The spool properties are shown in Table1.

Table 1. 12K carbon fiber Properties

	Fiber Properties
Tensile Strength	4900 Mpa
Tensile Modulus	230 Gpa
Filament Diameter	7 µm
Density	1.80 g/cm3
Sizing type and amount	60E with 0.3%

2.2. Tow Splitting line Arrangement

The proposed tow splitting system consist of three main components; tangential unwinding machine, splitting station and winding machine. The splitting process diagram is shown in figure 1. The 12k carbon spool was held in the tangential unwinding machine. The machine provides appropriate tension to the tow before splitting. The tow then approaches the "splitting station" which is the most critical part due to the delicacy of the process and requires optimisation. Considerable development in this research work has been conducted to improve this technique. The 12K carbon tow split into two equal halves and rewound into separate spools by the rewinding machine. Figure 2 illustrates virgin 12k tow and the split 6k tows which were produced after the splitting process.

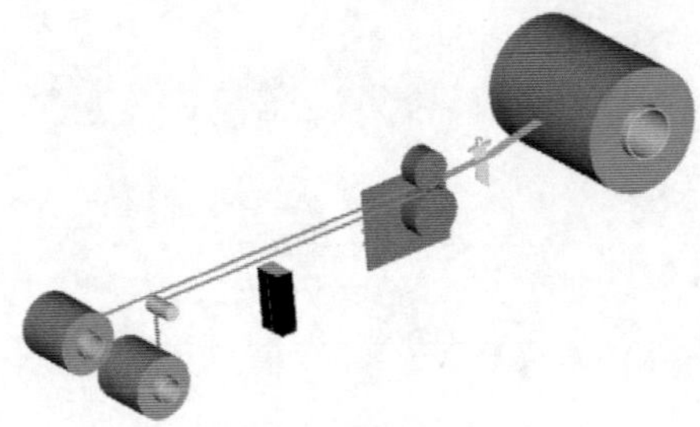

Figure 1. Tow Splitting process line.

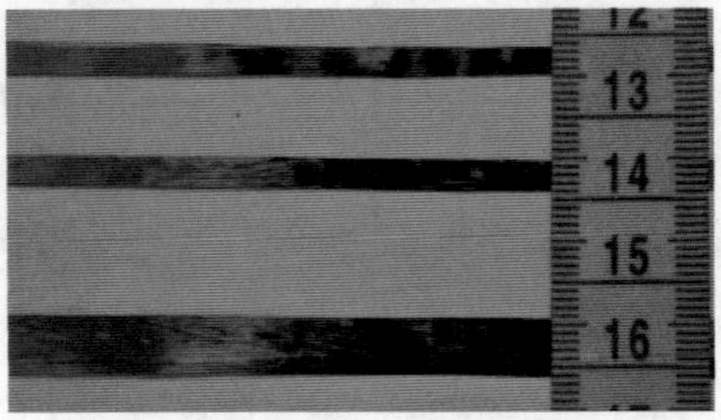

Figure 2. Virgin 12k and split 6k tows.

2.3. Development of Splitting Stations

Preliminary investigation revealed that when the filaments are sliding across the blade a twist is formed in front of the blade. It was noted that the filaments are not perfectly aligned but cross over to each other, instigating twist accumulation as can be seen in the figure 3 demonstration. Similar phenomenon was also observed by the *Peter A. Kiss* et al in his patent[5]. To reduce these process limitations, following splitting prototypes were developed and tested:

- Pneumatic splitting system
- Static blade splitting system
- Automated tow splitting

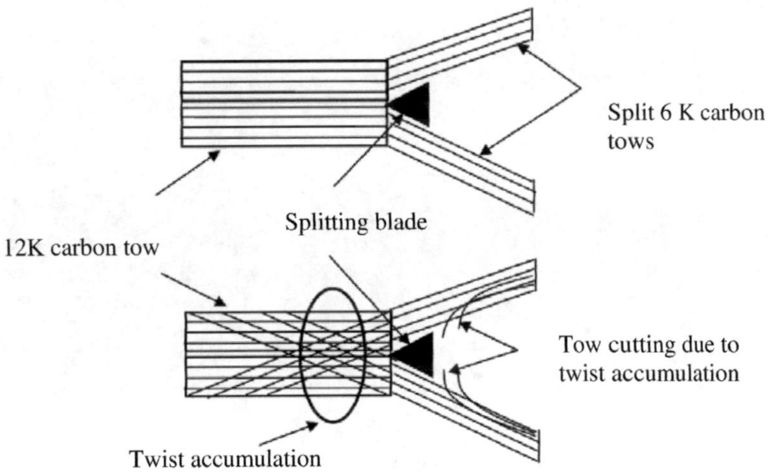

Figure 3. Splitting Phenomena of cutting and sliding.

2.3.1. Pneumatic splitting system. A combination of step and bulge roller were used to direct the tow in the centre of the splitting blade and spread the carbon tow respectively. The Splitting blades were attached to the pneumatic muscle which were controlled by the PLC (figure 4). During the splitting process, twist accumulated at the cutting edge of this blade which caused severe disturbance in the splitting process. As a solution the front blade was lifted to let the twist be cut by the rear lever-based blade. This technique did not produce the desired result. The rear blade made a sharp groove which trapped the filaments inside, causing the tow to wrap around the rubber roller. This also generated varied tension between the split tows; disrupting the splitting process and occasionally cutting off one tow.

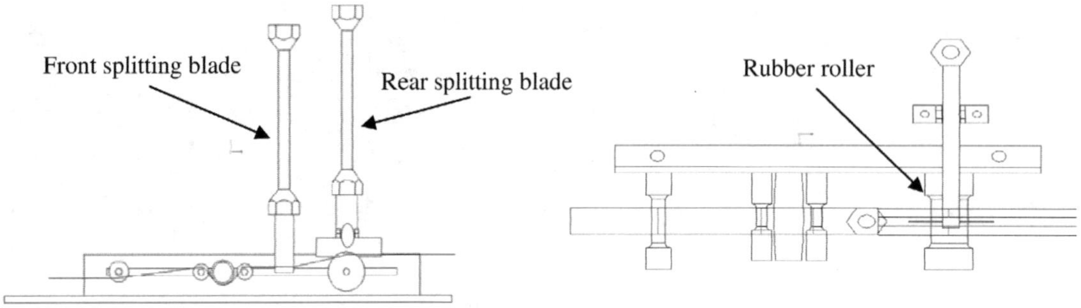

(a) Side view (b) Top view

Figure 4. Pneumatic Spiltting System.

2.3.2. Static blade splitting system. Another approach to splitting was considered by introducing a static blade next to the nip roller, which could be controlled manually in the event of uneven splitting (See figure 5 and 6). This procedure relied on the naked eye visualization and performed manually. The distance between the "spool unwinding machine" and splitting point was increased to reduce the twist intensity. A significant improvement was observed in the splitting process compared to the previous approach. With this approach, the issue of unbalanced tow tension was eliminated while the twist formation was reduced significantly. As the process was controlled manually there was a large variation observed which resulted in one spool weighing significantly higher than the other. This process concluded that this approach is applicable however the manual control of the splitting process is tedious, inconsistent and not appropriate for mass production.

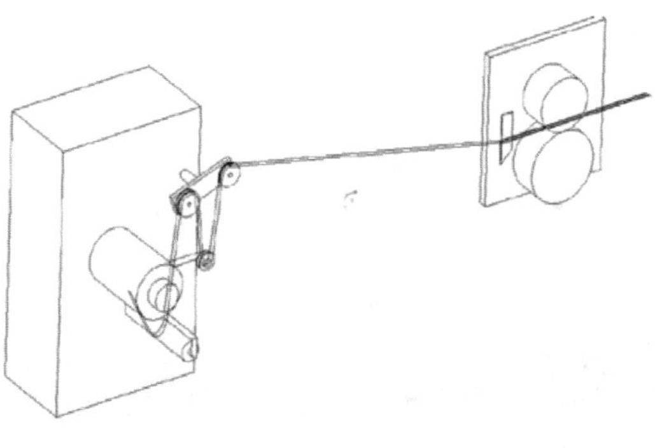

Figure 5. Static blade splitting system.

Figure 6. Nip roller and Static blade arrangement.

2.3.3. Automated Tow Splitting. Improvement on the manual system was made by replacing it with an automated PLC system which could control consistency of tow widths. The incoming carbon tow is split through the blade which is attached to the stepper motor. The split tows then approach the "laser and rollers assembly" station where they pass through the spreading and positioning rollers to ensure their correct position and widths with respect to the laser micrometre (see figure 7). The laser micrometers with a high frequency response are implanted between the set of rollers. It measures the split tow widths and their difference is used as a feedback to the stepper motor. If required, the PLC triggers a signal to the stepper motor to position the splitting blade (CW/CCW) accordingly to equalise the split tow widths. The process flow chart of the automated splitting process line is shown in figure 8.

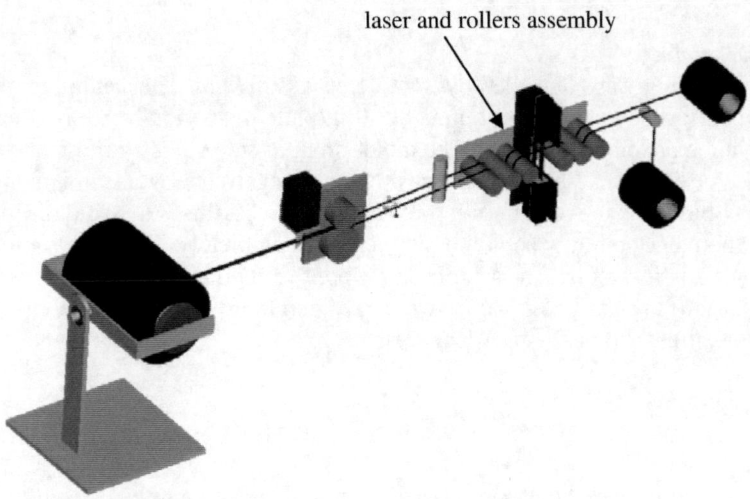

Figure 7. Automated tow splitting with feedback system.

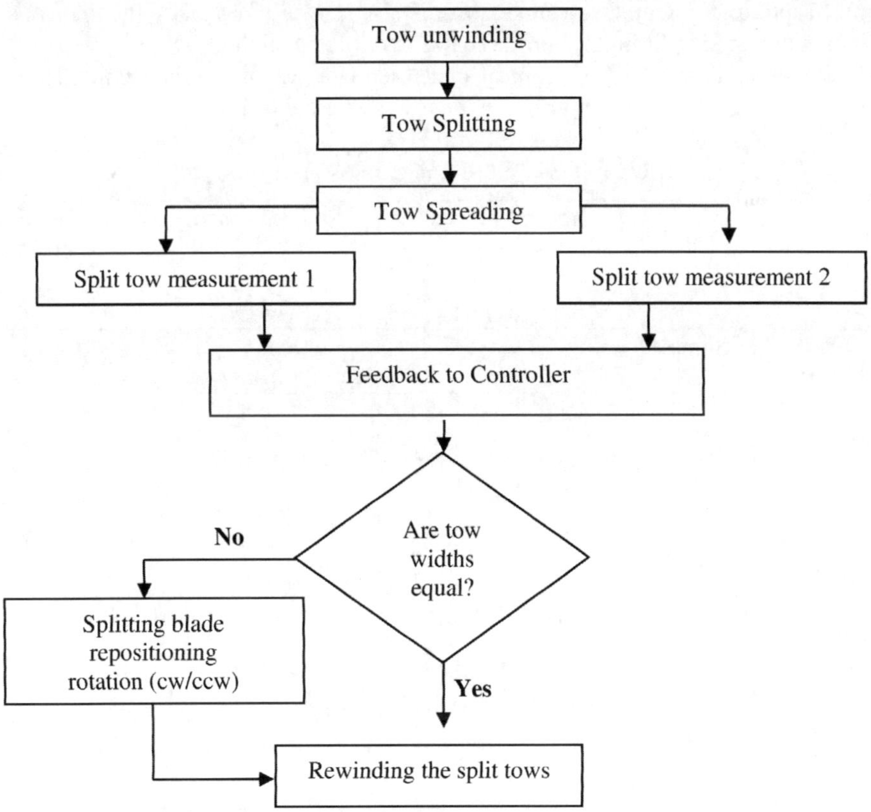

Figure 8. Process flow chart for automated tow splitting system.

2.4. Software Programming
The automated tow splitting was controlled by the CoDeSys software. The stepper motor and the laser micrometres were configured to the input-output (I/O) module of the PLC. At program initiation; the stepper motor was home positioned (centre). The motor angular step values (one degree on each signal) was assigned in the "position set table" for the positive and negative movement. An angular limit was also placed to restrict the blade movement of more than 20 degrees; this is to avoid the movement of the blade to the point where one split tow could possibly cut off completely. The quality of the split tows is determined by using statistical tools. The sample data were recorded at every second and the data values were saved in the form of arrays. These values were exported to an Excel file where they were used to find the standard deviations and other statistical data.

3. Results and discussions:

3.1. Quality of Split Tows
The quality of the split tows were considered for the static blade and automated splitting system only, as the pneumatic splitting system could not function for a reasonable length to be considered. The quality of split material was determined by the percentage weight difference between the split spools. Tow qualities for the automated splitting system were also determined statistically by recording the real time process data. Table 2 represents the weight difference of the split spools. It is observed that the split spools which are produced by the manual splitting have greater weight percentage difference compared to the automated splitting system. On average, this difference was 17.13 % for the manual splitting and 5.06 % for the automated splitting system. The reason of the higher weight difference, in manual splitting, was due to the lack of process control since the blade was moved manually. However, due to the continuous sensor feedback, the splitting on the tows were controlled effectively.

Table 2. Weight difference in 6k split spools.

		Actual split spool weight (grams)		Difference (grams)	Percentage difference (%)
	Sample	S 1	S 2	(S 1 – S 2)	
Manual splitting system	1	2114	1560	554	15.0
	2	3332	2324	1008	`17.8
	3	1641	1125	516	18.6
Automated splitting system	1	2809	2594	215	3.9
	2	3015	2717	298	5.1
	3	2211	2021	190	4.5
	4	1756	1562	194	5.8
	5	2747	2434	313	6.0

3.2. Statistical Analysis of Split Tows
The variation in the split tow is unavoidable. Hence, the overall quality of the split tows was determined statistically. The analogue values were recorded in real time as the split tows pass under the laser beam by PLC. These values were converted into the tow width by using the calibration curve equation. A total of 360 metres of tow length is recorded by the sensors at the speed of 0.6 metres per second. Hence, a total of 600 readings were taken. The proportion of split tows were used to determine the filament counts

as determined by equation 2 and 3. This is then multiplied by the total number of filaments as mentioned in equation 4 and 5.

Total Number of filament = 12000
Width of split tow 1 = $X1_i$
Width of split tow 2 = $X2_i$
Where i represent the number of recorded readings.

$$\text{Total width of split tow 1 and tow 2} = Y_i = (X1_i + X2_i) \quad (1)$$
$$\text{Proportion of split tow 1} = T1_i = X1_i / Y_i \quad (2)$$
$$\text{Proportion of split tow 2} = T2_i = X2_i / Y_i \quad (3)$$
$$\text{No of filaments in split tow 1} = T1_i \times (12000) \quad (4)$$
$$\text{No of filaments in split tow 2} = T2_i \times (12000) \quad (5)$$

Table 3 represent the statistical data of split tow in terms of filament counts. There is no significant difference in the mean value of the filament counts whilst the standard deviation is 0.68k. The filament counts distribution of both split tows can be assumed as normally distributed. The range of the split tows is approximately between 5k and 7K. The observed mean difference in filament counts is 227 and the standard deviation is 1.3k.

Table 3. Weight difference in 6k split spools.

	Split tow 1 Filaments	Split tow 2 Filaments	Difference
Mean	6113	5886	227
Standard Deviation	684	684	1368

4. Conclusion

The developed tow splitting system proved to be a very promising and successful technique to split carbon tows. This system provides an obvious cost reduction. The 12k carbon spool was split successfully by the automated system into 6k carbon spools. Almost no human interaction was required after setup. The splitting process was monitored by the laser feedback system. The sensitivity of the splitting blade was set to 0.02mm of the tow difference. Without the laser feedback system, on average, the difference of the split spools weight was 17%. However, laser feedback had drastically improved the performance of the splitting system to 4%. Currently, the split tow quality may not be up to the standard of the aerospace industry, but it has immense potential in automotive sector especially for short fibre preforming application.

References

[1] Liby A, Naskar A, Ott R and Jackson C 2016 Carbon Fiber Technology Facility, Advanced Manufacturing Office Program Review Meeting

[2] Rondeau R, Reeve S and Bond G 1999 The Effect of Tows and Filament Groups on the Properties of Discontinuous Fiber Composites *44th Int. SAMPE Symp. Exhib.*

[3] Johnson C F and Rudd C D 2000 Manufacturing Process Selection for Composite Components *Compr. Compos. Mater.* 1049–72

[4] Manson J-A E, Wakeman M D and Bernet N 2000 Composite Processing and Manufacturing—An Overview *Compr. Compos. Mater.* 577–607

[5] Peter A. Kiss, Jeffrey M.Deaton, Michael S. Parsons D C 2002 US 6,385,828 B1 United States Patent

IOP Conf. Series: Materials Science and Engineering **406** (2018) 012022 doi:10.1088/1757-899X/406/1/012022

Perspectives of web based composites from RCF material

G Stegschuster[1*], S Schlichter[1‡]

[1] Institut für Textiltechnik Augsburg gGmbH, Am Technologiezentrum 5, 86159 Augsburg, Germany

* georg.stegschuster@ita-augsburg.de, [‡]stefan.schlichter@ita-augsburg.de

Abstract. "Web Based Composites" consisting of nonwoven fabric made from recycled carbon fibres (rCF) have excellent economic and ecologic perspectives to find a place in industrial product design for cost-effective lightweight solutions in composites. However, the processing conditions of rCF are challenging and the industry is not yet aware of the full potential of the mechanical properties as well as the cost structure. The goal of the project "CaRinA" is to provide reliable material cards for the industry with crucial information on the mechanical properties to encourage the application of rCF nonwovens. Therefore a complete investigation of the parameters and boundary conditions of the production process to the final part is performed. Within the project CaRinA the ITA Augsburg is focused on understanding, explaining and enhancing the nonwoven process using an industrial sized nonwoven line from the company Dilo Systems GmbH, Eberbach, Germany. Nonwoven lines have i.e. over 40 parameters that influence the product quality. The most influential factors are determined using statistical experimental design to understand the relations and effects on the nonwoven and the composite properties. The approach of the ITA is explained in detail.

1. Introduction

Carbon fibres are expensive and energy consuming in manufacturing. With increasing quantities (> 9,000 components/year) even in small batches, material costs of the production of CFRP components have a disproportionately high impact on the unit costs (over 57% in the RTM process) compared to forming, semi-finished and finishing costs [1]. Against this background, the reduction of material costs is an essential lever for reducing component prices. Fibre composite approaches are in strong competition with other lightweight materials (ultra-high-strength steels, aluminium, magnesium, etc.). Nevertheless, they often only have a single life cycle and - unlike the other material classes - are not or only to a very small extent recycled. Even low-priced raw materials such as paper or glass have a significantly higher recycling rate. Other established construction materials, such as aluminium or steel, are several generations ahead of cost intensive carbon fibres in this regard.

1.1. State of the art for carbon fibre recycling

Considering the strong growth of the material class of carbon fibres, it is crucial to provide a suitable recycling strategy through a closed product life cycle in order to be able to compete with other materials and to exploit the full potential for lightweight construction economically and ecologically. The individual sub-steps for the realization of a complete recycling route are currently the subject of many research and development activities. Key advances have already been successfully demonstrated. The first important sub-process is the fibre-matrix separation for the treatment of matrix-containing waste streams. Currently various chemical methods (hydrogenation, solvolysis) and thermal processes (pyrolysis, partial oxidation, fluidized bed processes) are investigated. A variant of the partial oxidation is now implemented as the only method on an industrial scale by the project partner ELG Carbon Fibre Ltd., Birmingham, United Kingdom. In addition to matrix-containing waste, large amounts of dry fibre waste are produced in today's CF processes at cut rates of up to 40%. Further processing of the recycled carbon fibres is particularly suitable for the production of nonwovens due to the short nature of the fibres after the recycling process [2-7]. The established processes in this area are the carding process, the airlay and wet-laid process, which each differ considerably in processing parameters and nonwoven properties.

The wet-laid technique is based strongly on the classic paper production. In principle, the mostly shorter fibres are dispersed in water and continuously rinsed on a sieve. Due to the similarity with the processing of cellulose fibres, there are only minor adaptations of the process technology for the processing of carbon fibres.

In the field of dry web formation, the airlaid and carding processes are characterized by high productivity in the production of textile surfaces. For the price-oriented carbon fibre market this offers the advantage of low process costs. In addition, the nonwoven properties can be precisely developed for the respective applications.

1.2. Nonwoven production and challenges with the carding process

The main advantages of the nonwoven process besides the lowest level of production costs for the manufacturing of a textile fabric is the high level of universality of the achievable properties. The drapeability, the variation of thickness, the chance to vary isotropy and structure of the material and the chance to blend with all fibres with limited fibre length (natural and synthetic origin) are designable properties for a nonwoven semi-finished product.

In order to be able to process recycled carbon fibres successfully on a nonwoven line adjustments in terms of the opening, the air flow and the storage have to be made to reduce fibre damage and to increase the nonwoven quality. Due to the very compact and economical process, the carding process combines several transformation processes to produce a homogeneous, textile surface from fibres without any need of yarn formation. The opening and parallelizing process of the fibres is conducted between steel toothed rollers. The steel teeth are called (saw tooth) wires. The process enables the production of highly oriented nonwovens.

Nonwoven lines can be divided in three main process steps: fibre preparation, web formation and web bonding. The most important and universal line concept is based on the web formation principle of a roller card (Figure 1), which converts a mat of fibre tufts into a uniform web. A roller card manufactured by the company Dilo System GmbH, Eberbach, Germany, is installed at the facilities of the ITA in Augsburg and used for research purposes in the project CaRinA.

Particular attention is paid to the improvement of the doffing roller system, which is realized with a so-called hacker. Hackers are currently used on most of the carbon fibre roller cards for web doffing. The hacker allows a stable process with relatively short cleaning times of the doffer roller, but causes little hooks of the fibres during the web doffing leading to a worse orientation. In addition, the production speed is limited by the mechanical limits of the oscillating hacker. A roller based web doffing, commonly used in other nonwoven processes allows for higher production speeds and the possibility to influence the orientation and structure of the nonwoven fabric via the parameters warp and compression.

Industrial production lines based on nonwoven processes are available and first applications are known from research institutes but still only a few industrial lines are employed. Several research projects are addressing the topic of processing conditions [8, 9].

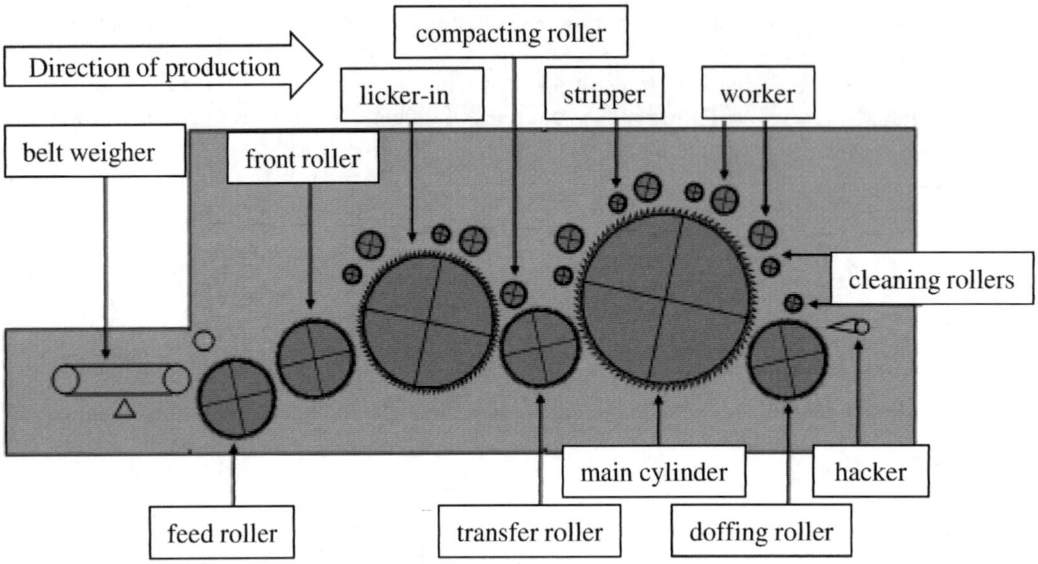

Figure 1. Schematics of a roller card.

2. Joint Project CaRinA and goals thereof

The aim of the project CaRinA is to meet the strong demand of the industry for a clear and easily accessible cost-specific performance spectrum for rCF nonwovens. Including all industrially relevant parameters, the nonwovens will be characterized by a comprehensive property profile for unambiguous traceability. Thus providing the required comparability both within the family of rCF nonwovens and in competition with other material classes. Furthermore the results allow for the first time to create a data model covering the entire process chain for the production of rCF products.

In order to achieve the set goals, the partners will produce, process and characterize several different nonwovens. The collected data is correlated with the process parameters and stored in a database to provide a complete data set of each individual material. In addition to processing parameters the data set will include thermal and mechanical characteristics of the respective materials as well as the history of the recycled and mixed fibres.

3. Work priorities of ITA

The focus of the ITA in the project is to gain knowledge about the processing parameters of the roller card and their respective interactions concerning the influence on the nonwoven and composite properties. For this purpose, the following investigations or developments are carried out by the ITA:

- Investigation of the processability of dry and pyrolysed secondary carbon fibres in the roller card process
- Improvement of fibre preparation, especially with the aim of a uniform mixing and gentle opening of the fibres
- Investigations of the influence and the improvement of the web bonding process "needling"
- Development of a roller based web doffing system for carbon fibres and the investigation of the influence on important nonwoven properties
- Investigation of different methods of application of thermoplastic matrices to rCF nonwovens
    - Investigation of powder spreading technology
    - Development, design and investigation of the incorporation of TP nonwovens in rCF nonwovens - Stacking
- Examination of various fibre blends and processing behaviour

- Investigation of the effects of modified PET fibres on the processability and nonwoven as well as composite properties

3.1. Approach for the investigation of opening, web formation and web bonding

The aim is to find the most influential factors beneath the more than 40 relevant parameters a roller card has to offer. The focus of the investigation is on the opening and mixing, the web formation and web bonding. Nine key factors are determined by an assessment of a panel of experts and then efficiently screened for their impact on four quality characteristics using a screening experiment design following a Plackett-Burman design [12].

Table 1. Factors that will be evaluated for screening by a panel of experts.

Section of the nonwoven line	Fibre preparation	Web formation	Web bonding
Factors	Degree of opening	Specification of wires	Warp
	Roller speeds	Roller speeds	Needle density
	Application of avivage	Amount of worker rollers	Needle penetration
	Roller spacing	Roller spacing	Needle type
	Material flow	Web weight	Area weight

A total of 12 plates is produced. The material combination tested consists of 40% pyrolysed carbon fibre and 60% PA6 fibres.

The quality features are:

The *regularity of the nonwoven area weight*, also called CV value. The CV value is the quotient of the standard deviation of an area weight measurement series (longitudinal, machine direction (MD) and transverse (cross directional (CD)) divided by the mean value of the respective measurement series.

The *intrinsic regularity of the nonwoven* (CV-I value) or the quality of the mixture of the carbon fibres within the nonwoven. For this purpose, the previously taken area weight samples are pyrolized, leaving the carbon fibre content remaining. Since the same test specimens are used this allows the measurement of the weight fraction of the carbon fibres. The determined CV-I value is the quotient of the standard deviation of the carbon fibre mass fraction of a measurement series (longitudinal (MD) and transverse (CD)) divided by the mean value of the respective measurement series. The CV-I value is a measure of the quality of mixing and distribution of carbon fibres within the nonwoven fabric.

The *flexural strength* and the *flexural modulus*, determined from mechanical bending tests in 0°, +/- 45° and 90° direction respectively to the machine direction of the nonwoven. The samples are tested in accordance with DIN EN ISO 14125 in a 4-point bending test.

The *orientation*, expressed in a polar diagram of the examined composites. The polar diagram is established from the mechanical characteristics of the four different angles, which each generated a data point. The polar diagram provides a more accurate orientation of the material than the typically in nonwoven industry employed MD/CD ratio. On the other hand the polar diagram allows to conclusions concerning the damage of the carbon fibres. Nonwovens with the same polar diagram orientation but reduced mechanical values will show those properties mainly due to a reduction of fibre length (Figure 2).

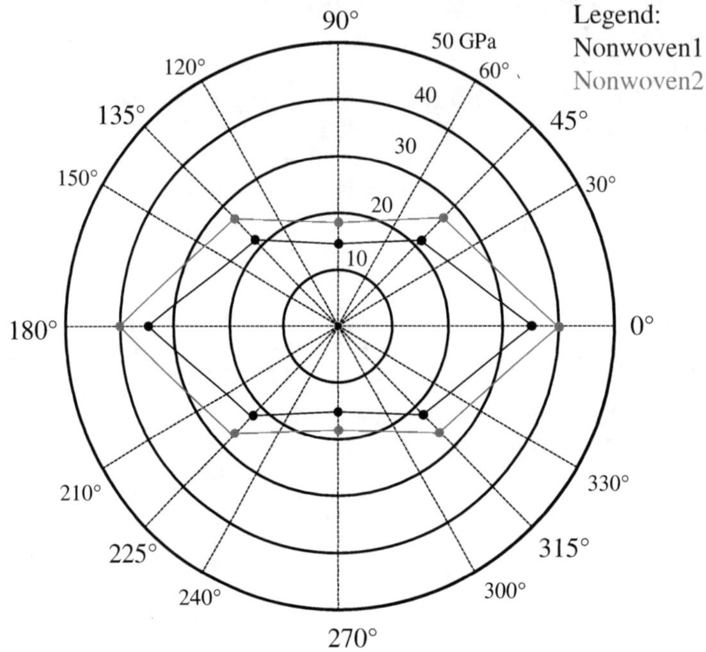

Figure 2. Schematic polar diagram of 2 nonwovens differing in fibre damage.

Based on the results of the screening experiments, the 2-4 most influential factors are determined. These factors are examined in a second statistical design plan for their exact relationship and interactions regarding the influence on the 4 quality characteristics. The procedure is the same as described for the screening experiments. The design is chosen from Central-Composite-Design, Box-Behnken-Design or Monte-Carlo-methods [12], enabling the detection of a cubic course of the factors influences. The results serve as a basis for the selection of the best operating point of the roller card for the production of nonwovens for material characterization and for further processing. Furthermore, at the end, the best degree of opening of the fibre mixtures and the settings for solidification are established.

4. Conclusion

Lightweight design is a key technology to enable the design and manufacturing of modern energy optimized cars especially for electro mobility. Although CFRP is the best available lightweight material available we have to face the challenge to reduce manufacturing costs and enable production scales of high volume production.

The combination of rCF and nonwoven processes shows high potential to contribute to solving this problem. The project CaRinA , which is presented in this lecture will help to understand the process key factors and create material cards which will help the rCF nonwoven industry as well as the parts manufacturer to apply this new technology in a much better way to use recycling materials for industrial purposes.

References

[1] Studie Wertschöpfungspotentiale im Leichtbau, 2014, Hrsg. Leichtbau BW.

[2] Stevenson A.: Recycling carbon fibre: state of the art and future developments. International textile conference, Dresden, 24.-25.11.2016.

[3] Dauner M. et al.: Rahmenbedingungen zur Verarbeitung von Carbonfasern durch Krempeltechnik. 31.Hofer Vliesstofftage, Hof, 9.-10.11.2016.

[4] Hengstermann M., Abdkater A., Cherif C.: Entwicklung von Prozessen zur Herstellung von Garnen aus recycelten Carbonfasern für die Wiederverwendung in lastragenden CFK-Bauteilen. 12. Kolloquium re4tex 2.-3.12.2015.

[5] Fuchs H., Albrecht W.: Vliesstoffe - Rohstoffe, Herstellung, Anwendung, Eigenschaften, Prüfung. Wiley-VCH Verlag, Weinheim, 2012.

[6] Thüringisches Institut für Textil- und Kunststoff-Forschung, Lützkendorf R.: Werkstoff- und Prozessentwicklung für thermoformbare hochleistungsfaserverstärkte Halbzeuge in Serienprozessen. Abschlussbericht zum Verwendungsnachweis zu FuE Vorhaben MF110118. 2013.

[7] Habers T., Ebel C., Drechsler K.: Highly Efficient Production and Characterizsation of CFRP Made from Recycled Carbon Fibres. Sampe Journal 50, No. 3, pp7-13, 2014.

[8] Schlichter S.: Processing Composites Waste, Status and Perspectives,1st International Composite Congress, Stuttgart 2015.

[9] Schlichter S.: Carbon Composites Recycling strategies and their realization for new products; 17th European Conference on Composite Materials ECCM17, Munich, 2016.

[10] Stegschuster G., Abbt M.: Results of a questionnaire concerning the market situation in the field of recycled carbon fibre and non-woven fabric produced thereof; CCeV.

[11] Schlichter S., Stegschuster G.: Web Based Composites – Potenziale vliesbasierter Verbundbauteile, Technische Textilien – Innovation, Technik, Anwendung; Deutscher Fachverlag GmbH 2017 (ISSN 0323 – 3243).

[12] Siebertz K., van Bebber D., Hochkirchen T.: Statistische Versuchsplanung – Design of Experiments (DoE), Springer-Verlag GmbH Deutschland 2010, 2017.

Prediction of elastic characteristics of spatially reinforced composite materials

A N Anoshkin[1], P V Pisarev[1*], D A Ermakov[1] and N A Merzlyakova[1]

[1]Department of Mechanics of Composite Materials and Constructions, Perm National Research Polytechnic University, Perm, Russian Federation,anoshkin@pstu.ru

* pisarev85@live.ru

Abstract. Numerical prediction of elastic characteristics of spatially reinforced composite materials (SRCM) with various reinforcement schemes is carried out in the work. Idealized SRCM structures constructed using TexGen, as well as parametrized geometric models constructed manually, are considered. It is revealed that, in spite of TexGen algorithms improved in recent times, it is impossible to achieve the required volume fraction of fibers corresponding to a real composite when constructing models of complex spatial structure of SRCM, and besides, self-intersection of weft and sewing threads often occurs. To eliminate the identified errors, idealized, parametrized geometric models in the language APDL were developed. In the framework of the developed geometric models, it was possible to realize a more dense arrangement of filaments in the frame, which allowed to maintain the percentage of the volume fraction of the reinforcing cage for the structures under consideration. The effective elastic characteristics of SRCM calculated using the developed numerical models and obtained as a result of mechanical tests were compared.

1. Introduction

One of the promising technologies for creating composite materials with high physical and mechanical properties is the technology of spatial weaving of reinforcing skeletons. Spatially reinforced composite materials (SRCM) are used to create highly loaded structures of aerospace industry. To develop such materials, corresponding to different operational requirements, it is necessary to use methods for predicting their mechanical properties [1].

At present, an approach based on computer modeling of their geometric structure is widely used to predict the mechanical properties of SRCM. Using the constructed geometric models, the micromechanics problem for predicting the mechanical properties of a SRCM reduces to a set of boundary conditions for either the periodicity cell or the structure fragments. From the solution of these problems, the stress and strain fields in the fibers and the composite matrix are determined, averaging and subsequent comparison of which makes it possible to find its effective mechanical properties.

To construct geometric models of SRCM, special software packages are developed, originally designed for modeling textile composites with two-dimensional weaving. The most famous packages are WiseTex [2-5] and TexGen [6, 7]. Using these packages, it is possible to construct an idealized composite structure, which leads to errors in predicting mechanical properties. To obtain real models of the three-dimensional structure of the composite, various corrections of model structures are used based on the results of optical microscopy or tomography of composite materials [8-10].

In this paper, we propose an algorithm for automatically correcting the model structure of SRCM obtained using the TexGen software package. It is shown that it is possible to improve the accuracy of predicting the elastic mechanical characteristics of SRCM using the proposed algorithm, which allows to describe the enveloping of the contacting fibers more accurate and to exclude their intersection.

2. Problem formulation

The object of the study was woven spatially-reinforced composite material with four variants of weaving (Figure 1): variants 1 and 2 have rectilinear fibers in three directions; 3 and 4 have straight-line fibers in two directions. The structure of weaving of SRCM fibers is a system of three threads: warp, weft and sewing yarns (Figure 1). The volumes between the reinforcing carcass filaments are filled with a matrix. The volume fraction of reinforcing yarns in the structures under consideration was determined within the framework of morphological studies of SRCM and constituted for weaving variants: 51.7%; 52.8%; 50.9%; 60.5% respectively.

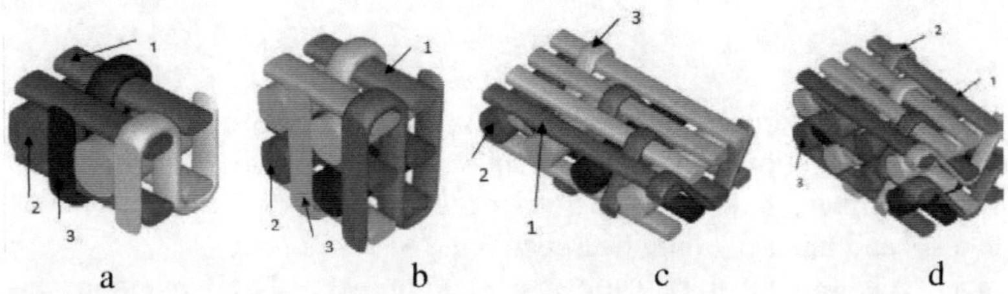

Figure 1. General view of geometric models of SRCM periodicity cells constructed with the help of TexGen, structures: a - 1; b - 2; c - 3; d - 4.

The technology of manufacturing SRCM includes the preparation of a woven fiber skeleton with the possibility of changing the parameters of the weave and subsequent impregnation with a reactive plastic binder under pressure in a rigid form (RTM). In the process of impregnation, the parameters of the fiber skeleton can be unevenly altered by the volume of the billet. All this leads to heterogeneity of mechanical properties of SRCM in the finished product. The solution of the problem of prediction of elastic characteristics makes it possible to evaluate the influence of the fiber framework parameters on the mechanical properties of SRCM.

For the prediction, geometric models of fragments of the spatial structures of SRCM were constructed using the TexGen software package (Figure 1) and corrected further using the developed algorithm. Further for the fragments, elastic problems with periodic boundary conditions were solved, and averaging was performed, obtained fields of structural stresses and deformations. By comparison of the averaged (macroscopic) stresses and deformations found, the effective elastic characteristics of the SRCM were found.

3. Construction of geometric models of the structure of SRCM

The initial data for constructing geometric models of the cells of periodicity were the cross-sectional area of the filaments and the distance between their centers. They were determined on the basis of the given parameters of the linear density of filaments and the density of laying on the base and weft. When constructing geometric models, the following assumptions were used: solid threads, the interphase layer between the matrix and the filaments is absent, the technological dispersion in the area of contact between the fibers of the warp, weft and sewing threads, a guaranteed interlayer of the matrix was set (Figure 2).

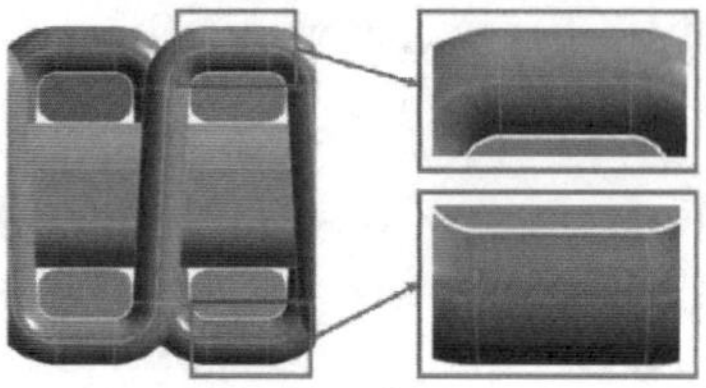

Figure 2. General view of the geometric model with a guaranteed interlayer of the matrix between the filaments.

Assuming an ideal packing of fibers in a complex filament, its conditional cross-sectional area S can be determined by specifying the linear density of the complex yarn T, the density of the yarn material ρ, and the degree of volume filling of the filaments in the yarn k can be determined using formula:

$$S = \frac{T}{pk} \tag{1}$$

Initially, cells of the periodicity of four different versions of idealized structures were constructed for the SRCM under consideration using the TexGen software package (Figure 1). The structures differ in the straightness of the fibers in the directions, in the number of layers, and in the volume fractions of the reinforcing carcass and threads along the directions

Subsequent studies have shown that, despite the improved TexGen algorithms recently, it is impossible to achieve a volume fraction of fibers corresponding to the real material when constructing geometric models of SRCM structures. In addition, self-intersections of weft and sewing threads appear in the construction of models (Figure 3), which in turn leads to errors when the geometric model is divided into finite elements, and also affects the accuracy of the results of numerical calculations.

To eliminate these shortcomings, an algorithms for constructing idealized parametrized geometric models of SRCM structures were developed. The algorithms took into account the deformability, density, volume fractions of threads along directions. With the help of developed algorithms implemented in the language APDL ANSYS, it was possible to implement a more dense arrangement of threads in the frame and eliminate their self-intersections. As a result, the SRCM periodicity cells (Figure 4) were obtained with a volume fraction of fibers corresponding to the real composite material. Further, predictions of the elastic characteristics of the SRCM with four variants of fiber weaving were made.

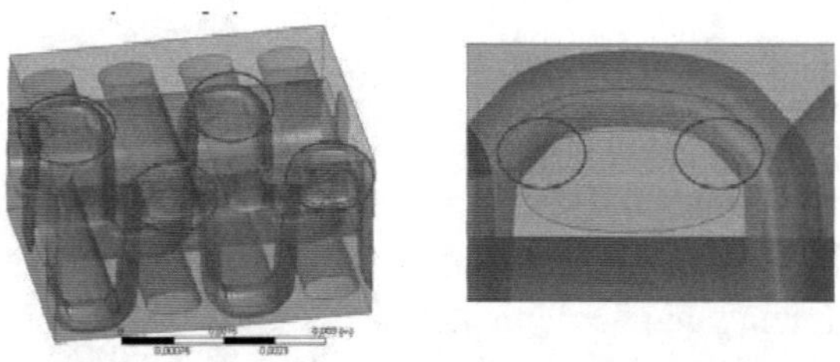

a b

Figure 3. Intersection of filaments at bending points: a - isometry; b - side view.

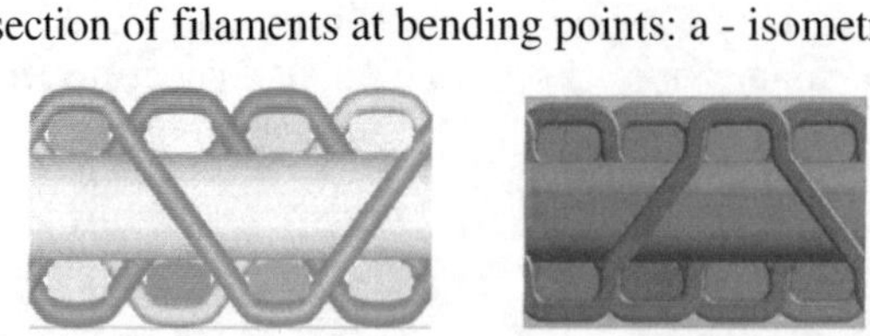

a b

Figure 4. A geometric model of structure 3 constructed using: a - TexGen; b - APDL.

Prediction of elastic characteristics was carried out for two types of periodic cells of SRCM with four variants of weaving of fibers, named further: 1, 2, 3, 4 (Figure3).

4. Peculiarities of the numerical solution of the problem for the fragment of the structure of SRCM

To predict the effective elastic characteristics of SRCM, the problems of elastic deformation of fragments of the constructed model structures with periodic boundary conditions were solved. The solution was implemented using the ANSYS software. Discretization of the investigated fragments was carried out using the spatial finite elements of SOLID186. The minimum size of the finite element, for all structures, was set equal to the thickness of the guaranteed interlayer of the binder between the filaments and was 0.01 mm. The degree of sampling was chosen on the basis of the estimated convergence of the numerical solution with respect to the maximum stresses. The total number of finite elements amounted to 4-8.5 million elements for structure 1-4. In Figure 5, as an example, a fragment of the finite element model of the textile framework and the representative volume of SRCM for the structure variant 3 is given.

a b

Figure 5. General view of the finite element model of the SRCM fragment (a) and the SRCM fiber skeleton (b) for the third variant of the structure 3.

Carbon filaments were modeled as a unidirectional fiber composite having a transversely isotropic symmetry system. In each finite element belonging to a carbon filament, local coordinate systems were assigned corresponding to the principal axes of symmetry of the material. Between the interacting threads in the model, the ideal contact condition was specified.

The SRCM structures under consideration are orthotropic and are characterized by nine independent elastic constants: Young's moduli E_X, E_Y, E_Z, Poisson's ratios ν_{XY}, ν_{YZ}, ν_{XZ}, and shear moduli G_{XY}, G_{YZ}, G_{XZ}. For each fragment, six boundary-value problems were solved numerically: stretching along the X, Y, Z axes to determine effective Young's moduli and Poisson's ratios and a net shift in the planes XY, XZ, YZ to determine the shear moduli.

The elastic characteristics of carbon filaments were calculated from the models for a unidirectional fibrous composite [11], the properties of the polymer binder were taken from experimental data (Table 1).

Table 1. Material properties used in the calculation.

Material	E_X, GPa	E_Y, GPa	E_Z, GPa	ν_{XY}	ν_{YZ}	ν_{XZ}	G_{XY}, GPa	G_{YZ}, GPa	G_{XZ}, GPa
Carbon thread	240	14	14	0.014	0.24	0.014	6.93	12.84	6.93
*		3.10			0.34			1.15	

5. Results

The resulting effective elasticity characteristics of SRCM for the considered variants of structural models constructed with the help of the TexGen software complex without correction and corrected according to the developed algorithm realized in the language APDL are presented in Table 2. Analysis of calculation results showed that the elastic moduli of SRCM for structure models, constructed using the corrective algorithm (APDL), are usually higher than those obtained based on the original TexGen models. This is due to a denser layout and the absence of fiber intersections in the model structure of SRCM, obtained as a result of the adjustment.

Table 2. Calculated effective elastic properties of SRCM.

Geometric model	E_x, GPa	E_y,GPa	E_z,GPA	ν_{xy}	ν_{yz}	ν_{xz}	G_{xy}, GPa	G_{yz},GPa	G_{xz}, GPa
Var. 1 *APDL*	59,98	44,28	6,931	0.074	0.260	0.273	3,39	2,83	2,71
Var. 1 *TexGen*	62,40	42,30	7,00	0,081	0,301	0,28	3,24	2,90	2,55
Var.2 *APDL*	61,01	43,88	7,23	0.079	0.254	0.260	3,33	2,84	2,76
Var. 2 *TexGen*	60,10	42,50	7,01	0,082	0,302	0,283	3,35	2,95	2,61
Var. 3 *APDL*	57,01	55,72	7,16	0.042	0.278	0.291	3,26	2,67	2,71
Var. 3 *TexGen*	48,70	46,82	6,55	0,054	0.313	0,326	2,63	2,45	2,45
Var.4 *APDL*	68,06	64,74	8,361	0.355	0.252	0.248	3,78	3,41	3,34
Var. 4 *TexGen*	68,01	63,50	8,05	0,356	0.243	0,226	2,91	3,35	3,45

The calculated effective elastic properties of SRCM were compared with the experimental estimates of the mechanical properties of the composite obtained from the OX stretching tests of the variants of structures 1 and 3 and the shear tests in the OXY plane of the variants of structures 2 and 4 (Table 3). The comparison showed that for models corrected using the developed algorithm (APDL), the error in predicting the elastic characteristics is several times smaller than in the initial models obtained with TexGen. The maximum prediction error is observed for the G_{xy} shear modulus and is no more than 10% for the developed geometric models (APDL), and for the TexGen initial models is 30%.

Table 3. Comparison of the calculated and experimental elastic characteristics of SRCM.

Geometric model	E_x^{calc}, GPa	ν_{xy}^{calc}	G_{xy}^{calc}, GPa	$\frac{(E_x^{calc} - E_x^{exp})}{E_x^{exp}}$,%	$\frac{(\nu_x^{calc} - \nu_x^{exp})}{\nu_x^{exp}}$,%	$\frac{(G_x^{calc} - G_x^{exp})}{G_x^{exp}}$,%
Var. 1 *APDL*	59,98	0.074	3,39	1,67-	5,4+	-“-
Var. 1 *TexGen*	62,4	0.081	3,24	2,24+	12,5+	-“-
Var.2 *APDL*	61,01	0.079	3,33	-“-	-“-	7,9+
Var.2 *TexGen*	60,10	0,082	3,24	-“-	-“-	5,27+
Var. 3 *APDL*	57,01	0.042	3,26	3,52+	4,76+	-“-
Var. 3 *TexGen*	48,70	0,054	3,26	14,58-	25,93+	-“-
Var.4 *APDL*	68,06	0.355	3,78	-“-	-“-	9,96-
Var.4 *APDL*	68,01	0,356	2,91	-“-	-“-	30,56-

6. Conclusion

When constructing geometric models of SRCM, it is important to ensure the volume fraction of fibers in general and in the directions of reinforcement corresponding to the values in a real composite material. The initial models obtained with the help of the TexGen software package may have inaccuracies in the laying and intersection of weft and sewing threads at the bending points. This leads to a decrease in the volume fraction of the fibers, to errors in the finite-element discretization of the geometric models of the SRCM, and reduces the accuracy of predicting the elastic properties. An algorithm for correcting the geometric structure of the SRCM for the TexGen software package implemented in the language APDL ANSYS is proposed. A selective comparison of the experimentally obtained elastic characteristics of SRCM with predictive results for geometric models of the structure constructed in TexGen and corrected by the proposed algorithm showed good agreement.

A further direction of research in the construction of geometric models and predicting the effective elastic properties of PACM is to take into account the deformation of threads in the composite at various technological stages of its production.

Acknowledgments

The work was carried out with the financial support of the state represented by the Ministry of Education and Science of Russia within the framework of the federal target program "Research and development in priority areas of development of the scientific and technological complex of Russia for 2014-2020" on the topic: "Development of experimental technologies for the automated production of large-sized details of advanced aviation engines made of thermoplastic composite materials". Unique identifier of applied scientific research RFMEFI57717X0261.

References

[1] Carey J.P. 2016, Handbook of Advances in Braided Composite Materials: Theory, Production, Testing and Applications, pp. 1-479

[2] Lomov S.V., Gusakov A.V., Huysmans G, Prodromou A and Verpoest I 2000, *Compos Sci. Technol.* **60**:2083-95

[3] Verpoest I and Lomov S.V. 2005,*Compos Sci Technol*. **65**:2563-74

[4] Lomov S.V., Ding X, Hirosawa S, Kondratiev S.V., Molimard J, Nakai H, et al. 2005, *SAMPE Europe International Conference*, p.28-33

[5] Sherburn M. N., Long A. C., Jones I. A., Robitaille F., Crookston J. J. and Ruijter W. 2005, *In 15th Int. Conf. on Composite Materials*

[6] El Said B., Ivanov D., Long A. C. and Hallett S. R., 2014, *ECCM-16*

[7] Drach A., Drach B., Tsukrov I., Bayraktar H., Goering J., *Realistic FEA modeling of 3d woven composites on mesoscale*, 2013

[8] Green S.D., Matveev M.Y., Long A.C., Ivanov D. and Hallett S.R., Mechanical modelling of 3D woven composites considering realistic unit cell geometry, vol. **118**, **№ 1**, pp. 284-293, 2014.

[9] Green S. D., Long A. C., El Said B. S. F. and Hallett S. R., *COMPOSITE STRUCTURES*, vol. **108**, pp. 747-756

[10] Liu Y., Straumit I., Vasiukov D., Lomov S.V. and Panier S. 2017, *COMPOSITE STRUCTURES*, vol. **179**, pp. 568-579

[11] Anoshkin A.N. 2003, *Mechanics of Composite Materials*. Vol. **39**. No. **5**, pp. 377-386

IOP Conf. Series: Materials Science and Engineering **406** (2018) 012024 doi:10.1088/1757-899X/406/1/012024

Characterisation and modelling of complex textile geometries using TexGen

L P Brown[*], A Endruweit, A Long, I A Jones

Composites Research Group, Faculty of Engineering, University of Nottingham

* louise.brown@nottingham.ac.uk

Abstract. TexGen is open source software developed at the University of Nottingham for the geometric 3D modelling of textiles and textile composites. It has a large number of users worldwide and underpins a significant number of research publications.

While many users make simplifying assumptions about the structure of a textile, in reality the internal geometry of a textile or textile composite is complex. Capturing this complexity is vital for the prediction of properties such as permeability and mechanical failure. Examples will be given of the characterisation of a material and how the complex features are captured and implemented in TexGen, making use of functionality such as the ability to vary the cross-sectional shape along the length of a yarn. The effect on prediction of properties as a model is refined will be demonstrated.

Recent additions to the software will also be highlighted. Laminated structures can be quickly and easily constructed from a selection of textiles and several nesting options are available. A new rotate textile option can then be used to create laminates with varying ply angles. Where the unit cell is also rotated, appropriate periodic boundary conditions have been implemented and are automatically generated in an ABAQUS input file.

A new feature is described which generates a TexGen model from a weave pattern file. Future developments of this may improve accessibility of the software to the weaving community. The generation of a pattern draft output from the TexGen model is also described.

1. Introduction

Textile composites are used in a wide range of applications, either in the form of laminates from 2D textiles, 3D textiles or increasingly complex net-shaped preforms. In order to speed up the design cycle for these materials it is essential to be able to create realistic geometric models so that accurate simulation results can be obtained for the prediction of properties such as permeability and mechanical properties. TexGen [1, 2] is open source software developed in the Composites Research Group at the University of Nottingham, enabling the creation of 3D geometric models of textiles and textile composites. There is built-in capability for the creation of standard textile forms, both 2D and 3D woven, and there are refinement algorithms built into the software for automatic reduction of intersections in the model and for creating different levels of compaction. This automation makes use of the versatility of the built-in modelling capability and can be recreated using the functions in the Python scripting API. This functionality is often overlooked by users and this paper seeks to highlight how this can be used to create more realistic textile models and how this can improve the final simulation results which use these models.

New features are also described, including an option to build up laminates using a variety of textile layers at varying rotations, and new input from weave pattern data as well as its output from a TexGen model.

2. Geometric modelling

The functionality used within TexGen to enable the creation and refinement of complex textile models is highlighted here and guidance is given on how these methods can be used for geometry model generation. The example used here is a 3D orthogonal weave shown in Figure 1a [3]. An idealized model of the geometry can be created using the 3D wizard in the TexGen graphical user interface (GUI), Figure 1b, in this case using the measured yarn widths and heights and a default power ellipse with a power of 0.6 as the constant cross-section. Nodes are created on a grid with one at each crossing of the warp and weft yarns. The binder yarns have extra nodes inserted around the top and bottom weft yarns in order to follow their shape, shown in Figure 1c. This uses a function ShapeBinderYarns which automatically calculates the yarn positions.

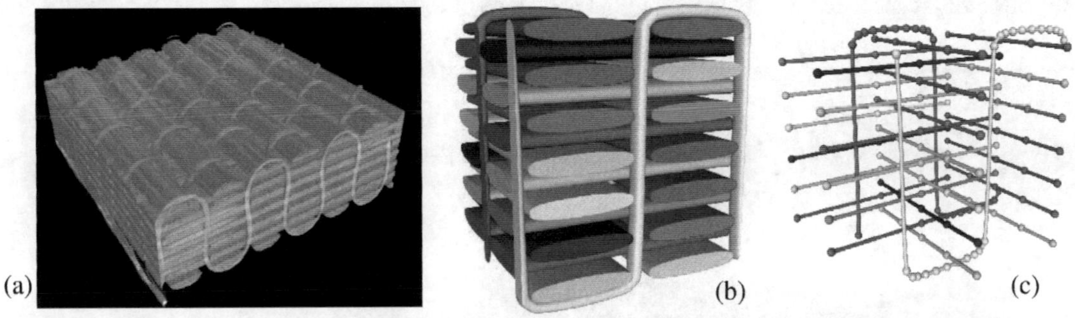

Figure 1. a) µCT image of 3D orthogonal textile b) Idealised model created using TexGen 3D wizard c) Nodes and paths in idealised model

2.1. *Use of complex cross-sections*

Figure 2a shows a slice from the µCT image of the orthogonal textile. It can be seen that the weft cross-sections, seen end-on in the image, show two distinct shapes. Where the binder yarns wrap around the wefts at the top and bottom of the textile the wefts form a curved section on the outer surface whilst remaining flat where they lie next to the warp yarn underneath. The weft yarns in the inner section of the textile form into a more rectangular shape which can be represented using a power ellipse section.

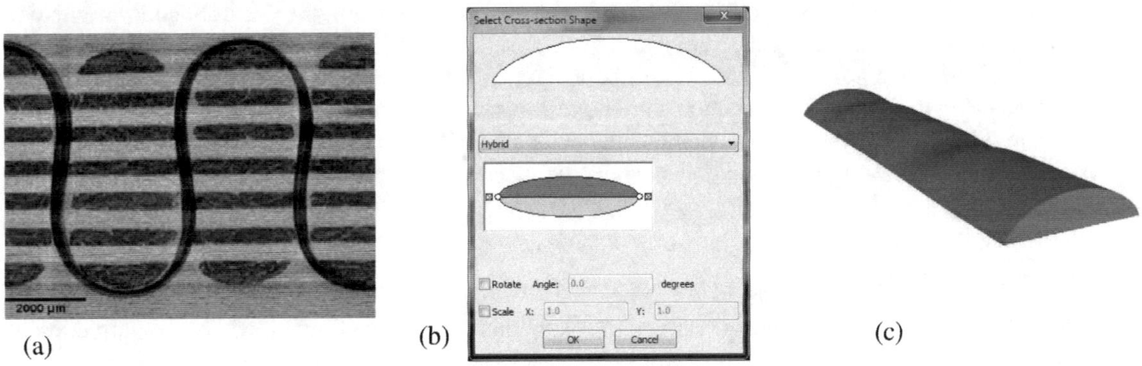

Figure 2. a) µCT image showing weft cross-sections b) Hybrid section dialog c) Yarn with hybrid sections

2.1.1. Hybrid sections. The asymmetric outer sections can be created using the hybrid section type. From the GUI the dialog in Figure 2b allows the shape to be divided into sections, here showing two but more can be chosen if desired. The size and section type of each division can then be entered to form the combined shape. Using Python the CSectionHybrid class is used followed by AddDivision and AssignSection to specify how the section is divided and which section type and size is assigned to each division. An example yarn using this cross-section is shown in Figure 2c.

2.1.2. Polygon sections. In some cases the required cross-section may not conform to a standard shape, or may be difficult to assemble as a composite shape using the hybrid option. In this case the CSectionPolygon class can be used. This option is only available via the Python scripting interface. In the case of the 3D orthogonal textile, as the textile is compacted crimp is induced in the top weft yarns which in turn affects the shape of the warp yarns underneath as shown in Figure 3a. During the refinement process implemented automatically in TexGen this effect is captured, Figure 3b, and the warp yarns shape is automatically changed to avoid intersections. The polygon section is used thus enabling sections such as that shown in Figure 3c to be produced. When the section is initialised a vector of points is specified which define the outer surface of the boundary.

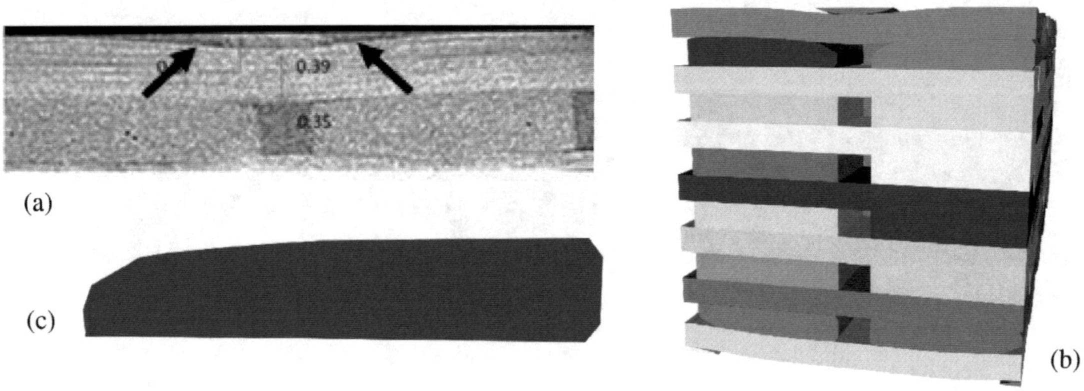

Figure 3. a) μCT image showing local geometry variations b) Refined geometry for compacted textile c) Polygon section

2.2. Variation of cross-section along length of yarn

When a yarn is created it will be assigned a constant cross-section by default using the CYarnSectionConstant class. In reality yarns will deform along their length and a realistic model should reflect these changes in cross-section. In TexGen it is possible to specify varying cross-sections along the length of a yarn either by specifying the sections at each node using the CYarnSectionInterpNode class or at specific points along the yarn using the CYarnSectionInterpPosition class. These classes are all inherited from the abstract CYarnSection base class. After these classes have been created sections are added using the AddSection function, either one section for each node or one section at each position as required. Once all of the sections have been added then the yarn section object is assigned to the yarn using the AssignSection function. These can also be selected in the GUI using the Select Yarn Section dialog given by the Modeller->Assign Section option. An example of a yarn with varying cross-sections can be seen in Figure 2c.

2.3. Local checking of volume fraction

It is important when adjusting cross-sections to ensure that the volume fraction within the yarn does not exceed a realistic limit. If yarn properties have been assigned to a yarn then the yarn volume fraction at a given section can be calculated.

$$Volume\ Fraction = \frac{Fibre\ Area}{Section\ Area}$$

where fibre area can be obtained using the yarn GetFibreArea function and the local yarn section can be extracted from the appropriate CYarnSection class described in section 2.2. This can be obtained using the yarn GetYarnSection function and then getting the actual cross-section at the required point using the GetSection function. From this the GetArea function will calculate the section area.

2.4. Permeability prediction

Flow through the 3D orthogonal textile was simulated using Computational Fluid Dynamics (CFD) software in order to determine the textile permeability [3]. A model of an orthogonal weave was created and then meshed using the voxel meshing built into TexGen. Incremental changes were made to the geometric model to assess the sensitivity of the permeability prediction to the level of geometric detail in the model using the techniques described in the previous sections. Changes were made to the binder cross-sections, surface weft yarn crimp and warp yarn cross-sections in successive models. The CFD simulation predictions are plotted with experimental data in Figure 4, demonstrating significant improvement in permeability prediction with inclusion of local variations in the geometric model.

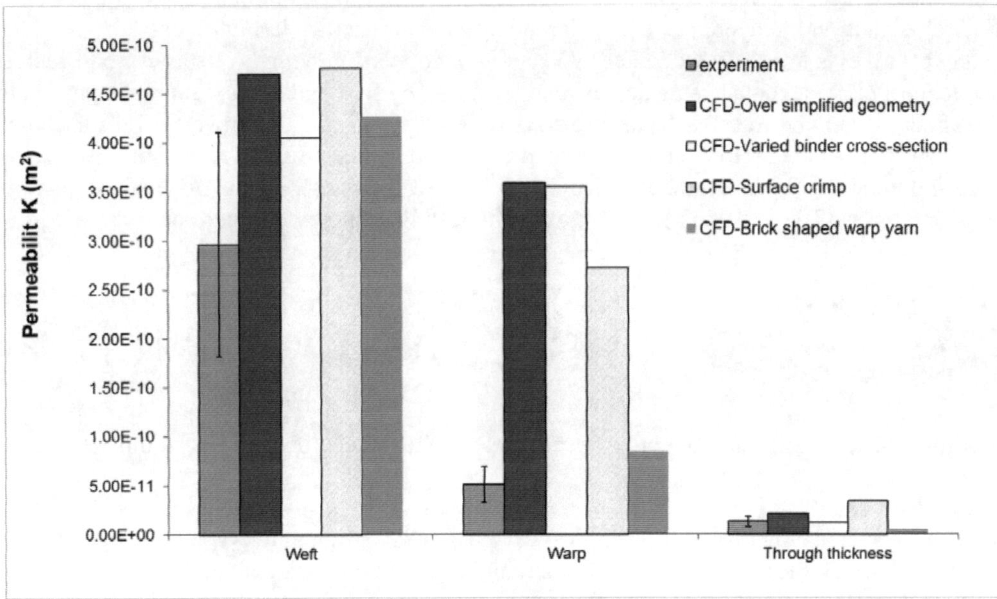

Figure 4. Permeability predictions with variations in model geometry

3. Constructing laminates

It is common for textiles to be built into laminates, often with varying rotations. The CTextileLayered class allows textiles to be combined into a single unit.

3.1. Layered textiles

In the GUI a layered textile can be created from any combination of textiles currently loaded. The Textiles->Create Layered option enables the selection of textiles and the ordering of the layers using the dialog shown in Figure 5a. The domain is created to be the size of the largest of those selected, therefore there may be more than one repeat of weaves with smaller unit cells. By default the textiles are stacked using the sizes of the domains in the z direction to govern the distance between the layers.

An example layered textile is shown in Figure 5b. In a Python script the CTextileLayered class is created and then textiles are added together with their x-y offsets using the AddLayer function.

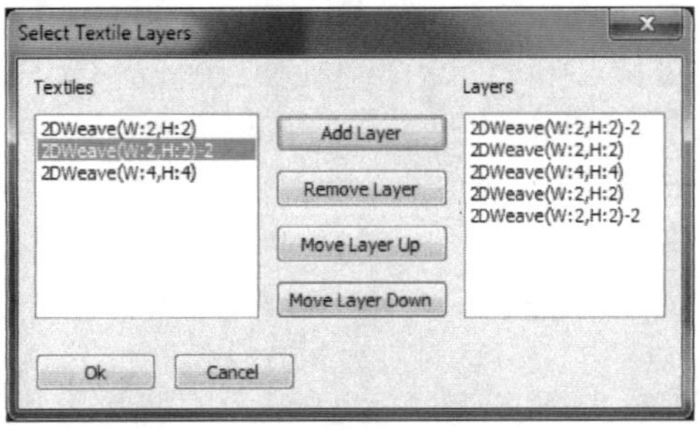

(a)

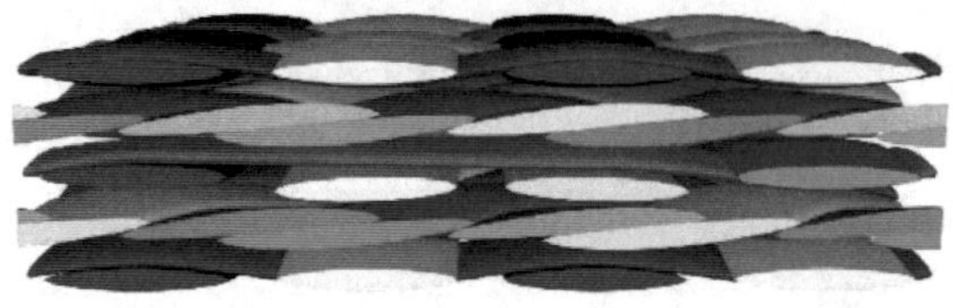

(b)

Figure 5. a) Dialog for selection of layers b) Layered textile

3.2. Layer-Offsets and Nesting

Layered textiles are created with an initial offset of zero in the x-y plane when created using the GUI. Using a script offsets can be specified for the individual layers as described in Section 3.1. Using the GUI offsets can either be specified as a constant offset between layers, random or specific offsets specified for each layer. A layered textile with random offsets is shown in Figure 6a.

Nesting of layers is implemented either by keeping the offsets, moving layers as close to each other as possible until the first contact is made between layers, or by finding the maximum nesting. In this case the offset is found between each pair of layers which gives the maximum vertical displacement of the layers towards each other to the point at which they make contact. In this case both the offset in the x-y plane and the vertical displacement are adjusted. Nesting retaining original offsets is shown in Figure 6b and the same layered textile with maximum nesting is shown in Figure 6c.

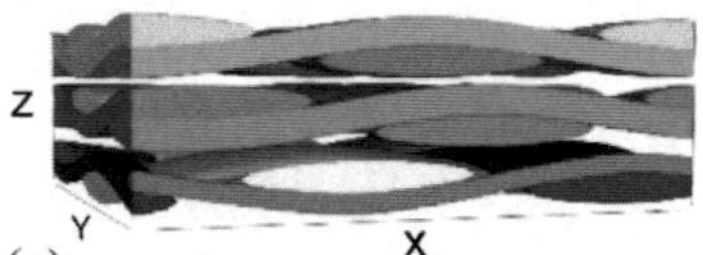

(a)

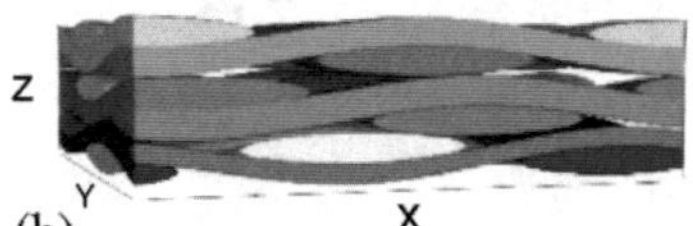

(b)

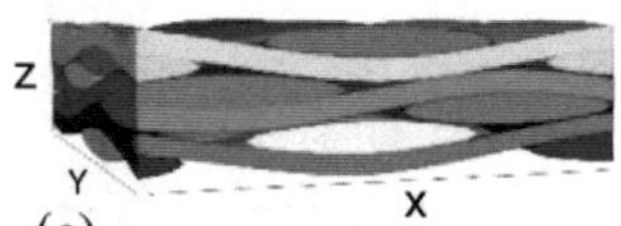

(c)

Figure 6. a) Textile with random offset b) Nested with same offset c) Maximum nesting

3.3. Rotated textiles

Textiles can be rotated either using the Textile->Rotate Textile option in the GUI or using the CTextile Rotate function which takes as parameters a quaternion, giving the axis and angle of rotation, and the centre of rotation. In the GUI the axis of rotation is limited to the x, y and z axes, however using the function in a script any axis of rotation can be selected. The domain can also be rotated if desired. Voxel file export has been modified for the rotated domain so that voxels are generated which are aligned to the domain axes and periodic boundary conditions have been implemented which account for the rotation.

4. Weave pattern input/output

4.1. Weave pattern import

Although it can be straightforward to create a 'standard' textile using the TexGen user interface, textile designers often design weave patterns using a weave pattern matrix. The ability to input these and automatically generate the corresponding textile model will speed up the modelling process significantly. In order to interpret a weave pattern in TexGen some extra information about the number of warp yarns in each stack is added as well as a keyword to indicate the type of weave generated.

Generation of general and 3D orthogonal weave patterns were reported in [4, 5]. This has now been extended to include 3D angle interlock weaves.

The weave pattern and layer-ID information are read from a text file in the format shown in Figure 7 using the WeavePattern Python script included in the TexGen installation. This is accessed from the GUI using the File->Import->WeavePatternFile option. The number of warp and weft yarns is extracted and the number of binder yarns deduced. From this the cell structure used to store weave patterns within TexGen is generated.

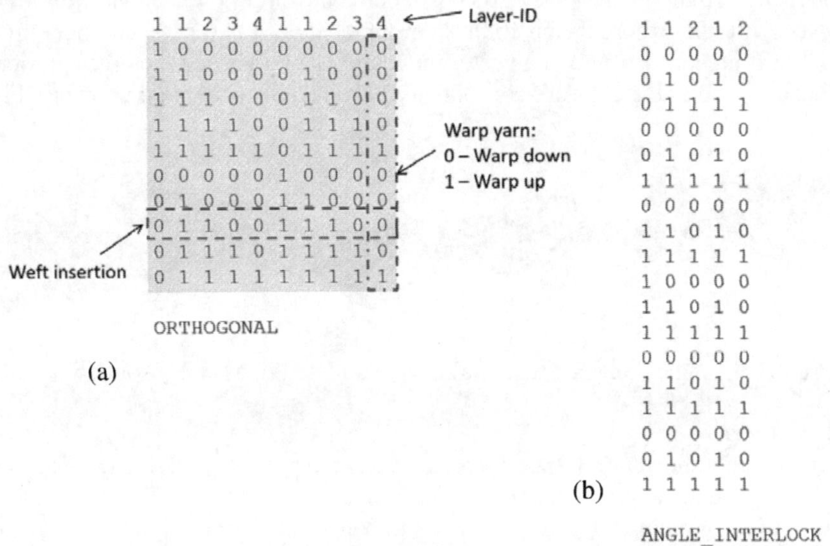

Figure 7. Weave pattern input file format a) Orthogonal weave b) Angle interlock weave

The appropriate class, CTextile3DWeave, CTextileOrthogonal or CTextileAngleInterlock, is used to create a textile each of which contain the cell array which describes the configuration of warp and weft yarns within the textile. The cell array is initialized by processing the weave pattern one row at a time, calling the SetupWeftRow function. This uses the layer-ID information to delimit the binder and warp stack information in the chart as illustrated by the dotted separators in Figure 8. For any stack of warp layers the weft will be positioned where the warp changes from being up to down.

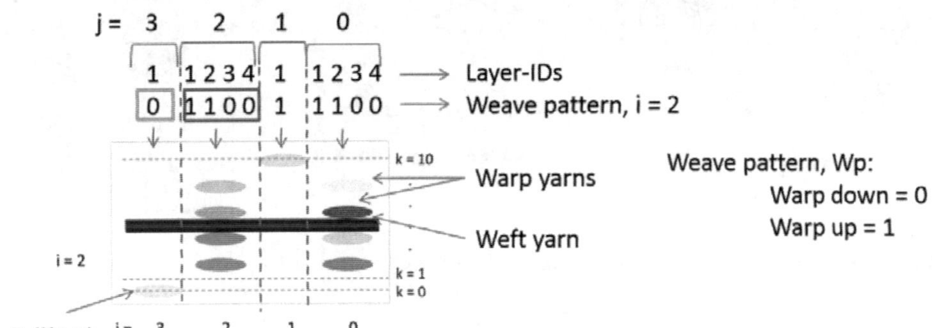

Figure 8. Creation of cell array entries

For an orthogonal weave, for each binder/warp stack, j, the cell array is populated as illustrated in Figure 8 which shows the row for i = 2 as highlighted in Figure 7a. In this case the binder yarn is assumed to be at either the top or bottom of the stack depending on whether the warp (ie binder) is up or down.

The angle interlock case is more complicated as the binder travels through the thickness of the textile at varying weft heights and the correct position needs to be retained in the cell array. In the first instance the binder position is set to one below the weft if the warp is down and to one above the weft if the warp is up. This results in a textile such as that shown in Figure 9a. At this point in the case of both types of weave a textile is formed with the warps stacked but the weft yarns spread out, one yarn for each cell row i.

A subsequent process, ConsolidateCells, moves the weft yarns into stacks where there is an empty cell in the previous row. For angle interlock yarns it can be seen from Figure 9a that the correct binder position for each stack is either where the position is constant between two rows or where it is at the top or bottom. This information is stored in a vector for each binder yarn and then the appropriate binder positions are assigned as the cells are consolidated. The resulting textile is shown in Figure 9b.

(a) (b)

Figure 9. Angle interlock textile a) unconsolidated b) after consolidation

4.2. Weave pattern export

For textiles which use either the 2D or 3D weave base classes the process described in Section 4.1 can be reversed in order to generate a weave pattern matrix from a TexGen model. For 3D textiles the ordering of the weft insertions, either stacking from top to bottom or bottom to top will change the weave pattern generated. In the user interface there is a dialog to select the weft insertion ordering (Figure 10b). From a Python script the ordering can be passed as a parameter to the weave ConvertToPatternDraft function.

A pattern draft is also generated based on a 1:1 tie up. This uses an adaptation of the algorithm defined for hand looms given by Griswald [6] and Glasner [7]. The pattern draft can either be printed or saved as a text file as shown in Figure 10c.

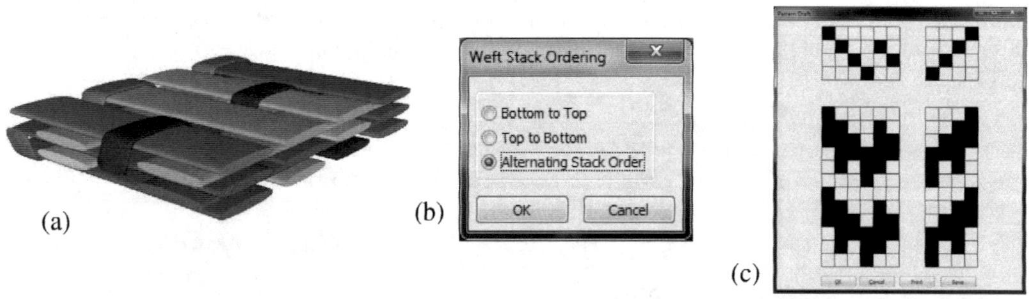

(a) (b) (c)

Figure 10. Weave pattern export a) Example weave b) Dialog to select weft stack ordering c) Pattern draft

5. Conclusions

Methods for increasing the accuracy of geometric textile models created using TexGen software have been described. The benefits of improved model accuracy were demonstrated for a case study where in-plane permeability was predicted using CFD simulation with models of increasing geometric accuracy.

Methods to quickly and easily create models of textile laminates using either the TexGen GUI or Python scripting are described. This includes nesting and new options for rotation of layers.

A method for importing a weave pattern matrix and converting it into a TexGen model is described enabling fast creation of models from a standard textile design format. The method has recently been extended to angle interlock textiles and this will be included in the next TexGen release. Creation of pattern drafts from a TexGen model is also described.

Acknowledgements
This work is funded by the Engineering and Physical Sciences Research Council RSE Fellowship [Grant number: EP/N019040/1].

References

[1] Brown, Louise P, & Sherburn, Martin. (2017, December 13). louisepb/TexGen: TexGen v3.10.0 (Version v3.10.0). Zenodo. http://doi.org/10.5281/zenodo.1115604
[2] A C Long and L.P. Brown, *Modelling the geometry of textile reinforcements for composites: TexGen*, in *Composite reinforcements for optimum performance* P. Boisse, Editor. 2011, Woodhead Publishing Ltd.
[3] Xuesen Zeng, Louise P. Brown, Andreas Endruweit, and A.C. Long. *Advanced Geometry Modelling for 3D Woven Fabrics and its Application in Permeability Prediction*. in *The 4th World Conference on 3D Fabrics and Their Applications*. 2012. RWTH Aachen, Germany.
[4] L P Brown, B Wendland, S Schaeffer, A C Long, and T. Gries, *3D-Woven Profile Design Tool for Users with a Non-Textiles Background*, in *10th Int. Conf. on Manufacturing of Advanced Composites (ICMAC 2015)*. 2015: Bristol, UK.
[5] L P Brown, S. Yan., X Zeng, and A.C. Long, *Mesoscale Geometric Modelling of Bifurcation in 3D Woven T-Beam Preforms*, in *12th Int. Conf. on Textile Composites*. 2015: Raleigh, NC, USA.
[6] Griswold, R.E. *Fabric Analysis - From Drawdown to Draft*. 2000 23/09/2014]; Available from: http://www.cs.arizona.edu/patterns/weaving/webdocs/mo/D/FabricAnalysis.pdf.
[7] Glassner, A., *Digital weaving.2*. Computer Graphics and Applications, IEEE, 2003. **23**(1): p. 77-90.

IOP Conf. Series: Materials Science and Engineering **406** (2018) 012025 doi:10.1088/1757-899X/406/1/012025

Virtual development and numerical simulation of 3D braids for composites

Yordan Kyosev[1,2], Matthias Hübner[3], Chokri Cherif[3]

[1]Hochschule Niederrhein, University of Applied Sciences, Research Institute for Textile and Clothing, Mönchengladbach, Germany
[2]TexMind UG –Mönchengladbach, Germany
[3]ITM – Institute of Textile Machinery and High Performance Material Technology, Technische Universität Dresden, Germany

Abstract. The 3D maypole braiding process is characterized with complex and often variable tracks for the carriers. During the last years smaller size (4x4 horn gears or larger) variable track machines become available on the market and the areas of industrial application of these increased. The design of the new products and the controlling of such machine is complex task, possible only with process simulation. The latest development of the algorithms and their implementation in the TexMind Braiding Machine Configurator software allows numerical simulation of any configuration of maypole braiding machine, with variable tracks. For the simulated geometry of the 3D braids an automated FEM mesh generator is developed, so that the input file for LS-Dyna is created automatically from the machine configuration. This speeds up significantly the preparation process of the FEM data and allow rapid numerical simulations of the mechanical properties of the different configuration of braided structures for composites

1. Introduction

The 3D braiding process allows production of complex profiles with larger changes in their dimensions and form [1]. Shell structures can be created during the overbraiding of complex mandrels and solid structures can be created if more carriers with several interlacing tracks are used. 3D Braiding allows as well development of mesh type structures and any combinations. There are different 3D braiding techniques, depending on the practical realisation of the carrier motion – row and column (cartesian) braiding [2], 3D braiding based on the lace principle [3–5], maypole 3D braiding [6][7]. The first two principle are implemented in a few single machines and the last one – the maypole 3D braiding principle become larger industrial application during the last time [8, 9]. The maypole 3D braiding has advantage against the two other principles in the productivity – the braiding process can (but should not) stop only, when the switches have to change their state, the remaining time they can run with 100 or more revolutions per minute of the horn gears, where the speed depends on the material and not on the machine limitations. The profiles, produced on such machines can have complex cross section without any stitching process – and because of this a lower danger of delamination is expected. Actually, the design of such profiles is complex task. This paper demonstrates the process of numerical simulation of the braiding process, generation of the idealized forms of the profiles and the relaxation of this geometry using FEM software LS Dyna.

IOP Conf. Series: Materials Science and Engineering **406** (2018) 012025 doi:10.1088/1757-899X/406/1/012025

2. Emulation of the braiding process and generation of idealized 3D braids

The motion of the carriers of any maypole braiding machine can be predicted numerically, as presented in [10]. All process related parts of the braiding machine – horn gears, carriers, switches – are modelled as objects in C++ language, and their motion laws are integrated in the time. The coordinates of each carrier in the x-y plane are obtained as result of this simulation. The idealized braid is received, as to each calculated position a z-coordinate is assigned

$$z(t) = V_{Take\_off} \cdot t \quad (1)$$

Where v_{Take_off} is the take-off velocity and t is the current time. The horn gears have significantly larger diameter $D_{HornGear}$ than the yarns, which means, that if the motion of the carrier is taken as basis for the yarn coordinates, the braid would be ticker than in the reality. To avoid this, x-y coordinates are scaled, based on the relation between the size of one ridge in the braid D_{Ridge} and the horn gear diameter [11]:

$$\xi = \frac{D_{HornGear}}{D_{Ridge}} \quad (2)$$

The scaling lead to the following set of coordinates for each point:

$$\left| \begin{array}{l} x_{scalled} = \dfrac{x_{Carrier}}{\xi} \\ y_{scalled} = \dfrac{y_{Carrier}}{\xi} \\ z_{scalled} = V_{Take\_off} \cdot t \end{array} \right. \quad (3)$$

All these coordinates are used as centre lines for the yarns, which are visualised in the form of tubes, sweeping circular cross section within the C++ and VTK based software TexMind Configurator. After the first report about the developed software in 2014 [12], its practical application for development of complex braided profiles for composites was demonstrate in [13, 14]. The latest development of the algorithms and the user interface allow simulation of braids with variable cross section [15], where the state of the switches and the time of their change is recorded and used in the simulation.

3. Application examples

Figure 1 demonstrates configuration of braid, similar to 6-track based solid braid, but with small differences in the configuration. All yarns in the central area of the braid are connected, but the yarns, working on the edges build separate small regions, which can be used for simpler connection to another parts. The edges are in this case thinner and allow for instance stitching with reduced resistance of the needle. Figure 2 demonstrates the simulated profile in top and side view.

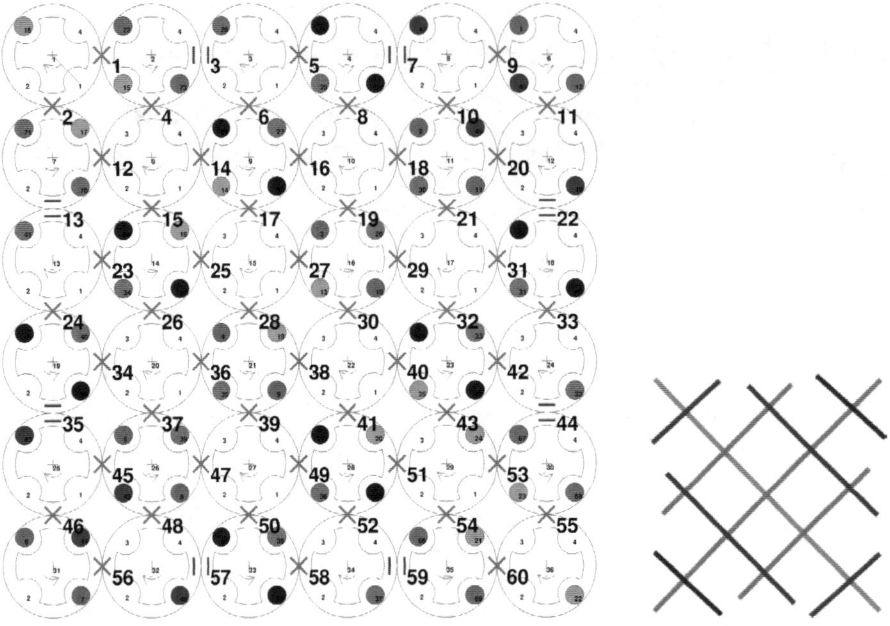

Figure 1. Braiding machine configuration for production of solid braided structure with not full connected 10 tracks

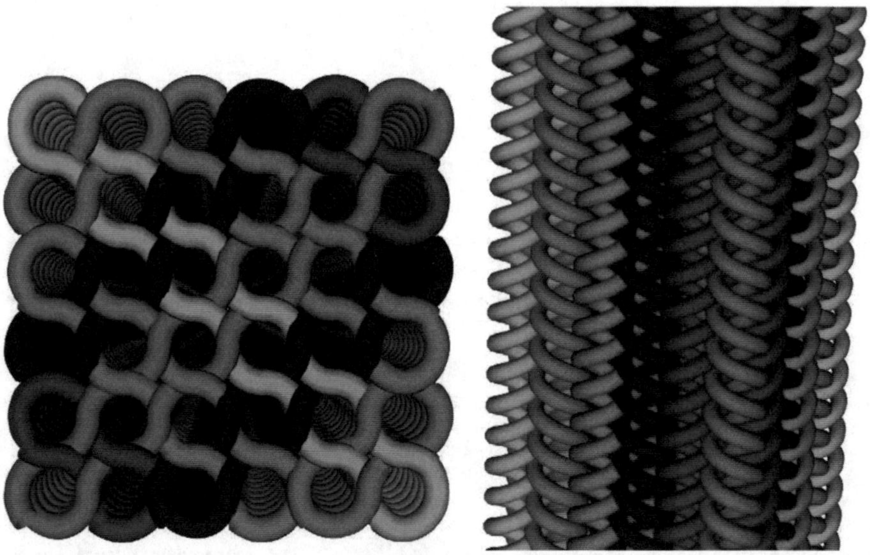

Figure 2. Idealized 3D simulation of the braided structure from the machine, presented in Figure 1

Figure 3 presents a configuration for the production of T-profile, based on one track. As found in [13], such configuration require not only gears with five slots on the turning points, as usual for flat braids, but as well five slots gear in the point, where the two tracks are crossing. In other case the arrangement of the carriers is not possible with that density. The 3D simulated structure in this case is based on less number of key-points. From the motion of one carrier along an angle, corresponding to one slot, only two points are taken into the structure. In this way the topology of the braid is preserved, but the simulation at the next steps can be performed faster, with less number of elements.

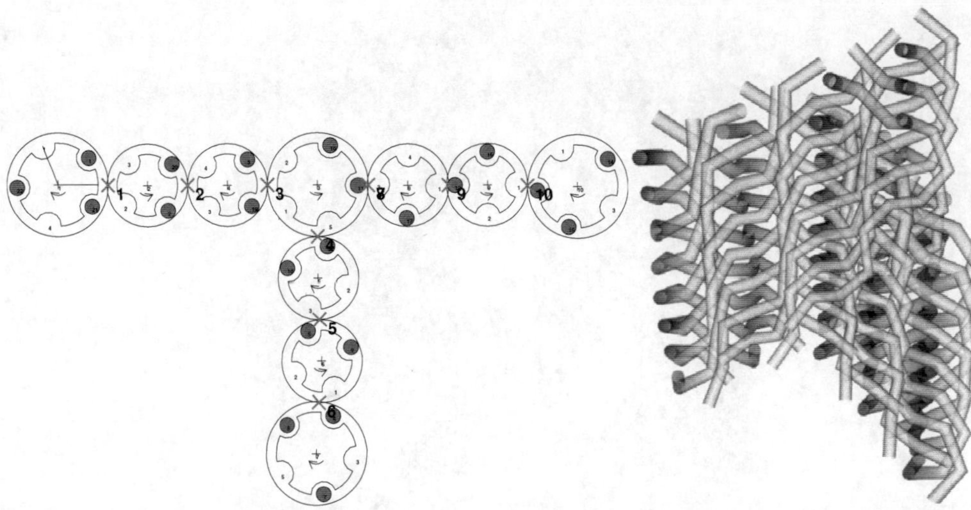

Figure 3. Configuration for production of short T-Profile with one track and the simulated 3D structure

4. From geometry to FE Mesh

The created braids with the software present idealized state of the real products only. The accuracy of the geometry can be improved by the use of different methods for mechanical or at least contact-detection algorithms. In the current case and automated export to LS-Dyna (LSTC) is implemented and tested (Fig.4). The yarn axis points from the 3D viewer are taken and the corresponding commands for LS-Dyna are created. The complete export is implemented as C++ library within the TexMind 3D Viewer.

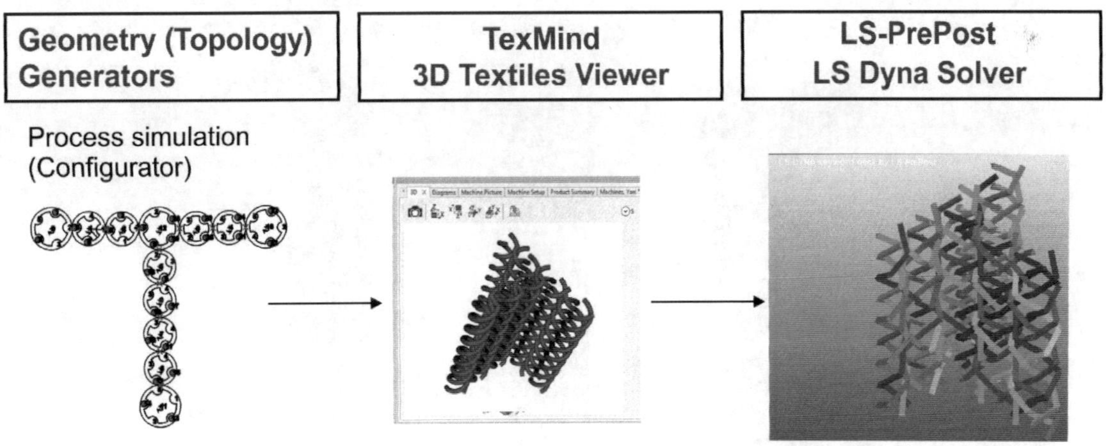

Figure 4. Simulation line from machine configuration, trough idealized geometry to FEM mesh

5. Export tests

The exporting interface is tested first with a tubular braid (Fig 5) with which as well tensile test simulation is performed. The deformed state of tubular braid are visualised in Fig. 6. The simulation results demonstrates the capability of the method and the correctness of its implementation in the software.

Figure 5. Tubular braid, generated with the TexMind software after import in the LS-Dyna as beam mesh.

Figure 6. Stages of the simulation of tensile test of the tubular braid

6. Complex profile example

Figure 7 demonstrates simulation of the deformation of the T profile, presented in figure 3. Such one deformation will happens for instance during the pultrusion process, where the product works under tension. The sample demonstrate that the outer areas of the structure have larger displacements in Y direction compared to the middle yarn pieces of the profile.

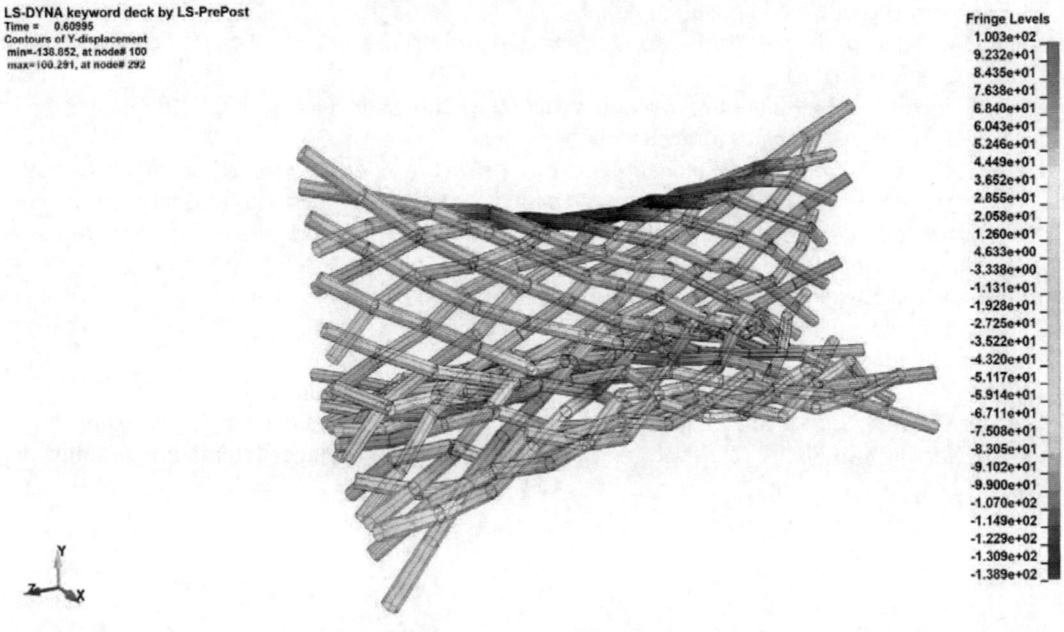

Figure 7. Deformation of braided structure with T-form of the cross section, simulated with the LS-Dyna

7. Conclusions

This work presents method for automated preparation of the simulation data of complex braided profiles for composites. The workflow starts with the common for the braider interface, where the machine and its carrier arrangement are defined and leads automatically to the input file for LS Dyna simulation software. The export is tested and verified with simple tubular structure. The developed tools can be used for any braid, created on 3D maypole braiding machine. The only disadvantage of the method is, that the generated mesh is based on not relaxed geometry, so initially a relaxation step with proper material data for the yarn has to be performed, if high accuracy of the simulation results is required.

References

[1] Bogdanovich A E 2016 An overview of three-dimensional braiding technologies *Advances in braiding technology: Specialized techniques and applications (Woodhead publishing in textiles* Number 177*)* ed Y Kyosev (Woodhead Publishing) pp 3–78

[2] Bilisik K 2016 Cartesian 3D braiding *Advances in braiding technology: Specialized techniques and applications (Woodhead publishing in textiles* Number 177*)* ed Y Kyosev (Woodhead Publishing) pp 107–45

[3] Bogdanovich A E 1993 *Composites Manufacturing* **4** 173–86

[4] Schreiber F 2016 Three-dimensional hexagonal braiding *Advances in braiding technology: Specialized techniques and applications (Woodhead publishing in textiles* Number 177*)* ed Y Kyosev (Woodhead Publishing) pp 79–88

IOP Conf. Series: Materials Science and Engineering **406** (2018) 012025 doi:10.1088/1757-899X/406/1/012025

[5] Schreiber F, Ko F K and et. al. 2009 Novel three-dimensional braiding approachand its products *ICCM-17 : 17th International Conference on Composite Materials (Edinburgh, 27.-31.07.2009)* (London: IOM Communications)
[6] Büsgen A 1993 *Neue Verfahren zur Herstellung von dreidimensionalen Textilien für den Einsatz in Faserverbundwerkstoffen: Dissertation*
[7] Lengersdorf M and Gries T 2016 Three-dimensional (3D)-maypole braiding *Advances in braiding technology: Specialized techniques and applications (Woodhead publishing in textiles* Number 177*)* ed Y Kyosev (Woodhead Publishing) pp 89–105
[8] Herzog Maschinenfabrik A 2011 *CAB Design: Computer Aided braid design* (Oldenburg: August Herzog GmbH & Co. KG)
[9] August Herzog 2013 *Product Catalogue* www.herzog-online.com (accessed 01.2013)
[10] Kyosev Y 2018 *Appl Compos Mater* **100** 416
[11] Kyosev Y 2018 *Topology based modelling of textile structures and their joint assemblies* (Springer)
[12] Kyosev Y K 2014 Machine configurator for braided composite profiles with arbitrary cross section *16th European conference on composite materials ECCM 16 16th European conference on composite materials ECCM 16 (22-26.06.2014)* (Seville-Spain)
[13] Kyosev Y and Küster K 2018 Development of Machine Configuration for T- and I-Profiles and Their Topological Modelling *Narrow and smart textiles* ed Y Kyosev *et al* (Cham, Switzerland: Springer) pp 81–9
[14] Kyosev Y K 2016 *TexMind Braiding Machine Configurator* (TexMind UG)
[15] Gleßner P and Kyosev Y 2018 Pattern Design with the Variation Braider VF of Company Herzog GmbH *Narrow and Smart Textiles* ed Y Kyosev *et al* (Cham: Springer International Publishing) pp 149–58

Meso-scale model for the forming process of biaxial reinforced weft-knitted fabrics

M Q Pham[1]*, O Döbrich[1], J Mersch[1], T Gereke[1], C Cherif[1]

[1] Technische Universität Dresden, Faculty of Mechanical Science and Engineering, Institute of Textile Machinery and High Performance Material Technology (ITM), 01062 Dresden, Germany

* minh_quang.pham@tu-dresden.de

Abstract. Numerical modelling of textile materials is important for developing new forming processes of textile reinforcements for composite parts, namely to decide on forming tool geometry and forming process parameters such as blank holder forces. A numerical model reduces the application of trial-and-error, which significantly reduces costs and material resources. The magnitude and distribution of blank holder forces play a decisive role in the quality of the formed textile. In general, no or even low blank holder forces will lead to wrinkles in the useful part of the preform. This reduces the mechanical quality of composite parts reinforced by textiles. In contrast, excessive blank holder force will cause damage to the textile. To observe these effects in complex forming situations, a material model has to be found which takes all forming modes of textile materials into account. A meso-scale model for biaxial reinforced weft-knitted fabrics is introduced, in which present yarns are simplified and discretized by beam elements. A single yarn is considered as a chain of multiple beam elements. Methods of modelling the complex geometry of the reinforced knitted structure are presented. The models are suitable for virtual mechanical tests and large scale forming simulations.

1. Introduction

The nature resources for economic development are limited. To achieve a sustainable development, the energy consumption, the emission of harmful substances and the use of material should be reduced. Fibre-reinforced polymer composites (FRP), which are increasingly used, are one of the innovative development to meet the mentioned needs. Continuous fiber-reinforced composites consist of a tensile load-absorbing textile reinforcement structure and a shaping, compressive-load-absorbing matrix material. FRPs have excellent properties, such as high specific strength and stiffness, good damping properties, chemical resistance and low thermal expansion. In comparison to conventional materials, particularly based on metals, they are distinguished by outstanding corrosion resistance and weight reduction. In order to fully exploit the potential of the fiber reinforcement in the composite, the fibers in the preform have to be arranged along the main load direction and suitably embedded into the matrix [1]. Vallons et al. [2] concluded that the strength of a biaxial carbon/epoxy non crimp fabric composite decreased approximately 20% under tensile stress due to a 5° misalignment of the fibre.

The forming of complex 3D components from 2D textile material with a load path optimized textile reinforcement structure requires innovative and reliable manufacturing processes at a reasonable price. To boost the usability of this material a reliable process design is necessary, which can only be achieved with the help of numerical methods. Many production parameters of the forming process have to be determined by "trial and error" experiments, for example the blank holder forces required for forming.

Excessive blank holder forces will pull the textile apart at a short travel distance of the stamp. Low blank holder forces lead to wrinkles on semi-finished product, which would reduce the mechanical properties of the composite part [3]. With the help of numerical simulation tools, the time and cost intensive "trial and error" experiments can be avoided.

There are some computing models available for analyzing the forming behavior of textile structures, such as woven fabrics [4–9],biaxial and multiaxial non-crimp fabrics [6,10], weft-knitted fabrics [11,12], and multiaxial warp-knitted fabrics [13]. The textile structure, in general, can be modelled at three different levels of objectivity: micro-scale (Figure 1), meso-scale (Figure 2) and macro-scale (Figure 3).

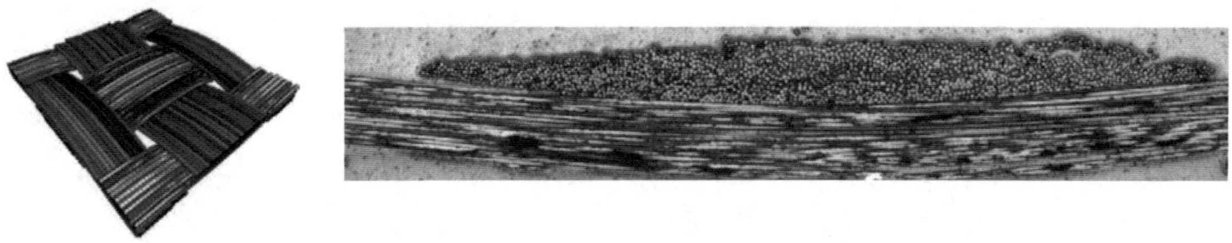

Figure 1. Micro-scale model of a plain woven fabric unit cell with 30 beam element chains for 1 yarn (left) and the cross-section of the real fabric (right) [14]

Figure 2. Meso-scale model of a twill woven fabric unit cell with shell elements [15]

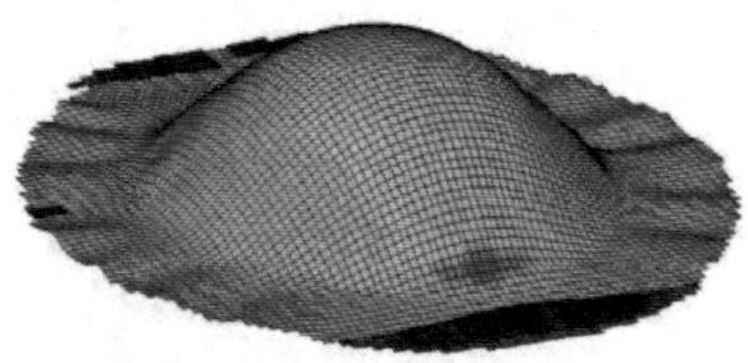

Figure 3. Macro-scale model of textile fabrics during forming process [16]

At the micro-scale, a single filament of the technical multifilament yarns can be modelled by beam or solid elements. But for the sake of simplification the quantity of representative beam or solid elements in the model is far less than the reality as shown in Figure 1. The models based on this method are closest to reality. The computational costs are highest compared to the other levels. For mechanical characterization of the filament, additional properties of the material such as the contact friction should be determined.

The models on the meso-scale level are simpler than the previous one. The smallest unit in these models is the yarn, not the filament itself. Depending on the diameter and the estimation of the behavior of yarn, beam elements [17], shell elements [15] or solid elements [18,19] are used. The computational cost of these models is lower than on the micro-scale but still relatively high compared to a continuum macro model. The contact between the single yarns, as well as the contact between the yarns and other objects like forming tools still has to be computed. For every new binding type, a new model for the textile is required. Figure 2 shows a meso-scale model of twill woven fabric unit cell with shell element [15]. Despite its simplicity, the meso-scale model can be used for mechanical [20,21], permeability [22,23] and structural analysis [24].

The macro-scale models have the least computational cost, which is significantly lower in comparison to the meso-scale and the micro-scale models. The textile structure is summarized in a continuum as shown in Figure 3. For a flat textile structure, shell elements can be applied for modelling. The textile structure plays no role while building the macro-scale models. The required parameters for the material model are the force-strain tensile behavior in warp and weft direction, the shearing behavior as well as the bending stiffness. Those properties have to be determined by physically or virtually testing the fabrics [25]. Additionally, according to Boisse et al. [26] the tensile behavior of the fabric, where the yarn systems are interwoven, is biaxial, i. e. the tension-deformation states in the warp or weft directions depend on the other direction. Therefore, this biaxial tensile behavior also needs to be tested on a biaxial tensile testing device. Despite its simplicity the macro-scale model can describe the shearing of the yarn system and typical failure modes during forming of textile such as wrinkle formation.

The development process of complex composite structures requires a higher level of objectivity, which the meso-scale model can offer. In this paper a simplified meso-model for biaxial reinforced weft-knitted fabrics is introduced, in which even yarn slippage and sliding are taken into account. Yarns are simplified and discretized by beam elements. A single yarn is considered as a chain of multiple beam elements. Methods of modelling, validation and the application of this model will be presented.

2. Material

The biaxial reinforced weft-knitted fabrics in this paper are made of commingled hybrid yarns of carbon fibre and polyamide 6.6. Folded yarn of glass fibre and polyamide 6.6 was used as the knitting yarn as shown in Figure 4. The fabrics were fabricated on a modified flat-bed knitting machine [27].

(a) (b)

Figure 4. Biaxial reinforced weft-knitted fabric: (a) Right side (b) Left side

3. Methods

3.1. Experimental testing of the textile material

To characterize the textile material, typical tensile experiments for yarns (ISO 3341 [28]) and textile stripes (DIN EN ISO13934-1 [29]) were carried out. Additionally, shear tests with a picture frame of size 200 mm × 200 mm and cantilever bending tests (DIN 53362 [30]) help to determine the shear resistance and bending rigidity of the fabrics accordingly. A summary of the textile characteristic is presented in Table 1.

The CF/PA hybrid yarn has a fineness of 1200 tex and is actual a folded yarn from four smaller commingled hybrid yarns of 300 tex fineness. It has a non-linear mechanical characteristic as shown in Figure 16. On the contrary the knitting yarn, which is a folded yarn of glass fibre of 68×2 tex and polyamide 6.6 of 31.2 tex, possesses a linear mechanical characteristic as shown in Figure 17. The result of tensile textile stripes test and shear test also shows non-linear characteristic as shown in Figure 19 and Figure 5 accordingly.

IOP Conf. Series: Materials Science and Engineering **406** (2018) 012026 doi:10.1088/1757-899X/406/1/012026

Table 1. Characteristics of the biaxial reinforced weft-knitted fabrics

Parameter	Value
Reinforcing yarn (commingled hybrid yarn and folded)	CF/ PA 6.6 (4 × 300 tex)
Distance between reinforcing yarns	3.6 mm
E-Modulus of reinforcing yarn	69.6 GPa
Knitting yarn (folded yarn)	GF (68×2 tex)/ PA 6.6 (31.2 tex)
E-Modulus of knitting yarn	68.4 GPa
Flat density of the fabrics	824.2 g/m^2
Thickness of fabric according to ISO 5084 [31]	1.9 mm
Max. tensile force of fabric according to DIN EN ISO 13934-1 [29]	7530 N
Fracture strain of fabric according to DIN EN ISO 13934-1 [29]	1%
Overhang length of fabric in Cantilever test according to DIN 53362 [30]	165 mm
Loop length	14.4 mm

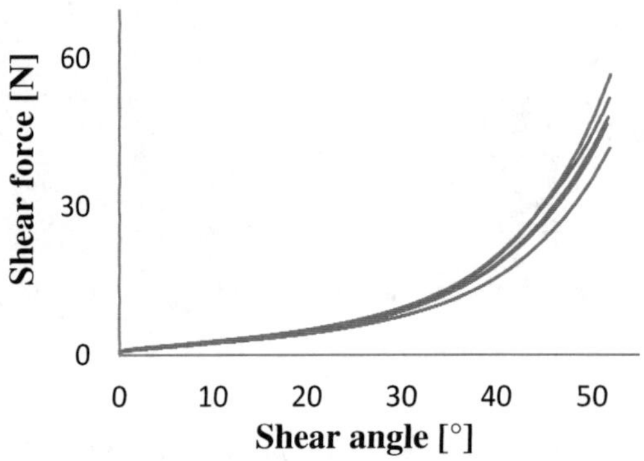

Figure 5. Shear force-shear angle curve for picture frame test of fabrics

Although the biaxial reinforced weft-knitted fabric is classified as a non-crimp fabric, the reinforcing yarns in the structure are actually not perfectly non-crimp. Due to the interaction between knitting yarn and reinforcing yarn during knitting process, the reinforcing yarns will be slightly crimped. To determine the crimp ratio of reinforcing yarn, the standard DIN 53852 [32] is applied. The crimp ratio is calculated as follows

$$E = \frac{l_f - l_w}{l_f} = 1 - \frac{l_w}{l_f} \qquad (1)$$

where E [-] is the crimp ratio, l_f [mm] is the length between two markings on the yarn after taking from the textile structure and stretching under conditions as in German Standard DIN 53 852 [32], l_w [mm] is the length between two markings on the yarn before stretching. The crimp ratio of warp and weft yarns are shown in Table 2.

The friction between the textile and metal part was tested on a biaxial tensile test machine. The principle of the test is shown in Figure 6. The textile stripe sample is pulled with force F_{pull} along its length while being pressed with a predefined force F_N in the vertical direction. The experimental result of the friction test is shown in Figure 7.

Table 2. Crimp ratio of warp and weft reinforcing yarns

	Crimp ratio E [-]
Warp yarn	0.003496
Weft yarn	0.002597

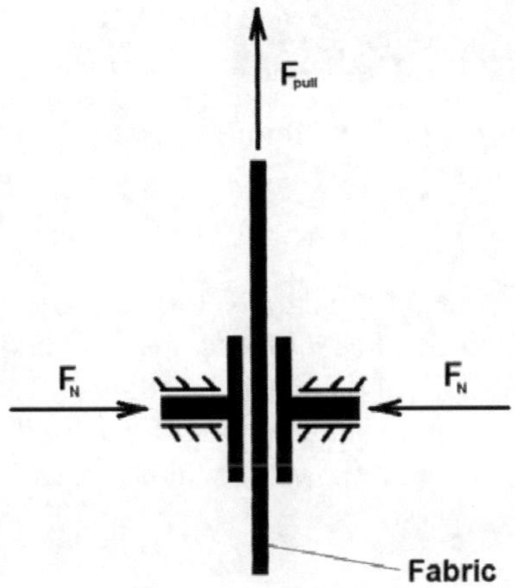

Figure 6. Testing principle of friction between textile and metal part on the biaxial tensile machine

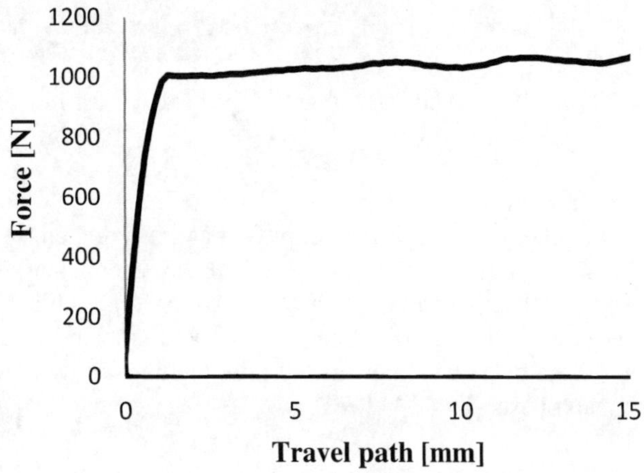

Figure 7. Experimental result of the friction test

Forming experiments of the textile fabric with L-profile were used to validate the usefulness of the model [33]. An overview of the stamp forming device with eight blank holder segments is shown in Figure 8. A formed sample with the L-Profile forming device for the biaxial reinforced weft-knitted fabric is shown in Figure 9. This sample was scanned with a laser device to determine the yarn orientation in the formed structure, which will be used as a comparison to the forming simulation using the textile meso-scale model.

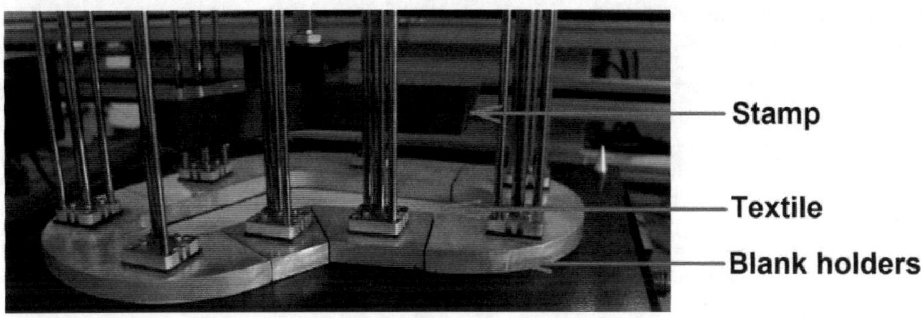

Figure 8. Forming device with L-shape [33]

Figure 9. Formed sample with the L-Profile forming device for the biaxial reinforced weft-knitted fabric

3.2. Modelling

In general the yarns are simplified and discretized by beam elements. A single yarn is considered as a chain of multiple beam elements. At the recent state of development, linear elastic material model is used. By the application of Belytschko-Schwer resultant beam formulation [34], the bending rigidity of the yarn can be described correctly. The diameter of the beam element can be chosen freely and does not affect beam bending rigidity. The methods of modelling the geometry of yarns will be presented in the next parts.

3.2.1. Modelling geometry of the knitting yarn system

The knitting yarn system in the biaxial reinforced weft-knitted fabric is a single jersey structure, which can be described mathematically by the equations of Choi and Lo [35]. The construction of one knitting loop is governed by the following equations:

$$x(t) = at^3 - 1.5at^2 + 0.5(a + w)t \tag{2}$$

$$y(t) = 0.5(c + 2e)(1 - \cos(\pi t)) \tag{3}$$

$$z(t) = 0.5(t_h - d)(1 - \cos(2\pi t)) \tag{4}$$

where x, y, z [mm] are coordinates of every single point on the knitting loop, w [mm] is the loop width, (c+2e) [mm] is the loop height, e [mm] is the adjacent loop overlapping distance, t_h [mm] is the fabric thickness, d [mm] is the yarn diameter, a [-] is a free parameter, t [-] (t ~[0,1]) is a parameter of the space curve. An example of a simple single jersey structure modelled by these equation is shown in Figure 10.

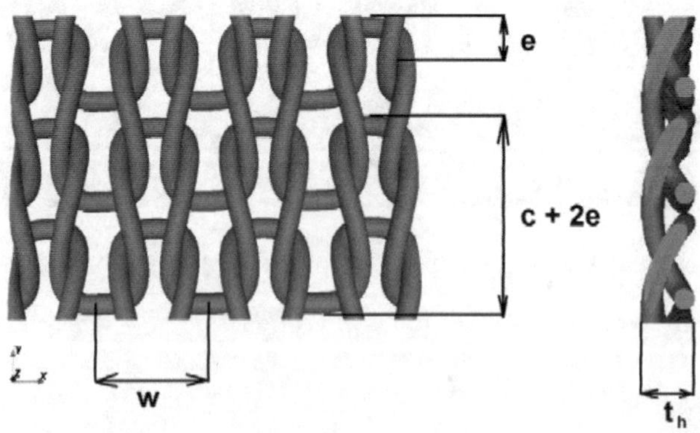

Figure 10. Single jersey weft-knitting structure according to Choi and Lo [36]

3.2.2. Modelling geometry of the reinforcing yarns
To implement the crimp of the yarns into the model, it is assumed that these yarns have only vertical waviness along its main axis in a harmonic manner as shown in Figure 11.

Figure 11. Reinforcing yarn within the fabric with waviness

The relationship between the coordinators of the yarn is described as follows:

$$z(y) = Acos(y) \tag{5}$$

where z [mm] is the z coordinate of a point on the reinforcing yarn, y [mm] is the y coordinate of a point on the reinforcing yarn, and A is a waviness parameter. The length of a cosine wave arc is from 0 to 2π and is determined with:

$$\int_0^{2\pi} \sqrt{1 + A^2 sin^2 y}\, dy \tag{6}$$

$$= 4 \int_0^{\frac{\pi}{2}} \sqrt{1 + A^2 sin^2 y}\, dy \tag{7}$$

Equation (7) is a Legendre complete elliptic integral of the second kind [36] and can be evaluated with functions available in the mathematical library of SymPy for Python. From Equation (1) we can calculate the crimp ratio of the beam chain in the model with the following formula:

$$E = 1 - \frac{4 \int_0^{\frac{\pi}{2}} \sqrt{1 + A^2 sin^2 y}\, dy}{2\pi} \tag{8}$$

By solving Equation (8) for each value of crimp ratio E of warp yarn and weft yarn, the waviness parameter A for each reinforcing yarn system can be determined.

4. Result
The diameter of yarns was chosen as 0.75 mm for reinforcing yarns and 0.25 mm for knitting yarns. An overview of the meso-scale model is shown in Figure 12 and Figure 13.

Figure 12. Cross-section of yarns in the fabric and its representative beams in the model

IOP Conf. Series: Materials Science and Engineering **406** (2018) 012026 doi:10.1088/1757-899X/406/1/012026

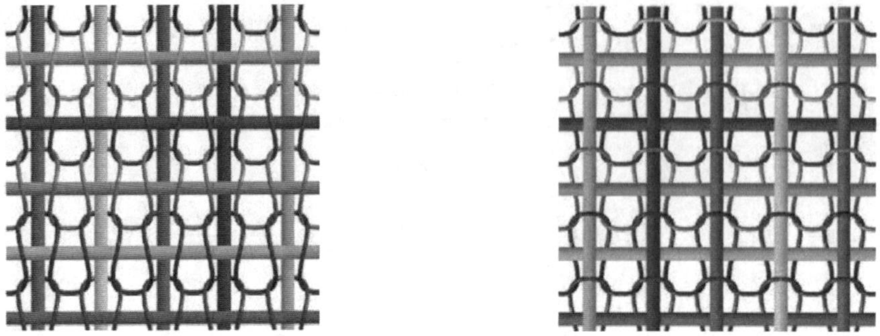

(a) (b)

Figure 13. Overview of the meso-scale model: (a) Right side (b) Left side

A simple simulation of the cantilever test for a single yarn helps to verify the bending rigidity of the beam element in the model as shown in Figure 14. For knitting yarn an analogous simulation was also made for the same purpose.

Figure 14. Simulation of cantilever test to verify the bending rigidity of beam element representative for reinforcing yarn in model

Another simulation of cantilever test for the textile fabric was used to verify the bending rigidity of the whole textile meso-scale structure in model as shown in Figure 15. Both bending rigidities – yarn and fabric – were verified with those simulations.

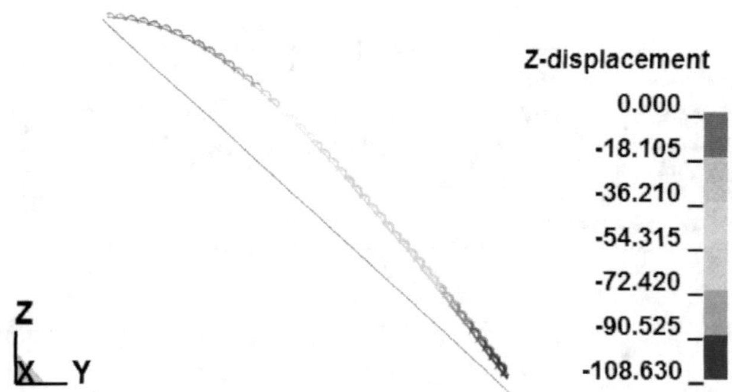

Figure 15. Simulation of cantilever test to verify the bending rigidity of the biaxial reinforced weft-knitted fabric in model

The simulation of tensile test helps to validate the force-strain behavior of the beam element with linear elastic material model. While this material model satisfied the requirement of yarn material with

IOP Conf. Series: Materials Science and Engineering **406** (2018) 012026 doi:10.1088/1757-899X/406/1/012026

linear characteristic such as knitting yarn (Figure 17), it causes a noticeable error by the yarn material with non-linear characteristic such as the commingled hybrid reinforcing yarn (Figure 16).

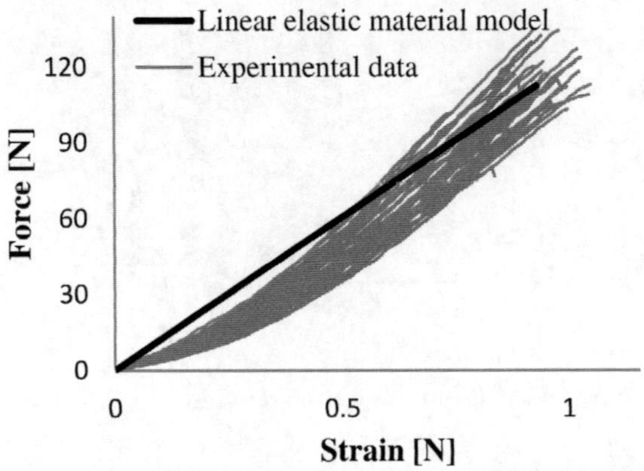

Figure 16. Comparison of force-strain curves of linear elastic material model and experimental data of reinforcing yarn

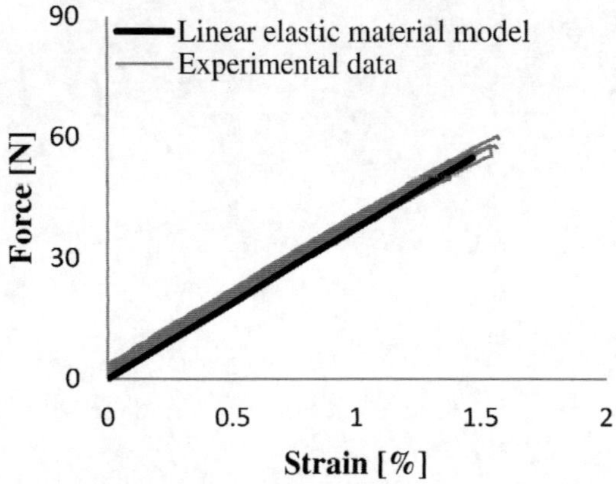

Figure 17. Comparison force-strain curves of linear elastic material model and experimental data of knitting yarn

A simulation of the fabric tensile test helps to check the correct mechanical behavior of the textile fabric model. Here, the fabric stripes with the dimension of 200 mm × 50 mm, consist of 14 yarns in the tensile direction. Due to the simplification of the yarn at the meso-scale, the main reinforcing yarn system in the model along the tensile direction takes most of the tensile stress. Figure 18 shows the axial force of the beam elements in the structure just before reaching fracture strain. Figure 19 shows the comparison of force-strain curves between simulation and experimental tensile tests.

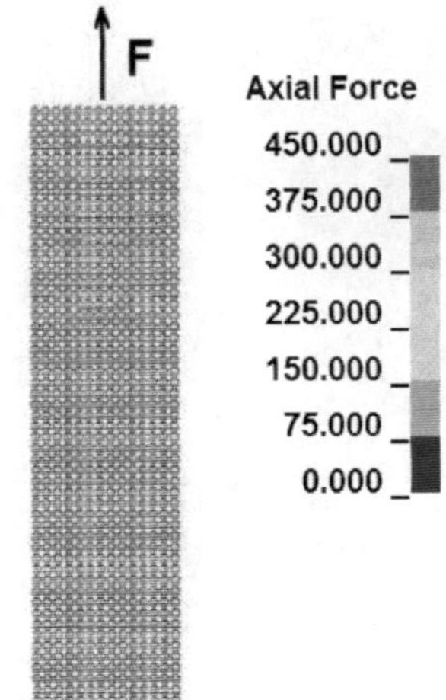

Figure 18. Axial force of the beam elements in structure just before reaching fracture strain

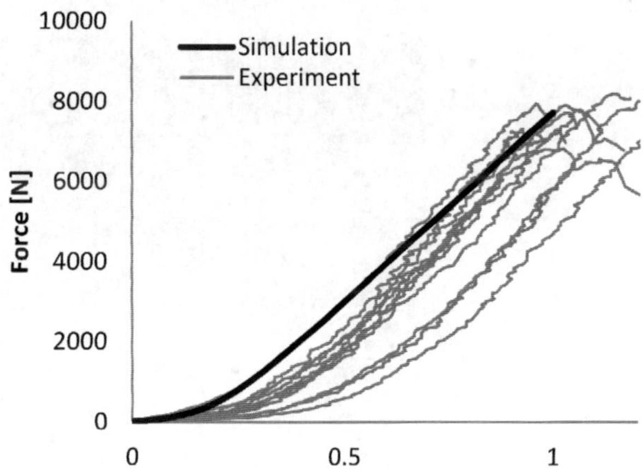

Figure 19. Comparison of force-strain curves between simulation of tensile test for textile stripes and experiment

A forming simulation using the L-profile was performed to check the usefulness of the textile meso-scale model as shown in Figure 20. The simulation was done with the same process parameters, which were used for the experiment (Figure 9), namely force of 44N was applied homogeneously for each

IOP Conf. Series: Materials Science and Engineering **406** (2018) 012026 doi:10.1088/1757-899X/406/1/012026

blank holder segment. The result of simulation shows the yarn run of the warp reinforcing system as in Figure 21. For model validation with experimental test, the experimentally formed sample as shown in Figure 9 was scanned with laser triangulation to record the yarn run (Figure 22). An optical comparison is shown in Figure 23.

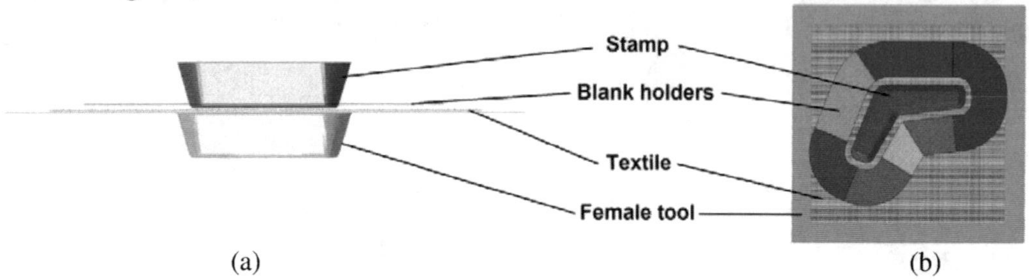

Figure 20. Forming simulation for L-Profile: (a) Front view (b) Top view

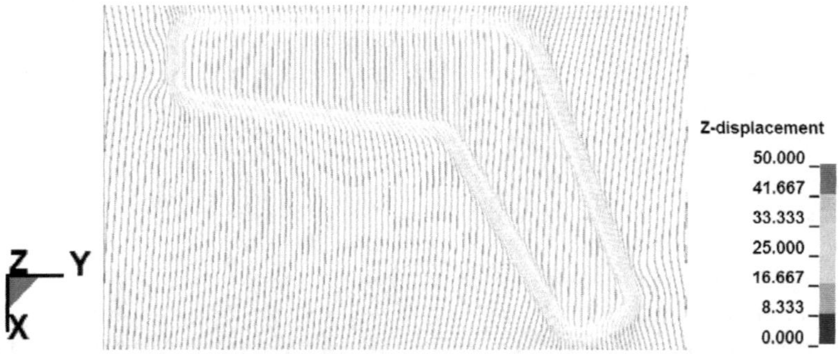

Figure 21. Yarn run of warp reinforcing system according to simulation result

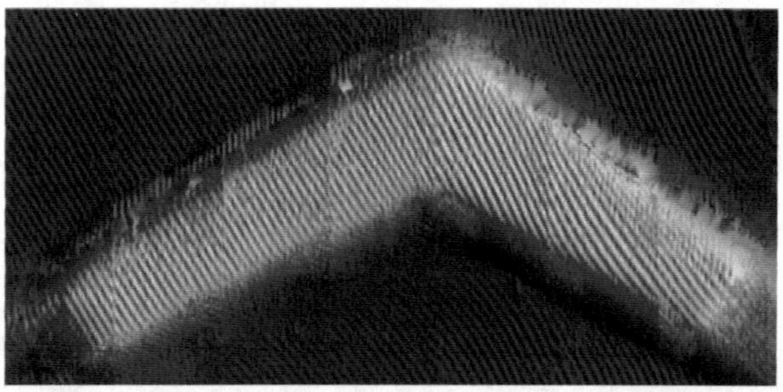

Figure 22. Scanned result of the formed sample with yarn run

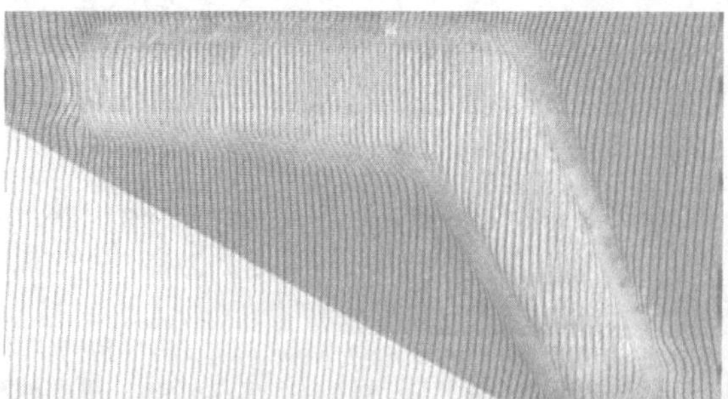

Figure 23. Optical comparison of yarn run between simulation result and scanned data

IOP Conf. Series: Materials Science and Engineering **406** (2018) 012026 doi:10.1088/1757-899X/406/1/012026

To quantitatively compare the yarn direction between model and experiment, the angle of each yarn run on top area of the formed part with the horizontal direction was measured as shown in Figure 24. The quantitative direction comparison of 70 warp yarns on the top side of the L-profile part is shown in Figure 25. At the left end of the L-profile, strong sliding of warp yarn system can be observed in both experiment and model as shown in Figure 26. Comparison between experiment and forming simulation using the meso-scale approach show very good agreement.

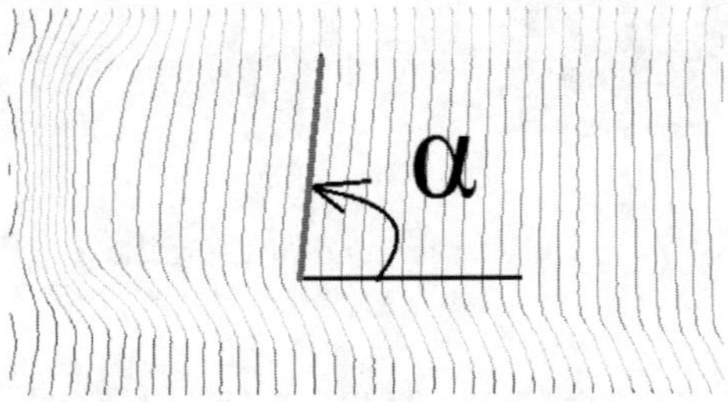

Figure 24. Measuring method of the direction of the warp yarn

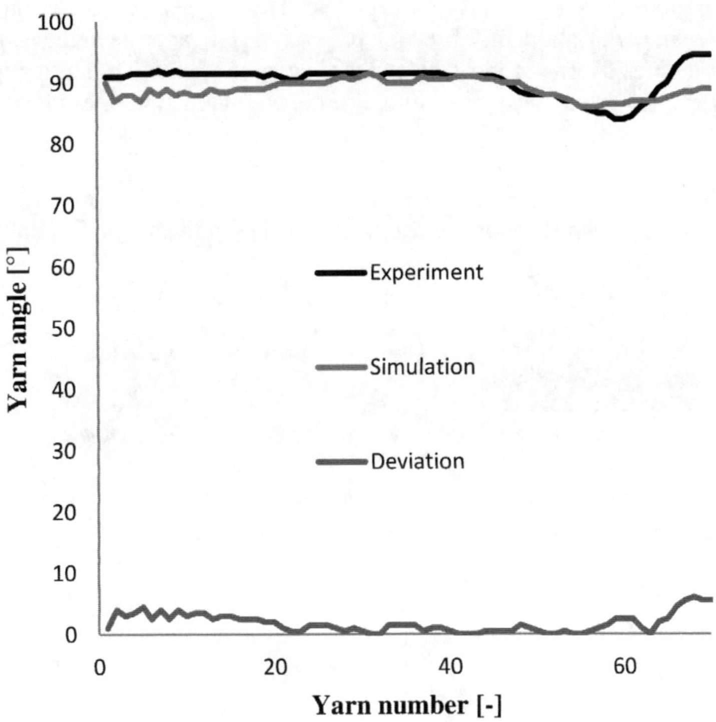

Figure 25. Quantitative direction comparison of 70 warp yarns

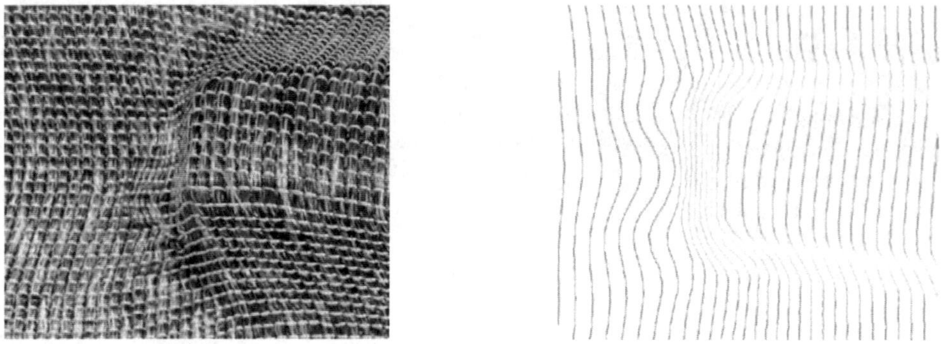

Figure 26. Strong sliding of warp yarns at left end of L-profile: (a) Experiment (b) Simulation

5. Conclusions

A meso-scale model of biaxial reinforced knitted fabric was introduced. The modelling and validating methods were presented. The model is simple and capable of taking into account most of the mechanical behaviours of the fabric. The usefulness of the model is proved by the application to determine the yarn misalignment during preforming process for an L-profile. The large-scale simulation predicted the direction of yarn with a slight deviation, this direction plays an significant role in the end properties of the composite part. The sliding of yarn is observable in the simulation results. The shear angle of reinforcing yarn system can also be validated, once the yarn run of the weft yarn system can be determined, which would be part of future studies.

Acknowledgements

This research was financially supported by the European Social Fund and the Federal State of Saxony (EFRE-Project "AMARETO Dresden").

European Union

Europe funds Saxony.
EFRE
European Regional Development Fund

The computations were performed partly on an HPC system at the Center for Information Services and High Performance Computing (ZIH) at TU Dresden.

ZIH
Zentrum für Informationsdienste und Hochleistungsrechnen

References

[1] Cherif C (ed.) 2016 *Textile Materials for Lightweight Constructions: Technologies -Methods – Materials - Properties* (Springer-Verlag Berlin Heidelberg)

[2] Vallons K, Duque I, Lomov S V and Verpoest I 2011 Loading direction dependence of the tensile stiffness, strength and fatigue life of biaxial carbon/epoxy NCF composites *Compos. Part A Appl. Sci. Manuf.* **42** 16–21

IOP Conf. Series: Materials Science and Engineering **406** (2018) 012026 doi:10.1088/1757-899X/406/1/012026

[3] Gereke T, Döbrich O, Hübner M and Cherif C 2013 Experimental and computational composite textile reinforcement forming: A review *Compos. Part A Appl. Sci. Manuf.* **46** 1–10

[4] Döbrich O, Gereke T and Cherif C 2012 Drape Simulation : Textile material model for correct property reproduction to improve the preform development process of fiber-reinforced structures *Proc. 12th Int. LS-DYNA® Users Conf.* (Dearborn, MI, USA) 1–10

[5] Khan M A, Mabrouki T, Vidal-Sallé E and Boisse P 2010 Numerical and experimental analyses of woven composite reinforcement forming using a hypoelastic behaviour. Application to the double dome benchmark *J. Mater. Process. Technol.* **210** 378–88

[6] Hamila N and Boisse P 2008 Simulations of textile composite reinforcement draping using a new semi-discrete three node finite element *Compos. Part B Eng.* **39** 999–1010

[7] Cherouat A and Billoët J L 2001 Mechanical and numerical modelling of composite manufacturing processes deep-drawing and laying-up of thin pre-impregnated woven fabrics *J. Mater. Process. Technol.* **118** 460–71

[8] Gelin J C, Cherouat A, Boisse P and Sabhi H 1996 Manufacture of thin composite structure by the RTM process: Numerical simulation of the shaping operation *Compos. Sci. Technol.* **56** 711–8

[9] Hamila N, Wang P, Vidal-Sallé E and Boisse P 2011 Simulation of forming and wrinkling of textile composite reinforcements *AIP Conference Proceedings* **1353** 942–7

[10] Hübner M, Gereke T and Cherif C 2014 Modeling of multiaxial non-crimp fabrics *Proc. 11th World Congress on Computational Mechanics WCCM XI* (Barcelona)

[11] Duhovic M and Bhattacharyya D 2006 Simulating the deformation mechanisms of knitted fabric composites *Compos. Part A Appl. Sci. Manuf.* **37** 1897–915

[12] Savci S, Curiskis J I and Pailthorpe M T 2000 A study of the deformation of weft-knit preforms for advanced composite structures Part 2: The resultant composite *Compos. Sci. Technol.* **60** 1943–51

[13] Hu J and Jiang Y 2002 Modeling formability of multiaxial warp knitted fabrics on a hemisphere *Compos. Part A Appl. Sci. Manuf.* **33** 725–34

[14] Döbrich O, Gereke T and Cherif C 2016 Modeling the mechanical properties of textile-reinforced composites with a near micro-scale approach *Compos. Struct.* **135** 1–7

[15] Hübner M, Staiger E, Küchler K, Gereke T and Cherif C 2016 Simulation of patched woven fabric composite structures under tensile load *Tekstilec* **59** 175–81

[16] Döbrich O, Staiger E, Andrews M-L, Gereke T and Cherif C 2016 A simulation model for metal textile hybrid material *Proc. 16th AUTEX World Textile Conference* (Ljubljana)

[17] de Araújo M, Fangueiro R and Hong H 2004 Modelling and simulation of the mechanical behaviour of weft-knitted fabrics for technical applications. Part IV: 3D FEA model with a mesh of tetrahedric elements *Autex Res. J.* **4** 72–80

[18] Bacarreza O, Abe D, Aliabadi M H And Kopula Ragavan N 2012 Micromechanical modeling of advanced composites *J. Multiscale Model.* **04** 1250005

[19] Fagiano C, Genet M, Baranger E and Ladevèze P 2014 Computational geometrical and mechanical modeling of woven ceramic composites at the mesoscale *Compos. Struct.* **112** 146–56

[20] Verpoest I and Lomov S V. 2005 Virtual textile composites software WiseTex: Integration with micro-mechanical, permeability and structural analysis *Compos. Sci. Technol.* **65** 2563–74

[21] Lomov S V., Gusakov A V., Huysmans G, Prodromou A and Verpoest I 2000 Textile geometry preprocessor for meso-mechanical models of woven composites *Compos. Sci. Technol.* **60** 2083–95

[22] Zeng X, Brown L P, Endruweit A and Long A C 2012 Advanced geometry modelling of 3D woven reinforcements in polymer composites: Processing and performance analysis *Proc. Fourth World Conf. 3D Fabr. their Appl.* (Aachen)

[23] Zeng X, Brown L P, Endruweit A, Matveev M and Long A C 2014 Geometrical modelling of 3D woven reinforcements for polymer composites: Prediction of fabric permeability and composite mechanical properties *Compos. Part A Appl. Sci. Manuf.* **56** 150–60

IOP Conf. Series: Materials Science and Engineering **406** (2018) 012026 doi:10.1088/1757-899X/406/1/012026

[24] Matveev M Y 2015 Effect of variabilities on mechanical properties of textile composites (University of Nottingham)

[25] Döbrich O, Gereke T, Diestel O, Krzywinski S and Cherif C 2014 Decoupling the bending behavior and the membrane properties of finite shell elements for a correct description of the mechanical behavior of textiles with a laminate formulation *J. Ind. Text.* **44** 70–84

[26] Boisse P, Gasser A and Hivet G 2001 Analyses of fabric tensile behaviour: Determination of the biaxial tension-strain surfaces and their use in forming simulations *Compos. Part A Appl. Sci. Manuf.* **32** 1395–414

[27] Großmann K, Bräunling S, Cherif C, Staiger E, Ulbricht E and Linse T 2013 Research report IGF-Forschungsvorhaben Nr. 377 ZBR 3D-Bauteile aus Blech und Textil durch umformende Verbundherstellung (Dresden)

[28] International Organization for Standardization ISO 3341:2000-05 Textile glass - Yarns -Determination of breaking force and breaking elongation

[29] International Organization for Standardization ISO 13934-1:2013 Textiles - Tensile properties of fabrics - Part 1: Determination of maximum force and elongation at maximum force using the strip method

[30] Deutsches Institut für Normung DIN 53362:2003-10 Testing of plastics films and textile fabrics (excluding nonwovens), coated or not coated fabrics - Determination of stiffness in bending - Method according to Cantilever

[31] International Organization for Standardization ISO 5084:1996 Textiles - Determination of thickness of textiles and textile products

[32] Deutsches Institut für Normung DIN 53852 Bestimmung von Garnlängenverhältnissen in Geweben und Maschenwaren

[33] Nosrat Nezami F, Gereke T and Cherif C 2013 Automated preforming of woven fabrics with segmented blank holders: a multiscale experimental approach *Proc. TexComp-11*

[34] Livermore Software Technology Corporation (LSTC) 2015 Theory Manual (California: Livermore software technology corporation)

[35] Choi K F and lo T Y 2003 An energy model of plain knitted fabric *Text. Res. J.* **73** 739–48

[36] Carlson B C 1987 A table of elliptic integrals of the second kind *Math. Comput.* **49** 595

IOP Conf. Series: Materials Science and Engineering **406** (2018) 012027 doi:10.1088/1757-899X/406/1/012027

Recent developments in "COMPOSITES DREAM" for meso-FE modelling of advanced materials

Tatsuo Sakakibara[1]*, Takuya Yamamoto[1], Gen Yamada[1], Aki Imaoku[1], Ryo Ohtagaki[1], Masaru Zako[2]

[1]Itochu Techno-Solutions Corporation (CTC), 3-1-3, Umeda, Kita-ku, Osaka, Japan
[2]Osaka University, 2-1, Yamadaoka, Suita, Osaka, Japan

* tatsuo.sakakibara@ctc-g.co.jp

Abstract. Composites Dream is a powerful software package of multi-scale finite element (FE) modelling and analysis for advanced composite materials. Composites Dream consists of WiseTex, MeshTex and SACOM which are originally developed by KU-Leuven and Osaka University respectively. In recent years, we are continuously developing some features to generate finite element model for several kinds of textile composites. In this paper, we present the recent developments in Composites Dream including meso FE-modelling of a cylindrical braided composites with inlay yarn, a bi-axial non-crimp fabric with stich modelled by beam element, a layered unidirectional non-crimp fabric etc. In addition, we also present a short fiber modelling capability associated arbitrary orientation tensor.

1. Introduction

Recently many kinds of composite materials and many type of composite sheets are produced for various industries. In order to design new products using such composite materials, a simulation software is strongly required to evaluate their characteristics. However, recent composite materials have complex structure so it is very difficult to generate numerical model for simulation. We have been developing the commercial software named "Composites Dream" which is modeling and simulation tool for textile composites. Composites Dream consists of WiseTex [1] [2], MeshTex [3] and SACOM [4] which are originally developed by KU-Leuven and Osaka University respectively. In this paper, we present the recent developments in Composites Dream to generate finite element models which are cylindrical braided composites, multi-ply stitched fabrics and random mat with short fiber reinforcements.

2. Modeling of Braided Tube

Figure 1 shows a schematic image of the braided fabric. It is very difficult to generate finite element model of cylindrical braided composites in especially when inlay yarn must be taken into account. The parameters which can be specified to generate braided tube are tube length, tube diameter, number of yarns, braided angle, consideration of inlay and braiding pattern which are shown in Figure 2. It is very easy to generate solid mesh for finite element analysis to simulate the complicated braided fabrics using Composites Dream. Figure 3 shows some examples by setting various generating parameters. Resin matrix can be taken into account by superimposed simulation. Figure 4 shows superimposed model of braided fabric and resin matrix. Braided yarn is modeled by solid element so that the deformation of the braiding yarn section along the braiding axis can be simulated [5].

IOP Conf. Series: Materials Science and Engineering **406** (2018) 012027 doi:10.1088/1757-899X/406/1/012027

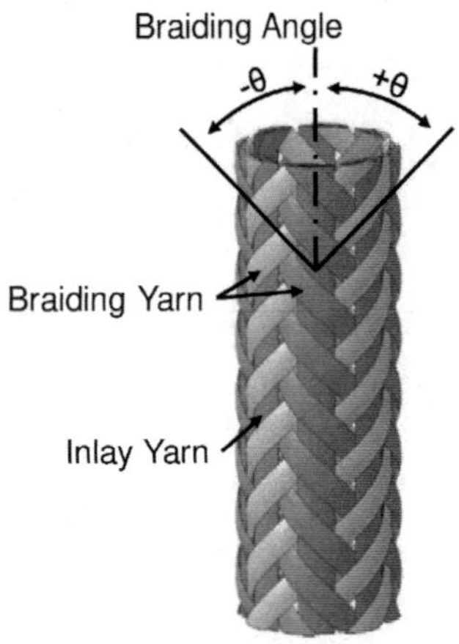

Figure 1. Schematic image of a braided fabric

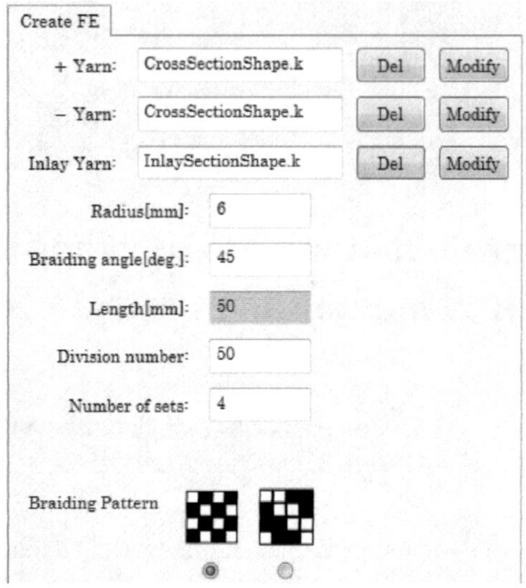

Figure 2. Input dialog of braided tube

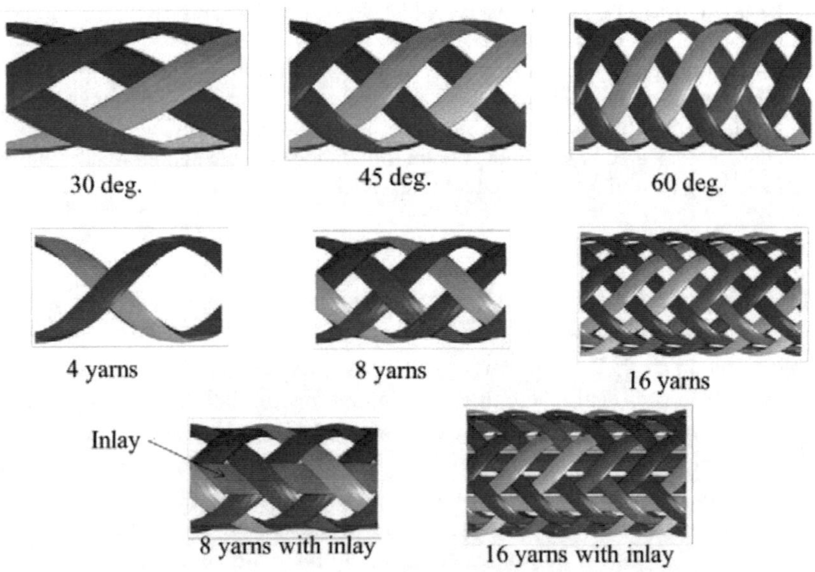

Figure 3. Example of various types of braided tube

Figure 4. Superimposed geometry of fiber and matrix

IOP Conf. Series: Materials Science and Engineering **406** (2018) 012027 doi:10.1088/1757-899X/406/1/012027

3. Modelling of Non-Crimp Fabrics

There is a growing need for using non-crimp fabric in automotive industry because of cost and draping deformability. Figures 5 and 6 show a biaxial non crimp fabrics and figures 7 and 8 show a unidirectional non crimp fabrics. Figure 9 shows numerical FE-model for biaxial NCF generated by Composites Dream. In Composites Dream, fiber yarns are modeled by solid element and stitching yarns are modeled by beam element (Figure 10). Unidirectional yarn and stitching yarn are discontinuous so that friction sliding and initial tension of the stitching yarn can be taken into account. It is also available to laminate biaxial fabrics and the direction of the yarn in each layer has arbitrary angle. Figure 11 shows a simple example of draping simulation. It can be seen gaps between straight yarns as shown figure 12.

Figure 5. Biaxial-NCF(front face)

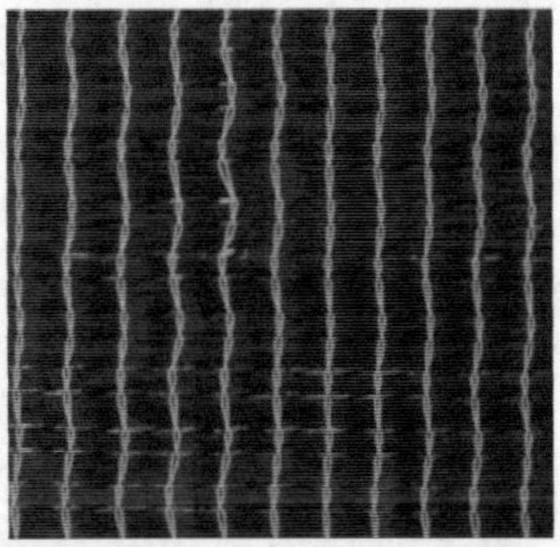

Figure 6. Biaxial-NCF(back face)

Figure 7. UD-NCF(front face)

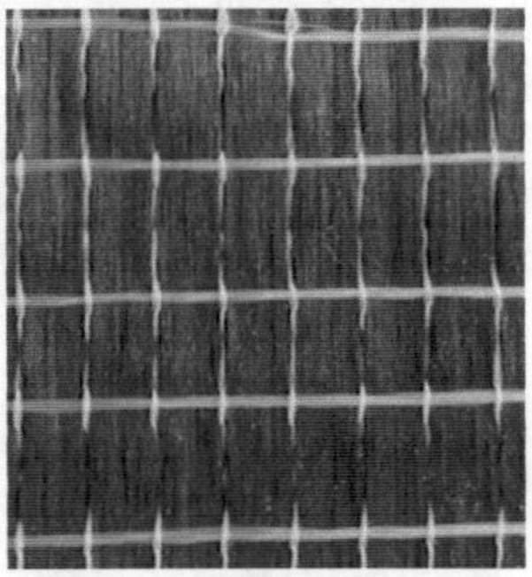

Figure 8. UD-NCF(back face)

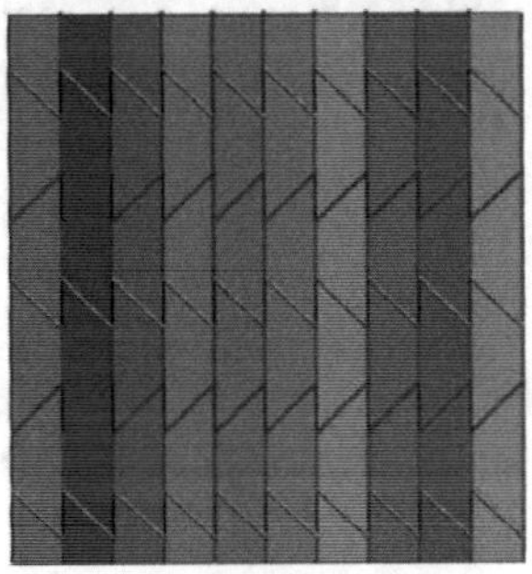

Figure 9. FE-model of Biaxial NCF

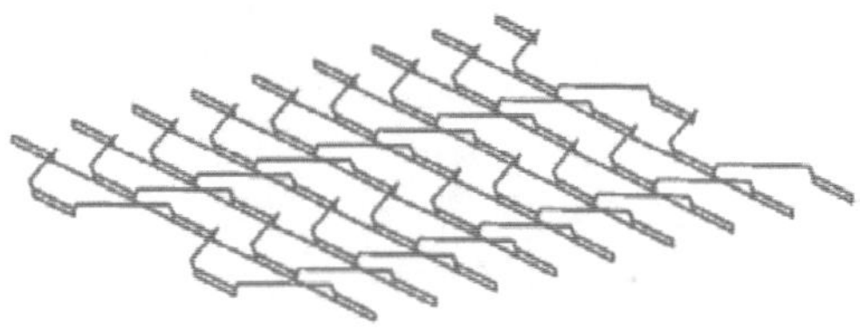

Figure 10. Beam elements of stitching

IOP Conf. Series: Materials Science and Engineering **406** (2018) 012027 doi:10.1088/1757-899X/406/1/012027

Figure 11. Draping deformation

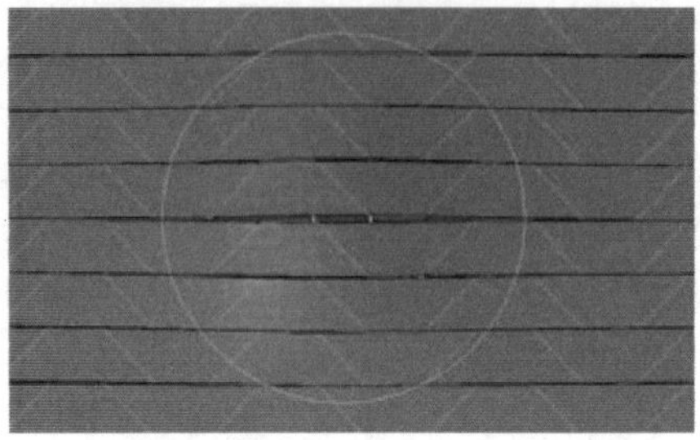

Figure 12. Gap opening between yarn

4. Modelling of fiber mat

A sheet just like non-woven fabric composed of the arbitrary oriented fiber is modeled by beam elements as shown in figure 13. The each fiber is assumed a straight line but has distribution in diameter and length. It is available to choose from uniform distribution, normal distribution and the Weibull distribution as a distribution function. The orientation distribution are specified by arbitrary orientation tensor. Multi-layer which has different orientation tensor respectively is available. The contacts between fibers are also generated automatically. Figure 14 shows input dialog for generating fiber mat. After generating fiber as beam element, it is possible to create voxel mesh. Each voxel element is distinguished whether it is fiber or space by the center coordinate of each element. The fiber element is automatically defined the anisotropy along the fiber axis.

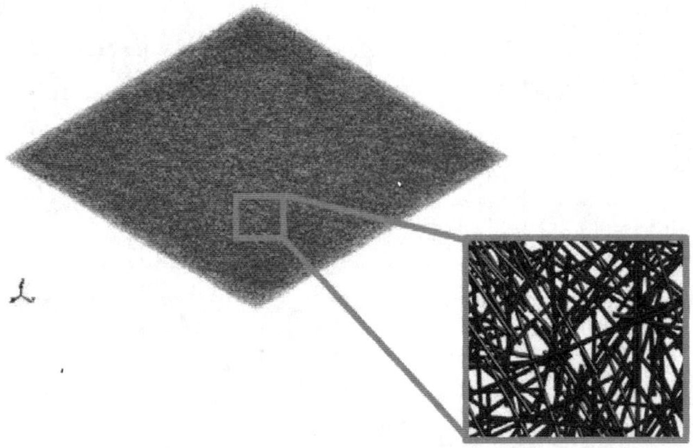

Figure 13. Example of fiber mat

IOP Conf. Series: Materials Science and Engineering **406** (2018) 012027   doi:10.1088/1757-899X/406/1/012027

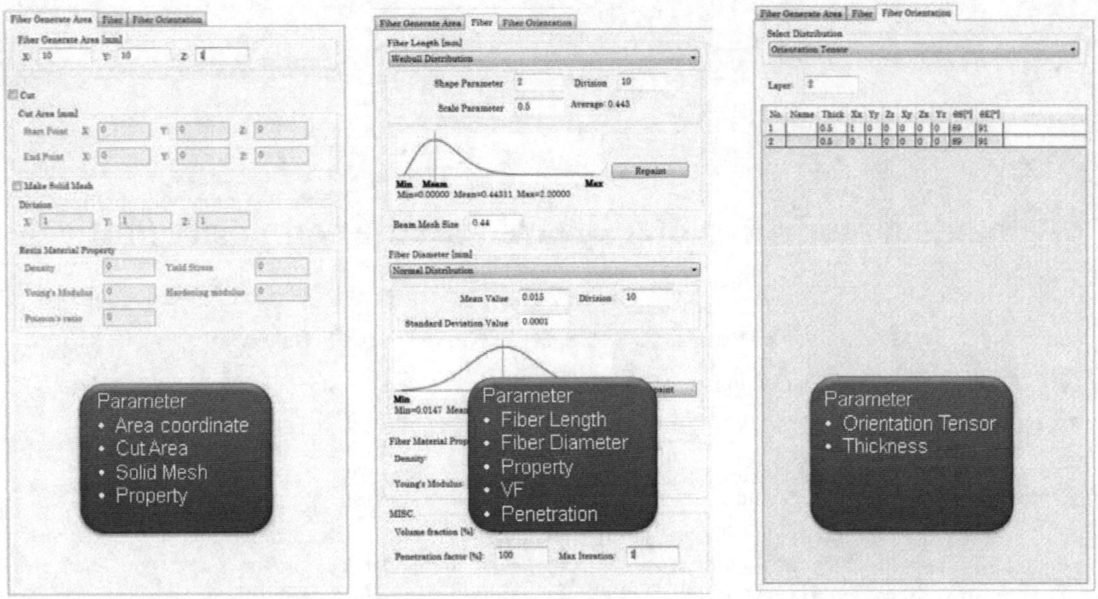

Figure 14. Input dialog for generating randomly oriented fiber mat

5. Conclusions

In this paper, we showed the recent developments in "Composites Dream" to generate finite element models for advanced composite materials such as braided tube, non-crimp fabrics and randomly oriented fiber mat. These developments are useful for the engineers of composite materials to simulate and evaluate the characteristics as stiffness, strength and deformability. These finite element models are exported for LS-Dyna or ABAQUS which are major finite element solver. We have been developing new modeling tools for more complicated composite materials like randomly oriented chopped carbon fiber tape reinforced thermoplastics sheet.

References

[1] Lomov S. V., Verpost I., Modelling of the internal structure and deformability of textile reinforcements: WiseTex software,ECCM, 2002

[2] Lomov S. V., Ivanov D. S., Verpost I., Zako M., Kurashiki T., Nakai H. and Hirosawa S., Composites Science and Technology, vol. 67, no. 9 (2007) pp. 1870

[3] Zako M., Lomov S. V. and Verpost I., "Finite Element and Damage modelings in textile composites", 22nd BEM-FEM Conference 2007

[4] Zako M., Kurashiki T., Nakai H. and Imura M., "A multi-scale analysis for textile composite materials to evaluate the mechanical properties", ECCOMAS Thematic Conference on Mechanical Response of Composite 2007

[5] Yamamoto T. Imaoku A. Sakakibara T. and Zako M., A proposal of conventional FE-modeling for layered braided composites, ICCM 2017

[6] Long Li et al, The effect of preforming quality on the permeability of non-crimp fabrics and the mechanical properties of their composites, ICCM 2015

Mechanical modelling of a sheared textile composite unit cell

M Song[1]*, A J Thompson[1], B El-Said[1], D Ivanov[1] and S R Hallett[1]

[1]Advanced Composites Centre for Collaboration and Science (ACCIS), University of Bristol, Queen's Building, University Walk, Bristol, BS8 1TR, UK

*Mengyi.Song@bristol.ac.uk

Abstract. Composites made using carbon fibre textile reinforcement give an anisotropic material with directional mechanical properties that are dependent on the fabric architecture. The directional properties of a 2D textile are primarily based on weave pattern: The direction of warp and weft yarns, undulation of the tows, and the change in fibre orientation during dry fabric forming process.

During dry fabric forming processes, shear is the dominant deformation mechanism as the 2D fabric conforms to 3D shapes. As the fabric changes its shape to match that of the tool, the fibres rotate away from the orthogonal axes as a function of the shear angle. The modified fibre orientation is carried through the curing process and ends up in the finished composite component. This has an influence on the mechanical properties of a cured composite in localised regions of high deformation. Currently the effect of shear during fabric forming is not usually taken into account when modelling the performance of the finished composite at the macro-scale. Through the development of a novel, robust virtual tool, it is possible to identify the change in mechanical properties for a given shear angle. This will allow the macro FE models to give more accurate results. High deformation regions can be identified from a forming simulation (e.g. a sheet of fabric draped over a tetrahedron tool and causes severe shear near the intrusion) and the change in mechanical properties of these local regions can be adjusted accordingly.

Using the digital element method previously developed in University of Bristol, the dry fabric architecture can be accurately predicted. The current work has created the ability to shear this virtual unit cell to update the unit cell geometry. This deformed woven geometry is then mapped onto a sheared voxel mesh and combined with an epoxy resin matrix to obtain the meso-scale Representative Volume Element (RVE) model. By applying periodic load cases to the model, the stiffness matrix and therefore the mechanical properties of the composite can be determined. This tool is applicable for any 2D woven resin-infused composite.

1. Introduction

The mechanical properties of 2D woven composites is affected by many parameters. Although the effect of individual constituent materials and weave definition is obvious, deformations induced during processing to achieve the final composite are part specific and can also affect final mechanical properties.

In-plane shear has been observed as the primary deformation when forming a textile to a 3D shape [1] [2], this is typically observed at regions of large double curvature. During this process the warp and weft yarns rotate away from their orthogonal axes, changing the anisotropy of the material. This makes the final composite stiffer in the direction where the fibres come together and more compliant at directions away from the fibres.

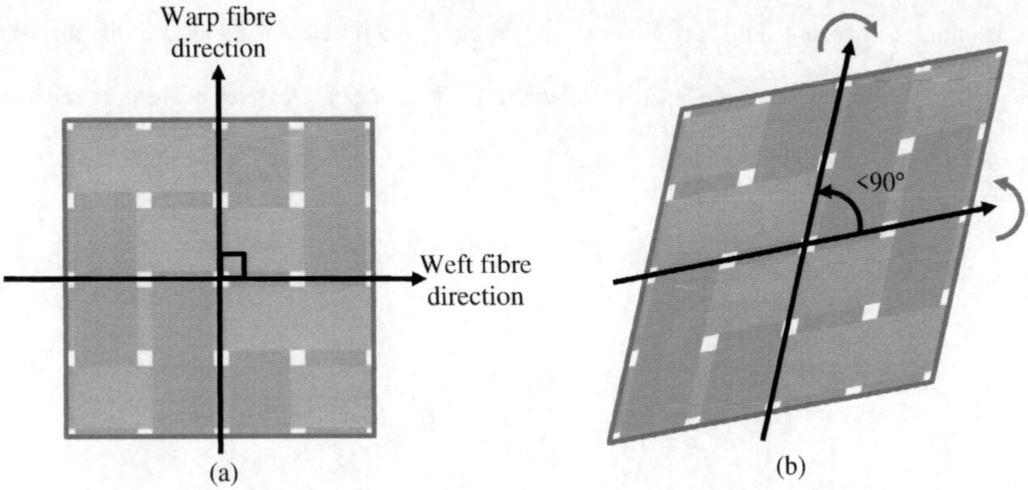

Figure 1. Fibre axis change during shear. (a) Before shear. (b) After shear.

A current challenge in composite manufacturing is the increase in part complexity as composites are seeing wider use in industries such as automotive and aerospace. Traditionally the effect of fabric shear during manufacture has not been considered in FE simulations due to the difficulty in determining shear angles, shear regions, mechanical property of the sheared region, and how to apply sheared properties in macro scale models.

This paper focuses on a methodology to address the latter two difficulties; to find the mechanical properties of the sheared material and application of these properties into macro-scale simulations. This process is written as a robust automated tool in Python to make it easy to use for FE analysts, with all simulations are performed in Abaqus. The processing effects of resin infusion and curing are not taken into consideration.

2. Extract shear angle

Shear angle can be obtained from either a forming/draping experiment or simulation. Figure 2 shows an example of a drape simulation conducted using Abaqus/explicit and an in-house user material developed at University of Bristol [3]. This is a hypothetical example of a swaged panel to show how the geometry can affect the shear angle over quite significant regions.

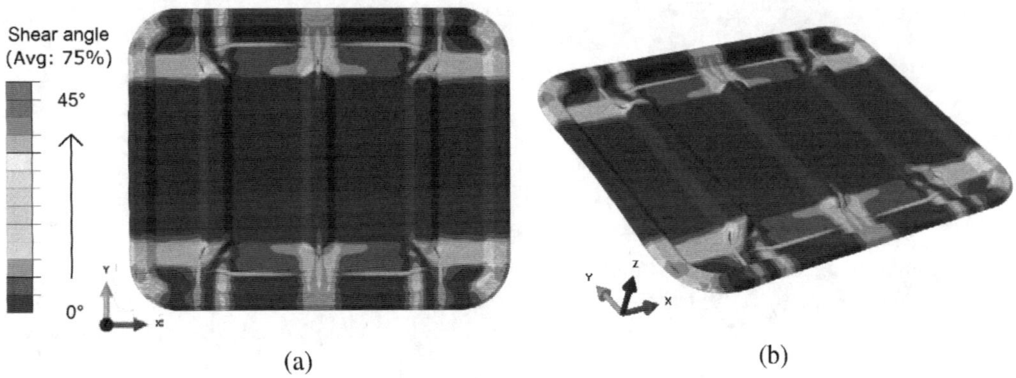

Figure 2. Shear angle distribution after draping simulation of a swaged panel. (a) Top down view. (b) Perspective view.

The shear angles are recorded for each element, and the change in fibre orientations as result of shear is also recorded.

Knowing the maximum shear angle, a set of discrete shear angles are selected in between the maximum and the minimum (0° is the unsheared state) where the mechanical properties are found for these discrete shear angles. A graph can be plotted (as in Figure 3) so the analyst can find the suitable mechanical property for any specified shear angle.

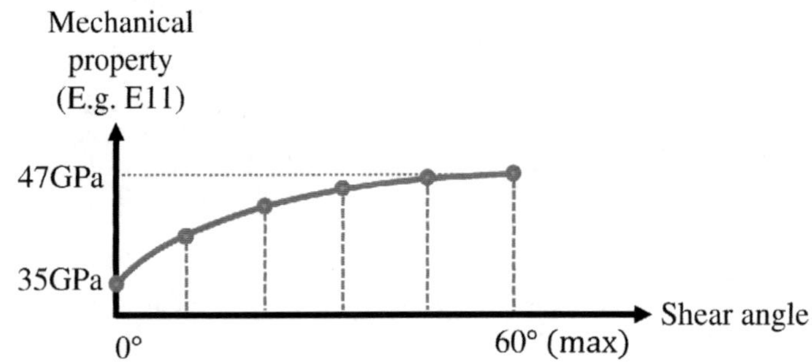

Figure 3. An example graph of material property vs shear angle. Where a total of six shear angles are investigated, with a maximum occurring shear angle of 60 degrees.

3. Digital element method

The compressed sheared fabric geometry (Figure 4) is generated using an in-house digital element solver from University of Bristol, which is based on the principles of [4] [5] [6] [7].

Figure 4. Sheared fabric geometry from in-house digital element solver. Presented via software TexGen. (a) Top down view. (b) Perspective view.

The solver models each tow as a set of 1D chain element, where weaving tension, compaction and shear are taken into consideration through external loads.

The sheared fabric geometry is extracted and then converted to csv files, which specifies the tow path nodes and cross-section profile as shown in Figure 5. This is repeated for the desirable discrete shear angles.

IOP Conf. Series: Materials Science and Engineering **406** (2018) 012028 doi:10.1088/1757-899X/406/1/012028

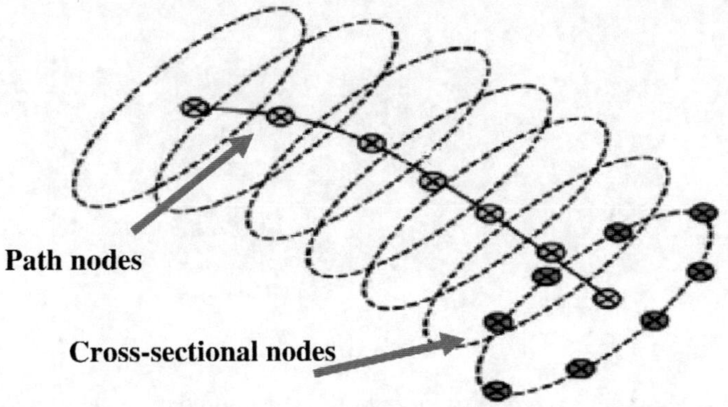

Figure 5. Structure of a tow. Where the path nodes describe undulation and direction of the tow; and the cross-sectional nodes outline the tow profile. A tow section is defined by two adjacent cross-sections [8].

4. Mapping and material cards

Knowing the unit cell dimension and the compacted thickness, a voxel mesh is generated in the unsheared state for the domain of the unit cell. The mesh is then sheared to the corresponding shear angle (Figure 6).

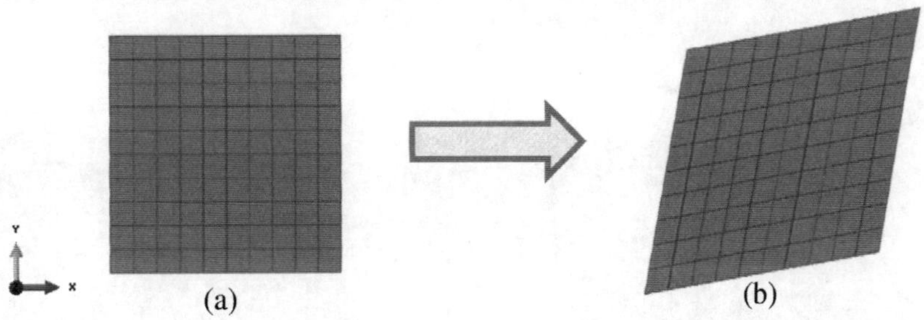

Figure 6. Sheared voxel mesh generation. (a) Unsheared mesh. (b) Sheared mesh.

To make sure that the fabric geometry is correctly mapped into the voxel mesh, the weave architecture is first tessellated to avoid any discontinuities in the tow profile. The mesh is then projected onto the tessellated fabric geometry and the centroid of each mesh element are queried whether it is inside or outside a tow (Figure 7).

IOP Conf. Series: Materials Science and Engineering **406** (2018) 012028 doi:10.1088/1757-899X/406/1/012028

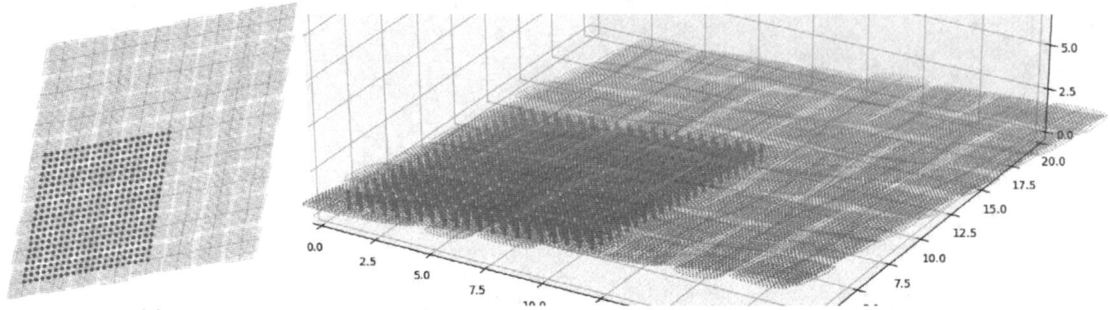

(a) (b)

Figure 7. Projecting the tessellated weave geometry onto the sheared voxel mesh, to cut out the mechanical model. Where the path nodes, the cross-section nodes and voxel mesh nodes are shown. (a) Top down view. (b) Perspective view.

These mesh elements are grouped according to which tow and which tow section they belong to. Figure 8 shows an example of the tow meshes of a sheared unit cell voxel model. The elements that do not belong to any tow sections represent the matrix.

Because of weaving tension, compression and undulation, each section of the tow will have a different orientation and tow cross-section, therefore the intra-yarn volume fraction also varies along the tows. The orientation of each element that belongs to a given section is assigned via the path node coordinates of the corresponding section.

To write material cards for elements belonging to each tow sections, the volume fraction is found and new elastic, orthotropic material cards are written for each corresponding section. The matrix is assigned as an elastic, isotropic material.

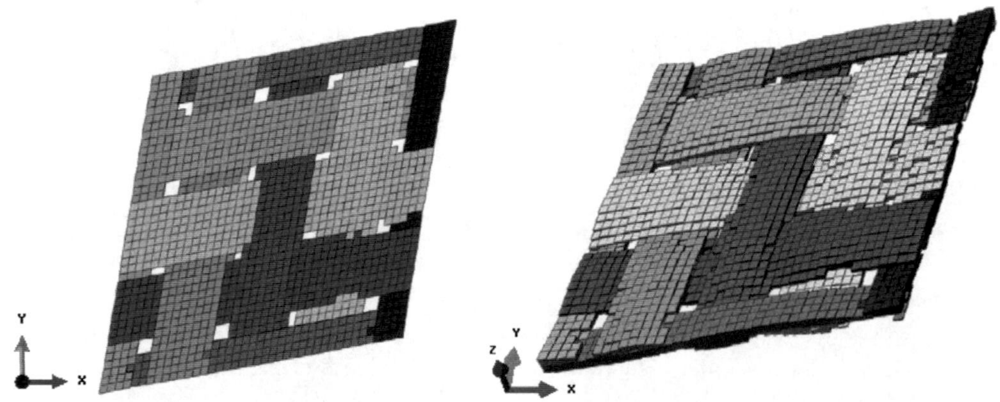

(a) (b)

Figure 8. Mechanical model of a sheared unit cell composite. Only showing elements that belong to the tows, for a 2x2 twill weave. (a) Top down view. (b) Perspective view.

5. Primary axis

Once the unit cell model is built, loads will be applied so the strains and stresses can be extracted, and a stiffness matrix can be assembled. This can be done in any position, for example in Figure 9a where one fibre axis is aligned to the global axis.

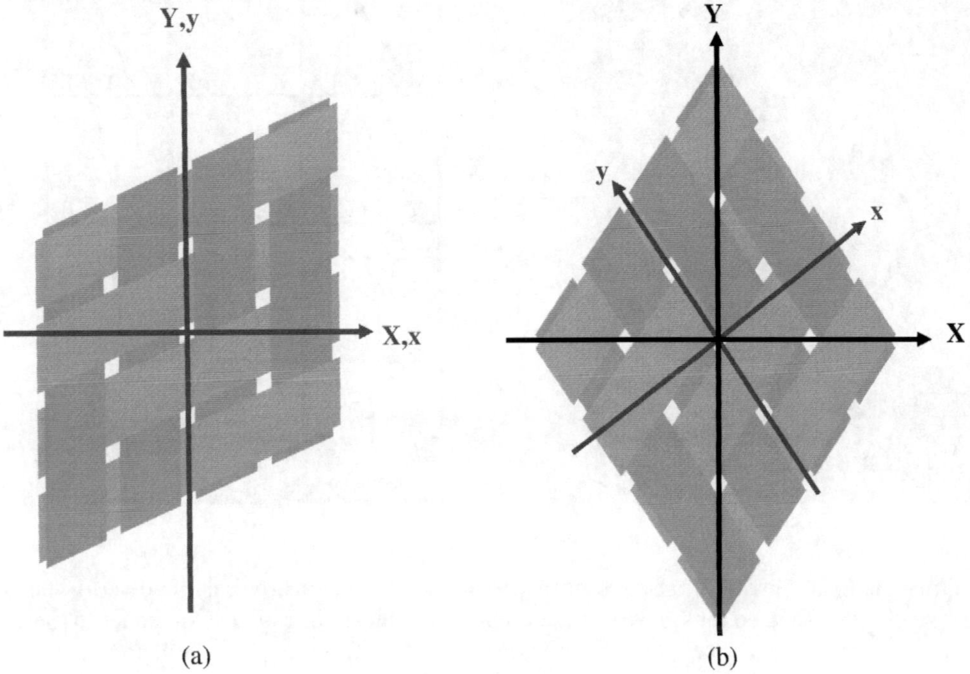

Figure 9. (a) An example unit cell axial orientation while extracting the stiffness matrix. (b) Rotate stiffness to match the global axis (from x, y to X, Y), to be used in macro simulations.

As the macro-scale shell and membrane model does not take through-thickness values, the stiffness matrix can be reduced to 2-dimensional plane stiffness matrix as shown in equation (1).

$$\begin{Bmatrix} \sigma_1 \\ \sigma_2 \\ \sigma_6 \end{Bmatrix} = \begin{bmatrix} Q_{11} & Q_{12} & 0 \\ Q_{12} & Q_{22} & 0 \\ 0 & 0 & Q_{66} \end{bmatrix} \begin{Bmatrix} \epsilon_1 \\ \epsilon_2 \\ \epsilon_6 \end{Bmatrix} \tag{1}$$

Once the reduced stiffness matrix is established, its components need to be rotated to find the effective mechanical properties in the global axis, as shown in Figure 9b. This can be achieved using a transformation matrix in (2) corresponding to the rotation direction between x and X.

$$[T] = \begin{bmatrix} cos^2\theta & sin^2\theta & 2sin\theta cos\theta \\ sin^2\theta & cos^2\theta & -2sin\theta cos\theta \\ -sin\theta cos\theta & sin\theta cos\theta & cos^2\theta - sin^2\theta \end{bmatrix} \tag{2}$$

As mentioned in section 2, the draping simulation is able to capture the rotation of the weft and warp tows. Therefore, the element orientations for each element belonging to the sheared region can be found, as shown by the white arrows in Figure 10a.

By orientating the material axis to align with the global X, Y axis as shown in Figure 9b, these axes are no longer dependent on shear angle. This allows the analyst to apply material orientation on a macro scale model where the global axis of the unit cell becomes local axis for each element, to assign mesh element orientation inside the macro scale simulation as shown in Figure 10b.

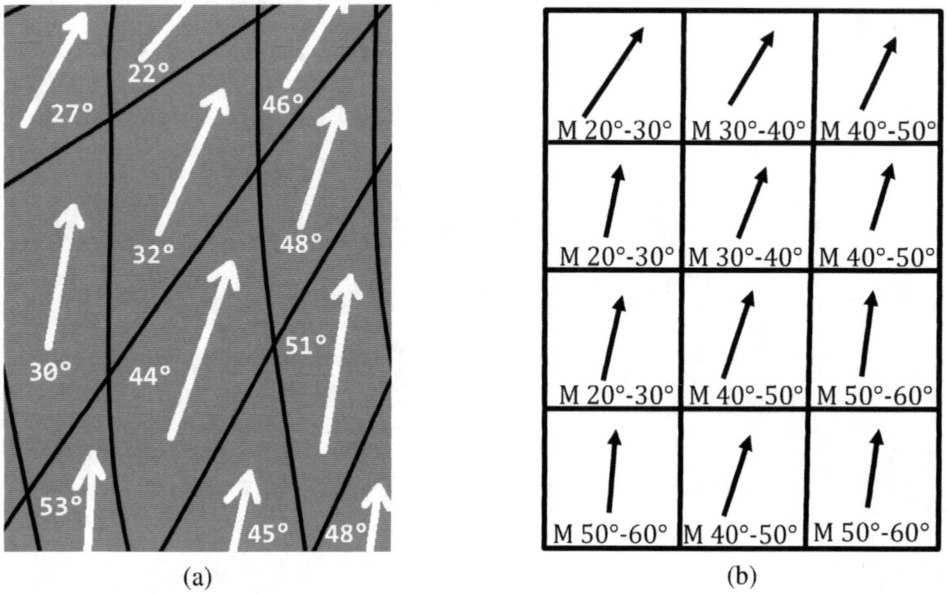

Figure 10. Example of how sheared properties are applied to macro scale model. (a) The unit cells from sheared region, where the orientation (thick white arrows) matches to the global axis of the unit cell, and shear angles are known from draping simulation. (b) Elements that corresponds to the sheared region. Where each element is assigned a local orientation that corresponds to the unit cell's global axis. Moreover, a set of material cards can be written for a set of shear angle ranges that are assigned to these 2D elements.

6. Conclusion

Shear deformations are inevitable as composite parts are becoming more complex. A numerical framework has been established here to account for the effect of composite fabric reinforcement shear due to forming processes on mechanical properties. A methodology to predict the mechanical properties of a sheared unit cell has first been proposed. A system for applying the sheared unit cell properties to a macro-scale simulation is then also defined.

The development of this tool is an essential part of determining the material properties as result of manufacturing deformations, and thus improving the accuracy of the structural scale simulations for mechanical performance.

Acknowledgements
This project has been funded by the EPSRC (grant no. EP/P510427/1) and Jaguar Land Rover (voucher no. 16000069).

References

[1] Long A C, Rudd C D, Blagdon M and Smith P 1996 Characterzing the processing and performance of aligned reinforcements during preform manufacture *Composites Part A* **27** 4 247-53

[2] Zouari B, Daniel J-L and Boisse P 2006 A woven reinforcement forming simulation method. Influence of the shear stiffness *Comput. Struct.* **84** 5-6 351-63

[3] Thompson A J, Belnoue J P-H and Hallett S R 2016 Numerical modelling of defect generation during preforming of multiple layers of 2D woven fabrics *ECCM17* Munich

[4] Thompson A J, El-Said B, Belnoue J P-H and Hallett S R 2014 High fidelity numerical simulations of textile preform deformation *Canada-Japan Workshop on Composites* Vancouver

[5] Zhou G, Sun X and Wang Y 2004 Multi-chain digital element analysis in textile mechanics *Comp. Sci. Technol.* **64** 239-44

[6] Green S D, Long A C, El-Said B and Hallett S R 2014 Numerical modelling od 3D woven preform deformations *Compos. Struct.* **108** 747-56

[7] Thompson A J, El-Said B, Ivanov D, Belnoue J P-H and Hallett S R 2017 High fidelity modelling of the compression behavior of 2D woven fabrics *Int. J. Solids Struct.* **0** 0 1-10

[8] El-Said B, Green S D and Hallett S R 2014 Kinematic modelling of 3D woven fabric deformation for structural scale features *Composites Part A* **57** 95-107

[9] Barbero E J 2013 Finite element analysis of composite materials using Ansys (2nd ed.) CRC Press

IOP Conf. Series: Materials Science and Engineering **406** (2018) 012029 doi:10.1088/1757-899X/406/1/012029

Development of methods for optimisation of complex 3D weave geometries

G H Spackman[1*], L P Brown[1] and I A Jones[1]

[1] Composites Research Group, Faculty of Engineering, University of Nottingham, Nottingham, United Kingdom

* george.spackman@nottingham.ac.uk

Abstract. The development of 3D weaves has resulted in the ability to produce near net shaped preforms, with the additional advantage over unidirectional lay-ups and 2D weaves of greater delamination resistance provided by through-thickness reinforcement. 3D weaving can allow the post-weave formation of bifurcations to form the web and flange of structural components. The mechanical properties of 3D woven components are highly dependent on the weave architecture, allowing the mechanical performance of the component to be tailored to its specific application. Given the number of design parameters to be varied, the design space is potentially infinite. This work focuses on the development of methods to find the optimum weave geometry of a unit cell based on the numerical evaluation of objective functions. This work demonstrates the development of methods to optimise 3D woven textile geometry, using the University of Nottingham's open-source software TexGen [1] to automatically generate each weave based on the input from a global optimisation algorithm. Methods of varying a number of the parameters will be reported alongside their geometric and physical constraints. Finally, the facility to automatically generate a wide range of weaves, with the ability to vary parameters as desired for input either directly into an optimisation algorithm or for further pre-processing is demonstrated.

1. Introduction

T and I joint beams are typically used in the aerospace industry as skin stiffeners, spar to skin joints and bulkhead joints. Typical in-service loading conditions include tensile and flexural loads [2–4]. These present conventional composite laminates with the problem of increased risk of delamination between the plies. It has been shown that composite 3D weaves have the advantage over laminates of greater through the thickness reinforcement which leads to greater impact resistance [5] and delamination properties [6]. Reduced fibre damage and crimp during the manufacturing process is an additional advantage of 3D weaving over z-pinning and stitching methods.

Due to material anisotropy, the mechanical performance of 3D woven structures is highly dependent on the meso-scale fibre architecture, allowing each structure to be tailored directly to its specific application. Some parameters that can be varied include the number of layers, the path of the reinforcing z-binders and the yarn cross-sections.

The design space of these complex 3D weaves can be near infinite which points to the use of direct search optimisation algorithms to aid optimum textile design. These employ computational algorithms to vary the design variables within a set of constraints and evaluate an objective function, typically through the use of finite element simulations, to score the performance of each textile. Within the field of composite design, the use of these optimisation algorithms has until recently been limited to the optimisation of laminates where the ply angles and number of layers are the design variables. The main algorithm employed in the field is the so-called 'genetic algorithm' which is known for its robustness, availability and ease of use as it has been implemented as an 'out of the box' function in MATLAB [7]. However, there is little information regarding its efficacy in comparison to other direct search algorithms.

An optimisation framework has been developed in [7, 8] to find optimum 3D weave architectures of flat woven pieces by obtaining the material moduli. Later in [10] Yan simulated several weave configurations by varying the weft yarn path of a woven T-joint in a tensile pull-off test, creating the weave variations manually. The aim of this work is to adapt the optimisation framework implemented to optimise the flat pieces and apply it to automatically generated FE models of T-pieces, replicating the tensile pull off test in Yan's work while being able to constrain the textile binder yarn paths to the feasible domain.

The use of optimisation algorithms relies on the ability to automatically generate a large number high fidelity finite element models with complex weave variations by creating a parameterised weave and then varying those parameters. The University of Nottingham's in-house geometric modelling software TexGen [1], can be used as a pre-processor for a finite element package, such as the commercial solver Abaqus, by creating the textile geometry and then automatically meshing with a voxel mesh before exporting to Abaqus for analysis under the specified loading conditions.

This study develops previous work by applying it to complex shape profiles with finite element simulations not limited to finding elastic moduli in unit cells. Following this, methods to find the optimum reinforcement architecture of a reduced model of the woven T-joint structure will be presented. Optimisation was by genetic algorithm for initial stiffness under an out of plane tensile load through the web of the joint. The developments in the optimisation framework will be presented along with a demonstration of a constraint handling method.

The novel aspects of this study include the application of the current optimisation framework to a post-deformation, idealised 3D model with a more complex objective function and the novel constraint handling algorithm.

2. Optimisation Problem and Constraints

An optimisation problem is defined as:

$$\begin{array}{ll} \text{minimise objective function} & f(x), \\ \text{subject to constraints} & g_i(x) \le 0 \\ \text{and} & h_i(x)=0 \\ \text{within bounds} & x_L \le x \le x_U \end{array} \tag{1}$$

where x is a vector of the design variables in the optimisation problem, $g_i(x)$ are the inequality constraints and the $h_i(x)$ are the equality constraint functions. The constraint functions are handled differently depending on whether they are linear or nonlinear. If there is constraint violation then a penalisation factor may be added to prevent the solution being considered as the minimum.

In this work the objective function is the initial stiffness of a reduced T-joint under tensile loading through the flange, a typical in service loading condition for this type of component. The T-joint is woven using crimp free yarns and the objective function is evaluated using the commercial finite element software, Abaqus [11]. In this case the path of the binder yarns at the junction were varied. This simple objective function was chosen to be able to demonstrate the ability of the optimisation framework to be applied to complex 3D woven structures. The genetic algorithm was used due to its ease of implementation.

Previous optimisation work [12] has focussed on weaves containing two binder yarns that are anti-symmetric to each other in order to make sure that binder yarns pass over and under each weft stack, meaning that they are mirror reflected copies of each other. However, when there are three or more binder yarns in the model there is no requirement that the binders be anti-symmetric so the following generalised constraints have been formulated:

1. One binder must pass over the top and another must pass under the bottom of each weft stack to ensure all the weft yarns are bound.
2. No straight, non-reinforcing binder yarns above or below weft stacks.
3. At least one binder yarn must cross all horizontal planes between weft layers in the unit cell to prevent separation of the textile.

The second constraint is easily implemented with a simple linear constraint. This, along with bounds, defines the region of feasible design solutions with respect to the linear constraints. In the genetic algorithm optimisation framework, MATLAB only generates feasible members with respect to the linear constraints, so the second constraint is automatically satisfied for each iteration.

Constraints 1 and 3 are difficult to write down in equation form, therefore, a novel algorithm was developed based on the binder yarn offsets from the top of the textile to determine whether or not a weave violates the constraints. This algorithm is used to add a penalty factor, post FE analysis, if constraint violation is found. This raises the effective objective function value and makes it unlikely that the optimisation algorithm will find that solution to be a minimum of the function f(x).

The objective function is calculated via finite element simulations, the load – displacement curve is plotted from the extracted data and the initial gradient calculated.

3. Optimisation Algorithms

The use of optimisation algorithms for finding the optimum composite design is well documented in the literature for laminates, with the genetic algorithm often chosen as the solver of choice [11-14]. Frequently, little justification is given for the choice of algorithm with the choice in practice often made on the basis of peer advice and/or prior familiarity.

3.1 Genetic Algorithm

The genetic algorithm is a population based solver based on the Darwinian principle of survival of the fittest. There are four stages:

1. Initialisation - An initial population of member solutions is generated with the set of design variables analogous to the genotype of that member. The linear constraints and bounds are satisfied here.

2. Selection - Each member solution is evaluated and given a fitness value. A penalty may be applied here if the member violates the nonlinear constraints. The population is then ranked with the best performing or elite solutions entered directly into the next generation.

3. Mutation - Of the remaining solutions, some of their design variables undergo a small alteration creating a slightly different combination of design variables which are then entered into the next generation. Analogous to gene mutations in biology.

4. Crossover - Two well performing textiles have their design variables combined to produce a new set of design variables. This process is analogous to breeding amongst a population. The rest of the next generation is made up of new initialised members.

By continuing this process, the optimum solution should be found that satisfies all the linear and nonlinear constraints.

4. Optimisation of a T-Joint

The optimisation workflow set out in [8], [9] was implemented to optimise the initial stiffness of the binder yarn paths for a reduced 3D woven T-joint structure. To apply this to a complex shape such as a

T-joint, the as woven geometry was modelled first in TexGen before the geometric transformations set out in [10] were employed to fold out the bifurcated part into the shape of a T. The model is then automatically meshed with voxel elements and submitted for finite element analysis in Abaqus. Post-analysis the binder yarn configuration is checked using a novel algorithm with a penalty factor added based on the constraint violation.

4.1 Optimisation Framework

The optimisation process is controlled via an algorithm implemented in Matlab which writes the design variables, in this case the binder yarn offsets, to a text file. Matlab then launches Abaqus and runs using its Python interface a script that uses the TexGen functions to set up the geometrical model, mesh it and apply the loads and boundary conditions before submitting the job for analysis. Once the job is finished, the objective function value is calculated from the extracted data and the penalty function added in the same Python script. The final value is written to a second text file from which Matlab reads in the final value. This process is repeated every time the algorithm produces a new set of design variables and continues until the algorithm's convergence criteria are met.

4.1.1 Varying Binder Yarns. To vary the binder yarn offsets and create different binder paths, the main Python script read in the final data line from the text file. Each data point in the data file is an offset from the top of the textile. For example, to define a binder yarn path in a textile with 6 weft yarns requires a data point each. Therefore, with 4 yarns there were 24 offset data points on each line of the data file, see figure 1. The TexGen function SetBinderPosition was then used to set the binder offset position.

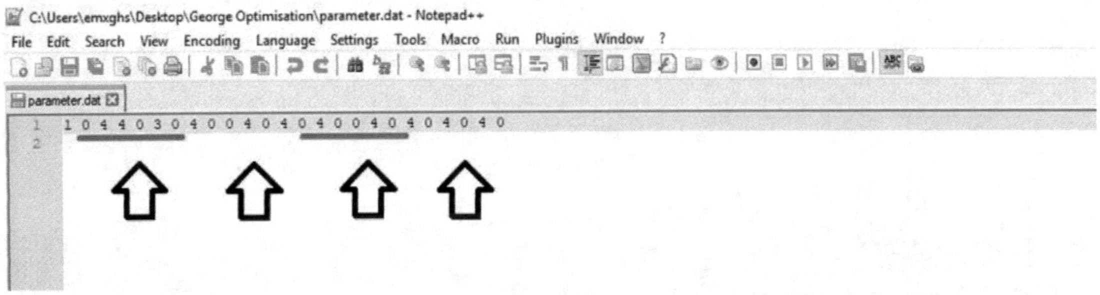

Figure 1. Data file containing the design variables for use in genetic algorithm optimisation. Each arrow points to the data points that define a single yarn in a textile with 6 weft stacks.

4.1.2 Penalty Algorithm. In order to ensure the optimum textile produced using the algorithm conforms to the constraints on the binder yarns set out in section 2, checking methods were developed to determine constraint violation for constraints 1 and 3.

The function to check constraint 1, that a binder must pass above and below each weft stack, takes all the binder offsets and orders them into lists; each list contains all the offsets in a weft stack. It then checks whether there is an offset value of 0, indicating a binder passing over the stack, and whether there is an offset value equal to the number of layers, implying a binder passing underneath the stack. If one or both of these are not present in the list a penalty is added. This process is repeated for each weft stack.

The algorithm to check constraint 3, that at least one binder must pass through all the layers in order to prevent the textile separating between the binders, is a recursive algorithm implementing the constraint that for any internal plane in the textile one of the binder yarns must have a binder offset position in the planes above and below that plane. A flow chart of the algorithm can be seen in Figure 2:

IOP Conf. Series: Materials Science and Engineering **406** (2018) 012029 doi:10.1088/1757-899X/406/1/012029

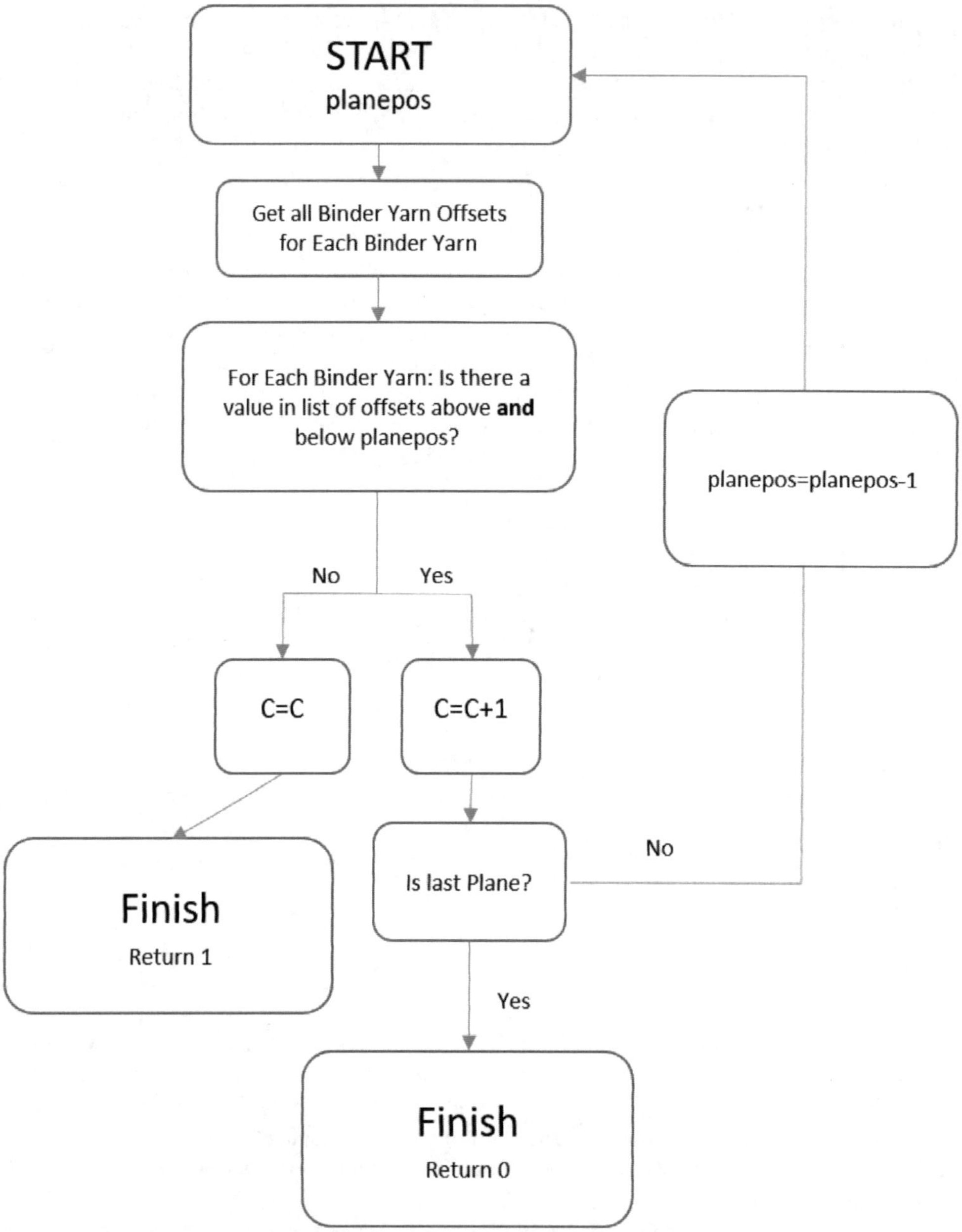

Figure 2. Flow chart of algorithm to implement constraint 3.

where planepos is the plane position of an internal plane.

To demonstrate the ability of the new algorithm and subsequent penalisation protocol to force the algorithm into the feasible domain, a simple flat piece model with three binder yarns was optimised using the genetic algorithm to find the binder yarn configuration with the lowest Von Mises stress in the structure. Results are shown in Figure 3:

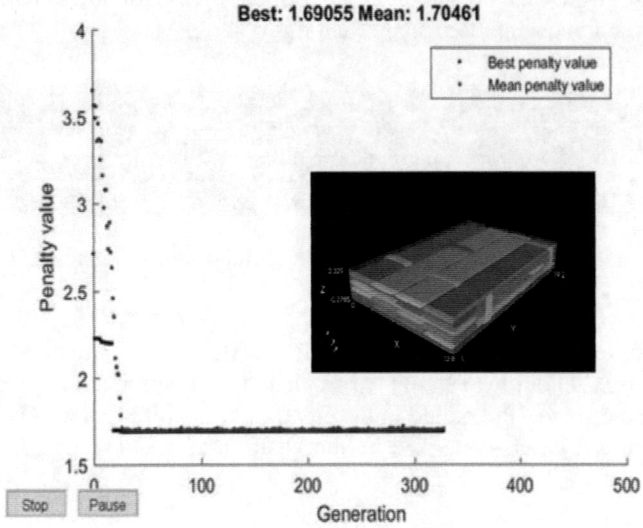

Figure 3. Graph of Penalty Value against Generation for optimisation showing the effect of the penalty parameter alongside the TexGen model of the optimum weave.

The textile produced conformed to all the constraints within around 20 generations. It can be seen that once the feasible domain has been reached, further variations in the weaving pattern causes little variation in the penalty value. This demonstrates the effectiveness of the constraint handling algorithm in forcing the genetic algorithm into the feasible domain.

4.2 Flat Woven Parameterised Weave

The T-joint is woven flat with the binder yarns constrained in different halves of the textile, binding only half the weft stack. This allows the creation of the bifurcation post weaving. The as-woven geometry is created as a parameterised weave using TexGen's CTextileLayerToLayer class with the parameters in Table 1.

Table 1. List of parameters used to create the as-woven flat model with 4 binder yarns

Parameters of CTextileLayerToLayer Class	
Number of X Yarns	10
Number Y Yarns	6
X Spacing	3.2
Y Spacing	3.2
Warp Height	0.35
Weft Height	0.25
Number of Binder Layers	1

The four binder paths were read in from the parameter file and set using TexGen's set binder position function. This results in an idealised model, with no yarn crimp. Since the reduced model focusses on the noodle or junction area, a simplification was made so that the binder yarns have to stay in the top half of the structure and were then copied and transformed to bind the lower half of the structure. This results in a textile that separates into two layers, see figure 4.

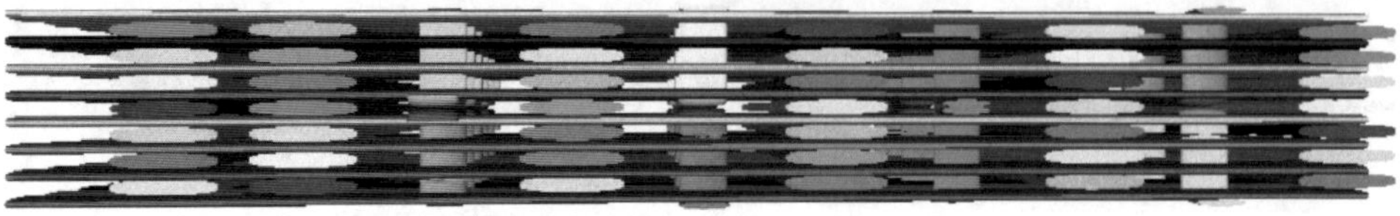

Figure 4. The flat as-woven geometry showing the binder yarns constrained in separate layers.

4.3 Bifurcated Model

Once the flat geometry has been created, a geometric transformation is applied to both the nodes and the yarn cross-sections, as set out by Yan in [10] and listed in equation 2. This creates the profile shape of the T and models the effect of the mould in the creation of the bifurcation. The angle the yarns make from a specified origin can be characterised by the position along the length of the weft yarn by the angle β as below.

$$\beta = \begin{cases} \frac{y'}{z'} & for\ \frac{\pi}{2}z' \geq y' > 0 \\ \frac{\pi}{2} & for\ y' > \frac{\pi}{2}z' \end{cases} \tag{2}$$

This is used to determine which of the transformations in Equation 3 to apply depending on the position of the node along the undeformed yarn.

$$\Delta y = \begin{cases} z' \sin\beta - y' & for\ \frac{\pi}{2}z' \geq y' > 0 \\ z' - y' & for\ y' > \frac{\pi}{2}z' \end{cases} \tag{3}$$

and

$$\Delta z = \begin{cases} z' \cos\beta - z' & for\ \frac{\pi}{2}z' \geq y' > 0 \\ (\frac{\pi}{2}z' - y') - z' & for\ y' > \frac{\pi}{2}z' \end{cases} \tag{4}$$

This results in the creation of the bifurcation, see figure 5.

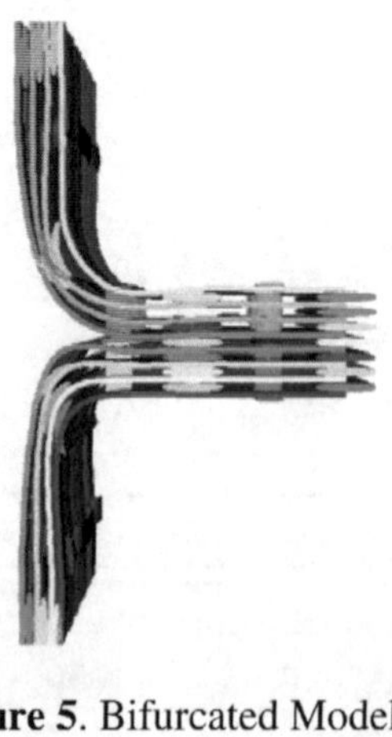

Figure 5. Bifurcated Model

4.4 Meshing

One of the main difficulties in the optimisation of weaving patterns is the automatic generation of a mesh, needed to allow the optimisation process to run without user interaction. It is difficult to mesh contacting yarns using conformal meshing. The solution is to employ voxel meshing [10] which involves the use of brick shaped hexahedral elements which are tessellated throughout the geometrical volume in a method analogous to the use of pixels in imaging.

Each voxel element is defined by its centre point and orientation. The voxel is assigned to be either matrix or yarn depending on which region its centre point lies within. In general, this method of meshing results in a jagged interface boundary which can lead to a greater interface stiffness or interlocking of the elements at the boundaries depending on the local stress condition [10]. However, the interface stiffness can be scaled following the procedures set out in the referenced work.

TexGen automatically meshes within a specified volume or domain set up by the user, either by assigning a default domain normally representing the unit cell or by specifying domain planes that enclose the volume. Currently, these domain planes must form a convex. For a complex shape with concave parts such as a T-joint the smallest shape that will enclose the T-joint is a cuboid.

When the model is meshed and exported using an Abaqus .inp file, the parts of the domain that are enclosed within the yarns are meshed with elements defined with the yarn properties while the rest of the domain is assigned the properties of the matrix. When the T-joint is meshed and exported, the result is a cuboid with surplus matrix elements with the yarn elements, forming the T-shape, embedded within. To represent the shape of the final composite component the extra matrix elements and nodes were deleted in Abaqus' part module using Python scripting. Once this was done the finite element model was set up.

4.5 Finite Element Model

The finite element model needs to be able to be automatically generated and submitted in Abaqus each time a new weave is created by the optimisation algorithm. The aim is to be able to create an accurate enough model each time while taking account of the computational expense of a large number of finite element simulations.

4.5.1 Modelling the Tensile Pull Off Test

The tensile pull off test involves applying a tensile force through the web of the T-joint [10]. The T-joint is clamped at the flanges while the machine head grips the top of the flange and applies a displacement controlled tensile force. The load – displacement graph is recorded and the stiffness is the gradient. In experimental and numerical simulations of the tensile pull out test, Yan found that the load-displacement curve rises linearly until a peak load is reached at a displacement of around 1mm where material damage has initiated and there is a drop in stiffness. Material damage comes mainly in the form of delamination and matrix cracking. The curve then rises further as the damaged component can still sustain a large load before catastrophic failure occurs between 5-7mm depending on the weaving pattern. A simplified diagram of the test is shown in Figure 4.

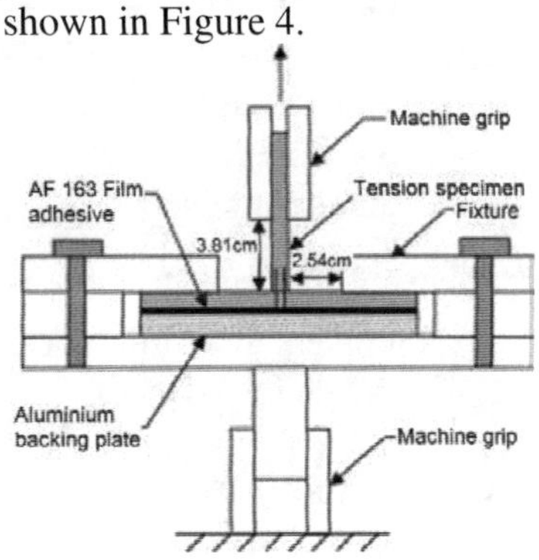

Figure 6. Diagram of the tensile pull off test, taken from [4].

Due to the large size of the model and the small element size needed to capture the geometry accurately, a reduced model focussing on the junction area was employed. This is an attempt to reduce the time and computational expense which finite element simulations of the full model would require. In addition to focussing closely on a smaller area of interest, the initial stiffness was chosen as the objective function.

In order to model the test, a displacement controlled loading scheme is applied in Abaqus to the nodes along the top surface of the web. The load is increased in a displacement controlled manner until the 1mm mark is reached. To model the clamping of the flange, encastre boundary conditions were applied to the bottom corners of the model. This would allow normal rotation of the noodle area while restricting rigid body motion, Shown in Figure 5.

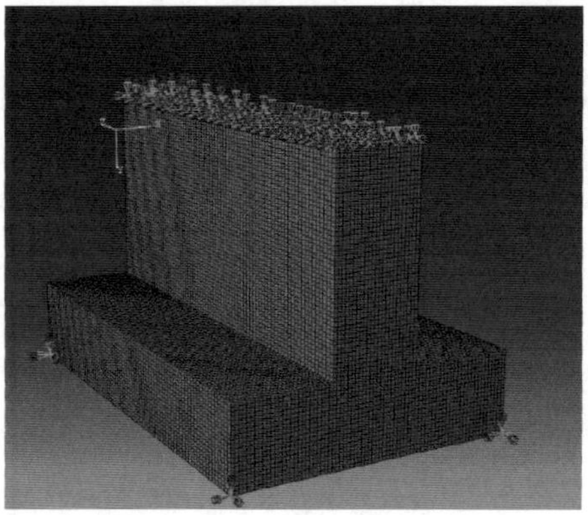

Figure 7. Finite Element model of T-joint with boundary conditions and loads prescribed.

The mesh was made up of 421,875 Abaqus C3D8R voxel elements, this mesh density was chosen as it best captured the yarn geometry which can include many gaps where it is difficult to distinguish between yarns and matrix if the mesh size is too large.

Once the model was set up and meshed and the loads applied, the models were submitted for analysis. The whole process of model set up and submission is automated using a Python script, following the process set out in previous optimisation work.

5. Discussion

Automatically meshing shapes with complex profiles such as the T-shape is currently difficult. TexGen generates the voxel mesh elements by dividing up a volume containing the model called the domain. The domain is specified by a set of planes. However, currently only domains that are convex in shape are able to be defined. This is due to the multiple plane intersections that would occur in concave shapes such as T-profiles, making it difficult to determine which intersection is required to form the domain. This works well for the vast majority of cases where unit cells are used to apply periodic boundary conditions as the woven cell is typically cuboidal in shape. However, for complex shapes this provides a significant barrier to automatic mesh generation. In previous work by Yan [16] the voxel mesh was automatically generated for a cuboid containing the embedded T-profile. This leads to an excess of matrix elements which Yan removed manually using the HyperMesh software which was also used to apply cohesive surfaces to the yarn-matrix interface.

Currently work is ongoing to solve this problem by writing a TexGen class specifically for meshing T and I-profiles by only writing the desired voxel elements to the Abaqus .inp file. These are selected based on whether or not the nodes of the element lie within the embedded T-profiles's volume. Once this has been done more results will be able to be generated.

IOP Conf. Series: Materials Science and Engineering **406** (2018) 012029 doi:10.1088/1757-899X/406/1/012029

6. Conclusions

Methods have been developed to optimise complex shapes such as T and I-profiles. First constraints on the feasibility of textiles were formulated and implemented in the existing optimisation framework. Results were shown that demonstrated the ability of the algorithm to converge to feasible weaves when this was applied.

Next modelling strategies from the literature were used to model T-profiles and lay the groundwork for their input into the optimisation of initial stiffness by the tensile pull off test using a genetic algorithm. Ongoing difficulties in automatically generating the voxel mesh was discussed which provides a barrier to use in optimisation algorithms. Further work will solve this problem and develop the finite element models so that they are more accurate, for example by incorporating delamination modelling.

References

[1] H. Lin, L. P. Brown, and A. C. Long, "Modelling and Simulating Textile Structures Using TexGen," *Adv. Mater. Res.*, vol. 331, pp. 44–47, Sep. 2011.

[2] F. Hélénon, M. R. Wisnom, S. R. Hallett, and R. S. Trask, "Numerical investigation into failure of laminated composite T-piece specimens under tensile loading," *Compos. Part A Appl. Sci. Manuf.*, vol. 43, no. 7, pp. 1017–1027, Jul. 2012.

[3] Q. . Yang, K. . Rugg, B. . Cox, and M. . Shaw, "Failure in the junction region of T-stiffeners: 3D-braided vs. 2D tape laminate stiffeners," *Int. J. Solids Struct.*, vol. 40, no. 7, pp. 1653–1668, Apr. 2003.

[4] P. . Stickler and M. Ramulu, "Investigation of mechanical behavior of transverse stitched T-joints with PR520 resin in flexure and tension," *Compos. Struct.*, vol. 52, no. 3–4, pp. 307–314, May 2001.

[5] H. Hong Hu, B. Baozhong Sun, H. Hanjian Sun, and B. Bohong Gu, "A Comparative Study of the Impact Response of 3D Textile Composites and Aluminum Plates," *J. Compos. Mater.*, vol. 44, no. 5, pp. 593–619, Mar. 2010.

[6] M. Pankow, A. Salvi, A. M. Waas, C. F. Yen, and S. Ghiorse, "Resistance to delamination of 3D woven textile composites evaluated using End Notch Flexure (ENF) tests: Experimental results," *Compos. Part A Appl. Sci. Manuf.*, vol. 42, no. 10, pp. 1463–1476, Oct. 2011.

[7] The Mathworks Inc., "Matlab and Statistics Toolbox Release 2017b." Natick, Massachusetts, United States.

[8] X. Zeng, A. C. A. Long, I. Ashcroft, and P. Potluri, "Fibre architecture design of 3D woven composite with genetic algorithms: a unit cell based optimisation framework and performance assessment," Jul. 2015.

[9] L. P. Brown, F. Gommer, X. Zeng, and A. C. Long, "Modelling framework for optimum multiaxial 3D woven textile composites," Sep. 2016.

[10] S. Yan, "Design optimisation of 3D woven reinforcements with geometric features. PhD thesis, University of Nottingham.," University of Nottingham, 2017.

[11] "Abaqus 2016 User Manual." Dassault Systemes, Providence, RI, USA.

[12] F. Gommer, L. P. Brown, X. Zeng, and A. C. Long, "Discovery of Optimum Textile Composites," 2016.

[13] R. Le Riche and R. T. Haftka, "Optimization of laminate stacking sequence for buckling load maximization by genetic algorithm," *AIAA J.*, vol. 31, no. 5, pp. 951–956, May 1993.

[14] G. Soremekun, Z. Gürdal, R. T. Haftka, and L. T. Watson, "Composite laminate design optimization by genetic algorithm with generalized elitist selection," *Comput. Struct.*, vol. 79, no. 2, pp. 131–143, Jan. 2001.

IOP Conf. Series: Materials Science and Engineering **406** (2018) 012030 doi:10.1088/1757-899X/406/1/012030

Utilizing numerical models to identify process-induced residual stresses in 3D woven carbon/epoxy composites

Igor Tsukrov[1,*], Kostiantyn Vasylevskyi[1], Borys Drach[2], Hilary Buntrock[1], Todd Gross[1]

[1]University of New Hampshire, Durham, NH 03861, USA
[2]New Mexico State University, Las Cruses, NM 88003, USA

* igor.tsukrov@unh.edu

Abstract. A combined experimental-numerical procedure is proposed to evaluate residual stresses in 3D woven carbon/epoxy composites. It is assumed that residual stresses are caused by the difference in thermal expansion coefficients of the constituents. The impact of residual stresses is quantified by drilling blind holes in the composite panels and mapping the resulting in-plane surface displacements by digital image correlation and electronic speckle pattern interferometry. Then mesoscale finite element models of the composite are used to determine the effective temperature drop ΔT^{eff} that results in the same predictions for hole drilling. The local distributions of residual stress are given by the finite element simulations of the cooling by this temperature drop ΔT^{eff}.

1. Introduction

3D woven carbon/epoxy composites develop curing-induced residual stresses that might limit their performance or lead to microcracking [1]. Distribution of the process-induced stresses can be quite complex due to morphology of the reinforcement and various residual stress formation mechanisms, e.g. inhomogeneity of the resin flow and temperature during injection and cure, chemical shrinkage, and mismatch between fiber and resin coefficients of thermal expansion (CTEs). There are no direct experimental techniques to determine this distribution. However, mesoscale numerical models of woven composites have been proven to provide a meaningful insight into the mechanics of the textile composites [2].

In this paper, we investigate whether recently developed finite element (FEA) models [3] can be utilized to determine the spatial variation of process-induced residual stresses from the surface displacements observed after drilling a hole. The surface displacements are measured by digital image correlation (DIC) and electronic speckle pattern interferometry (ESPI). The two methods are used because the displacements are sometimes on the lower end of the resolution for the DIC method and the displacement gradients near the hole are sometimes too steep to resolve the fringes obtained via the ESPI method.

There have been previous publication utilizing FEA or analytical solutions to evaluate residual stresses in the composites based on the hole drilling technique and interpretation of ESPI measurements, see [4, 5]. However, in those studies it was assumed that the residual stresses were uniform in the volume occupied by the hole which is not accurate for 3D woven composites. In our research, we assume that local distribution of residual stresses arises during cooling after curing and is mostly caused by the difference in CTEs between the fibers and the matrix. Thus, in our approach we compare the displacements from hole drilling with the results of FEA simulations to determine such an effective temperature drop ΔT^{eff} that the resulting distribution of stresses causes the same surface displacements. We then assume that the stress distribution caused by ΔT^{eff} is the actual residual stress field.

2. Experimental Methods

Experimental measurements were conducted on 4.1 mm thick composite panels fabricated by Albany Engineered Composites using Hexcel RTM6 resin and Hexcel 12K IM7 PAN-based carbon fibers. Several reinforcement architectures with different level of through-thickness reinforcement were considered but this paper presents results for only one of them, the so-called one-by-one orthogonal. This particular architecture was chosen because of its high through-thickness constraint preventing in-plane residual stress relaxation mechanism typical for laminates and low through-thickness constraint composites. The material had an overall fiber volume fraction of 55%. The in-plane dimensions of the repeating unit (as shown in figure 1) were 5.1 mm by 5.1 mm.

The composite sections containing at least nine unit cells were cut out from the panels and painted first with white, high heat spray paint. The black speckles were applied using an airbrush to get speckles that ranged in size from 5 to 10 μm. Then the surface was covered with a clear matte spray paint to protect the speckles from drilling debris deposits and prevent the speckles being removed with the water used for cooling during drilling. The sample was glued on a block mounted on a Thor Labs kinematic mount to allow precise repositioning of the "before drilling" and "after drilling" DIC images and interferograms. The apparent placement repeatability was on the order of 5 μm or less. The same specimen was probed on both the DIC setup and the interferometry system.

Residual stresses were studied by drilling a 1mm diameter blind hole to a depth of 0.5 mm and recording the resulting in-plane displacements on the surface of the specimen. Drilling was done with UKAM diamond coring tool. The depth was continuously measured with a dial indicator attached to the drilling head. A continuous flow of deionized water was manually applied during drilling using a squeeze bottle. The water was used to minimize the heat generated during drilling and to carry away the drilling debris. The sample was rinsed with more water after drilling and dried with a flow of warm air.

The in-plane displacements in the near field region around the hole were determined with a VIC-Micro3D Correlated Solutions digital image correlation system. The pixel size was 3.34 μm. The typical quoted maximum resolution is 0.01 pixels. The apparent displacement noise is ~0.05 pixels which corresponds to ~0.06 μm. The sampled region was large enough so that we subtracted any tilt and displacement from re-placing the sample by performing a 2D first order, least squares fit of a 10-pixel wide region on the outer border of the image. When the apparent displacement field extended to the border, the tilt correction was only performed on the borders not impacted by the displacement field of the hole.

The displacements outside the high displacement gradient region around the hole were measured using a custom-built electronic speckle pattern interferometry system similar to the one described by Diaz et al [6]. We used a 50 mW Melles Griot HeNe laser that had linear polarization. The angle between the normal to the specimen and the illumination beams was 45° which resulted in a 448 nm displacement for a phase difference corresponding to 2π. Our system exhibited phase noise of $<\pi/25$ which corresponds to a displacement of 9 nm which is roughly 7 times lower than the DIC measurements.

3. FEA modeling

This research utilized meso-scale FEA models of 3D woven composites described in publications [3, 7]. They were developed for the smallest repeating portion of the material – the so-called "unit cell" (UC). The entire composite can then be considered as a continuous assemblage of such unit cells. For this study we used the numerically simulated geometry of woven reinforcement obtained with Dynamic Fabric Mechanical Analyzer (DFMA), see [8]. Figure 1 presents an example of 3D finite element model for the orthogonally reinforced composite developed in MSC Marc Mentat [9].

Figure 1. Unit cell of one-by-one orthogonal composite panel: from the initial schematics of the reinforcement for DFMA simulations to the finite element mesh (reinforcement + matrix) in MSC Marc Mentat.

The matrix (fully cured HEXCEL RTM6 epoxy resin) was simulated as a linear isotropic material with constant Poisson's ratio $\nu_m = 0.35$, and temperature dependent Young's modulus E_m and thermal expansion coefficient α_m:

$$E_m = E_m^{0°C} - \beta_m T \quad (1)$$

$$\alpha_m = \alpha_m^{0°C} + \gamma_m T \quad (2)$$

where $E_m^{0°C} = 3{,}500\ MPa$, $\beta_m = 5.9\frac{MPa}{°C}$, $\alpha_m^{0°C} = 5 \cdot 10^{-5}\frac{1}{K}$, $\gamma_m = 1.05 \cdot 10^{-7}\frac{1}{K \cdot °C}$ are the material parameters, and T is the temperature in $°C$. The tows (12,000 IM7 carbon fibers impregnated with RTM6 epoxy) were simulated as transversely isotropic material with the properties obtained by micromechanical modeling as $E_{1t} = 221.38\ GPa, E_{2t} = 13.18\ GPa, G_{12t} = 7.17\ GPa, \nu_{12t} = 0.35, \nu_{23t} = 0.35, \alpha_{1t} = -2.29 \cdot 10^{-7}\frac{1}{K}, \alpha_{2t} = 2.23 \cdot 10^{-5}\frac{1}{K}$. In these expressions, direction 1 is parallel to the axis of the tow and directions 2 and 3 are transverse to the tow axis. Note that even though the properties of the matrix in the tows change with temperature as given by formulas (1) and (2), these changes will result in insignificant variations of the homogenized properties of the tows (see comparison in [7]), so in the numerical simulations the properties of the tows were assumed to be temperature independent.

The hole drilling was simulated in MSC Marc Mentat by deactivating elements belonging to the hole. First, the simulation of the cooling by the equivalent temperature drop ΔT^{eff} was performed. Then the elements corresponding to the position of the hole were manually selected to deactivate. Subtracting the displacements obtained after the simulation of curing from the displacements caused by removing the elements representing the hole produces the displacement field from the hole drilling.

In the simulations, upper and lower surfaces of UC were not constrained because the cell includes the entire thickness of the composite panel. Selection of the lateral constraints for hole drilling modeling, however, presented certain challenges. Periodic boundary conditions, traditionally used to evaluate the overall performance of woven composites, are not appropriate when the hole is located close to the boundary of the unit cell. For example, consider distribution of displacements due to hole drilling predicted by FEA for 2 different hole locations, see figure 2. Analysis of the lower right image shows that FEA predicts some unrealistically increased displacements far from the hole due to the imposed symmetry condition. On the other hand, FEA modelling of the entire composite panel with the same level of accuracy would require extremely large finite element models and excessive requirements for computer time and resources.

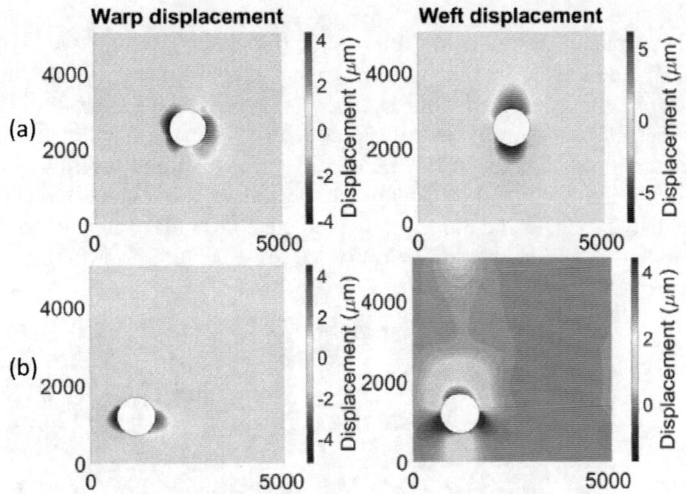

Figure 2. FEA-predicted displacement fields due to the temperature drop and subsequent removal of the elements simulating hole drilling: (a) displacements for centrally located hole; (b) displacements for hole near the corner of UC.

To select the most efficient way to model the structure we considered 3 possible approaches to selecting lateral boundary conditions. The most accurate one considered a portion of the structure consisting of 9 unit cells (3x3) with the sides that are required to be periodic but are allowed to move laterally to accommodate the overall shrinkage of the composite panel due to cooling. In the second approach, one unit cell is surrounded by the equivalent effective material and the periodicity is not imposed. In this case the required number of finite elements and, correspondingly, the simulation time was significantly lower. And, finally, the third approach was to consider one unit cell without the periodicity conditions. We calculated the displacements of the material due to cooling and normalized them with respect to 3x3 results, which we consider to be the most accurate. Based on the comparison presented in figure 3, we decided to perform our hole drilling simulations utilizing the model of a unit cell surrounded by the equivalent material (second approach).

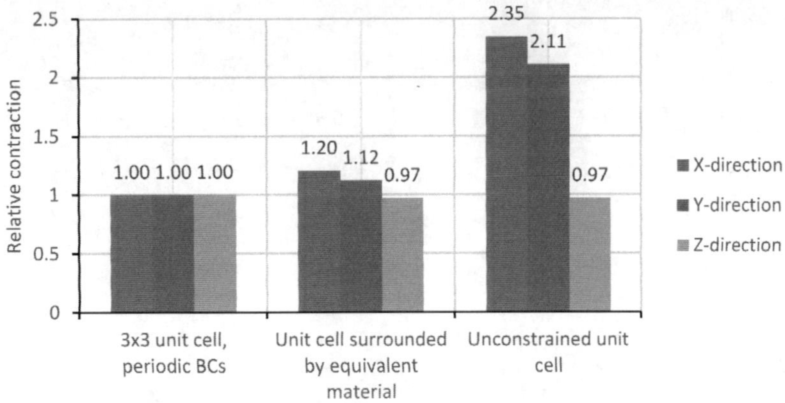

Figure 3. Comparison of predicted displacements in warp (x), weft (y), and through-thickness (z) directions for three choices of lateral boundary conditions (normalized with respect to 3x3 results).

4. Results and discussion

Our method is illustrated for the experimental measurements performed for a hole located as shown in figure 4. As can be deduced from the DIC images depicted in figure 5(b), the material removed by the hole was

IOP Conf. Series: Materials Science and Engineering **406** (2018) 012030 doi:10.1088/1757-899X/406/1/012030

mostly in the state of in-plane tension, so drilling resulted in opening of the hole. To determine the effective temperature drop ΔT^{eff}, we used the approach similar to that implemented in [4]. We matched FEA predictions for warp and weft displacements at the points located at a prescribed distance from the hole (in our case, we selected points one diameter away from the hole edge) with the experimental measurements. This resulted in the effective temperature drop $\Delta T^{eff} = -35^{o}C$. Figure 5(a) shows distributions of weft and warp displacements predicted by FEA for this temperature drop. Figure 5(c) presents displacements along the lines shown in figure 5(b), both experimentally observed and numerically predicted. Overall, good correspondence is observed with the exception of the weft displacements to the left of the lower left image, which was the area of significant decorrelation in experimental measurements.

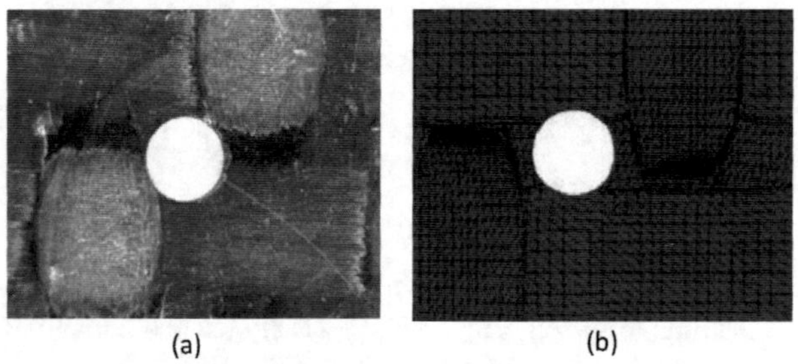

Figure 4. Microscope image of the drilled hole vs. FE model with matrix elements not shown.

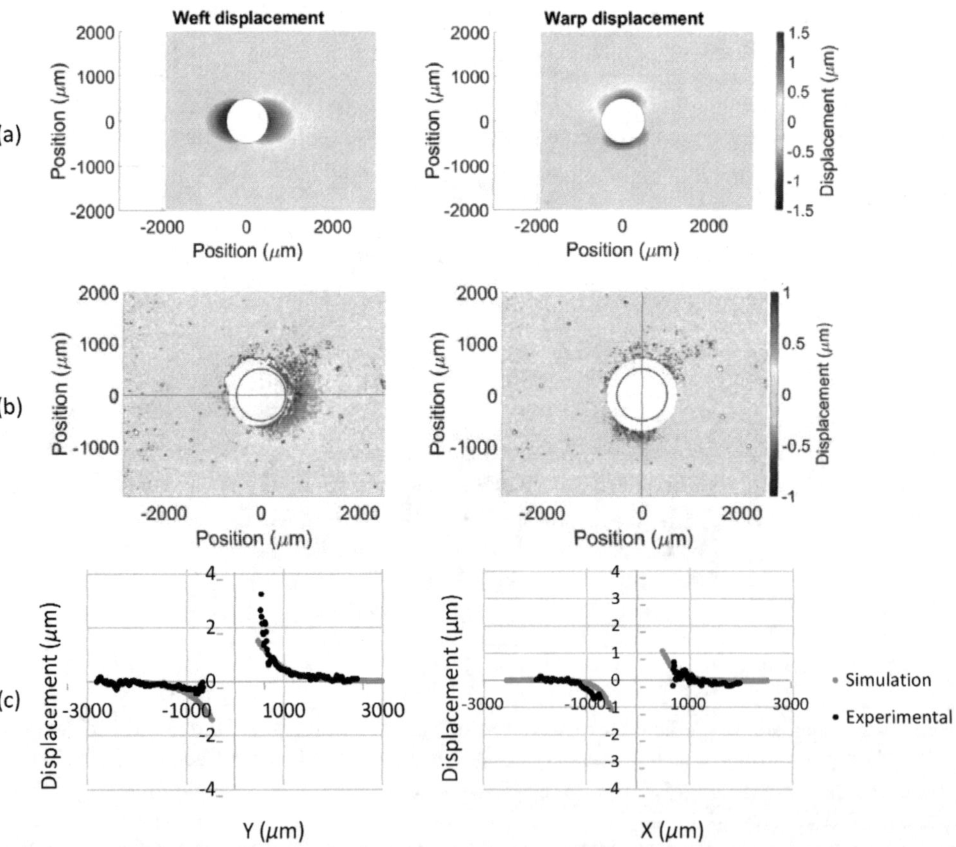

Figure 5. Displacement fields caused by hole drilling: (a) FEA predictions; (b) experimental DIC measurements; (c) slice plots along the lines shown in (b).

IOP Conf. Series: Materials Science and Engineering **406** (2018) 012030 doi:10.1088/1757-899X/406/1/012030

After the effective temperature drop ΔT^{eff} is determined and validated by comparison with the DIC and ESPI data, FEA simulations can be used to evaluate distribution of cure-induced residual stresses due to mismatch in CTE of constituents. Figure 6 presents distribution of residual stresses in both warp and weft directions (σ_{XX} and σ_{YY}), as well as the hydrostatic and von-Mises stress. The hydrostatic stress is of particular interest as it plays a major role in development of matrix microcracking for 3D woven composites utilizing thermoset resins, as discussed, for example, in [7].

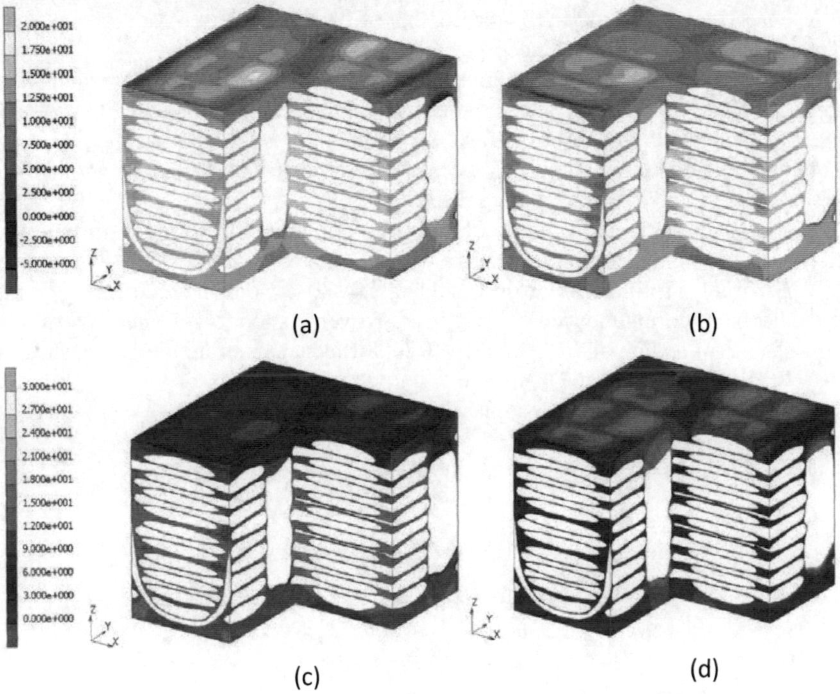

Figure 6. Distribution of cure-induced residual stresses in the matrix of orthogonally reinforced 3D woven composite: (a) warp; (b) weft; (c) hydrostatic; (d) equivalent von Mises.

5. Conclusions

Combination of meso-scale FEA modeling with experimental measurements of surface displacements caused by drilling blind holes in selected locations can be utilized to identify the full volume distribution of process-induced residual stresses in 3D woven composites. The proposed procedure involves determining the effective temperature drop ΔT^{eff} such that the surface displacements observed in the measurements are well reproduced by the numerical simulations of cooling by ΔT^{eff} and then removing the elements in the location of the drilled hole. In modelling, it is important to select the appropriate boundary conditions especially when the hole is drilled near the edge of the unit cell. The procedure can be used to evaluate levels of residual stresses in textile composites with various reinforcement architectures, and to optimize the material design by selecting the most beneficial reinforcement patterns and curing schedules.

Acknowledgements

This material is based upon work supported by the National Science Foundation through grant CMMI-1662098.

IOP Conf. Series: Materials Science and Engineering **406** (2018) 012030 doi:10.1088/1757-899X/406/1/012030

References

[1] Tsukrov I, Bayraktar H, Giovinazzo M, Goering J, Gross T, Fruscello M and Martinsson L 2012 Finite element modeling to predict cure-induced microcracking in three-dimensional woven composites *Int. J. Fracture* **172** 209-16

[2] Lomov S, Ivanov D, Verpoest I, Zako M, Kurashiki T, Nakai H and Hirosawa S 2007 Meso-FE modelling of textile composites: road map, data flow and algorithms *Compos. Sci. Technol.* **67** 1870–91

[3] Drach A, Drach B and Tsukrov I 2014 Processing of fiber architecture data for finite element modeling of 3D woven composite *Adv. Eng. Softw.* **72** 18–27

[4] Wu L F, Zhu J G and Xie H M 2015 Investigation of residual stress in 2D plane weave aramid fibre composite plates using Moiré interferometry and hole-drilling technique *Strain* **51** 429-43

[5] Pisarev V S, Eleonsky S I, Chernov A V 2014 Residual stress characterization in orthotropic plate by combining hole-drilling method and speckle interferometry *29th Congress of the Int. Council of the Aeronautical Sciences* (St. Petersburg, Russia)

[6] Diaz F V, Kaufmann G H and Galizzi G E 2000 Determination of residual stresses using hole drilling and digital speckle pattern interferometry with automated data analysis *Opt. Lasers Eng.* **33** 39-48

[7] Drach B, Tsukrov I, Trofimov A, Gross T and Drach A 2018 Comparison of stress-based failure criteria for prediction of curing induced damage in 3D woven composites. *Compos. Struct.* **189** 366-77

[8] Miao Y, Zhou E, Wang Y, and Cheeseman B 2008 Mechanics of textile composites: micro-geometry *Compos. Sci. Technol.* **68** 1671–8

[9] http://www.mscsoftware.com/product/marc

IOP Conf. Series: Materials Science and Engineering **406** (2018) 012031 doi:10.1088/1757-899X/406/1/012031

A proposal of conventional FE-modeling for layered braided composites: comparison of numerical results with experimental results

Takuya Yamamoto[1], Aki Imaoku[1], Tatsuo Sakakibara[1*], Masaru Zako[2]

[1]Itochu Techno-Solutions Corporation (CTC), 3-1-3, Umeda, Kita-ku, Osaka, Japan
[2]Osaka University, 2-1, Yamadaoka, Suita, Osaka, Japan

* tatsuo.sakakibara@ctc-g.co.jp

Abstract. A new FE modeling system for braided tube has been proposed. The input data are the shape of section of a braid, the braided angle, and the number of layers and the diameter of tube. The generated model has been verified by comparison with the experimental results and numerical ones. The mechanical properties for the tube have been analysed by the generated model and FEM code. On the other hand, the rigidity and the strength for the tube have been measured by three point bending and torsion tests. The experimental results have agreed with the numerical results. The agreement has encouraged that the proposed procedure can be applied to generate FE models for braided tubes with complex geometry.

1. Introduction

Braided tubes made of carbon fiber reinforce plastics (CFRP) have been widely used for various structures in automotive, civil engineering, sporting goods etc. due to their superior productivity and excellent features as performs [1, 2]. Moreover, they have a good feature that their mechanical properties can be controlled arbitrarily by several designing parameters such as braiding angles, tex of yarns, number of yarns and number of layers etc. (See Figure 1).

Controlling mechanical properties in practice, however, takes much effort because of many trial productions required under broad possibilities for the designing parameters. To ease the difficulty, finite element (FE) analysis has been invoked revealing relativity between the designing parameters and the mechanical properties such as stiffness, strength and internal stress etc. [3, 4]. We also have proposed a procedure to make braided models for FE analysis, whose structure of yarns is more precise than former models, presenting undulations by braiding and stacking of layers [5]. Further, we have implemented the procedure to software named "Composites Dream" which designers can operate easily.

This paper has reviewed the procedure for braided tubes and presented validation of a FE model generated by the procedure in virtue of comparing numerical results of the model and experimental results. A golf shaft with four layered triaxle braided has been modeled by the procedure. Cross-sectional observations of the model and the golf shaft have confirmed the similarity of structure of yarns. Further, comparing some tests (three-point bending and torsional tests) and FE analysis of them have validated the similarity of mechanical properties.

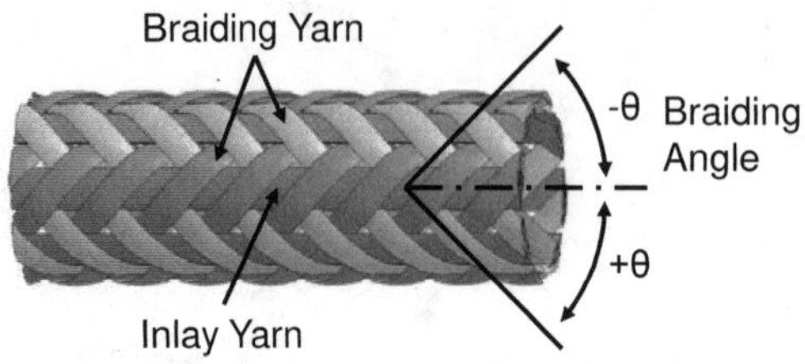

Figure 1. Structure of braided tube

2. Proposal model

2.1. Procedures

The modeling process is divided into three steps. The first step is to prepare shell elements and to determine the shape of cross-section of a bundle (Figure 2(a)). Width and thickness of the cross-section are estimated from thickness of each layer.

The next step is to duplicate the shell elements, and to generate solid elements for bundles by sweeping the shell elements along an axial direction of the tube with certain braiding angles (Figure 2(b)). In addition, the crimp can be represented by the sweeping trajectories whose radii change with a (periodic) spline curve with crimp height expected from thickness of each layer.

Finally, the generated FE-model packs up to the designed thickness of the tube (Figure 2(c)). The surfaces of solid elements contact with each other if some layers are applied. To preserve the tex of bundles in the calculation, the solid elements have an orthotropic material, and Poisson's ratios of them are 0.5. In order to remove redundant lengths generated by the compression of the packing, the thermal strain applies to reduce the lengths of bundles.

As example, Figure 3 shows several patterns of bundles that the procedure can generate. Figure 3(a) is for changing braiding angles, the degrees are 30, 45 and 60. And Figure 3(b) is for changing number of braiding bundles, the number are 4, 8 and 16.

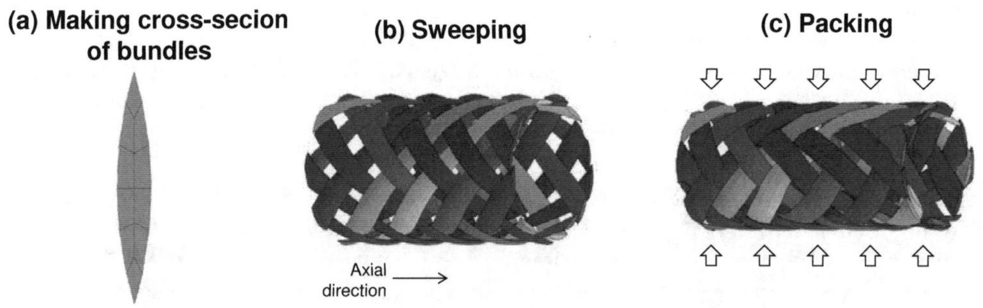

Figure 2. Proposal procedures to generate finite model of braided tube

2.2. Specimen

Figure 4 shows a golf shaft of braided CFRP tube used for validation. The length is 160 mm and the outer and inner diameters are 15.46 mm and 11.26 mm, respectively. The shaft consists of four layers and the specification is summarized in Table 1. A FE model of the tube generated by the proposed procedure is shown Figure 5.

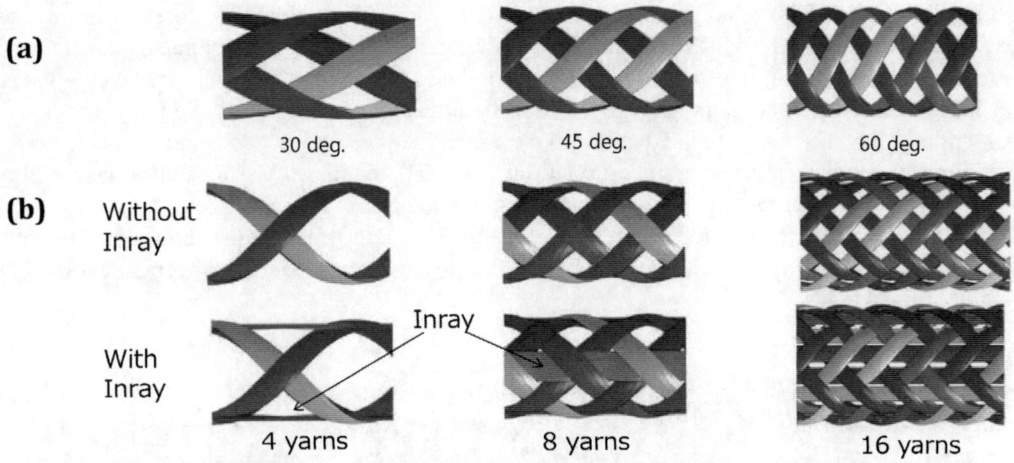

Figure 3. Examples generated by Composites Dream with some designing parameters (a)Changing braiding angle and (b)changing number of inlay

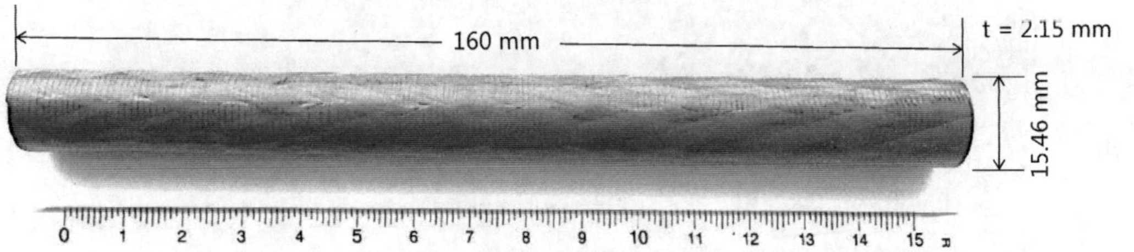

Figure 4. Golf shaft

Table 1. Specifications of the golf shaft

Position	No. of Ply	Tex	Number of Yarns		Braiding Angle
			Braiding Yarn	Inlay Yarn	
Inner layer	1	12K	16	0	46 deg.
⬇	2	12K	16	0	39 deg.
	3	12K	16	4	30 deg.
Outer layer	4	24K	16	8	19 deg.

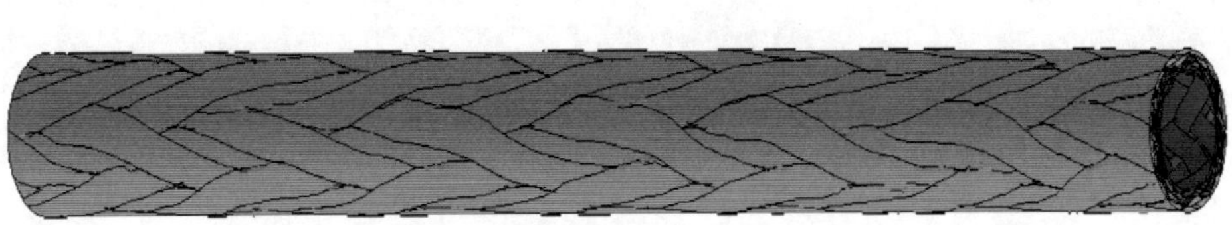

Figure 5. Generated finite model of the golf shaft

3. Results and discussions

3.1. Cross-sectional observation

In order to validate the modelling procedure, the shape of bundles in a cross-section of tube is compared. Figures 6(a) and (b) show the cross-section of the golf shaft. They have two types which are perpendicular and parallel sections to longitudinal direction, respectively.

The cross-sections for each direction of generated FE-model are shown in Figures 6(c) and (d).

IOP Conf. Series: Materials Science and Engineering **406** (2018) 012031 doi:10.1088/1757-899X/406/1/012031

From Figure 6(a), it is revealed that the cross-section is filled with bundles closely. Figure 6(c) shows the generated model and also reveals that the bundles are in existence without gaps. From these figures, it is recognized that the initial shape of section of bundle like a lens can't keep anymore after the packing, and that the trend of geometry change of each bundle deformed agree well. However, the explicit evaluation will be necessary to the quantitative evaluation.

Figure 6(b) indicates the section for axial direction. The bundles exist in section area without gap. Figure 6(d) is the generated model by the proposed procedure.

Comparing the golf shaft and FE model by the procedure confirms that the shape of section of bundle is similar. Consequently, we can conclude that the proposed procedure can generate FE models for layered braided tubes.

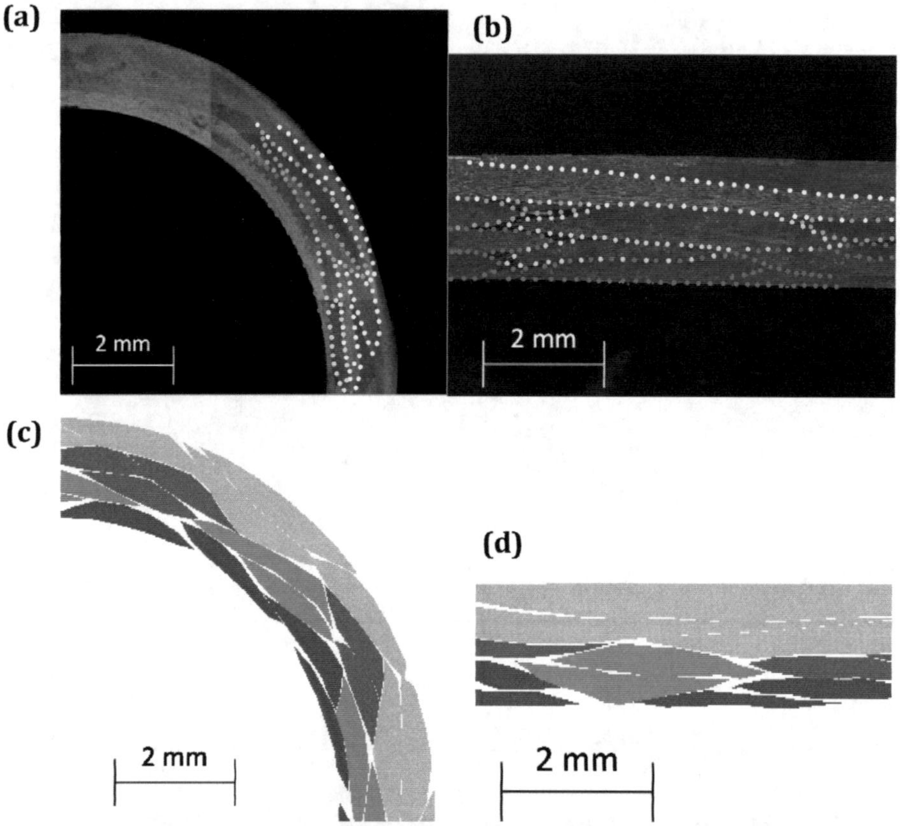

Figure 6. Cross-sectional images of the golf shaft and the finite model

3.2. Three-point bending test

Figure 7 shows the numerical and the experimental results for the three-point bending test. The span of bending test has been 120 mm, and the shapes of supporting and loading parts have been cylinder with 10 mm radius.

The analysis under three-point bending has been carried out by LS-DYNA. The engineering constants of CFRP bundle for analysis is shown in Table 2. The moduli have been estimated by Uemura's equation [6] with carbon fiber volume fraction $V_f = 54$ % (experimental) and epoxy resin. As the strain criteria of compression, $F_{\varepsilon c} = 0.25$ % has been employed.

The experimental result in the Figure 7 shows elastic response by 0.5 mm of displacement and failure response after that. The numerical result has a good agreement with it.

Figure 8(a) shows the top view of the specimen after test. The position of loading nose is the line connected to triangles in Figure 8(a). Figure 8(b) shows the numerical results concerning the failure elements and the displacement d is 1.0 mm. The damage propagations of both results have similar situation.

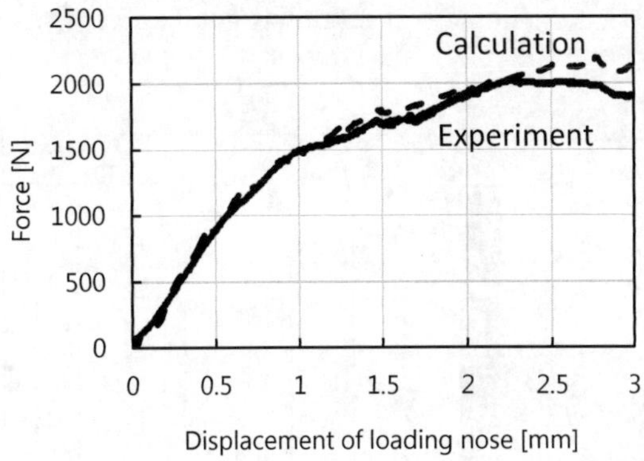

Figure 7. Load-displacement curves of the three-point bending test

Table 2. Engineering constants for bundles

Engineering constants		Description
E_1	140 GPa	Modulus of longitudinal (fiber) direction
E_2	6.91 GPa	Modulus of transverse direction
G_{12}	2.41 GPa	Shear modulus of longitudinal and transverse directions
G_{23}	2.74 GPa	Shear modulus of transverse directions
ν_{12}	0.427	Poisson's ratio of longitudinal and transverse directions
ν_{23}	0.263	Poisson's ratio of transverse directions

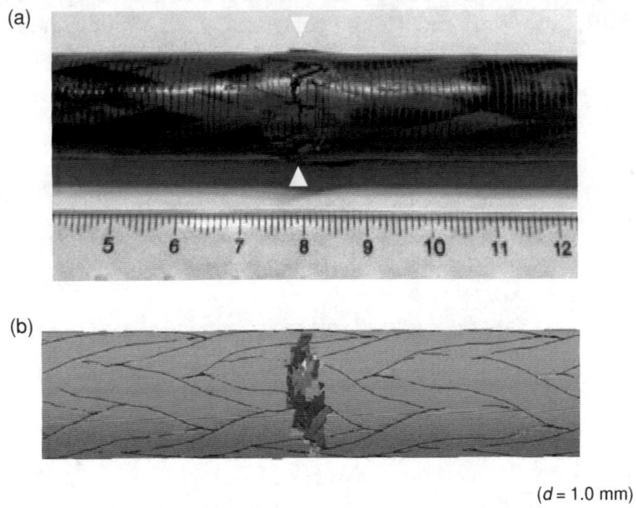

Figure 8. Status of damage propagation after the three-point bending test (a) Experimental result and (b) Numerical result

3.3. Torsion test

Figure 9 shows the torsional stiffness around the shaft axis obtained by torsion test and FE analysis. Four specimens are prepared. The scatter of experimental data is a little. But the numerical result is about 15% higher than the experimental ones. As the end of specimen must hold by the torsion testing apparatus, the both ends of specimens have been supported by GFRP. The support of GFRP did not consider to the numerical simulation. The torsional stiffness of GFRP is lower than that of CFRP. From these points, it is recognized that the numerical result has good agreements with the experimental ones.

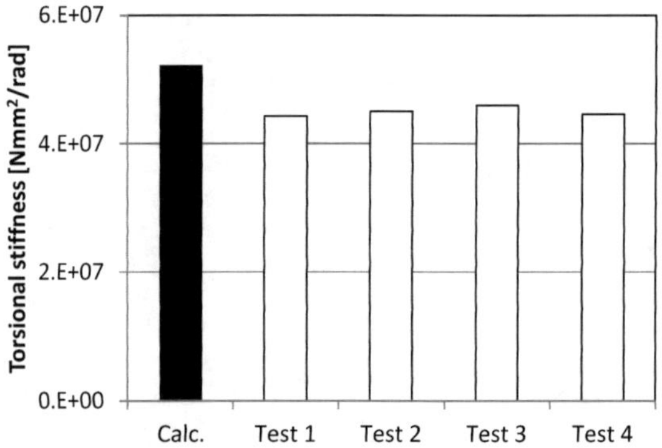

Figure 9. Torsional stiffness

4. Conclusions

The procedure to generate FE model of the layered braided tubes with complex architecture has been proposed. In order to verify the procedure, the similarity of shape of bundles has been confirmed by comparing the cross-sections of the real CFRP golf shaft and FE model generated by the proposed procedure. The proposed system is acceptable as a design tool, because the numerical results of mechanical properties such as the stiffness and failure behaviour with the FE model have agreed well with the experimental results

We have implemented the procedure to the software named "Composites Dream" in which some designing parameters for a braided tube can generate a FE model readily (See Figure 10). Providing the verified procedure with ease, the software should contribute designing of braided tubes.

Figure 10. Analysis procedures for braided composites tubes using Composites Dream

References

[1] T. Saito, H. Asada, M. Hirose, T. Uozumi, A. Ohtani and A Nakai, Effect of braided structure on mechanical properties of CFRP pipes, Transactions of the JSME, 82, 2016 (doi: 10.1299/transjsme.15-00354). (Japanese)

[2] T. Uozumi, M. Tanigawa, T. Nishida, T. Miura, A. Ohtani and A Nakai, Non-crimp tubular

IOP Conf. Series: Materials Science and Engineering **406** (2018) 012031 doi:10.1088/1757-899X/406/1/012031

preforming technology with high productivity, Proceedings of TEXCOMP-12 Conference, Raleigh, USA, May 26-29, 2015.

[3] A. Ohtani and A. Nakai, Effect of internal structure on mechanical properties of braided composites tubes, Proceedings of the 16th International Conference on Composite Materials, Kyoto, Japan, July 8-13, 2007.

[4] A. K. Pickett, J. Sirtautas and A. Erber, Braiding simulation and prediction of mechanical properties, Appl. Compos. Mater., 16, 2009, pp. 345-364 (doi: 10.1007/s10443-009-9102-x).

[5] T. Yamamoto, A. Imaoku, T. Sakakibara and M. Zako, A Proposal Of Conventional FE-Modeling For Layered Braided Composites, Proceedings of the 21st International Conference on Composite Materials, Xi'an, China, August 20-25, 2017.

[6] M. Uemura and N. Yamada, Elastic constants of carbon fiber reinforced plastic materials, J. Soc. Mater. Sci. Jpn., 24, 1975, pp. 156. (Japanese)

Tensile characterization of 3D nanostitched p-aramid/ phenolic MWCNTs composites

K Bilisik[1*], G Erdogan[1], N S Karaduman[2], G Kaya[3], E Sapanci[4], S Gungor[4]

[1]Nano/Micro Fiber Preform Design and Composite Lab., Faculty of Engineering, Erciyes University, 38039 Talas-Kayseri, Turkey
[2]Akdagmadeni Vocational High School, Bozok University, 66300 Akdagmadeni-Yozgat, Turkey
[3]Department of Textile Engineering, Faculty of Engineering and Architecture, Kahramanmaras Sutcu Imam University, Kahramanmaras, Turkey
[4]Advanced Material Development Unit, ROKETSAN Industries, 06780, Elmadag-Ankara, Turkey

* kadirbilisik@gmail.com

Abstract. The tensile properties of nanostitched and nanoprepreg three dimensional para-aramid/phenolic composites were studied. The tensile responses of para-aramid/phenolic composites showed three regions as initial inter-fiber friction, crimp exchange and yarn extension regions. Their tensile moduli in each region were defined. It was obtained that stitching and multiwall carbon nanotubes slightly increased the tensile strength of all stitched and stitched/nano composites. Tensile failure on all stitched structure showed brittle matrix and multiple ductile filament breakages in yarn after angular deformation in composite surface layers as well as local restricted delamination. Failure mechanism of carbon stitched composite was interestingly different compared to para-aramid stitched composite in where carbon stitching yarn was failed in the loop section. It was pulled out thickness direction of the composite and caused large local opening between layers.

1. Introduction

Nano/Micro filament preform composites are used in various fields such as aviation/aerospace, ballistics and civil engineering because of their high specific properties and exceptional damage tolerance behaviour [1, 2]. Z-directional reinforcement can be incorporated to textile preforms by using three-dimensional (3D) weaving, 3D braiding [4], and through-the-thickness stitching [5, 6]. In addition to these methods, recent studies have focused on the use of nanomaterials such as nanospheres, nanotubes, nano graphane platelets and nanofibers to obtain z-reinforcement. Techniques such as shear mixing, ultrasonication and transfer-printing have been used to disperse the nanomaterial in the resin [7, 8]. 3D textile reinforced composites generally exhibit low in-plane properties due to the presence of z-fibers which reduce the in-plane fiber volume fraction ratio. Nanomaterials incorporated in the composite structure, on the other hand, usually fail to provide an effective z-directional reinforcement because they generally have a discontinuous form and a random distribution rather than aligning in the thickness direction. In-plane tensile and impact strengths as well as flexural properties increased because of homogeneous topological arrangements of stitching fibers [9, 10].

Several studies have reported improvements in composite properties when nanomaterials are used in the structure. Tensile properties increased when multiwalled carbon nanotubes (MWCNTs) were used [11]. Incorporation of nanoclay enhanced the E-glass/vinyl ester composites [12, 13]. An increased amount of nanosilica from 2.5% to 7.5% significantly enhanced the principal and off-axis tensile properties of the E-glass/polyester composites [14]. Stitching also slightly increased the tensile characteristics of the E-glass/polyester nanosilica composites when compared to unstitched samples [15]. The damage was confined to a smaller area in stitched E-glass/polyester nanosilica composites [14, 15]. Composites reinforced with CNTs-coated carbon fibers displayed approximately 66% higher tensile strength compared with that of the base sample [16]. Addition of coated nanotubes noticeably improved the interface properties of the E-glass/vinyl ester composites [17]. Silane modification led to a more homogeneous dispersion of MWCNTs in epoxy matrix [18]. Amino functionalized carbon nanofibers (A-CNF, 1%) can considerably improve the in-plane mechanical strength of the laminated carbon/epoxy composite [19].

In this study, the tensile properties of stitched and MWCNTs-added para-aramid/phenolic composites were investigated.

2. Materials and methods

2.1. Nanostitched para-aramid/phenolic nanoprepreg based 3D nanopreform and composite

Para-aramid Twaron® basket (2/2) fabrics (CT736, Teijin, JP) with an areal density of 410 g/m^2 and warp/weft densities of 127 ends/cm were used to fabricate multistitched 3D nanopreform. The fabric was formed from 168 tex fibers and the fabric thickness was 0.62 mm. The average crimp of p-aramid fabrics was 10.35 % for basket (2/2). MWCNTs (Nanothinx, GR) used in this study were 15-35 nm in diameter, 10 μm in length and 1-2 nm in wall thickness. Their purity ratio was over 97%. Their average tensile strength and modulus values were 200 GPa and 1 TPa, respectively [20].

Principally, four types of p-aramid structures were developed: a) base (TBU), which was made up of six layers of Twaron basket (2/2) woven fabrics without stitch; b) stitched (TB-CS, TB-TS), the same as the base structure but the fabric layers are stitched through-the-thickness using carbon and Twaron stitching yarns. The stitches were applied in the longitudinal direction (0°) of the structure; c) base/nano (TBU-N) structures were obtained by adding MWCNTs to the base structures; d) stitched/nano (TB-CS-N, TB-TS-N) samples were obtained by adding MWCNTs to the stitched structures. Stitching was performed manually in the longitudinal direction of the structures using PAN based carbon and Twaron CT stitch yarns. The nanostitched and nanoprepreg composite samples are shown in figure 1. The stitch yarn properties are given in table 1. The preforms were consolidated via compression molding to obtain composite samples. The processing details for the mixing of the phenolic resin (Araldite EPN 1138, Biesterfeld Spezialchemie GmbH, DE) with MWCNTs and for the composite production were given in reference [21] and we would like not to repeat the processing steps and readers are advised to look through the above-mentioned reference.

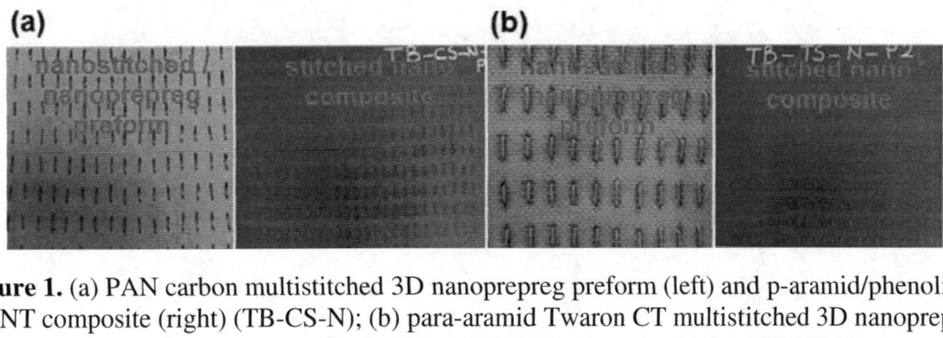

Figure 1. (a) PAN carbon multistitched 3D nanoprepreg preform (left) and p-aramid/phenolic MWCNT composite (right) (TB-CS-N); (b) para-aramid Twaron CT multistitched 3D nanoprepreg preform (left) and p-aramid/phenolic MWCNT composite (right) (TB-TS-N).

IOP Conf. Series: Materials Science and Engineering **406** (2018) 012032 doi:10.1088/1757-899X/406/1/012032

Table 1. Specifications of untwisted stitching yarns.

Fiber type	Tensile strength (GPa)	Tensile modulus (GPa)	Elongation at break (%)	Yarn linear density (dtex)
Twaron CT (Para-aramid fiber, Teijin, JP)	3.2	115	2.9	3360
Polyacrylonitrile (PAN) carbon (Carbon fiber, Aksaca, TR)	4.2	240	1.8	6 K[1]

2.2. Tensile test

The tensile tests of the composite samples were performed according to ASTM D3039-14 standard [22] on a Shimadzu AG-XD (JP) tester with a 5kN loading cell and equipped with Trapezium® data acquisition software (figure 2). The crosshead speed was 2 mm/min and the load was applied in the longitudinal (warp) direction of the structures. The tensile characteristics of the composite samples were calculated according to Eqs. 1-4.

$$F = P / A \tag{1}$$

$$\sigma_i = P_i / A \tag{2}$$

$$\varepsilon_i = \delta_i / L_g \tag{3}$$

$$E = \Delta\sigma / \Delta\varepsilon \tag{4}$$

Where F is the ultimate tensile strength (MPa), P is the maximum force before failure (N), A is the average cross-sectional area (mm^2), σ_i is the tensile stress at ith data point (MPa), P_i is the force at ith data point (N), ε_i is the tensile strain at ith data point, δ_i is the extensometer displacement at ith data point (mm), L_g is the extensometer gage length (mm), E is the tensile modulus of elasticity (GPa), $\Delta\sigma$ is the difference in applied tensile stress between the two stain points (MPa) and $\Delta\varepsilon$ is the difference between the two strain points.

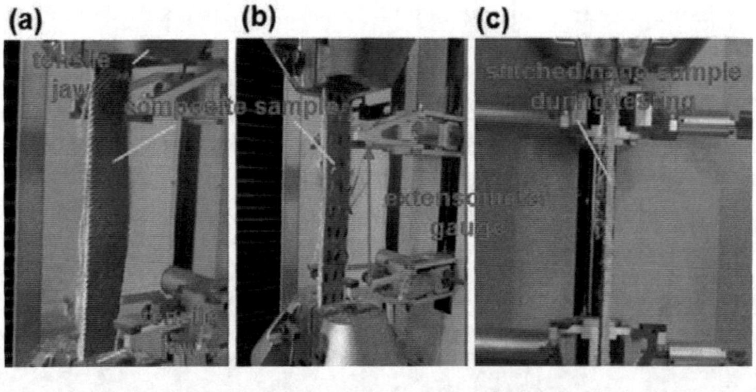

Figure 2. Tensile testing instrument with sample. (a) p-aramid/phenolic/MWCNT composite (TBU-N); (b) carbon multistitched p-aramid/phenolic composite (TB-CS); (c) p-aramid multistitched p-aramid/phenolic/MWCNT composite (TB-CS-N, digital image).

IOP Conf. Series: Materials Science and Engineering **406** (2018) 012032 doi:10.1088/1757-899X/406/1/012032

3. Results and discussion

3.1. Tensile results

The tensile test data of base (TBU), stitched (TB-CS, TB-TS), base/nano (TBU-N) and stitched/nano (TB-CS-N, TB-TS-N) composites are generated. Figure 3 shows the tensile stress-strain curves of some of the basket (2/2) fabric composites.

In figure 3, the tensile properties of stitched and stitched/nano composites were found to be generally higher compared with unstitched composites. Tensile stress-strain curves of the tested samples showed three distinct stages such as the inter-fiber friction, crimp exchange and yarn extension. In the first stage, yarn-yarn friction and micro shear took place. In the crimp exchange stage, warp crimp was removed leading to a large fabric extension. In the final i.e. yarn extension stage, the uncrimped yarns were strained to the point of final failure.

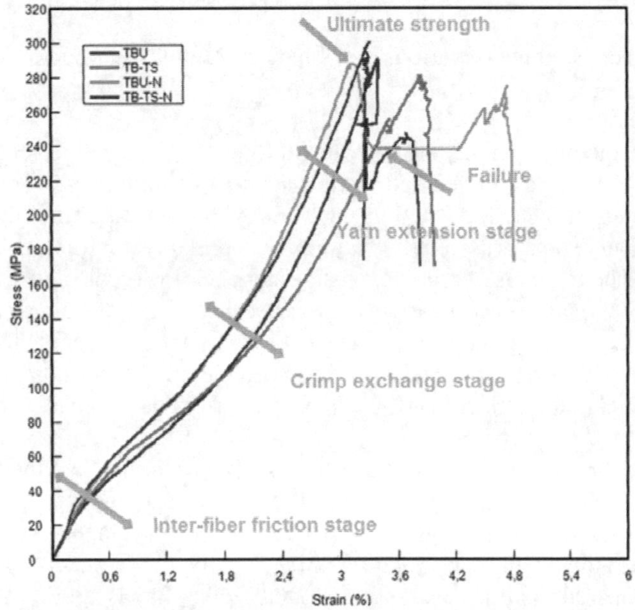

Figure 3. Stress-strain curves from tensile test for some of the multistitched p-aramid/phenolic MWCNT composites (based TBU, stitched TB-TS, base/nano TBU-N, stitched/nano TB-TS-N) [21].

3.2. Tensile strength

The tensile strength values of the developed structures are presented in figure 4. Stitched and stitched/nano structures generally displayed greater strength values (up to 7.73%) over the unstitched ones. There is no significant difference between the strength values of the stitched and stitched/nano composites and it is reasonable to state that the nanomaterial had only limited effect on the tensile properties. Stitching and nano addition slightly increased the strain values of the composites (up to 6.08%). The average tensile strain values were almost the same for the base and base/nano structures. However, the strain value of the stitched/nano sample was about 2.48% higher compared with that of the stitched structure.

IOP Conf. Series: Materials Science and Engineering **406** (2018) 012032 doi:10.1088/1757-899X/406/1/012032

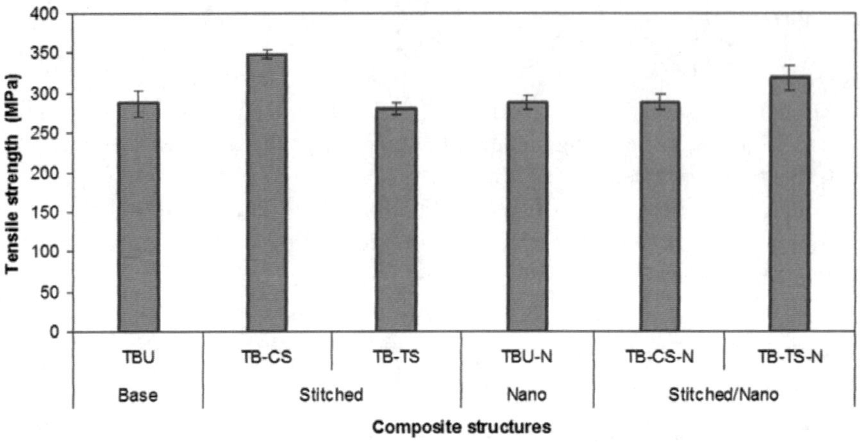

Figure 4. Tensile strength of various developed p-aramid/phenolic MWCNT composites.

3.3. Tensile modulus

The average tensile modulus values are shown in figure 5. Three different modulus values were calculated each corresponding to the three distinct load-elongation stages on the stress-strain curves (figure 3). The modulus of inter-fiber friction stage (E_{ff}) was calculated from the initial straight portion of the stress-strain curve and was relatively high due possibly to friction between fibers and the surrounding matrix. The modulus of crimp exchange stage (E_{ce}) was calculated from the following non-linear portion of the curve and it was relatively lower because of crimp extension at this stage. The final stage involved yarn extension in the form of tensile elongation of the yarns. The modulus at this stage was defined as the modulus of yarn extension stage (E_{ye}) and is higher compared with the E_{ce} due to decreased crimp ratio and to the fact that the extension took place as tensile elongation of low-twisted Twaron yarns (1-2 twist/meter).

As illustrated in figure 5, it was found that the E_{ff} was higher than the E_{ce} and E_{ye} due to initial inter-fiber friction stage of the stress-strain curve. The E_{ye} was also higher than the E_{ce} because of high extension in the crimp interchange stage of the composite. The average E_{ff} of all stitched composites was 7.20% higher than that of all base structures. Additionally, the E_{ff} of all base/nano composites were almost the same compared to all base structures. On the other hand, the E_{ff} of all stitched composites were slightly higher (18%) than the stitched/nano composites. It was realized that the stitching slightly increased the E_{ff} of all stitched composites. The E_{ce} and E_{ye} of all developed composites were almost near to each other.

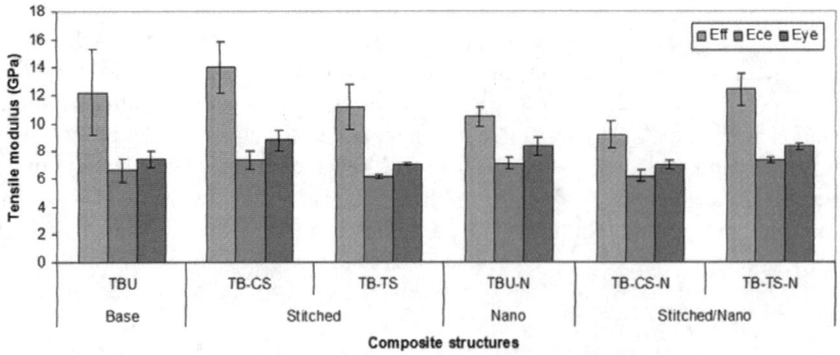

Figure 5. Tensile modulus of various developed p-aramid/phenolic MWCNT composites.

3.4. Failure characterization after tensile test

The post-tensile fracture characterization of the developed structures was performed. Figure 6 (a-c) presents the post-tensile failure pictures of base/nano (TBU-N) and stitched/nano (TB-TS-N, TB-CS-N) samples. TBU and TBU-N samples were subject to warp and lateral yarn splitting at the edge of the specimen as well as fiber and matrix failures due to the tensile load. Large delamination areas were also observed in these specimens (Figure 6 (a)). With regards to the stitched and stitched/nano structures i.e. TB-TS and TB-TS-N samples, delamination was largely constrained to smaller areas due to the stitch yarns. The failure mode of this specimen also involved warp yarn splitting, stitch yarn failure, warp/filling yarn breakages and matrix cracking as well as shear failures (Figure 6 (b)).

With regards to the carbon-stitched structures, i.e. TB-CS and TB-CS-N, fiber breakages and matrix cracking were the dominant failure mode whereas failure due to shear load was also observed. The carbon stitch yarn failed at the stitch loop section and pulled out of the lamina layers in thickness direction causing local delamination of the layers (Figure 6 (c)). It can be concluded that the stitch yarn largely limited the extent of delamination and reduced the total damage area when compared to the unstitched samples.

(a) (b) (c)

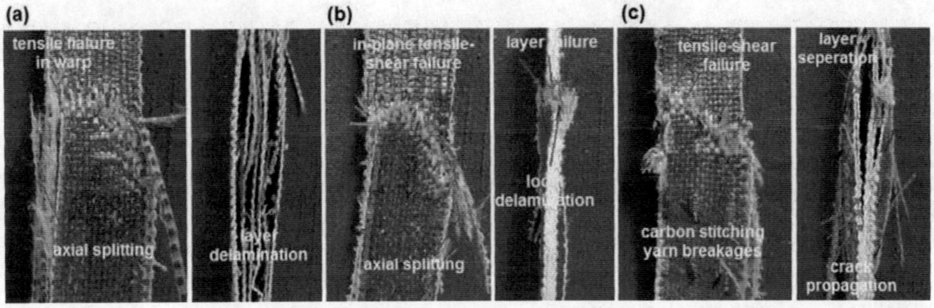

Figure 6. Warp directional top and side view (thickness) of tensile failure in various multistitched 3D p-aramid/phenolic MWCNT composites. (a) base/nano (TBU-N); (b) basket 2/2 p-aramid stitched/nano (TB-TS-N); (c) basket 2/2 carbon stitched/nano (TB-CS-N) (optical microscope, magnification x6.7) [21].

Figure 7 (a-b) shows the SEM photomicrographs of the failed samples TB-TS and TB-TS-N. TB-TS sample was subjected to multiple warp yarn failures on the top layer. Closer inspection of the failed region suggests warp yarn splitting and warps breakages (figure 7 (a)). Ductile multiple yarn end failure in the front face of the TB-TS-N structure was identified. Some of the failed filament has axial splitting probably due to weak cross bonding in the out-of-plane direction as exhibited in figure 7 (b).

(a) (b)

Figure 7. (a) SEM (LEO 440® model, UK) views of failed p-aramid stitching yarn filaments (TB-TS); (b) broken p-aramid stitching nano prepreg yarn in the delaminated failed surface of p-aramid/phenolic MWCNT composite (TB-TS-N) [21].

5. Conclusions

Tensile properties and failure pattern of the stitched and stitched/nano p-aramid composites were investigated. Stress-strain profile of the developed structures involved three different stages such as inter-fiber friction, crimp exchange and yarn extension. The modulus value in inter-fiber friction region was higher compared to those in the crimp exchange and yarn extension regions due to friction to the yarn in the matrix. Tensile strength and strain generally increased as a consequence of stitching and nanomaterial addition. The modulus value in inter-fiber friction region was slightly higher in stitched composites compared with that of the unstitched samples whereas other two moduli i.e. crimp exchange and yarn extensions were almost the same for all structures.

Ductile warp and weft yarn failures and matrix cracking were the primary failure mode in the structures. Para-aramid stitch yarn successfully constrained the layer-to-layer delamination in the samples whereas carbon stitch yarn failed at the stitch loop region and was pulled out of the structure in the thickness direction resulting in large delamination. It can be noted that the stitch yarn largely limited the extent of delamination and reduced the total damage area when compared to the unstitched samples. These results showed that stitching and multiwalled carbon nanotube addition improved the in-plane properties of the resulting p-aramid based composites.

Acknowledgements
This work was supported by Roketsan Industries Grant No. RS/ERCİYES DSM-76301-14-01N/R. The authors would like to thank Erciyes University Technology Research and Application Center (ERÜ TAUM) for tensile tests.

References

[1] Dow M B and Dexter H B 1997 Development of stitched, braided and woven composite structures *ACT Program at Langley Research Centre* NASA/TP-97-206234.
[2] Kamiya R, Cheeseman B A, Popper P and Chou T W 2000 *Compos. Sci. Technol.* **60** 33-47.
[3] Bilisik K 2012 *Text. Res. J.* **82** 725-743.
[4] Bilisik K 2013 *Text. Res. J.* **83** 1414-1436.
[5] Tong L, Mouritz A P and Bannister M K 2002 *Stitched Composites* In: 3D Fibre Reinforced Polymer Composites. New York: Elsevier B.V. 163-204.
[6] Bilisik K and Yolacan G 2014 J. *Compos. Mater.* **48** 2145-2162.
[7] Garcia E J, Wardle B L and Hart A J 2008 *Compos. A* **39** 1065-1070.
[8] Khan S U and Kim J-K 2011 *Int. J. Aeronaut. Space* **12** 115-133.
[9] Velmurugan R and Solaimurugan S 2007 *Compos. Sci. Technol.* **67** 61-69.
[10] Mouritz A P 2004 *Fracture and tensile fatigue properties of stitched fibreglass composites.* In: Proceedings of the Institution of Mechanical Engineers, Part L: Journal of Materials: Design and Applications **218** 87-93.
[11] Guo P, Chen X, Gao X, Song H and Shen H 2007 *Comp. Sci. Tech.* **67** 3331-3337.
[12] Chandradass J, Kumar M R and Velmurugan R 2008 *J. of Reinf. Plast. and Compos.* **27** 1585-1601.
[13] Bozkurt E, Kaya E and Tanoglu M 2007 *Compos. Sci .Tech.* **67** 3394-3403.
[14] Bilisik K and Yolacan G 2014 *Fiber. Polym.* **15** 1051-1061.
[15] Bilisik K and Yolacan G 2014 *Fiber. Polym.* **15** 614-624.
[16] Sharma S P and Lakkad S C 2011 *Compos. A* **42** 8-15.
[17] Zhu J, Imam A, Crane R, Lozano K, Khabashesku VN and Barrera EV 2007 *Compos. Sci. Tech.* **67** 1509-1517.
[18] Fukushima T, Kosaka A, Ishimura Y, Yamamoto T, Takigawa T, Ishii N and Aida T 2003 *Science* **300** 2072-2076.
[19] Srikanth I, Kumar S, Singh V, Rangababu B, Ghosal P and Subrahmanyam H 2015 *Bull. Mater. Sci.* **38** 1-9.
[20] http://www.nanothinx.com/raw-cnts-in-powder-form/; 2017 [accessed 17.02.17].
[21] Bilisik K, Erdogan G and Sapanci E 2018 *Polym. Compos.* https://doi.org/10.1002/pc.24847.
[22] ASTM D3039-14. 2014 Standard test method for tensile properties of polymer matrix composite materials.

Nonwoven fabrics with carbon nanotubes used as interleaves in CFRP

Paulina Latko-Durałek[1*], Kamil Dydek[1], Patryk Bolimowski[1], Emilia Golonko[1], Paweł Durałek[2], Rafał Kozera[1], Anna Boczkowska[1]

[1]Faculty of Materials Science and Engineering, Warsaw University of Technology, ul. Wołoska 141, 02-507 Warsaw, Poland
[2]TMBK Partners Sp. z o. o., ul. Pawińskiego 5A, 02-106 Warsaw, Poland

* paulina.latko@inmat.pw.edu.pl

Abstract. The goal of the present study was to implement thermoplastic nonwoven fabrics containing multi-walled carbon nanotubes as interlayers in Carbon Fiber Reinforced Polymers. These functional nonwovens were fabricated by a half-industrial scale melt-blown technique, starting with nanocomposite pellets of copolyamides doped with 3.5wt% of multi-walled carbon nanotubes. Three types of composite panels were fabricated using an out-of-autoclave technique (OoA): one without nonwovens and two with nonwovens. Incorporation of thermoplastic nonwovens doped with 3.5wt% of multi-walled carbon nanotubes increased the surface and volume electrical conductivity in direction K_z by about 2 and 3 orders of magnitude, respectively. Based on the images obtained from a Scanning Electron Microscope, it was found that melted nonwovens adhere well to the carbon fibers. It was also confirmed that carbon nanotubes are well dispersed in nonwovens, which results in an improvement of the overall electrical conductivity of the composite panels. The lack of homogenous layers of nonwovens between the carbon fiber layers decreased the interlaminar shear strength of the composite panels and affected the level of their electrical conductivity. Moreover, thermo-mechanical analysis showed an increase of the glass transition temperature of the resin in the presence of thermoplastic nonwovens and the appearance of an additional peak on the loss modulus curve caused by the polyamide 6 segments present in the copolyamides used.

1. Introduction

Lightweight non-metallic components made of Carbon Fiber Reinforced Polymer (CFRP) are commonly used in the automotive, aviation, defense and electronics industries, mainly due to their high strength and performance and low weight. On the one hand, CFRP offers increased manufacturing productivity, lower VOC emissions, better corrosion resistance, high durability and design flexibility. On the other hand, they suffer from insufficient electrical conductivity for lightning strike protection, static discharge, electrical bonding, grounding, interference shielding or current return through the structures required in many applications such as aircraft fuselages, wings, car bodies, satellites or antenna dishes [1][2].

Because of their low impedance and high reflectivity, carbonaceous structures like carbon fibers, graphite, graphene or carbon nanotubes (CNTs) are good candidates for increasing the electrical conductivity of CFRP. Moreover, in comparison to the copper meshes which are used nowadays in CFRP, they are much lighter but still show high mechanical strength [3][4]. Recently, much research has focused on incorporating single-walled (SWCNTs) and multi-walled carbon nanotubes (MWCNTs) into CFRP. This can be achieved by various methods: mixing with the neat epoxy resin used in liquid technologies, e.g. infusion or RTM [5][6]; application of thin films of CNTs (buckypaper) [7][8];

deposits of nanofillers on commercial prepregs [9][10]. Other approaches include modifying dry carbon fabrics by spraying epoxy and CNT on the carbon fabrics [11] or direct grafting of nanofillers onto carbon fabrics [12]. In each of the tested solutions, problems occur during fabrication and implementation into composites that in effect lead to insufficient enhancement of the electrical and mechanical properties of CFRP. For instance, excessively high viscosity of CNT-modified epoxy decreases its flowability and results in poor infiltration of CFRP layers [6].

In the composite industry thermoplastic nonwovens are commonly used as interlayers or surface finishing in CFRP. Applying a small amount of thermoplastic polymer makes the epoxy resin less brittle and improves the fracture toughness of the laminates [13][14]. Based on this, the new idea described in this paper involves fabrication of thermoplastic nonwoven fabrics containing CNTs which could be applied as interleaves in CFRP to improve not only the mechanical properties of composites, but also primarily their electrical conductivity. Such electrically conductive fabrics are of great interest nowadays due to their high potential for use not only in the composite industry, but also as electromagnetic absorbers [15], sensors [16], heating elements for aircraft de-icing [17] or in wearable electronics [18]. Recent studies have investigated various methods of producing nonwovens with conductive nanofillers, such as immersing the polymeric fabrics in a nanofiller solution [19][18], coating the fabrics with a thin layer of nanofiller [20], plasma treatment [21], ink printing [22], in an electrospinning process [23][24], by a melt-blown method [25] or by pressing thermoplastic fibers containing nanofillers [26].

Until now only a few papers have described the usage of thermoplastic nonwovens with CNTs as interleaves in CFRP manufactured by prepreg technology; however, they mainly describe the mechanical but not the electrical properties of CFRP. It was found that application of thin nonwovens containing CNTs produced by ultrasonication of polyethylene oxide and 15wt% MWCNTs resulted in an improvement of over 200% in Mode I and II interlaminar fracture toughness [27][28]. The authors of these papers suggested that this is due to increased adhesion between the epoxy and fibers of nonwoven fabrics. In other work, nanofibrous mats were produced from polysulfone by electrospinning and then kept in a carboxyl MWCNTs solution to cover the nanofibers [28]. Both Mode I and Mode II interlaminar fracture toughness were improved, but only by about 50%, and flexural strength was reduced due to the formation of microspheres of polysulfone and CNTs pulling out of the epoxy. The electrical conductivity of composite panels with inserted nonwovens containing CNTs has not been examined.

In the present study, we investigated the effect of thermoplastic nonwovens containing MWCNTs on the electrical, thermal and mechanical properties of CFRP. The nonwovens were produced by a half-industrial melt-blown pilot line by TMBK Partners company in Warsaw, Poland. In this technique, thermoplastic polymer is fed into the extruder, after which the melting material goes through the die orifices. Thin fibers are formed by application of high-speed hot air. These fibers are then deposited on a transport belt, thus creating the nonwoven web [29].

2. Materials and methods

Two types of thermoplastic copolyamides (coPA1 and coPA2) were selected as a polymer matrix; they have melting points of 108–118°C and 107°C, respectively. They were mixed with 3.5wt% of MWCNTs (NC7000) using a half-industrial twin screw extruder machine by Nanocyl, Belgium. Pellets containing 3.5wt% MWCNTs were processed directly using a melt-blown method. To decrease and homogenize the overall thickness of the nonwovens, they were thermally pressed below the melting point of copolyamides. The properties of the nonwovens are listed in Table 1, while macroscopic images are presented in Fig.1.

Table 1: Properties of nonwovens doped with MWCNTs.

Nonwoven designation	MWCNT content [wt%]	Areal weight [g/m^2]	Nonwoven thickness [mm]	Fiber diameter in nonwoven [μm]	Melting point measured by DSC [°C]
coPA1/3.5	3.5	30	0.12 ± 0.03	31.9 ± 10.7	113
coPA2/3.5	3.5	30	0.17 ± 0.05	40.9 ± 12.8	110

IOP Conf. Series: Materials Science and Engineering **406** (2018) 012033 doi:10.1088/1757-899X/406/1/012033

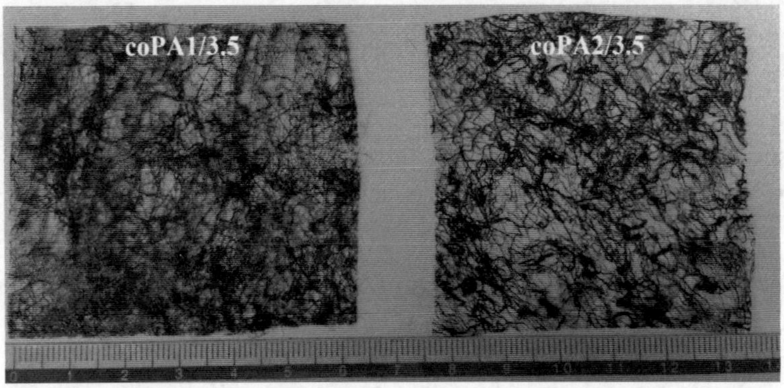

Figure 1: View of nonwoven coPA1/3.5 and nonwoven coPA2/3.5 produced by the melt-blown method.

The produced thermoplastic nonwovens with 3.5wt% MWCNTs were used as interlayers in prepreg technology. Three types of laminate were manufactured with the out-of-autoclave technique: a reference laminate without nonwovens (L1), a laminate with nonwovens made of coPA1+3.5wt% MWCNTs (L2) and a laminate with nonwovens made of coPA2+3.5wt% MWCNTs (L3). Each panel consisted of 13 layers of nonwovens placed between each layer of an unidirectional commercial prepreg with an areal weight 0.015g/cm^2. Laminates after lay-up were degassed, vacuum bagged on a glass tool plate and consolidated at 80°C for 60 minutes and then at 130°C for 120 minutes under a vacuum of 660–710 mmHg. After curing, composite plates were removed from the vacuum bag, trimmed, and cut into test specimen dimensions for electrical and mechanical measurements. The measured thicknesses of the laminates were as follows: 2.3mm (L1), 2.7mm (L2) and 2.7mm (L3). The calculated fiber weight fraction was 61% for L1, 55% for L2, and 54% 0.54% for L3.

Electrical conductivity of CFRP in directions K_z and K_x was measured. 5 samples with dimensions 85mm x 85mm were cut from different sections of the composite panels to measure the volume and surface electrical resistivity according to ASTM D-257 using a 6517B Electrometer/High Resistance Meter equipped with an 8009 Test Fixture. Before the test, each sample was degreased using acetone and the electrodes of the Test Fixture were cleaned with methanol. The sample was placed between stainless-steel electrodes, with a rubber pad which maintains perfect contact between the sample and electrodes. The voltage applied was 1V, time of the test was 15s, and the number of stored readings was adjusted to 5. Using this equipment, the electrical volume resistivity was measured throughout the laminate thickness. In order to measure the electrical volume conductivity along the carbon fibers (K_x), samples with 50mm length and 10mm width were cut from the composite panels. Before the test, each sample was degreased using acetone and covered with a silver paste tape to maintain good contact between electrodes and samples. The sample was placed between copper electrodes. Here, the variable current was set by a Keithley 6221 current source and the voltage change was read on a 2182A nanovoltmeter. In both tests, the final electrical conductivity was calculated, taking the dimensions and thickness of the samples into account.

Mechanical properties of the composite panels were analyzed in terms of the interlaminar shear strength (ILSS). A short beam shear test was conducted according to ASTM standard D2344 using an MTS Q/Test 10 universal testing machine. 8 specimens were cut from different places from each laminate in the direction of the carbon fibers, with dimensions of 25mm x 6.4mm. The test was performed at room temperature at a speed of 1 mm/min.

A Scanning Electron Microscope (SEM 3000, Hitachi, Japan) was used to analyze the microstructure and quality of the produced composite panels. For the observations, small pieces were cut from each laminate and finished using a polishing machine. Before the measurement, laminate cross sections were coated with a thin layer of gold. The applied voltage for SEM observations was 15kV. Using ImageJ software, the porosity in each panel was determined from a minimum 8 SEM images. The same microscope was used in the analysis of the composite panels after the ILSS test. In order to investigate

IOP Conf. Series: Materials Science and Engineering **406** (2018) 012033 doi:10.1088/1757-899X/406/1/012033

the dispersion and distribution of MWCNTs in the final composite panels, a high-resolution SEM SU70 (Hitachi, Japan) was used. For the observations, polished laminate samples were coated with an electrically conductive molybdenum layer of approximately 5 nm thickness. The acceleration voltage was 10kV.

Dynamic mechanical analysis (DMA) was performed using a Q800 DMA (TA Instruments, USA) in dual cantilever mode according to ASTM D4065-01. From each laminate, three rectangular specimens with 60mm length and 10mm width were cut and used in the test. The analysis was conducted from 0°C to 260°C with a heating rate of 2 °C/min at a frequency of 1Hz and amplitude of 30μm. The glass transition temperature (T_g) was determined from the peak value in the loss modulus curve.

3. Results

3.1. Electrical conductivity

Electrical properties of the CFRP panels were measured in terms of surface electrical conductivity and volume electrical conductivity along (K_x) and perpendicular (K_z) to the carbon fibers. As shown in Fig. 2a, it can be seen that implementation of nonwovens with 3.5wt% MWCNTs significantly improved the electrical conductivity throughout the laminate thickness. In comparison to the reference panel (L1), laminate L2 with nonwovens based on coPA1 has an electrical conductivity that is about 1882% higher. In the case of laminate L3, which has interlayers of nonwovens based on coPA2, this increase is much higher and was calculated as 8822%. The highest obtained electrical conductivity of CFRP with nonwovens containing 3.5wt% MWCNTs is 0.002 S/m through the laminate thickness. This was achieved for laminate L3, which incorporated nonwovens made of coPA2.

The in-plane electrical conductivity is higher than the out-of-plane electrical conductivity, which is obvious because of the parallelly laid carbon fibers. Incorporation of thermoplastic nonwovens with 3.5wt% MWCNTs leads to a tiny increase of about 47% and 63% for laminates L2 and L3, respectively, when compared to the reference panel (Fig. 2b). Such a small increase in electrical conductivity has already been reported in the literature: the increase of the in-plane conductivity by implementation of CNTs is only slight due to the presence of percolation pathways formed by carbon fibers [30]. Interestingly, the insertion of nonwovens with MWCNTs as interlayers also resulted in an increase of surface electrical conductivity, as shown in Fig. 2c. Electrical volume conductivity is dependent on the type of thermoplastic nonwovens used, but surface electrical conductivity is not. These differences will be explained in terms of the laminate microstructure in section 3.2.

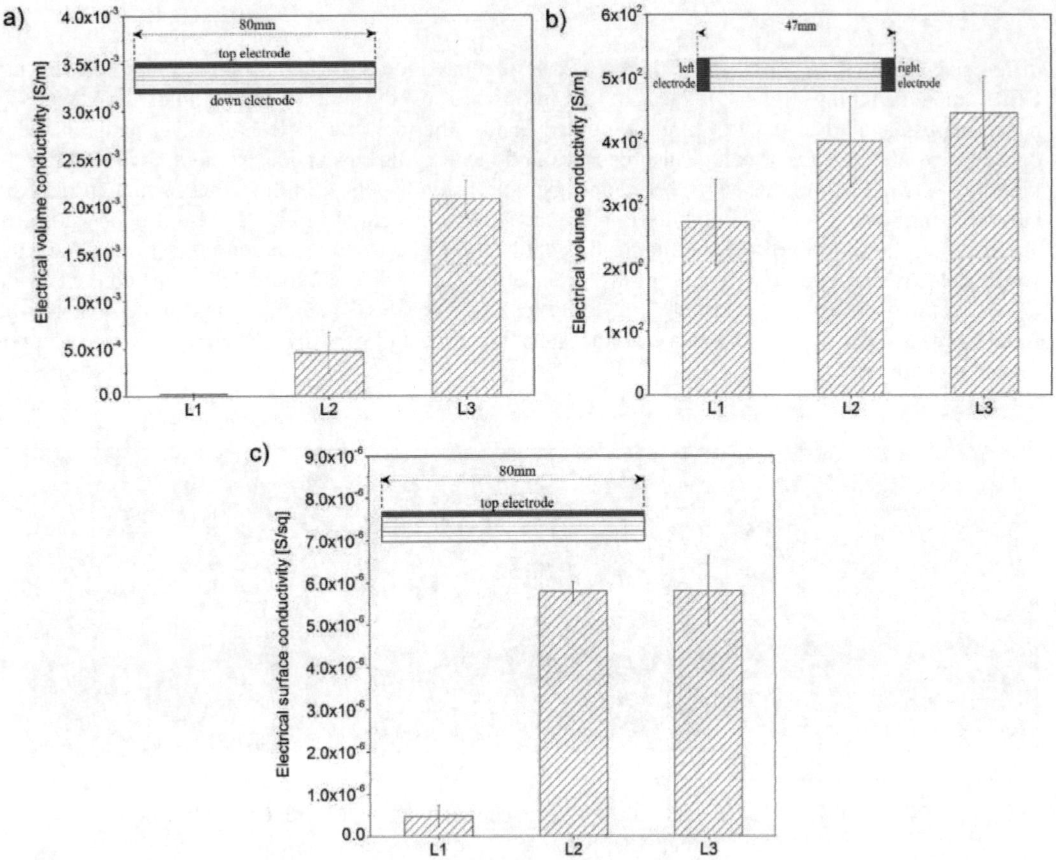

Figure 2: Volume electrical conductivity for CFRP panels in a) out-of-plane direction (throughout the thickness), b) in-plane direction (along the carbon fibers); c) surface electrical conductivity.

3.2. Microstructure

The typical microstructure of the CFRP reference panel cross-section is presented in Fig. 3a. There are visible layers of carbon fibers impregnated with the resin. The pore values determined by ImageJ, which are indicated as small black dots in the SEM images, equaled 0.53±0.25%, which proves the good quality of the panel. Microstructure images of the composite panel with inserted thermoplastic nonwovens are presented in Fig. 3b (L2) and Fig. 3c (L3). It can be observed that in both laminates the layers of melted nonwovens are not uniform in size through the whole panel cross-section. On one hand, this could be the result of insufficient compaction of the nonwoven layers. On the other hand, it could be due to the heterogenous structure of the nonwovens. If the pores between the fibers are too small, the resin is not able to penetrate easily between the fibers, it is not mixed with the melted polymer, and therefore the formed layer is discontinuous. Hence, in some places the nonwovens remain as droplets. The other possible reason is that the nonwovens formed from the copolyamides have a trace amount of water which causes swelling of the polymer during CFRP manufacturing. Obviously, there are also layers which are homogenous and thin, thus confirming the influence of the nonwovens' structural parameters on the quality of the layers. Such variances in the thickness of the nonwoven layers affect the overall thickness of the CFRP panels after curing, which are higher for the CFRPs with nonwovens. Laminates L2 and L3 are about 0.5mm thicker than the reference panel. The creation of a non-uniform layer of melted nonwovens led to the formation in the epoxy of free spaces without thermoplastic polymer and therefore more voids. Hence, the determined porosity in the composite panel L2 is 0.58±0.40%, while in L3 it is only 0.31±0.11%, which is smaller than that calculated for the reference panel. Moreover, in the composite panels the pore shapes were not only circular but also elongated. Such differences in laminate porosity and thickness can be related to the behavior of the nonwovens during CFRP manufacturing; this is related to the properties of the initial copolyamide + MWCNTs nanocomposites used to manufacture the nonwovens by the melt-blown technique. In particular, this could be an effect of the rheological properties and the varying dispersion of the nonwovens during melting. This means that the MWCNT cannot be uniformly dispersed in the CFRP, which affects the variances in the electrical conductivity improvement, as is presented in Fig. 2. Also, the non-uniform and local agglomeration of the thermoplastic copolyamides between the prepreg layers results in the varying electrical conductivity values of the laminates, which is difficult to explain in the context of the type of nonwovens used. Nevertheless, it indicates that the CFRP manufacturing process should be further improved to obtain laminates with the same thickness and porosity as well as uniform layers of melted nonwovens.

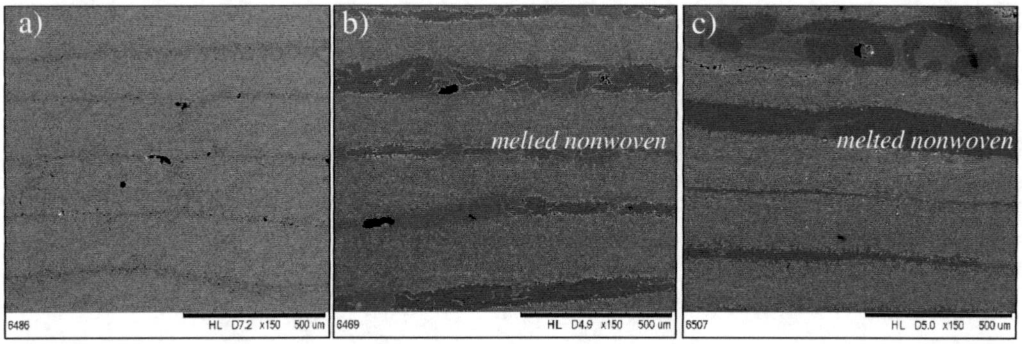

Figure 3: SEM images of cross sections: a) reference panel; b) panel with coPA1/3.5 nonwovens; c) panel with coPA2/3.5 nonwovens

In order to examine the dispersion and distribution of MWCNTs in CFRP, a high-resolution SEM was applied. The observations were performed for panel L2 and L3, but we did not find differences in the microstructure of these laminates. Therefore, the images presented in Fig. 4 show the microstructure of both L2 and L3 panels. Fig. 4a shows a melted layer of nonwoven that adheres perfectly to the carbon fibers. As has already been mentioned, it can be seen that the formed layers of nonwovens are not uniform in thickness. However, in some places the nonwovens occur as droplets after melting and the copolyamides used possess sufficiently high flow behavior to spread the polymer around the carbon fibers (Fig. 4b). The lack of voids in the boundary between the carbon fibers and the nonwovens confirms that the copolyamides used to produce the nonwovens have good adhesive properties and high compatibility with the carbon fibers. The dispersion of MWCNTs in the nonwovens after melting is shown in Fig. 4c. It can be seen that MWCNTs (white points) occur in a well-dispersed state and do not create agglomerates during the CFRP manufacturing process. Therefore, the electrical conductivity of these laminates can be improved because copolyamides doped with 3.5wt% MWCNTs are more conductive than the neat resin from the prepreg. MWCNTs create conductive pathways in the copolyamides and therefore the electric current which goes through the carbon fibers is easily conducted through the melted nonwovens. Obviously, to improve the overall electrical properties of the CFRP panel it is necessary to form continuous layers of the nonwoven throughout the whole laminate. In contrast, inhomogeneous layers of nonwovens introduce local agglomeration of the polymer without MWCNTs, which acts as an insulator. However, in such a situation single MWCNTs are pulled out of the thermoplastic polymer and have the tendency to create bridges, as presented in Fig. 4d. Thus, percolation pathways are created between neighboring nonwovens. Improving the electrical conductivity through the laminate thickness by formation of bridges between CNTs has been already suggested by Yamamoto et al. [31].

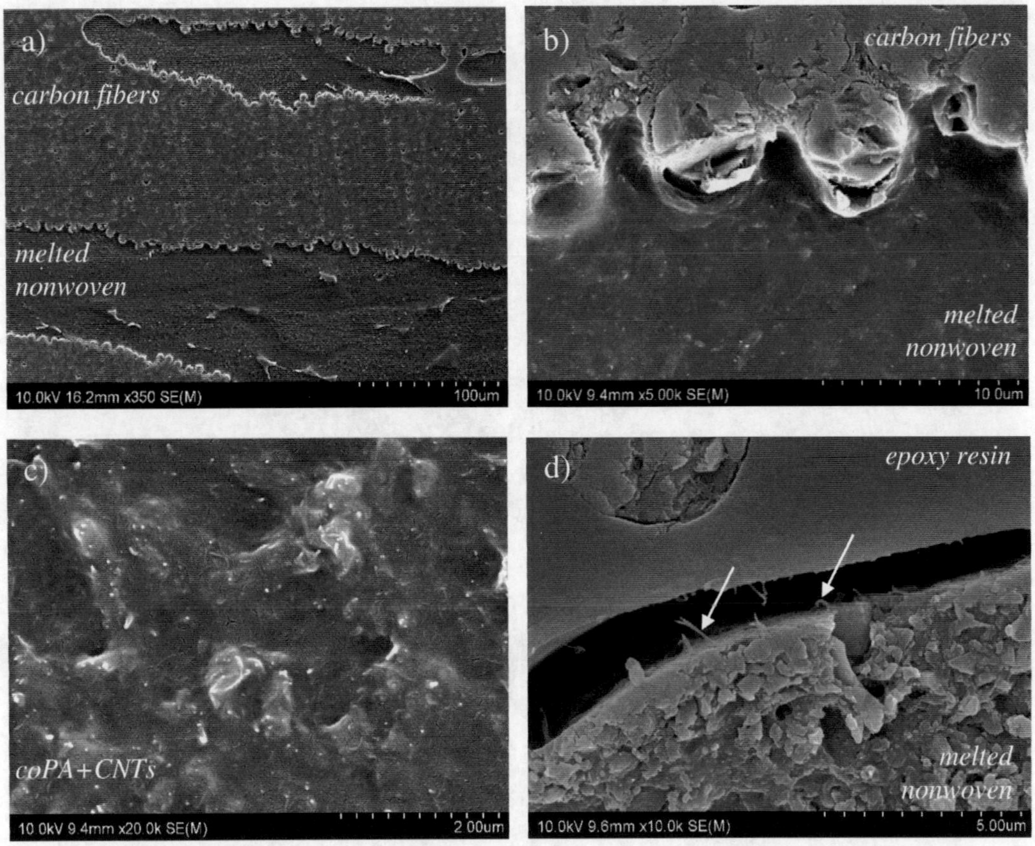

Figure 4: HRSEM images of CFRP panels containing nonwovens with CNTs: a) boundary nonwoven carbon fibers; b) melted nonwovens surrounded carbon fibers; c) well-dispersed CNTs in melted nonwoven; d) single CNTs (marked with arrows) between nonwoven layers.

3.3. Mechanical properties

In order to determine the resistance of the composite panels to delamination, an ILSS test was performed. It was expected that implementation of a small amount of thermoplastic polymer in the form of nonwovens between the carbon fabrics would prevent delamination of the CFRP. According to Fig. 5, the ILSS value determined for the reference panel L1 is 63.5±5.4 MPa. In the presence of thermoplastic nonwovens containing 3.5wt% MWCNTs there is a slight decrease in the shear properties of CFRP for both types of nonwovens. For laminate L2 with nonwovens coPA1/3.5, the ILSS value is 53.7±1.5 MPa, while for laminate L3 with nonwovens coPA2/3.5 it is–55.3±3.01 MPa. The differences between these two values are negligible, which indicates no effect of the type of nonwovens used. Beylergil at el.[32] fabricated CFRP panels with interlayers of 17 and 50gsm thermoplastic PA 66 nonwovens. They observed only a slight increase in the ILSS of the reference panel that was independent of the areal weight of the nonwovens used. In the tested CFRP, the quality of the manufactured CFRP had a more significant impact on the interlaminar shear stress. Discontinuity of the nonwoven layers, local agglomeration of the polymer, and the presence of pores and voids affected the delamination process. As presented in the SEM images of composite panels after the ILSS test (Fig. 6), the reference panel cracked in one place along the whole length of the specimen (Fig. 6a). In the case of the laminates containing nonwovens and MWCNTs (Fig. 6b, c), the delamination occurred in a few places, but not in each layer between carbon fabrics. Presumably, the delamination is pronounced in the layers in which the nonwovens created droplets of melted polymer instead of a continuous layer. Moreover, in those layers MWCNTs were also not well-dispersed and therefore the interlaminar shear stress decreased.

IOP Conf. Series: Materials Science and Engineering **406** (2018) 012033 doi:10.1088/1757-899X/406/1/012033

According to the literature, the introduction of CNTs into CFRP prevents the delamination process, but this depends on the amount of nanofiller [33].

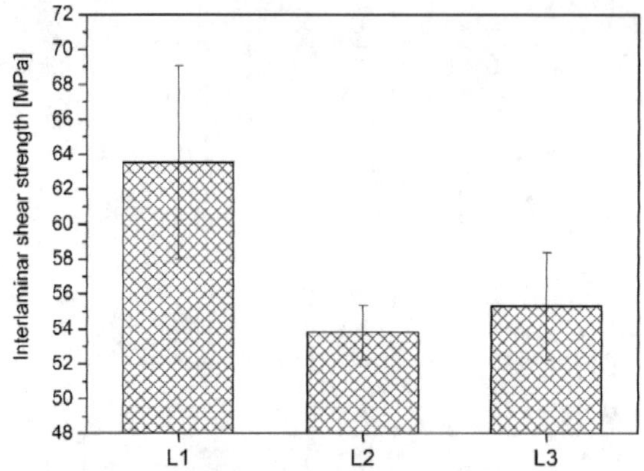

Figure 5: ILSS test results of fabricated CFRP panels without (L1) and with thermoplastic nonwovens containing 3.5 wt% MWCNTs (L2, L3).

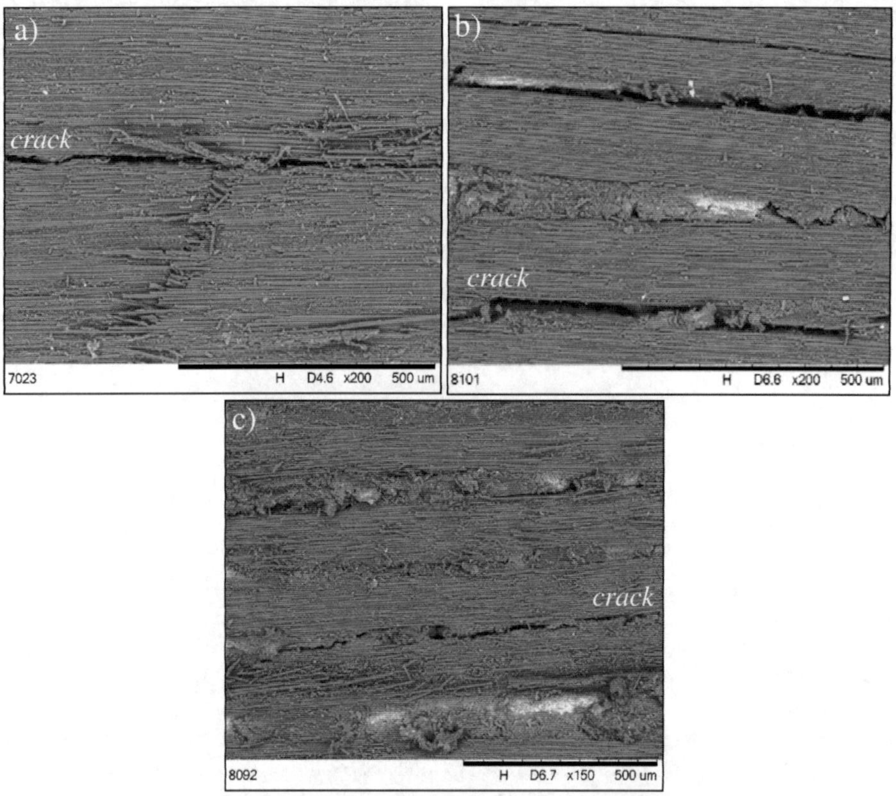

Figure 6: Microstructure of a) reference panel, b) panel L2 and c) panel L3 after ILSS test.

3.4. Dynamical Mechanical Analysis

The thermo-mechanical properties of the fabricated CFRP panels were measured using the DMA method as a function of temperature. From the loss modulus curves presented in Fig. 7, the changes in the glass transition temperatures as an effect of the addition of nonwovens was determined. The reference panel has T_g at 152.4°C. In the presence of thermoplastic nonwovens, the maximum of the T_g

peak increased. For laminate L2 containing nonwoven coPA1/3.5, it was around 161°C; for laminate L3 with nonwovens made of coPA2/3.5, it was around 170°C. The T_g values of the epoxy increased by a few degrees; this could be related to the presence of MWCNTs, as was also observed for CFRP fabricated with modified epoxy with CNTs [34]. The T_g temperature increase means that the composite panels containing copolyamide and MWCNTs should be stiffer than the reference panel. However, as presented in Fig. 7b, decreasing the storage modulus indicated that the introduction of thermoplastic polymer in the form of nonwovens increased the flexibility of the CFRP and therefore decreased their stiffness. This is also confirmed by ILSS results since laminates L2 and L3 have lower interlaminar strength (Fig. 5). An interesting phenomenon occurs in the loss modulus curves for the CFRP with implemented nonwovens. For both laminates the additional peak at around 41°C appeared, but this was not found in the CFRP panels reinforced by nonwovens made of PA66 [32]. This is due to the thermoplastic copolyamides used as a polymer matrix in nonwovens and indicates the presence of polyamide 6 segments in both copolyamides [35]. Presumably, the thermoplastic nonwovens used were too thick, which led to local agglomerations of the polymer. Hence, the overall mass of the introduced polymer was too high, as expressed by the additional peak in the loss modulus curve.

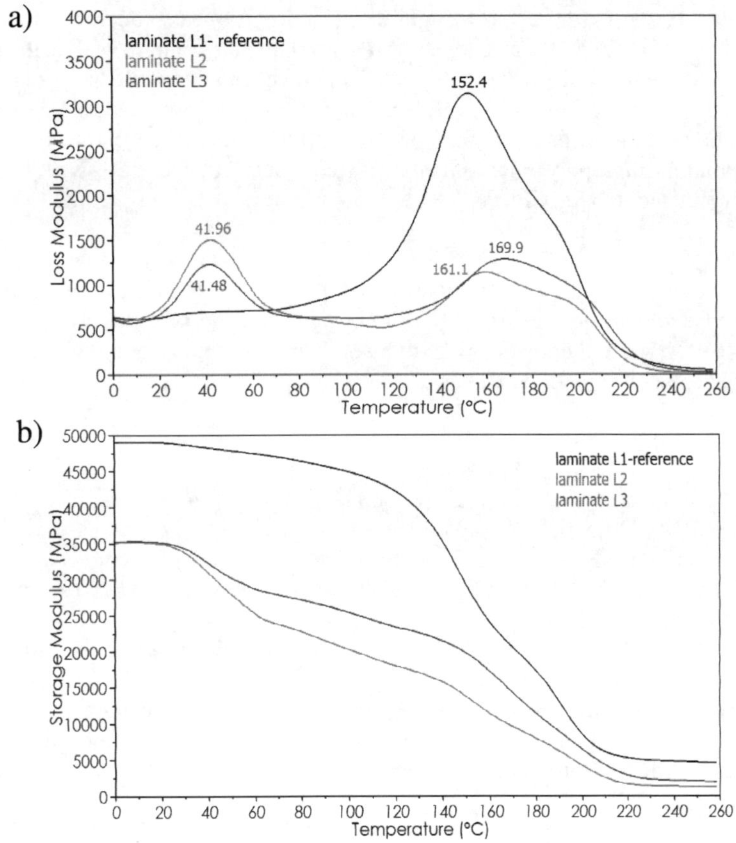

Figure 7: a) The loss modulus curves with marked Tg temperatures and b) the storage modulus curves obtained for CFRP composite panels by DMA test.

4. Conclusions

In this study, we demonstrated the effect of novel types of thermoplastic nonwovens containing 3.5wt% MWCNTs on the electrical and mechanical properties of CFRP. Two types of nonwovens based on different types of copolyamides (coPA1 and coPA2) and 3.5wt% of MWCNTs were fabricated directly from nanocomposite pellets by a half-industrial melt-blown method. Due to their high flexibility, implementation of the nonwovens was easily achieved by a standard prepreg method; this resulted in good quality laminates with less than 1% pore content. The electrical properties measured along the carbon fibers increased slightly due to the existing percolation pathways. However, both laminates containing nonwovens made of coPA1 and coPA2 had electrical volume conductivity through the laminate thickness which was about 1882% and 8822% higher than the reference panel, respectively. When the electrical surface conductivity of the composite panels was not affected by the type of nonwoven used, the electrical properties through the laminate thickness were dependent on the structural parameters of nonwovens. The SEM images showed that the formed layers of melted nonwovens differed in thickness and were discontinuous in some places. It was suggested that the excessively small pores in the nonwovens prevented the resin from flowing freely between the nonwoven fibers, which led to insufficient impregnation. This caused the inhomogeneity in the layers and the variances in the electrical and mechanical properties of the laminates. Nevertheless, the high-resolution SEM images confirmed that copolyamides+3.5wt% MWCNTs have sufficiently high flow behavior and compatibility to surround and adhere to the carbon fibers. Moreover, it was confirmed that MWCNTs were well dispersed in the nonwoven fibers after the melt-blown process. The analysis of the mechanical properties in terms of the interlaminar shear strength showed that in the presence of thermoplastic nonwovens containing 3.5wt% MWCNTs, there was a slight decrease in the shear properties of CFRP from 63.5 MPa for the reference to 53.7MPa and to 55.3MPa for laminates interleaved with coPA1/3.5 and coPA2/3.5, respectively. Introduction of thermoplastic nonwovens into CFRP also increased the resin glass transition temperature by about 9°C for L2 and 17°C for L3, as examined by DMA analysis. What is more, the additional peak in the loss modulus curves for laminates with nonwovens at 41°C was caused by the thermoplastic copolyamide and appears to be caused by the excessively high overall mass of the polymer introduced into the CFRP.

Acknowledgments

The studies were carried out within grant no. DOB-1-3/1/PS/2014 financed by the National Centre for Research and Development, and from the European Union Horizon 2020 Program under Grant Agreement n° 646307. The authors would like to thank Nanocyl from Belgium for manufacture of the nanocomposite pellets with carbon nanotubes.

References

[1] Yi X-S. Development of multifunctional composites for aerospace application. Elsevier Inc.; 2015. doi:10.1016/B978-0-323-26434-1.00011-8.

[2] Ma X, Scarpa F, Peng H, Allegri G, Yuan J, Ciobanu R. Design of a hybrid carbon fibre / carbon nanotube composite for enhanced lightning strike resistance. Aerosp Sci Technol 2015;47:367–77. doi:10.1016/j.ast.2015.10.002.

[3] Gagné M, Therriault D. Lightning strike protection of composites. Prog Aerosp Sci 2014;64:1–16. doi:10.1016/j.paerosci.2013.07.002.

[4] Gohardani O, Elola MC, Elizetxea C. Potential and prospective implementation of carbon nanotubes on next generation aircraft and space vehicles: A review of current and expected applications in aerospace sciences. Prog Aerosp Sci 2014;70:42–68. doi:10.1016/j.paerosci.2014.05.002.

[5] Gaztelumendi I, Chapartegui M, Seddon R, Florez S, Pons F, Cinquin J. Enhancement of electrical conductivity of composite structures by integration of carbon nanotubes via bulk resin and/or buckypaper films. Compos Part B Eng 2017;122:31–40. doi:10.1016/j.compositesb.2016.12.059.

[6] Quan D. Enhancing mode-I and mode-II fracture toughness of epoxy and carbon fibre reinforced epoxy composites using multi-walled carbon nanotubes 2018;143:81–92. doi:10.1016/j.matdes.2018.01.051.

[7] Lopes PE, van Hattum F, Pereira CMC, Novoa PJRO, Forero S, Hepp F, et al. High CNT content composites with CNT Buckypaper and epoxy resin matrix: Impregnation behaviour composite production and characterization. Compos Struct 2010;92:1291–8. doi:10.1016/j.compstruct.2009.11.003.

[8] Stahl JJ, Bogdanovich AE, Bradford PD. Carbon nanotube shear-pressed sheet interleaves for

Mode i interlaminar fracture toughness enhancement. Compos Part A Appl Sci Manuf 2016;80:127–37. doi:10.1016/j.compositesa.2015.10.014.

[9] Siddiqui NA, Khan SU, Ma PC, Li CY, Kim JK. Manufacturing and characterization of carbon fibre/epoxy composite prepregs containing carbon nanotubes. Compos Part A Appl Sci Manuf 2011;42:1412–20. doi:10.1016/j.compositesa.2011.06.005.

[10] Akcin Y, Karakaya S, Soykasap O. Electrical, Thermal and Mechanical Properties of CNT Treated Prepreg CFRP Composites. Mater Sci Appl 2016;7:465–83. doi:10.4236/msa.2016.79041.

[11] Fogel M, Parlevliet P, Olivier P, Dantras É. Composites : Part A Manufacturing of conductive structural composites through spraying of CNTs / epoxy dispersions on dry carbon fiber plies. Compos Part A 2017;100:40–7. doi:10.1016/j.compositesa.2017.04.020.

[12] Islam MS, Deng Y, Tong L, Faisal SN, Roy AK, Minett AI, et al. Grafting carbon nanotubes directly onto carbon fibers for superior mechanical stability: Towards next generation aerospace composites and energy storage applications. Carbon N Y 2016;96:701–10. doi:10.1016/j.carbon.2015.10.002.

[13] Beckermann GW, Pickering KL. Mode I and Mode II interlaminar fracture toughness of composite laminates interleaved with electrospun nanofibre veils. Compos Part A Appl Sci Manuf 2015;72:11–21. doi:10.1016/j.compositesa.2015.01.028.

[14] Hogg PJ. Toughening of thermosetting composites with thermoplastic fibres. Mater Sci Eng A 2005;412:97–103. doi:10.1016/j.msea.2005.08.028.

[15] Sano E, Akiba E. Electromagnetic absorbing materials using nonwoven fabrics coated with multi-walled carbon nanotubes. Carbon N Y 2014;78:463–8. doi:10.1016/j.carbon.2014.07.027.

[16] Miao M. Carbon nanotube yarns for electronic textiles. Elsevier Ltd.; 2015. doi:10.1016/B978-0-08-100201-8.00004-7.

[17] Kongahage D, Spinks GM, Wallace GG. Fabrication of a graphene coated nonwoven textile for industrial applications 2016;6:73203–9.

[18] Kongahge D, Foroughi J, Gambhir S, Spinks GM, Wallace GG. Fabrication of a graphene coated nonwoven textile for industrial applications. RSC Adv 2016;6:73203–9. doi:10.1039/C6RA15190F.

[19] Wu J, Ashraf SA, Wallace GG. Conducting textiles from single-walled carbon nanotubes 2007;157:358–62. doi:10.1016/j.synthmet.2007.04.010.

[20] Bonaldi RR, Siores E, Shah T. Characterization of electromagnetic shielding fabrics obtained from carbon nanotube composite coatings. Synth Met 2014;187:1–8. doi:10.1016/j.synthmet.2013.10.003.

[21] Haji A, Semnani Rahbar R, Mousavi Shoushtari A. Improved microwave shielding behavior of carbon nanotube-coated PET fabric using plasma technology. Appl Surf Sci 2014;311:593–601. doi:10.1016/j.apsusc.2014.05.113.

[22] Krucińska I, Skrzetuska E, Wiesława U-D. The use of carbon nanotubes in textile prinitng. J Appl Polym Sci 2011;121:483–90. doi:10.1002/app.

[23] Weng B, Xu F, Salinas A, Lozano K. Mass production of carbon nanotube reinforced poly(methyl methacrylate) nonwoven nanofiber mats. Carbon N Y 2014;75:217–26. doi:10.1016/j.carbon.2014.03.056.

[24] Mazinani S, Ajji A, Dubois C. Morphology, structure and properties of conductive PS/CNT nanocomposite electrospun mat. Polymer (Guildf) 2009;50:3329–42. doi:10.1016/j.polymer.2009.04.070.

[25] Krucinska I, Surma B, Chrzanowski M, Skrzetuska E, Puchalski M. Application of melt-blown technology for the manufacture of temperature-sensitive nonwoven fabrics composed of polymer blends PP/PCL loaded with multiwall carbon nanotubes. J Appl Polym Sci 2013;127:869–78. doi:10.1002/app.37834.

[26] Latko P, Kozera R, Salinier A, Boczkowska A. Non-Woven Veils Manufactured from Polyamides Doped with Carbon Nanotubes 2013;6:45–9.

[27] Lee S, Noguchi H, Kim Y, Cheong S. Journal of Composite Materials 2002. doi:10.1106/002199802026980.

[28] Zheng N, Huang Y, Liu H-Y, Gao J, Mai Y-W. Improvement of interlaminar fracture toughness in carbon fiber/epoxy composites with carbon nanotubes/polysulfone interleaves. Compos Sci

IOP Conf. Series: Materials Science and Engineering **406** (2018) 012033 doi:10.1088/1757-899X/406/1/012033

Technol 2017;140:8–15. doi:10.1016/j.compscitech.2016.12.017.

[29] Lewandowski Z, Ziabicki A, Jarecki L. The Nonwovens Formation in the Melt-blown Process. Fibers Text East Eur 2007;15:77–81.

[30] Pozegic TRR, Hamerton I, Anguita JV V, Tang W, Ballocchi P, Jenkins P, et al. Low temperature growth of carbon nanotubes on carbon fibre to create a highly networked fuzzy fibre reinforced composite with superior electrical conductivity. Carbon N Y 2014;74:319–28. doi:10.1016/j.carbon.2014.03.038.

[31] Yamamoto N, Guzman de Villoria R, Wardle BL. Electrical and thermal property enhancement of fiber-reinforced polymer laminate composites through controlled implementation of multi-walled carbon nanotubes. Compos Sci Technol 2012;72:2009–15. doi:10.1016/j.compscitech.2012.09.006.

[32] Beylergil B, Tano M, Akta E. E ff ect of polyamide-6 , 6 (PA 66) nonwoven veils on the mechanical performance of carbon fi ber / epoxy composites 2018;194:21–35. doi:10.1016/j.compstruct.2018.03.097.

[33] Xu H, Tong X, Zhang Y, Li Q, Lu W. Mechanical and electrical properties of laminated composites containing continuous carbon nanotube fi lm interleaves. Compos Sci Technol 2016;127:113–8. doi:10.1016/j.compscitech.2016.02.032.

[34] Korayem AH, Jian S, Hui Q, Yang C, Ling X. Failure of CFRP-to-steel double strap joint bonded using carbon nanotubes modi fi ed epoxy adhesive at moderately elevated temperatures. Compos Part B 2016;94:95–101. doi:10.1016/j.compositesb.2016.03.042.

[35] Greco R, Nicolais L. Glass transition temperature in nylons. Polymer (Guildf) 1976;17:1049–53. doi:10.1016/0032-3861(76)90005-7.

IOP Conf. Series: Materials Science and Engineering **406** (2018) 012034 doi:10.1088/1757-899X/406/1/012034

Influence of contents rate and dispersion of carbon nanotubes on interfacial adhesion between continuous fiber reinforced thermoplastics and injection resin

K Matsumoto[1], J Shiode[2], H Nishino[3]* and T Tanaka[4]

[1,2,3] Graduate School of Science and Engineering, Department of Mechanical Engineering, Doshisha University

[4] Faculty of Science and Engineering, Department of Mechanical Engineering, Doshisha University

* ctwc0552@mail4.doshisha.ac.jp

Abstract. Fiber reinforced plastics have been expected as an alternative of metal materials in order to reduce the lightweight of the automotive and the environmental loading. Carbon fiber reinforced thermoplastics (CFRTP) have been studied from the view point of reducing cycle time and recycling of polymer materials. Prepreg compression molding (PCM) enables to provide the excellent mechanical property. However, PCM is difficult to form complex shape. On the other hand, hybrid injection molding (HIM) has attracted attention because HIM makes it possible to form complex shape. However, the mechanical property of molded parts are affected by the interlaminar shear strength between continuous fiber reinforced thermoplastics and injection molded parts. Some researchers reported interlaminar shear strength improve by use of carbon nanotubes (CNTs). Therefore, we aimed at improving interfacial adhesion by preparing a film with CNT added in matrix and inserting the film at the interface of hybrid injection molded parts. We conducted short beam test on specimens inserted films. When CNT/PP film was used, the interfacial shear strength was improved compared to when no film was used. When the CNT was in the low dispersion state, the interfacial shear strength rose gently up to 1.0 wt% and dropped greatly at 3.0wt%. On the other hand, when the CNT was in the highly dispersed state, the interfacial shear strength rose up to 0.5 wt%, and thereafter dropped. When the CNT content was 1.0 wt%, the CF entered deep into interface at low dispersion. However, the interface was smooth and entry of CF could not be confirmed on the peeled surface at high dispersion. There is an optimum value for the content and dispersibility of CNT in the film.

1. Introduction

Continuous fiber reinforced thermoplastics have excellent mechanical property. Among reinforcing fibers, specific gravity of carbon fiber is 1/4 times as heavy as that of iron. Specific strength of carbon fiber is 10 times as strong as that of iron. Specific modulus of carbon fiber is 7 times as large as that of iron. CFRTP is excellent in specific strength and specific rigidity. It is also expected as a substitute material for metals from the viewpoint of ease of formability, high productivity and recyclability. PCM is one of the process to produce the CFRTP parts. However, PCM is difficult to form the complex shape and it requires long time for molding. On the one hand, Injection Molding (IM) is used for produce the complex shape parts which is reinforced by short fiber and long fiber (i.e. carbon fiber, glass fiber, and natural fiber). HIM process is promising new method which is combined these PCM and IM process for producing the hybrid parts with excellent mechanical property. In HIM process, CF/PP organosheets are heated by infrared (IR) heater to melt the resin of these sheets and making shape of melted organosheet is done by mold of IM. However, the mechanical property of molded parts is affected by the adhesive strength between continuous fiber reinforced thermoplastics and injection molded parts. Some researchers reported interlaminar shear strength is improved by use of CNTs in case of thermoset resin. It is known that the aspect ratio (fiber length / fiber diameter) of the reinforcing fiber contributes to the mechanical properties of the fiber reinforced composite material. Since CNTs have higher aspect ratios than general scaled reinforcing fibers and microfibers, high adhesive strength can be expected. However, the adaption of CNTs in HIM process for improving the adhesive strength of thermoplastics hybrid parts has not reported yet. In this research, therefore, we newly devised a method of connecting interfaces at the nano level as a new approach. We aimed at improving adhesive strength by preparing a film with CNT added in matrix and inserting the film at the interface of hybrid injection molded parts. We assumed that adhesive strength was affected by the content of CNTs added and dispersion state and investigated them.

2. Experimental

2.1. Material

NOVATEC PP (manufactured by Japan Polypropylene Corporation, part number MA 04 A) was used as a matrix. Multi-walled carbon nanotube MWCNT (manufactured by NANOCYL SA, NC 7000 TM) was used as the CNT. CF / PP organosheet was used as an insert material. PYROFIL pellet PP-C-20A (manufactured by Mitsubishi Chemical Corporation) was used as an injection resin. PP (manufactured by Prime Polymer, product number J 108 M) was used as a matrix of CF/PP films and glass fiber (average fiber diameter 0.28 μm, aspect ratio 36) was used as reinforcing fiber of CF/PP films.

2.2. Fabrication of CNT/PP granules

For fabrication of CNT/PP granules, co-rotating twin-screw extruder ZSK18 MEGAlab (manufactured by Coperion) was used. At first, we prepared masterbatch (MB) of PP/CNT nanocomposite with 5.0 wt% content ratio of CNT by co-rotating twin-screw extruder. The used configuration for compounding MB is shown in figure 1. PP granules are fed into the main hopper and CNTs are fed into side feeder by volumetric feeder. The compounds are extruded as strand under the cooling water and cut into granules by pelletizer.

Second, the MB pellets and PP pellets were dry-blended at room temperature so as to have content ratio (0 / 0.5 / 1.0 / 3.0 wt%), and the blends were charged from the main hopper and kneaded. The screws were shown in figure 2. In screw I, the rotation speed was 150 rpm and the throughput was 3.8 kg/h. In screw II, the rotation speed was 500 rpm and the throughput was 5.0 kg/h. The barrel temperature was 200 °C.

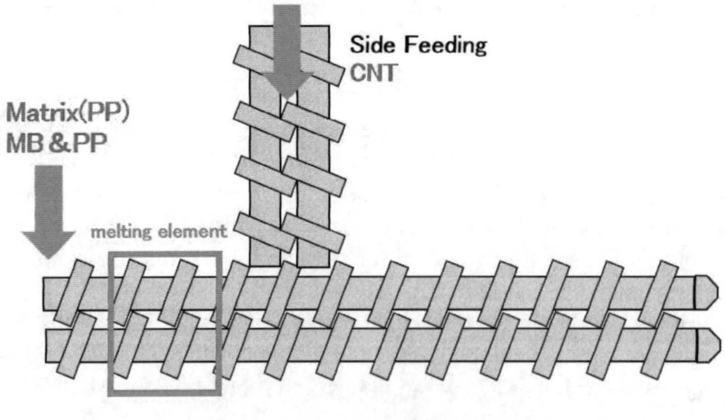

Figure 1. Screw configuration.

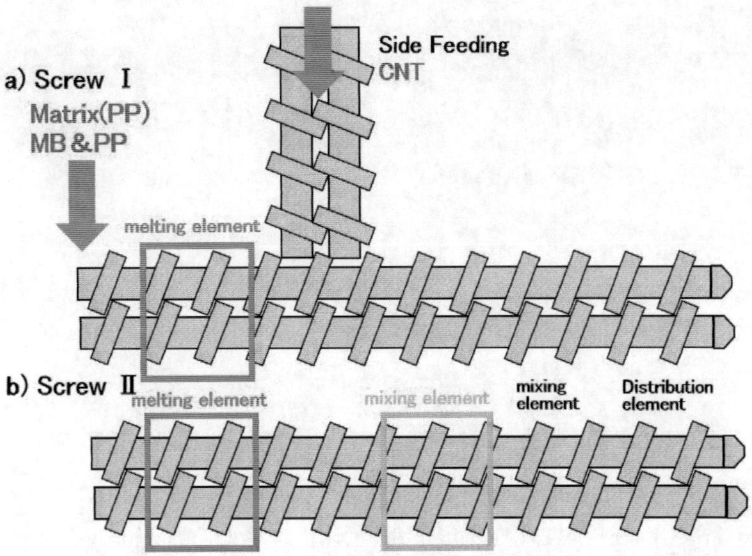

Figure 2. Screw configuration.

2.3. Fabrication of CNT/PP film

Equipment and process for producing CNT/PP films shown in figure 3. For fabrication of the film, granules were fed into a small twin-screw extruder (manufactured by Thermo Fisher Scientific, HAAKE Process 11) having a screw diameter of 11 mm. The thickness of the sheet was about 200 μm and width of the sheet was 16.5 mm. The film was taken up in a sheet winding device. The barrel temperature was 200 ° C, the rotation speed was 100 rpm, and the throughput was 1.0 kg/h.

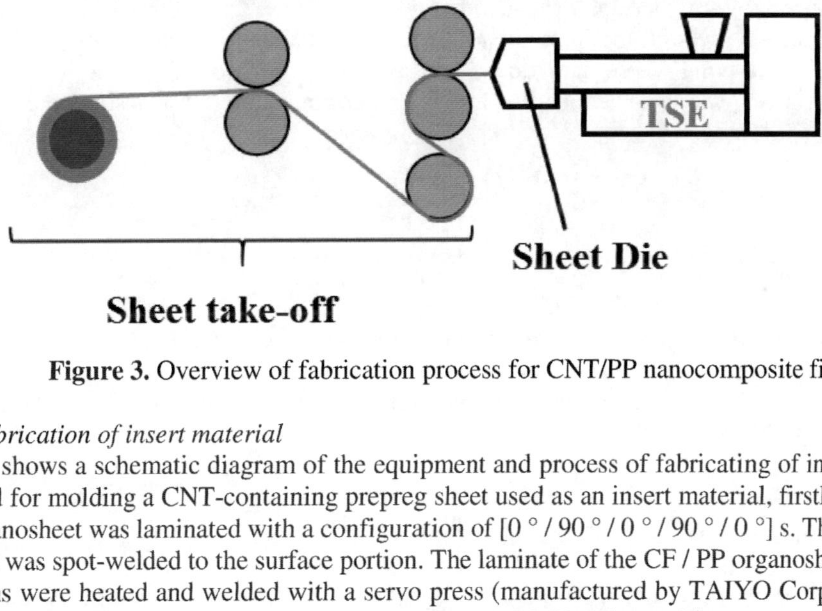

Figure 3. Overview of fabrication process for CNT/PP nanocomposite film.

2.4. Fabrication of insert material

Figure 4 shows a schematic diagram of the equipment and process of fabricating of insert material. As a method for molding a CNT-containing prepreg sheet used as an insert material, firstly 10 ply of a CF / PP organosheet was laminated with a configuration of [0 ° / 90 ° / 0 ° / 90 ° / 0 °] s. The CNT / PP film prepared was spot-welded to the surface portion. The laminate of the CF / PP organosheet and the CNT / PP films were heated and welded with a servo press (manufactured by TAIYO Corporation, product name: PQCS 2 - 60 kN - FC). The press pressure was 1.2 MPa. The molding temperature was 200 °C. The width of CNT/CF/PP insert material was 70mm. The length of that was 120mm. The thickness of that was 1.2mm.

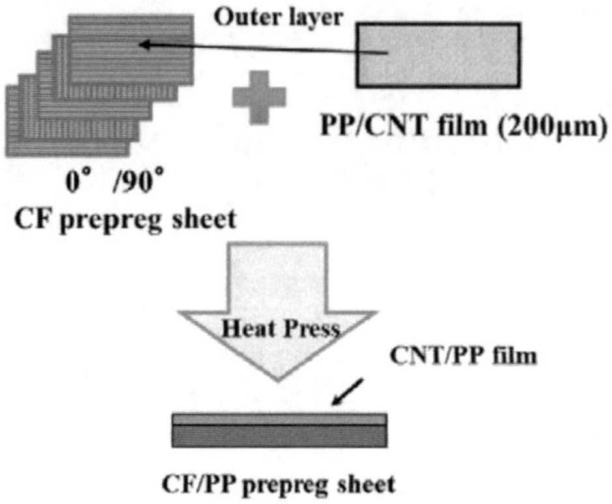

Figure 4. Overview of fabrication process of insert parts (CNT/CF/PP).

2.5. *Injection molding process*

The injection molding machine ET-40V (manufactured by Toyo Machinery Metals) was used. The thickness of the mold was 2.4 mm. The layer of CNT was molded at the center of thickness.

However, since the surface of the insert material is required to heat, a fully automatic vertical molding machine ET-80HR4 (manufactured by Toyo Machinery Metals) was used. Injection conditions were standard injection conditions (injection conditions: STD). The cylinder temperature was 240 °C, the holding pressure was 30 MPa.

2.6. *Evaluation of interlaminar shear strength*

Short beam three-point bending test was performed according to ASTM D 2344 standard. Autograph AG-1 (manufactured by Shimadzu Corporation) was used as a testing machine. Test speed was 1.0 mm/sec and distance between fulcrums was 9.6 mm. The insert side of the test piece was on the indenter side and the Injected side was on the fulcrum side. The interfacial shear strength τ was calculated by using equation (1). The point where the load was dropped slightly before fracturing of all materials during short beam testing was considered as the interfacial breaking load *F*. The test piece of width was *b*, that of thickness was *h*.

$$\tau = \frac{3F}{4bh} \tag{1}$$

2.7. *Dynamic viscoelasticity*

Dynamic viscoelasticity measuring device Rheogel-E 4000 (manufactured by UBM) was used. The CNT/PP film was cut into test pieces. The test pieces of thickness were 200 μm, that of length were 30 mm and that of width were 5.0 mm. The distance between fulcrums was 20 mm, the static load was 1.0 N, the displacement amplitude was 2.0 μm. The measurement temperature was 20 °C. The frequency range was 0.016 ~15.92 rad/sec.

3. Results and discussion

3.1. *Dispersion state of CNT*

The state of CNT dispersion in the CNT/PP pellets is evaluated from the viewpoint of melt viscoelasticity. Based on the results of the storage elastic modulus E ' in the melt viscoelastic properties, the state of dispersion of CNT in CNT/PP pellets is quantitatively evaluated. Elastic modulus E' at frequency 0.1 rad/s is shown in figure 5. In the case of the storage elastic modulus E ', it is considered that the screw II exhibits a higher value under all the conditions. This result suggested that CNT dispersion was good in screw II. From now on, CNT/PP pellets made with screw I are low dispersion and CNT / PP pellets made with screw II are made high dispersion.

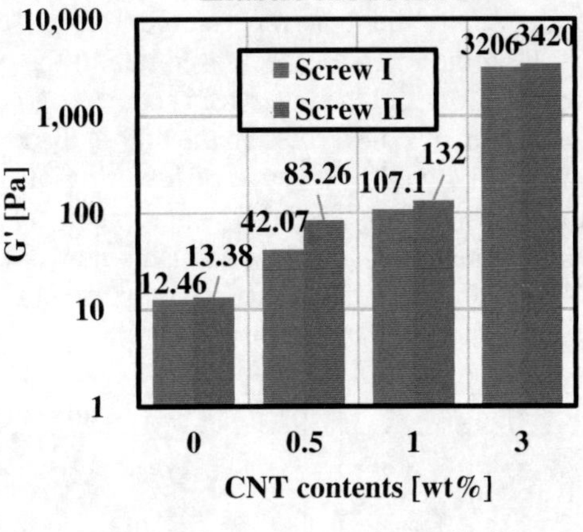

Figure 5. Comparison of elastic modulus E' between screw I and II process.

3.2. Influence of CNT content and dispersion state on interfacial shear strength

Figure 6 shows the relationship between the interfacial shear strength and the CNT content contained in the CNT/PP film. When CNT/PP film is used, the interfacial shear strength is improved compared to when no film is used. In addition, When CNT/PP film was used, the interfacial shear strength was improved compared to when no film was used. When the CNT was in the low dispersion state, the interfacial shear strength rose gently up to 1.0 wt% and dropped greatly at 3.0 wt%. On the other hand, when the CNT was in the high dispersion state, the maximum interfacial shear strength rose up to 0.5 wt%, and thereafter it dropped. From these results, it was revealed that the optimum content ratio exists to improve the interfacial shear strength. However, in the high dispersion state, the highest interfacial shear strength is shown at 0.5 wt%, but it is greatly reduced at 1.0 wt%.

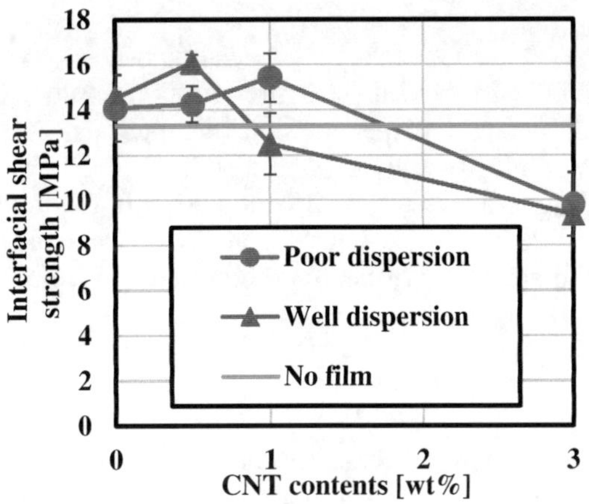

Figure 6. Correlation between CNT content and interfacial shear strength.

3.3. Adhesion effect of the CNT at peeled surface

SEM observation images of peeled surfaces of both the insert material side and the injection resin side after peeling under each condition are shown. The CNT content of 1.0 wt% is shown in figure 7. On the peeled surface at 1.0 wt%, a coarse interface can be confirmed at low dispersion showing the highest interfacial shear strength, and at high dispersion with reduced interfacial strength CNT has a two-dimensional network structure confirmed. Moreover, as a result of comparing with low magnification images, the CF entered deep into the interface at low dispersion, whereas at high dispersion the interface was smooth and entry of CF could not be confirmed. Therefore, one of the causes of the decrease in interfacial shear strength is considered to be that the number of CF at the interface decreased. Furthermore, as the number of CNTs at the interface decreased due to the construction of the network, it could not play the role of the wedge effect and the interface shear strength decreased. From this, it can be easily imagined that a network structure is constructed in the highly dispersed state, but it is presumed that network construction is probably involved as a cause of lowering of interfacial shear strength. In other words, it is presumed that the number of CNTs and CFs playing the role of wedge effect decreases and the interfacial shear strength decreases as the CNTs establish a network.

These results suggested that CNT content and dispensability may affect CNT 's network construction and that the network construction influences the interfacial shear strength.

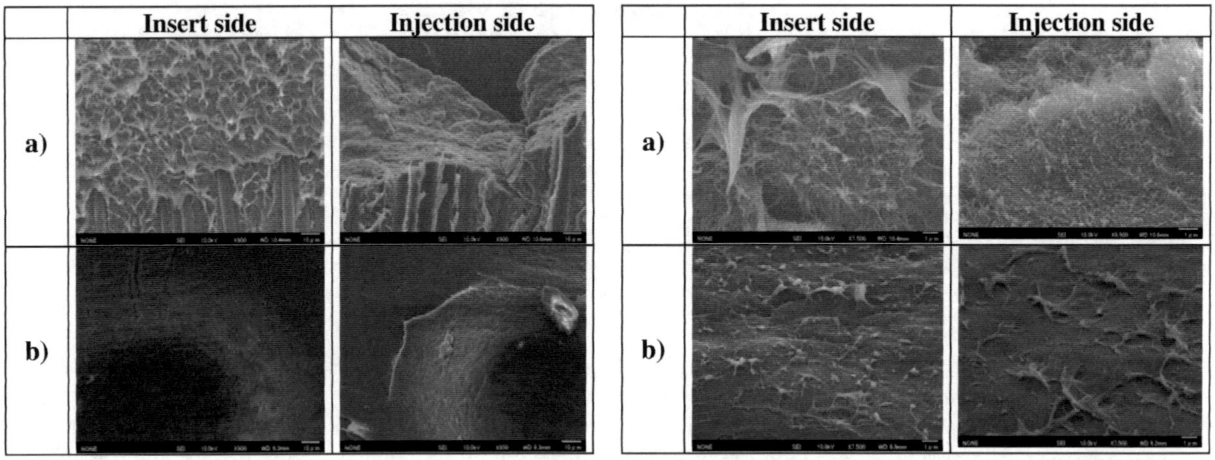

Figure 7. SEM micrograph of peeling surface between injection parts and insert parts at 1.0wt%, a) Poor dispersion and b) Well dispersion.

4. Conclusions

When CNT/PP film was used, the interfacial shear strength was improved compared to when no film was used. When the CNT was in the low dispersion state, the interfacial shear strength rose gently up to 1.0 wt% and dropped greatly at 3.0 wt%. On the other hand, when the CNT was in the high dispersion state, the interfacial shear strength rose up to 0.5 wt%, and thereafter it dropped. When the CNT content was 1.0 wt%, the CF entered deep into the interface at low dispersion. However, the interface was smooth and entry of CF could not be confirmed on the peeled surface at high dispersion. There is an optimum value for the content and dispensability of CNT in the film.

References

[1] Textile and Clothing Division Manufacturing Industries Bureau, Netsukaso CFRP ni kakaru kuni project ni tsuite, Journal of the Japan Society of Polymer Processing, Vol.27, No.3 (2015) pp.78–81 (in Japanese).

[2] Nakai, A., Hukugou zairyo jitsugen no key technology ha seikei kakou ni ari, Journal of Japan Society of Polymer Processing, Vol.27, No.3 (2015) pp.77 (in Japanese).

[3] Terada, K., Carbon Fiber Reinforced Thermo Plastics –Currently, Applications and Forecast–, Journal of the Japan Society for Precision Engineering, Vol.81, No.6 (2015) pp.485–488 (in Japanese).

[4] Tanaka, T. and Matsumoto, K., Tajikukonrenki oyobi kono tajikukonrenki wo mochiita nano

composite no seizou houhou, Japanese Patent Application No. 2017-030841 (2017).

[5] Zhou, H. W. and Mishnaevsky Jr., L., Yi, H. Y., Liu, Y. Q., Hu, X., Warrier, A., Dai, G. M., Carbon fiber/carbon nanotube reinforced hierarchical composites: Effect of CNT distribution on shearing strength, Composites Part B: Engineering, Vol.88 (2016) pp.201–211.

[6] Suetsugu, K. and Sakairi, T., Effect of Aspect Ratio on Flexural Modulus of Fiber-Reinforced Composite Material, KOBUNSHI RONBUNSHU Japanese Journal of Polymer Science and Technology, Vol.41, No.1 (1984) pp.57–61 (in Japanese).

[7] Ming-Chuen Yip and Yi-Chieh Lin, Chung-lin Wu, Effect of Multi-Walled Carbon Nanotube Addition on Mechanical Properties of Polymer Composites Laminate, Polymer & Polymer Composites, Vol. 19, No.2&3 (2011) pp.131–140.

[8] Tomioka, M. and Ishikawa, T., Tanaka, T., Shiode, J., Matsumoto, K., Seikeihinn oyobi sono seizou houhou, Japanese Patent Application No. 2017-034020 (2017).

IOP Conf. Series: Materials Science and Engineering **406** (2018) 012035 doi:10.1088/1757-899X/406/1/012035

Effect of addition of fine sub-micron sized and thick glass fibers into epoxy matrix on fatigue life of plain-woven textile carbon fiber composite

Kazuya Okubo[1*], Toru Fujii[1] and Tsubasa Kondo[2]

[1] Department of Mechanical Engineering, Doshisha University, Kyo-tanabe, Japan
[2] Former graduate student of Doshisha University, Kyo-tanabe, Japan

[*] kokubo@mail.doshisha.ac.jp

Abstract. This study investigated the mechanism of the effect of addition of fine sub-micron sized (close to nano-sized) and thick (11μm) glass fibers into epoxy matrix on the fatigue life of plain-woven textile carbon fiber composite. Static bending strength and fatigue life of the coupon shaped composite with the modified matrix were investigated by three-point static and cyclic loading test. The resistance of interlaminar delamination was characterized with double cantilever beam of Mode-I fracture test and ENF (end notch flexure) of Mode-II fracture test. The effect of the size of the added glass fibers on those mechanical properties was discussed. To explain the effect of the modification on the observed interfacial strength around the intertwining point of the reinforcing textile cloth, model specimens were prepared where an embedded carbon fiber bundle was put between other two carbon fiber bundles oriented to the perpendicular direction in the modified resin. Test results in current study showed that the addition of glass fibers with fine size was rather effective to dramatically improve the fatigue life. The fatigue life was dramatically increased 10-100 times by the addition of nano-sized glass fibers into the matrix in contrast to the static case. Fracture toughness in Mode-II was also improved by the modification of the matrix, while that in Mode-I was improved slightly. Macroscopic crack propagation in fatigue was totally prevented when the matrix was modified with fine glass fibers of 500nm in diameter. Those effects in improvement of mechanical properties of the textile carbon fiber composite were explained by the prevention of interfacial crack propagation with large resistance between carbon fiber bundles in Mode-II and modified matrix around the texture points where comparatively large strain energy would be released, if the epoxy matrix was adequately modified with the fine glass fibers.

1. Introduction

Textile carbon fiber composites with polymer matrix have superior advantages for many applications with attracting attentions ([1]). Generally, the interlaminar fracture toughness of the material with a layered structure is poor and cracks between the layers easily lead to fatal result. The interlaminar fracture toughness and interfacial debonding are important considerations at discussing the long-term reliability of the textile composites. Studies have been conducted to investigate the interlaminar fracture toughness and crack propagation of the composites ([2]-[9]). To improve the fatigue life and mechanical properties of the composites, various studies, such as the modification with micro and nano-sized fibers have also been conducted. It has been experimentally shown that the fatigue life and mechanical propertied of textile composite is improved by the addition of tiny fibers into the matrix ([10]-[14]). Although the effect of added fibers on fatigue life of the textile composite may depend on the diameter of the added fibers, the effect of the size on the fatigue life of the composite is not well be cleared. In addition, the behavior of crack propagation around carbon fiber bundle under Mode-I and Mode-II condition should be discussed to establish the efficacy of these fibers on the fatigue life of the composite. This study investigated the mechanism of the effect of addition of fine sub-micron sized (close to nano-sized) and thick (11μm) glass fibers into epoxy matrix on the fatigue life of plain-woven textile carbon fiber composite. Static bending strength and fatigue life of the coupon shaped composite with the modified matrix were investigated by three-point static and cyclic loading test. The resistance of interlaminar delamination was characterized with double cantilever beam of Mode-I fracture test and ENF (end notch flexure) of Mode-II fracture test. The effect of the size of the added glass fibers on those mechanical properties was discussed. To explain the effect of the modification on the observed interfacial strength around the intertwining point of the reinforcing textile cloth, model specimens were prepared where an embedded carbon fiber bundle was put between other two carbon fiber bundles oriented to the perpendicular direction in the modified resin.

2. Materials and testing machine

2.1 Materials

Plain-woven carbon fiber cloth with 3GPa of tensile strength and 240GPa of Young's modulus (Pyrofil TR-3110-MS: Mitsubishi Chemical Co., Ltd., Japan) was used as reinforcement. Epoxy resin (Epikote 828: Mitsubishi Chemical Corporation, Japan) and modified aliphatic polyamines (JER Cure 113: Mitsubishi Chemical Corporation), were used as matrix and curing agent, respectively. To enhance the matrix, in this study, conventional glass fibers with 11μm (EFH150-01: Central Glass Co., Ltd., Japan) (Figure 1) and thin 500 nm (FM1700: Nippon Muki Co., Ltd.) (Figure 2) in diameters were utilized as enhancing materials, respectively. The length of glass fiber with 11μm in diameter was 150μm and with 500 nm in diameter was 200~2,000μm. The weight fraction of the glass fibers was 0.2 wt%.

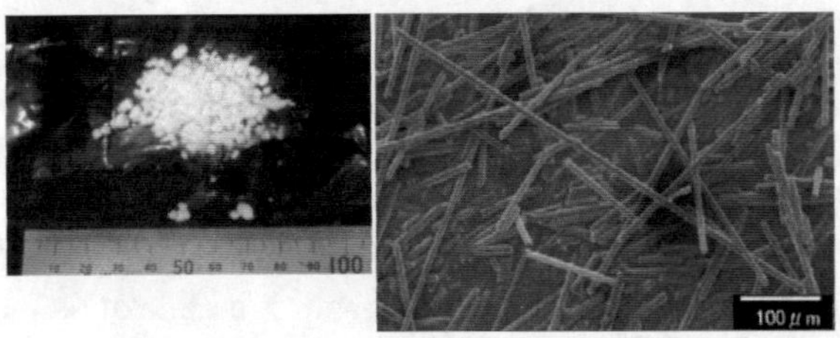

Figure 1. Conventional glass fibers (11μm).

Figure 2. Thin glass fibers (500nm).

2.2 Fabrication of composite plate

The composite plate was obtained after laminating with 8 cloths by conventional hand lay-up method and being cured at 80 degree-C for 1 hour and then at 150 degree-C for 3 hours under 5MPa on heating press machine (HB200HB: KAWANAKA SANGYO CO., LTD., Japan).Then heat was radiated slowly for 4 hours. The volume fraction of the fibers was set to 60%. After curing, the composite plate was cut into the dimension of specimen.

3. Test procedure

3.1 Static three point bending test and cyclic three point bending test

Static three point bending tests and cyclic three point bending tests were conducted to investigate the effect of the addition of two different sized glass fibers on the bending strength and fatigue life. The dimension of the composite specimen for measuring bending strength and fatigue life was 100×15×2mm (Figure 3). The three point bending strength of the specimen was determined using a universal testing machine (Servo-pulser EHF-EA5: SHIMADZU CORPORATION, Japan) under displacement controlled condition according to ASTM D790. The span length was 80mm and testing speed was 5.0mm/min. Test data were obtained for five runs under the same test conditions. The fatigue lives of the composite specimens were also measured using an electro hydraulic material testing machine

(AUTOGRPH AG-I: SHIMADZU CORPORATION, Japan) under load controlled conditions at 2Hz of frequency and 0.1 of stress ratio under 80% of maximum cyclic loading to the ultimate strength. The wave profile of cyclic loading was set to sinusoidal wave, and gage length was 80mm.

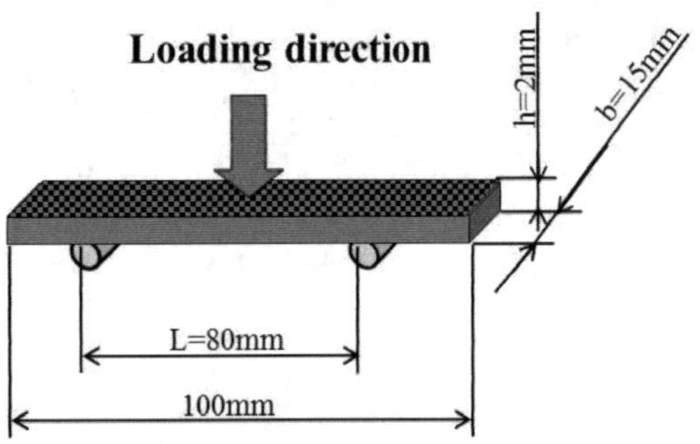

Figure 3. Specimen for static and fatigue bending test.

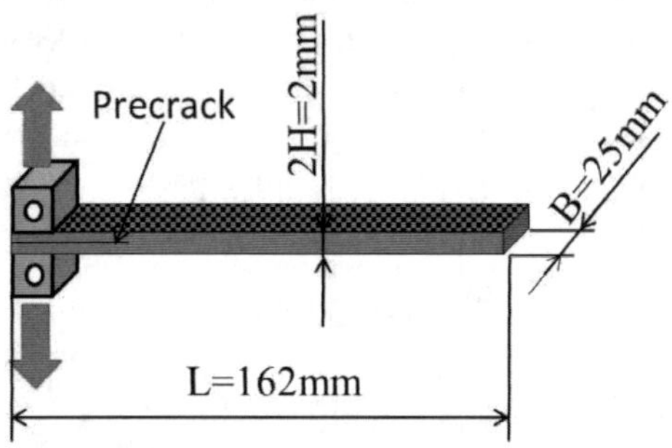

Figure 4. Specimen for DCB test.

3.2 Mode-I interlaminar fracture toughness test

The double cantilever beam (DCB) Mode-I fracture specimens were employed to characterize the resistance of interlaminar delamination according to ASTM D5528-01. The dimension of the DCB specimen was 162×25×2mm (Figure 4). The initial crack was introduced on interlayer of the DCB specimen by inserting a Teflon-sheet. Aluminium End-Blocks with the same width as the specimen were attached at the end of specimens by an epoxy adhesive to allow load application. The compact table-top universal tester (EZ test, load rating 500N: SHIMADZU CORPORATION) was used under displacement controlled test. The Mode-I interlaminar fracture toughness G_{Ic} was calculated by the following equation.

$$G_I = \frac{3}{2(2H)}\left(\frac{P}{B}\right)^2 \frac{(B\lambda)^{2/3}}{\alpha_1} \tag{1}$$

where $G_{Ic}=G_I$ at critical state , P, λ, B, 2H, and α_1 denote the Mode-I interlaminar fracture toughness, the critical load, the compliance corresponding to crack length, the specimen width, the thickness in equation, and α the end-block correction factor, respectively.

3.3 Mode- II *interlaminar fracture toughness test*

End-Notched Flexure (ENF) tests were applied to determine the Mode-II interlaminar fracture toughness of the specimens. The dimensions of the ENF specimen were 45×15×2mm (Figure 5). Pre-crack was introduced by inserting a Kapton film of approximately with 30μm in thickness (Kapton, Du Pont-Toray Corporation, Japan). The Mode II inter-lamina fracture toughness was calculate by

$$G_{IIC} = \frac{9a_1^2 F_c^2 C_1}{2B(2L^3 + 3a_1^3)} \tag{2}$$

$$a_1 = \left[\frac{C_1}{C_0} a_0^3 + \frac{2}{3}\left(\frac{C_1}{C_0} - 1\right) L^3\right]^{\frac{1}{3}} \tag{3}$$

where, a_0, F_c, C_0, C_1, L and B denote initial crack length, critical load, load point compliance at initial elastic part, load point compliance at critical load, span length and specimen width, respectively.

3.4 Measuring of crack propagation under fatigue loading

Notched coupon specimen was also prepared (Figure 6). Pre-crack with 2mm was initiated on the side surface of specimen by a razor after conventional mechanical cutting. In this test, cyclic simple tensile load was nominal applied to the specimen, where cyclic applied nominal load, cyclic stress ratio and loading frequency were 13kN, 0.1 and 5Hz, respectively.

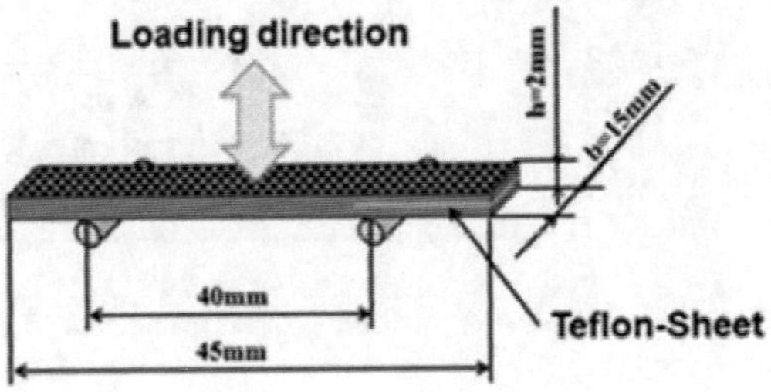

Figure 5. Specimen for ENF test.

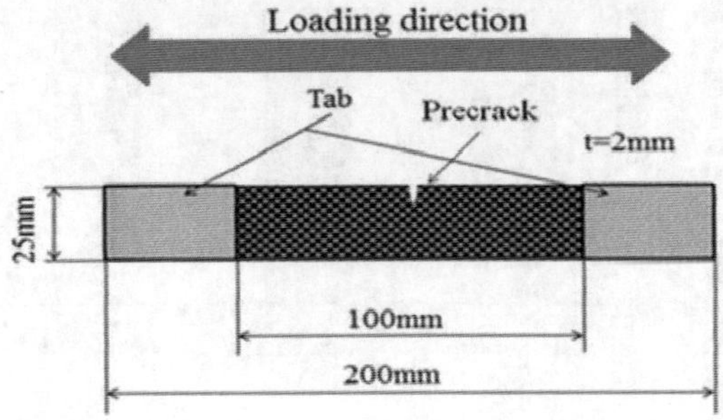

Figure 6. Notched specimen for measuring macro crack propagation

3.5 Estimating interfacial strength and frictional force around texture point.

Model specimen was originally prepared to estimate the interfacial strength and frictional force around texture point. One longitudinal carbon fiber bundle was passed between two transverse carbon fiber bundles which were all embedded into the unmodified and modified epoxy resin, respectively (Figure 7). Tensile force was applied to the longitudinal fiber bundle while the part of epoxy resin was hooked at an exit of tiny hole to fix the displacement.

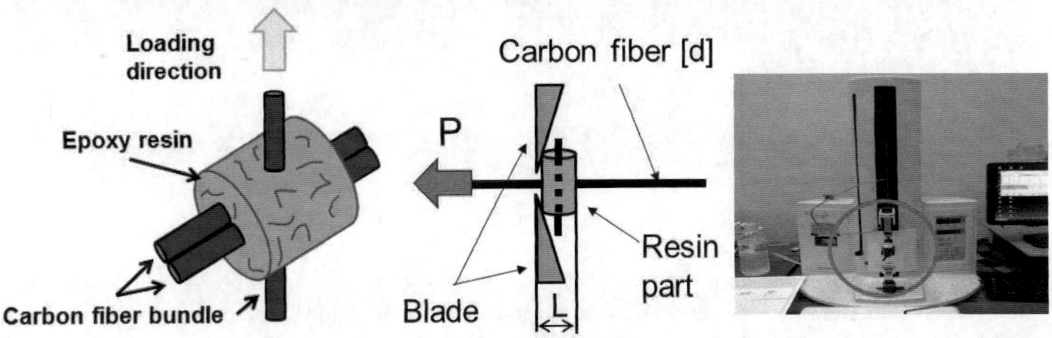

Figure 7. Model specimen and its testing machine.

4. Result and Discussion

4.1 Static strength and fatigue life

Figure 8 shows the test results of static three point bending strength of plain-woven fiber composite modified with nano-sized glass fibers with 11μm and 500nm in diameter. The static three point bending strength was about increased 17.8-29.9% by the addition of glass fibers into the matrix. These result shows that the static bending strength of composite was slightly increased with a reduction in the diameter of additional glass fibers. Figure 9 shows the change of the bending fatigue lives defined as the number of cycles to failure of the composite specimen subjected to 80% of cyclic maximum stress of static strength. The fatigue life was dramatically increased 10-100 times by the addition of nano-sized glass fibers into the matrix in contrast to the static case. After the cyclic three point bending test, the state of the side surface of the specimen was observed by microscope. Figure 10 shows an example of the damage state of the specimen after failure. It was shown that the fatigue crack of composite was grown along the carbon fiber bundle under cyclic bending loading.

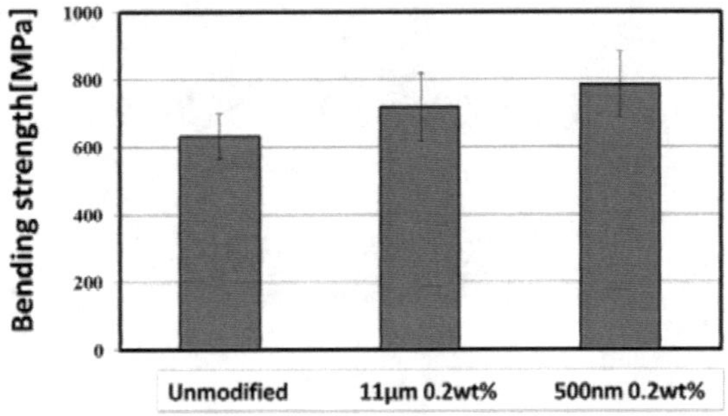

Figure 8. Change of static bending strength

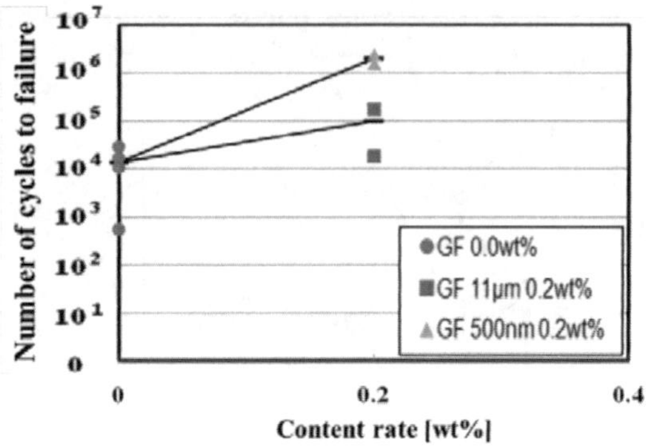

Figure 9. Change of fatigue lives.

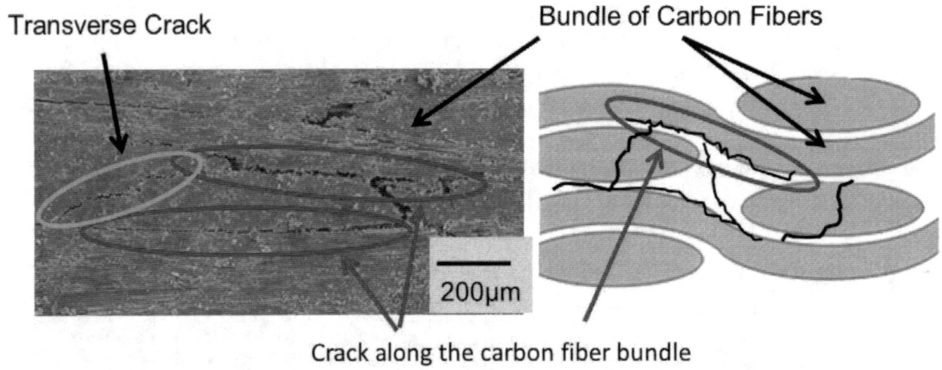

Figure 10. Example of the damage state of specimen after fatigue failure.

4.2 Mode-I interlaminar fracture toughness

Figure 11 shows the change of Mode-I interlaminar fracture toughness by the modification proposed in this study. When composite was modified with additional glass fibers with 11μm in diameter, the Mode-I internal fracture toughness (G_{IC}) was increased about 26%. The G_{IC} was also increased about 21%, when the composite was modified with glass fibers of 500nm in diameter. These result showed that the G_{IC} was basically increased due to addition of micro and nano-sized glass fibers into the matrix. However, the improvement in G_{IC} was comparatively slight with that in Mode-II shown in next session.

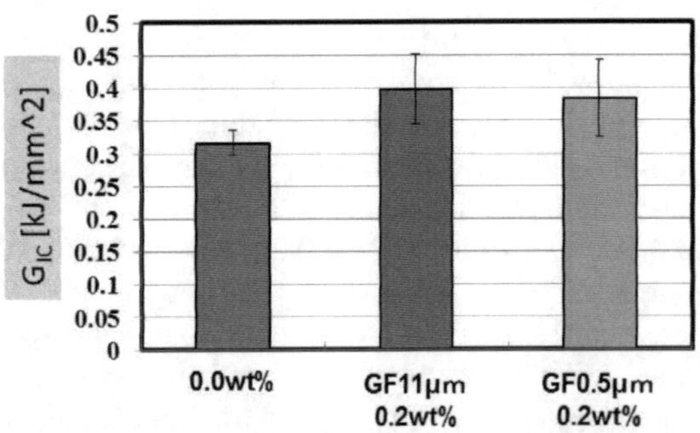

Figure 11. Change of Mode-I interlaminar fracture toughness.

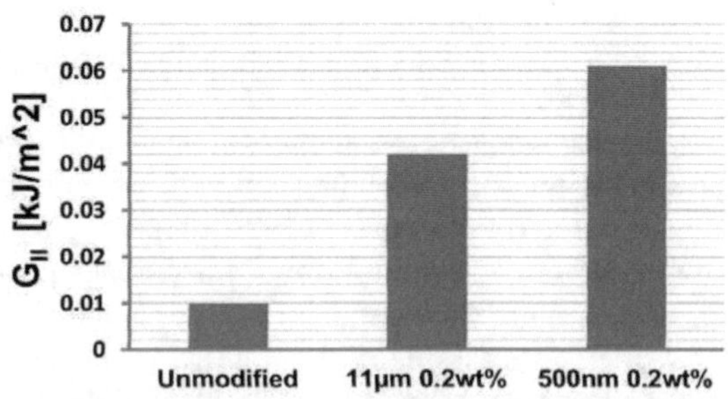

Figure 12. Change of Mode-II interlaminar fracture toughness.

4.3 Mode II inter-lamina fracture toughness

Figure 12 shows the change of Mode-II interlaminar fracture toughness by the modification. The Mode-II fracture toughness between the carbon fiber bundle and epoxy matrix was significantly improved about 4-6 times due to the addition of micro or nano-sized glass fibers compared with that of unmodified samples. The Mode-II fracture toughness of the specimen modified with fine glass fibers of 500nm in diameter was about 1.5 times higher than modified with 11μm in diameter. The fracture toughness in Mode-II was remarkably improved by the modification of the matrix, while that in Mode-I was improved slightly.

4.4 Macroscopic crack propagation under fatigue loading

Figure 13 shows the measurement result of the macroscopic crack length of the composite specimens with unmodified and modified matrix. Macroscopic crack propagation in fatigue was totally prevented when the matrix was modified with fine glass fibers of 500nm in diameter, while relatively high speed crack propagation was shown in the unmodified case especially in early stage of fatigue. This means that remarkable damage was prevented by the modification with fine glass fibers.

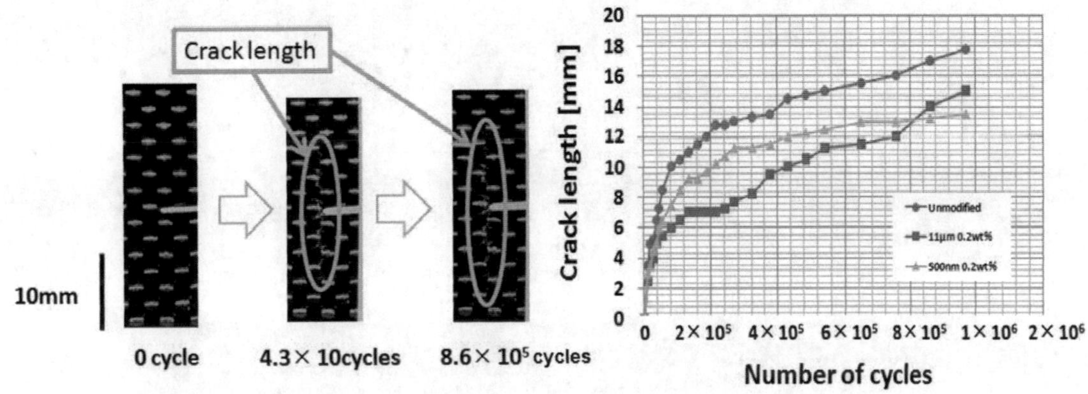

Figure 13. Propagations of macroscopic crack lengths of composite specimens

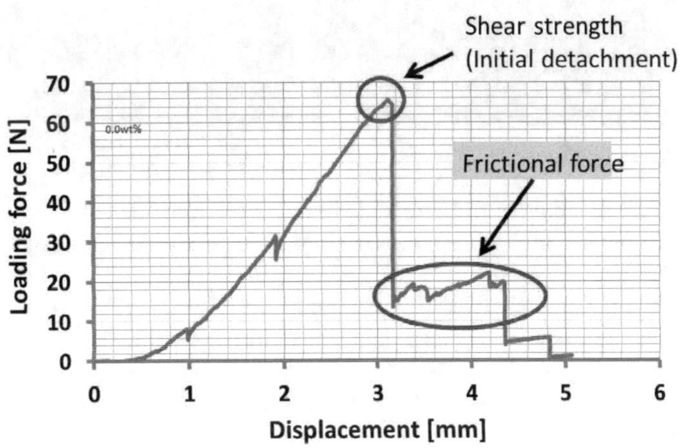

Figure 14. An example test result of loading force and displacement curve of model specimen.

4.5 Observed interfacial strength and frictional force around texture point.

Figure 14 shows an example test result showing the loading force applied to longitudinal carbon fiber bundle passing between two transverse carbon fiber bundles, as explained above. Loading force was linearly increased with the displacement of loading point, before a sharp peak of maximum, and then showed an almost plateau response with pulling behavior. These means the observed shear strength between carbon fiber bundle and matrix, and observed frictional force around the crossing point, respectively. Figure 15 shows the change of observed shear strength and frictional force by the modification. The observed frictional force was improved about 2.5 times due to the addition of fine glass fibers compared with that of unmodified samples, while the improved ratio was 28% in observed shear strength. Figure 16 shows the comparison of states of fracture and sliding surfaces of the tested longitudinal carbon fiber bundles embedded in the model specimen. Lots of particles of the resin were remained, if the matrix was modified with fine glass fibers. It was suggested that those effects in improvement of mechanical properties of the textile carbon fiber composite were explained by the prevention of interfacial crack propagation with large resistance between carbon fiber bundles in Mode-II and modified matrix around the texture points where comparatively large strain energy would be released, if the epoxy matrix was adequately modified with the fine glass fibers.

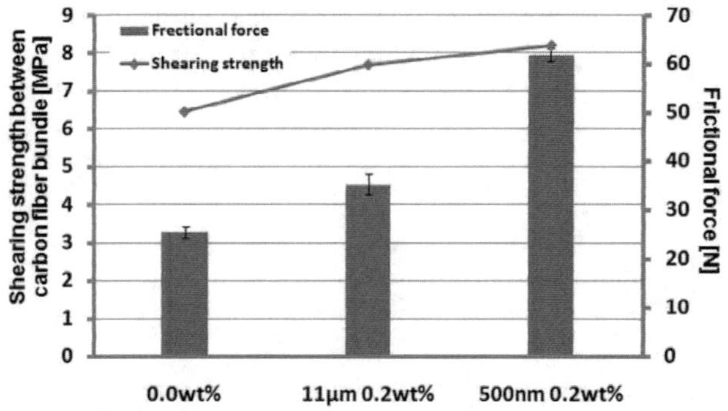

Figure 15. Change of observed shear strength and frictional force

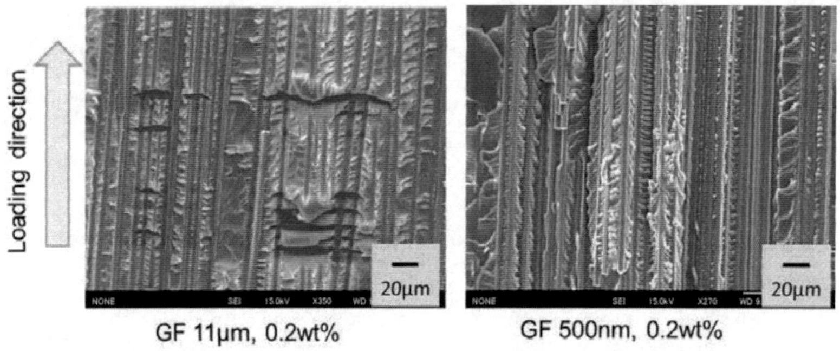

Figure 16. Comparison of states of fracture and sliding surfaces of tested longitudinal carbon fiber bundles embedded in model specimen.

5. Conclusion

5.1 Test results in current study showed that the addition of glass fibers with fine size was rather effective to dramatically improve the fatigue life. The fatigue life was dramatically increased 10-100 times by the addition of nano-sized glass fibers into the matrix in contrast to the static case.

5.2 Fracture toughness in Mode-II was also improved by the modification of the matrix, while that in Mode-I was improved slightly. The Mode-II fracture toughness between the carbon fiber bundle and epoxy matrix was significantly improved about 4-6 times due to the addition of micro or nano-sized glass fibers compared with that of unmodified samples. The fracture toughness in Mode-II was remarkably improved by the modification of the matrix, while that in Mode-I was improved slightly.

5.3 Macroscopic crack propagation in fatigue was totally prevented when the matrix was modified with fine glass fibers of 500nm in diameter

5.4 The observed frictional force of one longitudinal carbon fiber bundle passed between two transverse carbon fiber bundles embedded into the modified epoxy resin was improved about 2.5 times due to the

addition of fine glass fibers compared with that of unmodified samples, while the improved ratio was 28% in observed shear strength.
5.5 Those effects in improvement of mechanical properties of the textile carbon fiber composite were explained by the prevention of interfacial crack propagation with large resistance between carbon fiber bundles in Mode-II and modified matrix around the texture points where comparatively large strain energy would be released, if the epoxy matrix was adequately modified with the fine glass fibers.

Acknowledgment
This work was financially supported by a (advanced technological) research project on "Research and Development Center for Advanced Composite Materials" of Doshisha University and MEXT (the Ministry of Education, Culture, Sports, Science and Technology, Japan) -Supported Program for the Strategic Research Foundation at Private Universities, 2013-2017, the project S1311036.

References
[1] Yoshi, T., Okubo, K., Fujii, T. (2010) "Stiffness Reduction of Woven CFRP and CFRTP Spring under Ultra High Cyclic Fatigue for Vibration Conveyor", *Advanced Materials Research*, Nova Science Pub Inc, 123, pp.217-220. doi: org/10.4028/www.scientific.net/AMR.123-125.217.
[2] Hojo, M., Matsushita, Y., Tanaka, M., Adachi, T. (2012) "Interfacial fatigue crack propagation in microscopic model composite using bifiber shear specimens", *Composites, Part A*, Elsevier Limited, 43(2), pp.239-246. doi: org/10.1016/j.compositesa.2011.09.004.
[3] Adachi, T., Osaki, M., Araki, W., Kwon, S. (2008) "Fracture toughness of nano- and micro-spherical silica-particle-filled epoxy composites", Acta Materialia, Elsevier Limited, 56, pp.2101-2109. doi: org/10.1016/j.actamat.2008.01.002.
[4] Lin, T., Jia, D., He, P., Wang, M. (2010) "In situ crack growth observation and fracture behavior of short carbon fiber reinforced geopolymer", *Materials Science and Engineering, A*, Elsevier Limited, 527(9), 2404-2407. doi: 10.1016/j.msea.2009.12.004.
[5] Argüelles, A., Viña, J., Canteli, A.F., Castrillo, M. A. Bonhomme. J. (2008) "Interlaminar crack initiation and growth rate in a carbon-fibre epoxy composite under mode-I fatigue loading", Composites Science and Technology, Elsevier Limited, 68(12), pp.2325-2331. doi: org/10.1016/j.compscitech.2007.09.012.
[6] Kimura, H., Akiniwa, Y., Tanaka, K., Tanaka, H., Okumura, Y. (2006) "Electrical, thermal and mechanical properties of poly-etherimide epoxy-diamine blend", *International Journal of Fatigue*, Elsevier Limited, 28(10), pp.1147-1153. doi: org/10.1016/j.ijfatigue.2006.02.002.
[7] Arai, M., Sasaki, T., Hirota, S., Ito, H., Hu, N., "Quaresimin, M. (2011) "Mixed modes interlaminar fracture toughness of cfrp laminates toughened with CNF interlayer", *Acta Mechanical Solida Sinica*, Elsevie Limited r, 25(3), pp.321-330. doi: org/10.1016/S0894-9166(12)60029-9.
[8] Arai, M., Noro, Y., Sugimoto, K., Endo, M. (2008) "Mode I and mode II interlaminar fracture toughness of CFRP laminates toughened by carbon nanofiber interlayer", *Composites Science and Technology*, Elsevier Limited, 68(2), pp.516-525. doi: org/10.1016/j.compscitech.2007.06.007.
[9] Mohamed H., Elrahman, M. A., Okubo, K., Fujii, T. (2010) "Effect of microfibrillated cellulose on mechanical properties of plain-woven CFRP reinforced epoxy", *Composite Structures*, Elsevier Limited, 2(9), pp.1999-2006. doi: org/10.1016/j.compstruct.2009.12.009.
[10] Nishikawa, Y., Okubo, K., Fujii, T., Kawabe, K. (2006) "Fatigue crack constraint in plain-woven CFRP using newly-developed spread tows", *International Journal of Fatigue, Materials & Design* Elsevier Limited, 28(10), pp.1248-1253. doi: org/10.1016/j.ijfatigue.2006.02.010.
[11] Phong, N.T., Mohamed H. G., Okubo K., Chuong B., Fujii T. (2013) "Enhancement of mechanical properties of carbon fabric/epoxy composites using micro/nano-sized bamboo fibrils", *Materials & Design*, Elsevier Limited, 47, pp.624-632. doi: org/10.1016/j.matdes.2012.12.057.
[12] Higashino, M. Takemura. K, Fujii, T. (1995) "Strength and damage accumulation of carbon fabric composites with a cross-linked NBR modified epoxy under static and cyclic loadings", *Composite Structures,* Elsevier Limited, 32(1), pp.357-366. doi: org/10.1016/0263-8223(95)00027-5.
[13] Phong, N.T, Mohamed H. G., Okubo K., Chuong B., Fujii T. (2013) "Improvement in the mechanical performances of carbon fiber/epoxy composite with addition of nano-(Polyvinyl alcohol)

fibers", *Composite Structures*, Elsevier Limited, 99, pp.380-387. doi: org/10.1016/j.compstruct.2012.12.018.

[14] Mohamed H. G., Elrahman, M. A., Okubo, K., Fujii, T. (2010). *Composites Part A: Applied Science and Manufacturing*, Elsevier Limited, 41, pp.1263-1271. doi: org/10.1016/j.compositesa.2010.05.010.

IOP Conf. Series: Materials Science and Engineering **406** (2018) 012036 doi:10.1088/1757-899X/406/1/012036

Estimation of shear strength of FRP joints using adhesives with ceramics particles

Tetsusei Kurashiki*, Takumi Sakamoto, Koushu Hanaki, Kazutaka Mukoyama, Xingsheng Li

Management of Industry and Technology, Graduate School of Eng., Osaka Univ.
2-1, Yamadaoka, Suita Osaka, 565-0871, Japan

* kurasiki@mit.eng.osaka-u.ac.jp

Abstract. We proposed adhesives with thermosetting resin and ceramics particles for single lap joints of GFRP and tried to decompose by microwave irradiation. It was obvious that the optimum amount and radius of particles to satisfy high strength before irradiation and low strength after irradiation.

1. Introduction

Multi materials composed of metal and FRP with the high strength and the light weight are expected to reduce CO_2 and keep the energy saving for vehicles. The dissimilar material joining is one of the remarkable technologies for multi materials, however, it is difficult to decompose the joining structure for recycling. Therefore, the adhesive joining technology for metal and FRP with the high joining strength and easy decomposable function is needed.

There are several researches about decomposable adhesive using thermoplastic resin with filler [1][2], however, the adhesive based on thermosetting resin is more effective to apply the bonding structures such as FRP and metals to keep high bonding strength and easy decomposable function.

The purpose of this study is to develop adhesive materials based on thermosetting resin and ceramics filler for joining different kinds of materials such as FRP and metals, and to decompose the bonding structures conveniently. As the first step, we propose the adhesives with thermosetting resin and ceramics particles for a GFRP joint and try to decompose by microwave irradiation.

2. Single lap joints with thermosetting resin and ceramics particles

Test specimens were GFRP (E-glass: non-crimp fabric [0/90]s / Resin: vinyl ester), and the specimens were prepared as single lap joints in Figs.1, 2 based on JIS K 6850.

Test specimens are bonded with the adhesives based on thermosetting epoxy (Araldite®, Huntsman Japan Co.). The adhesive is included with SiC particles. The volume of the particles are 0, 10, 20, and 30%, and the diameter are 0.02, 0.3, 1.2, 80 μm. The thickness of adhesive layer is 0.1mm.

A tensile shear test is carried out with tensile testing machine (Shimadzu Co. Ltd.), and the tensile speed is 1.0mm/min. The microwave irradiation is controlled with the microwave apparatus (MOASTORE Co. Ltd.) and the frequency and power are 2.45GHz, 500W.

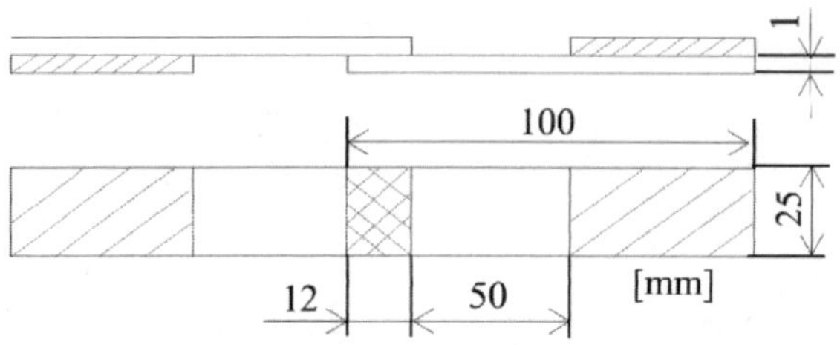

Figure 1. Shape of test specimen

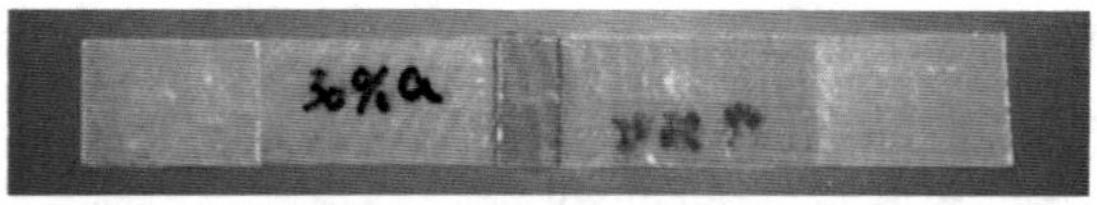

Figure 2. A sample of test specimen

3. Effects of volume fraction of particles on shear strength

A tensile shear test of GFRP joints with the proposed adhesive was carried out with/without microwave irradiation, and effects of the volume fraction of ceramics particles on the adhesive strength were investigated.

Figure 3 shows experimental results of shear strength of GFRP single lap joint before microwave irradiation in case of volume contents of particles 0, 10, 20 and 30% with 0.3μm diameter. The number of test specimen is 3. The average value is shown in the figure, and max. and min. values are shown in the scatter bar. The shear strength reduces with the increase of particles volume. In case of 30% volume, the strength shows 62.3% reduction compared with the strength of 0% volume.

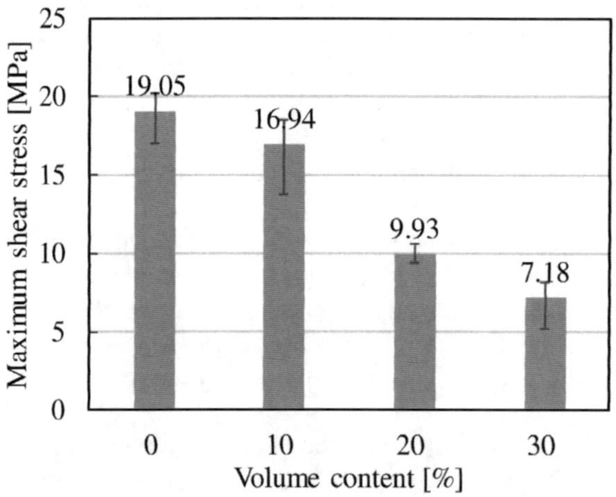

Figure 3. Results of shear strength of GFRP single lap joint before microwave irradiation (Volume contents of particle: 0, 10, 20, and 30%)

The temperature on the surface of test specimen is measured with thermometer (FT-H20, Keyence Co. Ltd.). Figure 4 shows the temperature change due to microwave irradiation. In case of 0% volume, the temperature tends to saturate 160°Cand the value is under the thermal decomposition temperature of thermosetting resin. On the other hand, the temperature increases 300°Cin case of over 10% volume due to the self-heating of ceramics particles.

Figure 5 shows the irradiation time of microwave until disappearance of adhesion, and the disappearance time of adhesion reduces with the increase of ceramics volume.

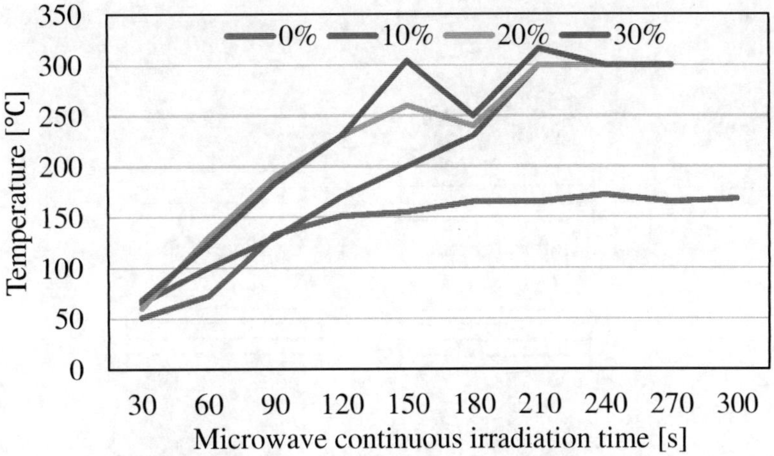

Figure 4. Temperature change due to microwave irradiation (Volume contents of particle: 0, 10, 20, and 30%)

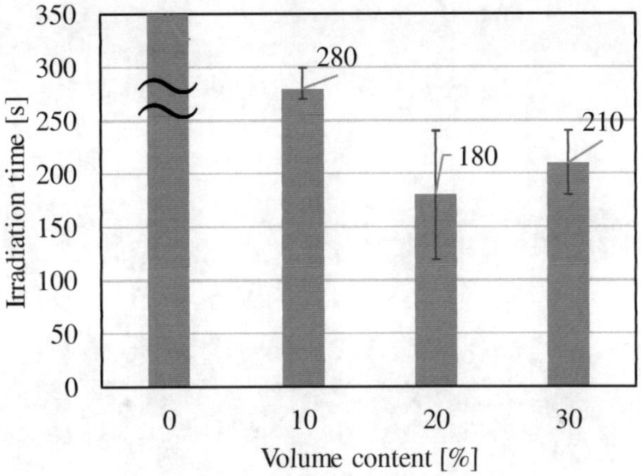

Figure 5. Irradiation time of microwave until disappearance of adhesion (Volume contents of particle: 0, 10, 20, and 30%)

4. Observation of fracture surface

After the tensile shearing test, the fracture surfaces of test specimens were observed with CCD microscope. The pattern of fracture on the cross section of specimens is clarified to 3 types (debonding in adhesives, interface fracture between adhesives and FRP, and fracture in FRP) in Fig.6.

Figure 7 shows the observational results with/without microwave irradiation in case of several volume contents of ceramics particles. Figure 7(a) shows the fracture surface after tensile shear test without microwave irradiation, and Figure 7(b) shows the fracture surface with microwave irradiation before tensile shear test. In case of 0% volume of particles, the fracture in FRP appeared after tensile shear test without microwave irradiation (Fig.7(a-1)), however, the specimen was not separated with only microwave irradiation (Fig.7(b-1)) because the temperature of adhesive resin was not reached to the thermal decomposition temperature (Fig.4).

IOP Conf. Series: Materials Science and Engineering **406** (2018) 012036 doi:10.1088/1757-899X/406/1/012036

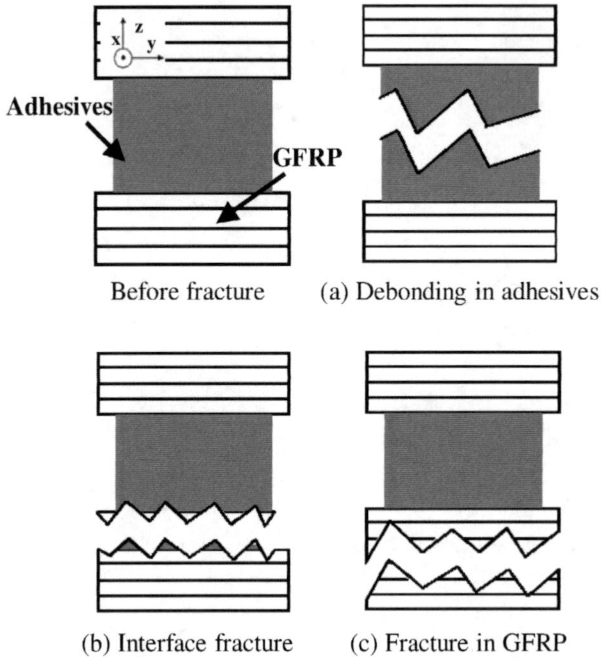

Figure 6. Pattern of fracture surface

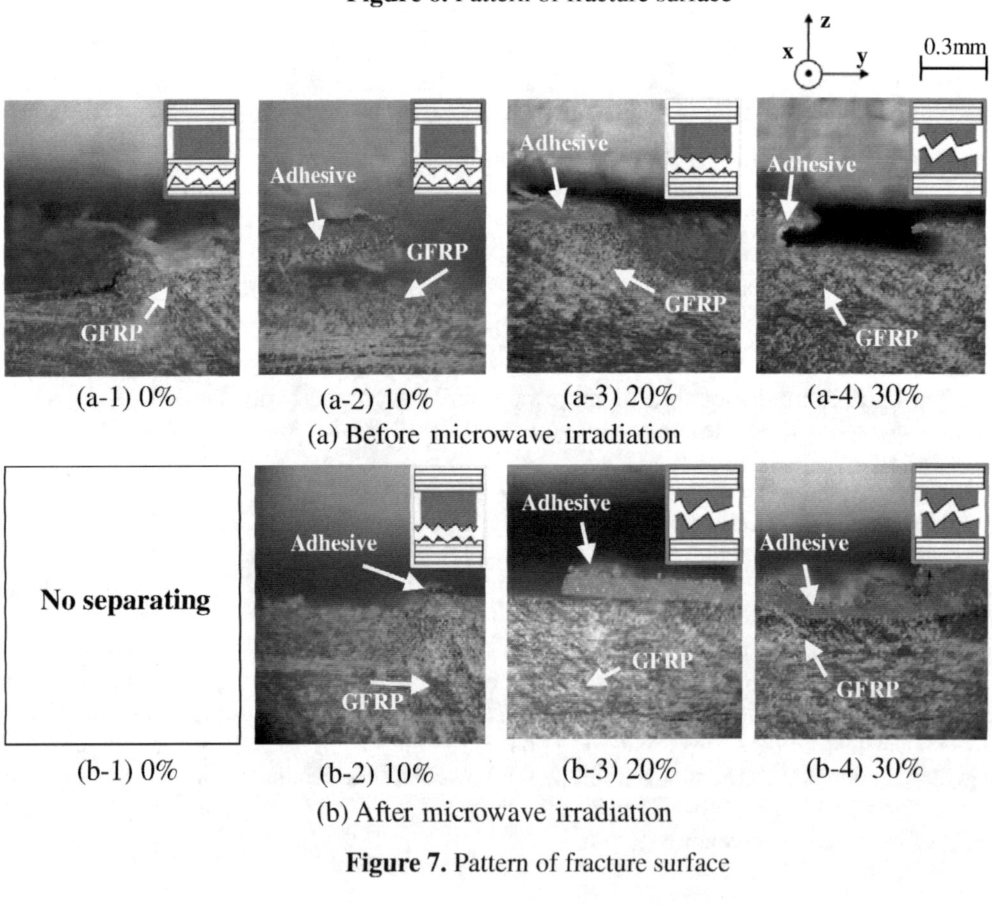

Figure 7. Pattern of fracture surface

In case of 10% volume, the fracture in FRP also appeared without microwave irradiation (Fig.7(a-2)), and the interface fracture appeared with microwave irradiation (Fig.7(b-2)). The pattern is effective for material joining which keeps the high bonding strength without microwave irradiation and easy decomposable function with microwave irradiation.

In case of over 20% volume, the interface fracture and fracture of debonding in adhesives appeared without microwave irradiation in Fig.7(a-3, a-4), and the max. shear strength are low in Fig.3. As the results, it was obvious that the optimum amount of volume fraction of ceramics particles to satisfy the high strength before microwave irradiation and l easy decomposable function after irradiation.

5. Effects of particles size on shear strength

The tensile shear test was carried out in case of various particle diameter with/without microwave irradiation and the irradiating time is 180 sec. Figures 8, 9 shows the experimental results of max. shear strength and elongation in case of various diameter of particles with 5.0% volume contents.

In case of 0.02μm diameter of particles, the shear strength and elongation before irradiation show a remarkable reduction compared with over 0.3μm diameter. Too small size of particles is not effective for reinforcement of adhesive resin.

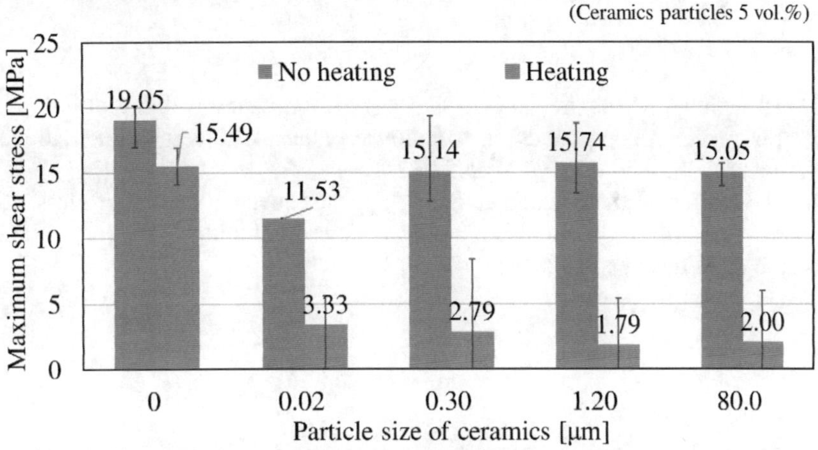

Figure 8. Results of shear strength of GFRP single lap joint with/without microwave heating (Case of particle size: 0.02, 0.3, 1.2, 80 μm)

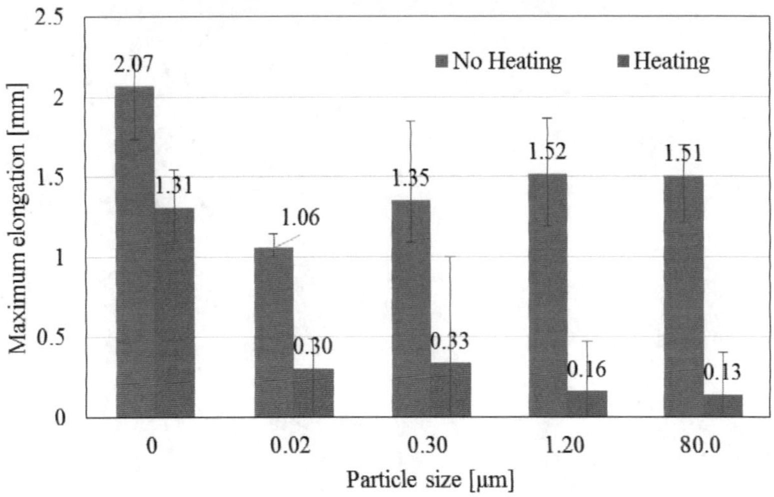

Figure 9. Results of max. elongation of GFRP single lap joint with/without microwave heating (Case of particle size: 0.02, 0.3, 1.2, 80 μm)

6. Conclusion

For bonding structures with different kinds of materials, we proposed the adhesives with thermosetting resin and ceramics particles for a GFRP joint and try to decompose by microwave irradiation. The single lap joints with GFRP were prepared, and a tensile shear test of GFRP joints with the proposed adhesive was carried out with/without microwave irradiation, and effects of the volume fraction and diameter of ceramics particles on the adhesive strength were investigated.

As the results, it was obvious that the optimum amount of volume fraction and radius of ceramics particles to satisfy the high strength before irradiation and low strength after irradiation. In case of the optimum volume contents, the observational results of fracture surface shows that the single lap joint keeps the high bonding strength without microwave irradiation and easy decomposable function with microwave irradiation.

Acknowledgements
The authors would like to express our thanks to Dr. Hideki HYUGA (National Institute of AIST, Japan) for the cooperation.

References

[1] M. Sano, H. Oguma, M. Sekine and C. Sato, "High-frequency welding of polypropylene using dielectric ceramic compounds in composite adhesive layers", International Journal of Adhesion & Adhesives 47 (2013) pp.57-62.
[2] M. Sano, H. Oguma, M. Sekine and C. Sato, "High-frequency welding of glass-fiber-reinforced polypropylene with a thermoplastic adhesive layer including SiC", International Journal of Adhesion & Adhesives 54 (2014) pp.124-130.

IOP Conf. Series: Materials Science and Engineering **406** (2018) 012037 doi:10.1088/1757-899X/406/1/012037

A promising reinforcement for textile composites: Study on the effect of cellulose nanofiber (CNF) content on the mechanical properties of CNF reinforced epoxy resin

Xingsheng Li[1*], K. Hanaki[1], X. Wang[1], T. Kurashiki[1] and K. Mukoyama[1]

[1]Management of Industry and Tchnology, Graduate school of Eng., Osaka Univ. 2-1, Yamadaoka, Suita Osaka, 565-0871, Japan

* xingsheng@mit.eng.osaka-u.ac.jp

Abstract. The reinforcement effect of volume fraction of CNF on different matrix resin (vinyl-modified epoxy ester resin emulsion; epoxy-modified alkyd emulsion) are investigated in this study. Tensile shear test is carried out following the JIS K 6850:1999 standard, by coating this colloid mixture on substrates. The results show that the maximum shear strength of both waterborne resin increase with the addition amount of CNF at relative humidity 32.78%. It is considered that there exits hydrogen bonding between fiber and matrix, and the stress concentration might be relieved by the 3-D network structure of CNF, thus the maximum stress is increased, and the crack generation and development is suppressed. But it leads to a loss of percentage elongation values. A peak of the deformation energy is found at 10wt.%, which is considered as a force balance point of 3-D network reinforcement and the aggregation of CNF at the interface. The reinforcement of CNF is sensitive to humidity, for epoxy ester matrix is found more easily influenced by humidity than alkyd matrix, which has bigger deformation energy than alkyd matrix in low humidity(32.78%RH), although deformation energy dramatically decreases in high humidity(97.30%RH), it still has compatible humidity as alkyd matrix.

1. Introduction

Fiber reinforced plastic(FRP) is widely used for lightweight structures in aerospace and automotive industry etc. However, as FRP is made by high strength fiber and low strength resin, FRP has high strength in fiber direction, but may weak in transvers direction and out-of-plane direction. Many efforts have been made to improve the transvers and out of plane properties. The bonding between fibers and matrix plays a crucial role in mechanical performance of FRPs. It is considered that to improve the surface condition by removing impurity on the surface or coating, which can improve the mechanical property of fiber and the wettability between fiber and matrix, as a result improve the mechanical property of FRP [1]. Plasma etching, and oxidation are widely used for surface modification. Research on coating functional group of an organic polymer which could reacting with matrix resin is carried out by Hojo M. etc. [2]. Effects of multi-scale modifications by adding nano-particales/ nano-tubes on surface CF on the mechanical property of FRP is studied by many researchers. For example, M. sharma et al. [3][4], interfacial strength has been increased by 113%, and failure was predominantly cohesive matrix failure, in contrast to the untreated fiber.

Epoxy resins are widely used in FRP. The three-dimensional cross-linked thermoset structures gives their high process ability and strength [5-7]. However, higher cross-link density contributes to lower fracture toughness [8-10]. This drawback can be overcome by modifying epoxy resins with integration of various nanofillers as a second microphase [11-14].

An initial study on the reinforcement effect of CNF on FRP is carried out. To investigate the reinforced effect of CNF on waterborne resin, specimen for dispersion status observation using scanning electron microscopy(SEM) and tensile shear test are prepared separately. Further, the effect of humidity on the reinforcement behavior of CNF is investigated.

2. Experiment

2.1. Materials

The stable suspension of cellulose nanofiber used in this study, is made by high pressure water jet method, which is provided by SUGINO MACHINE Ltd., Uozu, Toyama, Japan. General physical properties of cellulose suspension are shown as below. Concentration: 2 wt. %, viscosity: over 3000 MPa•s, specific surface area: 100~200 m2/g, degree of polymerization: 200~550 and the fiber is shaped in a diameter of about 20 nanometer and a length of a few micrometer. Two different waterborne resins were used as matrix material. (a) vinyl-modified epoxy ester resin emulsion EFD-5560 and (b) epoxy-modified alkyd emulsion WPW-601. Both of the waterborne resins are one-pack type and cure through polycondensation which provided by DIC Corp. . The detailed contents of the matrix waterborne resins are shown in table 1.

Table 1. Content of the matrix waterborne resin.

Content /wt.%	EE resin (EFD-5560)	EA resin(wPW-601)
Solid content	39.0~41.0	40.0~42.0
Amine	1.6	2.0
Water	58.4	50.0
Organic solvent	0	≤ 6.0 (Methoxybutanol & Dipropylene glycol dimethyl ether)

2.2. Methods

2.2.1. Mixture colloid preparation. Both the specimens for observation of dispersion state and tensile shear test, are prepared with same procedure. All mixture colloid used in this study are prepared using a mechanical mixer, 5min for stirring and degassing each. For shear tensile test, 6 kinds of mixture colloid are made with CNF Mass fraction ration (0, 2.5, 5.0, 10.0 and 20.0) for each type of resin.

2.2.2. Scanning electron microscopy. The dispersion state of CNF in epoxy ester resin with different content are studied using field emission scanning electron microscope JSM-7000F. The acceleration voltage is 10kV. Two specimens with neat CNF and CNF 20% reinforced epoxy ester resin are investigated. The specimens used for SEM observation are cured in the circumstance of 40°C for 2 days until the weight keeps still.

2.2.3. Tensile shear test. Tensile shear tests were performed on an Imada compact tensile/compression testing machine, to clarify the maximum strength of the material. Test pieces are prepared following the JIS K 6850:1999 standard, which is shown in Figure 1. The mixture colloid of CNF and resin were coated onto an aluminium film with a thickness of 2mm. Specimens are made by overlapping two aluminium film at 12.5±0.5mm length. To investigate the effect of humidity on the mechanical property of the coating, specimens are prepared under two environment humidity condition of 32.78RH% and

97.30RH%. According to the study of L. Greenspan [15], $MgCl_2$ and K_2SO_4 are used to control the environment humidity. The chemicals and the specimens are placed in a sealed container with a hydrometer. 8 specimens are prepared for each condition. The concentration and the solidification conditions of specimen are shown in table 2.

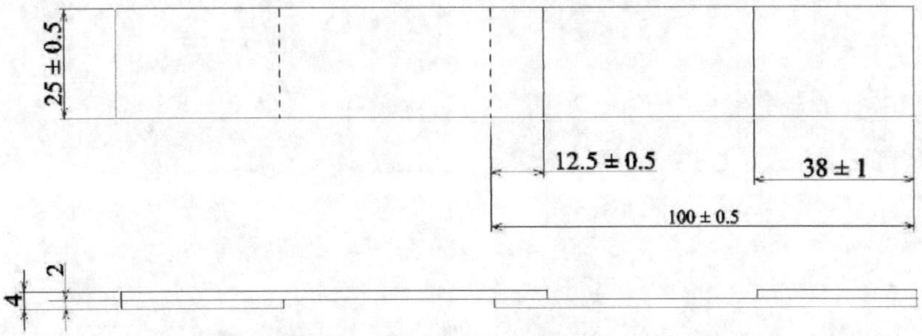

Substrate used in tensile shear test.(mm)

Figure1. Shape of the substrates used in tensile test (a) and tensile shear test (b).

Table 2. Conditions used in the preparation of test pieces.

Matrix	CNF content/ wt.%	Solidification conditions	Quantity of test pieces
Epoxy ester resin/ Alkyd resin	0	45×10Day + 25°C/32.78RH%×10Day /25°C/97.30RH%×10Day	8 pcs each
	2.5		
	5		
	10		
	20		

All the tensile shear tests were carried out in the room environment (at a temperature of 20~25 °C and a relative humidity that ranges between 25~40 %) and deformation speed was set to 0.015 mm/s.

3. Results and discussion

3.1. SEM observations of the dispersion state of CNF

The morphology of neat CNF and CNF reinforced waterborne epoxy ester with CNF content of 20wt.% were examined by SEM under 10kV accelerating voltage. Figure 2 and Figure3 shows the SEM image of at different magnifications respectively. In Figure2, severe aggregation of CNF is found and non-aggregated CNF with a diameter of several tens of nm is confirmed. High aspect ratio of CNF is observed, but the length of the fiber cannot be determined from these images. In the case of CNF composites film (20wt.%), fiber like morphology with nearly equal diameter is confirmed, from which CNF is supposed to be well distributed in the resin. The morphology is considered to be a fiber aggregation coated by the resin. As a result, more information could not be obtained above magnification of 2000×.

IOP Conf. Series: Materials Science and Engineering **406** (2018) 012037 doi:10.1088/1757-899X/406/1/012037

(a) (b) (c) (d)

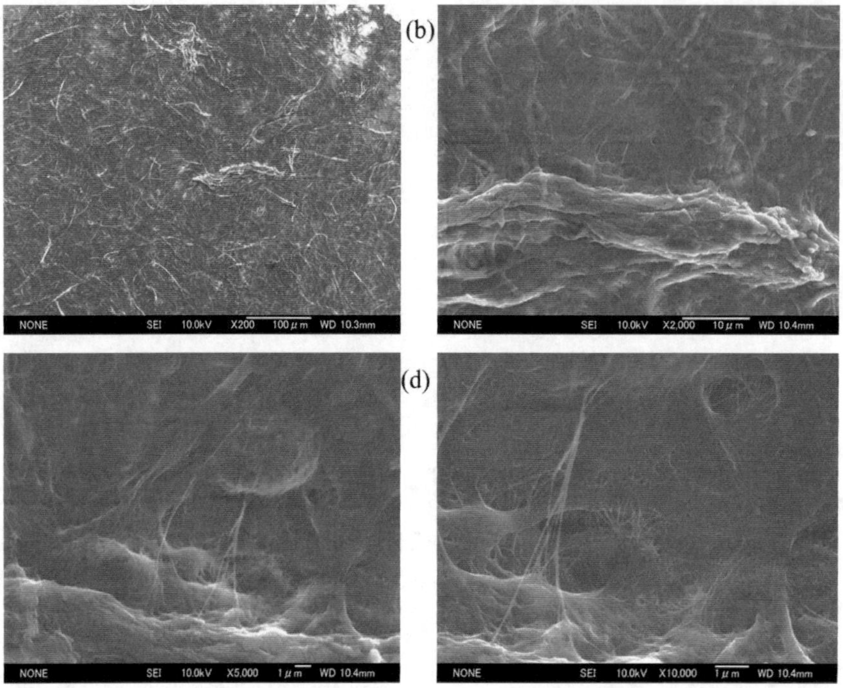

Figure 2. Surficial SEM photograph of neat CNF at magnifications: (a) 200×, (b) 2000×, (c) 5000×, (d) 10000×.

(a) (b) (c) (d)

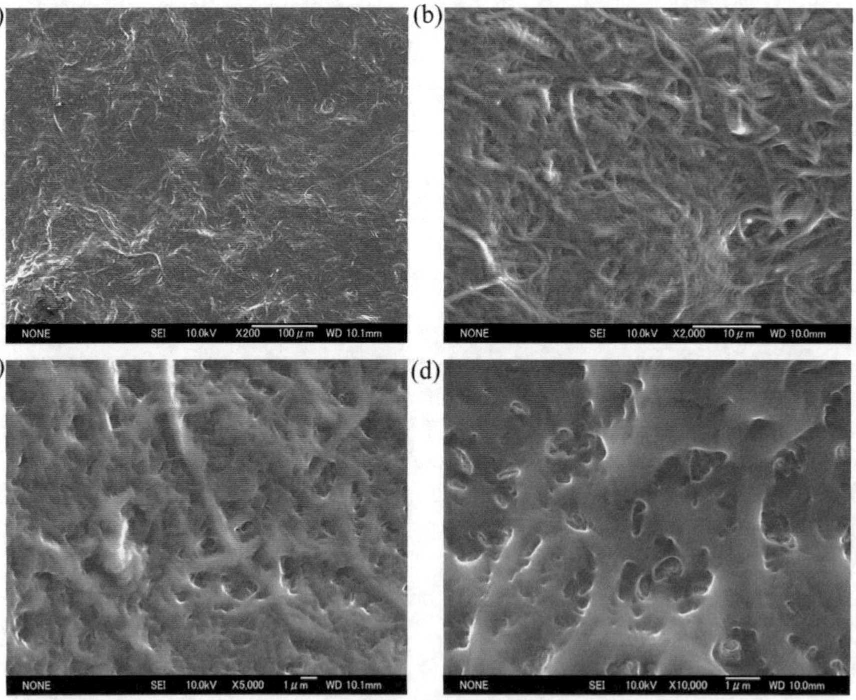

Figure 3. Surficial SEM photograph of 20wt.% CNF reinforced epoxy ester at magnifications: (a) 200×, (b) 2000×, (c) 5000×, (d)10000×.

3.2. Result of Shear tensile test

3.2.1. Stress-strain curve. Representative curves of shear stress-strain curve of Epoxy ester matrix and Alkyd matrix specimen at relative humidity 32.78%, are shown in Figure 4. In order to make further investigation, Maximum stress, strain at Max. stress and deformation energy are sort out in Figure 5 and 6 separately. The maximum stress in both kinds of matrix increase with the CNF addition level, and the effect on epoxy ester matrix seems is bigger than alkyd matrix. As regards to strain at Max. stress in epoxy ester, it is bigger than neat resin when CNF content is lower than 20wt.%. On the other hand, a monotonous decrease with the CNF content is found in alkyd matrix.

The CNFs are expected to be well compatible with epoxy ester matrix than alkyd matrix due to the fact that both the resin and the CNF surface contain hydroxyl functional groups and the hydrogen bonding between phases can occur. This interaction might induce an effective stress transfer from the matrix to CNF. But the stress improvement is at the expense of the elongation values. Tensile shear test of neat resin is very unstable at high humidity, which result in a boost in energy caused by irregular strain. Epoxy ester matrix has bigger deformation energy than alkyd matrix in low humidity, although deformation energy dramatically decreases in high humidity, also has compatible humidity as alkyd matrix. Deformation energy of epoxy ester matrix increases with CNF content, when CNF content is not more than 10wt.% at relative humidity 32.78% but decreases or not change very much at relative humidity 97.30%. In the case of alkyd matrix, CNF content 5wt.% shows the highest deformation energy without reference to humidity.

3.2.2. Observation of fracture surface. Example of fracture surface in tensile shear test bounded with CNF reinforced waterborne epoxy ester, is shown in Figure 7 in ascending order of CNF content (2.5, 5, 10, 20 wt.%). As the content is very small a relatively smooth resin like fracture was seen in CNF 2.5wt.%. At CNF content 5wt.%, CNF became in dominant, inside failure was observed due to the increase of the CNF. When CNF content increase further, there is more chance that CNF present in the interface of coating and the substrate, which result in interface failure and fiber pull out. This is consistent with the relation between deformation energy vs. CNF content in Figure 6 that deformation energy reaches peak when fiber pull out appeared at CNF10wt%.

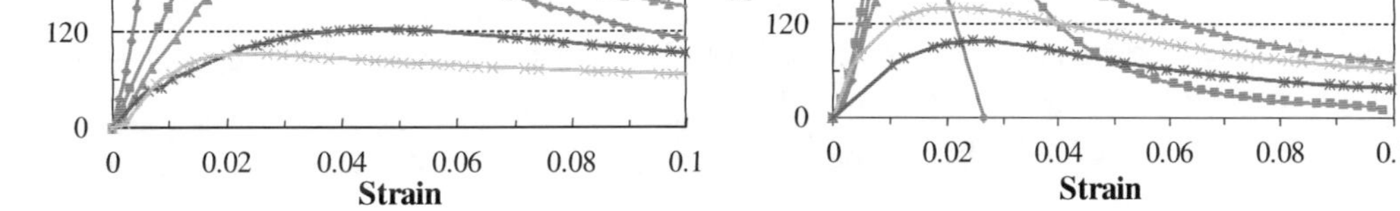

Figure 4. Stress-strain curve of shear tensile test (a) Epoxy ester matrix; (b) Alkyd matrix.

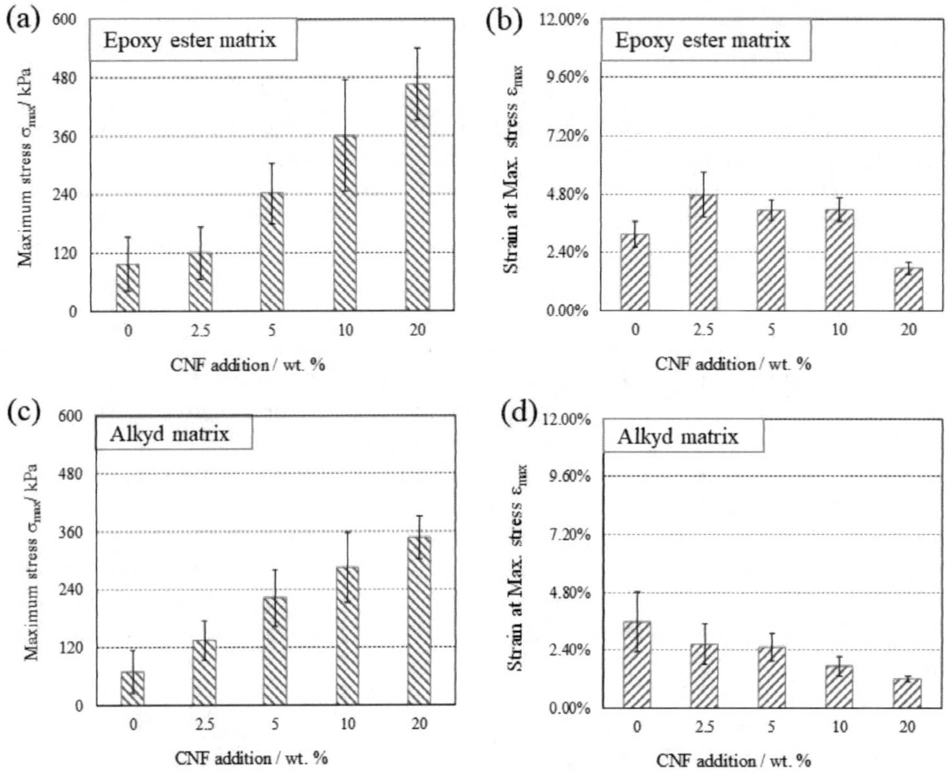

Figure 5. (a), (b) and (c), (d) are Max. strength and Strains at max. strength of Epoxy ester matrix and Alkyd matrix specimen at with various CNF respectively.

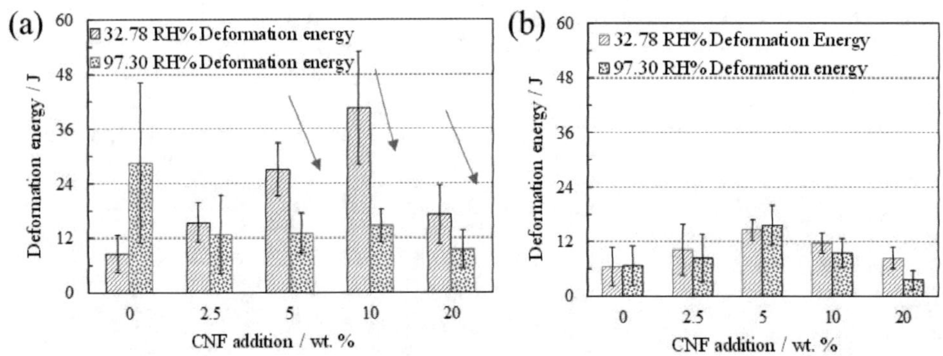

Figure 6. Deformation energy vs. CNF content of (a) Epoxy ester matrix; (b) Alkyd matrix at relative humidity 32.78% and 90%.

IOP Conf. Series: Materials Science and Engineering **406** (2018) 012037 doi:10.1088/1757-899X/406/1/012037

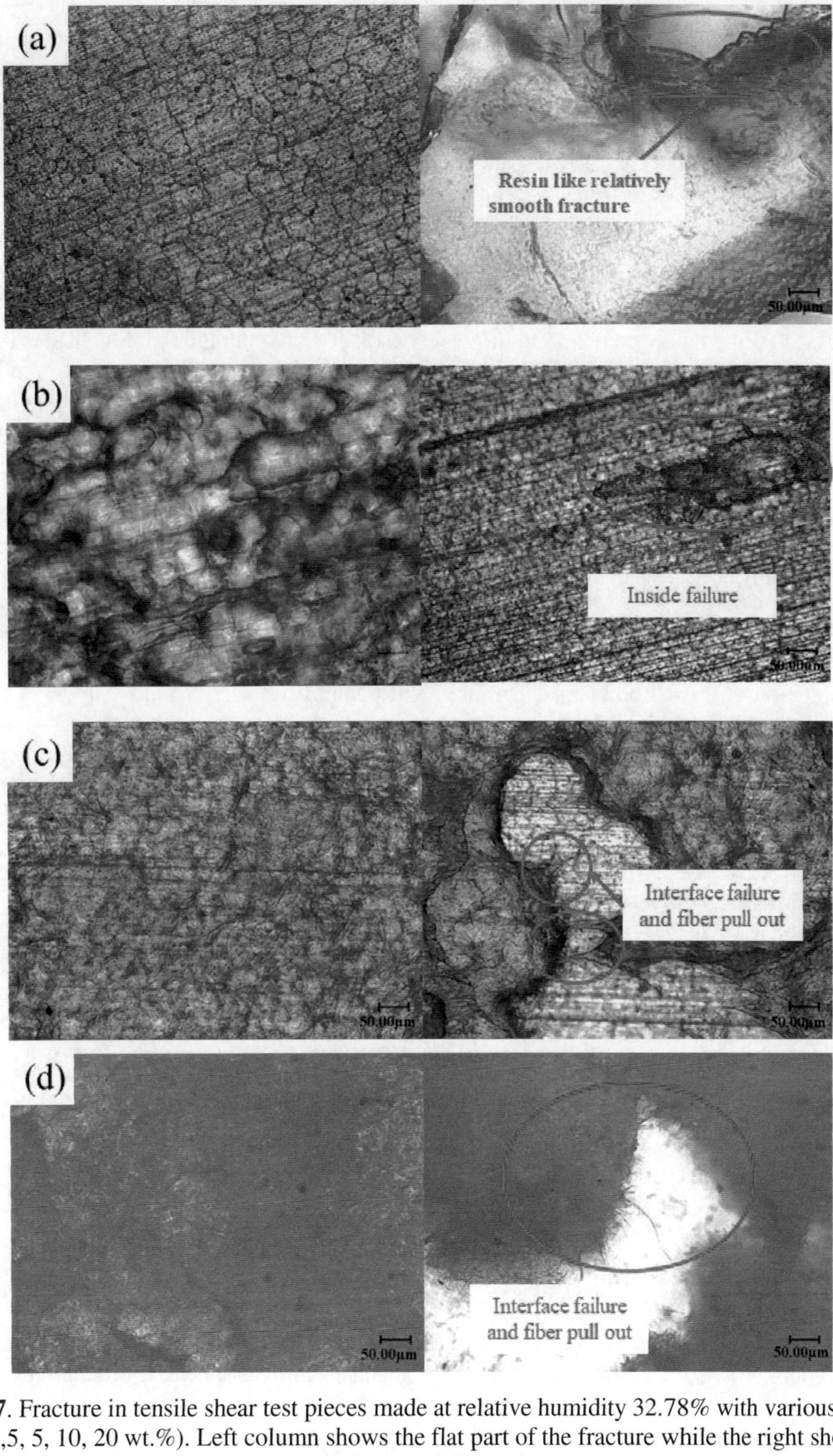

Figure 7. Fracture in tensile shear test pieces made at relative humidity 32.78% with various CNF content (2,5, 5, 10, 20 wt.%). Left column shows the flat part of the fracture while the right shows the part with intense deformation.

IOP Conf. Series: Materials Science and Engineering **406** (2018) 012037 doi:10.1088/1757-899X/406/1/012037

4. Conclusions

The aim of this study is to investigate the effect of CNF on different matrix resin (vinyl-modified epoxy ester resin emulsion and epoxy-modified alkyd emulsion). The study ends with the following observations and conclusions: (1). The maximum stress in both kinds of matrix increase with the CNF content at the expense of their percentage elongation values; (2). Deformation energy reach to the maximum at the content of CNF 10wt.% and 5wt.% for epoxy ester matrix and alkyd matrix respectively. The CNFs are expected to be well compatible with epoxy ester matrix than alkyd matrix due to the fact that it has more hydroxyl functional groups, thus has more hydrogen bonding between phases. This interaction might induce an effective stress transfer from the matrix to CNF respectively. On the other hand, more content of CNF increases the risk of exposure to the interface or sever aggregation occurred, which result in a strength decrease. With the increase of content of CNF, resin like failure, inside failure and interface failure mode are observed in sequence. (3). Relative humidity has considerable influence on the reinforcement effect of CNF in Epoxy ester matrix, but insignificant effect on alkyd matrix. Epoxy ester matrix has bigger deformation energy than alkyd matrix in low humidity, although deformation energy dramatically decreases in high humidity, also have compatible humidity as alkyd matrix.

References

[1] Morsyleide F. Rosa et al 2009 Effect of fiber treatments on tensile and thermal properties of starch/ethylene vinyl alcohol copolymers/coir *biocomposites Bioresource Technology* pp. 5196-5202

[2] Hojo M. et al 2001 Effect of interface control on Mode I interlaminar fracture toughness of woven c/c composite laminates. *JSME Int. J. Ser.A Solid Mech. Mater. Eng.* Vol.44 pp. 573-581

[3] Mohit S et al 2014 Carbon fiber surfaces and composites interphases *Composites Science and Technology* Vol.13 pp. 35-50

[4] Yirong L et al 2009 Increased interface strength in carbon fiber composites through a zno nanowire interphase *Advanced Functional Materials* Vol.19 No.16 pp. 2654-2660

[5] Kinloch A.J. Toughening epoxy adhesives to meet today's chanllenges *Materials Science of Adhesives:How to Bond Things Together* Vol.28 No.6 pp. 445-448

[6] Kinloch A.J. et al 2014 Improving the fracture toughness and the cyclic-fatigue resistance of epoxy-polymer bends *Polymer* Vol.55 pp. 6325-6334

[7] Wernik J.M. et al 2014 On the mechanical characterization of carbon nanotube reinforced epoxy adhesives *Mater. Des.* Vol.59 pp. 19-32

[8] T. D. Chang et al 1982 Studies of epoxy resin systems: Part: Fracture toughness of an epoxy resin: A study of the effect of crosslinking and sub-Tg aging *polym. Eng. Sci.* Vol.22 No.18 pp. 1228-1236

[9] R. A. Pearson et al 1989 Toughening mechanisms in elastomer-modified: Part 3 The effect of cross-link density *Mater. Sci.* Vol.24 pp. 2571-2580

[10] A. C. Garg, failure Mechanisms in toughened epoxy resins – Areview *Compos. Sci. Technol.* Vol.31 No.3 pp. 179-223

[11] A. Martone et al 2010 Reinforcement efficiency of multi-walled carbon nanotube/epoxy nano composites *Compos. Sci. Technol.* Vol.70 No.7 pp. 1154-1160

[12] Y. Zhao et al 2013 Simultaneously enhanced cryogenic tensile strength and fracture toughness of epoxy resins by carboxylic nitrile-butadiene nano-rubber *Composites Part A* Vol.66 pp. 178-187

[13] Y. Zhou et al 2006 Fabrication and evaluation of carbon nano fiber filled carbon/epoxy composite *Mater. Sci. Eno.* A Vol.426 No.1-2 pp. 221-228

[14] X. Wang et al 2013 An investigation of the mechanism of graphene toughening epoxy *Carbon* Vol.65 pp. 324-333

[15] Lewis Greenspan 1977 Humidity Fixed Pontes of Binary Saturated Aqueous Solutions *Journal of Research of the National Bureau of Standards- A. Phsics and Chemistry* Vol.8A No.1 pp. 89-96

IOP Conf. Series: Materials Science and Engineering **406** (2018) 012038 doi:10.1088/1757-899X/406/1/012038

Self-reinforced biobased composites based on high stiffness PLA yarns

Buyle G[1*], Van der Schueren L[1], Beauson J[2], Goutianos S[2], Schillani G[2], Madsen B[2]

[1] Centexbel Textile Research Center, Gent, Belgium
[2] Department of Wind Energy, Technical University of Denmark, Roskilde, Denmark

*guy.buyle@centexbel.be

Abstract. In self-reinforced polymer composites (SRPC) the same polymer material forms both the reinforcing fibre and the matrix phase. The current project aims to develop a biobased alternative for these composites using polylactic acid (PLA). Both the development of the reinforcing fibres, the modelling and production of the composites are being studied. The stiffness of the PLA filaments could be increased to 9 GPa by optimising processing parameters during extrusion. After consolidation to composites, promising results are obtained showing that the stiffness of the PLA SRPC can match the requirements of currently used commercial self-reinforced polypropylene composites (ca 4GPa).

1. Introduction

There is a worldwide increasing demand for replacing fossil-based with biobased raw materials for the production of polymers, leading to a significant growth of bioplastics in terms of technological developments. However, there are still some drawbacks which prevent their wider commercialization in many applications. This is mainly due to their low mechanical performance and durability when compared to conventional polymers. Enhancement of these properties remains a significant challenge for biobased polymers. Therefore, there is a need to develop biobased, sustainable polymeric materials with high stiffness, high impact and high durability without impairing recyclability and at a similar price level of non-biobased solutions.

The development of self-reinforced polymer composites (SRPC) is proposed as a means to enhance the mechanical performance of biobased polymers. In such SRPCs the same polymer material forms both the fibre reinforcing and the matrix phase. These materials use a highly-drawn polymer fibre to reinforce a matrix of the same polymer family. SRPCs offer many advantages compared to standard fibre reinforced composites such as: lightweight, high specific stiffness and strength, high impact resistance, excellent fibre-matrix adhesion, inherent thermoformability and a good recyclability thanks to the mono material composite. These materials have a high potential for a variety of applications like house appliances, automotive parts with a need of a high impact resistance, body armours or sports equipment.

Fossil-based SRPCs, mainly polypropylene based, are already available on the market. A biobased alternative, however, does not yet exist. Among biobased polymers, polylactic acid (PLA) is an ideal material for the preparation of SRPC, as it can be produced with controlled molecular configuration (molecular weight, molecular alignment, crystallinity, ratio between L- and D-lactic acid etc.), resulting in a wide range of mechanical and thermal properties, including different melting points.

The above described demand for high performance biobased materials is exactly the challenge being tackled within the H2020 project BIO4SELF which aims at fully biobased SRPCs. To produce such biobased SRPCs two PLA grades are combined: a low melting temperature (Tm) one to form the matrix and an ultra-high stiffness and high Tm one to form the reinforcing fibres. It was chosen to combine the low and high Tm PLA by producing hybrid yarns consisting of low Tm matrix fibres and high Tm reinforcement fibres. The latter remain intact during further processing at a temperature above the melting temperature of the low Tm PLA but below the one of the high Tm PLA, resulting in a fully biobased composite material (see Figure 1).

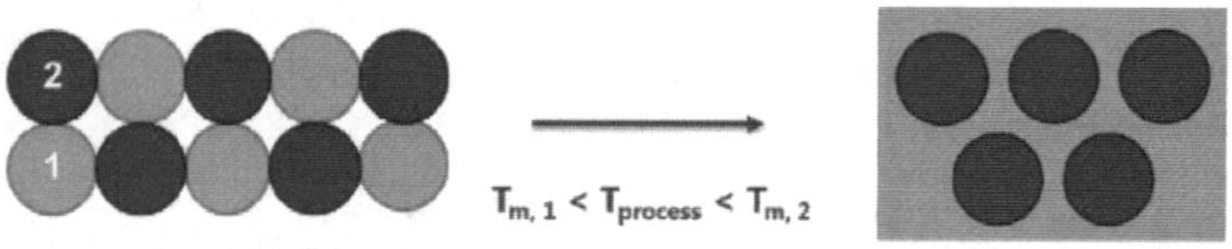

Figure 1. Production of self-reinforced composite (1: low Tm PLA, 2: high Tm PLA).

To successfully realise the BIO4SELF challenges, the complete value chain has to be actively involved. The consortium set-up within the project covers all required expertise and equipment going from the production of functionalised materials, over the melt spinning, finally resulting in composite end products. The current paper mainly focuses on the development of high stiffness PLA filaments and the modelling and production of simplified composite parts.

2. Description of materials and experimental procedure

2.1. Production of PLA filaments and yarns

The high Tm PLA material (with a Melt Flow Index of 10 g/10min at 190 °C) was purchased from Total-Corbion-PLA. The low Tm PLA material (with a Melt Flow Index of 15 g/10 min at 210 °C) was purchased from NatureWorks. At lab scale we used a multifilament extruder from Busschaert Engineering (named Spinmaster) at a throughput of 4 kg/hour. The extruder temperature was set at 240 °C. A yarn consisting of 48 filaments and a total titer of 240 dtex was produced. At industrial scale, we used a multifilament extruder in Svit (VÚCHV a.s) at a throughput of 18 kg/hour. The temperature was set at 245°C. A yarn consisting of 100 filaments and a total titer of 340 dtex was produced.

The high and low Tm PLA multifilaments, with a ratio 1:1, were then commingled in order to obtain a hybrid yarn consisting of both PLA grades.

2.2. Consolidation to composites

Unidirectional PLA composites were manufactured by winding the commingled yarns on a metal frame. The winding was dried overnight under vacuum chamber at 35°C and then press consolidated in a two-step process. First, the material is heated to 165°C under vacuum for 10 min. The heating is applied by the contact of two metal plates and controlled by two thermocouples. During heating, the low Tm PLA filaments are melted to form the matrix. Then, the winding is rapidly moved to the second section, where a pressure of 2 MPa is applied, and cooled down to 30°C for 1 minute. The produced composite plates have dimensions of 400 x 250 x 2 mm.

2.3. Mechanical characterization of composites

In order to determine the mechanical properties of the composites, the manufactured composites were tensile tested at different reinforcing filaments orientation, i.e. different high Tm PLA multifilaments orientation. Specimens for tensile tests, with dimensions of 180 x 20 mm, were cut from the manufactured unidirectional PLA composite plates. The specimens were cut at a range of angles to the winding direction ($\theta = 0^\circ$, 15°, 45°, 90°). Four specimens were prepared for each reinforcing filaments orientation. The static tensile tests were performed in an Instron tensile testing machine with a load cell of 25 kN. The cross-head speed was 10 mm/min. The strain was measured by two extensometers centred and aligned on each side of a specimen. The stiffness was determined in the strain range from 0.05 to 0.25 %.

To verify that the high Tm PLA multifilaments do not melt during the press consolidation, the measured values of composite stiffness, for $\theta = 0^\circ$, were compared to theoretical predictions using the rule of mixture:

$$E_c = V_f E_f + (1 - V_f) E_m \tag{1}$$

where E_c is the composite stiffness, V_f is the volume fraction of high Tm PLA multifilaments, E_f is the stiffness of the reinforcing filaments, i.e. the stiffness of the high Tm PLA multifilaments, and E_m is the stiffness of the matrix, i.e. the stiffness of the low Tm PLA filaments.

The composite stiffness of composites tested at an angle θ, is predicted by [1]:

$$\frac{1}{E_\theta} = \left[\frac{1}{E_{11}}\right] \cos^4\theta + \left[\frac{1}{G_{12}} - 2\frac{\nu_{12}}{E_{11}}\right] \cos^2\theta \sin^2\theta + \left[\frac{1}{E_{22}}\right] \sin^4\theta \tag{2}$$

where E_θ is the composite stiffness, θ is the orientation of the reinforcing filaments, i.e. the high Tm PLA multifilaments, ν_{12} is the major Poisson ratio, E_{11} is the composite stiffness for $\theta = 0^\circ$, E_{22} is the composite stiffness for $\theta = 90^\circ$, G_{12} is the shear modulus. The direction 1 refers to the winding direction and direction 2 refers to the perpendicular direction to the winding direction.

3. Results

3.1. Optimisation of the stiffness of PLA filaments

The mechanical performance of the PLA SRPC will mainly depend on the performance of the high Tm reinforcement fibres. Therefore, the challenge for the high Tm filament extrusion is to produce filaments with a maximum stiffness. Parameters investigated during the multifilament extrusion process were the influence of the capillary length over diameter ratio (L/D ratio) of the spinneret, the cold draw ratio and the comparison of a one- and two-step extrusion process.

When the extrusion and drawing was performed in one step, it was found that a high L/D ratio was beneficial for the filament's stiffness, going from 7.7 GPa for an L/D ratio of 2 to 8.7 GPa for an L/D ratio of 4. In addition, also a high cold draw ratio (up to 6 times drawing) resulted in a higher stiffness (Table 1). In general, the multifilaments, with a titer of 5 dpf (dtex per filament), reached a stiffness of approx. 9 GPa.

Table 1. Effect of L/D ratio and cold draw ratio on modulus of PLA yarn.

L/D ratio spinneret (-)	Cold draw ratio (x)	Modulus (GPa)
2	6.1	7.7
2.6	6.1	7.3
4	6.1	8.7
4	3.2	7.8
4	1.8	7.2

Next to the one-step process, also a two-step was investigated in which the extrusion was performed in a first step (at high take-up speed, 3000 m/min) followed by a separate drawing step (at 700 m/min). This also benefits the development of high stiffness PLA yarns. After successful parameterization at Centexbel, the extrusion process was upscaled at industrial scale by the company Fibrochem. This confirmed the advantages of a two-step versus a one-step process, resulting in a higher Young's Modulus (see Figure 2).

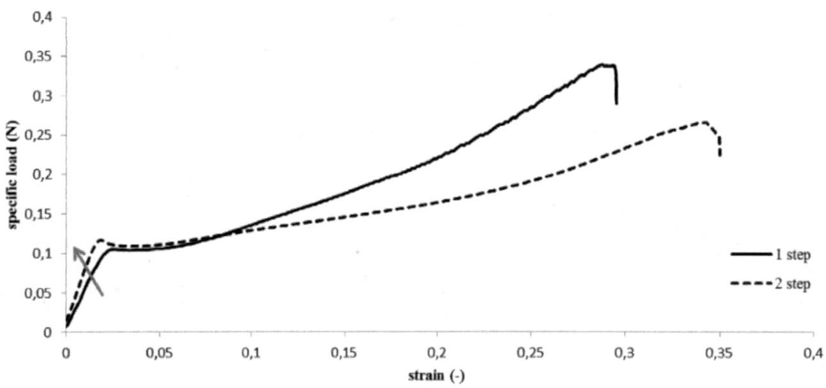

Figure 2. 1-step versus 2-step multifilament extrusion, arrow indicates increase in modulus.

Table 2. Filament properties produced with different spin finish.

	Oil based spin finish		Water based spin finish	
Properties	PLA high Tm	PLA low Tm	PLA high Tm	PLA low Tm
Stiffness [GPa]	7.6 ± 0.3	3.5 ± 0.1	7.5 ± 0.2	3.6 ± 0.1
Strength at break [MPa]	412 ± 67	66 ± 3	329 ± 51	28 ± 5

During multifilament extrusion, a spin finish needs to be applied on the yarn to regulate the friction among the filaments and the filament and the machine parts. That way, the built-up of electrostatic charges on the filament surface is prevented. In order to determine the effect of spin finish on the composite performance, composites were manufactured using multifilaments with two different spin finishes: an oil based one and a water based one. High and low melting temperature PLA multifilaments were produced with the two different spin finishes. As shown in Table 2, the strength at break is clearly influenced by the type of spin finish.

3.2. Consolidation to composites and the effect of spin finish on composite performance

In a composite material, the interphase region between the reinforcing filaments and the matrix material is crucial for the mechanical performance of the final composite. Some adhesion between the two constituents is required in order to obtain good composite mechanical properties. In a self-reinforced composite material one could argue that as the two constituents, the fibres and the matrix, are made of the same material, the adhesion between them should be strong. However, as spin finish is not specifically designed to promote adhesion between the reinforcing filaments and the matrix material, the presence of spin finish at the surface of the filaments might affect the fibre/matrix adhesion in a negative way.

The results of the static tensile testing are presented in Figure 3 and Figure 4. The composites are named "Oil" or "Water" depending on the spin finish used.

Regarding the composite stiffness, there are no large differences in the results obtained for the two types of composites "Oil" and "Water" for any θ angle. The experimental results obtained for $\theta = 0^{\circ}$ were compared to the theoretical estimation obtained with the rule of mixture (Equation (1)) and were found to be in good agreement assuming a 50% fibre volume fraction. The theoretical prediction was made using the filament properties presented in Table 1 and the composite stiffness is equal to 5.6 GPa. Equation (2) was used to predict the composite stiffness obtained for the composite tested at different angles. The Poisson's ratio was set to 0.25 and the shear modulus was adjusted to best fit the experimental results obtained at 15° and 45°. For the "Oil" composite the shear modulus was set to 1.3 GPa and for the "Water" composite it was set to 1.4 GPa. It can be seen on Figure 3 that the experimental stiffness obtained for the composites for $\theta = 15^{\circ}$ is clearly above the theoretical curves. This may be due to the sample preparation, where specimens with a lower filament angle could have been cut out. The stiffnesses obtained at 45° are located on the theoretical curves. In conclusion, it is not possible to detect any effect of the spin finish on the composite stiffness. This is in line with the results reported by Alcock et al., where it was demonstrated that the interface properties do not influence the composite stiffness [1].

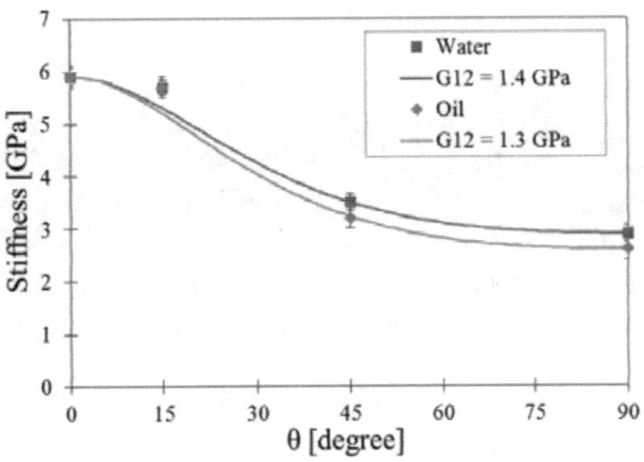

Figure 3. Stiffness of PLA composites tested at different loading directions.

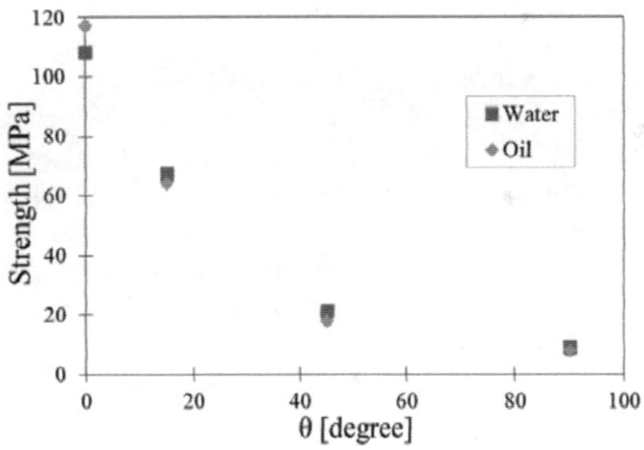

Figure 4. Strength of PLA composites tested at different loading directions.

Regarding the composite strength, similar conclusions can be drawn. All the results obtained are close to each other, except for the composite tested at $\theta = 0^{\circ}$. The strength values for the "Oil" composites are slightly higher than the "Water" composites. This is probably due to the difference in post yield response of the PLA_{HM} multifilaments. Spin finish has an effect on the manufacturing of the PLA filaments, which will result in filaments with different properties. Figure 5 and 6 show the stress-strain curve of the composite tested at $\theta = 0^{\circ}$, where it is possible to see the different behaviour around the yield region. Based on these results it is not possible to observe any effect of the different type of spin finish on the composite strength properties.

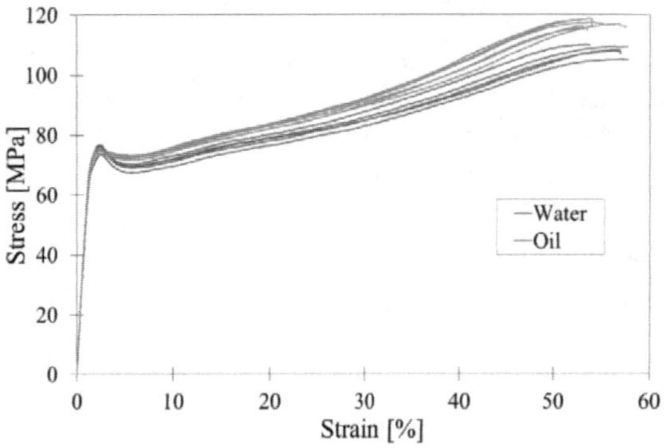

Figure 5. Stress-strain curve of the composite test for θ=0°.

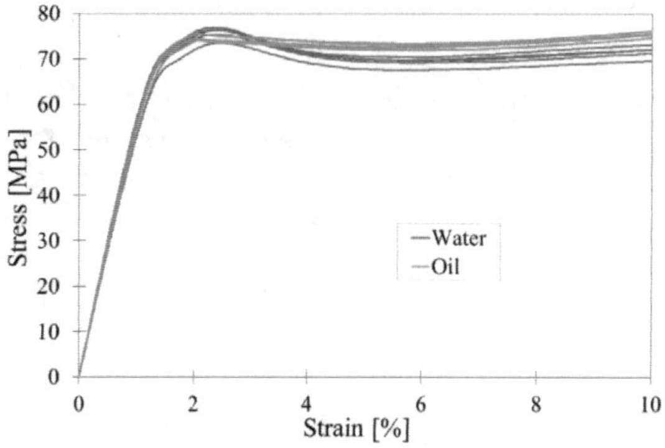

Figure 6. Detail view of the yield area of the stress-strain curve of the composite test for θ=0°.

In conclusion, no effect of the spin finish type on the composite performance was observed. The oil based spin finish was however selected, as PLA is sensitive to water and an oil based spin finish tends to result in higher mechanical properties of the PLA yarns.

Next to the production of unidirectional composites, the commingled yarns produced with the oil based spin finish were processed into a woven fabric which was consolidated to a bidirectional composite plate. Promising results were obtained that show that the stiffness of the PLA composite can match the requirements of currently used commercial self-reinforced polypropylene (ca 4GPa). As a next step, prototype composite parts for automotive and home appliances will be produced as demonstrators to illustrate the much broader range of industrial applications, e.g. furniture, construction and sports goods.

4. Conclusions and outlook

The results show that high stiffness PLA filaments can be produced on the standard multifilament extrusion equipment. These filaments were subsequently combined with low Tm PLA filaments into a hybrid yarn after which unidirectional and bidirectional SRPCs were manufactured. The effect of spin finish, needed to produce and process the multifilaments, on the composite properties was shown to be small. Overall composites with good mechanical performance, comparable to the current fossil based SRPCs, were obtained.

We will work on further optimisation of the PLA yarns, towards higher temperature resistance and higher stiffness e.g. by improvement of the 2-step process and the application of an additional heatsetting step. Future investigations will also include studying the adhesion between PLA filaments without any spin finish.

Acknowledgments
The BIO4SELF project has received funding from the European Union's Horizon 2020 research and innovation programme under Grant Agreement No 685614.

References

[1] B. Alcock, N.O. Cabrera, N.-M. Barkoula, J. Loos, T. Peijs, "The mechanical properties of unidirectional all-polypropylene composites", Composites: Part A 37 (2006) 716–726

Analysis of the natural fiber reinforced polymer: composite fiber rods (FRPCY)

M A El Messiry[1], R M El Deeb[2]*

[1,2]Textile Engineering Department, Faculty of Engineering, Alexandria University, Egypt

* Rania_eldeeb@yahoo.com

Abstract. Fiber reinforced polymer composites (FRPC) can be used in new or repaired reinforced concrete structures. Fiber reinforced polymer composite yarn (FRPCY) can be formed as rods of different diameter and length to be used in the formation of concrete structures due to their superior light weight, corrosion resistance, the lack of material ductility and energy absorbing capabilities. The formation of these FRPCY rods using natural fibers yarns is tried in this work. Due to the variation of the fiber volume fraction of the yarn along its length, it will directly reflect on the fiber volume fraction of the formed composite. In this work, the relation between the yarn volume fraction and the composite fiber volume fraction were studied, taken into consideration its variation along the yarn length. The formation of these rods using flax yarns is tried in this work and their mechanical properties are investigated.

Keywords: Yarn volume fraction, Composite fiber rod, Natural fiber composites

1. Introduction

Replacement of the steel reinforcement in concrete structures with more corrosion resistant substitutes such as composites is rapidly becoming a more economical option for construction facilities worldwide [1, 2]. The fiber Reinforced Polymer (FRPCY) composite can be applied in repaired reinforced concrete structures. The fibers can be used in form of single, plied or braided yarns, with different sizes starts from 0.5 mm [3]. The combination of braiding and pultrusion processes has been also named pull-braiding, braided-pultrusion and braid-pultrusion [4-7]. Civil engineering applications, namely for concrete internal reinforcement, composite rods were produced, varying the type of fibers used as a core reinforcement of a polyester braided structure [8]. To improve the adhesion between the FRP rods and the concrete several types of ribs are designed on the outer layer of the rods [9]. During this process the internal architecture of the reinforcement material is susceptible to large structural changes as it is forced to deform to allow for the reduction in its volume [10, 11].

Natural materials have many advantages when compared to traditional materials and are less harmful to the environment. The formation of natural fiber reinforced composite rods present a solution to deterioration problem. In the case of staple fiber yarns, theoretically, the fiber volume fraction varied between 0.76 – 0.9, depending on the fiber arrangement. The fiber obliquity in the twisted yarns increases the compactness of the yarn, which in its turn depends on the spinning technology [3]. It was revealed that the packing density of the fibers in the yarn cross section varied from radios to another with its highest value at the center of the yarn. The average packing density of the yarn depends of several factors: yarn count, spinning system, fiber properties, twist factor and yarn diameter. The distribution of the fiber packing density along the yarn affects the mechanical properties of the yarns and the fabric [11]. Consequently, The FRPCY fiber volume fraction and their mechanical properties are discussed in this work.

2. Materials and methods

Formation of the fiber reinforced polymer composite rod using different counts of flax yarns with following specifications given in table (1). The yarn composite formation through the extortion of the polymer in a paper tube under pressure with the yarn located at the middle of the tube with an inner diameter of 4 mm. the pressure was adjusted to high enough to insure the production of rod composite with no defects. All the specimens' rods were dried and cured for seven days in standard atmosphere before applying the mechanical tests. Figure (1) illustrates the shape of the rods after formation. After the composite formation the paper tube was removed to leaving a solid.

Table 1. Yarn properties

Yarn count Tex	Yarn diameter mm (CV%)	Yarn strength MPa	Young's Mod. (E) GPa
271	0.851(10.66%)	178.9	4.21
303	0.759(18.96%)	154.7	3.44
385	0.649(14.65%)	115.7	2.57
450	0.524(24.73%)	111.55	2.12
500	0.447(17.04%)	99.14	1.21

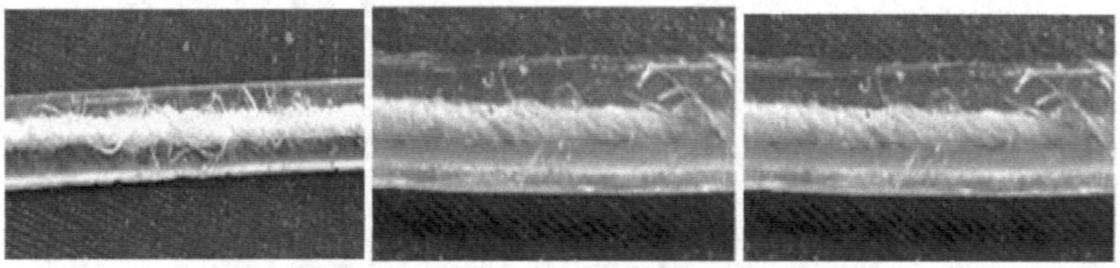

Figure 1. Composite Fiber Rods.

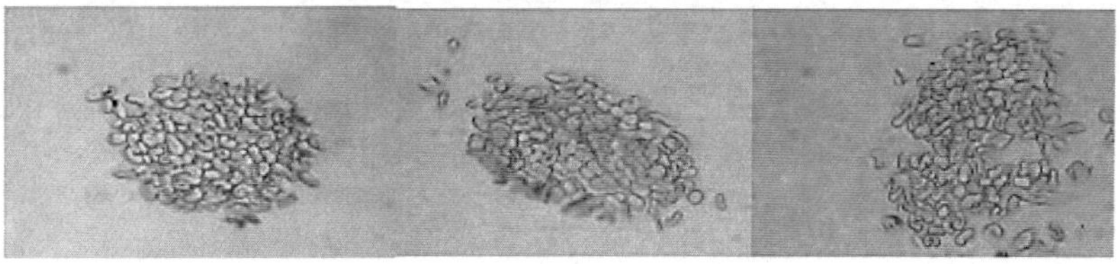

Figure 2. Various type of packing of spun yarns.

3. Composites mechanical properties

Tensile and bending tests were applied on the composite specimens. The tests were performed on Mecmesin (MultiTest5-xt) testing machine with a uniaxial load with load cell 5KN and crosshead speed 1 mm/min. The tensile test was according to ASTM D3039. The gauge length for the tensile test was 40 mm. The tensile results were analyzed to calculate the strength (MPa), strain, Young's modulus, and the work of rupture. Bending tests were applied with the 4 – point bending test was according to ASTM C348-8. The readings of bending forces and deflections were recorded with Newton and millimeter respectively. Five specimens were test for each type of composite rods and the average readings were recorded for both test.

4. Results and discussions

Due to the variability of the yarn diameter and the variability of the fiber radial packing with maximum value at the center of the yarn, its value depends on the spinning system levels of twist and the fiber parameters [3]. The yarn diameter is varied from section to section along the yarn length as well as the packing density. Generally, yarn is not circular and closer to oval shape which represents the highest likely hood as shown in Figure (2) [11]. The open packing arrangement has $V_{fy} = 0.76$. The packing of the spun yarns may be classified as highly packed to very low packing [3, 11]. Assuming "a" is the major axis and "b" is the minor axis of the yarn cross section, then the equivalent diameter can be expressed as, $D_{yarn} = D_{measured}\, \alpha^{0.5}$, $D_{measured}$ is the yarn diameter measured by one dimensional sensor.

For ring spun yarns: assume the ratio $\alpha = (b/a)$, 0.67-0.89, $D_{yarn} = 0.82 - 0.94\ D_{y\,measured}$. It was found that the average fiber volume faction in the yarn is given by [11], $V_{f\,yarn} = 0.64\ V_{max}$. In the case of FRPCY, the yarn is compressed during the composite formation.

The compactness of the yarn under the pressure will leads to the deformation of its cross section with increasing of the fiber packing density to reach its maximum value at the yarn center with reduction towards the outer layer [11]. In the case of rod type yarn reinforced polymer composite the outer diameter of the composite will be constant D_c and formed by pressing the yarn radially and its diameter $D_y(i)$ at any cross section will squeezed radially to D_m. Limit packing density is defined from closest arrangement of fibers in cross section empirical limit of fiber packing density is about 0.7- 0.8 at the core of yarn where fibers are squeezed by fibers in the adjacent fibrous layers. In the outer layers is packing density smaller and is decreasing towards the periphery. The average packing density V_{yarn} is in close connection with the yarn diameter which is not the case of spun yarns then at any cross-section i.

$$V_{yarn}(i) = 4\ tex_y /\ \pi \rho_f\ D_y(i)^2 \tag{1}$$

Figure (3) shows the change of the yarn diameter along the yarn length. In the case of apply pressure on the yarn during formation of natural fiber reinforced polymer composite rods the diameter is reduced according to the following equation:

$$\beta(i) = (D(i)_{y\ compact}\ / D(i)_{y\ initial}) = a\ P^{-b} \tag{2}$$

Where: $D(i)_{y\ initial}$ is the initial yarn diameter in free stat at i^{th} cross section, a, b are constants, p is the pressure on the yarn cross section. The experimental measurements indicated that, a =458861tex$^{-2.452}$, b = 4 tex$^{1.8856}$ 10^{-6} for flax yarns, which are used in this work. Figure (4) confirm the sounds of the calculation by equation (2).

5. The fiber packing density of the composite fiber rods

Based on the above, it can be assumed that measured mean diameter of the yarn D_y cannot be used to determine FRPCY volume fraction. Assume that the final composite yarn has a constant diameter D_c, then the yarn at any cross section $D_y(i)$ and the fiber packing density at that section is $V(i)_{f\ yarn}$. The fiber packing density of the composite yarn $V(i)_{fc}$ will be:

$$V(i)_{fc} = (D(i)y\ /Dc)^2\ V(i)_{f\ yarn} \tag{3}$$

Due to the compactness of the yarn the value of $V(i)_{fyarn}$ at any cross section reaches its limited value $V(i)max$ consequently,

$$V(i)_{fc} = (D(i)y\ /\ Dc)^2\ V(i)max \tag{4}$$

This indicates that the variation of yarn diameter directly reflected on the fiber volume fraction of the composite yarn. The fiber volume fraction of compacted yarns varied along its length can be assumed to be normally distributed [11]. Figure (5) illustrate the variation of the yarn fiber volume fraction along the FRP rod with deferent ratio $\beta = Dc\ /\ Dy$.

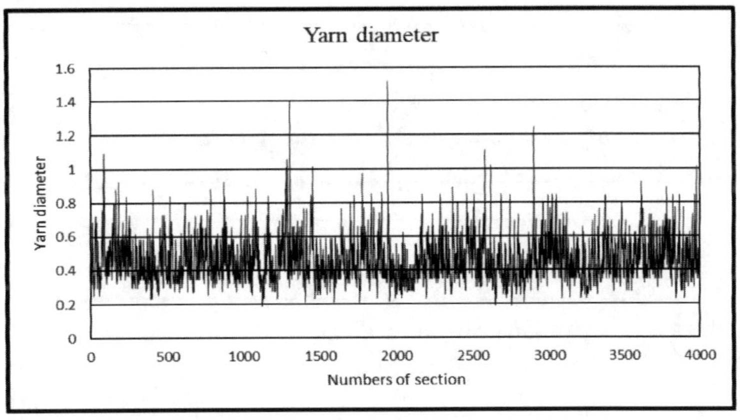

Figure 3. Measured yarn diameter versus its length.

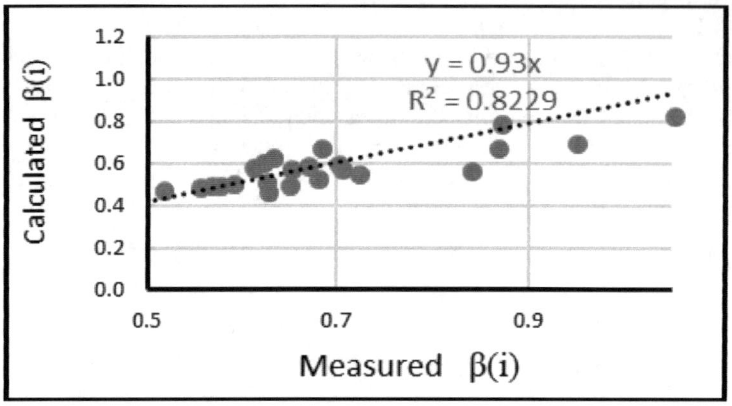

Figure 4. Measured and calculated ($D_y$compact/ $D_{yintial}$).

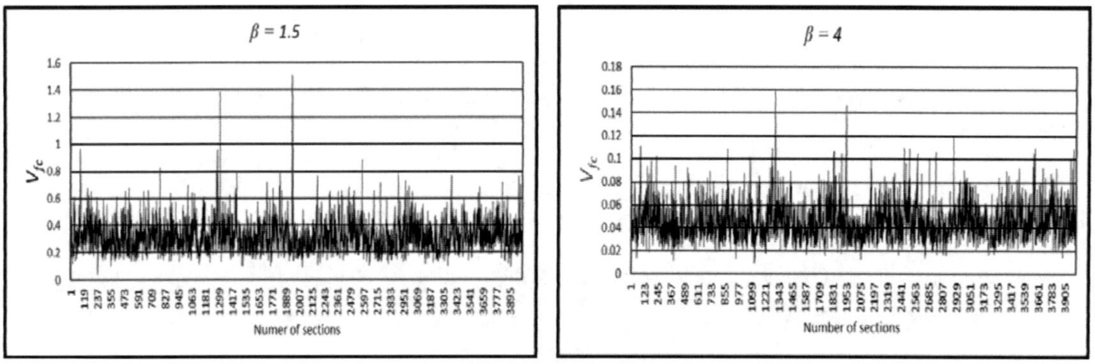

Figure 5. Fiber packing density of the composite rods V_{fc} along its length.

The change of the average value of fiber voluem fraction of the yarn will result in chang final FRPCY rod average fiber packing density as shown in figure (6). Figure (7) illustrated the relation between the average composite fiber voloum fraction (V_{fc}) and that of the yarn (V_{fy}), for diffrent values of (β).

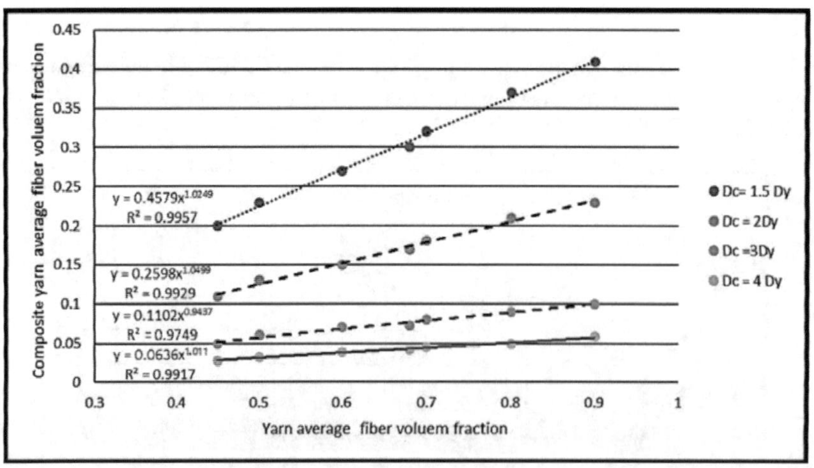

Figure 6. V_{fc} of FRPCY versus V_{fyarn}.

IOP Conf. Series: Materials Science and Engineering **406** (2018) 012039 doi:10.1088/1757-899X/406/1/012039

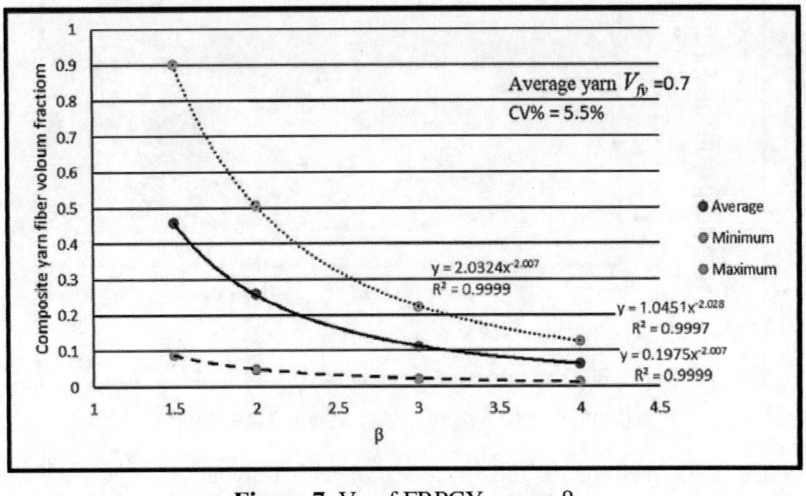

Figure 7. V_{fc} of FRPCY versus β.

Due to the variation of the fiber volume fraction of the yarn along its length, it will directly reflect on the fiber volume fraction of formed composite. In this work, the relation between the yarn fiber volume fraction and the composite fiber volume fraction was studied, taken into consideration its variation along the yarn length. The effect of the rod /yarn diameter ratio β, yarn fiber volume fraction variation was studied, taking into consideration the variability of yarn diameter. The relation between the average composite fiber volume fraction V_{fc} and that of the yarn (V_{fy}), for diffrent ratio β, indicated that the V_{fc} is inversely proportional to β^2 and the range of V_{fc} variation reduced as β increased. The value of average V_{fc} , in the case of V_{fy} 0.7, reaches 0.37 for β 1.5 and 0.048 for β 4. These values are function of yarn diameter variabily. Moreover for the value of β, the yarns compression during formation of the rod should take into consideration.

6. Mechanism of failure of the fiber reinforced polymer composite yarns

The FRPCY rods using different counts of flax yarns (271,303,385,450, 500 *tex*) were formatted through the extrusion of the polymer in a paper tube under pressure with the yarn located in the middle of the tube with an inner diameter of 4 mm. The pressure was adjusted to be high enough to insure the production of rod composite with no defects. Results on mean tensile strength, presented in Figure (8) show in most cases statrts with the failure of the polymr matrix followed with the straining of the core yarn till it breakes under the increasing loads as illustrated in figure (8). It was found that the pull off strength is given in figure (9).

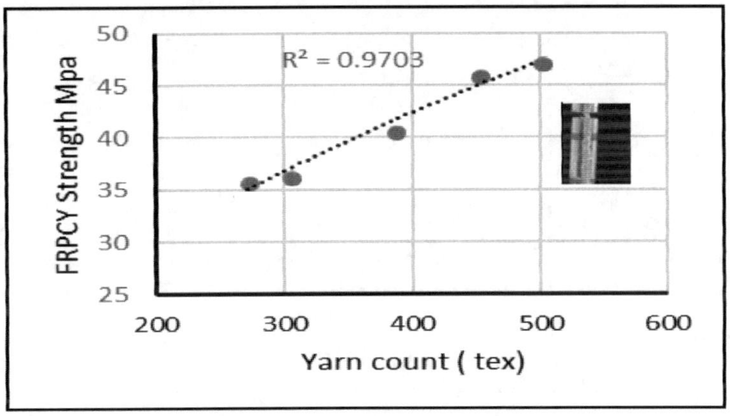

Figure 8. FRPCY strength versus yarn count.

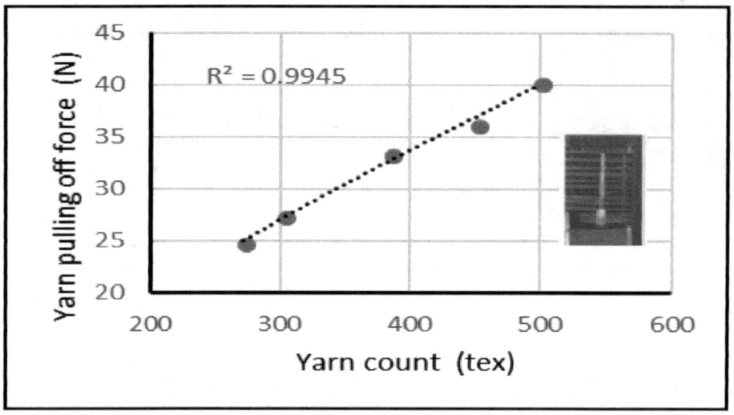

Figure 9. Yarn pulling force versus yarn count.

It was reveild that the strength od the rods increased as the yarn count (tex) increases due to the better anchering of the finers in the matrix that increase the interficial shear forces. The effect of the varation of count of the yarns used on the final FRPCY are given in figures (10-11).

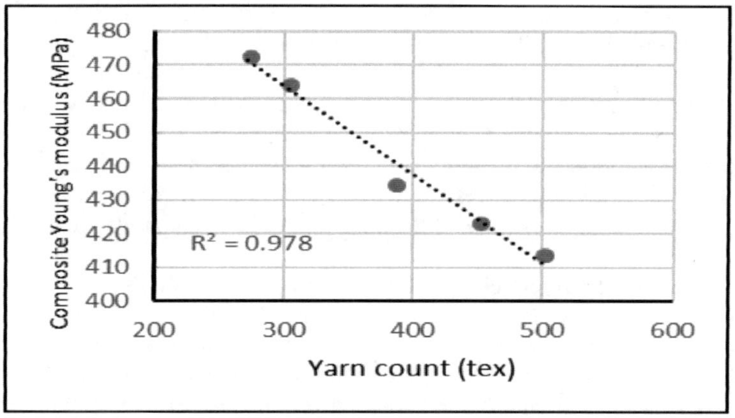

Figure 10. FRPCY modulus versus yarn count.

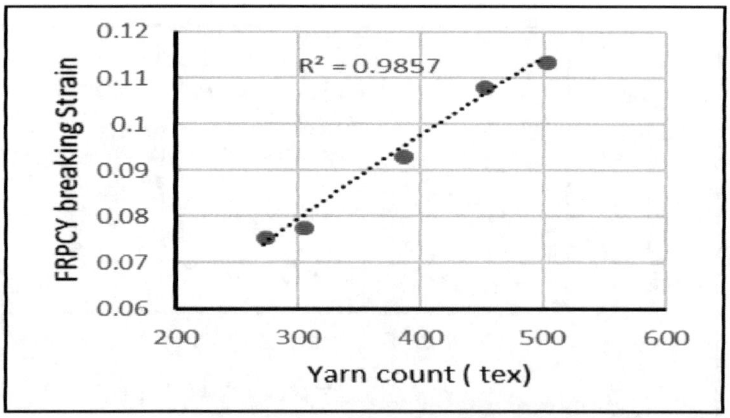

Figure 11. FRPCY breaking strain versus yarn count.

The work of rupture of the composite rods are found to be increased as yam count (tex) increase as shown in figure (12). It is well known that the bending stiffness of the yarns is very low. However, when the composite rod was tested it was revealed that the maximum deflection of the rod in 4-point bending test increased as a function of the yarn count (tex), which indicates that the FRPCY become more flexible with the increase of the yarn count (tex), figure (13).

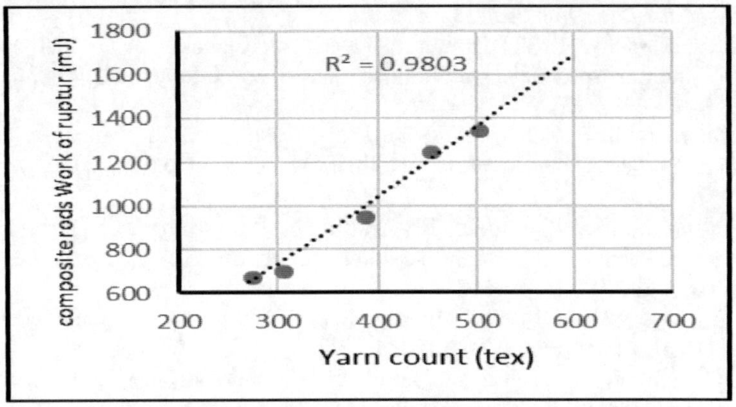

Figure 12. FRPCY work of rupture versus yarn count.

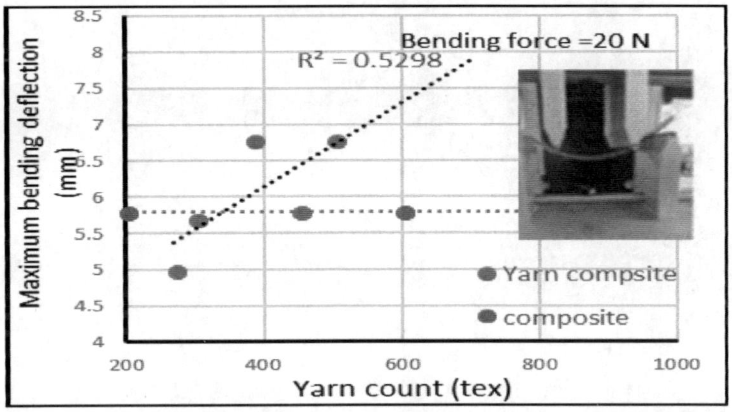

Figure 13. FRPCY bending deflection versus yarn count.

It was found that the mechanical properties of the flax yarn reinforced polyester polymer composite were improved by using yarns of coarse counts even at the low values of fiber volume fraction.

7. Conclusions

To enhance the resistance to bacterial and enzymatic degradation of flax fibers, a protection by low molecular weight polyester matrix was applied for its physical protection and bacteriostatic properties. The fabrication of FRPCY rod of flax yarns gives the possibilities to be used in concrete mixture. The effect of the rod /yarn diameter ratio, yarn fiber volume fraction on the FRPCY rod was studied, taking into considerations the variability of yarn diameter. The mechanical properties of the flax yarn reinforced polyester polymer composite was improved by using yarns of coarse counts even at low values of fiber volume fraction. Also, the enhancement of bending resistance of FRPCY makes it possible to be used for new or repaired reinforced concrete structures.

References

[1] Saadatmanesh H and Ehsani M R 1996 Jan. 15-17 Fiber composites in infrastructure Proc. First Int. Conf. on Composites in Infrastructure, Tucson, Arizona, Dept. of Civil Eng. and Eng. Mech., University of Arizona.

[2] Fangueiro R, Sousa G, Soutinho F, Jalali S and De Araújo M 2006 Application of Braided Fibre Reinforced Composite Rods in Concrete Reinforcement *Mat. Sci. Forum* **514-516** 1556-1560.

[3] ElMessiry M 2013 Theoretical analysis of natural fiber volume fraction of reinforced composites *Alex. Eng. J.* **52** **3** 301-306.

[4] Ahmadi M S1, Johari M S, Sadighi M and Esfande M 2009 An experimental study on mechanical properties of GFRP braid-pultruded composite rods eXPRESS Polymer Letters **3 no 9** 560–568.
[5] Byun J H and Chou T-W 1989 Modelling and characterization of textile structural composites – A review *J. of Str. Anal. for Eng. Des.* **24** 253–262.
[6] Hufenbach W, Bla ejewski W, Kroll L, Böhm R, Gude M and Czulak A 2005 Manufacture and multiaxial test of composite tube specimens with braided glass fiber reinforcement *J. of Mat. Proc. Tech.* **162-163** 65–70.
[7] Charlebois K M, Boukhili R, Zebdi O, Trochu F and Gasmi A 2005 Evaluation of the physical and mechanical properties of braided fabrics and their composites *J. of Reinf. Plas. and Comp.* **24** 1539–1554.
[8] Gonilho-Pereira C., Jalali S., Fangueiro R., Araújo M., Pina Marques P. Hybrid composite rods for concrete reinforcement. Structures and Architecture, –Cruz (Ed.) Taylor & Francis Group London ISBN 978-0-415-49249-2 1605-1612.
[9] Naujokaitytė L and Strazdienė E 2007 Investigation of Textile Fabrics Behavior under Compression *Mater. Sci. (MEDŽIAGOTYRA)* **13 4** 337- 342.
[10] Gutowski T and Dillon G 1992 The elastic deformation of lubricated carbon fiber bundles: comparison of theory and experiments *J. Compos. Mater.* **26** 2330–2347.
[11] El Messiry M. 2018 Theatrical investigation of the fiber volume fraction of natural fiber fabric reinforced polymer composite *J. Ind. Tex.*.

IOP Conf. Series: Materials Science and Engineering **406** (2018) 012040 doi:10.1088/1757-899X/406/1/012040

Manufacture and characterization of 3D warp interlock fabric made of flax roving

Henri Lansiaux[1]*, Damien Soulat[1], François Boussu[1], and Ahmad Rashed Labanieh[1]

[1] ENSAIT, GEMTEX, F-59100 Roubaix, France

* henri.lansiaux@ensait.fr

Abstract. The present study carries on the characterization of dry reinforcement structure for composite material. Yarn properties were analyzed to select the optimal level of twist. The influence of 3D-warp interlock structure's bending type was also investigated. Structures, made with 1000Tex flax roving, were woven on a dobby loom. Physical characterization and mechanical testing (tensile test) were conducted on the four different binding structures (AL, AT, OL and OT) to provide relevant data and identify the influence of the binding warp yarns' path inside 3D warp interlock structure.

1. Introduction

The number of publications about composites materials reinforced by natural fibre keeps increasing. Flax fibre is the strongest natural fibre, with a Young's modulus nearby 100 GPa [1]. Flax is a cellulosic fibre. Extracting the fibre from the plant is a long and specific process. The use of flax fibre as reinforcement structure became common in automotive and sport sectors [2] by the end of the 70's. Fibre architecture and fibre ratio have a preponderant role on the composite material's mechanical properties and failure mode [3, 4].

The main role of fabric inside a composite material is to resist to mechanical stress. Nowadays, most of the composites are made of multi-layers 2D woven fabrics. A 3D-warp interlock fabric is a multi-layer fabric in which layers are linked together by binding warp yarns [4]. This structure allows linking yarns on the z-direction, unlike 2D woven fabrics. Consequently, composite based on 3D-warp interlock structure shows higher through the thickness strength, delamination resistance and interesting improvement in impact damage tolerance [5]. 3D warp interlock fabrics are studied under many other aspect as characterization [6-11], or 3D modelling [8, 10-13].

Characterization of the roving can help us to optimise the process of weaving. Influence of twist value, for a given value of linear density of the roving, on tensile parameter was analysed. The most suitable value of twist was selected to proceed to the weaving. The aim of this study is to investigate the influence of 3D warp interlock structure's bending type, on dry preform. Structures, made with 1000Tex flax roving, were woven on a Dornier HTVS 4/S dobby loom machine. The weaving of 3D warp interlock structure requires weaving machine's adaptation [14-16]. For example, inserting weft yarn system must be adapted to thick fabric. The speed of weaving is significantly lower. This kind of fabric is also more subject to abrasion than traditional 2D fabric.

Physical characterization and mechanical testing have been conducted on four structures with different binding types to identify their influence on 3D warp interlock structure behaviour.

2. Materials and methods

Roving characterization has been realized first, and then 3D warp interlock fabric characterization.

2.1. Roving scale

Rovings used in this study were provided by Depestele Group, France. Two different linear densities have been compared. Roving is not twisted, only a cohesion agent maintains the filaments inside the yarn and contributes to 10% of the weight of the roving.

At this scale, the linear density, the twist level, and the tensile behaviour have been studied. Twist has been measured according to the ISO 2061 standard. Tensile test has been realized according to NF EN 2062 standard. The tensile machine is equipped of a 10kN cell. The length of the sample is 200mm. A preloading of 2N is applied before starting the test. The displacement rate during the test is 100mm/min. The tenacity, which allows comparing different linear density, is determined by

$$\boldsymbol{Tenacity} = \frac{\boldsymbol{Load}}{\boldsymbol{Linear\ density}} \quad (1)$$

2.2. 3D warp interlock scale

2.2.1. Definition

3D warp interlock fabric links the different layers of the weft yarns through-the-thickness thanks to binding warp yarns. Binding warp yarns, of this fabric, can be characterized by their path in the structure [4].

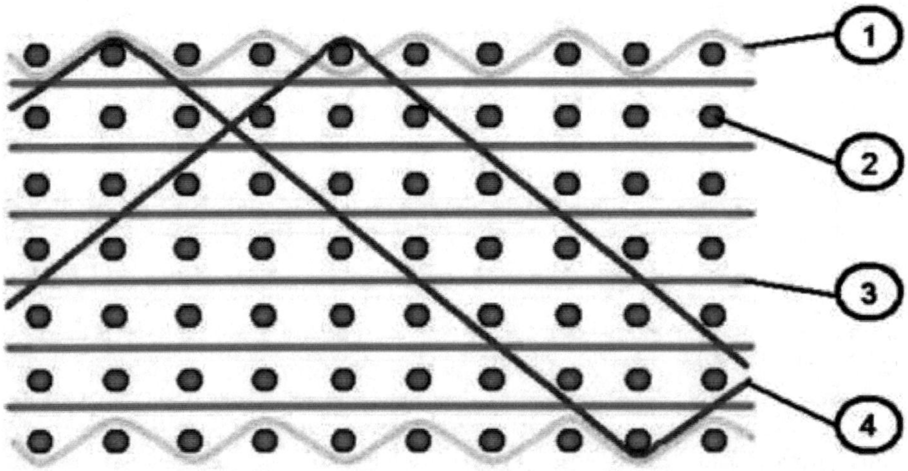

Figure 1: Scheme of 3D warp interlock fabric with the constitutive yarn

Surface weaver warp yarns (Figure 1, 1) give a specific surface aspect. Stuffer warp yarns (3) define longitudinal properties of the fabric. Binding warp yarns (4) define through the thickness properties of the fabric and ensure cohesion. Weft yarns (2) guarantee transverse properties of the fabric and define the number of layer of the fabric. Boussu et al. [3] sorted this fabric depending on the path of the binding warp yarn in the fabric. Binding warp yarn can be inserted diagonally (A, angle) or perpendicularly (O, orthogonal) to the surface of the fabric. The binding warp yarn can link a part of (L, layer to layer) or the whole (T, through the thickness) layer of the fabric. Figure 2 and 3 illustrate the four different types of woven diagrams. The placement of the binding warp yarns depends on the selected weave diagram.

(a) AL (b) AT (c) OL (d) OT

Figure 2: Scheme of 3D warp interlock woven patterns

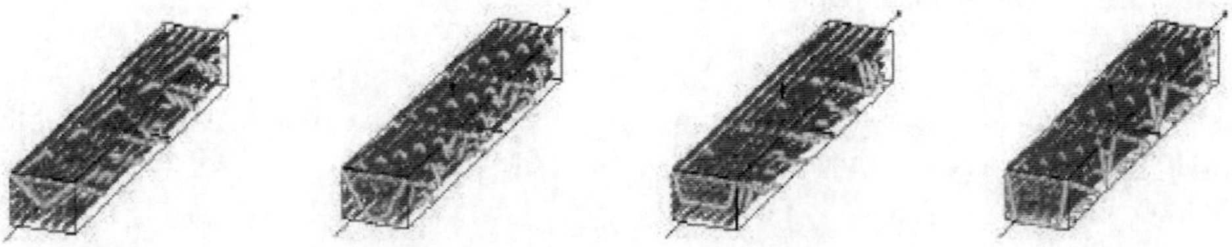

(a) AL (b) AT (c) OL (d) OT

Figure 3: 3D representation of the 3D weaves architectures

2.2.2. Parameters of production

A dobby loom (Figure 4) has been used to produce 3D warp interlock fabrics. Warp and weft yarns are 1000Tex flax roving. Four structures, corresponding to the four binding types, have been woven. The selected weave diagram is Twill 6, weft effect. The warp and weft densities are respectively 5 and 13 yarns/cm.

Figure 4: Production on dobby weaving loom

2.2.3. Characterization

The thickness of the manufacturing fabrics is determined according to the NF ISO 4603 standard. More than 30 samples have been done. Average values of mechanical test with standard deviation have been calculated. The areal density is determined according to the NF EN 12127 standard. Warp and weft densities are checked according to the NF ISO 4602. Warp crimp is determined according to the ISO 7211-3 standard. Binding and stuffer warp yarns crimp are determined separately and for each layer of the fabric.

Tensile tests have been realized according to the ISO 13934 standard. The size of the sample is wider than the one stated in the standard, in order to keep at least two complete woven patterns. A preloading of 10N is applied before starting the test. Then the displacement rate of the test is 100mm/min. Five samples are tested in warp and weft directions.

3. Result and discussion

3.1. Result of the roving characterization

3.1.1. Linear density

Figure 5 shows linear density of two different rovings (Tex 500 and Tex 1000) as a function of the twist. During twisting the linear density increases. Fibres lose their alignment, and twist counterbalance the gain of weight.

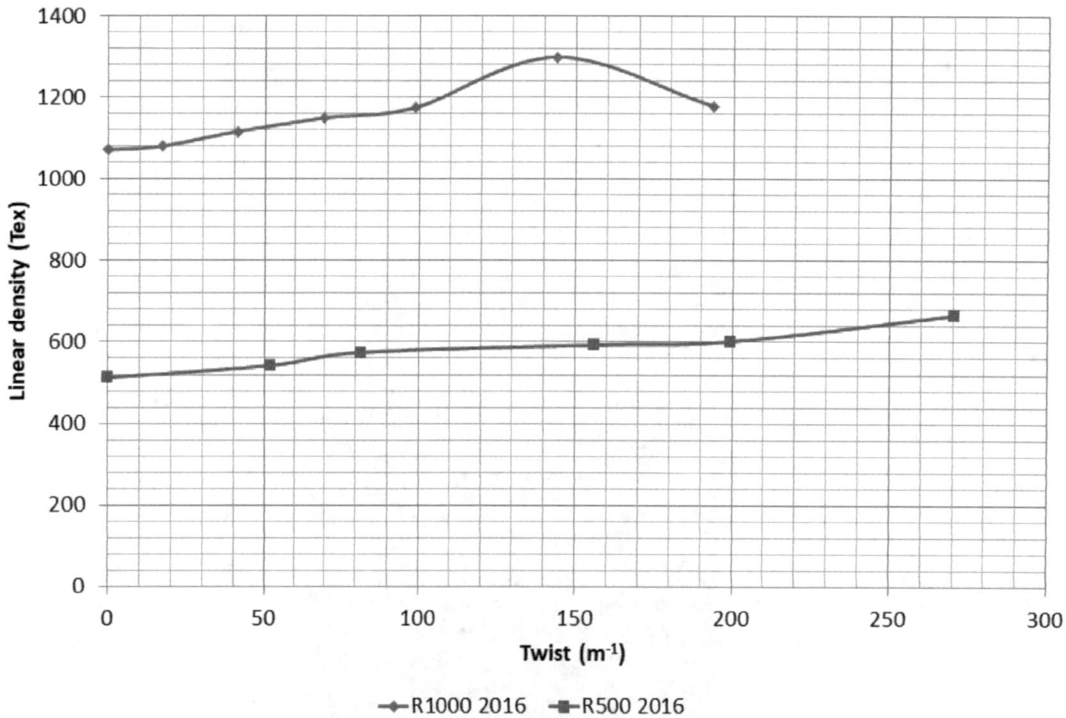

Figure 5: Linear density versus twist

3.1.2. Tensile test results

For the two linear densities, the tenacity and strain at break seem to have a similar evolution (Figure 6 and 7). The tenacity first increases until a maximum, and then decreases. First, the twist gives to the roving a higher mechanical performance by increasing the cohesion. After a specific number of twists, the loss of alignment of fibres in the roving is the cause of the decrease of properties. This specific number of twist is close to 75 rounds/m.

The strain at break increases as a function of the twist value. The number of fibres in the cross-section increases and the roving becomes stiffer.

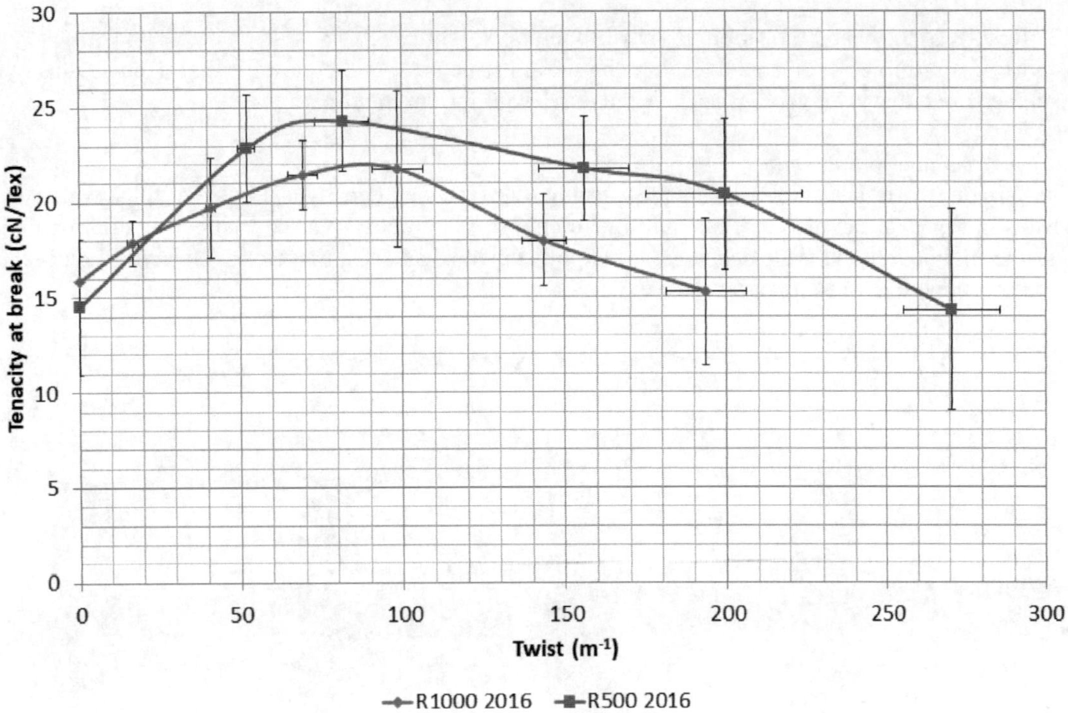

Figure 6: Tenacity of the two linear densities versus twist

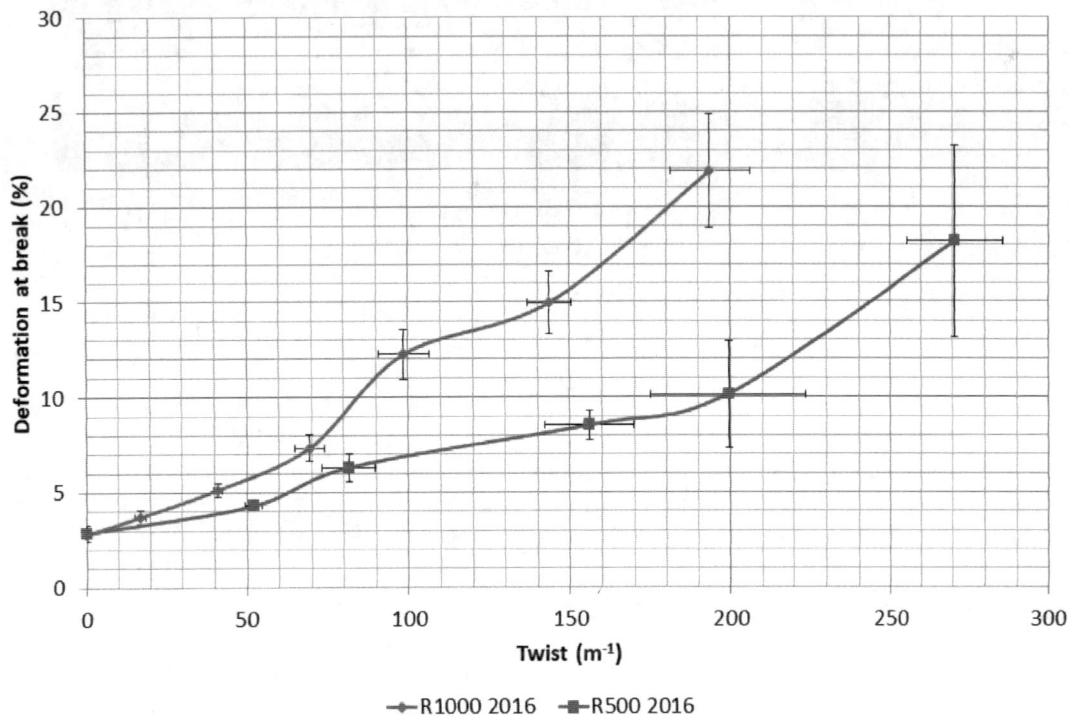

Figure 7: Deformation of the two linear densities versus twist

IOP Conf. Series: Materials Science and Engineering **406** (2018) 012040 doi:10.1088/1757-899X/406/1/012040

3.2. Results of the characterization of 3D warp interlock fabrics

A twist of 30 rd/m was selected to realize the woven structure. As the permeability decreases as twist level increases, this twist value was chosen to have equilibrium in properties.

3.2.1. Warp and weft densities

The warp density is determined during the warping step. The weft density is defined during the weaving. Results in Figure 8 can also be expressed by column by centimetre. Warp densities are slightly higher than the instruction and weft densities are slightly lower than the instruction. These differences are caused by the crimp of fabric after the weaving.

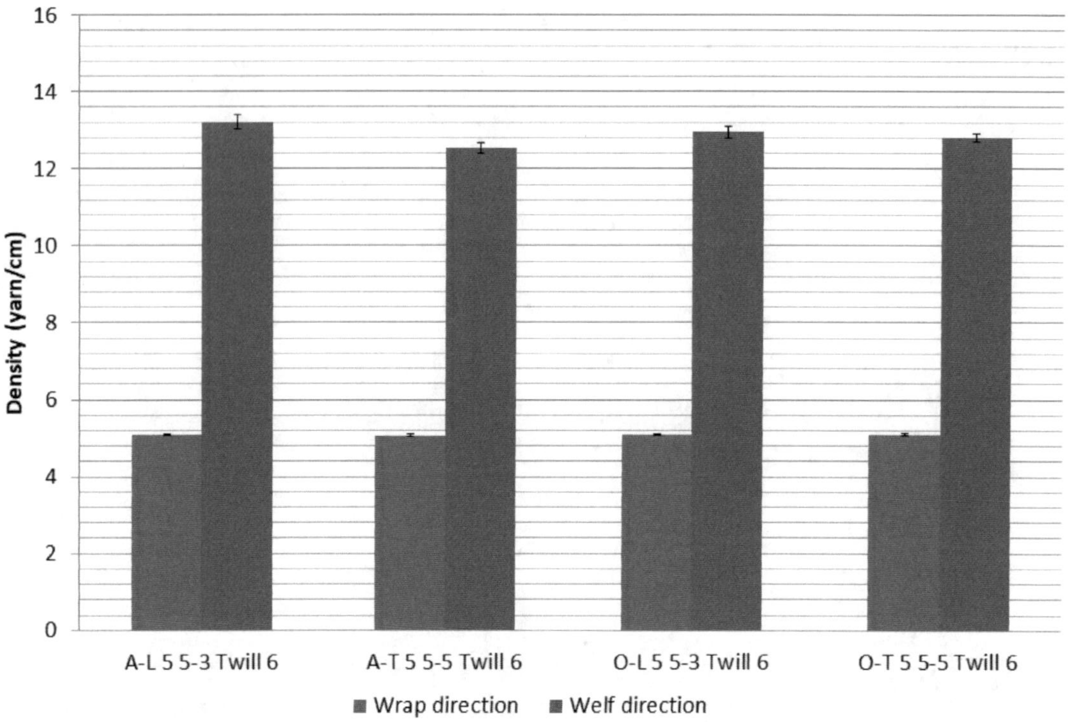

Figure 8: Warp and weft densities of the four manufactured 3D warp interlock fabrics

3.2.2. Warp crimp

The digit following term "Weaver" or "Stuffer" in Figure 9, is the minimum depth of the yarn in the 3D warp interlock structure. According to this definition, T structures which only have a binding yarn's path, don't show results in other depth. These structures present a higher binding warp crimp than L structures. The length of roving used to weave this type of structure depends on the geometry (cf. Figure 2). For similar reasons, O structures present a higher binding warp crimp than A structures.

The weft crimp was also studied. The stuffer warp crimp and the weft crimp evolve in order to stabilize the fabric after the weaving.

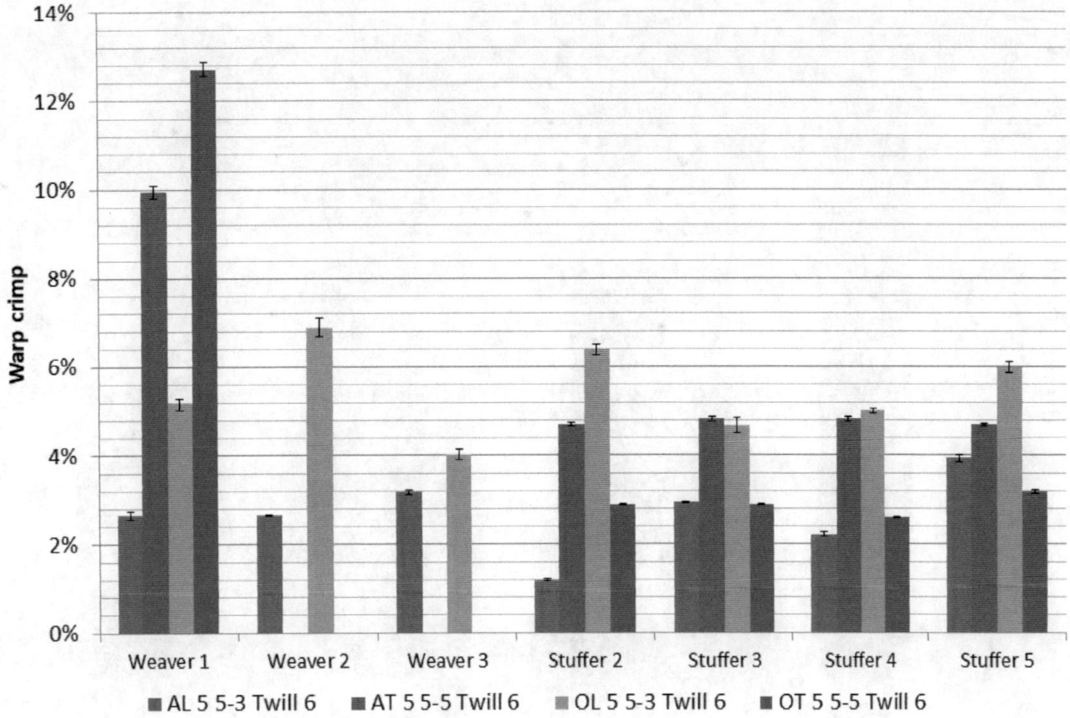

Figure 9: Warp crimp

3.2.3. Thickness and areal density
Figure 10 presents results of the thickness determination for the four fabrics. L structures present a higher thickness than T structures. The binding warp yarn, through the thickness of L structures, creates over-thickness.

Figure 11 presents results of thickness measurements. According to theses density values, the theoretical areal density must be 1.8kg/ m². Experimental values are close to this theoretical value. The difference between theses value is caused by the crimps of the fabric after weaving.

IOP Conf. Series: Materials Science and Engineering **406** (2018) 012040 doi:10.1088/1757-899X/406/1/012040

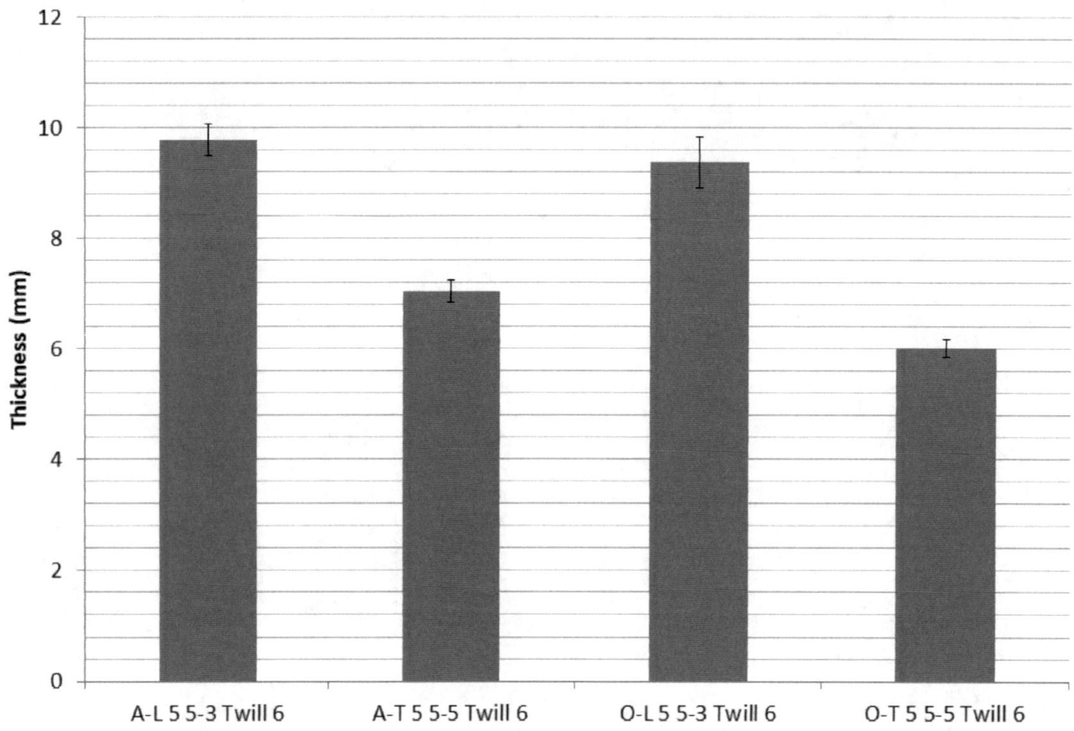

Figure 10: Thickness of the four manufactured 3D warp interlock fabrics

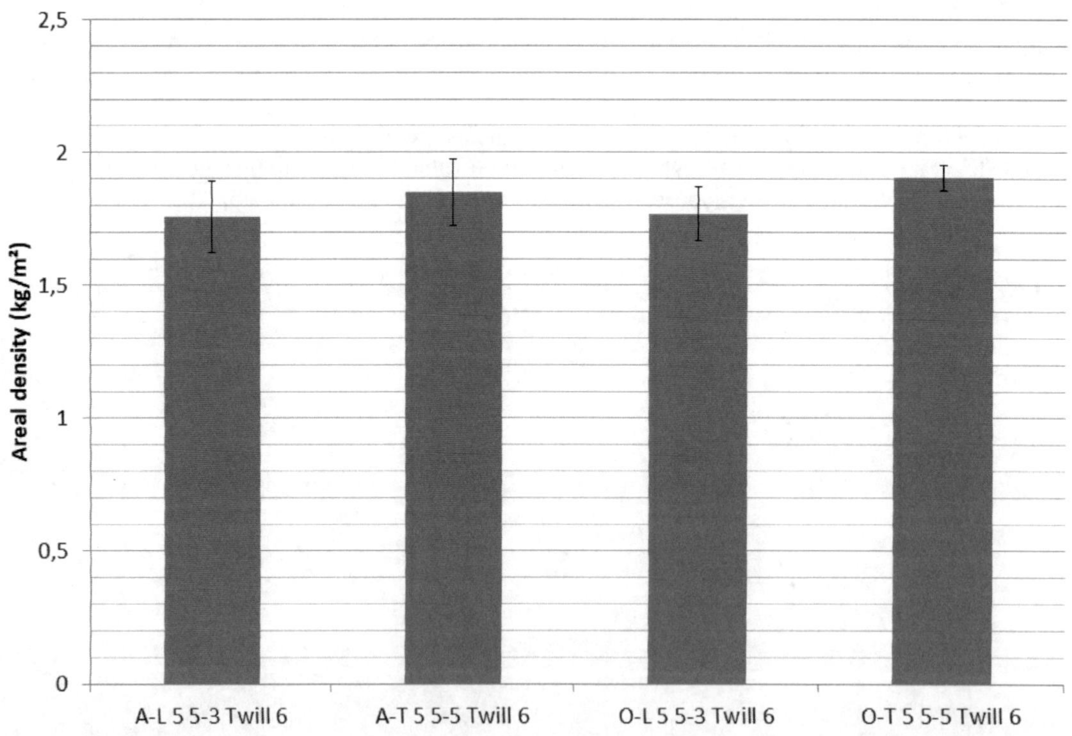

Figure 11: Areal density of the four manufactured 3D warp interlock fabrics

3.2.4. Tensile tests

Figure 12 displays a typical load-strain curve obtained during a tensile test, in the warp direction, of the structure A-T. The first peak is caused by the stuffer warp yarns' failure. The second peak is caused by the binding warp yarns' failure. This phenomena displays only on T structures. The distance between the peaks depends on the difference of length between stuffer warp yarn and binding warp yarn.

Figures 13 and 14 display averaged results of the tensile tests. Divided by the number of yarns in the tested section, the failure load is in the same order of magnitude for all tested fabrics. The stress at the first peak in warp direction and the deformation at break in weft direction are also in the same order of magnitude. In both case, stuffer warp yarn and weft yarn present the same failure load and deformation. The deformation failure of binding warp yarns depends on the crimp of the binding warps and the deformation failure of the stuffer warp yarns.

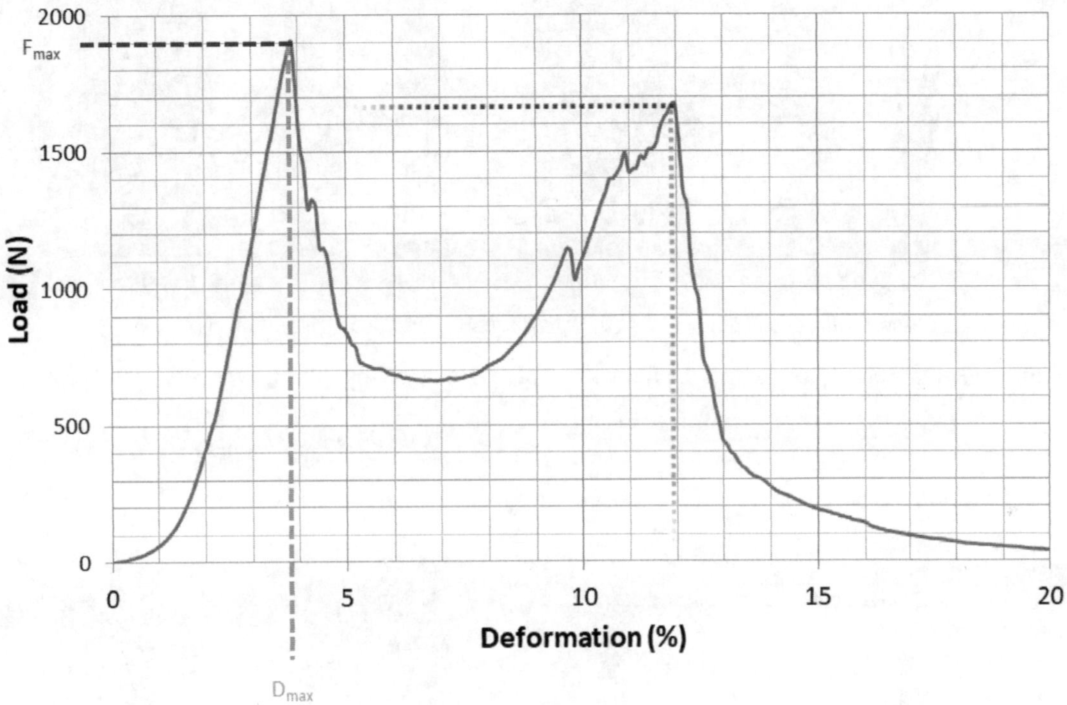

Figure 12: Load-strain curve for A-T structure in warp direction obtained on the tensile test machine

IOP Conf. Series: Materials Science and Engineering **406** (2018) 012040 doi:10.1088/1757-899X/406/1/012040

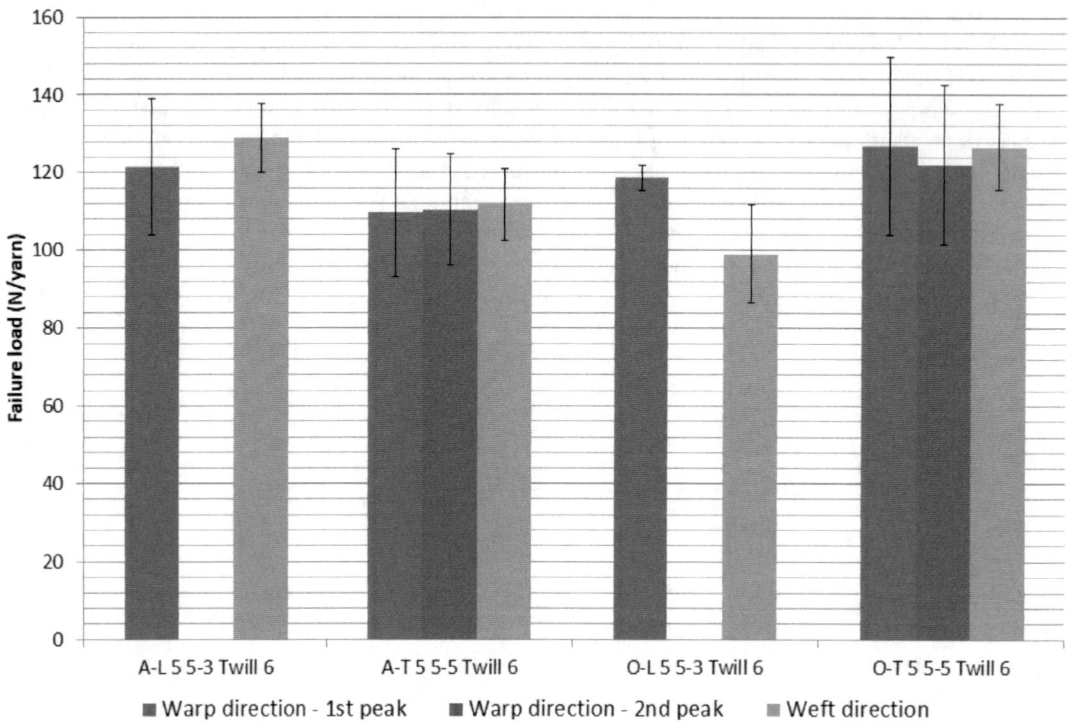

Figure 13: Failure load of the four manufactured 3D warp interlock fabrics (warp & weft directions)

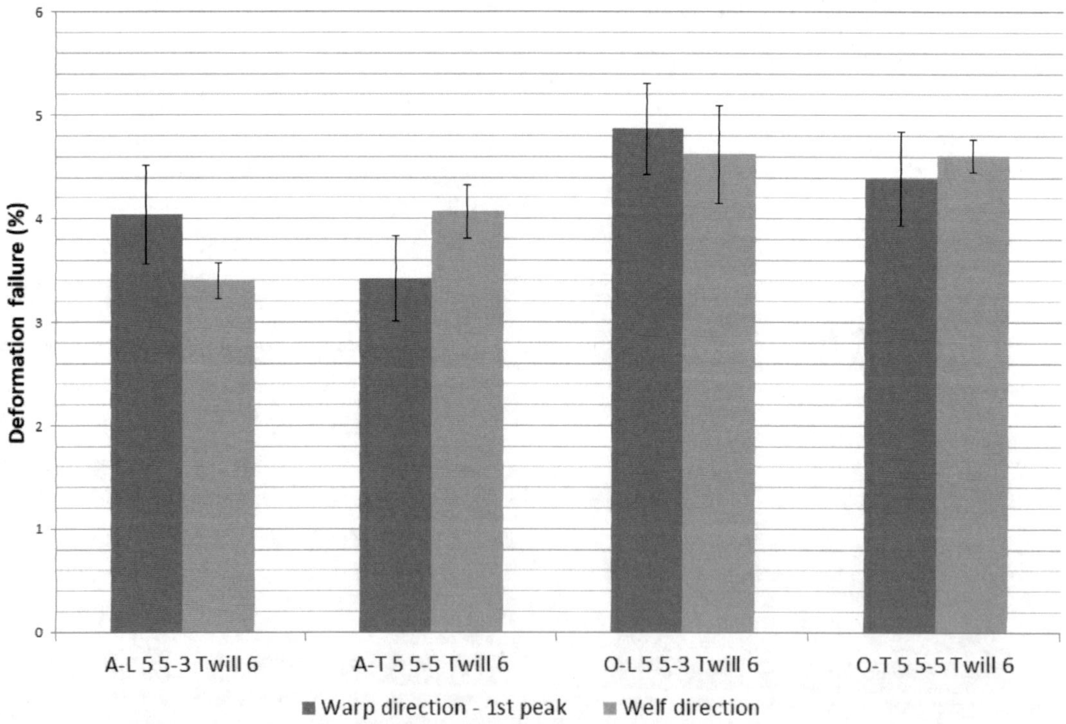

Figure 14: Deformation failure of the four manufactured 3D warp interlock fabrics

IOP Conf. Series: Materials Science and Engineering **406** (2018) 012040 doi:10.1088/1757-899X/406/1/012040

4. Conclusion

The optimized twist value in order to increase the mechanical performance of the 3D warp interlock fabric has been identified. Linear densities and deformation at failure are increasing function of the twist. Tenacities are also depending on the twist until a maximum value, after which the performance decreases.

The binding warp yarn's path of 3D warp interlock fabric affects different properties as the thickness, the density, the crimp or the tensile behaviour. The most remarkable phenomena is provided by T path structures with the appearance of a second peak during the tensile test. Bias and preforming tests will be realized to complete the study.

Acknowledgment

This study was fund by the Haut de France region, the Agence de l'environnement et de la maitrise de l'énegie (ADEME) and by Depestele group.

References

[1] D. U. Shah, Natural fibre composites: Comprehensive ashby-type materials selection charts, Materials & Design (1980-2015) 62 (2014) 21_31.
[2] L. Pil, F. Bensadoun, J. Pariset, I. Verpoest, Why are designers fascinated by flax and hemp fibre composites?, Composites Part A: Applied Science and Manufacturing 83 (2016) 193_205.
[3] F. Bensadoun, I. Verpoest, J. Baets, J. Müssig, N. Graupner, P. Davies, M. Gomina, A. Kervoelen, C. Baley, Impregnated fibre bundle test for natural fibres used in composites, Journal of Reinforced Plastics and Composites 36 (13) (2017) 942_957.
[4] F. Boussu, I. Cristian, S. Nauman, General de_nition of 3D warp interlock fabric architecture, Composites Part B: Engineering 81 (2015) 171 _ 188.
[5] J. Hu, 2 - applications of three-dimensional textiles, in: J. Hu (Ed.), 3-D Fibrous Assemblies, Woodhead Publishing Series in Textiles, Woodhead Publishing, 2008, pp. 33 _ 69.
[6] S. Adanur, C. Tam, On-machine interlocking of 3D laminate structures for composites, Composites Part B: Engineering 28 (5) (1997) 497 _ 506.
[7] C. Dufour, P. Pineau, P. Wang, D. Soulat, F. Boussu, 10 – threedimensional textiles in the automotive industry, in: X. Chen (Ed.), Advances in 3D Textiles, Woodhead Publishing Series in Textiles, Woodhead Publishing, 2015, pp. 265 _ 291.
[8] N. Vernet, F. Trochu, Analysis and modeling of 3D interlock fabric compaction behavior, Composites Part A: Applied Science and Manufacturing 80 (2016) 182 _ 193.
[9] A. Kumar Bandaru, Y. Sachan, A. Suhail, R. Alagirusamy, N. Bhatnagar, On the mechanical response of 2D plain woven and 3D angle-interlock fabrics, Composites Part B: Engineering 118 (2017) 135 _ 148.
[10] P. Hamelin, Polymeres renforces fibres (frp) identification et champs d'application, Techniques de l'ingenieur Les materiaux de construction base documentaire : TIB224DUO. (ref. article : c950)
[11] J. Moothoo, S. Allaoui, S. Ouagne, D. Soulat, A study of the tensile behaviour of flax tows and their potential for composite processing, Materials & Design 55 (2014) 764 _ 772.
[12] Buchanan, A. Grigorash, E. Archer, A. McIlhagger, J. Quinn, G. Stewart, Analytical elastic stiffness model for 3D woven orthogonal interlock composites, Composites Science and Technology 70 (11) (2010) 1597 _1604.
[13] S. Zhong Sheng, S. Van Hoa, Modeling of 3D angle interlock woven fabric composites, Journal of Thermoplastic Composite Materials 16 (1) (2003) 45_58.
[14] E. Capelle, Conception et fabrication de renforts tisses à base de fibres de lin pour la realisation de pieces composites à geometries complexes, phdthesis, Universite d'Orleans (2015).
[15] A.-C. Corbin, Élaboration et caractérisation de renforts 3D en lin, mathesis, ENSAIT (Jul. 2017).
[16] H. El-Dessouky, A. Snape, H. Tew, R. Scaife, D. Modi, K. Kendall, C. McHugh, Design, weaving and manufacture of a large 3D composite structures for automotive applications, in: 3D Fabrics & Their Applications, 2016.

IOP Conf. Series: Materials Science and Engineering **406** (2018) 012041 doi:10.1088/1757-899X/406/1/012041

Flexural behavior and damage extent in woven natural fibers/polypropylene composite laminates

Pietro Russo[1], Giorgio Simeoli[1], Vito Pagliarulo[2], Vittorio Bianco[1,2], Pietro Ferraro[2]

[1] Institute for Polymers, Composites and Biomaterials, National Council of Research, Pozzuoli (Na), Italy

[2] Institute of Applied Sciences and Intelligent Systems, National Council of Research, Pozzuoli (Na), Italy

Abstract. Recently, low environmental impact and potential use in a wide range of applications have been the main driving reasons supporting the rapid growth of the research interest toward the investigation and development of items based on natural fiber composites. This paper compares the performance and the extent of damage suffered by polypropylene (PP) composites reinforced with natural fiber fabrics and subjected to flexural loads. In particular, laminates including jute, flax and basalt woven fibers have been prepared by typical film stacking and compression molding techniques. A premodified matrix (PPC) including 2 wt% of a coupling agent was also considered. Measurements were carried out on at least 5 determinations for each kind of samples. All flexural parameters appeared to be significantly increased for specimens based on PPC with respect to the ones based on neat PP with effects particularly pronounced in presence of basalt fibers. Despite the use of the same content of coupling agent, this finding can be attributed the achievement of a different extent of interfacial adhesion, in turn, due to the different nature of reinforcing fibers. The influence of both the matrix modification and the nature of reinforcing fibers on the extent of induced flexural damage were assessed in non-destructive way by Electronic Speckle Pattern Interferometry and by Scanning Electron Microscopy observations of cryofractured surfaces.

1. Introduction

In the last years, the increased sensibleness about environmental concerns and depletion of fossil sources have stimulated a growing interest toward new-ecosustainable materials as the natural fiber composites (NFC). In fact, the advantages derive also in that natural fibers are cheap, have low density and may offer specific properties higher (eg. linen fibers) or, at least, superior to those of other synthetic ones conventionally used for reinforcements (i.e. glass, carbon fibers). On the other side, drawbacks as incompatibilities with the hosting matrices, their high sensitivity to processing conditions and the significant tendency to absorb moisture still reduce their potential applications. Therefore, special care must be devoted to improve the interfacial fiber-matrix adhesion through surface modification of the reinforcing fibers and/or with the aid of appropriate coupling agent to optimize the processing conditions and last an in-depth characterization is required with the aim to verify the meeting of suitable technical specifications [1]. Among these latter, a common requirement is the capability of the new material to support loads without failure for safety reasons.

For this purpose, the use of non-destructive testing techniques as ultrasonic methods [2], acoustic methods [3] and infrared thermography [4], that can be applied both remotely and in-line to monitor the behavior of reference materials during their loading, is increasingly common. In this frame, speckle interferometry is customary tool for precise measurements in engineering, displacement measurement, vibration analysis, velocity measurement, non-destructive testing [5-8] and it can be used on very different materials. In this paper, we have used the Electronic Speckle Pattern Interferometry (ESPI) as a tool to characterize in non-destructive way and at full field novel ecosustainable composite materials.

In particular, six different types of thermoplastic composite laminates were considered. All samples, including woven natural fibers were prepared by conventional film stacking and compression molding technologies. Tested specimens were investigated by ESPI to assess the effect of the main constituents 'nature on the detected extent of the damaged area. Moreover, results were also discussed in terms of morphological aspects by analyzing scanning electron microscopic observations of cryo-fractured surfaces taken in correspondence of the applied flexural load.

2. Experimental

2.1 Materials and sample preparation

Six different composite systems involving 4 matrices and three natural fiber fabrics were investigated. In details, samples were prepared from a polypropylene (PP) resin (Hyosung Topilene PP J640, MFI@230 °C, 2.16 kg: 10 g/10 min), also premixed with 2 wt% of a compatibilizing agent (PP-g-MA - Polybond 3000, MFI@190 °C, 2.16 kg: 405 g/10 min) (code PPC), a poly(lactic acid) (PLA – Ingeo 7001 D, ρ=1.24 g/cc, MFI@210 °C, 2.16 kg: 6 g/10 min, Tg=55-60 °C) and a polyamide 6 (PA6) (Lanxess Durethan B30S-000000 (MFI@260 °C, 5 kg: 102 g/10 min).

Films of all these matrices, obtained in pre-optimized conditions, were alternatively stacked with commercial reinforcing fiber fabrics. In particular, polypropylene, neat or modified, systems included a jute plain wave type fabric (J) with a specific mass of 250 g/m^2 by Deyute (Alicante, Spain), to make samples coded as PPJ and PPCJ, respectively, or a 2x2 twill flax fabric (L) supplied by Biotex (weight: 200 g/m^2), to prepare samples coded as PPL and PPCL, respectively. The PLA samples were reinforced with the above cited Jute fabric (PLAJ sample) while the polyamide 6 was reinforced by using a basalt plain wave type fabric (BS - weight: 210 g/m^2) from Incotelogy GmbH (Germany) (PA6BS).

Stacks of plastic films and reinforcing fabrics, made up of different number of layers, preliminarily estimated on the basis of the weight of the considered fabric and of previous experiences, to obtain laminates with thickness approximately equal to 3 mm, were compacted by a hot pressing step carried out under adequate temperature and pressure profile conditions.

2.2 Characterization techniques

Plates with dimensions of 100 mm × 12.7 mm were cut from the laminates and used for three-point bending tests on a Universal Testing Machine (Instron model 4505). Measurements were performed according to ASTM-D790 at a crosshead speed of testing equal to 2.47 mm/min, using a load cell of 1 kN and a span set at 70 mm at room temperature after being equilibrated under standard ASTM condition of 23 °C and 50% relative humidity for 24. Results are reported in terms of typical stress-strain curves and flexural parameters averaged on at least 5 determinations for each investigated sample.

A typical scheme of the system for recording speckle interferometric measurements can be found in [6]. The laser beam is split into a reference beam and an object beam by means of a Beam Splitter (BS). The reference beam is coupled into an optical fiber to reach a Charge-Coupled Device sensor (CCD), where also the light scattered from the object is addressed. Before reaching the object, the illumination beam passes through an electronically tilted etalon to shift the probing beam laterally in a parallel direction. In this way, it is possible to obtain multiple equivalent images of the speckle pattern, which are used to perform an average to reduce the speckle noise. Upon an external perturbation (i.e. thermal or mechanical), the plate is deformed and the reflected wavefront is slightly changed, while reference beam remains unperturbed. Thus,

IOP Conf. Series: Materials Science and Engineering **406** (2018) 012041 doi:10.1088/1757-899X/406/1/012041

the CCD camera records a new speckle pattern and the subtraction of the registered speckle patterns (deformed and non-deformed states) provides the correlation fringes. Analytically, the speckle pattern intensity for the unperturbed state is given by:

$$I_1 = I_0 + I_r + 2\sqrt{I_0 I_r}\cos(\theta)$$

while for the deformed state is given by:

$$I_1 = I_0 + I_r + 2\sqrt{I_0 I_r}\cos(\theta + \phi)$$

where I_0 is the intensity scattered from the object, I_r is the reference beam, θ is the phase angle and ϕ is related to the displacement, i.e. the deformation. The fringes pattern that represents the out-of-plane surface displacement is represented by the equation:

$$I_{1-2} = 2\sqrt{I_0 I_r}[\cos\theta - \cos(\theta + \phi)]$$

The correlation fringes, digitally acquired and treated for contrast enhancement and noise removal, allow obtaining the phase contrast maps, from which it is possible to measure the displacement field with high accuracy [5].

3. Results and discussion

Results of static flexural tests are summarized by the histograms depicted in Figure 1a, b and c, in terms of flexural modulus, flexural strength and strain at the maximum load, respectively.

The data demonstrate an evident improvement in the flexural performances in presence of the compatibilizing agent for polypropylene composite laminates, as a sign of a superior adhesion at the fiber-matrix interface and, consequently, of the enhancement of stress transfer between the polymeric and the reinforcing phases. Expectedly, this behavior occurs at the expense of the strain at maximum load which, in both cases, is reduced with a more pronounced effect detected for composite specimens containing woven jute fibers.

The latter, also used for the PLA, give rise to fully biodegradable PLAJ systems with further increased flexural modulus but which show a flexural strength slightly lower than that obtained for the more resistant polypropylene sample (PPCL).

Finally, the chemical affinity between the polar polyamide 6 and the basalt fibers as well as the superior mechanical properties of basic constituents with respect to other matrices and reinforcing natural fibers herein considered justify the extraordinary performances of the PA6BS composite laminate.

a)

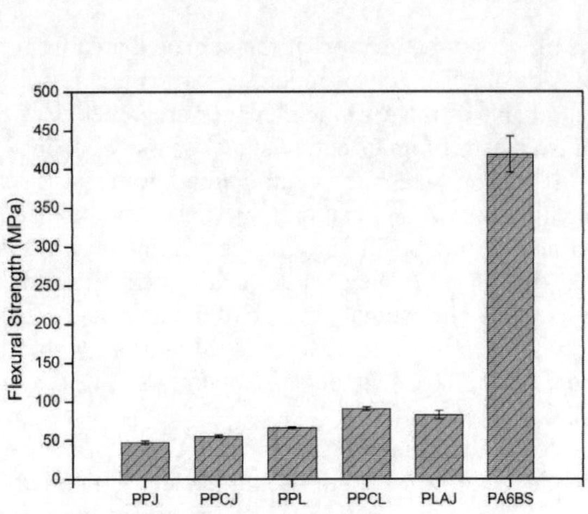

b)

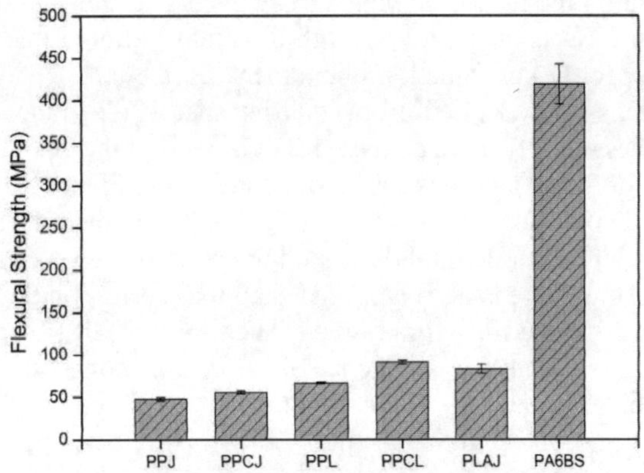

c)

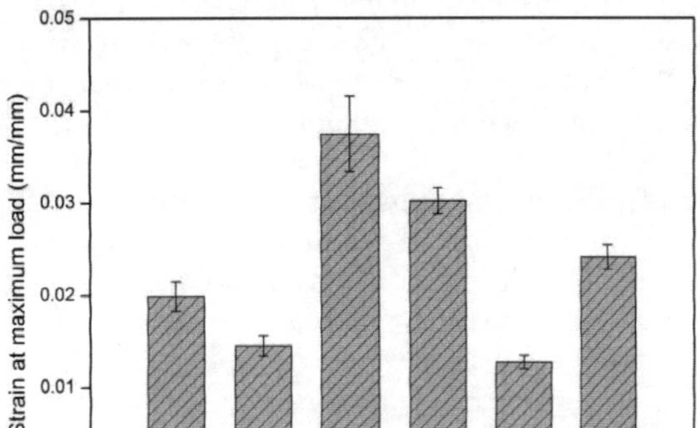

Figure 1. Flexural parameters of investigated composite systems

Figure 2 shows the process of estimation of the damaged area from the acquired fringe pattern in the specific case of the front side of a PA6BS specimen. In particular, the noisy phase map is reported in Figure 2(a). In order to obtain a reliable estimate of the damaged area, speckles need to be removed from the noisy map, which is achieved by means of an effective signal sparsity enhancement filter [9, 10]. The denoised map is shown in Figure 2(b), from which the central area of interest is selected (as shown in Figure 2(c)). Remarkably, numerical calibration is automatically performed in order to compensate for unwanted phase tilts eventually due to a non-perfect positioning of the specimen on the recording support. In this case, a second order polynomial fit was used. Hence, after calibration the estimated out-of-plane displacement is only due to the damage itself, i.e. the estimated value of the damaged area does not depend on the recording system. Figure 2(d) shows the area of interest after calibration, while the binary map in Figure 2(e) is obtained from (d) after applying image segmentation operators. This binary map is then used to measure the extent of the damaged area.

In this work, both sides of all the bended specimens were deformed by an external heating perturbation, focusing the attention on the section at which the flexural load was applied. Results showed that the damaged area is always higher on the back (tensile side) of the specimens with respect to the side where the force was applied (compression side). An exception is really the PA6/basalt system for which the extent of the damaged area evaluated on the front side seems to be even four times higher than the one recorded on the rear side. This behavior reflects the relatively high flexural rigidity of this system that induces a prevailing fiber breaking with respect to delamination phenomena on the tensile side of specimens and pronounced out-of-plane deformations on the loaded (front) side because of the hidden damages.

ESPI is able to assess the effective damaged area as well as the delaminated one by the measure of the specimen out-of-plane deformation due to a thermal stress and Figure 3 shows the extent of the damaged area for all investigated specimens. In this regard, it should be noted that inspections were carried out on specimens that have undergone a flexural deformation equal to 5%, as established by the reference ASTM Standard. In this cases, the value is always higher than that corresponding to the maximum load at which the incipient damage of included fibers is assumed. Therefore, the ESPI technique was used to compare the extent of the damaged area, due to the occurrence of various phenomena as delamination, fibers debonding, matrix cracking, fiber breaking and so on, for the different specimens, under the same flexural strain.

Clearly, the presence of the coupling agent in polypropylene composite laminates reinforced with jute fiber fabric, promoting a premature breaking of the reinforcing fibers, limits the delamination phenomena especially on the compression (front) side of the loaded specimen but the improved adhesion is not enough to prevent a certain sliding between adjacent plies on the back side, where the damage effects are particularly enhanced by tensile stresses. In this case, even because of the relatively low tensile properties of jute fibers in comparison with the other natural fibers considered in this research, the extent of the damaged area results to be comparable to that generated in presence of pure PP.

About composite samples including linen fibers, no damage seems to be verified on the front side, regardless the presence of the coupling agent. On the contrary, the damaged area on the tensile side is strongly affected by the improved interfacial adhesion: behavior reflecting the already mentioned effect of the PP-g-MA, more pronounced with respect to linen fibers with respect to jute fibers.

PLA/woven jute fibers specimens show damaged areas with significant extent only on the opposite side of the specimen with respect to the one where the load is applied.

Finally, the intrinsic good interfacial adhesion in PA6 systems including basalt fibers is responsible for the remarkable damage extent detected on both sides of the loaded specimens. However, surprisingly, it seems that the extent of the damage is wider on the side where the flexural load is applied. This unexpected consideration is still under investigation.

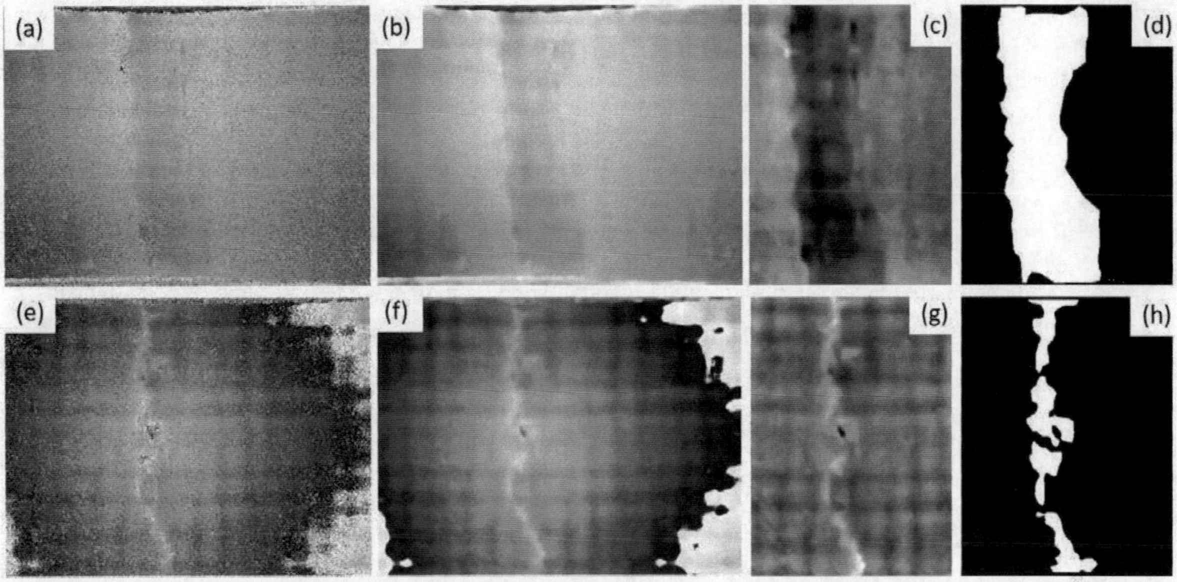

Figure 2. Example of numerical processing performed to estimate the damaged area. (a-d) The frontside of a PA6BS specimen is shown. (e-h) The backside of the same specimen is shown. (a, e) Acquired data. (b, f) Denoised data. (c, g) ROI after background calibration. (d, h) Damaged areas.

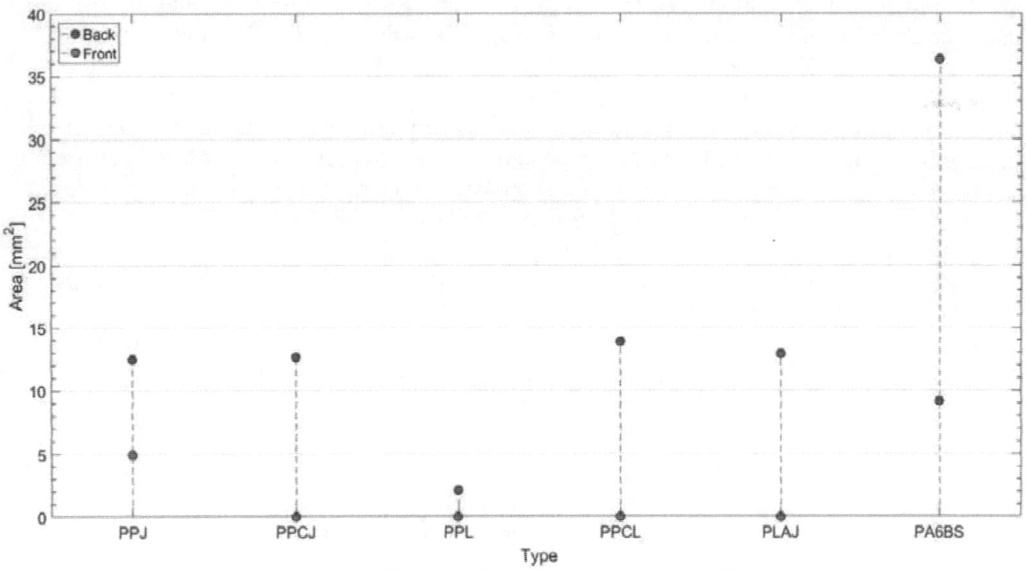

Figure 3: Damaged Area determined by ESPI.

4. Conclusions

The flexural damage of six different types of composites including woven natural fibers has been considered. In particular, the attention was focused on systems obtained by combining three thermoplastic matrices, polypropylene (apolar resin), poly(lactic acid) and polyamide 6 (polar resins), with also the inclusion of a coupling agent in the former case, with three kind of natural reinforcing fibers as linen, jute and basalt.

The damaged areas were investigated by means of ESPI to assess their actual extent. The results show that the interfacial adhesion significantly affects damage mechanisms and, consequently, the flexural parameters of the investigated composite materials, improving their performances with the increase of the fiber-matrix affinity. Moreover, the extent of the damage takes into account, among others, of the different type of stress at which plies on both sides of each specimens, with respect of the applied flexural load, are subjected. The damage, always detected in the side subjected to tensile stresses (backside), is generally negligible on the front side, in direct contact with the bending tool, subjected to compression stresses. The exception is the PA6BS system given its relatively high flexural rigidity.
Finally, with the awareness that SEM morphological observations, highlighting different damage mechanisms occurring on front and back side of tested specimens as well as the effect of the grafting agent, may support the detected flexural and damage results and provide a significant contribution about actual potentiality of studied composite systems, tests are still in progress.

References

[1] Pickering KL, Aruan Efendy MG and Le T A review of recent developments in natural fiber composites and their mechanical performance 2016 *Comp. Part A: Appl. Sci. Manufact.* **83** 98
[2] Papa I, Lopresto V, Simeoli G, Langella A and Russo P Ultrasonic damage investigation on woven jute/poly(lactic acid) composites subjected to low-velocity impact 2017 *Comp. Part B* **115** 282
[3] De Rosa IM, Santulli C and Sarasini F Acoustic emission for monitoring the mechanical behavior of natural fiber composites: A literature review 2009 *Comp. Part A: Appl. Sci. Manufact.* **40** 1456
[4] Meola C, Boccardi S, Carlomagno GM, Simeoli G and Russo P In-line monitoring of jute fiber-reinforced polylactic acid subjected to impact loading by using infrared thermography 2017 *J. Appl. Polym. Sci.* **134** 45579
[5] Georges MP, Thizy C, Languy F and Vandenrijt JF An overview of interferometric metrology and NDT techniques and applications for the aerospace industry 2016 Proc. of SPIE **9960** 996007
[6] Pagliarulo V, Rocco A, Langella A, Riccio A, Ferraro P, Antonucci V, Ricciardi MR, Toscano C and Lopresto V Impact damage investigation on composite laminates: comparison among different NDT methods and numerical simulation 2015 *Meas. Sci. Technol.* **26** 085603
[7] Pagliarulo V, Palummo R, Rocco A, Ferraro P, Ricciardi MR and Antonucci V Evaluation of delaminated area of polymer/carbon nanotubes fiber reinforced composites after flexural tests by ESPI 2014 *IEEE MAS Proc.* **6865922** 211
[8] Antonucci V, Caputo F, Ferraro P, Langella A, Lopresto V, Pagliarulo V, Ricciardi MR, Riccio A and Toscano C Low velocity impact response of carbon fiber laminates fabricated by pulsed infusion: A review of damage investigation and semi-empirical models validation 2016 *J Progress Aerospace Sci* **81** 26
[9] Bianco V, Memmolo P, Paturzo M, Finizio A, Javidi B and Ferraro P Quasi noise-free digital holography 2016 *Light: Science & Applications* **5(9)** 16142
[10] Bianco V, Memmolo P, Paturzo M. and Ferraro P On speckle suppression in IR digital holography 2016 *Opt. Lett.* **41(22)** 5226

IOP Conf. Series: Materials Science and Engineering **406** (2018) 012042 doi:10.1088/1757-899X/406/1/012042

Comparison of strength of 3D printing objects using short fiber and continuous long fiber

T Isobe[1], T Tanaka[2]*, T Nomura[3] and R Yuasa[3]

[1] Graduate School of science and Engineering, Doshisha University, Kyoto, Japan
[2] Faculty of science and Engineering, Doshisha University, Kyoto, Japan
[3] Kyoraku co., ltd., Kanagawa, Japan

* tatanaka@mail.doshisha.ac.jp

Abstract. In this research, composite materials were used to improve the strength of FDM 3D printed objects. The nanocomposites made from polylactic acid as matrix and multi-wall carbon nanotube as filler, short carbon fiber reinforced composite and continuous carbon fiber reinforced composite were prepared, and tensile test was carried out. As a result, the continuous fiber reinforced material exhibited tensile strength of about 7 times and elastic modulus about 5 times that of the other two materials. The strength was greatly improved by using the continuous fiber. The fracture surface after the test was observed using a scanning electron microscope. The result of observation shows that adhesion between the laminated layers and the relationship between the fiber and the matrix are bad, and improving these are necessary to increase strength. Comparing those materials, it is possible to improve the strength in some degree by using short fiber while maintaining ease of printing. On the other hand, by using continuous fiber it can be achieved significant strength improvement while printing was complicated.

1. Introduction

In recent years, 3D printers have attracted attention as multi-product small-volume production technology. 3D printing is one of Additive Manufacturing (AM) technology [1]. In modelling by 3D printers, the original three-dimensional shape (usually 3D CAD data) is divided into two-dimensional layers with constant thickness. The 3D printer makes and stacks the layers to create the original three-dimensional shape. By using a 3D printer, it is possible without metal mould to make complex parts which was difficult to shape with conventional processes. In other words, it is possible to reduce the time and cost to make metal mould so it is suitable for small lot production.

Depending on the forming method of each layer, 3D printers can be classified into various types. For example, there are Stereo Lithography (STL) which curing a photo-curable resin with laser, Selective Laser Sintering (SLS) which sintering a powder, Fused Deposition Modelling (FDM) which laminates thermoplastics, etc. Among them, we focused on FDM. The FDM uses thermoplastics formed into a wire shape which is called filament. The printer melts filament and extrudes it onto a forming table to create a three-dimensional object. The FDM has some advantages. Firstly, it have a simple mechanism that melt and extrude resin. It does not require expensive parts such as laser, so that the FDM system is relatively inexpensive. Secondly, thermoplastics can be used. The photo-curable resins used in the STL are inferior in strength and durability and the powder materials used in the SLS are difficult to handle. However, the thermoplastics used in FDM has enough strength and durability and it is easy to handle. Therefore, it is possible to manufacturing not only prototypes but practical parts easily.

However, the objects made by FDM does not have enough strength for mechanical parts. Therefore, we try to improve strength by adding carbon nanotube (CNT). The CNT has good mechanical property and can improve strength. It is also possible to impart functionality such as conductivity by adding CNT [2]. As another method, fiber reinforced composites are tried. In some examples, the strength of 3D printed objects is improved by using carbon fiber with excellent specific strength [3, 4]. Depending on the length of the carbon fiber, it can be divided into short fiber and continuous fiber. In this study, we compared 3D printed objects using MWCNT, continuous carbon fiber and short carbon fiber and consider strength and usefulness of them.

2. Introduction

2.1. Material

It is used that polylactic acid (PLA, REVODE 110, Zhejiong Hisun Biomaterials Co., Ltd.) as a matrix and multi-wall carbon nanotube (MWCNT, NC 7000 TM, NANOCYL SA.) as filler to make PLA/MWCNT nanocomposite. These materials were kneaded using a twin-screw kneading extruder (ZSK 18 MEGAlab, Coperion Gmbh) and formed into filament with 1.75 mm in diameter. The mass concentration of MWCNT were 0 (neat PLA), 0.5, 1.0, 2.0 and 3.0 wt%.

Commercial materials were used to prepare test pieces made from continuous carbon fiber reinforced thermoplastic (continuous CFRTP) and short carbon fiber reinforced thermoplastic (short CFRTP). The short carbon fiber reinforced composite (ONYX, Markforged), neat resin (NYLON, Markforged) and continuous carbon fiber reinforced composite (carbon fiber, Markforged) was used.

2.2. 3D printing

Tensile test pieces were made by 3D printing. SCOOVO X9 (Abee Co.) was used for printing of PLA/MWCNT nanocomposite and Markforged X7 (Markforged Inc.) was used for printing of continuous and short CFRTP. The test piece shape is shown in Fig. 1 (a) and (b). The specimen shape is different due to the limitation by the performance of the 3D printer. For PLA/MWCNT, smaller one was chosen to avoid warp and print quickly. The PLA/MWCNT has three outer walls and fills inside alternately at 45 degrees and -45 degrees. As shown in fig. 2, in the continuous CFRTP, carbon fibers oriented in the load direction were printed in the center and the outside was covered with neat resin (NYLON) or short fiber reinforced composite (ONYX). The short CFRTP was printed in the same way as the PLA composite using the short fiber reinforced composite (ONYX).

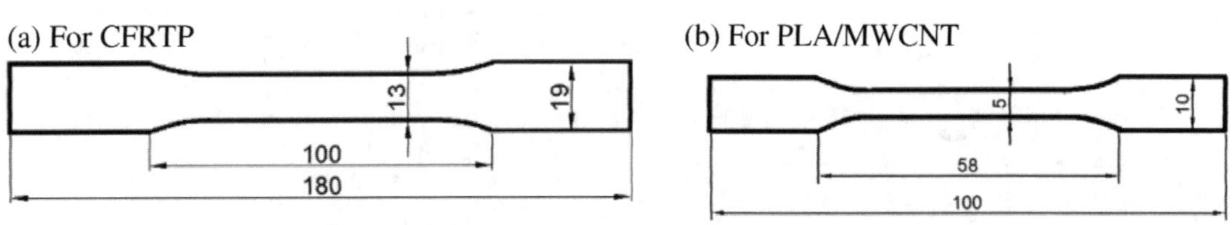

Figure 1. Dimension and shape of specimen (mm)

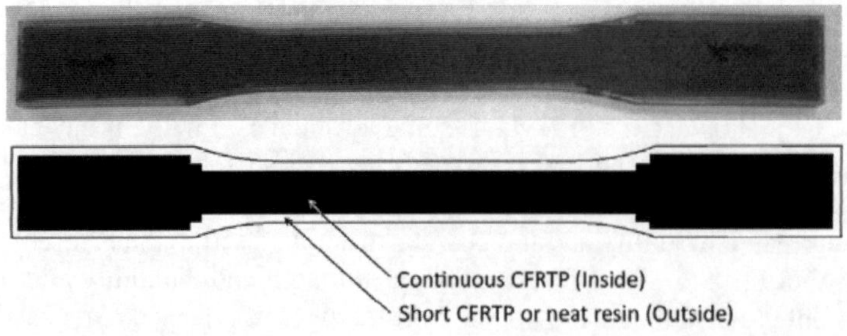

Figure 2. Continuous carbon fiber reinforced thermo-plastic

2.3. Tensile test

A tensile test was carried out using the prepared test piece. A universal testing machine (Auto graph AG-I 100 kN, Shimadzu Co.) was used for the test. The crosshead speed was set to 1 mm / min. In order to measure the strain, non-contact digital video extensometer (TRviewX 240S, Shimadzu Co.) and cross head displacement was used.

2.4. Image observation

Image observation of the fracture surface was carried out. The scanning electron microscope (JSM-7001FD, JEOL Ltd.) was used. Trimming, marking and colour adjustment were performed on the image using painting software.

2.5. Observation of carbon fiber

The 3D printed object using carbon fiber were heated in an electric muffle furnace to remove the matrix. Heating was carried out at 550 ° C for 30 minutes. The volume contents of the fiber were measured by comparing the weights before and after. The remaining carbon fiber was taken out and the fiber length was measured by an optical microscope

3. Results and discussion

3.1. Mechanical properties

The stress-strain diagram of the PLA/MWCNT composite is shown in the fig. 3. In PLA/MWCNT, the stress increased almost linearly until fracture. The breaking strain was about 1 ~ 2%, and no stress reduction was occurred. Compared with neat PLA, the elastic modulus was not greatly improved but the tensile strength was improved and increased by 48% when 1wt% of MWCNT is added. In that case, the tensile strength was 53 MPa and the Young's modulus was 3 GPa. Until 1 wt%, the tensile strength was improved as more CNT is added, but strength was decreased when 3wt% was added. It is because the aggregation of MWCNT. The aggregations are considered to act as internal defects of the material. The cross section of the PLA/MWCNT specimen is shown in the fig. 4. Using a microtome (RMD-5, Nihon Microtome Laboratory, Inc.), a thin piece was cut out from the cross section of the test piece and photographed with a microscope. The white spots are voids and the black spots are aggregates. As a result of measuring the area using the image processing software, both the area and the number of the aggregates increased with the increase of the addition amount of MWCNT. The image of the specimen fracture surface is shown in the fig. 5. In the SEM image, it was confirmed that the MWCNTs were dispersed, but there are some aggregates. Several voids were generated. These were made when overlaying. There was no crack between the laminated layers, and the interlayer adhesion may be good.

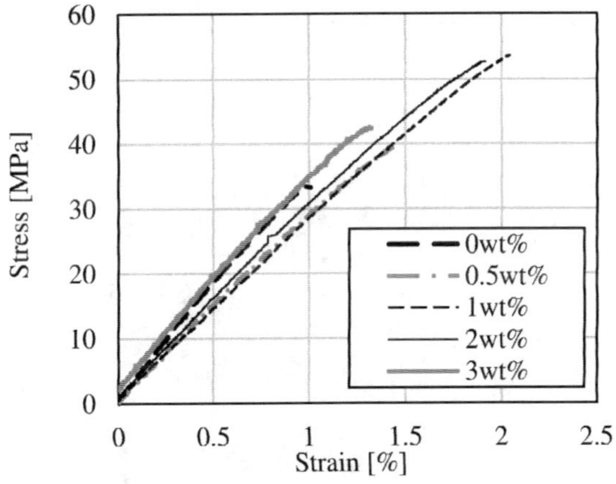

Figure 3. Stress-strain curve of PLA/MWCNT

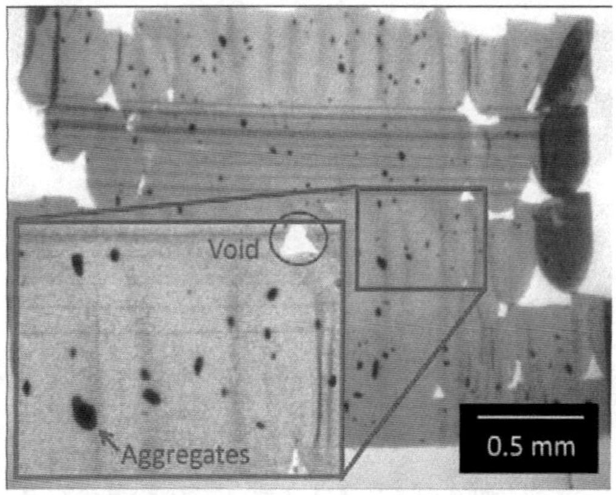

Figure 4. Aggregates and voids

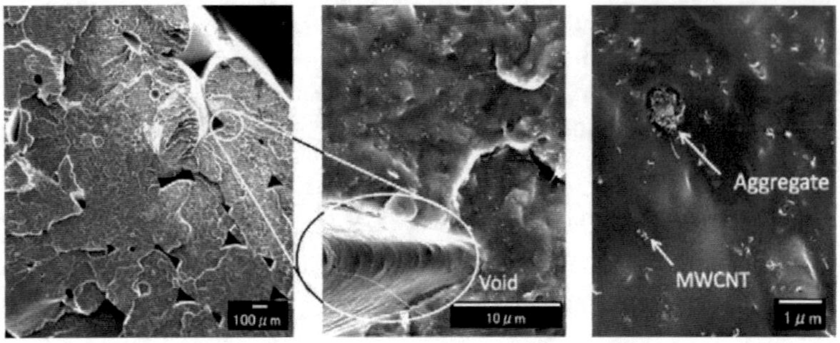

Figure 5. The fracture surface of PLA/MWCNT

As a result of the measurement, the weighted average fiber length of the short carbon fiber was 0.13 mm. The stress strain diagram of short CFRTP (ONYX) is shown in the fig. 6 (dashed Line). At first, it showed a linear relationship and the elastic modulus was 2 GPa. But the gradient decreased from about 1% strain, and finally the breaking strain reached 10%. The tensile strength was 45.5 MPa. This is lower than the neat resin material (NYLON) that is described in the catalogue. The fracture surface is shown in the fig. 7 and fig. 8. There were some fiber which was pulled out of the matrix in fig. 8. This indicates poor adhesion between the fiber and the matrix. It is thought that this is the cause of the deterioration of the mechanical properties of the short CFRTP. In addition, as shown in fig. 7, the fracture surface has a step for each laminated layer. This is because adhesion between the laminated layers was poor. It also affected the mechanical properties. With the short carbon fiber reinforced material, it was seemed that the fiber orientation was different for each laminated layer in fig. 8, and the fiber were oriented along the resin extrusion direction. It is because the short fiber dispersed in the material are oriented in one direction through the process of forming filaments and extruding from the nozzle.

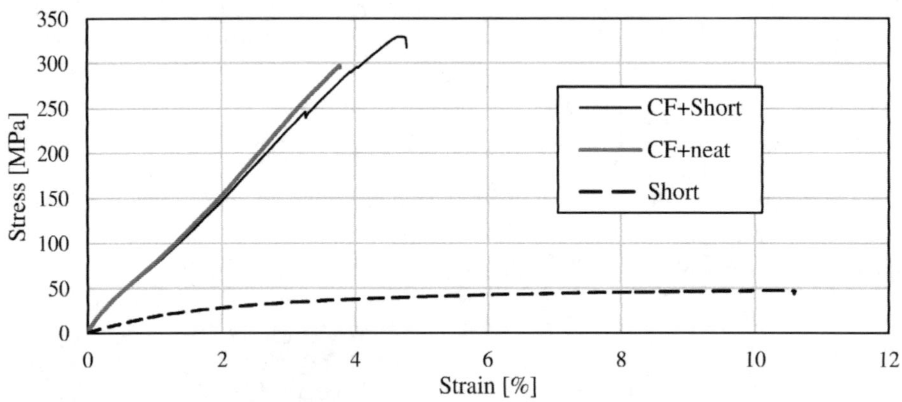

Figure 6. Stress-strain curve of CFRTP

Figure 7. The fracture surface of short CFRTP

Figure 8. Short carbon fiber pulled out

The stress-strain diagram of the continuous fiber reinforced material is shown in the fig. 6 (solid line). Looking at the stress-strain diagram, the slope varies at 0.5% strain. Also, the curve was disturbed around 3 ~ 4% strain. In the continuous CFRTP, the continuous fiber reinforced material is covered with a neat resin or a short fiber reinforced material to improve surface roughness and so on. It is thought that the change and disorder of the inclination occurred due to breakage of the interface between the fiber and the matrix and peeling of the interface between the outer layer and the strong central layer. The test piece after the test is shown in the fig. 9. The specimen did not break completely and the center continuous reinforcing fiber part and the outer peripheral part peeled at the boundary. This material had tensile strength of 341 MPa and elastic modulus of 10 GPa. These were much higher than that of short CFRTP. However, as mentioned above, the adhesion between the layers are weak, and destruction has also occurred from there. By improving this, it is considered that the mechanical properties of this material can be further improved.

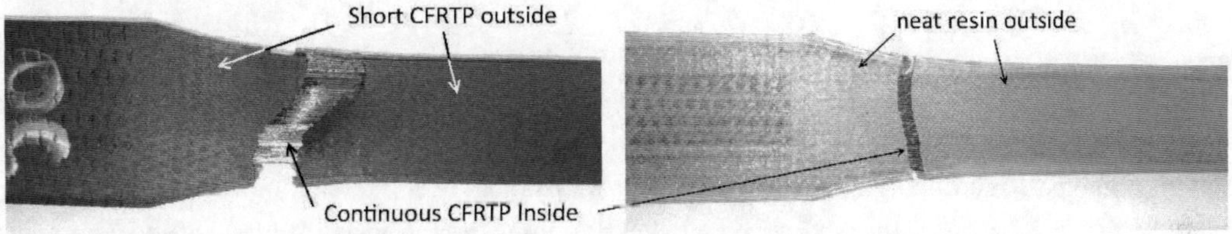

Figure 9. The broken specimen (continuous CFRTP)

3.2. Comparison by material

The tensile strength and elastic modulus of each material prepared in this study are shown in the fig. 10. The continuous CFRTP has the strongest elastic modulus and tensile strength. The FDM 3D printer extrudes the material little by little while moving the nozzle, so it can control the internal structure. For example, it is possible to place a lot of fiber where the stress concentrates or align fiber along load direction. Therefore, using in the FDM, parts with high mechanical properties can be manufactured. However, to use continuous fiber reinforced material in 3D printer, modified equipment must be used. It require special nozzle to extrude fiber and mechanism to cut fiber.

The short CFRTP and PLA/MWCNT are inferior in mechanical properties compared to the continuous. But they can be printed with conventional 3D printers without special modifying. Especially the nanocomposites demonstrate its effect by adding a small amount. The mass concentration of fiber was 35.7 wt% for continuous CFRTP and 14.3 wt% for short CFRTP, but MWCNT was 3wt% or less. Generally, the smaller the amount of reinforcement, the more easy to print. In fact the PLA/MWCNT nanocomposite can be printed with commercially available 3D printer without special modified in this study. Continuous fiber and short fiber material should each have merits and demerits and should be used properly.

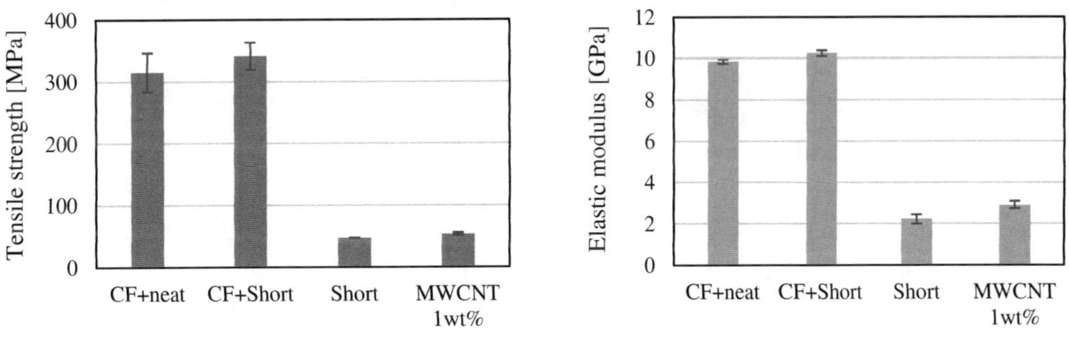

Figure 10. The tensile strength and elastic module

4. Conclusion

In this paper, mechanical properties of 3D printed object using PLA/MWCNT nanocomposite, short CFRTP and continuous CFRTP were evaluated. Nano fiber and short fiber were mixed in the filaments and replace the conventional neat filament. The continuous fiber was mixed with the resin at the nozzle. Experimental results show that the continuous fiber reinforced material has a tensile strength of 341 MPa, a short fiber of 47 MPa and a nanocomposite of 53 MPa. The elastic modulus was about 10 GPa for continuous fiber, about 2 GPa for short fiber and about 3 GPa for nanocomposite. By observation of the fracture surface, it was shown that adhesion between layers and interfacial strength between fiber and matrix affects strength in fiber reinforced materials. And in the nanocomposite, strength decreases due to formation of aggregates. Continuous fiber reinforced materials are superior in strength, but it require special equipment to print. However, nanomaterials improve strength and can be replaced from neat materials without modification. By properly using both, it leads to the manufacture of better parts.

References

[1] Sano Y, Yagyu J, Keishi T and Kawashima I, *Tokoton yasashii 3Dpurinta no hon (very easy, book of 3D printer)*, The Nikkan Kogyo Shimbun(2014)A reference

[2] Chenfeng K, Hsuchiang K, Chenchi M. M, Chiahsun C, Mechanical and electrical properties of multi-wall carbon nanotube/poly(lactic acid) composites, Journal of Physics and Chemistry of Solids, *Journal of Physics and Chemistry of Solids*, Vol.**69** (2008), pp.1395-1398

[3] Nanya L, Yingguang L, Shuting L, Rapid prototyping of continuous carbon fiber reinforced polylacticacid composites by 3D printing, *Journal of Materials Processing Technology*, Vol.**238** (2016), pp.218-225

[4] Vinayak Chandrakant G, Pravin R. K, Hrushikesh B. K, Mechanical and thermomechanical properties of carbon fibre reinforced thermoplastic composite fabricated using fused deposition modelling (FDM) method: a review, *International Journal of Mechanical and Production Engineering Research and Development*, Vol.**8** (2018), pp.1161-1168

IOP Conf. Series: Materials Science and Engineering **406** (2018) 012043 doi:10.1088/1757-899X/406/1/012043

Thermoplastic vs. thermoset epoxy carbon textile composites

Hironori Nishida[1,3*], Valter Carvelli[2], Toru Fujii[1], Kazuya Okubo[1]

[1] Department of Mechanical Engineering, Doshisha University, Tatara Kyotanabe, Japan
[2] Department A.B.C., Politecnico di Milano, Milan, Italy
[3] Hiroshima Prefectural Technology Research Institute, Kure, Hiroshima, Japan

* nishidacfrtp@gmail.com

Abstract. The main advantage of thermoplastic resins compared to thermoset counterpart are: increased toughness, better recyclability, and mainly the ability to deliver fast manufacturing processes. However, available thermoplastic resins (TP) have higher melt viscosity than thermoset ones. Due to high viscosity, the infusion process with conventional thermoplastic resins could lead to inappropriate impregnation of the fiber bundles. Recently, a thermoplastic epoxy resin (TP-EP) was developed with both advantages of thermoset and thermoplastic resins. The TP-EP couples the workability of thermoset, formability and recyclability of thermoplastic. The purpose of the present research is to comparatively assess the mechanical features of thermoplastic and thermoset epoxy carbon textile composites. As main outcome, the composite with highly polymerized thermoplastic epoxy has better mechanical performance than the conventional thermoset epoxy textile composite.

1. Introduction

Carbon fiber reinforced plastics (CFRP) have been studied because of excellent specific strength and specific rigidity. CFRPs have been widely used in aerospace and automotive industries. To take advantage of the recyclability and fast manufacturing process, carbon fiber reinforced thermoplastics (CFRTP) has been extensively studied ([1]-[5]). However, thermoplastic resins (TP) have higher melt viscosity than the thermoset ones. Hence, the infusion process with conventional thermoplastic resins could lead to inappropriate impregnation of the fiber bundles ([6], [7]). Recently, a new resin coupling the advantages of thermoset and thermoplastic was developed ([8], [9]). It has the good workability of thermoset and the post-formability and recyclability of thermoplastic. The CFRTP produced with in-situ resin had a lot of attention ([10], [11]).

In previous investigations, the effect of the weight-average molecular weight (Mw) on mechanical properties of thermoplastic epoxy carbon textile composites was measured. The result highlighted the better properties of high Mw (Mw>60,000) CFRTP ([12], [13]). However, to better understand its advantages for real industrial applications, it is necessary to assess and compare the effects of thermoplastic epoxy (TP-EP) and thermoset epoxy (TS-EP) on the main mechanical properties of CFRTP and carbon fiber reinforced thermoset (CFRTS) counterpart.

The purpose of the present research is to comparatively assess the mechanical features of CFRTP-EP and CFRTS-EP carbon textile composites. Micro droplet, quasi-static tensile, Izod impact, mode I and mode II inter-lamina fracture and inter-laminar shear tests were conducted.

2. Materials and Manufacturing

2.1. Reinforcement and Matrix

Plain weave carbon fiber fabric (Mitsubishi Rayon TR3110MS) was used as reinforcement (yarn TR30S 3L, linear density 1.79 g/cm^3, pick and end counts 12.5 per inch, areal weight 200 g/m^2). Thermoplastic epoxy resin (Denatite XNR 6850A, Accelerator XNH 6850B; supplied by Nagase ChemteX Corporation, Japan) [9] was used as matrix (T_g was approximately 100 degree-C). Thermoset epoxy resin (JER828, Mitsubishi Chemical Corporation) and amine (JER113, Mitsubishi Chemical Corporation) were used as resin and curing agent, respectively.

2.2. Manufacturing of CFRTP laminates

Plain weave CFRTP prepreg was made by the following procedure. The resin, 'XNR6850A', was heated by an electric oven at 120 degree-C. When the temperature of the resin reached 105 degree-C, the accelerator 'XNH6850B' was added to the resin with stirring. The plain weave carbon fabric was impregnated with the thermoplastic epoxy resin by hand lay-up. The molecular weight of prepreg was finally controlled by a predetermined time and temperature sequence.

CFRTP prepreg impregnated with the thermoplastic epoxy resin in the state of oligomer was polymerized at a given temperature in an electric oven. The obtained prepreg was dried at 50 degree-C for 12 hours. CFRTP laminates were prepared by press molding at 175~195 degree-C and 6~12 MPa. The carbon fiber volume fraction of the CFRTP laminates was approximately 45 %.

2.3. Fabrication of CFRTS laminates

The same plain weave carbon fiber fabric was used as reinforcement. The CFRTS plates were laminated by hand lay-up impregnation. The mold was cured in a hot press at 80 degree-C for 1 hour and then at 150 degree-C for 3 hours. The CFRTS laminates had approximately 45 % fiber volume fraction, as the CFRTP ones.

3. Experimental Activities

3.1. Measurement of weight-average molecular weight

The weight-average molecular weight of the TP-EP matrix was measured for each batch by the gel permeation chromatography (GPC) adopting a CLASS-LC10 (Shimadzu Corporation) and a GPC column (Styragel HR4E, Styragel HR5E: waters). Tetrahydrofuran (THF) was used as solvent. The calibration curves were drawn based on the retention time and the Mw of standard polystyrene.

3.2. Micro-droplet tests

Figure 1 shows a scheme of the micro-droplet test setup. Both ends of a single carbon fiber were fixed on a sheet of paper using an epoxy-based adhesive. The droplets were attached to the single carbon fiber by a soldering iron. The interfacial shear strength was determined with a pullout loading speed of 0.12 mm/min. The average strength was obtained using at least 20 specimens for each resins. Equation (1) was used to calculate the interfacial shear strength (τ).

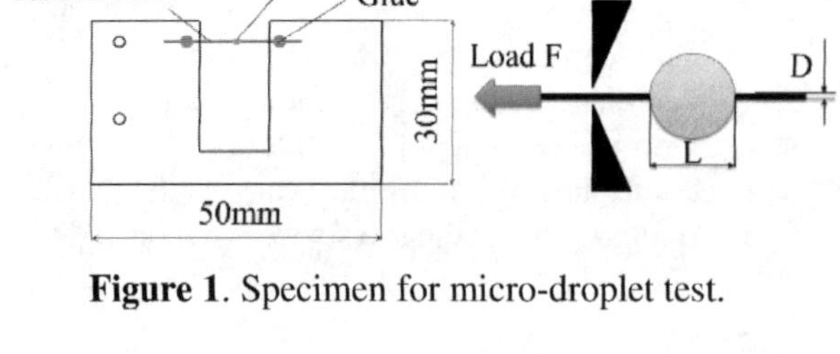

Figure 1. Specimen for micro-droplet test.

$$\tau = \frac{F}{\pi D L} \tag{1}$$

where τ is the interfacial shear strength; *F* is the pull-out load; *D* is the fiber diameter; *L* is the adhesion length.

3.3. Quasi-static tensile tests

Tensile tests in the warp direction of CFRTP and CFRTS were conducted according to the standard ISO 527 at 1 mm/min (cross head speed) by using a universal material testing machine. Dimensions of the specimen were: length 200 mm, width 25 mm and thickness 2 mm. The distribution of strain on the surface of the specimen was estimated by the DIC (Digital Image Correlation) technique. The surface was painted with white and speckled with black acrylic paints for a length of about 4 cm (see Figure 2). During loading, digital images were recorded with a frequency of 1 Hz by a digital camera. Processing of images was performed by the software VIC-2D™.

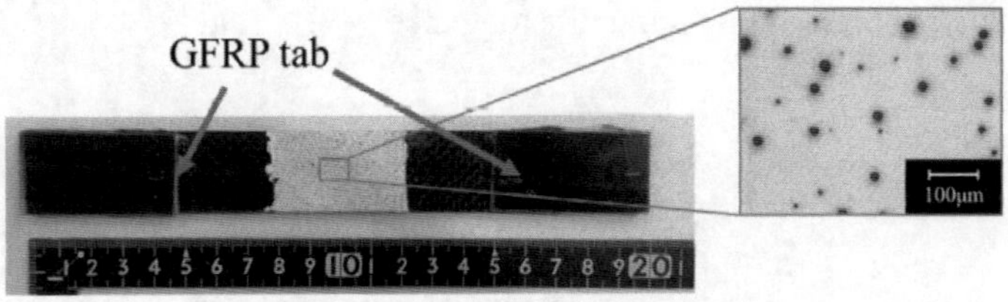

Figure 2. Specimen for quasi-static and fatigue tensile tests.

3.4. End Notched Flexure (ENF) test

Figure 3 shows geometry of specimen for ENF test. Laminates had 20 layers of plain weave carbon fabric. The quasi-static three-point bending loading at 0.5 mm/min of cross-head speed allowed for the estimation of the Mode II inter-lamina fracture toughness. The length of pre-crack was 50 mm. Kapton film of approximately 30 μm thick (Kapton, Du Pont-Toray Corporation) was inserted between 10th and 11th ply of the laminate. Equation (2) was used to calculate the Mode II inter-lamina fracture toughness.

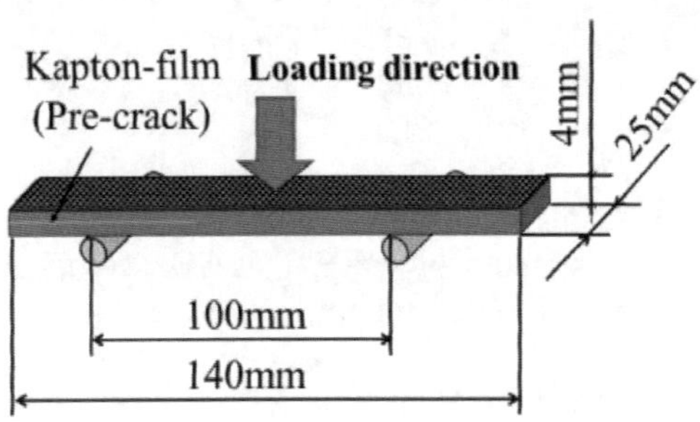

Figure 3. Specimen for ENF test.

$$G_{IIC} = \frac{9a_1^2 P_c^2 C_1}{2B(2L^3 + 3a_1^3)} \tag{2}$$

where

$$a_1 = \left[\frac{C_1}{C_0}a_0^3 + \frac{2}{3}\left(\frac{C_1}{C_0} - 1\right)L^3\right]^{\frac{1}{3}} \tag{3}$$

In Eqns. (2, 3), G_{IIC} is mode II inter-lamina fracture toughness; a_0 is initial crack length; P_c is initial critical load; C_0 is load point compliance of the initial elastic part; C_1 is load point compliance at initial critical load; a_1 is estimated crack growth length at initial critical load; L is supports span; B is specimen width.

3.5. Double Cantilever Beam (DCB) test

Specimen with the size of 160 (length) x 25 (width) x 4 (thickness) mm (Figure 4) was loaded to measure the Mode I inter-lamina fracture toughness at 0.5 mm/min cross-head speed by a universal material testing machine (EZ-test, Shimadzu Corporation). The pin-blocks for loading were attached as shown in Figure 4. The length of pre-crack was 50 mm. Kapton film of approximately 30 μ m thick (Kapton, Du Pont-Toray Corporation) was inserted between 10th and 11th ply of the laminate. Equation (4) was used to calculate the Mode I inter-lamina fracture toughness.

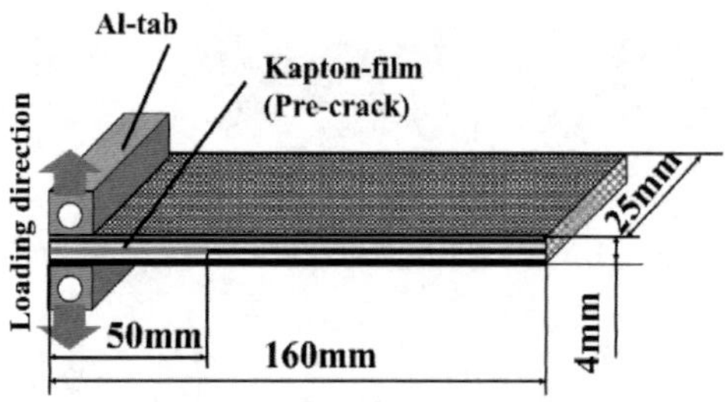

Figure 4. Specimen for DCB test.

$$G_{Ic} = \frac{3}{2(2H)}\left(\frac{P_c}{B}\right)^2 \frac{(B\lambda_0)^2}{\alpha_1} \tag{4}$$

where

$$\frac{a}{2H} = \alpha_1(B\lambda)^{\frac{1}{3}} + \alpha_0 \tag{5}$$

Symbols in Eqns. (4, 5) have the meaning: G_{IC} is mode I inter-lamina fracture toughness; $2H$ is specimen thickness; P_c is initial critical load; B is specimen width; λ_0 is COD compliance of the initial elastic part; α_0 is slope of straight line; α_1 is intercept of the straight line; a is crack length

3.6. Inter-laminar shear strength (ILSS) test

Three point bending tests of short beam specimens were conducted by a universal material testing machine, following the recommendations of ISO 14130. Figure 5 shows geometry of specimen for ILSS test. The specimen had dimensions of 20 (length) x 5 (width) x 2 (thickness) mm. The span length was 10 mm. Bending load was applied with a displacement rate of 5.0 mm/min.

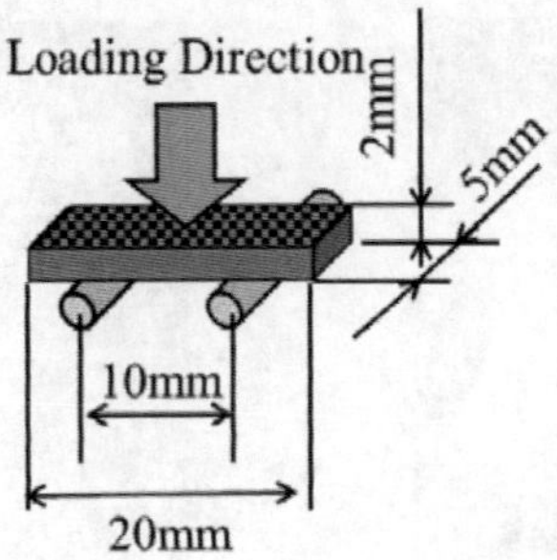

Figure 5. Specimen for ILSS test.

3.7. *Izod impact strength test*

Izod impact tests were conducted using an impact pendulum (Yonekura seisaku-sho, Co., Ltd., A1040). The izod impact test was according to ISO 180. Figure 6 shows geometry of specimen. The impact direction was edgewise and samples had dimensions of 80 (length) x 10 (width) x 2.5 (thickness) mm.

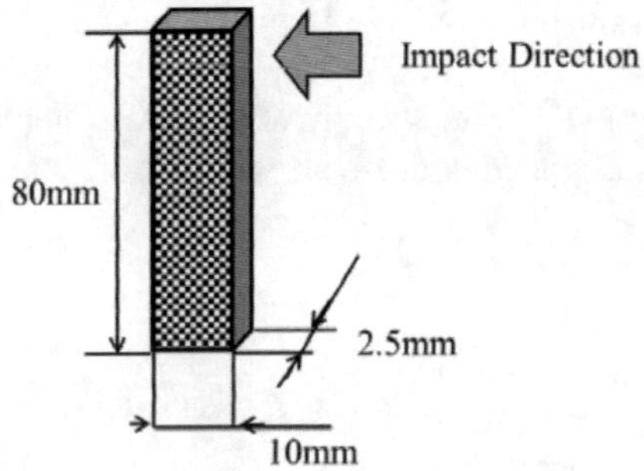

Figure 6. Specimen for Izod impact strength test.

4. Results and discussion

4.1. *Micro-droplet test*

The micro-droplet tests provided the interfacial shear strength compared in Figure 7. The interfacial shear strength of CFRTP depends on the weight-average molecular weight (Mw) of the matrix. The better strength was measured for Mw>60,000 by previous study ([12], [13]). For comparison, in Figure 7, the composite with Mw=90,000 was adopted. The interfacial shear strengths of CFRTS and CFRTP are comparable and in the same experimental scatter band.

Figure 8 shows the SEM observations of carbon fiber surface after micro-droplet tests. Some resin still in contact demonstrate the good adhesion with carbon fiber of both resins.

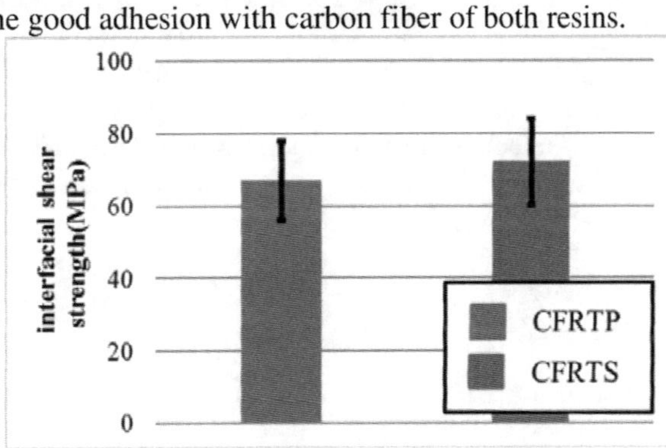

Figure 7. Interfacial shear strength.

IOP Conf. Series: Materials Science and Engineering **406** (2018) 012043 doi:10.1088/1757-899X/406/1/012043

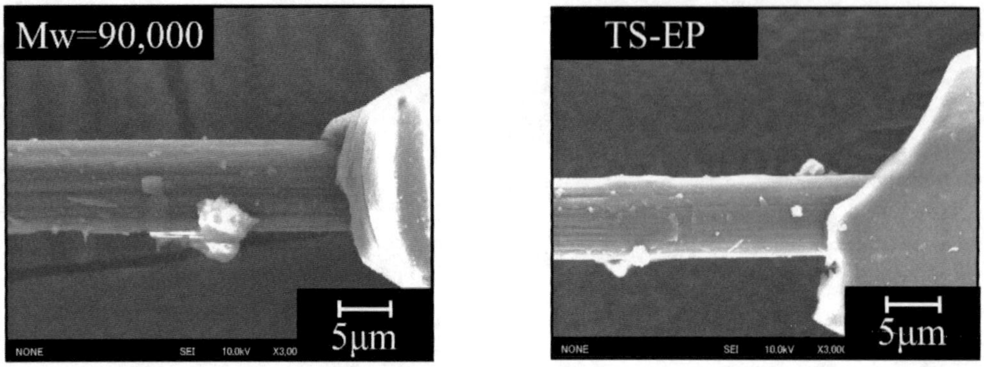

Figure 8. Micro-droplet test: surface of carbon fiber after adhesion failure with (left) TP-EP (Mw=90,000) and (right) TS-EP.

4.2. Quasi-static tensile tests

Figure 9 shows the quasi-static tensile test results in term of elastic modulus and strength. Previous study [13] highlights that the tensile strength of CFRTP depends on weight-average molecular weight (Mw) of the matrix and the better properties were for Mw>60,000. Therefore, here for comparison, the composite with of Mw=84,000 was considered. As a result, tensile strength and elastic modulus do not have considerable variations.

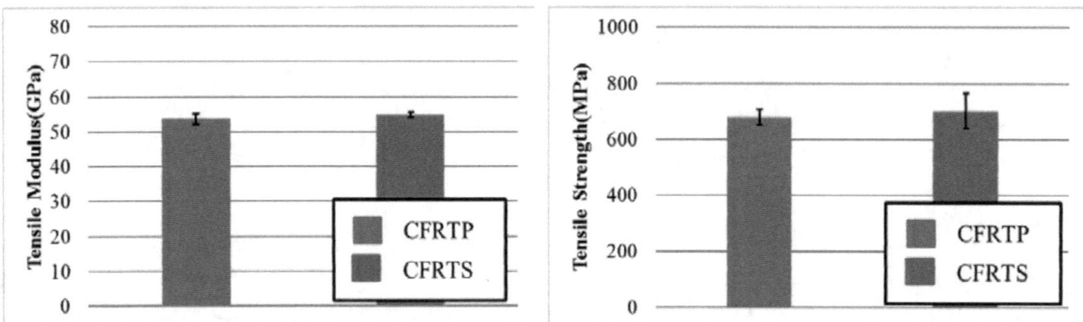

Figure 9. Quasi-static tensile properties: (left) Elastic modulus and (right) Strength.

4.3. Inter-laminar properties

One of the main mechanical weaknesses of laminated composite materials is the inter-lamina fracture toughness. It is measured here by three tests: ENF, DCB and ILSS.

4.3.1. End Notched Flexure (ENF) test. Figure 10 shows the results of the Mode II inter-lamina fracture toughness. The Mode II fracture toughness of CFRTP depends on the weight-average molecular weight (Mw) of the matrix [13]. The better properties is for Mw>60,000. Here, for comparison, the TP-EP with Mw=108,000 was considered. The comparison shows a considerable higher, almost double, Mode II inter-lamina fracture toughness of CFRTP than CFRTS.

IOP Conf. Series: Materials Science and Engineering **406** (2018) 012043 doi:10.1088/1757-899X/406/1/012043

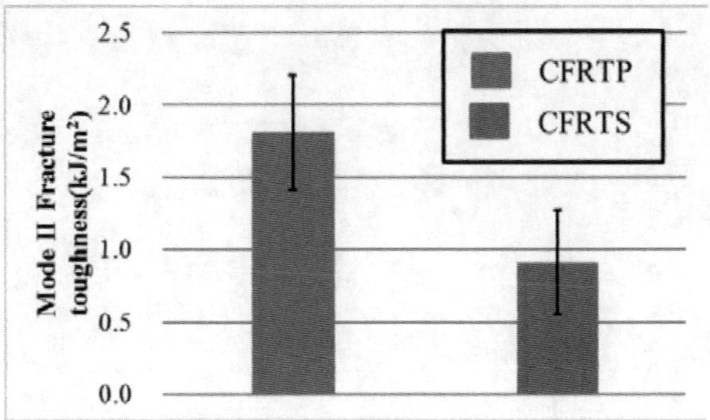

Figure 10. Mode II inter-lamina fracture toughness.

4.3.2. Double Cantilever Beam (DCB) test. Figure 11 shows the results of the Mode I inter-lamina fracture toughness. For comparison, the TP-EP with Mw=70,000 was adopted. As for Mode II, the Mode I inter-lamina fracture toughness of CFRTP is almost double than the CFRTS counterpart.

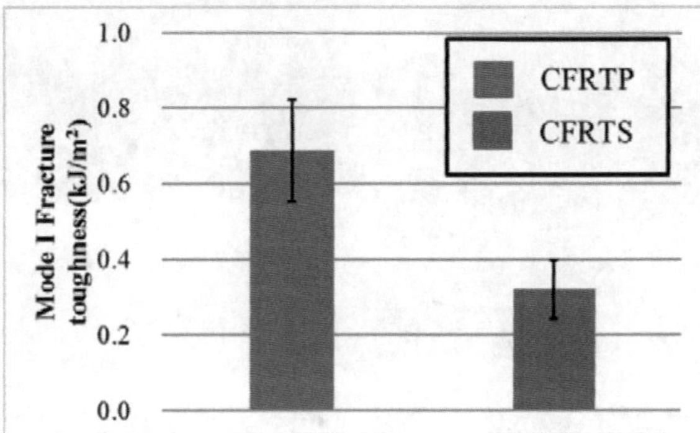

Figure 11. Mode I inter-lamina fracture toughness.

4.3.3. Inter-laminar shear strength (ILSS) test. The inter-laminar shear strength of CFRTP was ~20% higher than that of CFRTS laminate (Figure 12).
Figure 13 shows the load - mid span displacement curves, which had similar maximum load but different evolution. CFRTP curve had bi-linear shape with 'hardening' behaviour, while CFRTS curve had a sudden drop after the maximum load, meaning a fast inter-laminar crack development. The difference is visible observing the damage imparted after the ILSS tests. Figure 14 shows the failure modes. The CFRTS specimen had several and longer inter-laminar clacks than the CFRTP counterpart.

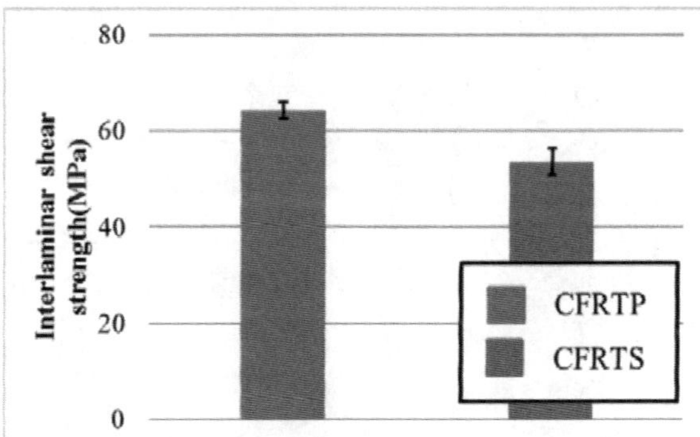

Figure 12. Inter-laminar shear strength.

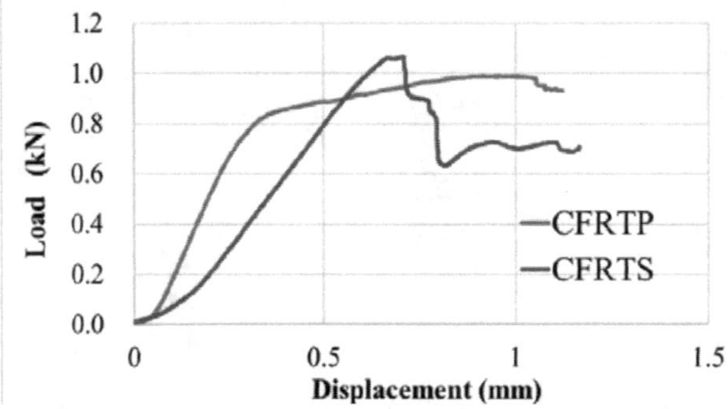

Figure 13. Inter-laminar shear strength: load - mid span displacement curve.

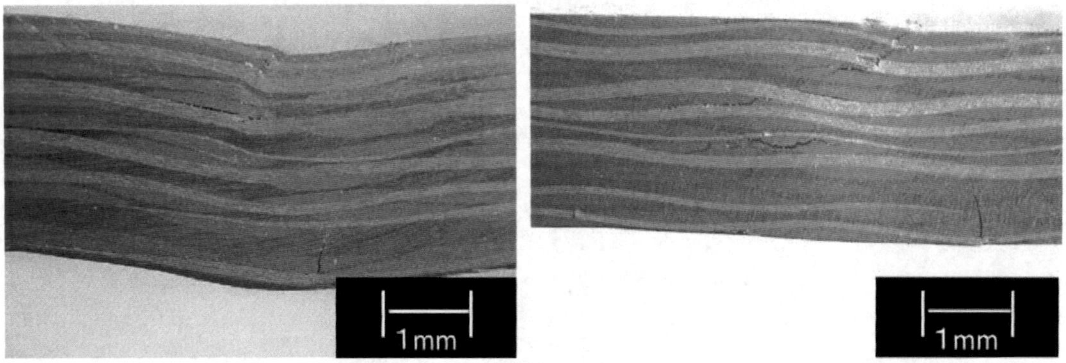

Figure 14. Inter-laminar shear strength: failure mode of (left) CFRTP (Mw=103,000) and (right) CFRTS.

4.4. Izod impact strength test

The Izod impact strength of CFRTP was ~10% higher than CFRTS (Figure 15).

Figure 16 shows the SEM observation of the fracture surface of CFRTP and CFRTS, respectively. The 'plastic' deformation of the matrix with good energy absorption was recorded for the TP-EP matrix of CFRTP (Figure 16 left). The 'brittle' behavior of the TS-EP matrix allowed lower energy absorption. Consequently, the impact strength was improved due to the energy dissipated for 'plastic' deformation of the matrix.

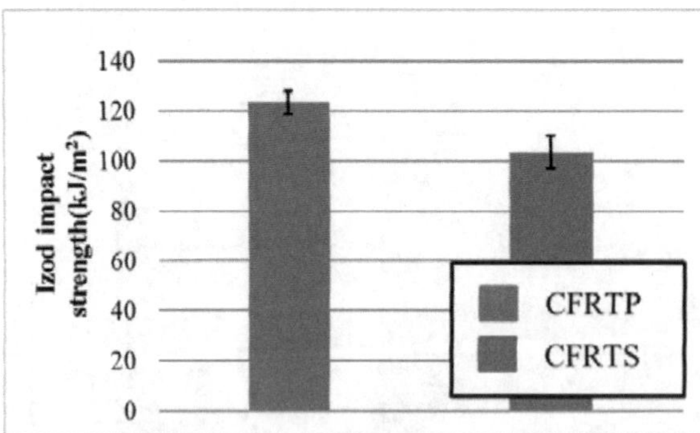

Figure 15. Izod impact strength.

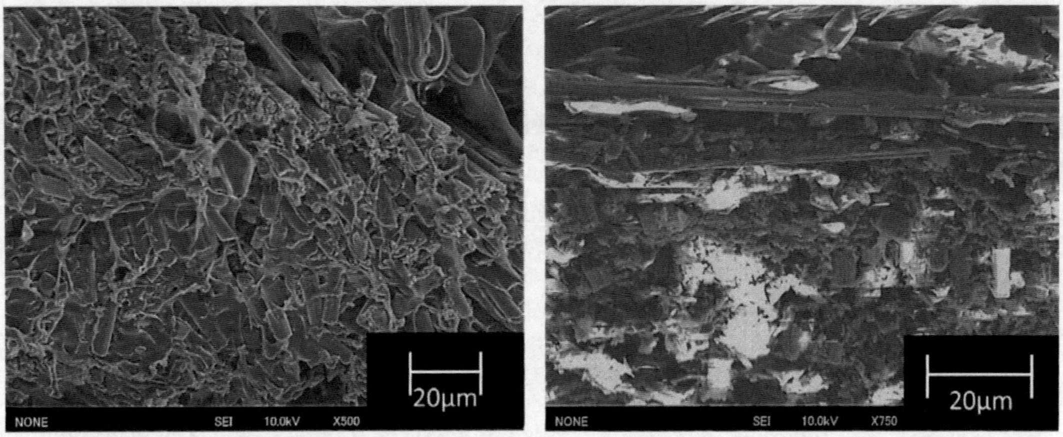

Figure 16. Izod impact tests: failure mode of (left) CFRTP and (right) CFRTS.

5. Conclusion

A summary and comparison of the complete set of experimental results is shown in Figure 17. As main outcome, the composite with highly polymerized thermoplastic epoxy has better mechanical performance than the conventional thermoset epoxy textile composite.

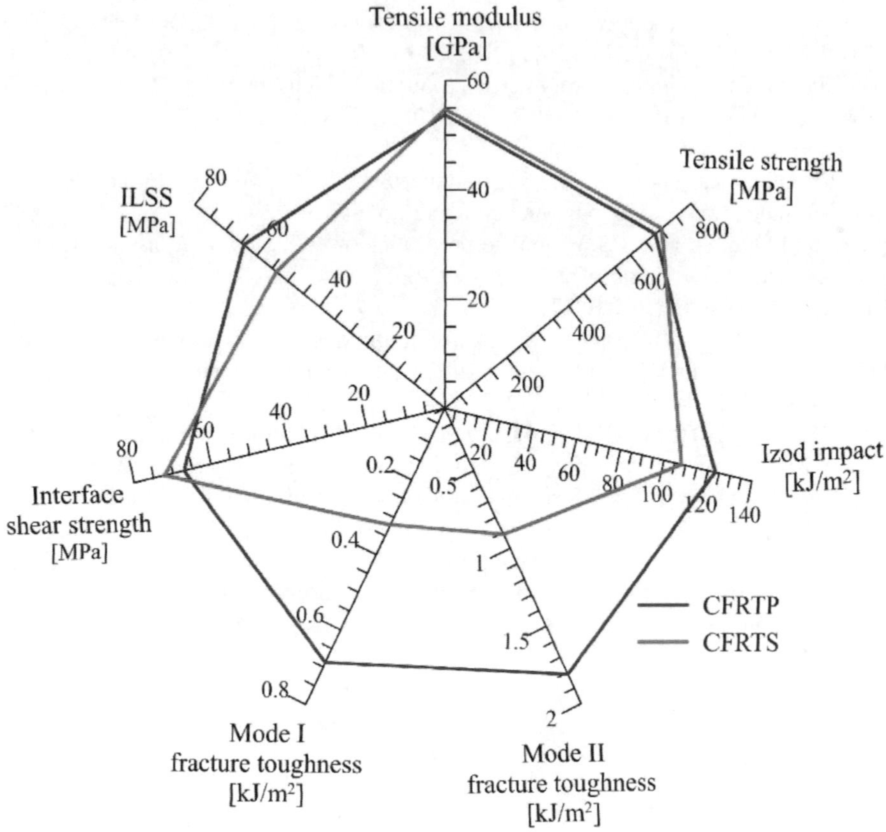

Figure 17. Comparison of CFRTP and CFRTS mechanical properties.

Acknowledgement
The authors are grateful for support of Doshisha University Research & Development Center for Advanced Composite Materials and Nagase Chemtex Corporation, Japan. Keisuke Nagai and Souichiro Imagawa are gratefully acknowledged for contribution in the experimental tests.

References

[1] Erber, A. and Spitko, S. (2014) 'Expanded role for thermoplastic composites', *Reinforced Plastics*. Elsevier Advanced Technology, 58(4), pp. 29–33. doi: 10.1016/S0034-3617(14)70178-X.

[2] Stewart, R. (2010) 'Automotive composites offer lighter solutions', *Reinforced Plastics*. Elsevier Advanced Technology, 54(2), pp. 22–28. doi: 10.1016/S0034-3617(10)70061-8.

[3] Marsh, G. (2013) 'Composites poised to transform airline economics', *Reinforced Plastics*. Elsevier Advanced Technology, 57(3), pp. 18–24. doi: 10.1016/S0034-3617(13)70088-2.

[4] Klimke, J. and Rothmann, D. (2010) 'Carbon composite materials in modern yacht building', *Reinforced Plastics*. Elsevier Advanced Technology, 54(4), pp. 24–27. doi: 10.1016/S0034-3617(10)70140-5.

[5] Mike and Brady, P. (2010) 'Technology developments in automotive composites', *Reinforced Plastics*. Elsevier Advanced Technology, 54(6), pp. 25–29. doi: 10.1016/S0034-3617(10)70213-7.

[6] Halawani, N., J.L. Augé, H. Morel, O. Gain, and S. Pruvost.. (2017) 'Electrical, thermal and mechanical properties of poly-etherimide epoxy-diamine blend', *Composites Part B: Engineering*. Elsevier, 110, pp. 530–541. doi: 10.1016/J.COMPOSITESB.2016.11.022.

[7] Dogan, A. and Arikan, V. (2017) 'Low-velocity impact response of E-glass reinforced thermoset and thermoplastic based sandwich composites', *Composites Part B: Engineering*. Elsevier, 127, pp. 63–69. doi: 10.1016/J.COMPOSITESB.2017.06.027.

[8] Nagase ChemteX Corporation, Process for Production of Thermoplastic Cured Epoxy Resin with Transparency to Visible Light, and Thermoplastic Epoxy Resin Composition. United States Patent Application Publication, Pub. No. US 2014/0194590 A1; 2014

[9] Nagase ChemteX Corporation. In situ-polymerizing Thermoplastic Epoxy Resin, Technical information sheet. http://www.nagasechemtex.co.jp/en/

[10] Naito, K. and Oguma, H. (2017) 'Tensile properties of novel carbon/glass hybrid thermoplastic composite rods', *Composite Structures*. Elsevier, 161, pp. 23–31. doi: 10.1016/J.COMPSTRUCT.2016.11.042.

[11] Ben, G. and Sakata, K. (2015) 'Fast fabrication method and evaluation of performance of hybrid FRTPs for applying them to automotive structural members', *Composite Structures*. Elsevier, 133, pp. 1160–1167. doi: 10.1016/J.COMPSTRUCT.2015.07.093.

[12] Nishida, H., Okubo, K., Fujii, T., and Carvelli, V. (2017) 'Improvement of Bending Strength of Carbon Fiber/Thermoplastic Epoxy Composites-Effects of Molecular Weight of Epoxy on Carbon Fiber/Matrix Interfacial Strength and Connection of Cracks in Matrix', *Open Journal of Composite Materials*, 7(4), pp. 207–217. doi: 10.4236/ojcm.2017.74014.

[13] Nishida, H., .Carvelli, V., Fujii, T., and Okubo, K. (2018) 'Quasi-static and fatigue performance of carbon fibre reinforced highly polymerized thermoplastic epoxy', *Composites Part B: Engineering*, 144, pp. 163–170. doi: 10.1016/j.compositesb.2018.03.002.

IOP Conf. Series: Materials Science and Engineering **406** (2018) 012044 doi:10.1088/1757-899X/406/1/012044

Using micro-CT to quantitatively validate the textile compaction simulation

W Wijaya[1], Y X Lim[1], P A Kelly[1*], S Bickerton[1], M A Ali[2], R Umer[2,3], K A Khan[2]

[1] Centre for Advanced Composite Materials, University of Auckland
[2] Department of Aerospace Engineering, Khalifa University of Science and Technology
[3] Centre for Future Materials, University of Southern Queensland

* pa.kelly@auckland.ac.nz

Abstract. The results of any dry textile compaction simulation have to be validated. Micro-CT is a perfect tool to facilitate such a task. By performing an equivalent dry textile compaction experiment inside a micro-CT, the actual deformation and placement of fibrous tows can be captured within a sample as it deforms. The resulting micro-CT images can readily be used to qualitatively validate a compaction model. This work presents a set of methodologies to facilitate a straight forward quantitative comparison between a micro-CT sample and an equivalent compaction model. The images from the micro-CT and the compaction models are subjected to various image processing techniques that are implemented in MATLAB. The program is designed to allow a user to process the images in a consistent and semi-automated manner while minimising the manual labour time. Various cross-sectional parameters of the tows at each image slice are computed automatically using an algorithm that can handle any arbitrarily shaped cross section. By presenting the computed data as a series of Excel files and 3D plots, this work concludes that micro-CT can be used to conveniently quantitatively validate the result of compaction models.

1. Introduction

The permeability of textile reinforcements is a critical model parameter for any mould filling simulation software. A complete simulation chain to predict the textile permeability can be broadly broken down into three logical steps: textile modelling, textile compaction simulation, and flow simulation. Due to the dominant influence of textile flow channel geometry on the predicted permeability, a textile compaction model that can capture realistic tow deformation behaviour is of utmost importance. Such a model needs to be validated both qualitatively and quantitatively. Over recent years, micro-CT has emerged as a reliable tool to perform qualitative validations because of its ability to capture the actual geometry of textile stacks at various compaction levels [1-3]. Its potential for quantitative validation has been hindered by the insufficient image contrast for automatic image segmentation. Bale et al. [4] coated carbon fibre preforms with SiC by chemical vapour infiltration technique before scanning the samples under micro-CT to have better observations of the tow boundaries. Template contours were drawn manually along the tow boundaries before fitting ideal ellipses that allowed various measurements of tow parameters. Pazmino et al. [5] analysed a single layer of E-glass non-crimp 3D orthogonal woven reinforcement under micro-CT. The cross-sectional shape of the tows was outlined by six points that allowed various geometric parameters of the tows to be measured using a convex envelope algorithm. Barburski et al. [6] studied the micro-CT images of twill-weave carbon fabric. The tow boundaries were outlined by a free-hand drawing operation followed by fitting ideal ellipses to these outlines. All of these

IOP Conf. Series: Materials Science and Engineering **406** (2018) 012044 doi:10.1088/1757-899X/406/1/012044

works, however, did not show how explicit quantitative comparison between the micro-CT images and the compaction models can be made. Furthermore, most works relied on having nearly elliptical tow cross sections so that ideal ellipses can be fitted to extract width and thickness information. This work addresses these issues and also means of minimising manual labour time.

In this work, a series of image processing techniques are executed in MATLAB to analyse the images of both micro-CT and compaction model. A convenient free-hand drawing technique is used to outline the tow boundaries from the micro-CT images. The tow boundaries from the compaction model are outlined automatically using MATLAB's inbuilt edge detection function. An in-house algorithm allows both the width and thickness information to be extracted from any arbitrarily shaped tow cross-section. Finally, all of the outlined cross sections and the associated geometric parameters are summarized in a series of Excel files and 3D plots for aid of comparison.

2. Material and micro-CT scanner

The material used in this study was an 800 GSM E-glass plain weave. Several 40 mm diameter preform samples were prepared and scanned with a GE® Phoenix Nanotom device available in the Khalifa University of Science and Technology [7]. The device has a 40 mm diameter compaction fixture with a load cell capacity of 5 kN. The circular samples were compacted in this fixture at a crosshead speed of 5 mm/min to a certain compaction level and then paused to relax. For each sample, after sufficient stress relaxation was achieved, scanning was done at 25 μm voxel size. This installation allows the same compacted sample to be further compacted, relaxed and scanned, thus enabling the deformed shape of the same sample to be observed at various compaction levels.

3. Processing micro-CT images

The raw micro-CT images were subjected to various image processing techniques, automatically executed within MATLAB. At the beginning of the program execution, a graphical user interface (GUI) allowed the user to enter basic inputs which include the initial image, the final image, the total number of images, and the total number of fibrous tows to be analysed. As will be explained later, the user has the flexibility to choose which tows in the images to be processed. The general processing methodology is presented below. The program was designed to process a series of images along one tow direction (weft or warp) at one time instant (the geometry at different times can be analysed in the same way).

3.1. Image enhancement

The raw grey-scale micro-CT images needed to be enhanced to allow accurate identification of the tow boundaries. Based on the initial input from the user, the program automatically chose the set of images to be processed and ordered them based on their names. Each image was complemented such that the dark regions become lighter and the light regions become darker. Following this, the image contrast was enhanced by saturating the bottom 1% and the top 1% of all pixel values. This percentage was determined by trial and error and can be adjusted as wished. An example of raw and processed images is presented in figure 1(a) and 1(b), respectively. Notice how the tow boundaries in the raw image are very difficult to discern and how the image enhancement process improved their visibility and sharpness significantly.

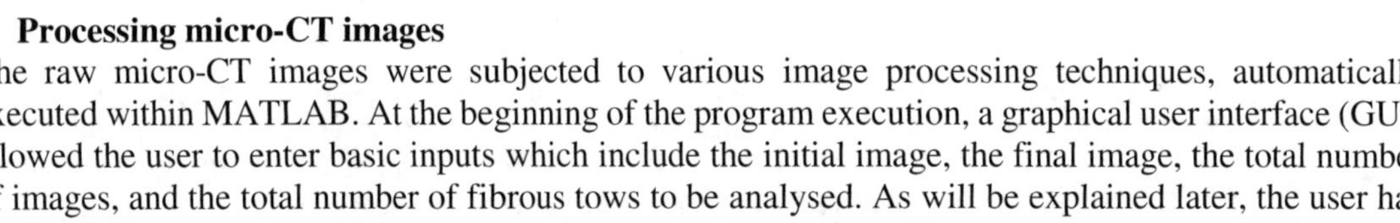

(a) (b)

Figure 1. Comparison of the (a) raw and (b) processed micro-CT image.

3.2. Identification of tow boundaries

Following the image enhancement process, the tow boundaries were ready to be located. An interactive GUI was created to allow the user to perform free hand drawing to outline the tow boundaries at every chosen image. The order in which the user outlines the tows defines the tow IDs (numbered sequentially) and this is established when processing the very first image. Therefore, when outlining the tow boundaries of subsequent images, the user must be consistent with this established order. The GUI allows zoom and undo operations, allowing for a robust and accurate outlining process. Following this outlining process, both the centroid location and the area of each tow cross section of each image were determined and recorded automatically. Figure 2 shows an example of the result of these procedures, showing the drawn tow outlines and the identified centroid locations.

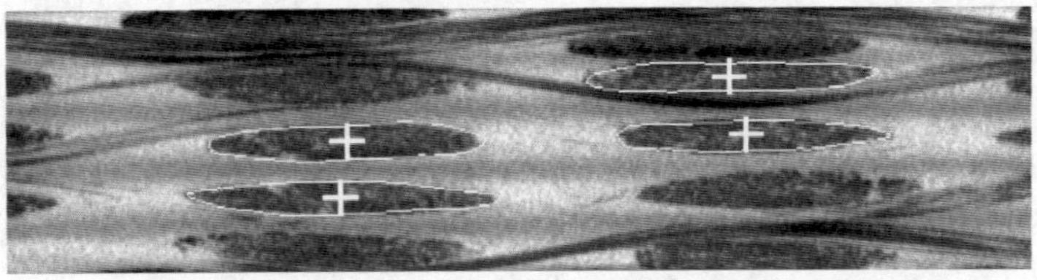

Figure 2. Drawn tow outlines and the identified centroids.

3.3. Geometric measurements

From the boundary location process described previously, the locations of the pixels that made up the tow boundaries were recorded which allowed various cross-sectional parameters (width, thickness, and angle of rotation) to be computed automatically. The algorithms to compute these geometric parameters must be able to handle any arbitrary cross-sectional shape in a consistent manner. The geometric data were then stored for later 3D visualization and written to a series of Excel file. In the following, the methodologies to compute these geometric parameters are outlined.

3.3.1. Width and angle. To compute the width of a particular tow cross section, the coordinates of the most left and right pixels were identified. It was found in some cases that there was more than one pixel along the same vertical line at these extreme positions. To address this issue, the y-coordinates of these pixels were averaged to give a unique pixel coordinate at each extreme position. The width was then computed as the magnitude of the line connecting the (y-averaged) coordinates at the left and right-most position of the tow boundary, by Pythagoras' theorem. An example of such a connecting line is shown in figure 3. Following this, the angle between the connecting line and the global horizontal axis gave the measure of cross sectional rotation.

3.3.2. Thickness. To compute the thickness of a particular tow cross section, the midpoint of the connecting line described previously was computed. From this midpoint, a perpendicular line (refer to figure 3) was drawn out in both directions until it hit the identified tow boundary. The distance between the two points of intersection gave the measure of tow thickness.

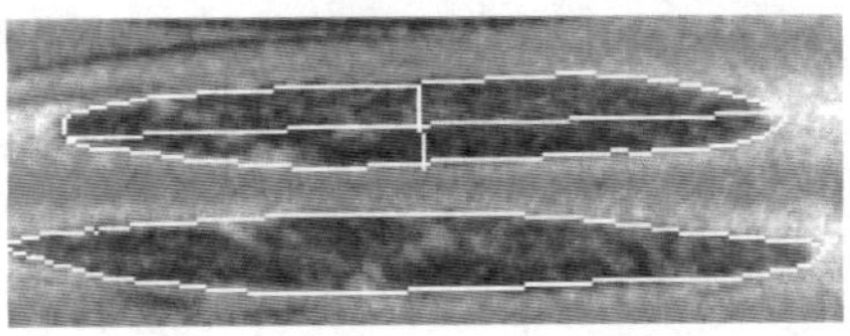

Figure 3. The width and thickness lines

Figure 4. Raw cross sectional image captured by HyperView

4. Processing images from the compaction model

The dry textile compaction simulation was done in ABAQUS. To allow both quantitative and qualitative comparisons with the micro-CT images, it is desirable to extract the same geometric parameters from the deformed textile model as for the micro-CT images presented in section 3. The general processing methodology is presented below.

4.1. Image extraction

The ABAQUS output (.odb) file contained the information of the textile deformed geometry. This file was imported into HyperView™. HyperView™ allows the current view at its viewport to be exported as an image file. This way the cross-sectional pictures at various locations within the deformed textile model were extracted. The user has the flexibility to control the number of cross-sectional pictures along the direction of interest (weft or warp) by using the in-built options within HyperView™. Once this direction of interest is chosen, the deformed tows running in the other direction (warp or weft) were removed manually by the user for easier analysis in the subsequent steps. Figure 4 illustrates an example of a cross sectional image captured by HyperView™ from a compacted 4-layer non-nested plain weave stack. The images extracted from HyperView™ were then inputted to MATLAB to be processed.

4.2. Image preparation

Since the extracted images came from a simulation, the colours of the tows and the background could easily be made distinct. This allowed for automatic image segmentation to isolate individual tow cross section as can be seen in figure 5(a). Minor modification to the binarised images were made to remove the incomplete tow cross section connected to the image boundaries. An algorithm was employed to detect for any object connected to the image boundaries and to delete them. The final cleaned image can be seen in figure 5(b).

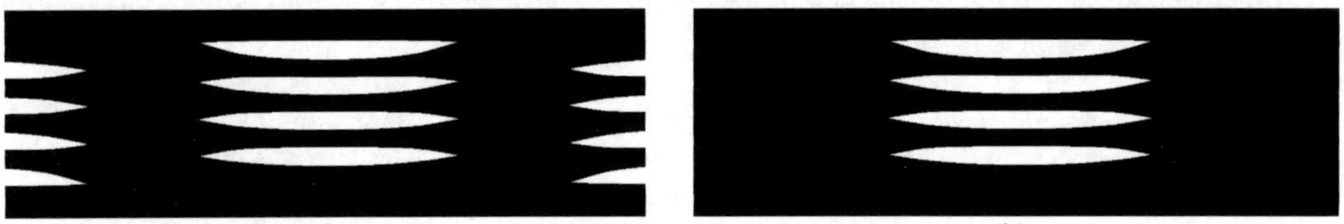

(a) (b)

Figure 5. The binarised images from the compaction model (a) before and (b) after boundary cleaning.

4.3. Identification of tow boundaries

Referring to figure 5, the identification of tow boundaries was simpler than for the micro-CT analysis. Instead of applying the free hand drawing algorithm as done for micro-CT images, Canny's edge detection algorithm [8] was employed to identify the tow boundaries. Following the edge detection procedure, information such as the tow centroids, the tow areas, and the minimum rectangles enclosing individual tow cross section were readily computed and recorded. Using a GUI, the user assigns IDs to each tow by clicking on the tow cross sections in the very first image, from the preferred first to last cross section in an orderly manner. For the subsequent images, the program can keep track of the tow IDs automatically. For example, consider a tow cross section with a tow ID=1 in the ith image. For the (i+1)th image, the program computes the distance between the centroid of image i, ID=1 and the centroids of all tow cross sections in the (i+1)th image; the cross section which results in the minimum distance is considered to be image i+1, ID=1. Figure 6 shows an example of the identified tow boundaries from the edge detection algorithm, super imposed on the original grey-scale image. The tow edges and centroids were accurately identified. The numbers located near the centroids corresponds to the user-assigned tow IDs.

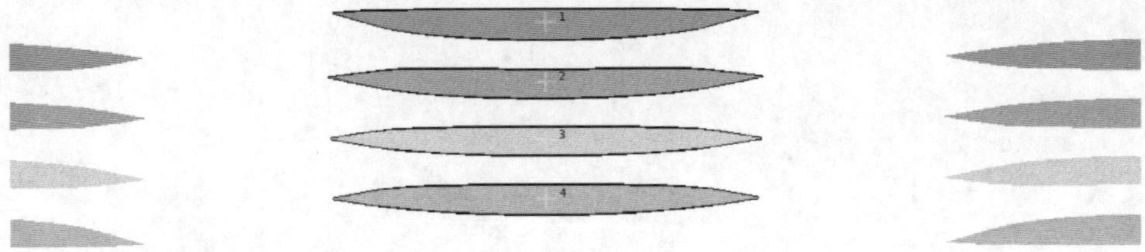

Figure 6. Drawn tow outlines and the identified centroids from compaction simulation.

The boundary pixels identified from the edge detection algorithm did not distinguish which tow cross section the tow-slices belongs to. Hence, an extra assignment step was necessary. The minimum bounding rectangles obtained from the edge detection algorithm were used to isolate/crop each tow cross section from the cleaned binary image (figure 5(b)). Each of the cropped binary images contained the information of a complete tow cross section (from which the minimum bounding rectangle was calculated) and (possibly) some incomplete tow cross sections from the adjacent tows in its close proximity. These unwanted cropped parts were cleaned using the same algorithm that transformed the figure 5(a) to figure 5(b). By performing this extra step, the boundary pixels that corresponded to a particular tow cross section could be assigned to that cross section.

4.4. Geometric measurements

Following the previous steps, the location of the pixels that made up the tow boundaries at each image are available which allowed the measurements of various cross-sectional parameters (width, thickness, and angle of rotation). To understand the methodologies to compute these parameters, the reader is referred back to section 3.3. The whole computation process was completely automated.

5. Results

In the previous sections, slightly different approaches were adopted to process the images from the micro-CT and the compaction model. At the end, both approaches resulted in the same data type that could be processed and presented in an identical manner. This way, direct comparisons between a real sample and a compaction model can be carried out rationally. The results are gathered and can be presented as a series of 3D plots and excel files. Figure 7 shows an example of such a 3D plot from processing the micro-CT images. The plot shows the 3D centreline path of a particular tow from the micro-CT sample. The discrete user free-hand drawn tow boundaries distributed along the tow centreline are also shown. The centroid coordinates are also presented for completeness sake. For convenience and ease of visualisation, the variation of the geometric parameters (width, thickness, and angle of rotation) can be displayed using colour transition along the tow centreline, which can be interpreted from a colour bar/legend besides the plot. In this example, the colour transition refers to the variation of tow width. Figure 8 shows the equivalent 3D plot of a tow from the compaction model. In this case, the discrete tow boundaries along the tow centreline correspond to the boundary pixels determined by the edge detection algorithm. Both figures can be opened in MATLAB simultaneously and rotated for better visualization. In addition to these plots, a series of similar 3D plots that contain all of the tows are readily produced.

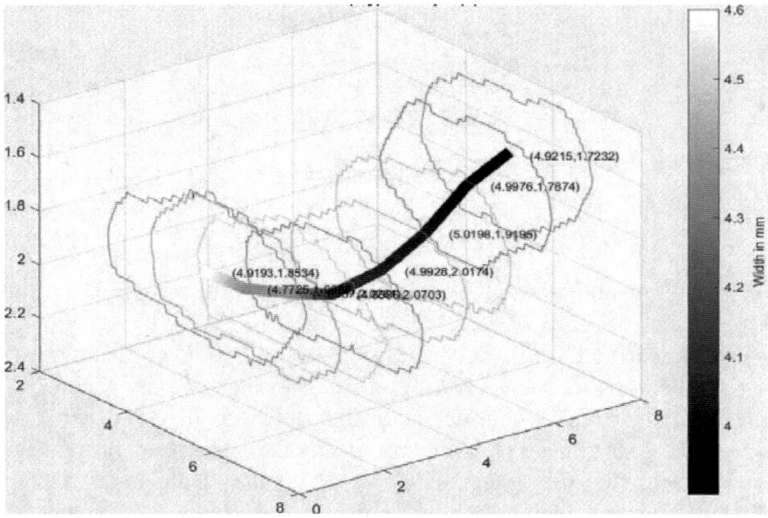

Figure 7. 3D plot of a tow from the micro-CT sample.

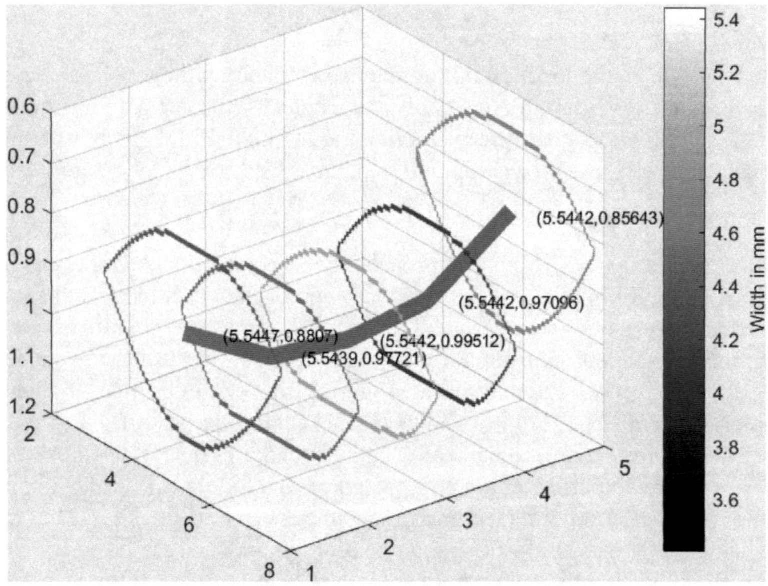

Figure 8. 3D plot of a tow from the compaction model.

6. Conclusions

This work has presented a set of methodologies to extract the geometric parameters of fibrous tows from micro-CT images and textile compaction models. These methodologies are coded and implemented in MATLAB to create a consistent and semi-automated image processing program that minimises the manual labour time. Due to the insufficient contrast available in micro-CT images, the boundaries of tow cross sections are difficult to discern automatically, and so are captured using a free hand drawing algorithm that allows zoom and undo operations. The tow colours from the compaction model can be freely adjusted which allows image segmentation and edge detection algorithms to be executed automatically. The program allows important tow cross sectional parameters (width, thickness, and

IOP Conf. Series: Materials Science and Engineering **406** (2018) 012044 doi:10.1088/1757-899X/406/1/012044

angle of rotation) to be computed for any arbitrarily shaped cross sections, in a consistent fashion. The geometric data of each tow cross section at each image are then stored. A series of 3D plots are readily produced using MATLAB. These plots show the 3D tow paths along with the identified tow cross sections. The variation of the geometric parameters can be added by having a colour transition along the tow paths. The same computed geometric parameters are also stored in a series of Excel files should the user wish to perform further analysis and of the data. Using these approaches, it has been demonstrated that it is possible to make a qualitative and a quantitative comparison between a micro-CT sample and a textile compaction model.

References

[1] Green S D, Long A C, El Said B S F and Hallett S R 2014 Numerical modelling of 3D woven preform deformations *Composite Structures* **108** 747-56

[2] Naouar N, Vidal-Salle E, Schneider J, Maire E and Boisse P 2015 3D composite reinforcement meso F.E. analyses based on X-ray computed tomography *Composite Structures* **132** 1094-104

[3] Thompson A J, El Said B, Ivanov D, Belnoue J P and Hallett S R 2017 High fidelity modelling of the compression behaviour of 2D woven fabrics *International Journal of Solids and Structures*

[4] Bale H, Blacklock M, Begley M R, Marshall D B, Cox B N and Ritchie R O 2012 Characterizing Three-Dimensional textile ceramic composites using synchrotron X-Ray Micro-Computed-Tomography *J Am Ceram Soc* **95** 392-402

[5] Pazmino J, Carvelli V and Lomov S V 2014 Micro-CT analysis of the internal deformed geometry of a non-crimp 3D orthogonal weave E-glass composite reinforcement *Composites Part B: Engineering* **65** 147-57

[6] Barburski M, Straumit I, Zhang X, Wevers M and Lomov S V 2015 Micro-CT analysis of internal structure of sheared textile composite reinforcement *Composites Part A: Applied Science and Manufacturing* **73** 45-54

[7] Aziz A R, Ali M A, Zeng X, Umer R, Schubel P and Cantwell W J 2017 Transverse permeability of dry fiber preforms manufactured by automated fiber placement *Composites Science and Technology* **152** 57-67

[8] Canny J 1986 A computational approach to edge detection *IEEE Transactions on Pattern Analysis and Machine Intelligence* **PAMI-8** 679-98

Improvement of mode-II interlaminar fracture toughness of carbon textile composites with modified matrix of thermoplastic and thermoset epoxy -addition of glass fibers

Soichiro Imagawa[1], Hironori Nishida[2], Kazuya Okubo[2], Toru Fujii[2]

[1] Graduate School of Doshisha University, Kyotanabe, Kyoto 610-0394, Japan
[2] Department of Mechanical Engineering, Doshisha University, Tatara Kyotanabe, Japan

Abstract. Interlaminar fracture toughness of a material with a layered structure such as CFRP is poor and cracks between the layers easily lead to fatal result. Interlaminar fracture toughness still plays an important role in damage propagation of CFRP. In this study, mode II inter-lamina fracture toughness of CFRP was investigated by ENF test for clarifying effect on weight-average molecular weight of matrix and matrix modification with glass fiber. Micro glass fibers whose diameter was 0.5μm with 0.3wt% of weight content were added into thermoplastic epoxy resin and thermoset epoxy resin, respectively, for the matrix of the textile carbon fiber composite prepared by our laboratory. State of the fracture surface was also changed accompanying with the change of the weight-average molecular weight of the matrix. In these results, Mode II interlaminar fracture toughness of the composite was improved by preventing the crack propagation with fiber bridging of added glass fibers on the interface regardless of the types of matrix of thermoplastic and thermoset epoxy resin.

1. Introduction

Carbon fiber reinforced thermal polymer (CFRTP) is one of the most important materials for structural applications, particularly in aviation industries owing to its high strength to weight ratio.[1]They show benefits over their thermoset counterpart, such as an increased toughness, a better recyclability, and shorter production cycle times. In an effort to reduce costs and to bring CFRTPs to their full potential, many developments were made. However, common thermoplastic resins have a higher melt viscosity than the thermosetting ones. This does not allow an easy infusion process and good impregnation of fibers. Therefore, various impregnation methods were studied. In [2], a method was proposed for reducing the viscosity of the resin with a solvent. However, the solvent must be removed during the manufacturing of the composite. Powdered and commingled methods were studied to reduce the impregnation length, which is the path of the matrix to complete the impregnation process [3]. A fine matrix powder was combined with reinforcing fibers in [4, 5], but the powder can be easily dislodged from the filaments. Commingled fabrics were proposed in [6, 7]. Recently, CFRTPs using in-situ resin received a lot of attentions [8], for the development of thermoplastic epoxy resin (TP-epoxy) with linear crosslinked structure. The weight-average molecular weight (Mw) of TP-epoxy depends on the polymerization temperature and the polymerization time [9]. It is also known that the mechanical properties of TP-epoxy depend on the Mw of resin, see e.g. [10]. In previous study, it is also known that mechanical performance of CFRP was improved by addition of glass fiber. However, effect of the addition of glass fibers as extra modification into highly polymerized thermo-plastic epoxy resin on

IOP Conf. Series: Materials Science and Engineering **406** (2018) 012045 doi:10.1088/1757-899X/406/1/012045

mechanical properties of the composite has not been experimentally discussed, comparing the effect on that with thermoset-plastic. In this study, mode II inter-lamina fracture toughness of CFRP was investigated by ENF test for clarifying effect on weight-average molecular weight of matrix and matrix modification with glass fiber. Micro glass fibers whose diameter was 0.5 μ m with 0.3wt% of weight content were added into thermoplastic epoxy resin and thermoset epoxy resin, respectively, for the matrix of the textile carbon fiber composite preparedby our laboratory.

2. Materials and testing machine

2.1 Materials

Plain weave carbon fiber fabric (Mitsubishi Rayon TR3110MS) was used as reinforcement (yarn TR30S 3L, linear density 1.79 g/cm3, pick and end counts 12.5 inch, areal weight 200 g/m2). Thermoplastic epoxy resin (DENATITE XNR 6850A, ACCELERATOR XNH 6850B; supplied by Nagase ChemteX Corporation, Japan) was used as matrix (Tg was approximately 100 °C). Thermoset epoxy resin (supplied by Mithubishi Chemical Corporation, Japan) was used as matrix. Glass fiber (FM1700) were used as modified matrix. Glass fiber with a diameter of 0.5μm and length of 150μm. Figure 1 shows a scheme of thermoplastic epoxy resin. Figure 2 shows a scheme of glass fiber (Nippon Muki Corporation, Japan).

Figure 1. Thermoplastic epoxy resin.

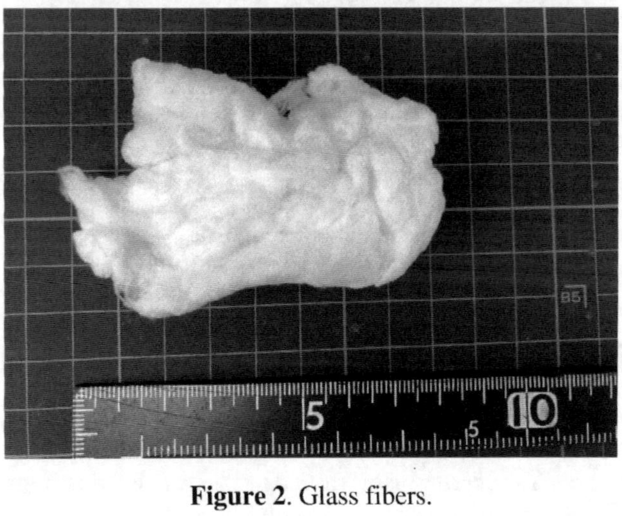

Figure 2. Glass fibers.

2.2 Condition of highly polymerization of thermo-plastic epoxy for matrix and measurement method of weight average molecular weight.

Figure 3 shows example of conditions for increasing the molecular weight Mw vs. Polymerization time of each temperature and Figure 4 shows measurement method of weight average molecular weight. The weight-average molecular weight of the thermoplastic epoxy matrix were measured by the gel permeation chromatography (Gel Permeation Chromatography; GPC) (CLASS-LC10: Shimadzu Corporation) and a column (Styragel HR4E, Styragel HR5E: waters). Tetrahydrofuran (THF) was used as the solvent. The calibration curves were drawn based on the retention time and the molecular weight of standard polystyrene (Polystyrene Polymer Laboratories Ltd.).

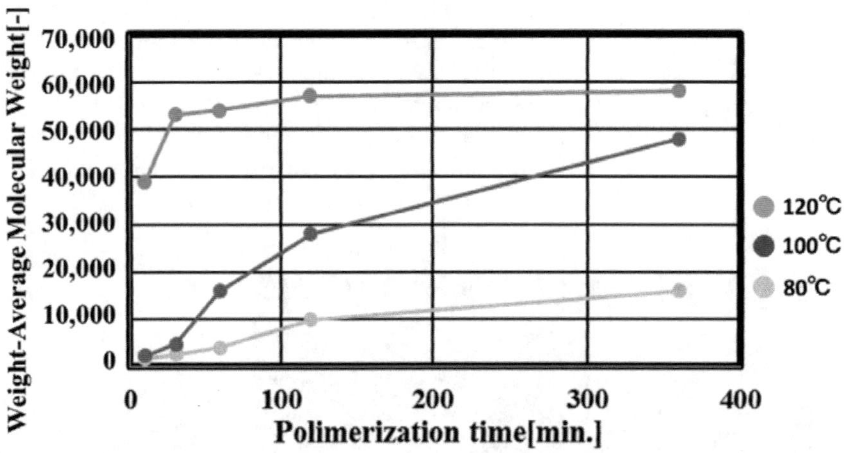

Figure 3. Relationship between Mw and polymerization time on temperature.

Figure 4. Measuring equipment of weight average molecular weight of resin. (Gel Permeation Chromatography; GPC)

2.3 Fabrication procedure of matrix plate specimen of modified matrix

The resin, 'XNR6850A', was heated by using an electric oven at 120 °C. Next, when the temperature of the resin reached 105 °C, the accelerator 'XNH6850B' was added to the resin or modified resin with stirring. Viscosity of thermoplastic epoxy resin was reduced using electric furnace. After that, glass fiber was added on thermoplastic and thermoset epoxy resin in according to Figure 5 procedure. Matrix of resin were prepared in which the weight content of glass fiber were 0.3wt%. Glass fiber was added into thermoplastic and thermoset epoxy resin using process homogenizer under the condition of 5,000 rpm 10min.

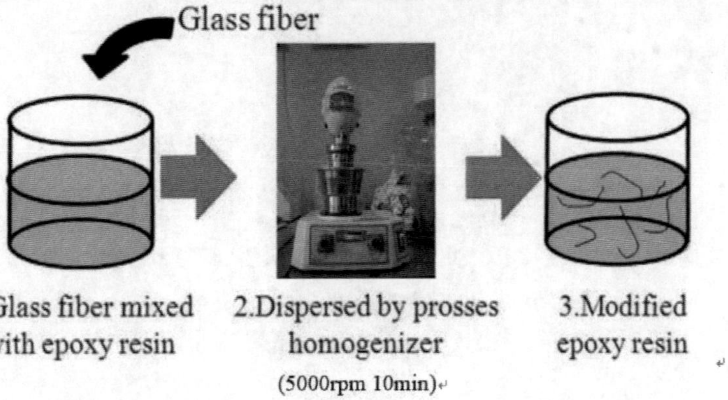

Figure 5. Mixed glass fibers with epoxy resin.

2.4 Fabrication of plain-woven CFRTP and CFRTS laminate

Plain weave CFRTP prepreg was made by the following procedure. The plain weave carbon fabric was impregnated with the thermoplastic epoxy resin by hand lay-up. The molecular weight of prepreg was finally controlled by a predetermined time and temperature sequence. CFRTP prepreg impregnated with the thermoplastic epoxy resin in the state of oligomer was polymerized at a given temperature in an electric oven. The obtained prepreg was cut into 245x245 mm and dried at 50 °C for 12 hours. CFRTP laminates were prepared by press molding with 10 layers (for quasi-static three point bending test) and 20 layers (for mode II tests) of dry prepreg at 175~195 °C and 6~12 MPa on a heat-press device. The carbon fiber volume fraction of the CFRTP laminates was approximately 45 %. In the case of CFRTP which added glass fiber on thermoplastic epoxy resin was the same of proceduer.

2.5 Fabrication of plain-woven CFRTS laminate

Plain weave CFRTS prepreg was made by the following procedure. The plain weave carbon fabric was impregnated with the thermoplastic epoxy resin by hand lay-up. The CFRP plates were cured at 80 degree-C for 1 hour and then at 150 degree-C for 3 hours. After curing, the plain-woven CFRP plates were cut into the dimension of specimen in orthotropic direction of fiber cloths. In the case of CFRTS which added glass fiber on thermoplastic epoxy resin was the same of proceduer.

3. Test procedure

3.1 Three point bending test

The resin which wrote Section 2/3 was poured into a glass plate and they were insert in electric oven. After that, modification of epoxy resin were polymerized and use it for three point bending test.
Dimensions of the specimen were 80 (length) x 10 (width) x 4 (thickness) mm. Three point bending test were conducted by using machine shown in Figure 6.

IOP Conf. Series: Materials Science and Engineering **406** (2018) 012045 doi:10.1088/1757-899X/406/1/012045

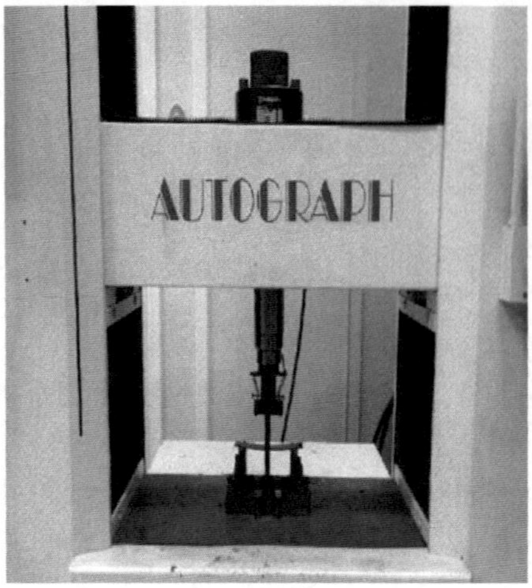

Figure 6. Test device for bending test.

3.2 ENF test

The CFRTP laminate, prepared for ENF test, had 20 layers of plain weave carbon fabric. Five specimen, for each considered Mw (see Table 2), with the size of 140 (length) x 20 (width) x 3 (thickness) mm were subjected to quasi-static three-point bending load. The Mode II inter-lamina fracture toughness of CFRTP was determined at 0.5 mm/min of cross-head speed. The length of pre-crack was 50 mm. Kapton film of approximately 30 μ m thick (Kapton, Du Pont-Toray Corporation) was inserted between 10th and 11th ply of the laminate. The Mode II inter-lamina fracture toughness was calculate by (1)

$$G_{IIC} = \frac{9a_1^2 F_c^2 C_1}{2B(2L^3 + 3a_1^3)} \tag{1}$$

where

$$a_1 = \left[\frac{C_1}{C_0} a_0^3 + \frac{2}{3}\left(\frac{C_1}{C_0} - 1\right) L^3\right]^{\frac{1}{3}} \tag{2}$$

Variables of GIIC, a0, Fc, C0, C1, a1, L, B denote the mode II interlaminar fracture toughness, initial crack length, critical load, compliance of the initial elastic region at loading point, compliance at critical load, estimated crack growth length at critical load, supports span, specimen width, respectively. Figure 7 shows a scheme of specimen of ENF. ENF test were conducted by using machine shown in Figure 6.

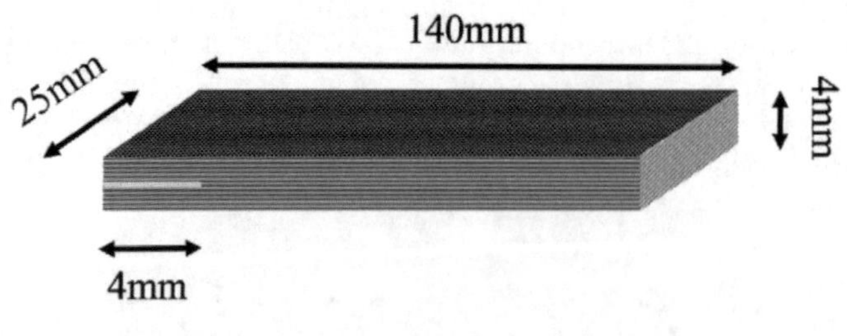

Figure 7. Geometry of ENF specimen.

4. Result and discussion

4.1 Monolithic matrix strength

Figure 8 shows the monolithic resin strength with respect to the weight average molecular weight. For comparison, the strength of TS-Epoxy and TSGF0.3wt% were also plotted. Test results showed that the monolithic resin strength of TP-Epoxy was improved with increase of its molecular weight. Moreover, when the molecular weight of TP-Epoxy exceeded about Mw=60,000, its monolithic strength was greater than that of TS-Epoxy. Test results also showed that the addition of glass fibers was effective to improve the monolithic resin strength regardless the type of epoxy. However, the degree of improvement in monolithic resin strength was significant when glass fibers were added into TS-Epoxy compared to that of TPGF0.3wt%.

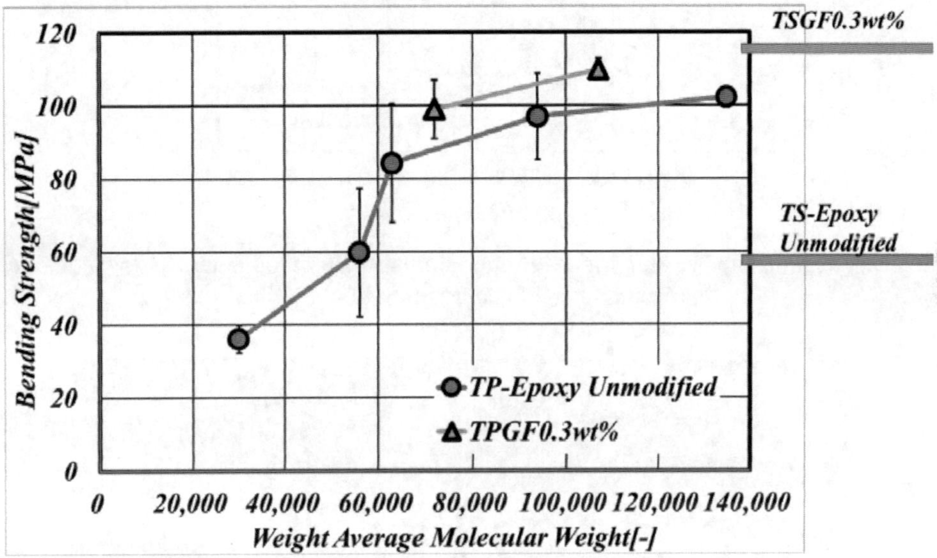

Figure 8. Relationship between bending strength and weight-average molecular weight on epoxy resin plate.

4.2 Mode II inter-laminar fracture toughness

Figure 9 shows the Mode-II interlaminar fracture toughness with respect to the weight average molecular weight. For comparison, the fracture toughness of TS-Epoxy and TSGF0.3wt% were also plotted. Test results showed that the Mode-II interlaminar fracture toughness was improved with almost linearly along its molecular weight increases. When the molecular weight of TP-Epoxy exceeded about Mw=60,000, the fracture toughness of CF/TP-Epoxy composites was greater than that of CF/TS-Epoxy composites. Figure 10 shows the fractured surfaces of ENF specimen observed by SEM. The observation of fractured surfaces revealed that the manner of fractured surface was changed with change of molecular weight of TP-Epoxy. At low molecular weight condition (Figure 10-a), the rough fractured surface was observed, while high molecular weight condition (Figure 10-b) showed smooth fractured surface. The addition of glass fibres was effective to improve the fracture toughness regardless the type of matrix. However, when TS-Epoxy was used as matrix, the degree of improvement in fracture toughness by adding the glass fibres was depend on its molecular weight. At low molecular weight TP-Epoxy (MW=45,000) was used, the improvement in fracture toughness by adding the glass fibres was about 25%, while high molecular weight condition (MW=60,000), its improvement in fracture toughness was about 140%. Figure 11 shows the magnified observation of fractured surface of TPGF0.3wt%. The existence of added glass fibres at its fractured surface was confirmed and residual resin on fibre surface was also confirmed. These results suggested that the Mode-II interlaminar fracture toughness of the composites was improved by fibres bridging of added glass fibres to preventing the crack propagation.

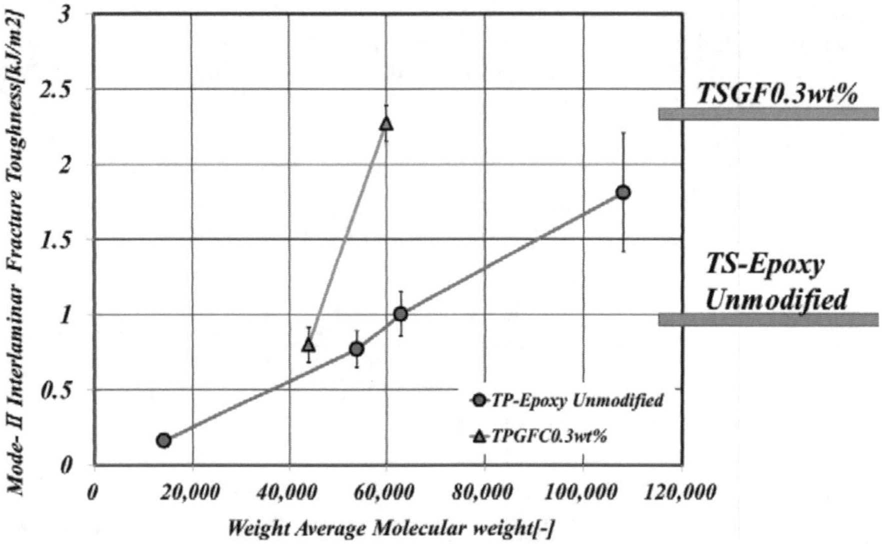

Figure 9. Relationship between Mode- II interlaminar fracture toughness and weight-average molecular weight and addition of glass fiber.

Figure 10. State of the fracture surface for Mw: (a) 14,000 and (b) 108,000.

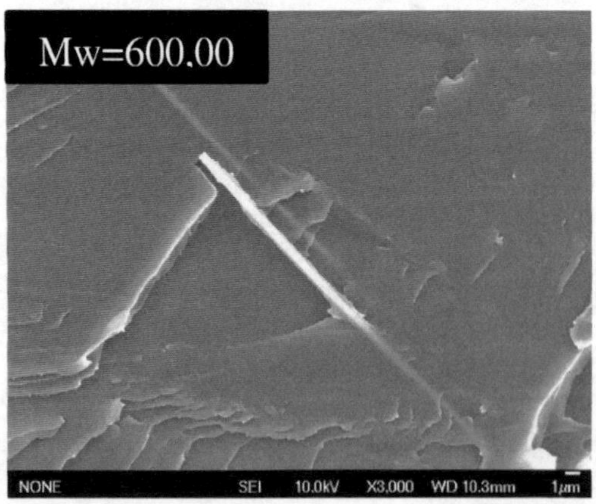

Figure 11. State of the fracture surface on addition of glass fiber on thermoplastic epoxy.

5. Conclusions

5.1 Monolithic resin strength was improved with increase of weight average molecular weight of thermo-plastic epoxy and also improved by addition of glass fibers to the thermoplastic and thermoset epoxy resin.

5.2 When addition of glass fiber on thermoplastic epoxy resin, fiber bridging was caused on CFRTP. Mode II interlaminar fracture toughness of the composite was improved by preventing the crack propagation with fiber bridging of added glass fibers on the interface regardless of the types of matrix of thermoplastic and thermoset epoxy resin.

Acknowledgements

The authors gratefully acknowledge the support of Doshisha University Reserch & Development Center for Advanced Composite Materials and Nagase ChemteX Corporation, Japan. And also, this work was financially supported by a (advanced technological) research project on "Research and Development Center for Advanced Composite Materials" of Doshisha University and MEXT (the Ministry of Education, Culture, Sports, Science and Technology, Japan) -Supported Program for the Strategic Research Foundation at Private Universities, 2013-2017,

References

[1] CFRTP and stainless steel laser joining: Thermal defects analysis and joining parameters optimization Junke Jiao Zifa Xu Qiang Wang Liyuan Sheng Wenwu Zhang doi: org/10.1016/j.optlastec.2018.01.023

[2] B. Liu, A. Xu, L. Bao, Preparation of carbon fiber-reinforced thermoplastics with high fiber volume fraction and high heat-resistant properties, Journal of Thermoplastic Composite Materials, 2015, DOI: 10.1177/0892705715610408. doi: org/10.1177/0892705715610408

[3] A. Xu, L. Bao, M. Nishida, A. Yamanaka, Molding of PBO Fabric Reinforced Thermoplastic Composite to Achieve High Fiber Volume Fraction, Polymer Composites, 34, 2013, pp. 953-958. doi: org/abs/10.1002/pc.22501

[4] J.H. Chen, E. Schulz, J. Bohse,G. Hinrichsen, Effect of fibre content on the interlaminar fracture toughness of unidirectional glass/fibre polyamide composite, Composites: Part A, 30, 1999, pp. 747-755 doi: org/10.1016/S1359-835X(98)00188-2

[5] A. K. Mohanty, L. T. Drzal & M. Misra, Engineered natural fiber reinforced polypropylene composites influence of surface modifications and novel powder impregnation processing, Journal of Adhesion Science and Technology, 16, 2002, pp. 999-1015 doi: org/10.1163/156856102760146129

[6] A.C. Long, C.E. Wilks, C.D. Rudd, Experimental characterization of the consolidation of a commingled glass/polypropylene composite, Composites Science and Technology, 61, 2001, pp. 1591-1603. doi: org/10.1016/S0266-3538(01)00059-8

[7] Phong, N.T, Mohamed H. G., Okubo K., Chuong B., Fujii T. (2013) "Improvement in the mechanical performances of carbon fiber/epoxy composite with addition of nano-(Polyvinyl alcohol) fibers", *Composite Structures*, Elsevier Limited, 99, pp.380-387. doi: org/10.1016/j.compstruct.2012.12.018.

[8] G. Ben, K. Sakata, Fast fabrication method and evaluation of performance of hybrid FRTPs for applying them to automotive structural members, Composite Structures, 133, 2015, pp. 1160-1167 doi: org/10.1016/j.compstruct.2015.07.093

[9] H. Nishida, The development of thermoplastic epoxy resin and continuous fiber reinforced thermoplastics using it, Journal of the Adhesion Society of Japan, 51, 2015, pp. 516-523. doi: org 10.11618/adhesion.51.516

[10] T. Imanishi, H. Nishida, N. Hirayama, N. Tomomitsu, In situ polymerizable thermoplastic epoxy resin and high performance FRTP using it and fiber fabrics, Proceedings of 16TH International conference on composite materials, Kyoto, Japan, 2007.

[11] Hironori Nishida, Kazuya Okubo, Toru Fujii, Valter Carvelli, Improvement of bending strength of carbon fiber/ ther-moplastic epoxy composites- Effects of molecular weight of epoxy on carbon

fiber/matrix interfacial strength and connection of cracks in matrix -,Kyoto, Japan,2017 doi: 10.4236/ojcm.2017.74014.

[12] Nishida, H., .Carvelli, V., Fujii, T., and Okubo, K. (2018) 'Quasi-static and fatigue performance of carbon fibre reinforced highly polymerized thermoplastic epoxy', Composites Part B: Engineering, 144, pp. 163–170. doi: 10.1016/j.compositesb.2018.03.002. doi: 10.1016/j.compositesb.2018.03.002.

[13] Nishida, H., Okubo, K., Fujii, T., and Carvelli, V. (2017) 'Improvement of Bending Strength of Carbon Fiber/Thermoplastic Epoxy Composites-Effects of Molecular Weight of Epoxy on Carbon Fiber/Matrix Interfacial Strength and Connection of Cracks in Matrix', Open Journal of Composite Materials, 7(4), pp. 207–217. doi: 10.4236/ojcm.2017.74014. doi: 10.4236/ojcm.2017.74014.

IOP Conf. Series: Materials Science and Engineering **406** (2018) 012046 doi:10.1088/1757-899X/406/1/012046

Novel high performance auxetic fibrous structures for composite reinforcement

C Gonçalves[1*], R Magalhães[2], S Rana[2], R Fangueiro[3], J Nunes[1] and G Dias[1]

[1] Institute for Polymers and Composites (IPC), University of Minho, Campus Azurém, 4800-058 Guimarães, Portugal

[2] Centre for Textile Science and Technology (2C2T), University of Minho, Campus Azurém, 4800-058 Guimarães, Portugal

[3] Department of Mechanical Engineering, University of Minho, Campus Azurém, 4800-058 Guimarães, Portugal

* a68476@alunos.uminho.pt

Abstract. This paper reports the development of novel auxetic fibrous structures from high performance yarns using knitting technology for application as reinforcement in composites. Two types of high performance yarns, namely, para-aramid (p-AR) and polyamide 6.6 (PA), were used to produce three distinct auxetic structures: purl, re-entrant lozenge and re-entrant star structures. It was observed that all structures produced could achieve maximum negative Poisson's ratios (NPR) of -0.60, -0.54 and -0.47 in the course direction of purl and lozenge and in wale direction of star structures, respectively. The auxetic textile structures also presented better tensile strength and energy absorption (except re-entrant lozenge structure) than plain knitted textiles produced by using the same type of fibres and parameters.

1. Introduction

In recent years, the use of textile technology to fabricate auxetic materials has attracted more and more attention due to their superior fracture toughness and impact and indentation resistance, besides other interesting characteristics [1,2]. Auxetic reinforcements are now used in composite materials to improve the properties that are desired to achieve, for example, in crash helmets, sport clothing, sponges, ropes, filtration and other shock absorbing products. They are also used in the biomedical, automotive and defence industries [3,4].

Several studies revealed that auxetic effect can be achieved by developing structures that involve the rotation of units such as squares, rectangles, triangles, rhombi and parallelograms [5,6]. The auxetic fabric structures were manufactured by using specially designed fibrous arrangements that were thoroughly investigated. The present study evaluates the auxetic behaviour of NPR weft-knitted fabrics made from PA and p-AR fibres and polymeric matrix composites using them as reinforcing layers. Wherein an epoxy resin has been selected as matrix due to its excellent toughness, adhesion, thermal and chemical resistance [2].

2. Experimental work

2.1. Materials

Knitted fabrics were produced from p-AR fibre yarns Standard Twaron and high-tenacity PA fibre yarns, supplied by Teijin Company (Arnhem, Netherland) and Rhodia Company (Lisboa, Portugal), respectively. An epoxy resin Biresin® CR83 combined (100:30 ratio) with hardener Biresin® CH83-2, both from Sika Deutschland GmbH/Germany, were used as matrix of the composites produced in this work.

2.2. Production of fibrous structures and their composites

In order to compare the behaviour of auxetic structures with non-auxetic ones, jersey plain knitted fabrics were also produced using the same parameters and the schematic geometries provided in Figure 1. Weft-knitted textile structures were manufactured on a 10-gauge Stoll CMS 320 TC flat knitting machine (Stoll GmbH & Co., Reutlingen, Germany). Weft-knitting refers to the knitting direction: the stitches are knitted successively in the weft direction, also called course direction, and the loops are interconnected in the wale direction.

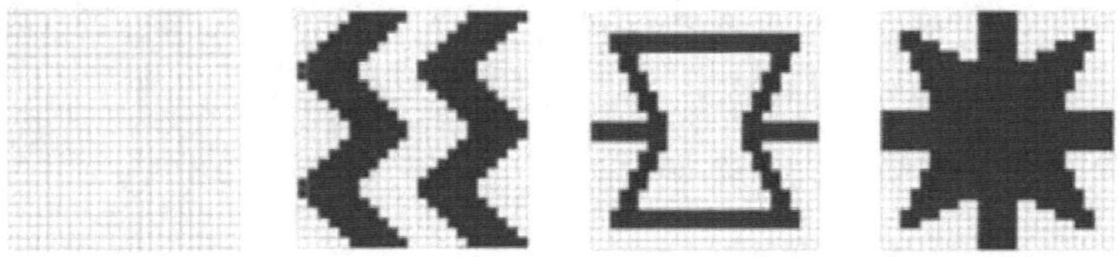

(a) (b) (c) (d)

Figure 1. Schematic geometries regarding high performance fibrous structures produced using knitting technology: a) Jersey plain, b) purl (Aux 1), c) re-entrant lozenge (Aux 2) and d) re-entrant star (Aux 3.1, 3.2).

Flat knitting technology provides a simple, but highly effective way for manufacturing auxetic fabrics from conventional yarns. All knits under study were produced for each type of fibres using the same knitting parameters in order to get equivalent stitch sizes, given by the geometrical parameters listed in Table 1.

Table 1. Parameters used to produce jersey plain and auxetic knitted fabrics.

Knitted fabric	Fibre type	Linear density (Tex)	No. of yarns	Loop length*	Take down load*
Jersey plain	Polyamide	97			
Aux 1	Polyamide	97			
Aux 2	Polyamide	97	1	11	8
Aux 3.1	Polyamide	97			
Aux 3.2	Para-aramid	172			

* in machine units

The produced fabric structures were subsequently impregnated with an epoxy resin by using the vacuum infusion technique. One layer of fabric was used and the average weight gain after resin impregnation was approximately 50% for all composites specimens.

2.3. Auxetic, physical and mechanical properties

In this work, the digital image correlation (DIC) method was used to determine the Poisson's ratio and, therefore, to assess the auxetic behaviour of the developed textile structures.

The measurement setup was fixed to a universal mechanical testing machine (see Figure 2) in which the structures were vertically subjected to tensile tests after being attached to the clamps of the equipment with regard to measure their displacements in course and wale directions. The specimens were marked at specific points in both directions, as shown in Figure 2. Images taken from fabrics and composites under tensile deformation were recorded, analysed and treated by using the MatLab® high-performance software for assessing their Poisson's ratio. The image recording was performed using 3840x2160 pixels and a frame rate of 24 images/second.

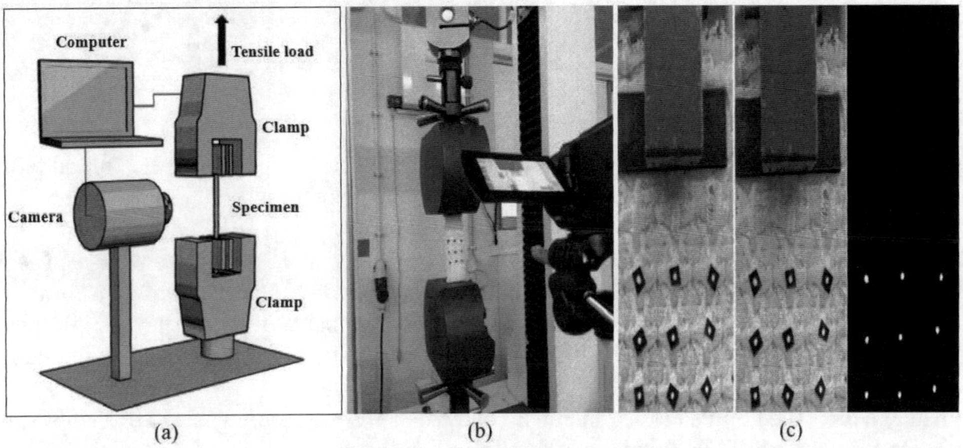

Figure 2. A schematic diagram and photograph of the tensile testing system Poisson's ratio measurement: (a, b) setup of testing and (c) image processing steps.

Then, the course and wale strains in the specimens were determined using the following equations:

$$\varepsilon_x = \frac{x_n - x_0}{x_0} \tag{1}$$

$$\varepsilon_y = \frac{y_n - y_0}{y_0} \tag{2}$$

Where x_n and y_n are the distances between the points marked in the wale and course directions, respectively, in each deformation step, i.e., 1 cm. x_0 and y_0 are the initial intervals (first frame) between the reference points. The Poisson's ratio was then calculated through the relation between the deformation in the wale and course directions in each deformation step. It can be seen these deformations in a schematic representation shown in Figure 3a.

The tensile tests were also used to determine the tensile strength, strain and energy absorption until failure in the developed fabrics and composites. The tests on fabrics were performed according to the standard ISO 13934-2 Textiles (Tensile properties of fabrics – Part 2 Determination of maximum force using the grab method), in an universal testing machine, *Hounsfield H 100 KS* with a load cell of 10 kN. Whereas, the mechanical properties of the composites were made according to the ASTM D3039/D3039M-00 standard (Standard Test Method for Tensile Properties of Polymer Matrix Composite Materials)in the same machine at constant crosshead displacement rate of 5 mm min^{-1}, with a 2.5 kN load cell but using different clamps, (see Figure 3 c). In both cases, five specimens of each configuration were tested.

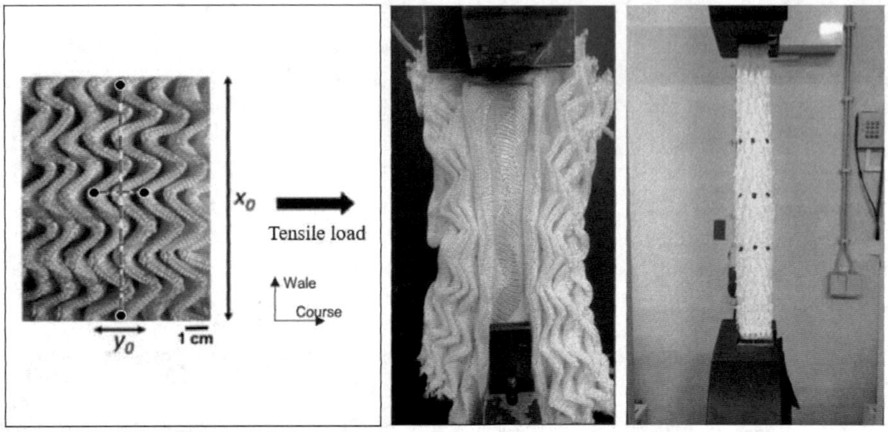

(a) (b) (c)

Figure 3. Tensile testing setup of (a) distance between the marked points on the knitted fabrics (b) knitted fabrics and (c) composite materials.

To assess the impact behaviour of the developed composites, their specimens were also subjected to drop weight impact testing accordingly to the ASTM D7136/D7136M standard in a *Fractovis Plus®* drop weight impact testing equipment from Ceast. An hemispheric impactor having 20 mm diameter and a mass of 10.044 kg was made to fall from a height of 500 mm in order to impact the sample at the speed of 3.13 ms^{-1}.

The thickness of the knitted fabrics and their composites were measured using NP EN ISO 5084:1999 Textiles (Determination of the thickness of textile and textile products) standard.

3. Experimental results and discussion

3.1. Mechanical behaviour of the knitted fabrics

The results of thickness, areal density and NPR measurement of the fabrics produced are listed in Table 2.

Table 2. Thickness, areal density and NPR of the knitted fabrics produced [a].

Knitted fabric	Thickness (mm)	Areal density (g/m^2)	Maximum NPR values	
			Wale	Course
Jersey plain	0.88 (2)	2.46 (3)	-	-
Aux 1	10.45 (1)	14.24 (4)	-0.15	-0.60
Aux 2	7.38 (4)	5.98 (5)	-0.23	-0.54
Aux 3.1	10.48 (2)	8.96 (2)	-0.47	-0.32
Aux 3.2	13.90 (4)	14.54 (7)	-0.08	-0.29

[a] The values of coefficient of variation are given between parentheses

It may be observed that the auxetic fabrics exhibited much higher thickness and areal density as compared to the jersey plain one made using the same parameters. Aux 3.2 had also more thickness and areal density than all the other auxetic structures made from PA fibre yarns due to the higher linear density of p-AR fibres. Furthermore, the p-AR yarn fabrics presented much lower NPR than those produced from the PA yarns. This could be due to the higher flexibility of the PA fibre yarns, which may result in better structural rearrangement under tensile loading leading to superior auxetic behaviour.

Almost all auxetic fabrics exhibited maximum NPR value in the course direction. The exception was the Aux 3.1 knitted fabric which present a maximum NPR of -0.47 in wale direction.

The mechanical properties determined on the fabric structures produced are listed in Table 3. It is clear that in general the auxetic fabrics exhibited superior maximum force, elongation as well as energy absorption than conventional fabrics (except Aux 2). This indicates the positive influence of the auxetic design in improving the performance of fabric structures.

Table 3. Mechanical testing results of knitted fabrics [a].

Knitted fabric	Maximum Force (N)	Elongation at Maximum Force (mm)	Energy absorption (kN.mm)
Jersey Plain	1304.0 (6)[a]	104.7 (8)	30.3 (22)
Aux 1	1480.5 (*)	169.4 (*)	44.9 (37)
Aux 2	1070.0 (19)	140.98 (*)	22.4 (19)
Aux 3.1	1465.5 (10)	177.4 (7)	41.9 (14)
Aux 3.2	1610.0 (10)	234.0 (2)	68.0 (17)

[a] The values of the coefficient of variation (CV) values are given between parentheses
* CV ≥ 20

3.2. Behaviour of knitted fabric reinforced composites

The results of thickness, areal density and NPR measurement of the composites are listed in Table 4. It is expected that this study could help us understanding the deformation mechanism of auxetic composites under axial tension load.

Table 4. Thickness, areal density and NPR of the composites produced [a].

Composites	Thickness (mm)	Areal density (g/m^2)	Maximum NPR	
			Wale	Course
Jersey plain	1.57 (12)[a]	4.04 (5)	-	-
Aux 1	15.10 (3)	25.67 (4)	-[b]	-0.1
Aux 2	9.70 (9)	9.26 (10)	-0.35	-[b]
Aux 3.1	14.20 (6)	13.80 (11)	-0.17	-0.05
Aux 3.2	13.25 (7)	19.77 (8)	-0.10	-0.20

[a] The values of the coefficients of variation are given between parentheses
[b] Clamp slippage before fabric reinforced composite break

The same trends were also observed for the composites produced from the auxetic fabrics, which presented much higher thickness and areal density than the conventional ones made from jersey plain fabrics. This was caused by the different arrangement of the knitted loops in the auxetic structures, which increase the thickness and mass per unit area not only of the fabrics themselves but also of their reinforced composites.

Anyway, the auxetic composites showed lower NPRs than their predecessor fabrics. This might be explained by the much higher restriction to movements and rearrangements of the composites, resulting from the addition of a solid, continuous and stiffer epoxy matrix, which strongly contributed for the reduction of their lateral expansion. It may be also seen that composites and their prior fabrics seemed to present maximum NPR values in the same direction (course and wale directions). Aux. 2 was the only exception but that seemed to be caused by the slippage at clamps that happened in the tensile tests made in the course direction.

Also like their previous fabrics, composites reinforced with p-AR fabrics had lower NPR than those produced with PA ones. The highest NPR values obtained in the composites were -0.35 and -0.2, in wale and course directions, respectively.

As it may be seen in Table 5, the composites made from auxetic fabrics demonstrated to absorb higher energy than those from non-auxetic ones (jersey plain). While the p-AR (Aux 3.2) and polyamide (Aux 3.1) auxetic knitted fabrics showed to withstand very a similar load (see Table 3), the composite derived from the p-Ar (Aux 3.2) has shown to resisted to a much higher load and presented a considerable lower elongation than the composites based on both polyamide auxetic and jersey plain fabrics. Hence, it may be concluded that the material of the fibre reinforcement seemed to play more important role on the mechanical properties obtained in the final composites produced from the auxetic knitted fabrics.

Table 5. Mechanical testing results of knit fabric reinforced composites [a].

Composites	Maximum force (N)	Elongation at maximum force (mm)	Energy absorption (kN.mm)
Jersey plain	758.6 (6)	224.8 (4)	68.3 (8)
Aux 1	1689.0 (9)	191.0 (*)	170.8 (12)
Aux 2	1125.4 (3)	113.9 (3)	60.3 (8)
Aux 3.1	699.2 (8)	278.2 (5)	87.0 (9)
Aux 3.2	2681.5 (8)	51.0 (18)	78.7 (14)

[a] The values of the coefficients of variation (CV) are given between parentheses
* $CV \geq 20$

Figure 4 depicts the representative energy–time–force curves obtained from the low-velocity impact tests made on non-auxetic and auxetic textile composites using an impact energy of 49.2 J. Among the four different structures tested, the non-auxetic presented the highest peak force which is 1035 N, while auxetic textile composite exhibited slightly lower peak force under the same impact energy. The peak force for Aux 1, 2, 3.1 and 3.2 composites were 805 N, 711 N, 856 and 995 N, respectively. Moreover, the impact duration for the non-auxetic and auxetics Aux 1, 2, 3.1 and 3.2 textile composites were 14, 25, 13, 14 and 10 ms, respectively. Despite the slightly higher peak force presented by the non-auxetic composite, the auxetic textile composites exhibited greater absorption of energy (34.8 J for Aux 1) and better force reduction than the non-auxetic under low-velocity impact at the energy of 49.2 J, which seems to indicated that the auxetic textile composites may have superior impact protective performance than the non-auxetic one.

It may be also seen that the impact energy absorption of all knitted fabric reinforced composites can be divided into three distinct stages. In the first stage, the value of absorbed energy is relatively low due to small deformation of the matrix and insignificant compression of textile structure. In a second stage, the energy–time curve increases rapidly and reaches the peak value ($E_{máx}$), representing the deformation of the reinforced auxetic textile structure and the increase in vertical displacement of the composite structure. At the last stage (from $E_{máx}$ to the end), the absorbed energy gradually decreases until the impact process finishes.

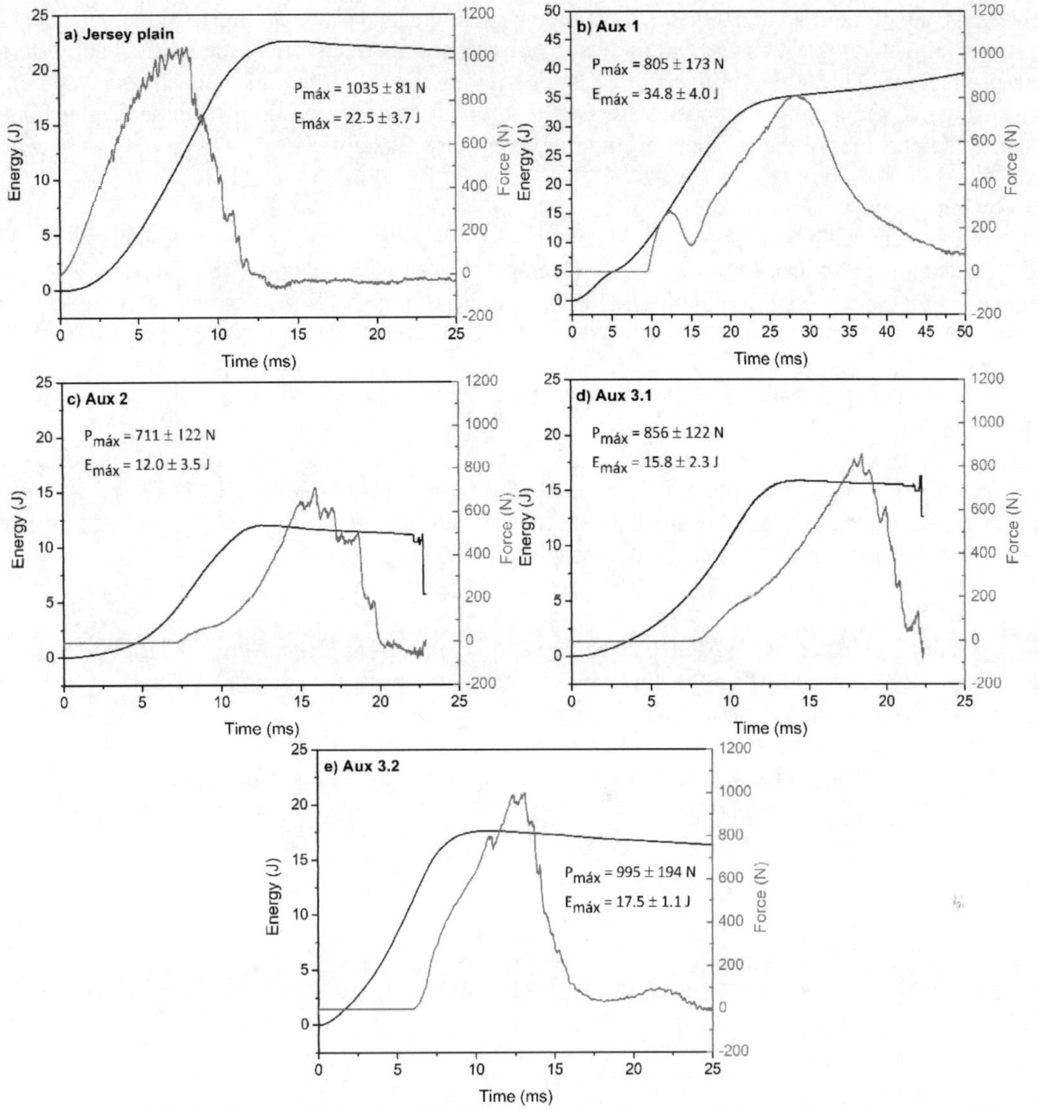

Figure 4. Low velocity impact behaviour of the five composite materials configurations tested (Jersey plain, Aux 1, 2, 3.1 and 3.2) with the peak force ($P_{máx}$) and impact energy ($E_{máx}$) average.

4. Conclusions

In the present research, all fabrics and composites produced using the developed auxetic knitted yarn structures exhibited negative Poisson's ratio (NPR) whose value, however, was strongly dependent on the type of structure (i.e., fabric configuration or composite) as well as on type of fibre yarns used. The maxima NPR values obtained were 0.47 -0.60 in the wale and course directions in auxetic knitted fabrics, respectively, and -0.35 and -0.20 in wale and course directions of their composites, respectively. The auxetic knitted fabrics themselves and their composites exhibited higher thickness, areal density, elongation and resistance under tensile loads as well as energy absorption than non-auxetic jersey plain knitted fabrics and its composite. The auxetic composites reinforced with p-AR knitted fabrics was the one that was able to support higher tensile load. Under stretching, the composites reinforced with auxetic knitted fabric also proven to present better energy absorption capability than the composite reinforced with the non-auxetic structures. However, under impact test, the difference in mechanical behaviour between the auxetic and non-auxetic textile composites diminished probably due to reduction in negative Poisson's ratio effect that was verified in the auxetic composites relatively to the knitted fabrics used in their production. To better optimise the design of the auxetic structures and enhance the impact performance of the auxetic composites further theoretical study is required to better understand the mechanical behaviour of the composites under low-velocity impact.

Acknowledgments
This work was funded by the Operational Program of Competitiveness and Internationalization (POCI) with financial support from the European Regional Development Fund (FEDER) and Portuguese Foundation for Science and Technology (FCT), under the project "Development of Auxetic Fibrous Composite for Personal Protection" (ref. POCI-01-0145-FEDER-016893 and PTDC/CTM-POL/5814/2014).

References

[1] S K Krishna, D Raj and C Emilio 2016 Three decades of auxetics research−materials with negative Poisson's ratio: a review. *Advanced Engineering Materials* **18** 1847-1870
[2] S Fernanda, R Sohel and F Raul 2016 Development of novel auxetic textile structures using high performance fibres *Materials & Design* **106** 81-89
[3] H M A Kolken and A Zadpoor 2017 Auxetic mechanical metamaterials *RSC Advances* **7** 5111–5129
[4] D Rant, T. Rijavec and P Alenka 2013 Auxetic Textiles *Acta Chim. Slov.* **60** 715–723
[5] H Hu, Z Wang and S Liu 2011 Development of auxetic fabrics using flat knitting technology *Textile Research Journal* **81** 1493–1502
[6] J N Grima, P Farrugia, R Gatt and D Attard 2008 On the auxetic properties of rotating rhombi and parallelograms: A preliminary investigation P*hys. stat. sol.* **33** 521–529

IOP Conf. Series: Materials Science and Engineering **406** (2018) 012047 doi:10.1088/1757-899X/406/1/012047

Influence of process induced defects for biaxial carbon fiber braids

R Czichos[1,*], O Bareiro[2], A K Pickett[1], P Middendorf[1] and T Gries[2]

[1] Institute of Aircraft Design (IFB), University of Stuttgart, Pfaffenwaldring 31, 70569 Stuttgart, Germany

[2] Institut für Textiltechnik (ITA), RWTH Aachen University, Otto-Blumenthal-Straße 1, 52074 Aachen, Germany

* czichos@ifb.uni-stuttgart.de

Abstract. The influence of different machine settings for the manufacture of biaxial carbon fiber braids is investigated using numerical models and compared with experimental results. For these studies different process parameters like yarn pre-tension and rotational speed of the bobbins have been varied to generate cylindrical biaxial braids with an average braid angle of ±45°. From a test program the non-uniform distribution of the braid angle and cover factor is observed using image processing techniques of the produced samples. In parallel, ongoing work is using numerical simulations of the braiding process to identify defects, and investigations have been able to correlate numerical results with experimentally observed defects.

1. Introduction

Rotary braiding is an effective manufacturing technique to produce near net-shape preforms in a cost- and time-efficient manner, and is a widely used textile preforming process in the automobile, aerospace, sports and other industries. This process uses two sets of bobbins that move in sinusoidal paths in the radial braiding machine; one set rotates clockwise, whereas the other set rotates counter-clockwise. The final shape of the preform and fiber architecture is determined by the size of the machine, the mandrel geometry and machine settings such as mandrel take-up speed, yarn pre-tension and rotational speed of the bobbins. However, one drawback in this process is the interaction of these parameters, which can cause different kinds of defects like those depicted in Figure 1.

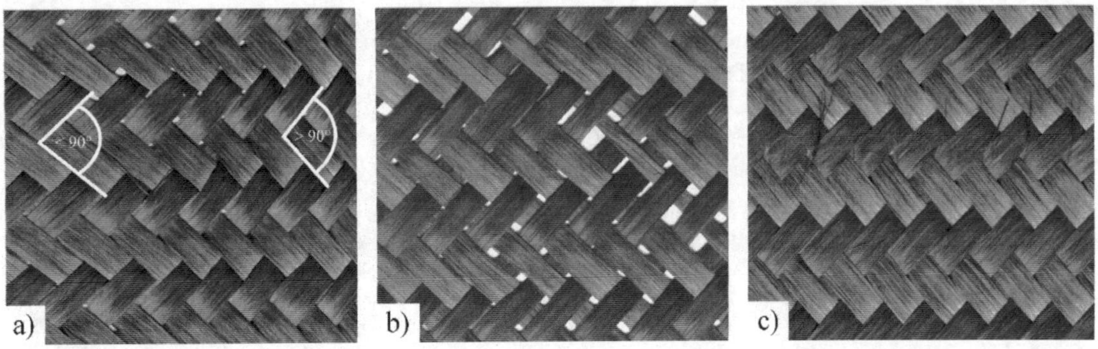

Figure 1: a) Varying braid angle; b) Varying cover factor; c) Fiber damage in form of broken filaments.

1.1. Braid angle
The braid angle is defined as the angle between the longitudinal axis and the direction of the off-axis yarns. For a biaxial preform produced over a cylindrical mandrel this angle can be calculated from simple kinematic relation [1] as

$$\tan \alpha = \frac{2\pi\omega d}{vn} \tag{1}$$

where ω describes the rotational speed of the bobbins in revolutions per seconds, d the mandrel diameter in millimeter, v the mandrel take-up speed in millimeter per seconds and n the number of horn gears of the braiding machine. However, in practice this angle may not be constant as seen in **Figure 1a** and can vary due to irregular mandrel take-up speed, a varying geometry, different yarn pre-tension or stick-slip effects during the braiding process.

1.2. Cover factor
The cover factor describes the ratio of the yarn cover on a surface to the whole surface of the mandrel geometry for a single layer. It has a high influence on the in-plane mechanical properties, and therefore most applications require a fully closed braid for best mechanical properties [2]. The low cover factor shown in Figure 1b is caused by variations in the yarn width, which is mainly influenced by friction causing a compaction of the yarn during the rewinding or braiding process. Alternatively, a large mandrel diameter, insufficient bobbins or excessive mandrel/bobbin velocities will also produce preforms with a low cover factor.

1.3. Fiber damage
The fiber damage mentioned in the present work is shown in Figure 1c, which has resulted in significant breakage of single filaments. This is induced by excessive friction and will result in some loss of composite mechanical properties. Furthermore, fiber breakage during manufacture is undesirable in the workplace since this can cause health problems and is dangerous to other electrical equipment which are likely to short circuit.

Experimental testing is very costly and time-consuming and the influence of different process parameters like yarn pre-tension or machine speeds cannot be tested for every possible machine configuration. A purely analytical approach is unable to account for the afore-mentioned parameters, consequently the use of the finite-element method is investigated here as a means to undertake a holistic simulation of the process chain that includes the effects of imperfections. The main goal of this paper is therefore to transfer the results from a process simulation, with included defects, to a structural mechanical calculation which enables effective stiffness to be calculated.

For this study biaxial 2D carbon fiber braids with a twill weave pattern have been manufactured on a radial braiding machine. Different process parameters were used to generate biaxial ±45° braids, from which quantitative analyses of fiber angle, cover factor and yarn width are made. From simulations of the braiding process it is possible to correctly capture imperfect yarn architecture and to estimate fiber damage. These results can then be used to construct FE models of the imperfect architecture for stiffness analysis, and two different approaches are demonstrated here.

2. Experimental test program and defect characterization
This section gives a brief overview of the test program and characterization of the defects mentioned in section 1. After a short description of the different machine settings used, the methods of characterization are explained, followed by measurement results at the end of this section.

2.1. Test program & setup
The investigated braids are produced on a fully mounted radial braiding machine Herzog RF 1/64-120 at the Institut für Textiltechnik (ITA), see Figure 2a. Braids were produced over a cylindrical mandrel of 85 mm diameter with Tenax®-E HTS40 F13 24K 1600tex yarns. Since yarns are delivered on spools

IOP Conf. Series: Materials Science and Engineering **406** (2018) 012047 doi:10.1088/1757-899X/406/1/012047

it is necessary to do a rewinding operation. This step is done with a Herzog SP 280 as shown in Figure 2b. A constant tensile force is required to ensure a proper rewinding, which is done with a pre-tensioner as shown in Figure 2c, where the yarn is guided through two eyelets and a normal force is applied with a spring. One potential problem of this operation is that high friction forces can create fiber damage, before the actual braiding process is started.

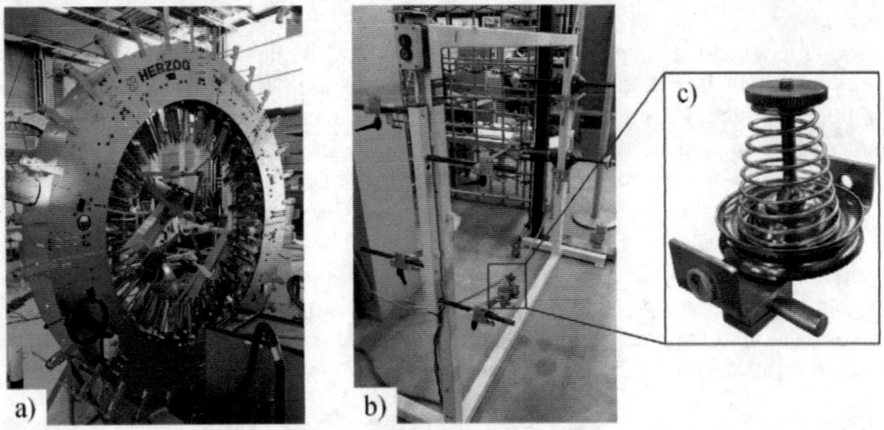

Figure 2: a) Radial braiding machine Herzog RF 1/64-120; b) Rewinding machine Herzog SP 280; c) Pre-tensioner to apply a constant rewinding force.

The investigated process parameters as well as the calculated mandrel take-up speed according to equation (1) are listed in Table 1.

Table 1: Investigated process parameters.

Reference	Yarn pre-tension [g]	Bobbin rotation speed [rpm]	Mandrel take-up speed [mm/s]	Calculated mandrel take-up speed [mm/s]
A	100	50	14	13.91
B	100	150	40	41.72
C	350	100	30	27.82
D	700	50	14	13.91
E	700	150	32	41.72

For every configuration a total number of three braids were manufactured and subsequently cut from the mandrel for further investigation of variabilities and defects. This characterization is performed with a CanoScan LiDE 200 flatbed scanner from Canon as shown in Figure 3.

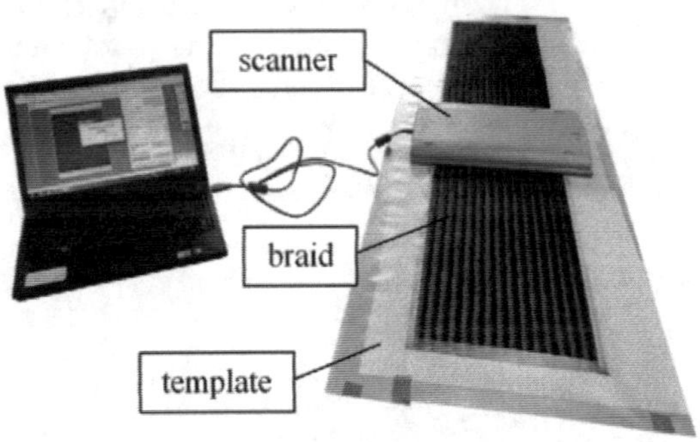

Figure 3: Experimental setup for the characterization of the braided preforms.

The braid is surrounded by a template which contains the dimensions as well as a reference coordinate system that is also scanned during the characterization process. The main advantage of this approach is the possibility to identify the fiber architecture for the future coupons, as the braids are not moved anymore until they are infiltrated with resin and fully cured.

2.2. Test evaluation

The evaluation of the scanned braids is mainly done automatically. For the calculation of the braid angle a software tool from the Faserinstitut Bremen e.V. (FIBRE) is used, that was originally developed within the scope of the EU project IMac-Pro [3]. Each preform has a length of 1300 mm and is divided into 10 single scans. A part of the scanned raw data can be seen in Figure 4a. In the next step, the scanned image is segmented into 6 x 10 subcells for which the fiber angle is calculated as shown in Figure 4b. The braid angle for every layer contains about 600 measuring points. The mean as well as the standard deviation for all three layers are listed in Table 2.

For determination of the cover factor a MATLAB-script is used. The scanned raw data is taken and converted into a grey-scale image which only contains black and white pixels as depicted in Figure 4c. For a correct allocation of the pixels a threshold is set manually and black and white pixels are counted, whereby black indicates cover of the mandrel surface and white specifies gaps in the braid.

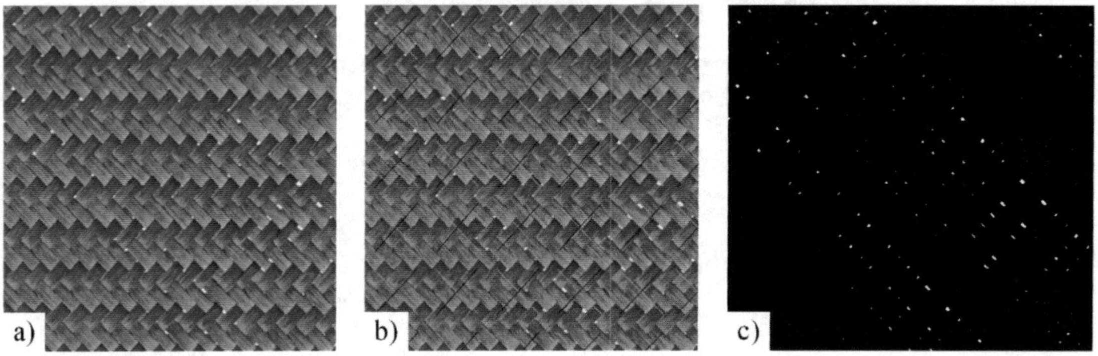

Figure 4: a) Scanned raw data; b) Analysed braid angle; c) Grey scale image for automated cover factor determination.

The measuring of yarn width is done manually, supported by the open source software GIMP. Hereby, a scale is scanned which is taken as a reference to convert the pixels into a length unit. As every braid consists of 10 single scans, a total of 10 x 40 measuring points are recorded for a single layer. An overview of the results of the characterization can be seen in Table 2. Considering the process parameters of Table 1, a comparison of RefB (150 rpm, 100 g) and RefE (150 rpm, 700 g) shows that a higher process speed leads to a wider distribution of the braid angle. The cover factor is not influenced by the process speed, however, the yarn width shows the lowest value for those two configurations. The yarn pre-tension does not have a significant influence on the yarn width, as demonstrated by the comparison from RefA (50 rpm, 100 g) and RefD (50 rpm, 700 g) which show nearly equivalent results.

Table 2: Results of the characterization of the dry braids.

Reference	Braid Angle [°]	Cover Factor [%]	Yarn Width [mm]
A	44.87 ± 0.64	99.89 ± 0.05	5.59 ± 0.47
B	43.66 ± 1.47	99.66 ± 0.34	5.45 ± 0.56
C	42.69 ± 0.27	99.72 ± 0.09	5.64 ± 0.40
D	44.65 ± 1.11	99.86 ± 0.07	5.60 ± 0.41
E	44.76 ± 1.48	99.76 ± 0.12	5.52 ± 0.41

3. Numerical investigations

This section describes the numerical investigations made to virtually produce the braids using the same process parameters as listed in Table 1. In a second step, the results are transferred to a structural analysis containing the imperfect yarn architecture.

3.1. Process simulation

The software code used for the braiding simulation is the Virtual Performance Solution 13.0 from the ESI group [4]. The numerical model is built-up following methods discussed in [5], using the same configuration as the real machine. The 64 braiding yarns are therefore moving on a sinusoidal path, whereas the mandrel translates with constant speed. In order to consider the effect of different yarn pre-tensions the bobbins are modelled as spring elements that apply a constant tensile force on every yarn. The mandrel and the braiding ring are modelled as rigid bodies while the discretization is done with shell elements. The yarns are also discretized with shell elements, as it has been shown that this approach is best suited for process simulation [6] since yarn width can correspond to the width of the shell elements. The physical distance between the yarns is controlled by the contact properties and should be defined as the subsequent yarn thickness. A constant friction coefficient of $\mu = 0.2$ is applied to all contact interfaces. The virtual braiding process is shown in Figure 5.

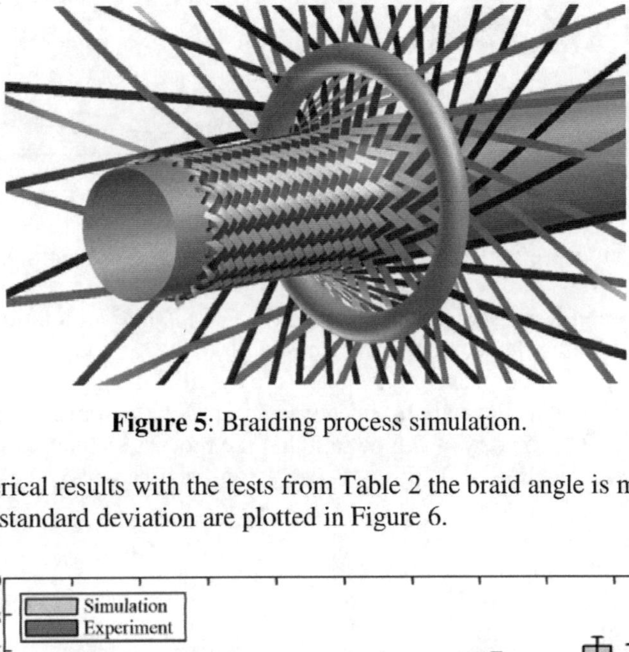

Figure 5: Braiding process simulation.

To compare the numerical results with the tests from Table 2 the braid angle is measured and the mean values, as well as the standard deviation are plotted in Figure 6.

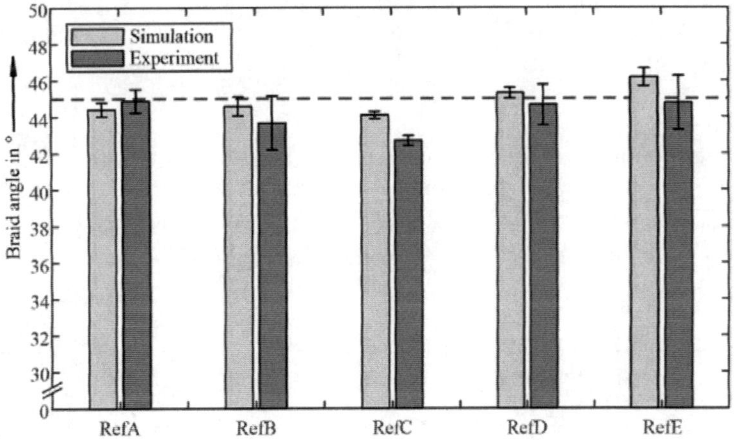

Figure 6: Comparison of the braid angle from the numerical process simulation and the test results.

It can be seen from Figure 6 that the simulation fits the experimentally determined braid angle quite well. For all cases, except RefA, a process-dependent tendency can be observed and the influence of the investigated process parameters is reflected in the numerical results. The remaining deviation of numerical and experimental results can be caused by several factors such as the yarn material model or simplification of the constant friction coefficient. These influences will be further investigated in ongoing work.

As mentioned in section 1.3, fiber damage is mainly caused by excessive friction. In order to estimate the influence of the different process parameters on this particular defect, the dissipated frictional energy from simulation models is depicted in Figure 7.

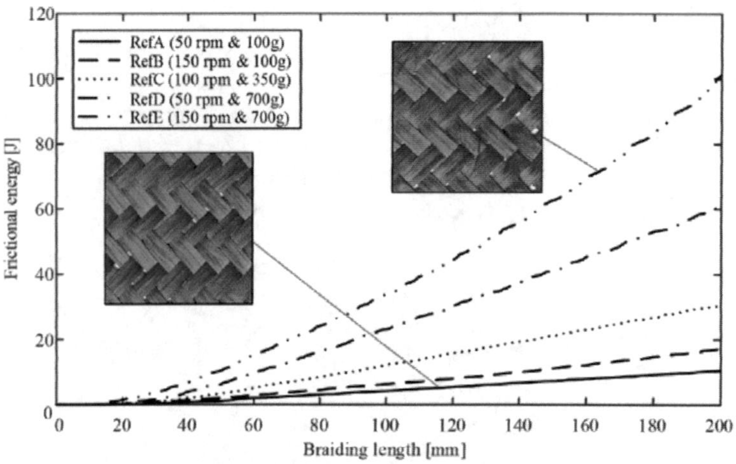

Figure 7: Frictional energy of different process parameters in consideration of fiber damage.

A comparison of RefA and RefB shows a minor influence of the bobbin rotational speed respectively the process speed. On the other hand, comparing RefA with RefD reveals that the yarn pre-tension has a strong influence on contact friction energy and therefore the fiber damage. This can also be seen in the graphs of Figure 7. The braid angle is approximately constant for RefA and RefE, but increasing fiber damage can be observed. In summary, it can be said that the process simulation is capable of capturing the fiber architecture as a function of the process parameters and can also be used as an indicator for fiber damage.

3.2. Structural simulation

This section describes two methods to transfer the results from the process simulation to a numerical analysis for prediction of elastic stiffness including the effect of irregular braid angle.

3.2.1. Voxel method

The first method uses volumetric pixels or short voxels, where the meso-scale composite is represented by a regular 3D mesh. In contrast to other works [7], [8] and [9], the presented method aims to generate a three-dimensional yarn based on the two-dimensional mesh with an irregular yarn architecture. This procedure is depicted schematically in Figure 8 and detailed in Figure 9.

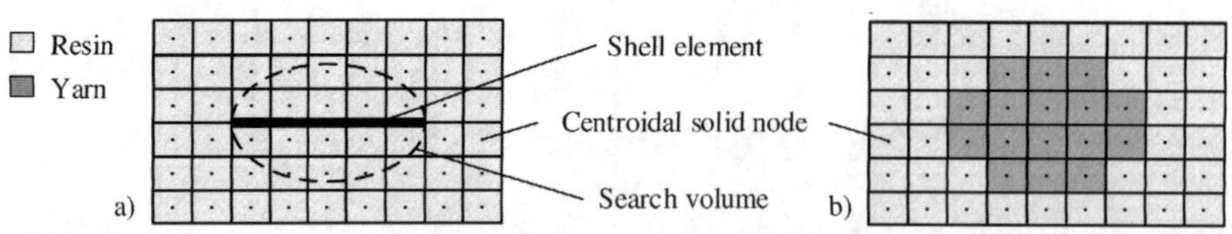

Figure 8: Schematic procedure of the voxel method. a) Shells overlaid with solids. b) Final yarn segmentation based on the selected search volume.

The algorithm to set-up the voxel analysis is implemented in a python script and is undertaken as following. In a first step, the region of interest is extracted from the process simulation results according to Figure 9a. The shell mesh is then superimposed with solid elements as shown in Figure 9b. These two steps are performed manually in the used CAE environment. Thereafter a search area is defined that characterizes the yarn cross-sectional area and is expanded in the third dimension by using the shell-element length. This generates a three-dimensional search volume. As this is done, the script automatically calculates a centroidal node for every solid element. This serves to speed up the algorithm as it checks whether the centroidal node is inside the search volume or not. If so, a new part identification number as well as the material principal axes are assigned to the corresponding element. By iterating through all elements the fiber architecture is segmented. The final voxel mesh is shown in Figure 9c, as well as a section cut of the yarn architecture. As can be seen the specified ellipsoidal cross-section is well developed. In a last step the user defines the material properties, boundary conditions and applied loading manually. By computing the resistance to deformation the elastic properties of the composites volume are determined.

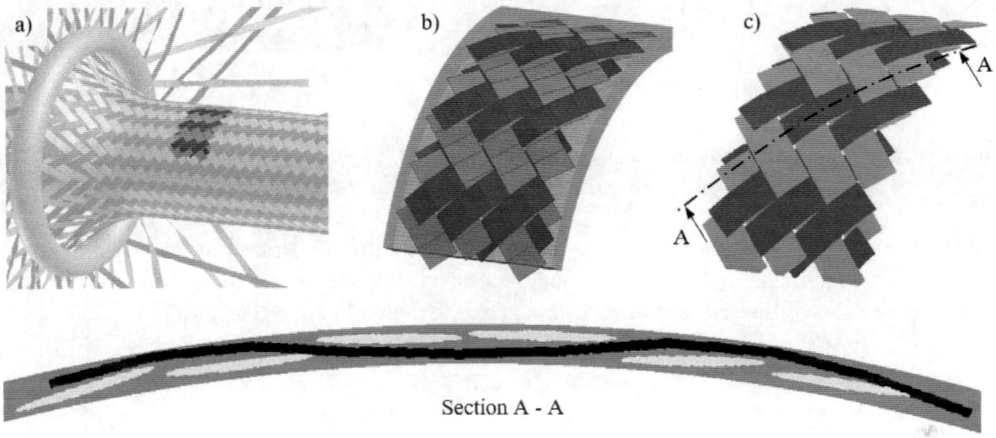

Figure 9: Voxel method: a) Cut-out of the sample; b) Overlay with solid elements; c) Final yarn architecture discretized with volume elements.

Due to the step-wise discretization of the yarn surface (see Figure 8b), stress concentrations do occur at the connection of fiber and matrix parts due to the high mismatch of modulus and the meshing. This is unimportant for determination of the homogenized elastic stiffnesses, since these properties are not influenced by local stress concentrations. However, for a prediction of damage onset and failure this method can be expected to lead to premature failure due to the unrealistic stress concentrations [7] and is therefore not suitable. In order to overcome this limitation an alternative approach is described in the following section.

3.2.2. Embedded elements technique

This second method uses a mesh superposition technique, in which an embedded region constraint is imposed via an option available in Abaqus [10]. The underlying concept is to overlay two meshes where a group of elements is embedded in a set of host elements. The response of the host elements is then used to constrain the translational degrees of freedom (DOF) of the embedded elements. The main benefit of this approach is the independent meshing for the yarns and the resin, while a full model often leads to a poor element quality due to very thin gaps between the single yarns. However, one of the main challenges of this technique is the volume redundancy, which causes an added stiffness that must be taken into account. An interpenetration of the yarns must also be avoided, as this is not covered automatically by the embedded region constraint in Abaqus. In [11] a methodology is presented to overcome these problems and different cases are compared with full models, where a good correlation has been shown. The basic idea of this technique is schematically depicted in Figure 10a, where the embedded nodes must not necessarily lie on the host edges, but can be located inside or even outside the host element. If an embedded node lies outside the host element, a geometric tolerance is used to define how far it can lie outside the host region [10].

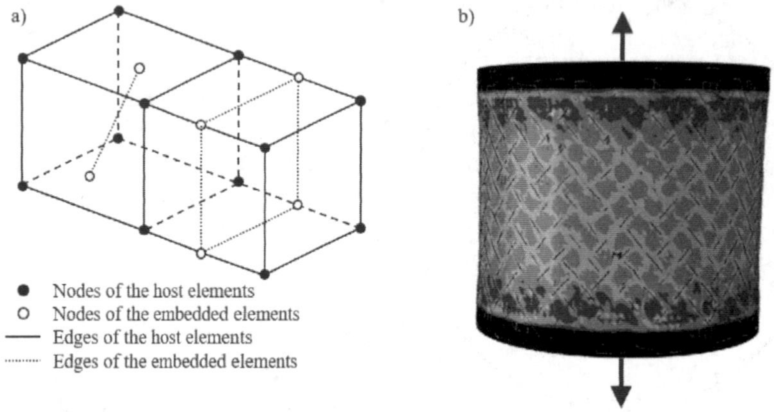

Figure 10: a) Embedded elements in accordance with [10]; b) Axial tension loaded cylinder of an embedded shell-braid in a solid mesh.

In Figure 10b an axial tension loaded cylinder is shown, where the shell-mesh of the process simulation is embedded in a solid-mesh that is used to model the resin. In order to get the correct overall stiffness the material properties of the reinforcement region are calculated in accordance to [11] as

$$C_R^* = C_R(x) - C_M, \quad (2)$$

where $C_R(x)$ designates the position-dependent stiffness matrix of the reinforcement and C_M the stiffness matrix of the resin. In order to avoid an interpenetration of the yarns while loaded under tension, a general contact is used with normal behaviour of *hard contact* and tangential behaviour of *rough* in accordance to [10]. So once the yarns are in contact, no slip will occur; this can be seen as a perfect adhesion condition. Reference [11] also presents an approach to calculate the correct stress- and strain fields for the reinforcement region which can then be used to predict damage onset and failure. A limitation of the embedded elements technique is the inability to represent fiber-matrix debonding, or delamination between layers. Since there is no suitable interaction criteria this effect has to be further investigated, as well as the influence of the non-constraint rotational DOF's for the case that shell elements are embedded in solid elements.

3.2.3. Comparison of the presented methods with WiseTex

In order to evaluate the forecast quality of the presented numerical methods, the results are compared with WiseTex which is an established meso-scale analysis tool to determine elastic stiffness of textile composites [12]. As the embedded shell elements are not able to represent the assumed elliptical yarn cross-section, their geometrical thickness is chosen to have the same cross-sectional area as the elliptical yarns. The models are loaded in the longitudinal direction and the corresponding stiffness is described by $E_{45°}$. The WiseTex result is calculated with the Mori-Tanaka scheme and a comparison with the presented numerical methods can be seen in Table 3.

Table 3: Results of the stiffness comparison from WiseTex and the presented numerical methods.

	WiseTex	Voxel method	Embedded elements
$E_{45°}$	13689 MPa	11206 MPa	13969 MPa

These results show that the embedded element technique corresponds well to the WiseTex results with only a small overestimate of stiffness of about 2 %. The voxel method underestimates stiffness by about 18 % which can be caused by a too low fiber volume fraction. Considering section A-A from **Figure 9** it can be seen that the yarns are not in contact with each other which does not reflect reality and does lead to resin rich zones between the yarns. Further investigations are necessary to overcome this problem.

4. Results and outlook

In this work biaxial carbon fiber braids with an average braid angle of ±45° have been manufactured with different process parameters. A characterization of the preforms is done using image processing techniques. These results show that the braid angle and the yarn width are dependent on process parameters. For example, higher process speed leads to a wider standard deviation of the braid angle and a smaller mean of the yarn width. Also different yarn pre-tension does not significantly effects the braid angle, but can cause significant fiber damage.

For a virtual characterization the braiding process has been simulated and braid angles have been compared with experimental results. The influence of the process parameters is demonstrated in the simulation as well as the experimentally captured standard deviation of the braid angle. A comparison of the dissipated frictional energies shows a strong influence of the yarn pre-tension, which leads to higher fiber damage as observed in the experiments. A voxel method, as well as an embedded element technique, was presented to transfer the results from the process simulation into a structural analysis, which were shown capable to predict composite stiffness compared to the analytical results from WiseTex.

For future work the preparation of test coupons and their mechanical testing is planned with characterization of the strain-field to be undertaken using digital image correlation (DIC). Further investigations are also planned to quantify fiber damage using laser sensors.

Acknowledgments
The present work is funded by the Deutsche Forschungsgemeinschaft (DFG, German Research Foundation) – Project number: 323019910

References

[1] Kyosev Y 2015 *Braiding Technologies for Textiles* (Cambridge: Elsevier) chapter 6 pp 188 - 191
[2] Heieck F, Hermann F, Middendorf P and Schladitz K 2017, *Influence of the cover factor of 2D biaxial and triaxial braided carbon composites on their in-plane mechanical properties* (*Composite Structures* vol 163) pp 114 – 122
[3] Miene A 2008, *IMac-Pro Industrialization of Manufacturing Technologies for Composite Profiles for Aerospace Applications*, annual report of the Faserinstitut Bremen e.V. (FIBRE) 2008/2009
[4] Virtual Performance Solution 2017, *Solver Reference Manual*, ESI Group
[5] Pickett A, Erber A, Reden T von and Drechsler K 2009, *Comparison of analytical and finite element simulation of 2D braiding* (*Plastics, Rubber and Composites* vol 38) pp 387 – 395
[6] Böhler P 2018, *Einzelfadenbasierte Modellierung von textilen Preform-Prozessen*, Dissertation, University of Stuttgart, not published, pp 127-130
[7] Doitrand A, Fagiano C, Irisarri F-X and Hirsekorn M 2015, *Comparison between voxel and consistent meso-scale models of woven composites*, (*Composites: Part A*, vol 73), pp 143 – 154

[8] Hello G, Schneider J and Aboura Z 2014, *Numerical simulations of woven composite materials with voxel-FE models*, ECCM 2014 – 16th European Conference on Composite Materials, Seville, Spain
[9] Kim HJ and Swan Colby 2003, *Voxel-based meshing and unit-cell analysis of textile composites* (*International journal for numerical methods in engineering*, vol 56), pp 977 – 1006
[10] Abaqus (2017), *Abaqus Documentation*, Dassault Systèmes
[11] Tabatabaei S A 2016, *Meso-Finite Element Modelling of Textile Composites using Mesh Superposition Method*, Dissertation, KU Leuven
[12] Verpoest I and Lomov S V 2005, *Virtual textile composites software WiseTex: Integration with micro-mechanical, permeability and structural analysis*, (*Composites Science and Technology* vol 65) pp 2563 – 2574

IOP Conf. Series: Materials Science and Engineering **406** (2018) 012048 doi:10.1088/1757-899X/406/1/012048

Tetraxial textiles: assessment of mesoscale mechanical modelling by experimental measurements

M Ghazimoradi[1], N Naouar[2], V Carvelli[1], P Boisse[2], R Frassine[3]

[1]Department A.B.C., Politecnico di Milano, Milan, Italy
[2]LaMCoS UMR 5259, INSA-Lyon, Université de Lyon, Villeurbanne, France
[3]Department of Chemistry, Materials and Chemical Engineering, Politecnico di Milano, Italy

Abstract. Tetraxial technical textiles were recently manufactured by a new loom developed to weave yarns in four directions. The textile has warp, weft, and two diagonal yarns oriented at symmetrical angles (typically ±45°) with respect to the warp direction. The peculiar yarns distribution could enhance the mechanical response of the textile in multiple directions aiming to almost isotropic in-plane behaviour. For the prediction of the mechanical performance of such tetraxial textiles, a reliable and accurate predictive model is of relevant importance. The present investigation aims to adopt the finite element numerical approach at the mesoscopic scale to predict the mechanical response for any in-plane loading of tetraxial textiles. An accurate modelling of the constitutive behaviour of the fibrous yarns was adopted considering a hyperelastic model. The modelling of the tetraxial unit cell allowed to have the mechanical behaviour for uniaxial and biaxial tensile and for shear loading conditions. The assessment of the accuracy of the numerical model was performed considering a huge experimental campaign dedicated to several hybrid tetraxial textiles. The comparison highlights the accuracy of the numerical model to predict the nonlinear behavior of the fabric for any loading condition and to provide the proper mechanical model for further optimization of tetraxial textiles supposed for different industrial applications.

1. Introduction

Woven fabrics with special architecture have received considerable attention in literatures [1], [2], due to their mechanical behaviour allowing for peculiar tear resistance as well as improved formability characteristics. Two-dimensional (2D) and three-dimensional (3D) fabric architectures can be considered in classification of technical textiles, which have more than two weaving directions. For instance, triaxial braiding produces a 2D architecture with three interlaced yarns; such fabrics have axial yarns in the longitudinal direction and the braided yarns tilted at an angle ranging from 30 to 60 with respect to the previous ones [3]. Advantages of triaxial braided fabrics are geometry flexibility and good tear resistance. Increasing the number of weaving direction to four, a 2D multiaxial textile is obtained, named tetraxial. The fabric architecture combines a traditional orthogonal interlacement with two additional threads (bias), which can be oriented at different angles with respect to the warp/weft direction (Figure 1). It has warp, weft, and two diagonal yarns oriented at symmetrical angles (typically 45°) with respect to the warp direction.

This study has two main parts; experimental and numerical. The first part, an extensive experimental campaign was conducted on tetraxial fabrics made of different yarns. Tensile and friction properties of the yarns were determined, as well as uniaxial and biaxial in-plane tensile tests of the fabric.

In the second part, a numerical model was adopted to predict the mechanical behaviour of tetraxial textiles varying several parameters, namely: material and spacing of the yarns. The meso-scale model takes into account the realistic geometry of woven structure, and the interaction between yarns. Furthermore a hyperelastic model has been adopted [5] considering four different deformation modes. The mesoscopic FE modelling of the tetraxial representative volume (RV) allowed simulating the mechanical behaviour for uniaxial, biaxial, and in-plane shear deformations. Finally, the comparison highlights the accuracy of the meso-scale FE model to predict the nonlinear behaviour of the fabric for any loading condition.

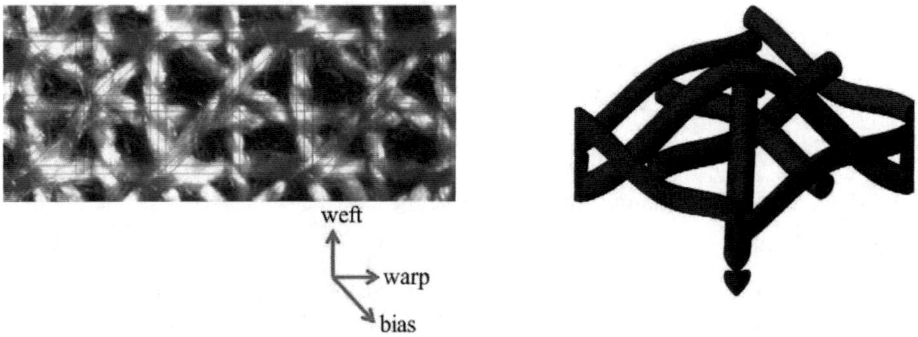

(a) (b)

Figure 1. (a) Tetraxial woven fabric, (b) sketch of the interlacement in the unit cell.

2. Textile features

The tetraxial machine, depicted in Figure 2, is a combination of a traditional woven machine (lower part, Figure 2a), and a specific automatic tension system (upper part, Figure 2b).

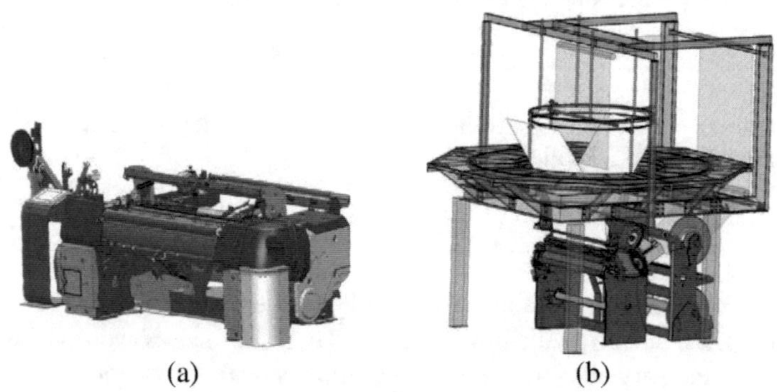

(a) (b)

Figure 2. Scheme of (a) lower and (b) upper part of the tetraxial loom.

The fabric can be considered as a combination of a traditional orthogonal fabric with two additional threads (bias), which can be oriented at different angles with respect to the warp directions (see Figure 1b).

Table 1. Some features of the tetraxial textile

Material	Count [tex]	Nominal strength [cN/tex]
Polyester (PET)	63	20
Aramid	130	219

In the investigation two types of tetraxial technical textile were considered: PET/PET/PET and PET/Aramid/PET hybrid tetraxial textile. Table 1 lists the features of the yarns (PET and Aramid). Additionally, the fabric construction of the tetraxial technical textiles are summarised in Table 2. In the present investigation, yarn material has been varied only in the weft direction.

Table 2. Features of the tetraxial textiles

Yarns [Warp / Weft / Bias]	Areal density [g/m2]	Warp [ends/cm]	Weft [picks/cm]	Bias [threads/cm]
PET / PET / PET	136	6	7	5
PET / Aramid / PET	159	6	7	5

3. Overview of the experimental investigation

3.1. Yarns

Yarn tensile tests (extracted from fabric in warp, weft, and diagonal directions) were performed by INSTRON device with 500 N load cell. The gauge length was 150 mm and testing speed 5 mm/min. Five tests were performed for each yarn direction (warp, weft, diagonal).
The tensile force vs. strain curves of yarns were depicted considering the free length as base length for strain calculation. PET shows a nonlinear behaviour (a), which, since the early loading, is related to rearrangement of fibres and their alignment in the load direction. On the other hand, Aramid has almost linear behaviour until failure point (b).

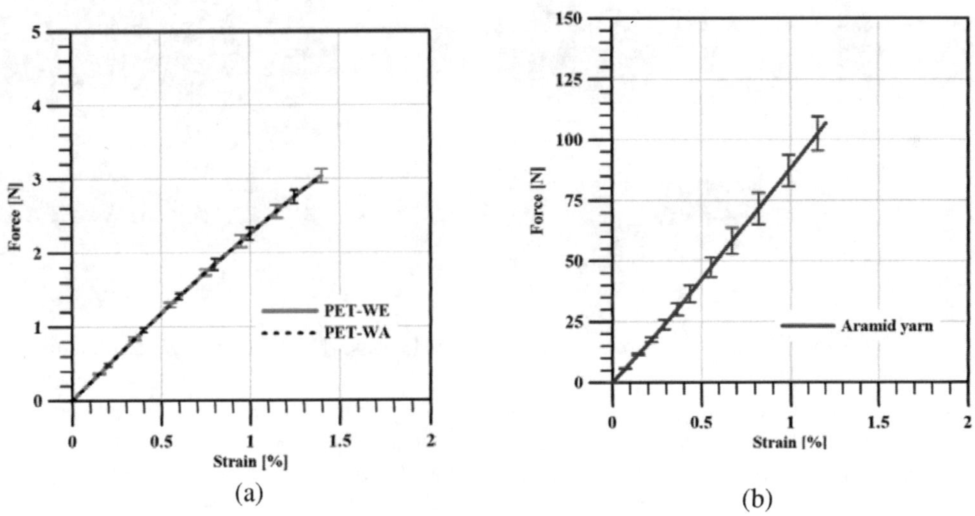

Figure 3. Force vs. strain curves for yarns extracted from tetraxial fabrics: (a) PET in warp and weft directions, and (b) aramid. Bars indicate standard deviation.

Table 3 collects the average specific strength in warp, weft and diagonal directions for all considered yarns. Special weaving process used to manufacture the tetraxial hybrid fabric has considerable effect on tensile behaviour of some yarns. The comparison of unwoven nominal yarns strength in Table 1and measured in Table 3 of Aramid yarn extracted from weft direction of tetraxial illustrated a significant reduction of strength, while it is slightly changed for weft PET yarn (Figure 3). This is probably due to the intrinsic properties of these materials and the damage induced in the weft path of the weaving loom. Moreover, the weaving loom creates some damage in the warp and diagonal tracks, as well, resulting in

IOP Conf. Series: Materials Science and Engineering **406** (2018) 012048 doi:10.1088/1757-899X/406/1/012048

a relative reduction of the PET yarns strength. An in-depth investigation is necessary to better understand the correlation of the yarn material properties and the imparted degradation in the weaving process.

Table 3. Average specific strength of yarns [cN/tex]. (± indicates standard deviation)

	PET	Aramid
Warp yarn	17.56 ± 1.8	-
Weft yarn	20.53 ± 0.82	119.81 ± 0.09
Diagonal yarn	17.61 ± 0.6	-

3.2. Fabrics

Biaxial and uniaxial tensile tests have been considered to investigate the mechanical properties of tetraxial technical textiles. A home design biaxial device was adopted to evaluate the biaxial tensile behaviour. The displacement is generated in each direction of cross-shaped specimen (Figure 4a), which is set in the centre of the machine. Arms of the specimens had width of 100 mm to create a biaxial loaded portion of 100 × 100 mm^2 with an almost uniform distribution of strain on centre part. All tests were performed at room temperature and speed of 1mm/min.

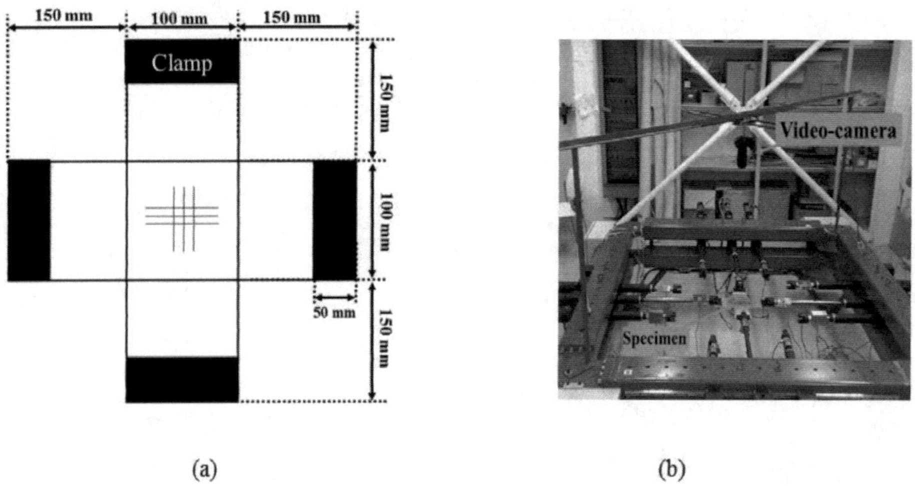

Figure 4. (a) Cross-shaped fabric specimen dimensions, (b) device for biaxial tensile tests.

In order to evaluate the in-plane shear behaviour of tetraxial textiles, bias-extension tests were performed. A tensile test involves rectangular specimen of material such that the warp and weft directions of the yarns are orientated initially at ±45° to the direction of the applied tension load Tabs 3 mm thick, 140 mm wide, and 60 mm long were glued at ends of the specimens, having 280 mm free length between the grips, which give a ratio $\lambda = H_0/W_0$ of 2. Tests were conducted by an INSTRON device with a speed of 10 mm/min. During test, a digital camera records photograms at a frequency of 1 Hz for image correlation analysis. Post processing by Digital Image Correlation (DIC) [6] allowed estimation of the in-plane shear deformation on the reinforcement surface.

4. Finite element model

To define the constitutive law, it is essential to describe the specific deformation modes of the yarn by a phenomenological approach based on experimental observations. According to [5] four deformation modes, which are related to the behaviour of the yarns and their interaction, are considered: the elongation in the fibre direction, the compaction of the transverse section of the yarn, the transverse section and the longitudinal shear (Figure 5).

4.1. Hyperelastic constitutive equation

In order to define a hyperelastic material, an elastic strain energy potential per unit volume w must exist which depends on the current strain state. To ensure that this constitutive equation fulfils the principle of material frame indifference, a function $\tilde{w}$ of the right Cauchy Green deformation tensor $\underline{\underline{C}}$ must exist.

It has been shown [8] that an isotropic function $\hat{w}$ exists which depends on the right Cauchy-Green strain tensor $\underline{\underline{C}}$ and structural tensors $\underline{\underline{G}}_i$ defining the symmetry group, such as:

$$\tilde{w}\left(\underline{\underline{C}}\right) = \hat{w}\left(\underline{\underline{C}}, \underline{\underline{G}}_i\right) \tag{1}$$

The representation theorems for transversely isotropic functions of scalars, vectors and/or symmetric tensor variables define a finite number of scalar functions ($I_1, ..., I_n$) and a function w such as:

$$\hat{w}\left(\underline{\underline{C}}, \underline{\underline{G}}_i\right) = \bar{w}\left(I_1, ..., I_n\right) \tag{2}$$

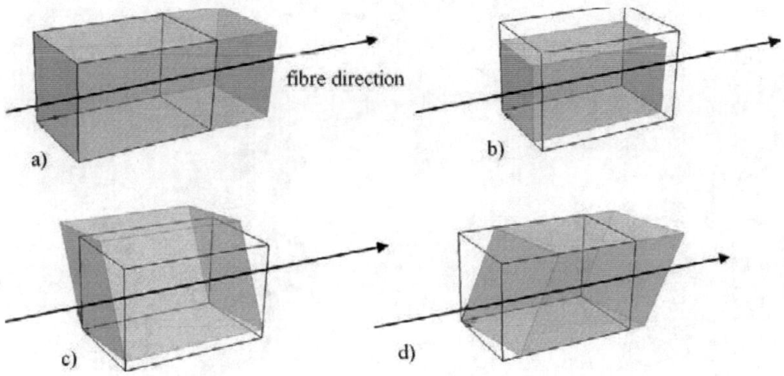

Figure 5. Four different deformation modes of single yarn: (a) elongation, (b) compaction of the cross section, (c) distortion of the cross section and (d) longitudinal shear.

Number n depends on the number and type of arguments of the isotropic function under consideration. The symmetry group of a transversely isotropic material is defined, in the initial configuration with a preferred unit direction $\underline{A}$. This allows the definition of a structural tensor $\underline{\underline{A}} = \underline{A} \otimes \underline{A}$. Then, the strain energy potential become:

$$w^{it} = w^{it}\left(I_1, I_2, I_3, II_4, II_5\right) \tag{3}$$

where I_1, I_2, I_3 are the classical invariants of $\underline{\underline{C}}$. II_4 and II_5 are mixed invariants defined from the structural tensor $\underline{\underline{A}}$ by:

$$II_4 = \underline{\underline{C}} : \underline{\underline{A}} = \underline{A} \cdot \underline{\underline{C}} \cdot \underline{A} \quad and \quad II_5 = \underline{\underline{C}}^2 : \underline{\underline{A}} = \underline{A} \cdot \underline{\underline{C}}^2 \cdot \underline{A} \tag{4}$$

4.2. Constitutive equation for a yarn

If $\underline{A}$ is a unit vector, the elongation of the yarn in the direction of the fibres is directly given by invariant II_4 defined in (4). Consequently, this quantity can be chosen as the representative value of the length change of the yarn:

$$I_{elong} = \underline{\underline{C}} : \underline{\underline{A}} \tag{5}$$

IOP Conf. Series: Materials Science and Engineering **406** (2018) 012048 doi:10.1088/1757-899X/406/1/012048

Yarn compaction, distortion and longitudinal shear can be characterized using the following expressions [9]:

$$I_{comp} = \sqrt{\frac{I_3}{I_4}} \quad I_{dist} = \sqrt{I_1 - \frac{I_4}{I_5} - 2I_{aire}} \tag{6}$$

The constitutive equation is the summation of all contributions:

$$\underline{\underline{S}} = 2\frac{\partial w}{\partial \underline{\underline{C}}} = 2\left(\frac{\partial w_{elong}}{\partial I_{elong}}\frac{\partial I_{elong}}{\partial \underline{\underline{C}}} + \frac{\partial w_{comp}}{\partial I_{comp}}\frac{\partial I_{comp}}{\partial \underline{\underline{C}}} + \frac{\partial w_{dist}}{\partial I_{dist}}\frac{\partial I_{dist}}{\partial \underline{\underline{C}}}\right) \tag{7}$$

More details can be find in [5]. The constitutive equation has been implemented in a user material subroutine VUMAT for ABAQUS/Explicit finite element code, developed at INSA Lyon [5].

5. Results and comparisons

For each deformation mode (see Figure 5), a strain energy density function based on the experimental behaviour of the technical textile [1], is defined. Afterwards, the strain energy density is used to determine the model parameters by means of a sum of differences squared algorithm, which optimizes the calculated and experimental energies.

5.1. Mesh sensitivity

In Figure 6, two types of mesh are proposed to check the mesh sensitivity. The force vs. logarithmic strain for both meshes is depicted in Figure 7. This comparison shows that improvement of mesh does not have any substantial effect on the numerical prediction.

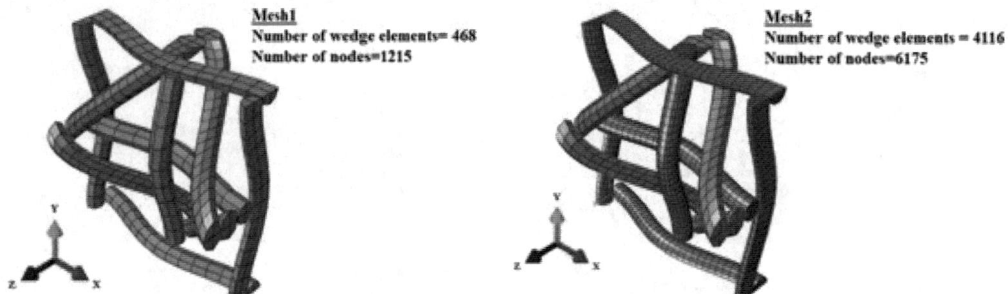

Figure 6. Two different meshes for PET/PET tetraxial unit cell.

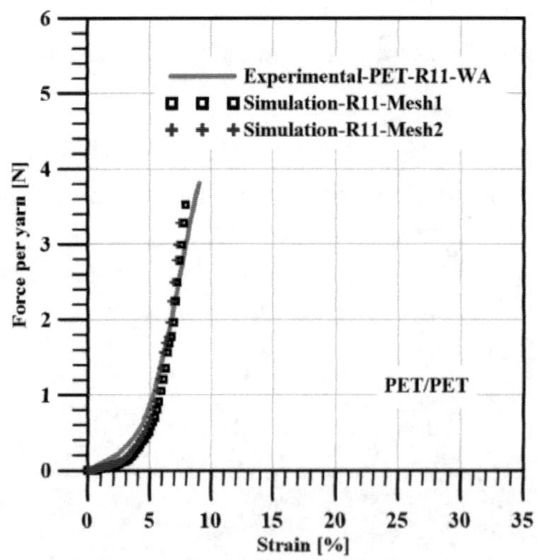

Figure 7. Effect of the refinement of mesh in PET/PET tetraxial hybrid textile.

5.2. Tensile loading

The numerical results were validate by comparison with experimental results of bi-axial tensile loadings. Comparison of bi-axial tensile test for PET/PET and PET/Aramid tetraxial textiles with different warp to weft displacement ratio *R* is depicted in Figure 8. The good agreement between the experimental and numerical curves for "R11", "R12" and "R21" validated the proposed hyperelastic constitutive law.

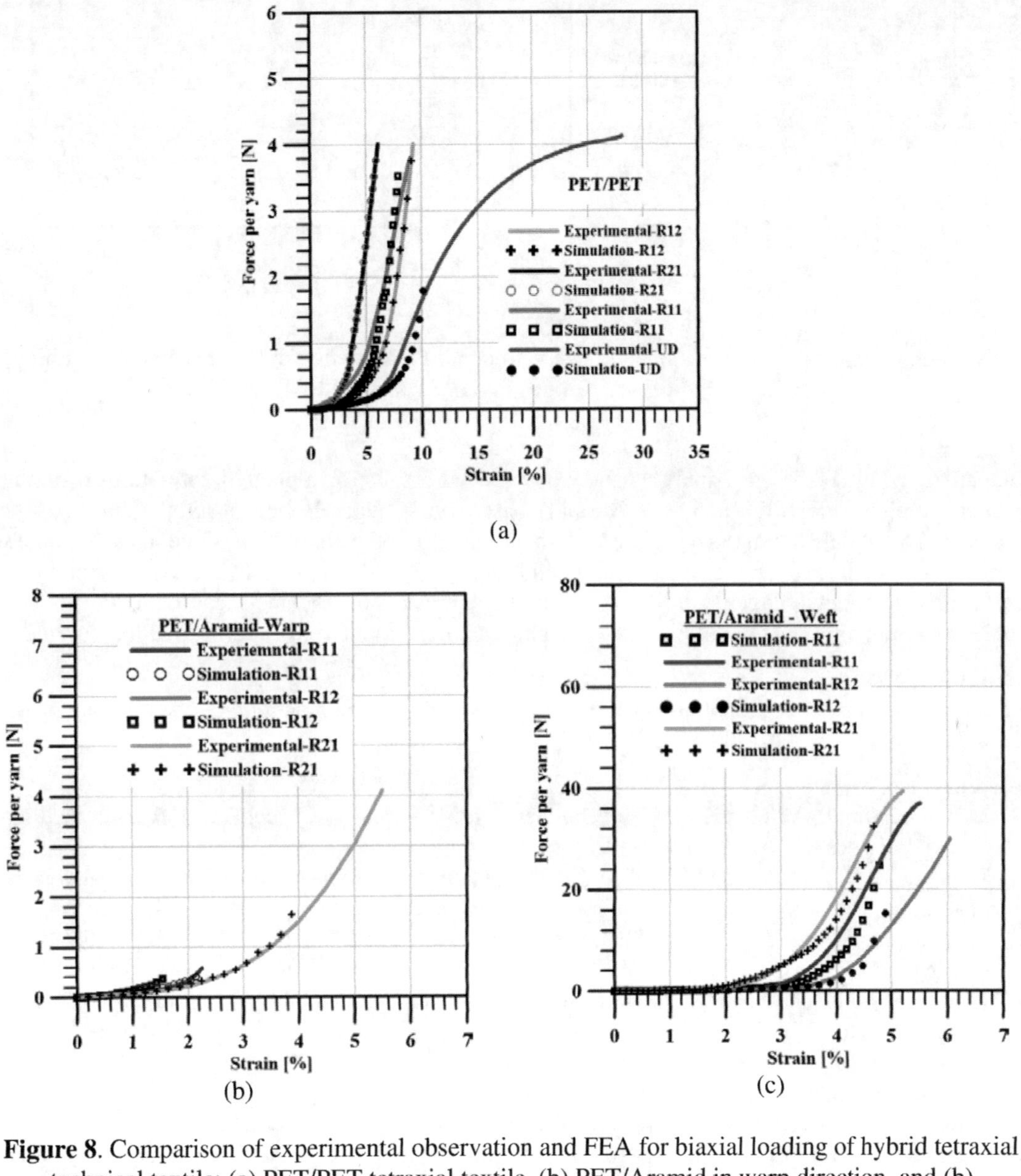

Figure 8. Comparison of experimental observation and FEA for biaxial loading of hybrid tetraxial technical textile; (a) PET/PET tetraxial textile, (b) PET/Aramid in warp direction, and (b) PET/Aramid in weft direction.

5.3. Shear Loading

Figure 9 presented the comparison of FEA and experimental observation for bias extension test. The force vs. in-plane shear deformation curves show a significant agreement between experimental and numerical results. Comparisons in Figure 9a,b illustrated that changing the yarn material in weft direction has not any significant effect on in-plane deformation of hybrid tetraxial textiles, and this point is evident with the numerical results too.

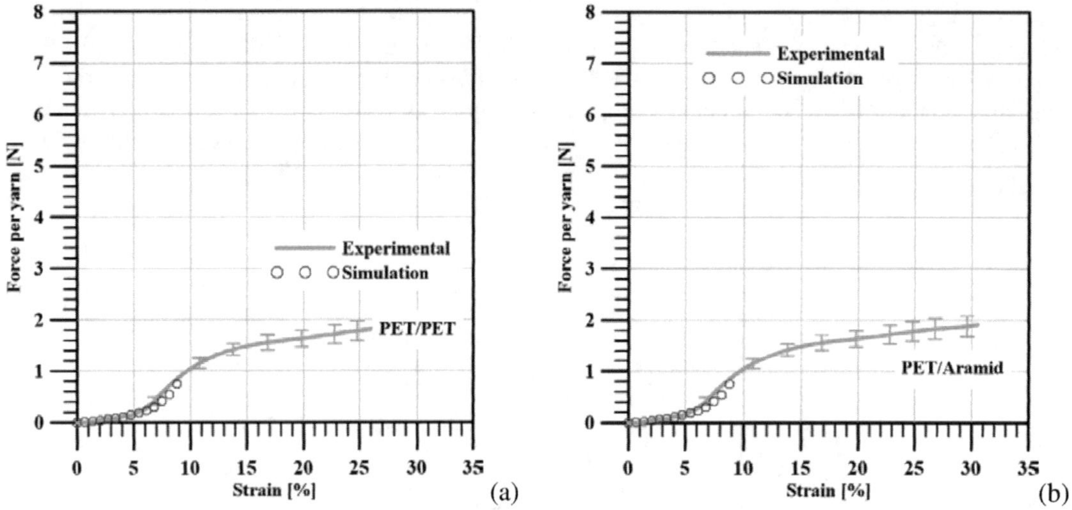

Figure 9. Comparison of experimental observation and FEA results for bias extension loading of tetraxial technical textiles: (a) PET/PET and (b) PET/Aramid.

6. Conclusions

Comparison of meso-scale FE analyses and experimental results of in-plane deformations of tetraxial technical textiles (PET/PET and PET/Aramid) based on a hyperelastic constitutive law has been presented. To validate the meso-scale finite element modelling the simulation of the unequal and equi-biaxial tensile loading of a representative unit cell is performed. The numerical and experimental results have a substantial good agreement. Moreover, numerical results of in-plane shear loading of PET/PET and PET/Aramid tetraxial textiles have also a significant agreement with experimental measurements.

Acknowledgements

The work was supported by the Italian Ministry of Economic Development under the research project 'Tetraxial' (Grant No. MI01_00202).

References

[1] M. Ghazimoradi, V. Carvelli, M. C. Marchesi, and R. Frassine, "Mechanical characterization of tetraxial textiles," *J. Ind. Text.*

[2] J. Pazmino, V. Carvelli, and S. V. Lomov, "Formability of a non-crimp 3D orthogonal weave E-glass composite reinforcement," *Compos. Sci. Technol.*, vol. 61, pp. 76–83, 2014.

[3] S. C. Quek, A. M. Waas, K. W. Shahwan, and V. Agaram, "Analysis of 2D triaxial flat braided textile composites," *Int. J. Mech. Sci.*, vol. 45, no. 6–7, pp. 1077–1096, 2003.

[4] F. Loix, P. Badel, L. Orgéas, C. Geindreau, and P. Boisse, "Woven fabric permeability: From textile deformation to fluid flow mesoscale simulations," *Compos. Sci. Technol.*, vol. 68, no. 7–8, pp. 1624–1630, 2008.

[5] A. Charmetant, E. Vidal-Sallé, and P. Boisse, "Hyperelastic modelling for mesoscopic analyses of composite reinforcements," *Compos. Sci. Technol.*, vol. 71, no. 14, pp. 1623–1631, 2011.

[6] "Correlated Solutions, Inc., VIC-2D, Irmo, SC, USA," 2012. [Online]. Available: http://www.correlatedsolutions.com.

[7] A. Charmetant, J. G. Orliac, E. Vidal-Salle and P. Boisse, "Hyperelastic model for large deformation analyses of 3D interlock composite preforms," *Compos. Sci. Technol.*, vol. 72, no. 12, pp. 1352–1360, 2012.

[8] Q.-S. Zheng, "Theory of Representations for Tensor Functions—A Unified Invariant Approach to Constitutive Equations," *Appl. Mech. Rev.*, 1994.

[9] J. C. Criscione, A. S. Douglas, and W. C. Hunter, "Physically based strain invariant set for materials exhibiting transversely isotropic behavior," *J. Mech. Phys. Solids*, 2001.

IOP Conf. Series: Materials Science and Engineering **406** (2018) 012049 doi:10.1088/1757-899X/406/1/012049

Micro-scale model for a multi-scale modeling approach of thermoplastic fiber reinforced polymers

Stefan Hesseler[1*], Scott E Stapleton[2], Valeriy Yakovlev[1], Lars Appel[1] and Thomas Gries[1]

[1]Institute of Textile Technology, RWTH Aachen, Otto-Blumenthal-Straße 1, 52074 Aachen, Germany

[2]Department of Mechanical Engineering, University of Massachussetts Lowell, 197 Riverside St., Lowell, MA 01854, USA

* stefan.hesseler@ita.rwth-aachen.de

Abstract. Thermoplastic fiber reinforced composites (TPFRCs) are becoming more important due to the potential of short part cycle times and recyclability. One major problem of these materials is process stability. During processing, unwanted distortions appear due to residual stresses on different scales. The prediction of these distortions and stresses would reduce the cost of TPFRCs and would assist in achieving a higher market share. For the prediction of the residual stresses, a new multi-scale simulation approach is required. The research presented in this paper is a contribution to the micro-scale model. A method for the representation of the thermoplastic matrix in a 2D micro-scale model is developed. With this, the a solid as well as a semi-liquid state of the polymer which appear during thermoforming can be represented.

1. Introduction

The demand for lightweight materials is rising due to resource scarcity and environmental awareness. Fiber reinforced polymers (FRPs) are among the strongest lightweight materials, which combine high mechanical properties with low density. The FRPs with the highest mechanical properties contain thermoset polymer matrices, which are cured irreversibly through a chemical reaction. Even though thermoplastic composites have inferior mechanical properties, they are becoming more popular due to the potential of shorter part production cycle times, ease of storage and handling, increased toughness and recyclability [1][2].

Thermoplastic fiber reinforced composites (TPFRCs) often come in the form of blanks which are processed by applying heat and pressure in a thermoforming process. Despite the short cycle times and easy handling, TPFRCs do not have a high market share, which is around ten percent [2]. One major issue for these materials is the process stability. Residual stresses occur due to several effects on different scales, which cause unwanted deformations [3]. Currently, attempts to mitigate unwanted residual stresses rely on trial and error, which can be cost intensive and time consuming [4].

Computational models of the thermoforming process which can predict shape distortions and residual stresses are needed to replace trial and error. A multi-scale approach is required due to the multi-scale nature of the residual stresses [1]. Typical scales for modeling of TPFRCs are micro-, meso- and macro-scale [1]. On the micro-scale, the discrete element method (DEM) will be utilized to include the solid as well as the semi-liquid state of the TPFRCs during the thermoforming process. For this, an existing code for the calculation of dry carbon fibers introduced in [5] will be expanded by adding a matrix, which is

IOP Conf. Series: Materials Science and Engineering **406** (2018) 012049 doi:10.1088/1757-899X/406/1/012049

temperature dependent. Thus, with the model contacts mechanics between single filaments will be represented. Additionally, the solid as well as the semi-liquid state of the matrix will be considered. For this, springs representing the matrix are implemented in the DEM model shown in [5]. These springs are dependent on geometrical and material property factors. With this, a temperature dependent material model will be developed and applied to the springs. The basic rules for the implementation of the matrix springs in the DEM model are explained in the following chapters.

2. Model description

The matrix model is based on springs with a stiffness dependent on the material properties and geometry factors. For this, an independent continuum mechanics model is created. The model calculates the forces for each fiber solely based on the matrix springs. For the application of these springs, rules to implement the matrix are developed. These rules are introduced hereinafter.

The model is based on two types of connections: fiber-fiber and fiber-matrix connections (Figure 1). These connections are meant to resemble the influence of the matrix on fibers. The fiber-fiber connection is used to represent the interaction between any two adjacent fibers. It is assumed that the two fibers are connected by a conical portion of the matrix. The fiber-matrix connection is used to represent interaction between fibers and matrix. These portions are attached to the given fiber, fiber *i* and to the portion of the matrix between any two of the neighboring fibers, fiber *j* and fiber *k*. The exact point of the attachment for the fiber-matrix connection will be explained below. To establish any of these connections it is required to know the neighbors of a given fiber. The method used to find neighboring fibers for any of these fibers is presented in the following section.

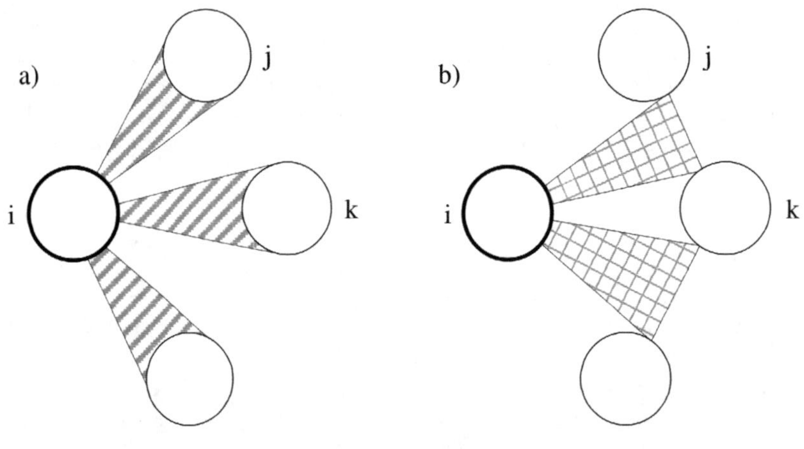

Figure 1. a) Matrix portion between two fibers, b) matrix portion of a fiber and matrix without an adjacent fiber

2.1. Neighbor search

The connections must be established between appropriate fibers and must be attached to appropriate regions of the matrix. Thus, finding correct neighbors for each fiber is a crucial step. Neighbor selection is controlled by three parameters: search radius, maximum overlap angle and maximum relative fiber angle (Figure 2). The neighbor search for a given fiber starts with finding all fibers within the search radius. Next, all these fibers are sorted in ascending order with respect to the distance from the given fiber. After that, all sorted fibers, starting from the closest one, are checked. When a fiber is within the set conditions imposed by a maximum overlap angle and a minimum relative fiber angle, it is considered to be a neighbor of the given fiber. After finding the neighboring fibers the stiffness of the matrix portions need to be calculated, which is done in the following.

IOP Conf. Series: Materials Science and Engineering **406** (2018) 012049 doi:10.1088/1757-899X/406/1/012049

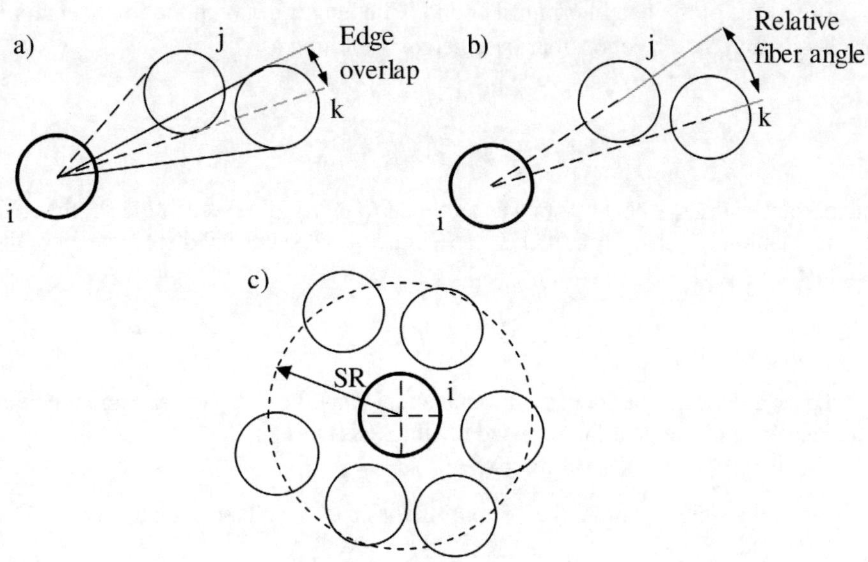

Figure 2. Parameters considered in the neighbor search: a) edge overlap, b) relative fiber angle, c) search radius

2.2. *Stiffness Calculation*

After all neighboring fibers are found, the stiffness of the matrix portions needs to be calculated. For this, the assumption is made, that each connection has solely normal stiffness in the direction of the connection. For now, all connections are assumed to be tapered rods made of linear elastic material under plane stress. Thus, normal stiffness of these connections can be found. The tangential displacements are assumed to be small compared to the normal displacement. Therefore, a tangential stiffness is not included in the calculation.

For the calculation of the matrix stiffness the geometry of the matrix is a key part. The geometry is approximated by a frustum shown in Figure 3. The frustum has the length *L*, the frustum angle *θ*, a base height h_0 at the given fiber and the height h_1 at the neighboring fiber.

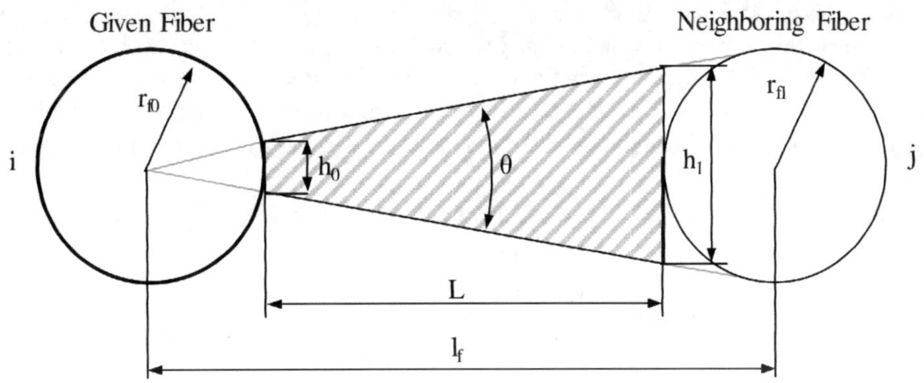

Figure 3. Geometrical parameters used in stiffness calculation

Considering the geometry, the frustum angle, *θ*, can be found as

$$\sin\left(\frac{\theta}{2}\right) = \frac{r_{f_1}}{l_f} \tag{1}$$

where r_{f_1} is the radius of the neighboring fiber and l_f the length between the fiber centers. The given fiber side base height and the neighbor fiber base height are found as

$$h_0 = 2\tan\left(\frac{\theta}{2}\right) r_{f_0} \quad (2)$$

$$h_1 = 2\tan\left(\frac{\theta}{2}\right)\left(l_f - r_{f_1}\right) \quad (3)$$

To find the normal stiffness, the frustum was assumed to be rigidly fixed at the neighbor fiber's side and loaded by an axial force at the given fiber's side (Figure 4). The normal displacement of the frustum, δ_n, is calculated as

$$\delta_n = \int_0^L \frac{F}{EA(x)} dx \quad (4)$$

where F is the axial load, E the Young's modulus and $A(x)$ the area, which can be calculated with the height in dependence of the position $h(x)$ and the thickness t as

$$A(x) = t \cdot h(x) \quad (5)$$

Assuming a general spring equation, the normal stiffness, k_n, can be calculated as

$$k_n = \frac{Et(h_1 - h_0)}{\ln\left(h_1/h_0\right) L} \quad (6)$$

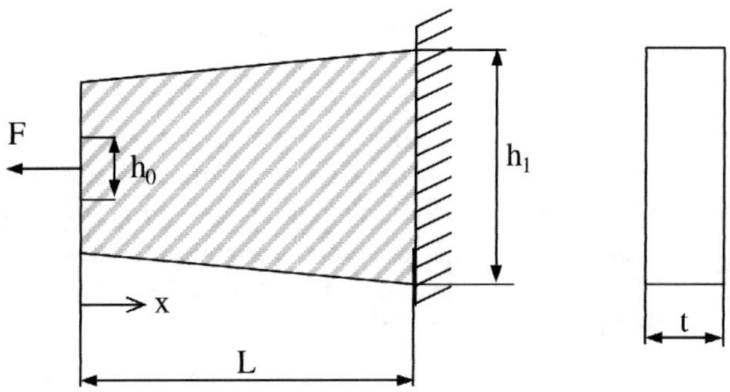

Figure 4. Calculation model for the matrix stiffness

2.3. Calculation of the fiber-matrix connections

The calculation of the stiffness k_n for fiber-matrix connections is done equally as for the fiber-fiber connections. The geometric parameters are calculated in a different way, which is shown in Figure 5. The frustum angle θ is calculated by subtracting the angle between two fibers θ_{jk} by the half of each fiber's angle θ_{ik} and θ_{ij}. The position vector of the given fiber, fiber i is $\underline{r}_{A0}$, the position vector of the neighboring fiber, fiber j is $\underline{r}_{B0}$ and the position vector of the neighboring fiber, fiber k is $\underline{r}_{C0}$. The position of the matrix between two fibers is described as $\underline{r}_{D0}$. The vector from the given fiber to the matrix position $\underline{r}_{D0}$ is $\underline{a}$.

$$\underline{r}_{D0} = \frac{\underline{r}_{B0} + \underline{r}_{C0}}{2} \quad (7)$$

$$\underline{a} = \underline{r}_{D0} - \underline{r}_{A0} \quad (8)$$

The length of the frustum is calculated with the radius of the given fiber R_i and the positions of the matrix as

$$L = \left\|\underline{a}\right\| - R_i \quad (9)$$

IOP Conf. Series: Materials Science and Engineering **406** (2018) 012049 doi:10.1088/1757-899X/406/1/012049

With the length of the frustum and the angle between the fibers, the base height and the height can be found as

$$h_0 = 2\tan\left(\frac{\theta}{2}\right) R_i \tag{10}$$

$$h_1 = 2\tan\left(\frac{\theta}{2}\right) L \tag{11}$$

With the deformation for each connection, the forces acting on each fiber can be calculated.

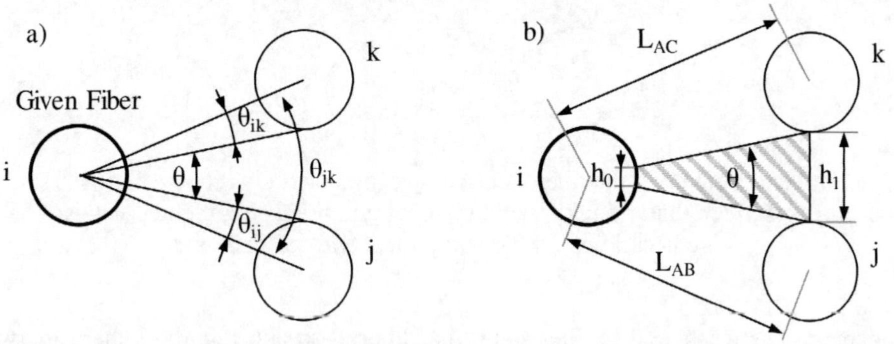

Figure 5. Geometric parameters of fiber-matrix connections. a) angles between fibers, b) frustum between fibers

2.4. Calculation of the deformation

After the determination of the stiffnesses the deformations during one time step of the model need be calculated. To calculate the deformation, the given and its neighboring fiber are assumed to have initial positions $\underline{r}_{A_0}$ and $\underline{r}_{B_0}$ respectively. These positions define the initial configuration. During a load step the given fiber moves the distance $\underline{u}_A$, while the neighboring fiber moves the distance $\underline{u}_B$ (Figure 6).

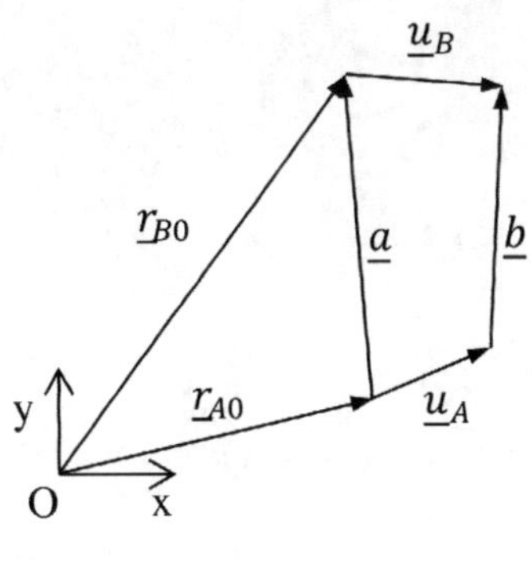

Figure 6. Relative movement of fibers described with the fiber position vectors $\underline{\boldsymbol{r}}_{A_0}$ and $\underline{\boldsymbol{r}}_{B_0}$, the fiber displacements $\underline{\boldsymbol{u}}_A$ and $\underline{\boldsymbol{u}}_B$ and the relative fiber vectors a and b.

The final positions after displacement can be calculated with the initial position $\underline{r}_{A_0}$ and $\underline{r}_{B_0}$, the displacement $\underline{u}_A$ and $\underline{u}_B$ and the relative fiber vectors $\underline{a}$ and $\underline{b}$. Finally, the change in length Δl_n of the connections can be calculated with the relative fiber vectors $\underline{a}$ and $\underline{b}$ as

$$\Delta l_n = \|\underline{\mathrm{b}}\| - \|\underline{\mathrm{a}}\| \tag{12}$$

2.5. Deformations of the fiber-matrix connections

For the calculation of the change in length at a fiber-matrix connection, three fibers are considered: The given fiber, fiber *i* and the two neighboring fibers, fiber *j* and fiber *k* (Figure 7). The calculation is equal to the calculation of the fiber-fiber connection.

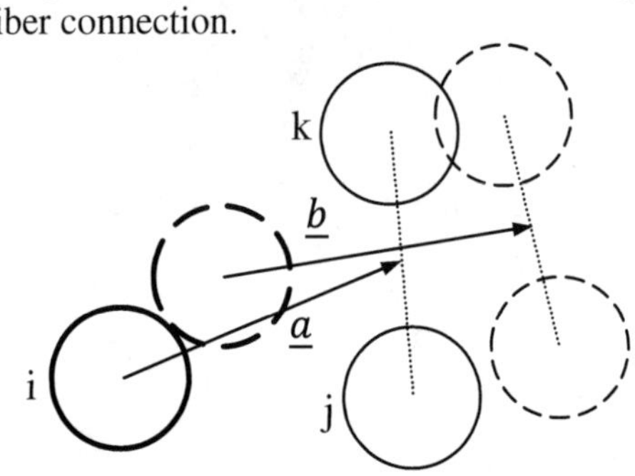

Figure 7. Relative movement of the fiber-matrix connection including the given fiber, fiber i and the neighboring fibers, fiber j and fiber k. The initial configuration is highlighted with solid lines, the configuration after displacement with dashed lines

2.6. Calculation of the forces

For the connections between a given fiber and its neighboring fibers (Figure 8), there are two types of connections: Fiber-fiber connections established between the given fiber and its neighboring fibers (Figure 8 a), and fiber-matrix connections established between the given fiber and the middle portion of the matrix between two of the neighboring fibers (Figure 8 b). The middle portion of the matrix between two of the neighboring fibers is also connected to the neighboring fibers with a rigid rod (Figure 8 c).

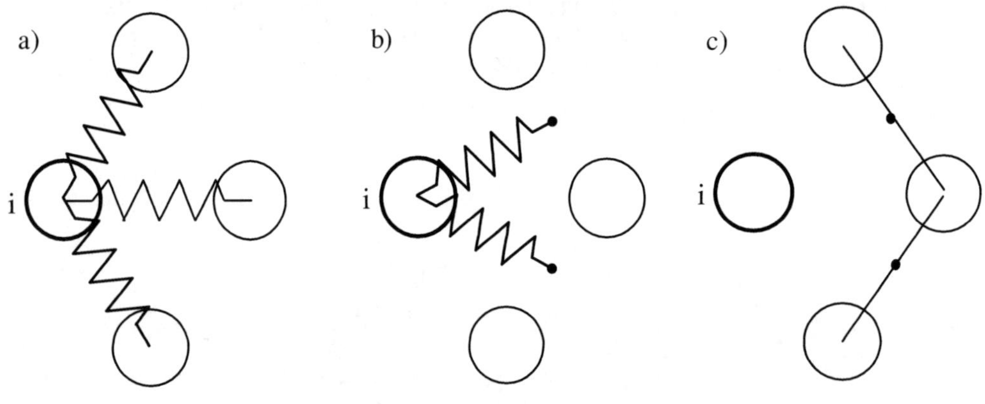

Figure 8. Connection types: a) fiber-fiber connections, b) fiber-matrix connection, c) matrix to fiber rod connection

Forces acting on the given fiber are calculated in the deformed configuration. Deformations and directions of all connections relevant to the given fiber are calculated. With stiffness, deformations and directions of all direct and indirect connections, the resulting force acting on the given fiber can be calculated.

Before the calculation of the forces acting on fibers, all connections and stiffnesses of the connections are defined. To calculate reaction forces acting on a given fiber, all connections relevant to the fiber should be found. In Figure 9 all relevant connections for a given fiber example are shown. The next step is to calculate the deformation of each connection, which is done as explained in the previous sections. The forces can be calculated with $\underline{F_n} = (\Delta l_n \cdot k_n) \cdot \hat{n}_n$, whereas $\underline{F_n}$ is the connection force for the connection *n*, Δl_n the change in length of the connection *n*, k_n the stiffness of connection *n* and $\hat{n}_n$ the unit vector which defines the direction of the connection *n*.

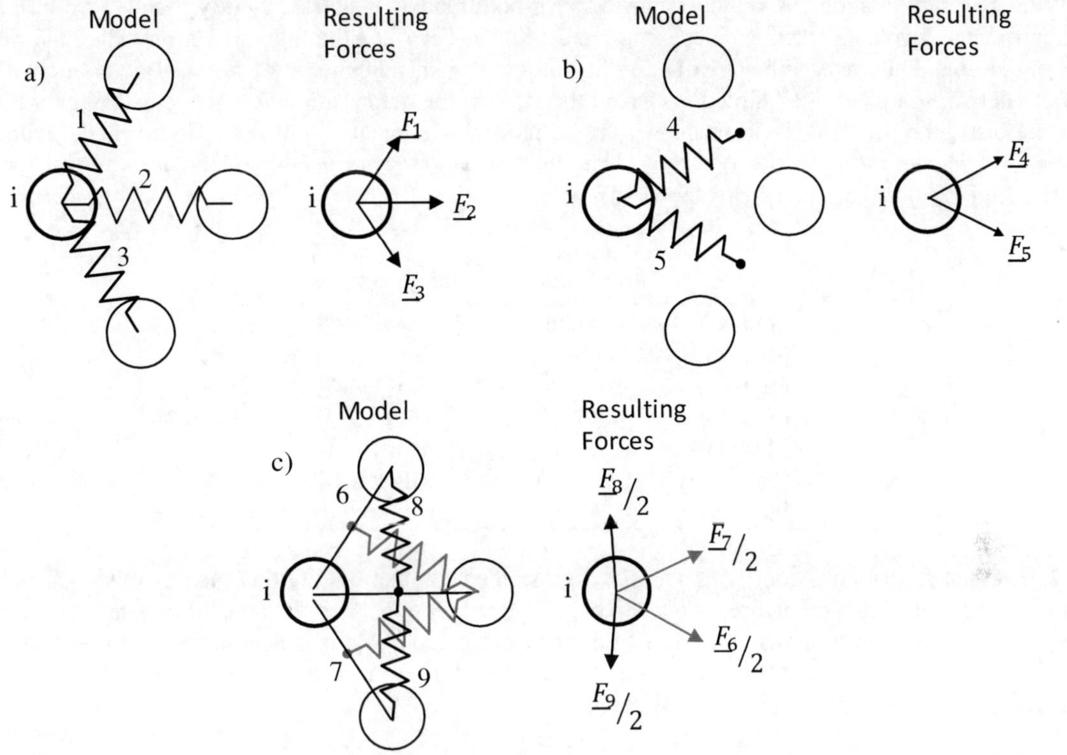

Figure 9. Force directions of: a) Fiber-fiber connections, b) fiber-matrix connection and c) indirect fiber-matrix connections.

3. Validation

The validation of the reduced spring model is done by comparing the reduced model to an equal FE model created in Abaqus unified FEA from Dassault Systèmes, Vélizy-Villacoublay, France. [6] This is done for a model a size of 0.4 mm times 0.4 mm in which the fiber radius is 0.01 mm. The search radius for neighbors was determined with a convergence study which is shown in Figure 10 and set to eight times the fiber radius (0.08 mm).

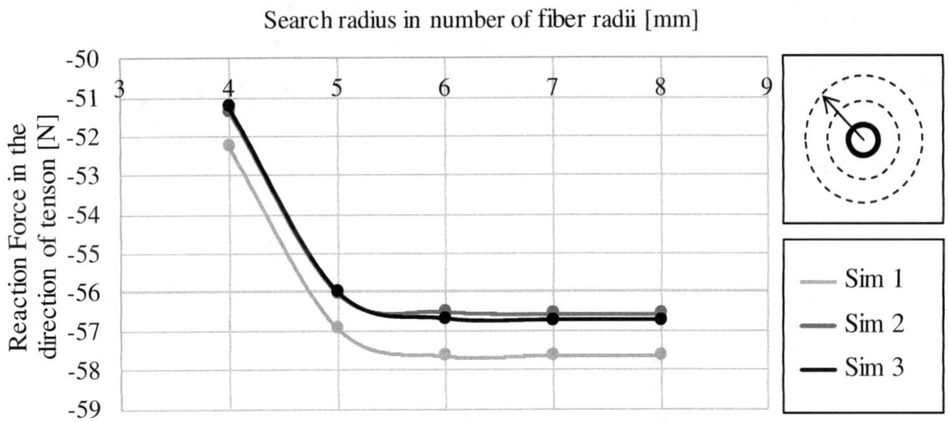

Figure 10. Search radius convergence study

IOP Conf. Series: Materials Science and Engineering **406** (2018) 012049 doi:10.1088/1757-899X/406/1/012049

The maximum edge overlap angle was set to 5° and the minimum relative fiber angle was set to 25°. With this configuration, all neighboring fibers for both models were sufficiently found. Two different approximated volume fractions were studied: 20 % and 48 %. At the top and bottom edge, the fibers were arranged in a row with a small gap of 5 % of the fiber radius to apply boundary conditions. The bottom row was fixed in all directions and at the top row the deformation was applied. The other fibers were distributed at random positions within the model. A comparison of the reduced model with FE model is shown in Figure 11 a) and b). The material parameters for both calculations as well as the specific FE model parameters are shown in Table 1.

Table 1. Simulation parameters

Matrix Young's modulus	2000 MPa
Matrix Poisson's ratio	0.3
Fiber Young's modulus	73000 MPa
Fiber Poisson's ratio	0.22
Mesh size	0.001 mm
Element type	CPE4R
Step	Nonlinear static

The first validation is done with a tensile displacement simulation. For this, the top row of fibers was moved by 0.01 mm in the *y*-direction and fixed in the *x*-direction. A comparison of the undeformed and the deformed model is shown in Figure 11 c) and d). For each volume fraction, three different random fiber distributions are used for calculation. The reaction forces in loading direction of both models for each random fiber distribution are shown in Figure 12.

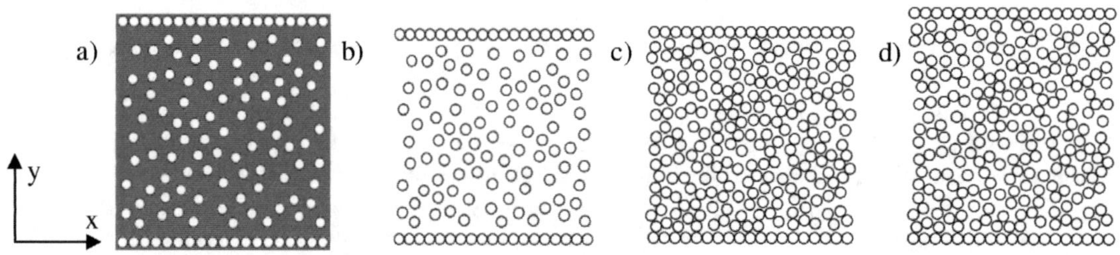

Figure 11. Comparison of the FE model a) and the reduced model b) for a low volume fraction and comparison of the undeformed c) and deformed d) model for a higher volume fraction

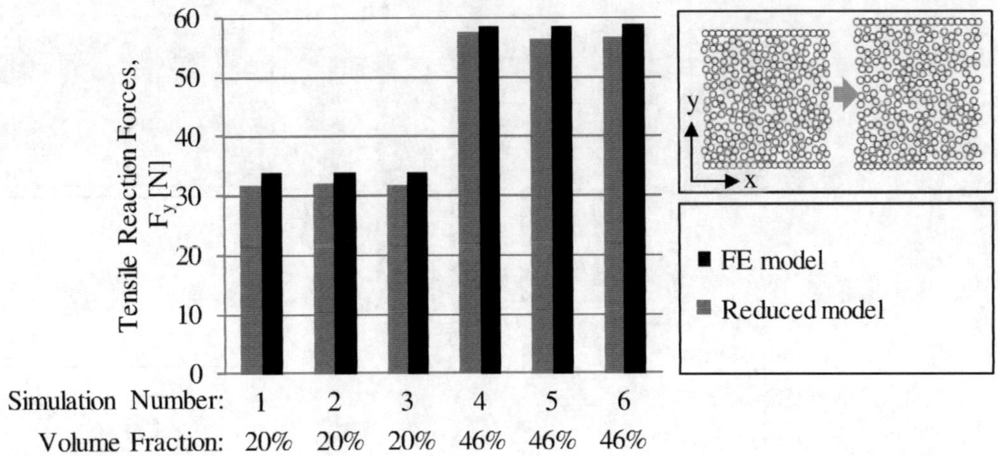

Figure 12. Comparison of reduced model and FE model reaction forces for tensile loading

The reaction forces predicted by the reduced model in the direction of deformation are close to the reaction forces calculated with the FE model. The forces in x-direction are insignificantly small and therefore not discussed anymore. The standard deviation of the different random distributions is smaller than 1 % difference. The mean results of the model are within 10 % deviation from the FE model results.

The second validation is a compression loading simulation. For this, the top row of fibers is moved by 0.01 mm in negative y-direction and fixed in x-direction. The reaction forces in loading direction of the bottom row are shown in Figure 13.

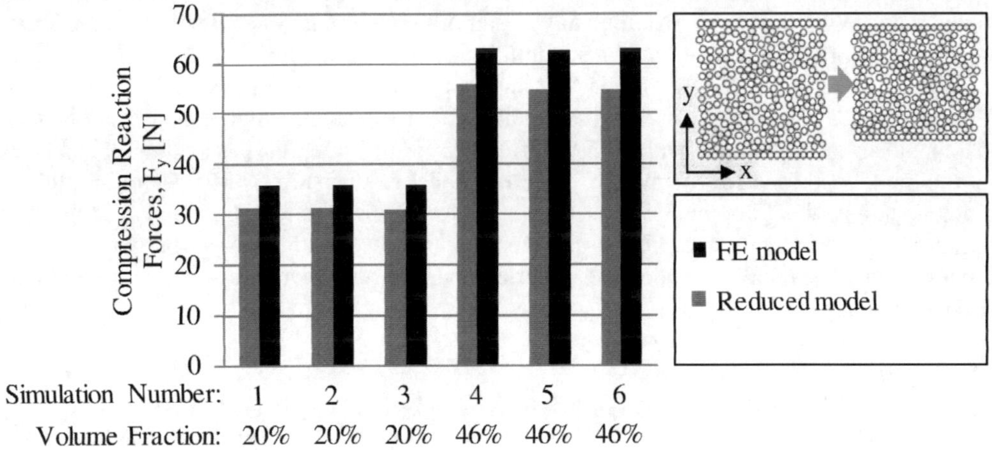

Figure 13. Comparison of reduced model and FE model reaction forces for compression loading

The reaction forces predicted by the reduced model in the direction of deformation are also close to the reaction forces calculated with the FE model. The forces in x-direction are again insignificantly small and therefore not discussed anymore. The standard deviation of different random distributions is less than 1 % difference. The mean results of the model are within 15 % deviation from the FE model results.

The third validation is a shear loading simulation. For this, the top row of fibers is moved by 0.01 mm in positive x-direction and fixed in y-direction. The reaction forces in x-direction of the bottom row are shown in Figure 14.

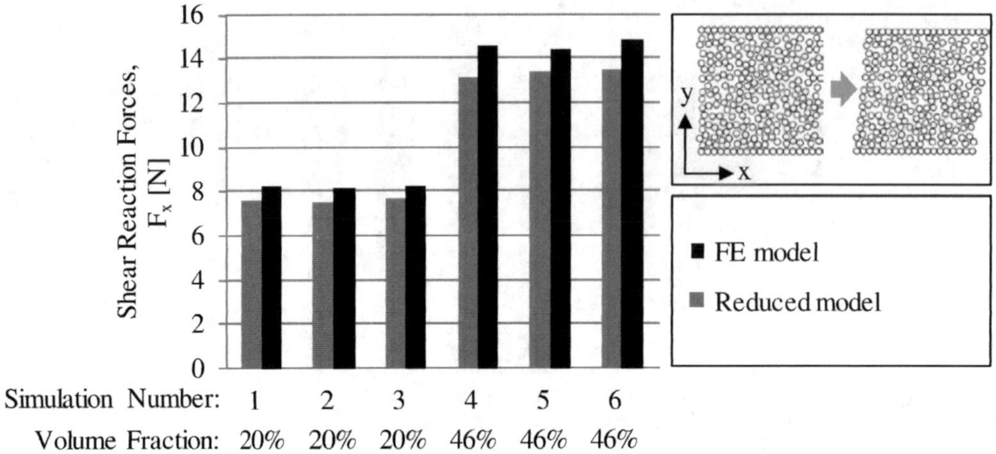

Figure 14. Comparison of the reduced model and FE model reaction forces for shear loading

The reaction forces predicted by the reduced model for the shear load are close to the reaction forces in FE model. The forces in y-direction are insignificantly small and therefore not discussed anymore. The standard deviation of different random distributions is less than 10 % difference. The mean results of the model are within 10% deviation from the FE model results.

4. Conclusions

In summary, a matrix model for a 2D micro-scale simulation is introduced. The reduced model is dependent on material properties as well as geometrical factors. With these, neighboring fibers are identified and two different types of connections are established: fiber-fiber and fiber-matrix connections. With this discretization, any material model (e.g. one which is more accurate and temperature dependent) can be implemented in the micro-scale model.

For the validation, the reduced model is compared to an equivalent FE model. Different calculations are carried out. The results are within a 10 % difference of the reduced model and the FE model for the tension and shear loading and within 15 % difference for the compression loading. It can be concluded that the model approach for the matrix fit the FE model results sufficiently. Therefore, the matrix can be integrated in the fiber model presented in [5]. After the integration and testing of this solid configuration, a matrix in semi liquid state needs to be observed. This can be done by lowering the material properties. Finally, an adequate material model for the thermoplastic matrix will be introduced for the calculations.

References

[1] Parlevliet P P, Bersee H E and Beukers A 2006 Residual stresses in thermoplastic composites – A study of the literature – Part I Formation of residual stresses *Composites Part A* **37** pp 1847-1857

[2] Stack R and Lai F 2013 Developments in Thermoforming Thermoplastic Composites *Thermoforming Quarterly* **32** pp 48-53

[3] Chapman T J, Gillespie JR J W, Pipes R B, Manson J –A E and Seferis J C 1990 Prediction of Process Induced Residual Stresses in Thermoplastic Composites *Journal of Composite Materials* **24** pp 616-643

[4] Kho G S 2000 An investigation of temperature control issues related to intelligent thermoforming (Bethlehem: Lehigh University, PA, USA)

[5] Stapleton S E, Appel L, Simon J-W and Reese S 2016 Representative volume element for parallel fiber bundels: Model and size convergence *Composites: Part A* **87** pp 170-185

[6] Smith M 2009 ABAQUS/Standard User's Manual 6.9 Simulia Providence RI USA

IOP Conf. Series: Materials Science and Engineering **406** (2018) 012050 doi:10.1088/1757-899X/406/1/012050

Novel textile preforming for optimised fibre architectures

Mikhail Matveev[1*], Vivek Koncherry[2], Sree Shankhachur Roy[2], Prasad Potluri[2], Andrew Long[1]

[1] Composites Research Group, Faculty of Engineering, University of Nottingham, Nottingham, UK

[2] Robotics & Textile Composites Group, School of Materials, University of Manchester, Manchester, UK

* Mikhail.Matveev@nottingham.ac.uk

Abstract. Development of novel textile preforming concepts driven by multi-objective optimisation techniques will be reported here. One of the main goals here is to completely relax the manufacturing constraints imposed by conventional textile machinery. Several examples towards this goal are presented in this paper. Robotic/mechatronic concepts have been employed in creating optimised fibre architectures.

1. Introduction

Conventional textile preforming techniques, such as 3D weaving and braiding, are capable of producing a wide range of reinforcement architectures but these techniques impose significant constraints in terms of local tow densities, tow orientations and degree of through-thickness reinforcement. As a result, composites manufactured with traditional textile structures seldom achieve their true potential in terms of light-weighting and damage tolerance. However, the design of novel preforms not constrained by conventional textile machines and processes is a complex problem owing to an infinite number of possible design. Therefore, the design needs to be guided by numerical modelling which can predict properties of a composite before it has been manufactured.

It has been shown that by optimizing of yarns widths and spacing in a 3D woven reinforcement it is possible to achieve up to 20% weight-saving for structural parts [1, 2]. An optimization of the weave style of 3D woven reinforcements has been performed by Zeng et al. [3] to improve the buckling properties of a tubular section. Most of the previous studies were constrained to optimization of existing types of reinforcements e.g. orthogonal 3D woven or similar. However, various preforming techniques make it possible to create more complex reinforcements such as multi-axial 3D preforms [4] and even extended towards the less conventional preforms which are currently not produced. Such designs can be explored by using reinforcement optimization framework where the reinforcements have less manufacturing constraints than in previous studies. This paper aims to extend the existing approaches by adopting and extending TexGen schema along with multi-objective optimization.

The present framework relies on the unit cell modelling to optimise properties at the mesoscale (scale of a unit cell) [5]. The workflow consists of several steps. First, the geometry of a unit cell is automatically generated in TexGen via a Python script. Then the unit cell is discretised and the finite element model is used to obtain its properties. The results of the analysis are then used in an optimisation procedure based on the genetic algorithm (GA). Such modular approach can be modified and extended by replacing some of the steps/blocks without major changes to the others. This paper addresses two of

IOP Conf. Series: Materials Science and Engineering **406** (2018) 012050 doi:10.1088/1757-899X/406/1/012050

the steps of the approach aiming to improve and extended each of them. In particular, the paper outlines a novel meshing approach and a multi-objective optimisation procedure. Results of the multi-objective optimisation of reinforcements with various manufacturing constraints are presented along with the steps towards its validation.

Manufacturing techniques for optimised fibre preforms are developed in two directions – 3D woven preforming and cylindrical preforming. Both of the techniques make it possible to place the off-axis yarns into preforms. These techniques will be used for the validation of the optimisation framework. Moreover, optimisation of fibre preforms with less manufacturing constraints will be driving further development of these techniques.

2. Numerical framework

Multi-scale analysis of fibre reinforced composites is most often based on the unit cell approach when in the first step of the analysis a representative volume element of a periodic composite structure is used to derive the homogenised elastic properties of the composite [5]. These properties are then used for a macro-scale analysis of a composite part and then a subsequent sub-modelling of a critical location can be performed using the unit cell again. In some cases, this sequence of the multi-scale analysis can be reduced to a single analysis of the unit cell e.g. in the case of a uniaxial tensile or compressive load [3].

The geometry of a unit cell, dimensions of yarns, spacing between yarns and the binder path, can be parametrised which make it possible to perform the optimisation of these parameters in an optimisation framework. However, this requires robust and automated procedures to generate a unit cell geometry and a corresponding mesh. Furthermore, a fast computation of the elastic properties is also desired since these properties are needed for evaluation of both constraints and objective functions.

2.1. Model generation

A unit cell geometry of a conventional 3D woven textile can be parameterised within TexGen by a range of parameters such as number of the warp/weft yarns in the unit cell, number of layers, dimensions of the yarns, spacing between the yarns, number of the binder yarns and their path in between the layers. The latter can be coded for each of the binder yarns as a vector which defines the position of the binder yarn within a stack of layers at each intersection of the binder and weft yarns. A unit cell geometry can be made more complex by introducing e.g. off-axis yarns which also can be parameterised.

TexGen pre-processor Python library allows the geometries to be created via a Python script. Examples of the geometry are given in Figure 1.

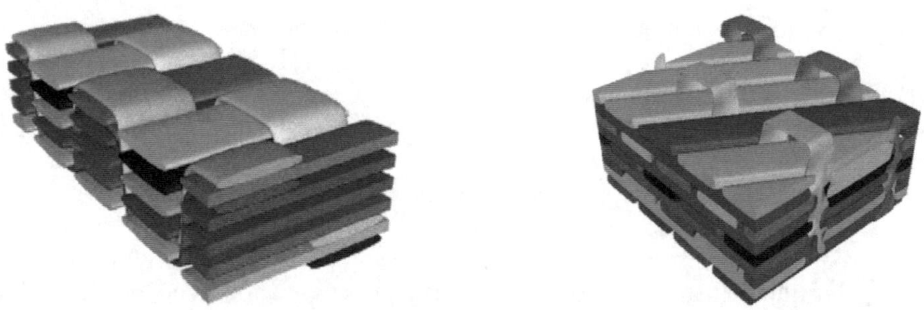

(a) (b)

Figure 1. (a) Orthogonal 3D woven preform; (b) 3D preform with off-axis yarns (top yarns removed)

2.2. Mesh generation

Evaluation of the mechanical properties and local stresses within a unit cell requires a good quality finite element mesh of the geometry. However, a conformal finite element mesh for complex textiles is difficult to generate. Voxel mesh has been used in many studies showing good results for relatively simple geometries [6] but exhibiting spurious stress concentrations in more complex geometries [7].

IOP Conf. Series: Materials Science and Engineering **406** (2018) 012050 doi:10.1088/1757-899X/406/1/012050

A novel meshing techniques, a combination of two existing techniques, has been implemented within TexGen to generate meshes for the geometries of arbitrary complexity. The meshing starts with creating a uniform voxel mesh with the element size being the maximum desired size in the mesh. This is then followed by an iterative octree refinement of each voxel, a procedure when a voxel is split into eight equally size voxels if the voxel consists of more than one material as defined by its corners. The octree refinement is continued until no more voxels need to be refined or the maximum allowable level of the refinement achieved. The refined voxel mesh is not conformal and contains hanging nodes which must be constrained to the neighbouring nodes of larger elements using linear constraints [8].

The refined voxel mesh does not resolve the problem of artificial stress concentrations caused by sharp angles in the voxel mesh. This can be resolved by applying Laplacian smoothing to all the surface nodes of the refined mesh. Examples of the refinement and smoothing are shown in Figure 2.

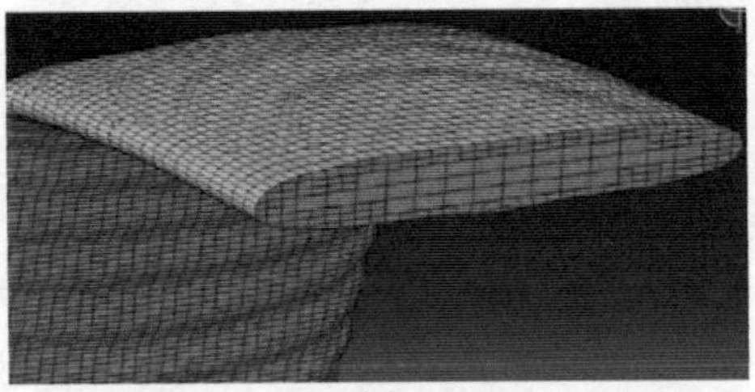

(a)

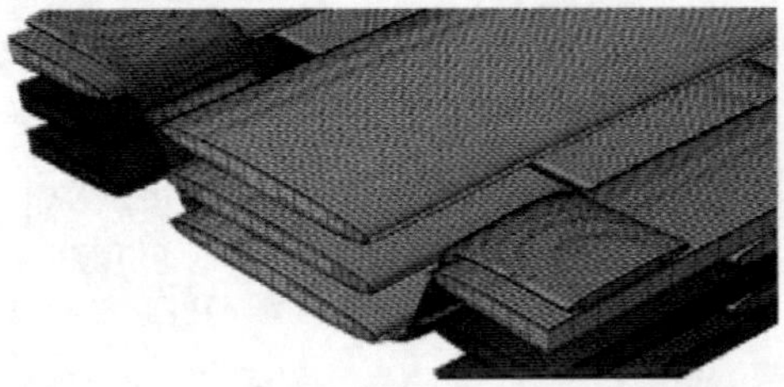

(b)

Figure 2. (a) Refined and smoothed binder yarn; (b) Refined and smoothed 3D orthogonal weave

The proposed meshing technique was validated against conformal meshing using a unit cell of a plain weave textile composite. Stress distributions in one of the cross-sections of the unit cell is given in Figure 3. The difference between the maximum Mises stresses is about 3.5%.

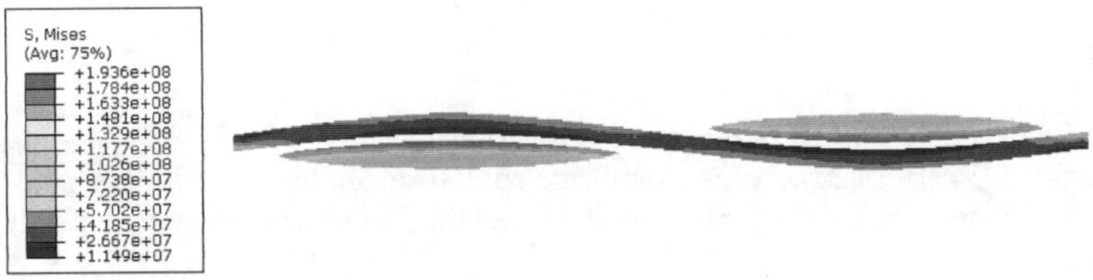

(a)

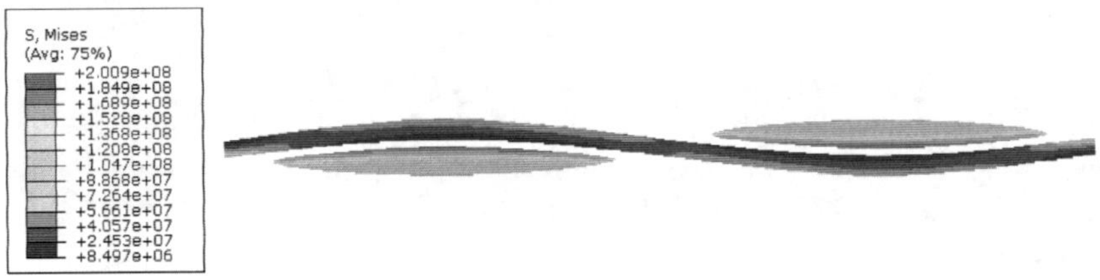

(b)

Figure 3. (a) Stress distribution in the conformal mesh; (b) Stress distribution in the smoothed octree mesh

2.3. *Orientation averaging*

The optimisation algorithm requires multiple evaluations of the constraints and objective functions. Therefore, any speed up in these evaluations will translate in a significant speed up of the overall optimisation procedure. The elastic properties can be evaluated using the orientation averaging (OA) approach, an analytical approach based on the iso-strain assumption for the yarns and matrix within a unit cell.

The OA has been implemented within TexGen framework via a Python script which extracts local yarn orientations and volumes of sub-sections. Comparison with the FE analysis showed that the properties predicted with the OA are within 5% for the in-plane Young's modulus while the Poisson's ratios are up to 20-30% different as it can be seen from Table 1.

Table 1. Comparison of the elastic properties of a plain weave (VF=37%)

	FE	OA
E1 / GPa	32.9	28.5
E2 / GPa	32.9	28.5
E3 / GPa	5.6	7.6
G12 / GPa	6.2	4.76
v12	0.1	0.065

When compared to an FE analysis the OA procedure was up to 7 times faster and took only 1 minute in order to create a reinforcement and compute the elastic properties using high discretisation. The time saving was achieved not only by avoiding the FE analysis but also by avoiding the mesh generation.

2.4. Optimisation algorithm

The optimisation of the parameterised unit cell geometry was carried out using the GA implemented within Matlab. The single-objective optimisation was performed using the standard implementation of the GA. The multi-objective optimisation was performed using a non-dominated sorting GA (NSGA-II) [9] which can provide faster convergence towards the optimum solution.

The range of the parameters defining the binder path in a reinforcement has only one constraint which ensures that the modelled preform is held together by binder yarns interlacing either entire stack or subsequent parts of it.

3. Case studies

3.1. Single-objective optimisation of the buckling resistance: comparison of FE and OA approaches

A previous study [3] performed optimisation of the composite landing gear in buckling. The problem has been reduced to the analysis of a dimensionless buckling coefficient given by a combination of the elastic constants:

$$\beta = \frac{D_{12} + 2D_{66}}{\sqrt{D_{11}D_{22}}} \tag{1}$$

where D_{ij} are flexural stiffnesses of a homogenised unit cell.

The optimisation framework with the FE analysis found that a 3D woven composites with the reinforcement having two binder yarns separated by a layer of weft yarns and running parallel have the buckling coefficient almost 50% higher than a conventional orthogonal 3D woven composite. It was reported that the optimisation procedure took up to 420 minutes on a desktop with 5 CPUs employed and assessed about 300 possible designs.

An identical optimisation was performed using the OA approach which yielded the identical design of the binder path. However, the overall optimisation time was about 70 minutes which is 5 times faster than FE based optimisation. The resulting binder path is shown in Figure 4.

Figure 4. Reinforcement with the binder path optimised to achieve maximum buckling coefficient

3.2. Multi-objective optimisation using OA

The OA approach has been utilised for the multi-objective optimisation of in-plane stiffness and the buckling coefficient (1) of 3D preforms. These two objective functions can be viewed as competing targets as they depend on the amount of the in-plane and through-thickness reinforcement, respectively.

A multi-objective optimisation provides a collection of the solution, so-called 'Pareto front' shown in Figure 5, with each of them being an optimum. The point of the maximum buckling coefficient corresponds to the reinforcement described in the previous section and the point of the maximum in-plane stiffness represents a trivial solution with straight binder yarns. The points in between these two maxima can be selected according to the design preferences. Some of the examples of the reinforcements are shown in Figure 5.

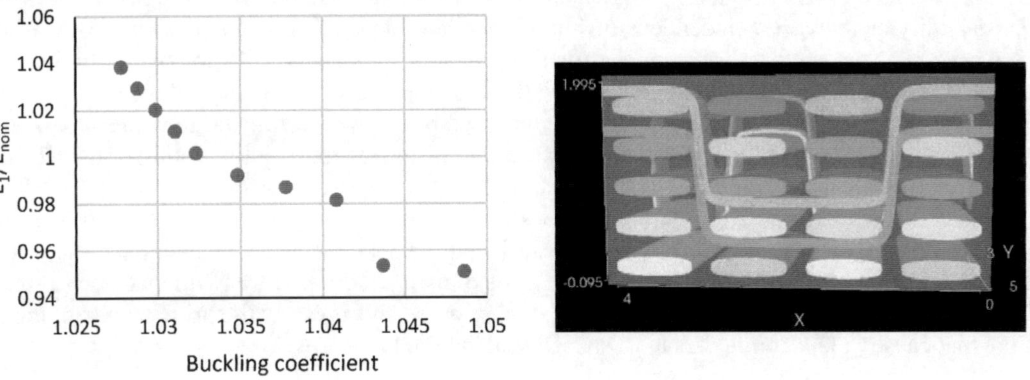

(a) (b)

Figure 5. (a) Pareto front for the multi-objective optimisation; (b) Example of a reinforcement from the Pareto front

3.3. Multi-objective optimisation of the multi-axial composites

Optimisation of the buckling coefficient and the transverse Young's modulus was performed for multi-axial preforms. The design variables were rotations of the layers while the binder path was fixed to follow an orthogonal pattern. The layer orientation was limited to the range between 20° and 60° with a step of 5°. The lower and higher orientations were not considered owing to a large size of the unit cell required for the analysis. Properties of the composites with the multi-axial reinforcements were calculated using OA approach.

The multi-objective optimisation results in the Pareto front shown in Figure 6. As in the previous example, it can be seen that the two objectives are "competing" against each other. The resulting front can be used to select a solution which is the most suitable for a particular design criterion.

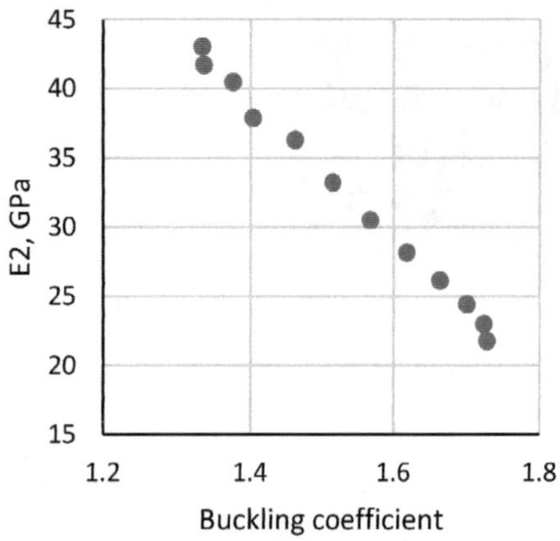

(a) (b)

Figure 6. (a) Pareto front for the multi-objective optimisation of multi-axial preforms; (b) Example of a reinforcement with β=1.56 and E_2 = 30.5 GPa ([30°/0°/90°/0°/-30°/0°/30°/0°/90°/0/-30°])

4. Optimised 3D woven preforming with novel manufacturing technologies

A certain degree of optimisation is possible on the conventional weaving technologies, such as varying density, varying thickness, binder density, but it is not possible to place stuffer tows in off-axis direction or weave near-net preform. The preforming approach in this study will not be constrained by the architecture that can be manufactured by using the conventional weaving technologies.

Traditionally, in order to weave a T-shape the 3D fabric is woven flat and then the web is lifted up post weaving. This post lifting of the web creates fibre distortion in the joint section of the web and the flange as well as resin rich area. Resin rich/delta-fillet can be seen in Figure 7 [10]. 3D Weaving machines that are used currently in the composites industry was originally developed for weaving 2D fabrics. These machines work well for the yarns that are flexible and stretchable. In the traditional machines shed is changed for every weft. The issue of using traditional weaving machine to weave 3D carbon fibre is the filamentation during the weaving process; this is due to the brittle nature of the carbon fibre that causes a lot of fibre damage, especially in the shedding area.

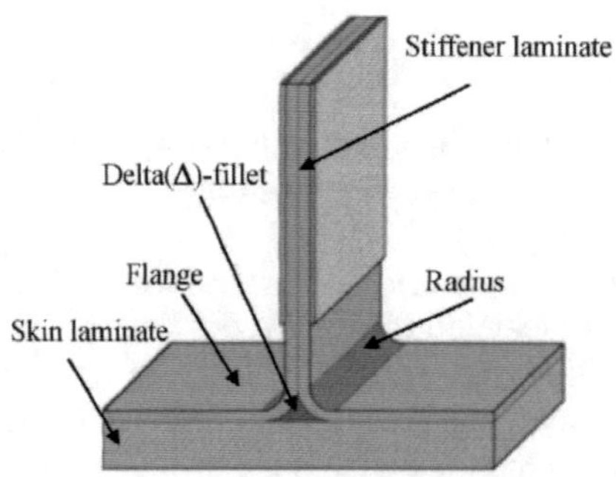

Figure 7. T-structure with delta-fillet/ resin rich area [10]

Two novel preforming prototype machines are being developed and are driven by optimisation. In the first approach, a multi-weft insertion 3D weaving machine for manufacturing near-net ultra-thick composite was developed, as seen in Figure 8 [11]. In this new approach, angle-interlock binders can be used to place off-axis fibres. In this multi-weft weaving system, the fabric is woven near-net in the shape

IOP Conf. Series: Materials Science and Engineering **406** (2018) 012050 doi:10.1088/1757-899X/406/1/012050

of T. This weaving technology also has the advantage of salvage on either side, preventing the tows from falling out of the structure. In the multi weft weaving system, the damage in the shedding area is considerably less because multiple sheds are opened and less movement of carbon fibre.

Current multi weft system developed for this study can insert twenty-two wefts simultaneously, and one beat-up push all twenty-two wefts into the fabric fell. Multiple weft systems give higher output with less weaving motion compared to that of the conventional loom.

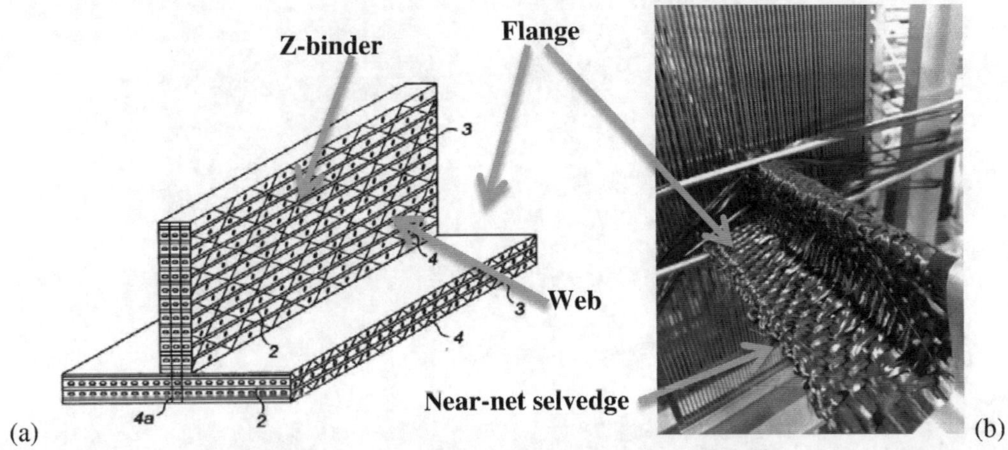

Figure 8. (a) Near-net 3D preform concept [11] and (b) Near-net 3D woven T shape fabricated using a novel weaving equipment developed at the University of Manchester

In the second preforming process the prototype machine can place stuffer in any off-axis angle, as seen in Figure 9. To study this 3D orthogonal fabric, samples are being developed by changing the number of layers, tow size and off-axis angles.

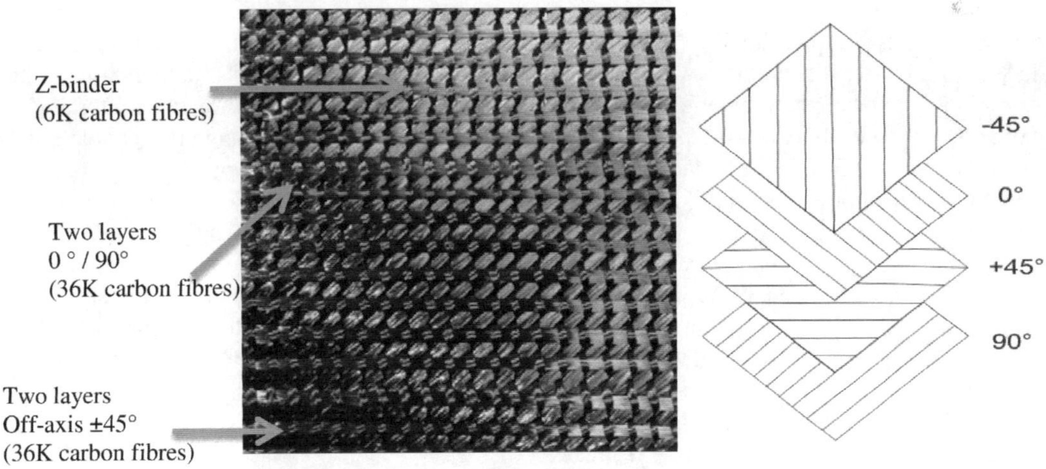

Figure 9. Multiaxial fabric with orthogonal binders

5. Cylindrical preforming for optimised fibre architecture

Braiding and filament winding are widely used for manufacturing cylindrical and closed shape structures using continuous fibres [12]. However, these technologies have their own limitations in achieving desired fibre path for an optimum design of composites. Since this study aims to develop preforms based on the optimisation framework that uses numerical modelling to predict composite properties, individual constraints of both braiding and filament winding are addressed in this section. In addition, other available technologies for producing 3D preforms were reviewed. A concept of a novel technology for manufacturing cylindrical preforms is also proposed to eliminate the design constraints of braiding and filament winding.

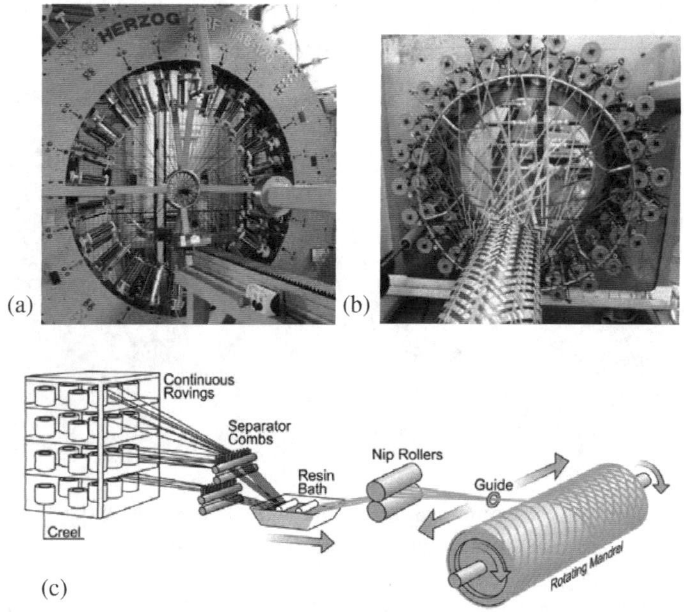

Figure 10. Cylindrical composite preforming technologies (a) Radial braiding (b) Maypole braiding (c) Filament winding [13]

Conventional two-dimensional (2D) circular braiders - maypole and radial (Figure 10) both can produce cylindrical textile preforms with fibres along three different axes. Braid structures usually have fibres running along helical paths and fibres from bobbins rotating in opposite directions interlace with each other to produce a biaxial structure. The bias fibre orientations in a braid range from about ±10° to ±85° and the third set of fibres are placed between the interlacing fibres and along the cylinder axis (0°) to generate a triaxial braid structure [14]. High braid angle often generates very high crimp and higher crimp affects the in-plane properties of the composite as observed by previous studies [15, 16]. Hence achieving higher angle or hoop (near 90°) orientation without compromising the mechanical properties is a limitation of conventional braiding. The undulation of fibres can contribute to impact damage tolerance when compared with filament wound composites [17]. However, the undulation effect is limited to a single braided layer and often composite is manufactured using multi-layer of 2D braided structures. Hence, 2D braiding method lacks the capability of producing a thick multi-layer cylindrical preform with through-thickness fibres which is one of the design requirements for improved impact resistance.

Filament winding (Figure 10c), on the other hand, can conveniently generate hoop and helical fibre orientations. However, through thickness fibre path is not possible to achieve as the structure is mostly non-crimp. Also, placing fibres along the cylinder axis (0°) is not convenient as it requires additional arrangements such as pinwheels [18, 19]. Overall the filament winding process on its own is capable of producing a triaxial structure ($\pm\alpha/90°$) with small undulations similar to a non-crimp architecture.

Three dimensional (3D) braiders, as opposed to 2D braiders, can produce a structure with through-thickness reinforcement eliminating one of the constraints mentioned above. However, often 3D braiding machines produce solid preform structures those are not suitable for over moulding on a core

IOP Conf. Series: Materials Science and Engineering **406** (2018) 012050 doi:10.1088/1757-899X/406/1/012050

or even produce large cylindrical structures. Using the concept of interconnecting tracks, 3D braiding equipment was developed in vertical deck arrangements [20, 21]. In these equipments, the through-the-thickness fibre paths are often at an angle as these paths are generated by braider bobbins running along helical directions and also fibre undulations are likely to be high.

At the initial steps of the novel manufacturing concept development, two unique types of equipment were developed in Manchester as part of the feasibility studies. One of the concepts was a multi-axis filament winding with an increased degree of freedom for winding complex shape preform (Figure 11a). In order to increase the number of fibre axis in a preform over-winding of a triaxial braid can achieve a quadriaxial structure [19]. In the second concept, an equipment was developed that can produce a preform with two layers of helical reinforcement and the axial reinforcement passing through the thickness (Figure 12) [22].

(a)

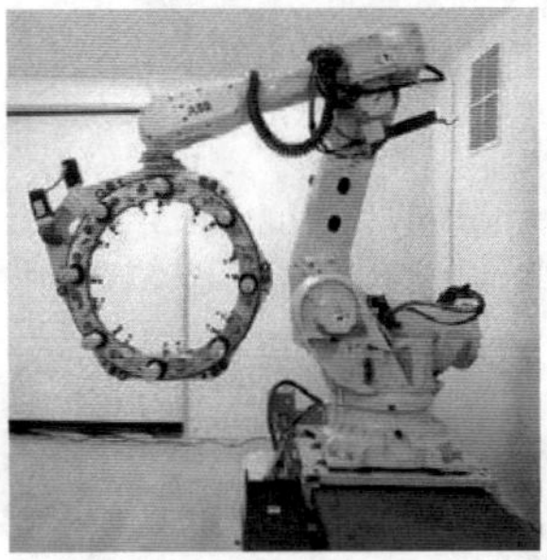

(b)

Figure 11. (a) Prototype multiaxis winding developed at the University of Manchester (b) The multi-axis winding concept used in commercial equipment developed by Cygnet Texkimp UK Limited

In order to optimise a preform based on the numerical modelling requirement, the constraints of traditional equipment need to be overcome. As part of the initial investigations, two multiaxis winding types of equipment provide the necessary foundation to develop novel equipment that can produce a multi-axis reinforcement that none of the existing equipment can produce. This equipment will be capable of combining multiple layers of $\pm\alpha$ fibre orientations with 0°/90° orientation as well as through-thickness reinforcement. In addition, since fibre at $\pm\alpha$ orientation have crimp, in the new concept, the fibres at $\pm\alpha$ orientation will not be interlaced thus have the potential to improve in-plane properties by minimising fibre undulation. Since individual tows will be inserted from different bobbins, local fibre densities can be changed by changing the fibre linear density.

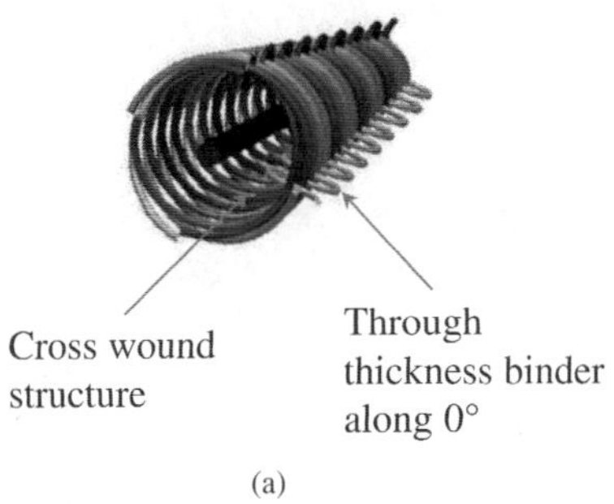

(a)

(b)

Figure 12. (a) Schematic of a 3D cylindrical preform concept (b) Novel winding concept developed at the University of Manchester incorporating through thickness fibre reinforcement [22]

6. Conclusions and Future Work

The paper presented the extended framework for the optimisation of 3D preforms. It was shown that approximate OA technique for the computation of the elastic properties can be used to accelerate the optimisation procedure. OA approach was used in the multi-objective optimisation of the reinforcement where two objective functions are essentially "competing" against each other. The generated Pareto fronts can be used as design guidelines in the selection of the appropriate reinforcement design.

The novel meshing technique has been presented but not yet fully utilised in the framework which is now limited only to the optimisation of the elastic properties. The further work will include multi-objective optimisation which includes failure onset as one of the objective functions.

Finally, the validation of the generated Pareto front for the multi-axial preforms is now enabled by the development of the multi-axial preforming technologies such as 3D multiaxial cylindrical preforming and 3D woven preforming.

Acknowledgements

This work was supported by the Engineering and Physical Sciences Research Council [grant number EP/P006701/1]; through the EPSRC Future Composites Manufacturing Research Hub.

References

[1] Fu X W, Ricci S and Bisagni C 2015 Minimum-weight design for three dimensional woven composite stiffened panels using neural networks and genetic algorithms *Composite Structures* **134** 708–15

[2] Tao W, Liu Z, Zhu P, Zhu C and Chen W 2017 Multi-scale design of three dimensional woven composite automobile fender using modified particle swarm optimization algorithm *Composite Structures* **181** 73–83

[3] Zeng X, Long A C, Ashcroft I and Potluri P 2015 *20th International Conference on Composite Materials* Copenhagen, Denmark.

[4] Labanieh A R, Liu Y, Vasiukov D, Soulat D and Panier S 2017 Influence of off-axis in-plane yarns on the mechanical properties of 3D composites *Composites Part A App. Science and Manufacturing* **98** 45–57

[5] Lomov S V, Ivanov D S, Verpoest I, Zako M, Kurashiki T, Nakai H and Hirosawa S 2007 Meso-FE modelling of textile composites: Road map, data flow and algorithms *Composites Science and Technology* **67** 1870–18

[6] Zeng X, Brown L P, Endruweit A, Matveev M and Long A C 2014 Geometrical modelling of 3D woven reinforcements for polymer composites: Prediction of fabric permeability and composite mechanical properties *Composites Part A Applied Science and Manufacturing* **56** 150–60

[7] Green S D, Matveev M Y, Long A C, Ivanov D and Hallett S R 2014 Mechanical modelling of 3D woven composites considering realistic unit cell geometry *Composite Structures* **118** 284–93

[8] Panesar A, Brackett D, Ashcroft I, Wildman R and Hague R 2017 Hierarchical remeshing strategies with mesh mapping for topology optimisation *International Journal for Numerical Methods in Engineering* **111** 676–700

[9] Deb K, Pratap A, Agarwal S and Meyarivan T 2002 A fast and elitist multiobjective genetic algorithm: NSGA-II *IEEE Transactions on Evolutionary Computation* **6** 182–197

[10] Bigaud J, Aboura Z, Martins A T and Verger S 2018 Analysis of the mechanical behavior of composite T-joints reinforced by one side stitching *Composite Structures* **184** 249–55

[11] Potluri P, Jetavat D and Sharma S 2017 Method and apparatus for weaving a three-dimensional fabric *Google Patents*

[12] Potluri P 2012 *International Encyclopedia of Composites* ed Nicolais L, Borzacchiello A and Lee S M, Wiley, New York 138–153

IOP Conf. Series: Materials Science and Engineering **406** (2018) 012050 doi:10.1088/1757-899X/406/1/012050

[13] Allnex Composites *Filament winding (Available at www.nuplex.com/composites/processes/filament-winding) Access date: June 25, 2018.*
[14] Potluri P and Nawaz S 2011 *Developments in braided fabrics. Specialist Yarn and Fabric Structures* Cambridge Woodhead Publishing Limited 333-353.
[15] Selver E, Potluri P, Hogg P and Soutic C 2016 Impact damage tolerance of thermoset composites reinforced with hybrid commingled yarns *Composites Part B: Engineering* **91** 522–38
[16] Stig F and Hallström S 2013 Influence of crimp on 3D-woven fibre reinforced composites. *Composite Structures* **95** 114–122
[17] Beni Z T, Johari M S and Ahmadi M S 2014 Comparison of the post-impact behavior of tubular braided and filament wound glass/polyester composites under compression *J Eng Fibre Fabr* **9** 140–5
[18] Rousseau J, Perreux D and Verdiere N 1999 The influence of winding patterns on the damage behaviour of filament-wound pipes *Composites Science and Technology* **59** 1439–49
[19] Roy S S, Potluri P and Soutis C 2017 Tensile Response of Hoop Reinforced Multiaxially Braided Thin Wall Composite Tubes *Applied Composite Materials* **24** 397–416
[20] Brookstein D S, Skelton J, Dent J R, Dent R W and Rose D J 1994 Solid braid structure *Google Patents*
[21] Némoz G, Dréano L and Bigaud D 2005 *Proceedings of the European Conference on Spacecraft Structures, Materials and Mechanical Testing* 3D braided preforms for shaped composite material
[22] Jan K 2015 *Development of 3D Braiding Concept for Multi-Axial Textile Preforms* in *School of Materials* The University of Manchester: Manchester, UK.

IOP Conf. Series: Materials Science and Engineering **406** (2018) 012051 doi:10.1088/1757-899X/406/1/012051

Two different modelling approaches for fabric composites subjected to ballistic impact

R Scazzosi, A Manes*, G Petrone, M Giglio

Politecnico di Milano, Department of Mechanical Engineering, Via La Masa 1, 20156 Milan, Italy

* andrea.manes@polimi.it

Abstract. Fabric composites are widely applied in different fields and, in particular, have become a valid option in the realization of protective structures due to their low weight combined with high strength and stiffness. Finite element numerical models are an established method for the assessment and the optimization of the design of engineering structures. Two main approaches for the analysis of impacts on fabric composites are present in the literature. A macro-homogeneous approach considers the materials as an equivalent homogeneous medium with no distinction between the constituents. A meso-heterogenous approach reproduces the material microstructure and therefore its components are considered as different entities. The capacity of these two numerical models to model the ballistic impact of a real bullet against a panel made of plain weave Kevlar® 29 embedded in an epoxy matrix is compared in this study. The outcome of the comparison of the two numerical model is assessed using experimental data from actual ballistic tests. Quantitative results, such as the residual velocity of the bullet, as well as qualitative results, such as the damage morphology, are compared. The latter is especially important in case the impacted structure is multifunctional, aimed to protect as well as to carry load.

1. Introduction

Fabric composites are widely applied in different fields and, in particular, have become a valid option in the realization of protective structures due to their favourable combination of low weight and high strength [1]. The design of protective structures is mainly based on experimental tests and experience. However, the use of simulation tools as a predictive method of the dynamic behaviour of textile-based composites is growing. The impact on protective structures is a complex phenomenon which involves large deformation and failure and for this reason simplified models are not able to catch the whole phenomenon in its complexity. Finite element numerical models are an established method for both the assessment and the optimization of the design of engineering structures. Two main modelling approaches for the analysis of impact on fabric composites are present in the literature. A macro-homogeneous approach considers the materials as an equivalent homogeneous medium with no distinction between its constituents. A meso-heterogeneous approach reproduces the material microstructure and therefore its constituents are considered as different entities. In this approach, not only the material behaviour, but also the interaction between the constituents are fundamental in determining the response to a determined impact. This approach is expected to provide a more detailed description of the problem and therefore should lead to more accurate results, however it requires more computational resources.

In the macro-homogeneous approach, the material behaviour is modelled using the classic orthotropic elasticity. Different failure criteria have been used in the literature, including Chang-Chang [2], [3] and Tsai-Wu [1], [4] which consider the different failure modes of composites. The composite plate can be considered as a uniform body [1], [5] or as a layered body where each layer is considered as a separate part. Zhang et al. [5] compared the two approaches and found that both deliver similar quantitative results such as, ballistic limit velocity and residual velocity in agreement with experimental tests. The layered model was shown to reproduce the damage morphology with higher precision while requiring more computational time [5]. The layered model allows the reproduction of delamination by using a tiebreak interface between the layers [3], [4].

A mesoscale reproduction of the material geometry was adopted for both woven fabrics [6], [7] and fabric composites [8]–[11]. Modelling of fabric composites adds complexity to the problem due to the reproduction of the matrix and its interaction with the yarns. The meso-heterogeneous approach can be used to reproduce all of the target [10], [11], or only in the region directly involved in the impact [4]. In the latter case the surrounding regions of the target are modelled using the macro-homogeneous approach. Different strategies can be adopted in the reproduction of the unit cell geometry. Kudryavtsev and Sapozhnikov used shell elements for the yarns and solid elements for the matrix to reproduce a simplified geometry of the unit cell [10]. Bresciani et al. reproduced an accurate geometry of the unit cell using solid elements for both the yarns and the matrix [4]. For the analysis of aramid fiber different solutions were adopted to characterise the material behaviour of the yarns: isotropic [4] or orthotropic along the central line of the yarn [7]. In both cases the material was considered to be linear elastic until failure which may be based on von Mises [4] or maximum principal [5] stress threshold.

The aim of the present work is to exploit both the macro-homogeneous and meso-heterogeneous approach to develop two numerical models of the ballistic impact on fabric composites. The paper aims also at establishing a methodology of the calibration and validation of the two numerical model by means of characterization based on comparison and further validation with experimental ballistic impact tests on composites made of Kevlar® 29 plain-woven fabrics embedded in an epoxy matrix. The main objective is to compare the accuracy of the two modelling approaches to predict the residual velocity and the damage morphology. The most meaningful mechanical properties of the composite, which are the elastic modulus and the tensile strength in the principal directions, were calibrated by means of simple tensile tests. The static and dynamic mechanical properties were assumed to be identical. This may be an oversimplifying assumption but, it has already been successfully used for the development of a macro-homogenous model of ballistic impact on aramid fiber-reinforced composites [3]. Furthermore, a dynamic characterization of the dynamic properties of the material is beyond the scope of this study. The developed numerical models were validated considering the highly complex impact conditions of a soft-core projectile subjected to large deformation during the penetration process. Large deformations of the projectile not only increase the complexity of the phenomenon but also require an accurate description of the material behaviour of both the projectile and the target. The outline of the paper is therefore the following: in section 2 the experimental test performed for the calibration and the validation of the numerical models are described; in section 3 the numerical models developed following either the macro-homogeneous and meso-heterogeneous approach are described; in section 4 the results of the experimental tests and the prediction of the numerical models are discussed and the accuracy of the two approaches is compared; finally some conclusion are drawn in section 5.

2. Experimental Tests

Experimental tests for the calibration and the validation of the numerical models were performed on textile-based composites. These composites were manufactured from several layers of plain woven Kevlar® 29 embedded in an epoxy matrix. Kevlar® 29 is an aramid fabric manufactured by DuPont® with a low density and high strength, and is commonly used in ballistic shield design. The elastic modulus of Kevlar® 29 is stated to be 83 GPa and a tensile strength of 3600 MPa by the manufacturer [12]. However, research on the mechanical properties of Kevlar® 29 showed different values depending on the test conditions [13]–[16]. The datasheet of the epoxy used as the matrix states a flexural modulus of 2.9 GPa and a flexural strength of 120 MPa. Tensile tests were performed according to the ASTM D3039 [17] for the calibration of the numerical models while the ballistic tests were performed according to EN 1522 [18] for validation.

2.1. Tensile Tests
Tensile tests were performed according to the ASTM D3039 [17] which is the standard test method for tensile properties of polymer matrix composite materials. Tensile specimens consisted of 5 fabric layers, had a nominal thickness of 2.5 mm and a dimension of 25x250 mm. The specimens were pulled in displacement control at a constant speed of 2 mm/min until failure. The axial load was measured by the machine load cell. The axial strain was measured using a 50 mm extensometer positioned in the center of the specimen and the strain range for the calculation of the elastic modulus was between 1.8 and 2.2 %. Five specimens were tested to obtain an elastic modulus of 10.06 ±0.65 GPa and a tensile strength of 405.24±18.03 MPa.

2.2. Ballistic Tests
Ballistic tests were performed according to the EN 1522 [18] which is the European standard for the classification of bullet resistance for windows, doors, shutters and blinds. Panels for the ballistic test had a nominal thickness of 6.5 mm, a dimension of 270x270 mm and consisted of 14 layers of fabric. The panels were assessed against EN 1522 bullet resistance class FB3 which requires a .357 Magnum bullet with a nominal mass of 10.2 g and a velocity of 430 m/s. The experimental apparatus is shown in Figure 1, including the .357 Magnum bullet which has a conical tip and consists of a soft lead core and a brass jacket. According to the standard requirements, the gun was positioned at a distance of 5 m from the target. The support system consisted of two steel plates with a central square hole of 160x160 mm which were held together with the target by means of four screws. Only one shot with an angle of incidence 90° was fired at the center of the target. Two velocity screens were positioned 2.5 m before and 2.5 m after the target to measure respectively the impact and residual velocity of the target. Two repetitions of the test were performed, and the bullet always completely perforated the panel with high residual velocity.

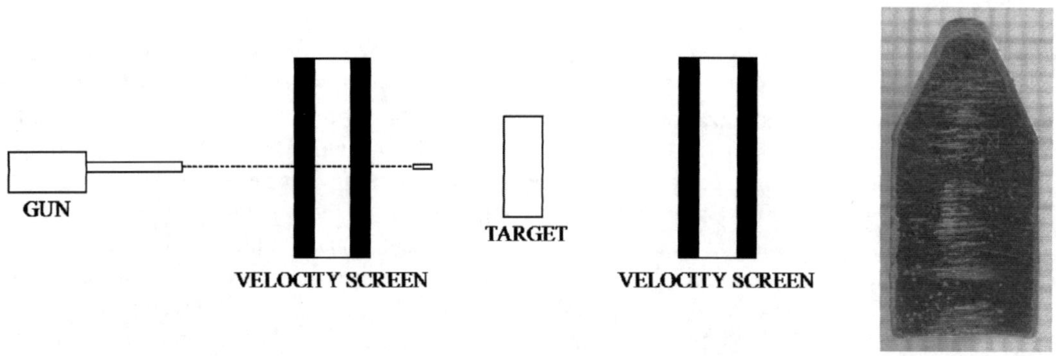

Figure 1. Experimental apparatus of the ballistic tests (left) and section of .357 Magnum (right).

3. Numerical Model
The two numerical models were based on finite element method and were implemented using the commercial software LS-DYNA. In both the macro-homogeneous and meso-heterogeneous approach a three-dimensional model was implemented, and the double symmetry of the problem was exploited to decrease the computational cost of the analysis. Considering the configuration of the support system, only the material of the panel which is free to deform during the impact, that is the portion of the material in the region of the plate square holes of 160x160 mm, was considered in the numerical model and all the degrees of freedom were constrained in the boundary regions. Symmetry constraint were indeed

IOP Conf. Series: Materials Science and Engineering **406** (2018) 012051 doi:10.1088/1757-899X/406/1/012051

applied on the faces cut by symmetry planes. Therefore, the two numerical models implemented are shown in Figure 2. In the case of meso-heterogeneous approach, the yarn and the matrix were only reproduced in the regions near the impact point. The remaining part of the target was modelled using the macro-homogeneous approach.

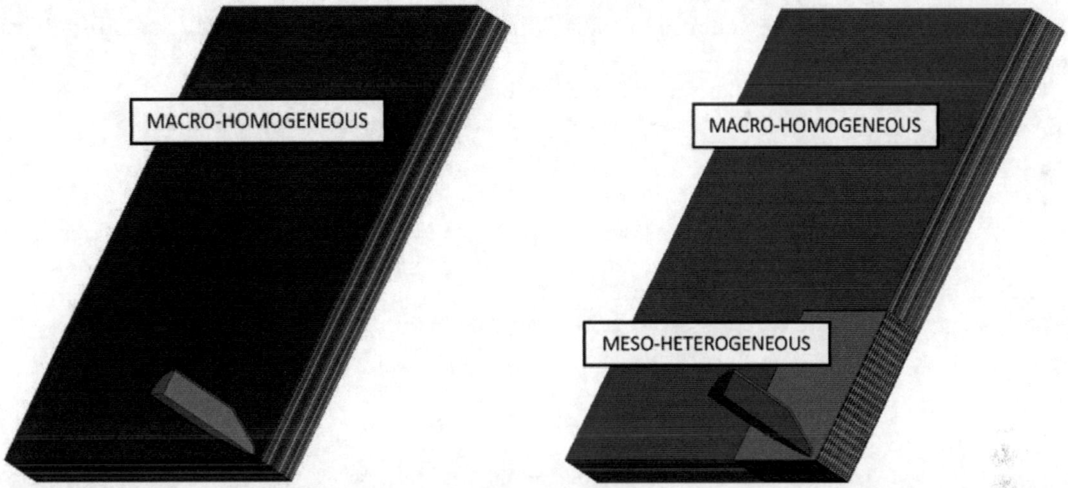

Figure 2. Numerical model with macro-homogeneous (left) and meso-heterogeneous (right) approach

3.1. Model of the Projectile

The geometry of the lead core and the brass jacket of the projectile was accurately reproduced and meshed with constant-stress solid elements. The Modified Johnson-Cook and Cockcroft-Latham failure criterion (MAT_107) were used for modelling the material behaviour of lead and brass, and the material parameter used are reported in Table 1: E and ν are respectively the elastic modulus and the Poisson's ratio, A, B and n are the Johnson-Cook strain hardening parameters, C is the strain rate sensitivity parameter and $\dot{\epsilon}_0$ the reference strain rate, m is the thermal softening parameters and T_m is the melting temperature and W_{cr} is the Cockcroft-Latham parameter. This is an established approach for metallic material under high-velocity impact [19], [20]. Johnson-Cook is a material model for metals which take into account strain hardening, strain-rate sensitivity and temperature softening [21]. For a detailed description of the material model the reader is referred to the software manual [22].

Table 1. Material parameters for projectile lead core and brass jacket. [23]

Material	E [MPa]	ν [-]	A [MPa]	B [-]	n [-]	$\dot{\epsilon}_0$ [s^{-1}]	C [-]	T_m	m	W_{cr} [MPa]
Lead	16000	0.42	0	36.62	0.0987	72.108	0.1593	525	1	175
Brass	115000	0.21	111.69	504.69	0.42	1	0.0085	1288	1.68	914

3.2. Macro-Homogeneous Approach

In the macro-homogeneous model each layer of the panel was modelled separately and meshed using constant stress solid elements with a dimension of 0.25x0.25x0.25 mm. This element size was chosen to match as closely as possible the element size of the meso-heterogeneous model which is bound by the geometry of the yarns. The material model used for the composite layers was MAT_054 which is based on orthotropic elasticity and the Chan-Chang failure criterion. The option 2WAY was used so that material have bilinear behaviour in direction 1 and 2, which are the warp and fill direction on the layer. The tensile failure mode in direction 1 is given by Eq. 1

$$\left(\frac{\sigma_{11}}{X_t}\right)^2 + \beta\left(\frac{\sigma_{12}}{S_c}\right)^2 - 1 \geq 0 \tag{1}$$

where X_t is the tensile strength in direction 1 and S_c is the shear strength on plane 12 and β is the shear weighting factor.

The tensile failure criterion in direction 2 is given by Eq. 2

$$\left(\frac{\sigma_{22}}{Y_t}\right)^2 + \beta\left(\frac{\sigma_{12}}{S_c}\right)^2 - 1 \geq 0 \tag{2}$$

where Y_t is the tensile strength in direction 2.

The compressive failure criterion in direction 1 is given by Eq. 3

$$\left(\frac{\sigma_{11}}{X_c}\right)^2 - 1 \geq 0 \tag{3}$$

where X_c is the compressive strength in direction 1.

The compressive failure criterion in direction 2 is given by Eq. 4

$$\left(\frac{\sigma_{22}}{Y_c}\right)^2 - 1 \geq 0 \tag{4}$$

where Y_c is the compressive strength in direction 2.

The matrix tensile and compressive failure criterion is given by Eq. 5

$$\frac{\sigma_{22}^2}{Y_c Y_t} + \left(\frac{\sigma_{12}}{S_c}\right)^2 + \frac{(Y_c - Y_t)\sigma_{22}}{Y_c Y_t} - 1 \geq 0 \tag{5}$$

When one of the failure criterion is met inside an element, the relevant stiffness parameters are degraded (for a detailed description the reader is referred to the software user manual [22]). Element are deleted from the analysis when the tensile failure criterion is met. MAT_054 also allows to consider other strain-based element deletion criteria. Therefore, also a tensile strain threshold equal to the ratio between the tensile strength and the elastic modulus (in direction 1 or 2) was considered. The material parameters for Kevlar29/Epoxy which were mostly obtained from the literature are reported in Table 2. The elastic moduli in warp and fill directions were assumed to be the same and equal to the average elastic modulus obtained in the tensile tests. The tensile strength in the warp and fill direction was obtained by an inverse calibration method as described in section 3.4. The interaction between the layers was modelled using the tiebreak interface: dethatching and sliding of each couple of adjacent layers is inhibited until the interface failure criterion, given by Eq. 6, is met.

$$\left(\frac{\sigma_n}{S_n}\right)^2 + \left(\frac{\tau}{S_s}\right)^2 - 1 \geq 0 \tag{6}$$

where S_n is the interfacial normal stress threshold and S_s is the interfacial shear stress threshold which were considered respectively to be equal to 34.5 MPa and 9 MPa [4].

Table 2. Material parameters for Kevlar29/Epoxy (macro-homogeneous approach).

Symbol	Property	Value
ρ [kg/m^3]	Density	1025
E_1 [GPa]	Elastic modulus 1	10.06
E_2 [GPa]	Elastic modulus 2	10.06
E_3 [GPa]	Elastic modulus 3	6 [4]
G_{12} [GPa]	Shear modulus 12	0.77 [4]
G_{13} [GPa]	Shear modulus 13	5.43 [4]
G_{23} [GPa]	Shear modulus 23	5.43 [4]
ν_{12} [-]	Poisson's ratio 12	0.25 [4]
ν_{13} [-]	Poisson's ratio 13	0.33 [4]
ν_{23} [-]	Poisson's ratio 23	0.33 [4]
X_t [MPa]	Tensile strength 1	425
Y_t [MPa]	Tensile strength 2	425
X_c [MPa]	Compressive strength 1	185 [4]
Y_c [MPa]	Compressive strength 2	185 [4]
S_c [MPa]	Shear strength plane 12	77 [4]
β	Shear weighting factor	0 [4]

3.3. Meso-Heterogeneous Approach

In the meso-heterogeneous approach the yarn unit cell geometry was accurately reproduced using the software TexGen which is a textile geometric modeller developed by the University of Nottingham [24]. The input dimensions for the geometry reproduction were measured by high-resolution pictures of the test specimen. The geometry of the yarn unit cell was subsequently imported in the commercial software Abaqus for the generation of the mesh. TexGen delivers also the geometry of the matrix which however is too complex to generate a mesh. Therefore, a simplified geometry of the matrix was reproduced starting from the geometry of the yarn unit cell. The unit cells were then used to create the whole part for the numerical model and imported in LS-DYNA. The meso-heterogeneous approach was used in the regions surrounding the impact point.

For local meso-heterogeneous modelling of the target two main approaches are used in the literature which are equally valid for the reproduction of the phenomenon. Bresciani et al. reproduced the primary yarns, i.e. the yarns in direct contact with the projectile, in their entire length [4]. Barauskas and Abraitiene [6] reproduced only a squared region in the center of the impact point whose width is approximately three times the diameter of the projectile. The second approach was followed and considering the diameter of the projectile used in this study a square of about 52x52 mm was reproduced with the meso-heterogeneous approach which means 18 unit cells per side. Constant stress solid elements were used for the mesh of both the yarns and the matrix. The size of the element was bound by the geometry of the yarns and was chosen as a trade-off between the accuracy in the reproduction of the geometry and the computational time. The average element size of the macro-homogeneous and meso-heterogeneous model is similar, as visible in Figure 3, in order to compare consistent results. The other regions of the target were modelled using macro-homogeneous approach (described in section 3.2). The connectivity between the two regions of the model was guaranteed by a tied interface.

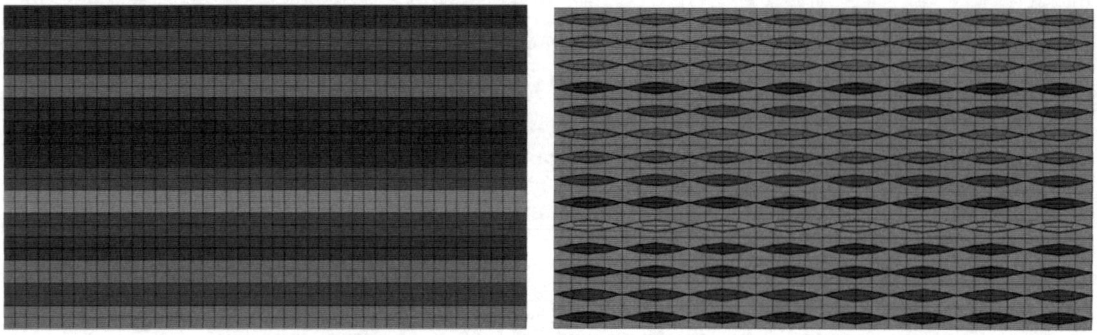

Figure 3. Target mesh for the macro-homogeneous (left) and meso-heterogeneous (right) approaches

Both the yarns and the matrix were considered to be isotropic and linear elastic until failure (MAT_001). Maximum principal stress threshold (σ_f) was defined for the element deletion (MAT_000). The material parameters for Kevlar® 29 yarns and epoxy resin are reported in Table 3. The properties used for epoxy were obtained from the material data sheet while the properties for Kevlar® 29 were obtained by an inverse calibration method as explained in section 3.4. The Poisson's ratio of the two materials was obtained from the literature.

Table 3. Material parameters for Kevlar® 29 and Epoxy (meso-heterogeneous approach).

Symbol	**Property**	**Kevlar® 29**	**Epoxy**
ρ [kg/m^3]	Density	1.44	1.18
E [GPa]	Elastic modulus	62	2.9
ν [-]	Poisson's ratio	0.44 [4]	0.35 [4]
σ_f [MPa]	Failure stress	3000	120

The definition of the contact in the meso-heterogeneous approach is far more complex than in the macro-homogeneous approach leading to longer computational times. Friction between the yarns is a relevant feature for the capability of the target to absorb the kinetic energy of the bullet. Interaction between the yarns of each layer was modelled using automatic surface to surface contact considering the static friction coefficient, the dynamic friction coefficient and the exponential decay respectively equal to 0.23, 0.19 and 10^8 [4]. The interaction between the matrix and the yarns was modelled using a tiebreak interface considering S_n and S_s respectively equal to 34.5 MPa and 9 MPa [4].

3.4. Material Parameter Calibration

The Kevlar29/Epoxy tensile strength in direction 1 and 2 (respectively X_t and Y_t) for the macro-homogenous approach and the Kevlar® 29 elastic modulus and maximum principal stress threshold (respectively E and σ_f) for the meso-heterogeneous approach were calibrated using an inverse method. Therefore, numerical models of the tensile test were developed following the same approach used for the model of high velocity impact. The simulations were run iteratively, changing the value of the parameters to be calibrated, until obtaining a match between the experimental and numerical stress-strain curves. The geometry of the specimen was fully reproduced and only the portion of the material outside the apparatus clamps was taken into consideration. Thus, the size of each layer was 25x150 mm with a thickness of 0.5 mm. The meshes used in the model of the tensile test have to be identical to the meshes used in the high-velocity impact model to assure consistency of the material parameters obtained following this procedure. As shown in Figure 4, one side of the specimen was constrained in the longitudinal direction while to the other side a longitudinal displacement was imposed to cause tensile strain inside the specimen. The displacement of the nodes which lie on the extensometer application points and the reaction forces were hence calculated to obtain a stress-strain curve which can be compared with the experimental one. The stress-strain curves obtained with the numerical models were compared with the experimental one as shown in Figure 4. The optimized value of X_t and Y_t of Kevlar29/Epoxy was 425 MPa slightly higher than the experimentally measured value of 405 MPa. This mismatch of 5% may be caused by the imperfect uniaxial state of stress inside the specimen and the mesh dependency of the failure criterion. The Kevlar® 29 elastic modulus and maximum principal stress threshold obtained were respectively 62 GPa and 3000 MPa. These values are inside the experimental ranges measured by different authors for Kevlar® 29 elastic modulus and tensile strength [13]–[16]. The stress-strain curve obtained with meso-heterogeneous approach show a noticeable change from the linear behaviour at strain level of 0.02 which corresponds to onset delamination between the yarn and the matrix. This behaviour of the stress-strain curve is also observed for the experimental stress-strain curve but, at a higher level of strain.

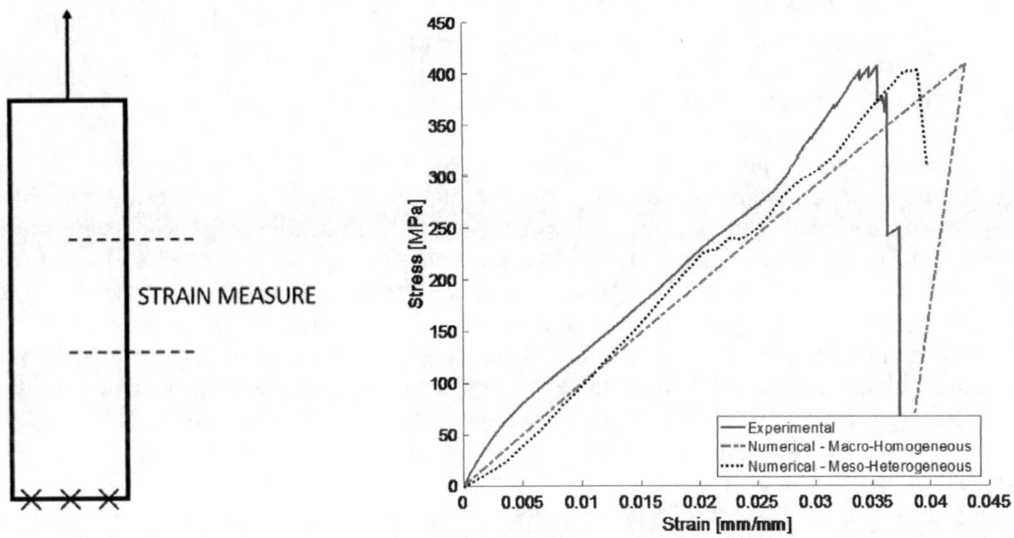

Figure 4. Scheme of the tensile test numerical model (left) and inverse calibration results (right)

4. Discussion

The most meaningful result of the ballistic test, in case of complete perforation of the target is the residual velocity of the projectile. The residual velocity of the projectile predicted by the numerical models was compared with the experimental results in Figure 5. The meso-heterogeneous approach predicts the average experimental residual velocity with higher accuracy. Indeed, the percentage error of the macro-homogenous and meso-heterogeneous approaches are respectively 17% and 11%.

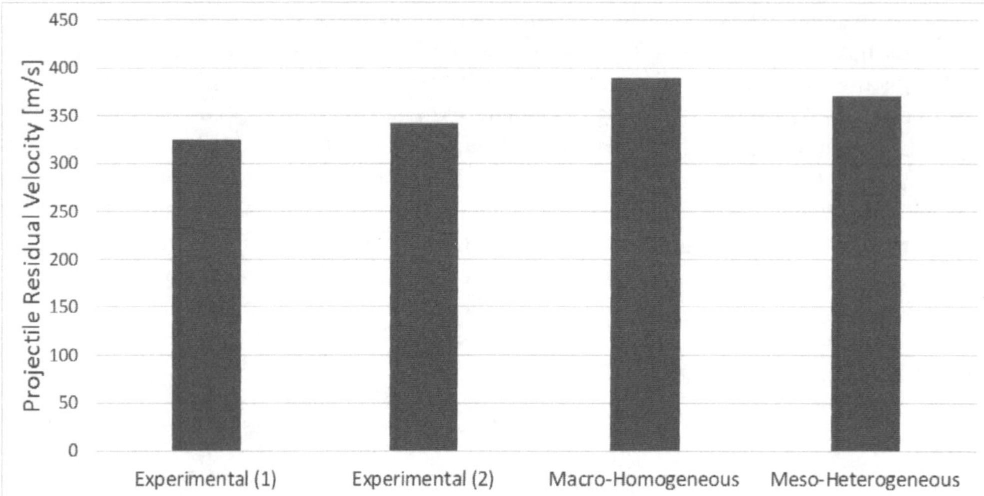

Figure 5. Comparison of experimental and numerical projectile residual velocities

This discrepancy may be due to several reasons related to the modelling of the target and of the projectile. First of all, the projectile used in the experimental tests is a soft-core bullet which is subjected to a mushroom deformation during the penetration process. Deformation of the projectile plays an important role in its kinetic energy dissipation and consequently its correct reproduction in the simulation is essential for the accurate prediction of the residual velocity. The projectile material parameters were obtained from the literature, and a complete characterization of the projectile material, which would require resources beyond the scope of this study, would increase the accuracy of the model. These considerations are also valid for the material parameters related to the target which were obtained from the literature. On the other hand, the tensile strength of the material (either Kevlar29/Epoxy tensile strength for the macro-homogeneous approach, or Kevlar® 29 tensile strength for the meso-heterogeneous approach) were calibrated using data form a static test. This parameter is fundamental in the characterization of the ballistic performance of the target. Since high-velocity impact is a phenomenon which involves high strain rates further experimental test should be performed to evaluate the dynamic strength of the target material and the dependence on strain rate of the material response.

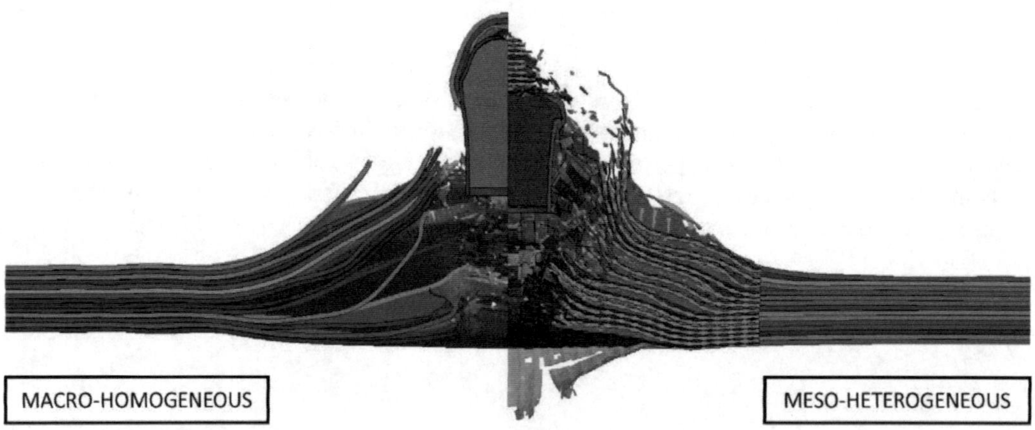

Figure 6. Target deformation predicted by the numerical model (at 9 μs)

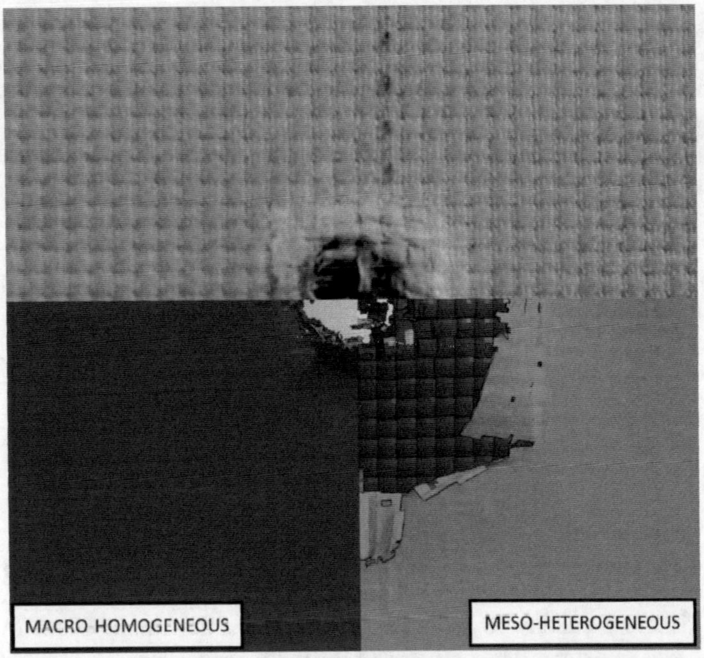

Figure 7. Comparison of experimental and numerical front face damage morphology

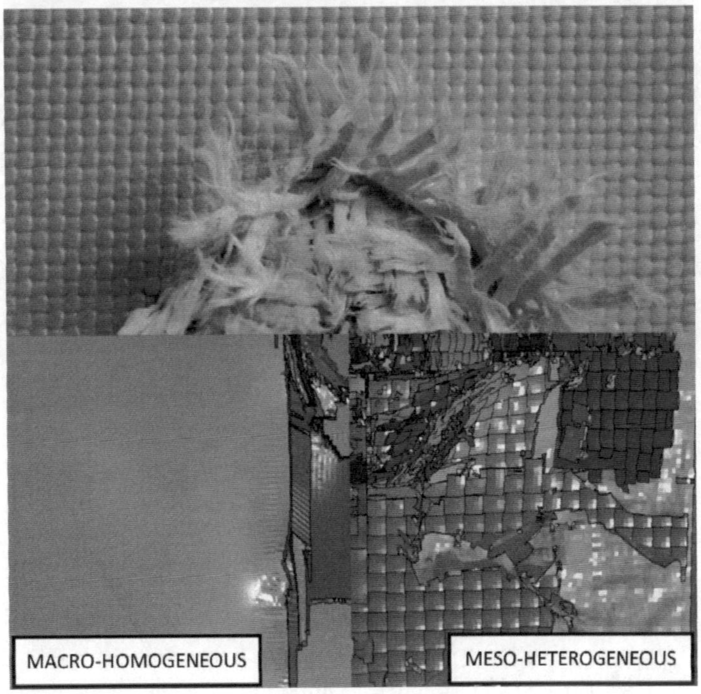

Figure 8. Comparison of experimental and numerical back face damage morphology

The target deformation predicted by the numerical models is shown in Figure 6. The macro-homogeneous approach predicted a higher extension of delamination whereas the meso-heterogeneous approach predicted failure of the tiebreak interface between the yarns and the matrix rather than failure of the matrix itself. Both approaches were able to replicate fiber crushing in the first layers, tensile failure in the back layers and delamination. In Figure 7 the target front face damage morphology is compared with the predictions of the numerical models. Both the two approaches correctly predict the size of the hole generated by the projectile. Only meso-heterogeneous approach is capable to reproduce matric cracking and deboning of the matrix from the yarn even though the size of the predicted damaged area is greater than the experimental. It is thought that more accurate calibration of the matrix material model and of the tiebreak interface criterion would lead to more accurate results. In Figure 8 it is reported the comparison of experimental and numerical back face damage morphology. The macro-homogeneous approach predicted a damage area of elliptical shape which was not observed experimentally. The meso-heterogeneous approach, instead, predict the right damage morphology even if, as in front face damage morphology, it overestimates the area of the region of failed matrix. Therefore, the meso-heterogeneous approach was shown to reproduce the experimental damage morphology more accurately as it simulated the physical phenomenon more precisely due to the more detailed modelling of the material. The higher accuracy of the meso-heterogeneous approach was however obtained at three times higher computational time which is due to the higher number of contact interfaces of the model.

5. Conclusions

The macro-homogeneous and meso-heterogeneous approaches were exploited to develop two numerical models of a ballistic impact against fabric composites. The accuracy of the two models in predicting the projectile residual velocity and the damage morphology was compared. For this reason, ballistic tests according to the EN 1522 standard were performed on panels manufactured from Kevlar 29 plain weave fabrics embedded in epoxy matrix. The bullet used for these tests was a .357 Magnum with an impact velocity of 430 m/s. Tensile tests were also performed to calibrate the tensile properties both for the macro-homogeneous and meso-heterogeneous approach. This calibration was performed using an inverse method and thus numerical models of the tensile tests were developed. Both two approaches were shown to be capable of reproducing the fundamental features of composite failure: fiber tensile failure, fiber crushing and delamination. The meso-heterogeneous approach delivers more accurate results both in terms of prediction of the residual velocity and reproduction of the damage morphology. On the other hand, the advantage of the macro-homogeneous approach was a three times lower computational time required to complete the simulation.

References

[1] N. Nayak, A. Banerjee, and T. R. Panda, "Numerical Study on the Ballistic Impact Response of Aramid Fabric- epoxy Laminated Composites by Armor Piercing Projectile," *Procedia Eng.*, vol. 173, pp. 230–237, 2017.

[2] E. Palta, H. Fang, and D. C. Weggel, "Finite element analysis of the Advanced Combat Helmet under various ballistic impacts," *Int. J. Impact Eng.*, vol. 112, no. August 2016, pp. 125–143, Feb. 2018.

[3] B. Berk, R. Karakuzu, and A. K. Toksoy, "An experimental and numerical investigation on ballistic performance of advanced composites," *J. Compos. Mater.*, vol. 51, no. 25, pp. 3467–3480, Oct. 2017.

[4] L. M. Bresciani, A. Manes, A. Ruggiero, G. Iannitti, and M. Giglio, "Experimental tests and numerical modelling of ballistic impacts against Kevlar 29 plain-woven fabrics with an epoxy matrix: Macro-homogeneous and Meso-heterogeneous approaches," *Compos. Part B Eng.*, vol. 88, pp. 114–130, Mar. 2016.

[5] X. Zhang, T. Liu, N. He, and G. Jia, "Investigation of two finite element modelling approaches for ballistic impact response of composite laminates," *Int. J. Crashworthiness*, vol. 22, no. 4, pp. 377–393, Jul. 2017.

[6] R. Barauskas and A. Abraitienė, "Computational analysis of impact of a bullet against the multilayer fabrics in LS-DYNA," *Int. J. Impact Eng.*, vol. 34, no. 7, pp. 1286–1305, Jul. 2007.

[7] T.-L. Chu, C. Ha-Minh, and A. Imad, "A numerical investigation of the influence of yarn mechanical and physical properties on the ballistic impact behavior of a Kevlar KM2 ® woven

fabric," *Compos. Part B Eng.*, vol. 95, pp. 144–154, Jun. 2016.

[8] A. Manes, L. M. Bresciani, and M. Giglio, "Ballistic Performance of Multi-layered Fabric Composite Plates Impacted by Different 7.62mm Calibre Projectiles," *Procedia Eng.*, vol. 88, pp. 208–215, 2014.

[9] L. M. Bresciani, A. Manes, and M. Giglio, "An analytical model for ballistic impacts against plain-woven fabrics with a polymeric matrix," *Int. J. Impact Eng.*, vol. 78, pp. 138–149, Apr. 2015.

[10] O. Kudryavtsev and S. Sapozhnikov, "Yarn-Level Modelling of Woven and Unidirectional Thermoplastic Composite Materials Under Ballistic Impact," *PNRPU Mech. Bull.*, vol. 3, no. 3, pp. 108–119, 2016.

[11] S. Sapozhnikov and O. Kudryavtsev, "Modeling of Thermoplastic Composites Used in Protective Structures," *Mech. Compos. Mater.*, vol. 51, no. 4, pp. 419–426, 2015.

[12] DuPont, "Kevlar Aramid Fiber Technical Guide." .

[13] Y. Ou, D. Zhu, M. Huang, and H. Li, "The effects of gage length and strain rate on tensile behavior of Kevlar® 29 single filament and yarn," *J. Compos. Mater.*, vol. 51, no. 1, pp. 109–123, Jan. 2017.

[14] J. A. Bencomo-Cisneros *et al.*, "Characterization of Kevlar-29 fibers by tensile tests and nanoindentation," *J. Alloys Compd.*, vol. 536, no. SUPPL.1, pp. S456–S459, Sep. 2012.

[15] M. R. Roenbeck *et al.*, "Probing the internal structures of Kevlar ® fibers and their impacts on mechanical performance," *Polymer (Guildf).*, vol. 128, pp. 200–210, Oct. 2017.

[16] V. P. W. Shim, C. T. Lim, and K. J. Foo, "Dynamic mechanical properties of fabric armour," vol. 25, pp. 1–15, 2001.

[17] ASTM International, "ASTM D3039/D3039M - 17: Standard Test Method for Tensile Properties of Polymer Matrix Composite Materials." 2017.

[18] CEN, "EN 1522 - Windows, Doors, Shutters and Blinds - Bullet Resistance - Requirements and Classification." 1998.

[19] T. Børvik, S. Dey, and A. H. Clausen, "Perforation resistance of five different high-strength steel plates subjected to small-arms projectiles," *Int. J. Impact Eng.*, vol. 36, no. 7, pp. 948–964, Jul. 2009.

[20] E. A. Flores-Johnson, M. Saleh, and L. Edwards, "Ballistic performance of multi-layered metallic plates impacted by a 7.62-mm APM2 projectile," *Int. J. Impact Eng.*, vol. 38, no. 12, pp. 1022–1032, Dec. 2011.

[21] G. R. Johnson and W. H. Cook, "A Constitutive Model and Data for Metals Subjected to Large Strains, High Strain Rates and High Temperatures," in *Proceedings 7th International Symposium on Ballistics*, 1983, pp. 541–547.

[22] Livemore Software Technology Corporation (LSTC), "LS-DYNA Keyword User's Manual - Volume II - Material Models." 2017.

[23] S. G. Nunes, R. Scazzosi, A. Manes, S. C. Amico, W. F. de Amorim Junior, and M. Giglio, "Influence of Projectile and Thickness on the Ballistic Behavior of Aramid Composites: Experimental and Numerical Study (under revision)."

[24] R. Wang and H. Liu, "Modelling and Simulatng Textile Structures Using TexGen," *Adv. Mater. Res.*, vol. 331, pp. 44–47, 2011.

A numerical study examining the formation of consolidation induced defects in dry textile composites

A. J. Thompson*[1], J. P-H. Belnoue[1] and S. R. Hallett[1]

[1]Advanced Composite Collaboration for Innovation and Science (ACCIS), University of Bristol, United Kingdom

* adam.thompson@bristol.ac.uk

Abstract. A key factor in the performance of composite materials is the quality of the final manufactured part. Regardless of the material form or process there are two common steps to produce this; the forming of an initially planar material into a three-dimensional shape and the consolidation of the material to achieve the required thickness and fibre volume fraction. During these processes the material is required to deform to achieve the final geometry. Defects which are induced by these processes can severely degrade the performance of the final structure. While a large body of research has been dedicated to understanding and modelling forming induced defects, consolidation induced defects have gone relatively unexplored. In this paper a simple approach is introduced to include the high transverse compliance of textile materials into conventional forming simulations, using 2-d finite elements, so that consolidation induced defects during forming can be analysed. This approach allows for the high stiffness of the fibre direction, non-linear shear stiffness, out-of-plane bending stiffness and through thickness compressibility of textiles to be fully present within the model. Using this approach, the consolidation of a large numbers of plies over curved tooling is simulated to examine the various parameters that contribute to the formation of defects.

1. Introduction

The consolidation of textile reinforcement materials is an integral process in composite manufacture. Its purpose is to increase the fibre density in the composite to ensure a high volume fraction (Vf) is achieved. This is crucial for high performance composites as the Vf corresponds directly to the mechanical properties of the composite. The maximum achievable volume fraction is dependent on both the internal architecture of the textile and the applied pressure.

Fibrous materials are highly compliant in the plane transverse to the fibre direction (i.e. the application of small pressures can significantly reduce their thickness). This is due to initial gaps which exist between the individual fibres, only small pressures are required to close these gaps which force the fibres to rearrange through sliding and bending. As the fibres come into contact and the number of contact points increase, rapid stiffening occurs until the inter-fibre contacts fully constrain the fibrous network.

For consolidation on a simple flat surface, the distortions induced by the reduction in thickness typically occur at the meso scale, where the architecture of the textiles distorts to accommodate the change in shape through migration of crimp, increase crimp angle and change in yarn cross section [1]. When consolidated over curved tooling, severe defects can form such as ply waviness, wrinkles and thickness variation. Tools with even single curvature can be a catalyst for the formation of such defects. One example of this is when multiple layers are laid up and consolidated over a radius. During this process the external radius of the lay-up is reduced producing an excess of material. If the material is constrained in-plane, via inter-ply friction or geometric features of the tool, the excess material in the radius is unable to migrate and hence will be accommodated through buckling of the fibres out of plane, causing wrinkles, or in-plane, causing severe fibre misalignment [2]. An example of such a defect can be seen in Figure 1.

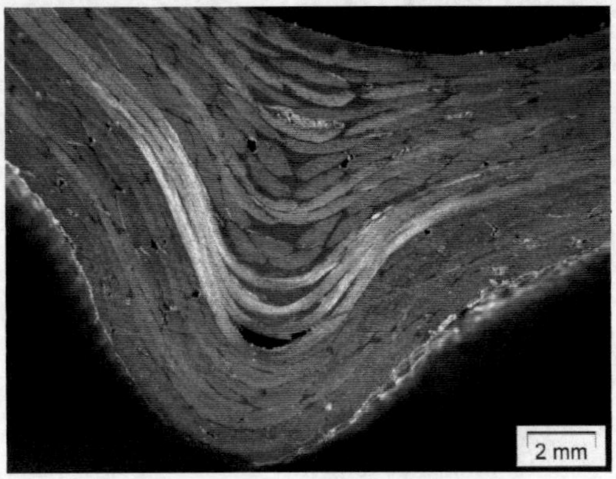

Figure 1. example of a consolidation induced wrinkle, formed in the corner radios of a non-crimp fabric composite part

Research into consolidation induced deformations has increased in recent years for pre-impregnated fibrous materials, with mechanisms behind their formation and models to predict their occurrence being presented [2-4]. For textile materials, consolidation-driven defects have had little attention as their magnitude is often superseded by deformations caused during the forming process. In this paper a simple approach for including the high transverse compliance of textile materials into conventional forming simulations, using 2-d finite element analysis, is introduced. Implemented into the commercially available Abaqus Explicit finite element software, the method allows for the high stiffness of the fibre directions, non-linear in-plane shear stiffness, out-of-plane bending stiffness and through thickness compressibility of textiles to be present within the model. With this approach a large number of layers consolidated over curved tooling can be simulated.

2. Modelling approach

Deformations induced during the consolidation of the material are dependent on the change in path length of the material and how much excess length is generated. This is determined by the geometric features of the tool and the change in thickness of the layered material. The way the excess length is accommodated is dependent on the material behavior and the constraints enforced by the geometry. Simulating this phenomenon therefore necessitates the inclusion of the through thickness compliance of the material into a model, as well as the other characteristic material behaviors typically included in forming simulations, i.e. stiffness along the fibre direction, in-plane shear stiffness, out of plane bending stiffness etc.

The starting point of this method is a model typically used for forming simulations where the fabric is represented as a continuous planar material. The mechanical behavior of the meso-scale structure is represented implicitly through constitutive models. A hypo-elastic material model [5] is used to ensure the directions of the high stiffness fibres are tracked during the simulation and to implement the non-linear in-plane shear stiffness of the material. Both shell and membrane elements are used to represent the material, which are superimposed and their nodes mutually constrained. The hypo-elastic material model is assigned to both element types, the membrane elements are assigned the in-plane material properties (tensile modulus of fabric fibre directions and the non-linear shear stiffness) the shell elements are assigned with just the flexural modulus of the material.

This approach has been successful at modelling the forming of textiles over complex geometries, capturing both the shear and wrinkle deformations accurately as well as the interaction of multiple layers, as shown in Figure 2.

Figure 2. Comparison of experiment with prediction for two layers of plane woven fabric [+/-45 0/90] formed over tetrahedron tool using double diaphgram forming process – experiment (left), simulation using shell/element forming approach (right)

A limitation of 2-D elements, both shell and membrane, is that they have no through thickness compliance, so are unable to include the high transverse compliance of the fibrous materials, preventing them from modelling consolidation processes. A compliant contact was therefore introduced to include the through thickness compliance of the material into the shell/membrane forming approach.

A commonly used contact in finite element analysis is the penalty contact algorithm. This method introduces a force at contact detection points which have penetrated across a defined surface. The purpose of this force is to eliminate the penetration by pushing the two surfaces out of contact. The method uses a simple formula:

$$F_c = k_c D_p \tag{1}$$

where k_c is the contact stiffness, often referred to as the penalty stiffness, and D_p is the depth of the penetration. The magnitude of the force required to remove the penetration is typically unknown, consequently there is often a finite ammount of penetration at the end of each time step. The value of k_c is a fictional stiffness which must be large enough to minimise the penetration while at the same time not be too large as to cause instability. The contact stiffness is defined as the contact pressure between two surfaces as a function of the overclosure. The overclosure of the two surfaces is the depth of the penetration. By manipulating this relationship, to capture the non-linear through thickness stiffness of the textile material, through thickness compliance can be included in the forming simulations as inter-penetration between adjacent layers.

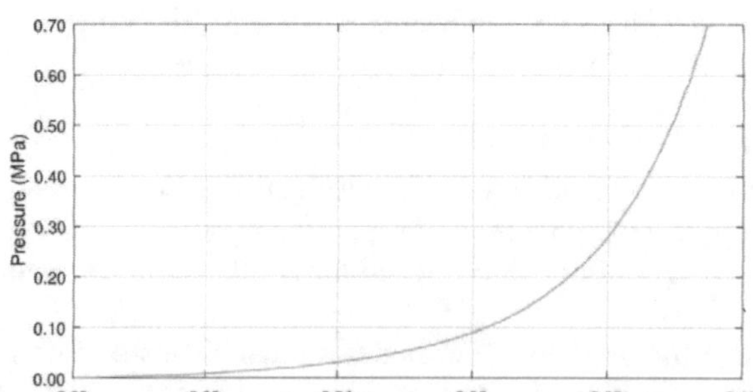

Figure 3. Pressure-overclosure relationship derived from compaction results of a plain woven fabric.

Figure 3 shows the pressure overclosure relationship for a single layer of a carbon fibre plain woven fabric, with 6000 fibres per tow, areal weight of 320gsm and 7 micron fibre diameter. This relationship was deduced from a compression test on samples of the material, where the overclosure is considered as the change in material thickness and the contact pressure is the externally applied pressure.

2.1. Verification

A simple compaction test on a stack of 2D elements was first performed to verify the approach. Single square shell elements with a length of 1 mm were stacked up, with each element representing a layer of material and given a thickness of 0.4mm. A displacement boundary condition was applied to all of the stacked elements, constraining their displacements to just vertical translations. The stacks were placed on a rigid surface prescribed with a fully fixed boundary condition. The pressure-overclosure relationship shown in Figure 4 was applied to the general contact definition in tabular form. Three models were created, each with a different number of layers, to examine the performance of the contact. A schematic of the verification model for two layers is presented in Figure 4.

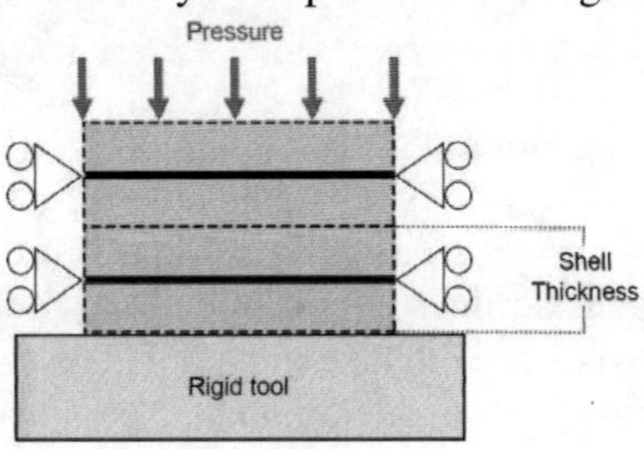

Figure 4. Schematic of consolidation model verification study for the compression of two shell elements onto a flat plate. Each shell element representing a layer of material

Figure 5 shows each of the models before and after compaction. As expected, thickness reduction is achieved through inter-penetrations of contacting elements. The thickness-pressure response output of each model is shown in Figure 6. The model with just two layers matches well with the expected values, however as the number of layers is increased the models become more compliant and diverge from the input values. This is likely due to the build up of small numerical errors which will increase with the number of layers. The difference is still relatively minor, with less than 3% error in the 24 layers at 0.1MPa.

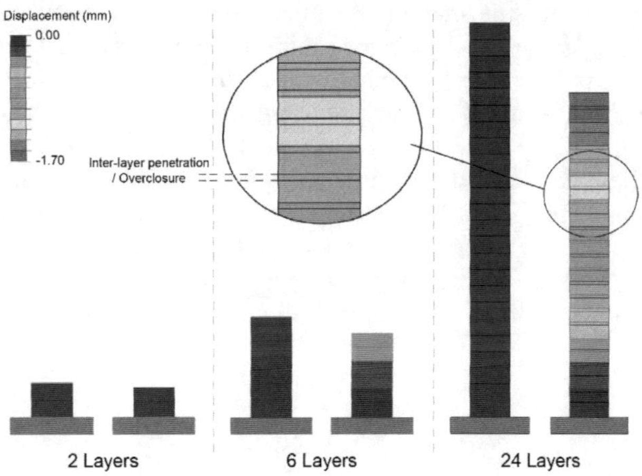

Figure 5. Verification of compliant contact – pre and post consolidated models

IOP Conf. Series: Materials Science and Engineering **406** (2018) 012052 doi:10.1088/1757-899X/406/1/012052

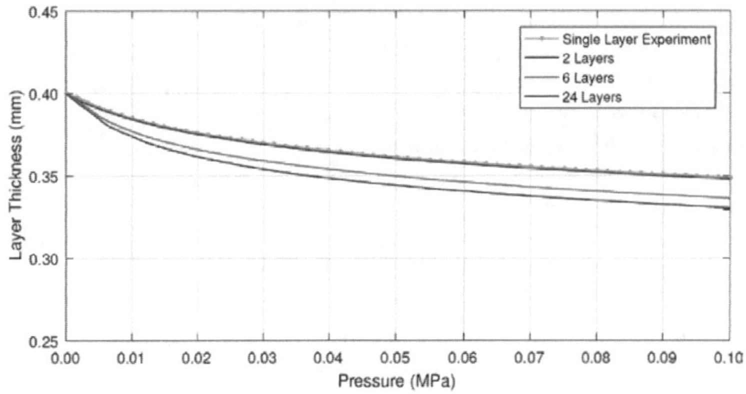

Figure 6. Mechanical out-put of compliant contact verification study showing the normalised Thickness-Pressure relationship for each model

3. Consolidation of C-section geometry

Consolidation-induced wrinkles form as a result of multiple plies being compacted over three dimensional surfaces as the planar materials is forced to assume the shape of the tool surface. Taking a simple male radius for example, during the consolidation process the external radius of the layup is reduced, forcing the material into a tighter geometry. If the material is constrained in-plane and the layers are unable to shear relative to one another, then excess material becomes present in the region of the radius. To accommodate this excess material the individual layers are forced to deform out-of-plane, causing wrinkles, or in plane, resulting in severe fibre misalignment. Depending on their severity, these deformations can compromise the structural integrity of the final part, so developing an understanding of how these defects form and the parameters which affect them is of major importance.

The magnitude of excess length created through the consolidation of plies over a radius can be determined using Equation 2.

$$L_{ex} = \frac{2\pi\Delta t}{4} \tag{2}$$

The change in thickness Δt is the only physical parameter which affects the amount of excess length, with L_{ex} increasing with increasing magnitudes of Δt. The change in thickness is dependent on the compliance of the material as well as the number of layers being consolidated. It is evident from Figure 6, for example, that there is a greater change in thickness in the 24 layers than in the consolidation of 2 layers, this is due to there being the same strain but applied to a much thicker sample, therefore compaction of 24 layers is likely to lead to more severe defects than thinner samples.

The formation of these defects is dependent on details of the geometry and the deformation modes of the material. The compliant contact is used here to analyse the effect of consolidating multiple layers of the plain woven fabric over a simple, male C-section tool. In order to do this, the compliant contact has been combined with the shell-membrane forming approach. The through thickness compliance, high tensile stiffness, in-plane shear and out-of-plane bending behaviour are therefore all present within the model.

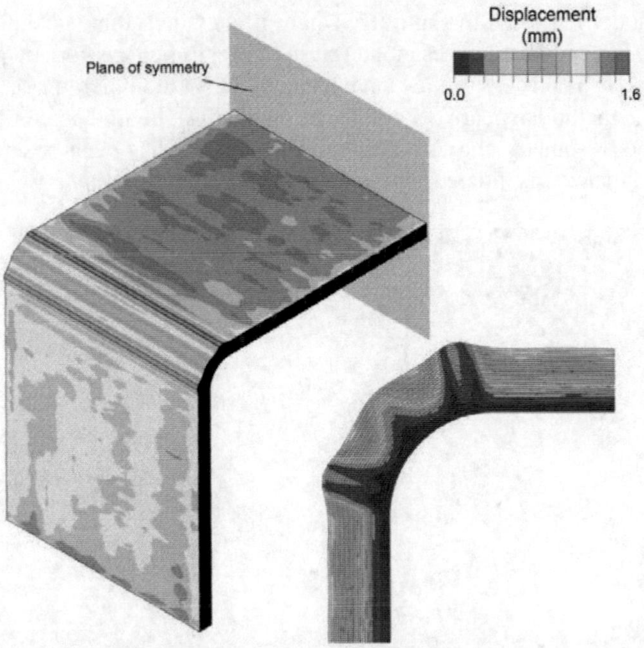

Figure 7. Deformed C-section geometry

The male C-section tool considered has a web length of 300mm, flange length of 150mm and 15mm radius. The width of the part is 150mm. The tool was modelled as a rigid surface. A stack of 24 plies, each with 0.4mm thickness was created above the tool. The layup was quasi-isotropic with a layup of [0/90\ +/-45\ 0/90\ +/-45\ 0/90\ +/-45\ 0/90\ +/-45\ 0/90\ +/-45\ 0/90\ +/-45]$_s$. As the tool only exhibits single curvature it was assumed that the forming of the geometry caused no deviation of fibre angle and formed perfectly to the tool. Furthermore, it was assumed that the pre-consolidated formed part had a constant thickness of 9.6mm (0.4mm x 24 plies) across the profile of the tool. Only half of the C-section was modelled and a symmetry boundary condition was applied along the centre of the web. The input properties for the model are presented in Table 1. The pressure-overclosure relationship in Figure 3 was used for the contact between layers. Coulomb friction was used to model the tangential interactions between the plies, with a coefficient of 0.28. This value was taken from experimental results of the frictional interactions between layers of plain woven carbon fibre fabric in [6]. A ramped pressure load of 0.1MPa was applied to the top ply to simulate vacuum consolidation.

Table 1. Properties for material model. E represents the stiffness in the two fibre directions and G_{12} is the shear stiffness as a function of the shear angle γ. Membrane elements include the in-plane material properties and shell elements the flexural stiffness

Element Type	E_{f1} (MPa)	E_{f2} (MPa)	G_{12} (MPa)
Membrane	40,000	40,000	$32.37\gamma^4 - 59.31\gamma^3 + 36.36\gamma^2 - 7.384\gamma + 2.37$
Shell	33	33	0

The deformed model is shown in Figure 7, where it is evident that substantial thickness reduction has occurred, which has produced excess material in the region of the corner radius. To accommodate this excess material, out-of-plane wrinkles have formed; two at the junctions of the radius and a smaller one in the centre. The method appears to capture the mechanism behind consolidation induced wrinkles well and the deformed geometry shows a similar trend to what has been observed in literature for the case of a prepreg laminate consolidated over a C-section, Figure 8.

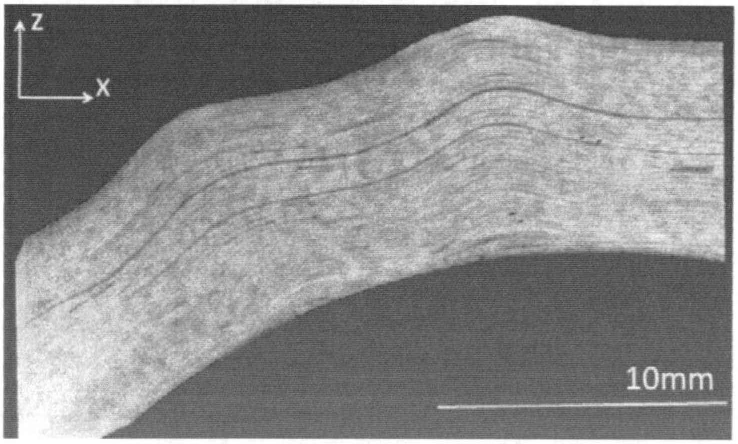

Figure 8. Consolidation induced wrinkles in prepreg C-section geometry, taken from Dodwell et al. [2]

4. Consolidation of different lay-up configurations

Four different lay-ups have been modelled, the lay-up orientations for each model is presented in Table 2. It was assumed that the compressive stiffness and interface friction was independent of ply orientation, the only varying parameter was the orientation of the fibres in each layer.

Table 2. Layer orientations for each model

	Lay-up
LU1	$[\ 0/90\ ]_{24}$
LU2	$[\ +/-45\ ]_{24}$
LU3	[0/90 +/-45 0/90 +/-45 0/90 +/-45 0/90 +/-45 0/90 +/-45 0/90 +/-45]s
LU4	[0/90 0/90 +/-45 +/-45 0/90 0/90 +/-45 +/-45 0/90 0/90 +/-45 +/-45]s

The thickness variation of each of the consolidated models in Figure 9 shows that in each layup, the presence of the radius causes some non-linearity in thickness around that corner region. The aligned configurations LU1 and LU2 vary significantly, the magnitude in LU2 is higher and varies across part width as well as length. When ply orientation is varied through thickness of the laminate, the presence of two large wrinkles becomes evident at both the web radius and flange radius junctions. There is little difference evident between LU3 and LU4.

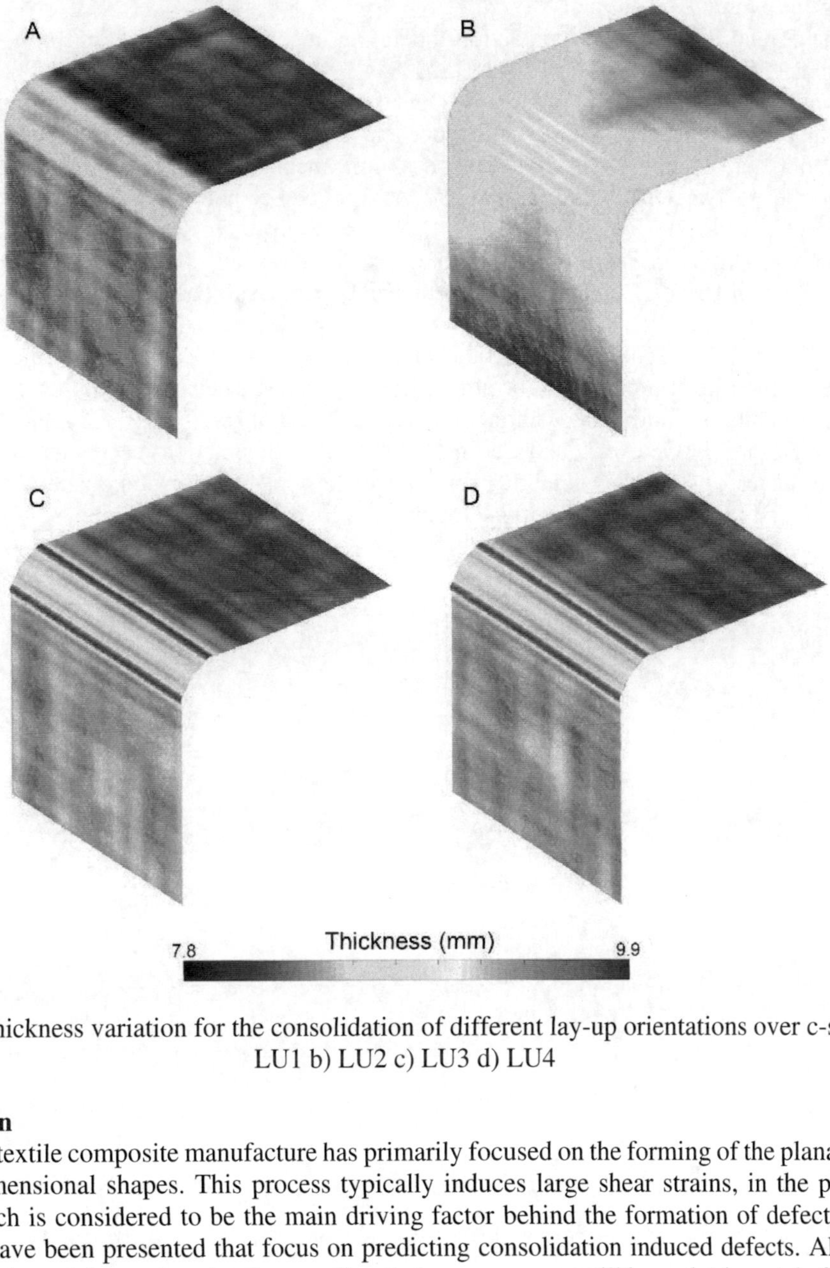

Figure 9. Thickness variation for the consolidation of different lay-up orientations over c-section, a) LU1 b) LU2 c) LU3 d) LU4

5. Conclusion

Modelling of textile composite manufacture has primarily focused on the forming of the planar materials into three dimensional shapes. This process typically induces large shear strains, in the plane of the material, which is considered to be the main driving factor behind the formation of defects. As such, few models have been presented that focus on predicting consolidation induced defects. Although the magnitude of these deformations is often smaller, their presence can still have detrimental effects on the performance of the final composite component. A simple numerical approach has therefore been developed to include the non-linear through thickness compliance of textiles into modelling approaches designed for forming simulations. This method allows for the compliance of the material to be captured through surface interactions, rather than the material behaviour, by manipulating the pressure-overclosure relationship of the penalty contact. As this method is simple and fully seperated from the material definition it can be easily applied to different approaches for modelling fabric forming behaviour. The approach has been used here to examine the consolidation of multiple layers of fabric over a C-section tooling showing the effects of consolidating different lay-up configurations.

IOP Conf. Series: Materials Science and Engineering **406** (2018) 012052 doi:10.1088/1757-899X/406/1/012052

References

[1] Potluri P, and Sagar T. V. Compaction modelling of textile preforms for composite structures. Composite Structures, 86(1-3):177–185, 2008.

[2] Dodwell T. J, Butler R, and Hunt G. W. Out-of-plane ply wrinkling defects during consolidation over an external radius. Composites Science and Technology, 105:151–159, 2014.

[3] Lightfoot J. S, Wisnom M. R, and Potter K. A new mechanism for the formation of ply wrinkles due to shear between plies. Composites Part A: Applied Science and Manufacturing, 49:139–147, 2013.

[4] Belnoue J. H, Nixon-Pearson O. J, Thompson A. J, Ivanov D. S, Potter K. D, Hallett S. R. Consolidation-Driven Defect Generation in Thick Composite Parts. ASME. J. Manuf. Sci. Eng. 2018;140(7)

[5] M. A. Khan, T. Mabrouki, E. Vidal-Salle, and P. Boisse. Numerical and experimental analyses of woven composite reinforcement forming using a hypoelastic behaviour. application to the double dome benchmark. Journal of Materials Processing Technology, 210(2):378–388, 2010.

[6] Nosrat Nezami F, Gereke T, and Cherif C. Analyses of interaction mechanisms during forming of multilayer carbon woven fabrics for composite applications. Composites Part A: Applied Science and Manufacturing, 84:406–416, 2016.

IOP Conf. Series: Materials Science and Engineering **406** (2018) 012053 doi:10.1088/1757-899X/406/1/012053

Assessment of process-induced microcracks in 3D woven composites using meso-scale model simulations

Nikolay V Timoshchuk[*], **Kent D Kasper and Harun H Bayraktar**

Albany Engineered Composites Inc, 216 Airport Drive, Rochester, NH 03867, USA

* nikolay.timoshchuk@albint.com

Abstract. One of the advantages of 3D woven composites is the ability to create unique fiber architectures in which tows can be individually arranged to optimize performance to meet specific structural requirements. This flexibility may lead to a design with unintended mismatch of thermal properties in areas of the composite, creating the potential for manufacturing issues such as process-induced microcracking. This paper presents results of a case study in which various architectures were analyzed, including some that led to intra-tow microcracks in the molded composite and others that did not. The objective was to develop an approach using meso-scale finite element modeling at the unit cell level to evaluate the propensity of these architectures to microcrack. Analysis results showed that the transverse principal stresses which developed in tow elements during cooling from cure to room temperature were significantly different in the various architectures. Comparison to the micro-computed tomography scans of the molded composites established strong correlation between regions with microcracks and areas of high predicted transverse principal stresses in the simulations. This study suggests that meso-scale finite element modeling of 3D woven architectures can be used in the design cycle to reduce or eliminate microcracking in the finished composite parts.

1. Introduction

Superior fatigue resistance and damage tolerance of 3D woven composites over traditional 2D laminates [1] made this class of composites an obvious choice for a number of applications, particularly in the aerospace arena. Some examples where 3D composites provide a significant advantage include: integrally woven T- and Pi-joints, cruciforms, outer guide vanes, and various 3D woven fittings. The most widespread applications are the fan blades and fan cases used in the LEAP turbofan engines, which currently power the Boeing 737 MAX and Airbus A320neo series aircraft [2].

An ability to create unique fiber assemblies (i.e. architectures) where tows can be individually arranged for optimal performance coupled with a possibility to tailor thickness by adding or dropping layers offers great flexibility in designing 3D woven preforms. Such flexibility, however, heightens the risk of creating a design with potentially large mismatch in thermal properties within the composite, which can lead to process-induced microcracks in the finished part. The cost associated with evaluating multiple part designs for the propensity to microcrack by means of experimental testing can be high and the turnover rates low. Therefore, a validated simulation approach which ties the meso- and macro-scale modelling of 3D woven composites and takes into account preform level design features such as layer drops could provide a great practical advantage in evaluation of architecture designs prior to manufacture.

IOP Conf. Series: Materials Science and Engineering **406** (2018) 012053 doi:10.1088/1757-899X/406/1/012053

Understanding process-induced microcracks in 3D woven composites was attempted by researchers in the past. Tsukrov et al [3] utilized micro-scale numerical modeling of a representative as-woven textile geometry to predict microcracking of carbon fiber/epoxy composites during resin curing. They relied on the parabolic failure stress criterion to study two typical 3D woven architectures, orthogonal and ply-to-ply, to assess their propensity to microcrack by comparing parabolic stresses developed within resin pockets. They found that the orthogonal architecture exhibited higher parabolic stresses, both in cumulative sense across the entire unit cell as well as locally, thus having a higher potential to produce microcracking during curing. They were able to correlate the regions of high parabolic stresses with the locations of microcracks as observed in the scans obtained by micro-computed tomography (μCT).

The microcracks presented in this paper appear to be different in nature compared to those analyzed by Tsukrov et al, and found not only within resin pockets but primarily through the tows as shown in micrograph in figure 1(a) below.

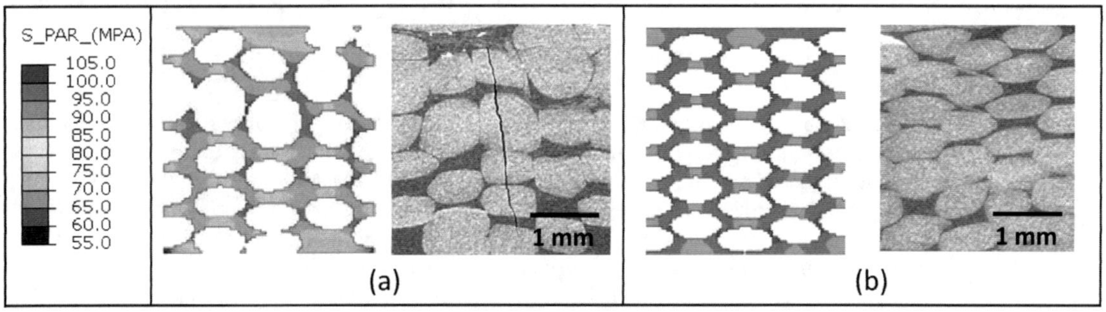

Figure 1. Two 3D woven architectures showing (a) microcracks and a lower level of the predicted parabolic stress, and (b) no microcracks and a higher level of the predicted parabolic stress.

By applying the parabolic stress failure criterion, no correlation was established between the levels of parabolic stresses as predicted by meso-scale FEA modeling and intra-tow microcracks as detected in μCT scans. In fact, the parabolic stresses were found to be higher for the architecture that did not exhibit microcracking as shown in figure 1(b). This suggests that for the type of intra-tow microcracks discussed in this paper a different approach is necessary in order to predict this type of failure in 3D woven composites.

2. VERVE™ property predictions

Albany Engineered Composites Inc. uses a proprietary method to assess the performance of 3D woven composites using meso-scale modeling of realistic 3D woven architectures on the unit cell level. This method relies on conducting Virtual Experiments on Representative Volume Element (VERVE™) to extract the equivalent stiffness matrix [4], which makes it possible to predict effective elastic constants of an RVE. VERVE™ can also be used to extract the effective thermal and electrical constants. With the application of periodic boundary conditions (PBCs), the performance of a unit cell can represent the performance of an entire part or coupon which makes VERVE™ a powerful virtual surrogate technique to study property changes due to process and material variations [5]. Additionally, models generated by VERVE™ allow one to visualize the spatial variations in stress/strain developed within a unit cell, a level of detail often necessary to identify regions and/or certain features of an architecture responsible for stress concentrations and/or failure.

An example of a meso-scale finite element unit cell model of a 3D woven composite visualized by Abaqus/CAE software (Version 6.14-1, Dassault Systèmes Simulia Corp., Providence, RI) is shown in figure 2. This example plots the stresses developed along the longitudinal directions of a tow, after applying load in the global X (warp) direction. Elements representing the surrounding resin were removed for clarity.

IOP Conf. Series: Materials Science and Engineering **406** (2018) 012053 doi:10.1088/1757-899X/406/1/012053

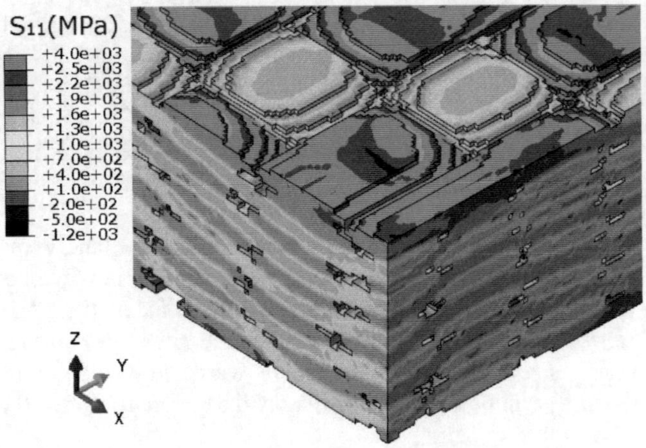

Figure 2. VERVE™ model of a 3D woven architecture.

By conducting a series of virtual tests, all nine independent material constants of a unit cell representing an orthotropic material can be determined. These effective properties can be used as input to the finite element analysis of a part on the macro-scale level, where tows and resin are no longer modeled exclusively due to computational time and machine capacity limitation considerations. This method of meso-scale modeling can also be applied to applications that have a more specific set of load and boundary conditions, as it was used in this study to understand the effects of residual stresses on microcracking in 3D woven composites.

2.1. Selected Architectures

Four 3D woven fiber architectures considered in this paper are presented in figure 3. They have attributes of either (or both, in case of A3) ply-to-ply or orthogonal topologies, which are two common types in 3D woven composite designs. Topology is a general term that is used to describe the weave pattern without specifying any of the parameters associated with it such as number of layers, warp and weft column spacing, tow types, etc. Architecture is considered to be a specific case of a topology with all those parameters defined.

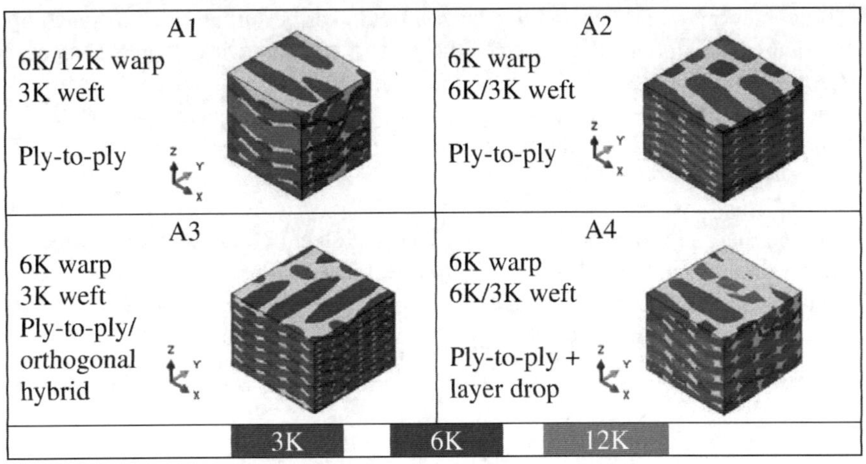

Figure 3. 3D woven architectures created using meso-scale model simulations.

A1 and A2 were two ply-to-ply architectures in which all fibers provide layer-to-layer interlocking (i.e. first layer interlocks with second, second with third, etc. until all layers are interlocked).

Architecture A1 also used two different tow sizes, 6K and 3K, in the X (warp) direction which was not the case in any other architecture where the warp tow size was kept constant. Architecture A3 had attributes of both ply-to-ply and orthogonal topologies. In orthogonal topologies, there are two sets of fibers: stuffers, which are essentially straight yarns, and weavers which dive between top and bottom surfaces interlocking all layers. A3 had a combination of stuffers (i.e. orthogonal feature) as well as layer-to-layer interlocking (i.e. ply-to-ply feature). A4 was a special case of a ply-to-ply architecture with a layer drop in the Y (weft) direction, a design feature most commonly thought of as a preform-level detail. The layer drop was implemented by exiting a yarn immediately under the surface yarn.

All architectures were modeled using the unit cell approach. In 3D woven composites, a unit cell is considered to be a building block for the whole textile, such that the preform can be created by assembling the unit cells in warp and weft directions, sometimes in the through-thickness direction as well. A unit cell can vary in size depending on how many warp and weft columns are necessary to define the architecture but generally can be as little as 2 by 2 and as large as 10 by 10 columns.

2.2. Material properties and material orientations
All 3D composites considered in this paper utilized Toho's HTA carbon fiber tows in 12K, 6K, and 3K sizes, depending on the architecture. Physical coupons that were used for validation were injected by resin transfer molding (RTM) using Tactix 123 resin system. A homogenization technique based on composite cylinder assemblage (CCA) and upper/lower bounds on the elastic constants as given by Hashin [6] was used to obtain tow properties with a tow packing factor in the range of 0.7-0.8. The transversely isotropic tow properties were mapped onto the tow elements with the axis of material symmetry lined up with the longitudinal direction determined from known positions of centroid points for each tow. The matrix was treated as an isotropic material with the temperature-independent Young's modulus, Poisson's ratio and coefficient of thermal expansion.

2.3. Boundary conditions and loadcases
VERVE™ employs periodic boundary conditions as a way to study the behavior of an entire part or a coupon by considering just a single representative volume element or a unit cell of that structure. PBCs exist in a form of equations relating the displacement on one part of the boundary to those on another [7]. Full periodic boundary conditions (i.e. periodicity in X, Y and Z directions) were applied to all architectures considered in this study, except for architecture A4 which, due to a layer drop, maintained periodicity in X direction only.

All property predictions were calculated based on the linear perturbation analysis completed within Abaqus/CAE. Simulations intended to extract the elastic constants involved six steps necessary to construct the equivalent stiffness matrix: three axial and three shear load cases in each of the three primary directions and on three primary planes, respectively. Thermal expansion coefficients were determined from the change in RVE dimensions due to a uniform temperature drop of 1°C.

2.4. Effective properties
Effective elastic and thermal properties for each of four architectures as predicted by meso-scale model simulations are summarized in table 1. A significant difference in fiber volumes used in these four cases as well as in values of material constants suggests a different internal state of stress that these composites would likely to experience during the curing cycle.

Table 1. Effective properties of architectures A1 – A4.

	A1 (*Vf=0.56*)	A2 (*Vf=0.58*)	A3 (*Vf=0.64*)	A4 (*Vf=0.58*)
E_1 (GPa)	62.8	61.4	88.9	50.9
E_2 (GPa)	13.7	42.7	32.9	25.1
E_3 (GPa)	8.4	9.8	10.7	9.2
G_{12} (GPa)	3.1	3.4	4.2	3.3
G_{13} (GPa)	4.4	3.2	4.3	3.9
G_{23} (GPa)	2.9	3.2	3.7	3.6
ν_{12}	0.19	0.13	0.16	0.24
ν_{13}	0.81	0.51	0.55	0.59
ν_{23}	0.48	0.53	0.5	0.55
α_1 (με/°C)	-1.46	1.66	0.489	0.353
α_2 (με/°C)	17	3.38	5.3	5.15
α_2 (με/°C)	37.2	42	35.6	38.6

These values were used as the material properties for the 3D woven core in the macro-scale analysis as discussed in Section 3.

3. Cure-induced microcracking in 3D composite core

3D woven preforms were molded with 2D cover plies attached to top and bottom surfaces (figure 4), which possess different elastic and thermal properties compared to those of the 3D composite core. The constraint due to the addition of cover plies yielded a different state of residual stress experienced by the core from a uniform temperature drop, compared to an unconstrained case. This paper does not discuss in detail the cooldown analysis on the macroscopic level, however, the strain output from these models was used to define the boundary conditions in meso-scale level simulations as described in Section 3.1.

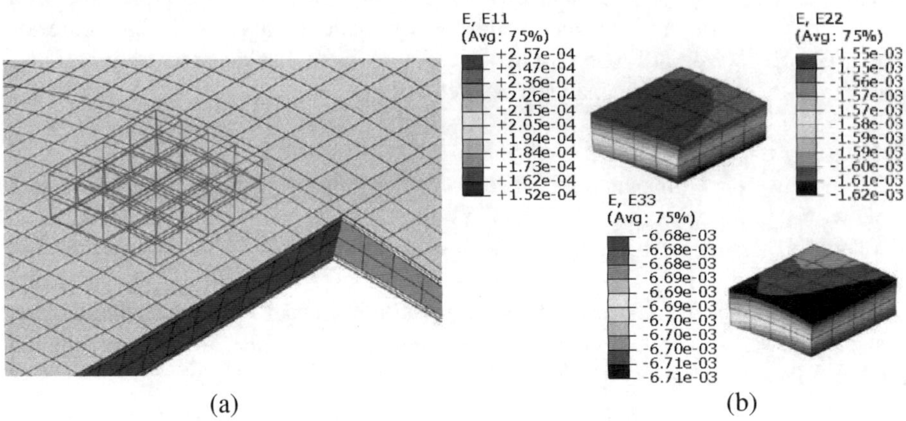

Figure 4. Partial view of the macro-scale model showing (a) 2D cover plies on both sides of the 3D woven core with the area of interest corresponding to an approximate area where most microcracks were observed in molded parts, and (b) the amount of strain experienced by the core in that region.

The comparison charts that demonstrate the difference in strains as predicted in constrained versus unconstrained cases are shown in figure 5 for (a) direction-2 (i.e. weft direction), and for (b) direction-3 (i.e. through-thickness direction). Strains in direction-1 (i.e. warp direction) for both meso- and macro-scale models were found to be negligible and thus were not shown.

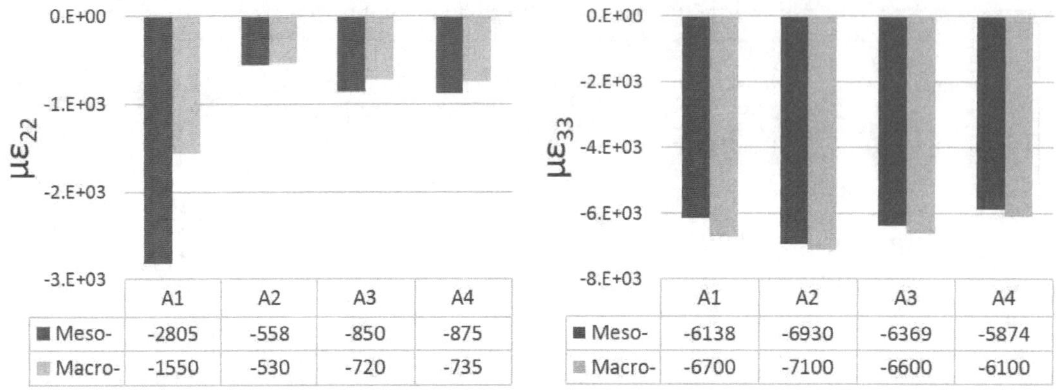

(a) (b)

Figure 5. Comparison of microstrain in (a) direction-2 (weft direction), and (b) through-thickness direction between constrained and unconstrained cases.

In most cases the difference was less than 10%, particularly for microstrain in the through-thickness direction. The biggest difference was observed in $\mu\varepsilon_{22}$ for architecture A1 which suggests that in order to predict a more accurate state of stress in a 3D woven core it is necessary to take into account the contribution of 2D cover plies that effect the overall response. Failure to do so could lead to wrong conclusions as it relates to microcracking in those cases. To account for the contribution of 2D cover plies in meso-scale thermal analysis, the microstrains $\mu\varepsilon_{22}$ and $\mu\varepsilon_{33}$ were included as direct boundary conditions.

3.1. Meso-scale thermal analysis

The meso-scale unit cell models for all four architectures presented in Section 2 that were used to obtain the effective elastic and thermal constants were modified to apply an uniform temperature drop from cure to room temperature. Two major adjustments that were made included: (1) changing the temperature drop from 1°C to 165°C, and (2) prescribing a fixed amount of displacement in directions 2 and 3, to match the results of the macro-scale analysis for each architecture.

3.2. Transverse principal stresses

The type of microcracks discussed in this paper appear to occur not only in resin pockets, but more so through the tows of 3D woven composites, as was shown in figure 1. Presumably, the stresses acting in the transverse plane of a transversely-isotropic tow were significant enough to split a tow, forming a microcrack. The maximum and minimum values attained by the normal stress in the plane of isotropy could provide an insight into this type of failure. The meso-scale models of four architectures considered in this paper were post-processed to calculate the principal normal stresses (i.e. transverse principal stresses) according to the following expression:

$$\sigma_{TPS1}, \sigma_{TPS2} = \frac{\sigma_{22} + \sigma_{33}}{2} \pm \left(\left(\frac{\sigma_{22} - \sigma_{33}}{2} \right)^2 + \tau_{23}^2 \right)^{1/2} \quad (1)$$

where 2 and 3 are tow directions in the plane of transverse isotropy.

3.3. Validation by micro-computed tomography scans

The hypothesis that transverse principal stresses can be used to evaluate the propensity of an architecture to exhibit the intra-tow microcracks due to cure-induced stresses was validated by comparing the results of the meso-scale model simulations with the micro-computed tomography scans as shown in figure 6. The μCT images correspond to the same warp and weft columns for which transverse principal stresses

IOP Conf. Series: Materials Science and Engineering **406** (2018) 012053 doi:10.1088/1757-899X/406/1/012053

are shown. There appears to be a strong correlation between architectures with high TPS1 and whether or not those architectures exhibited microcracking. Architecture A1 and A4 both have considerably higher TPS1 compared to A2 and A3, reaching in excess of 115-120 MPa. The locations of microcracks tend to correlate with areas of stress concentration.

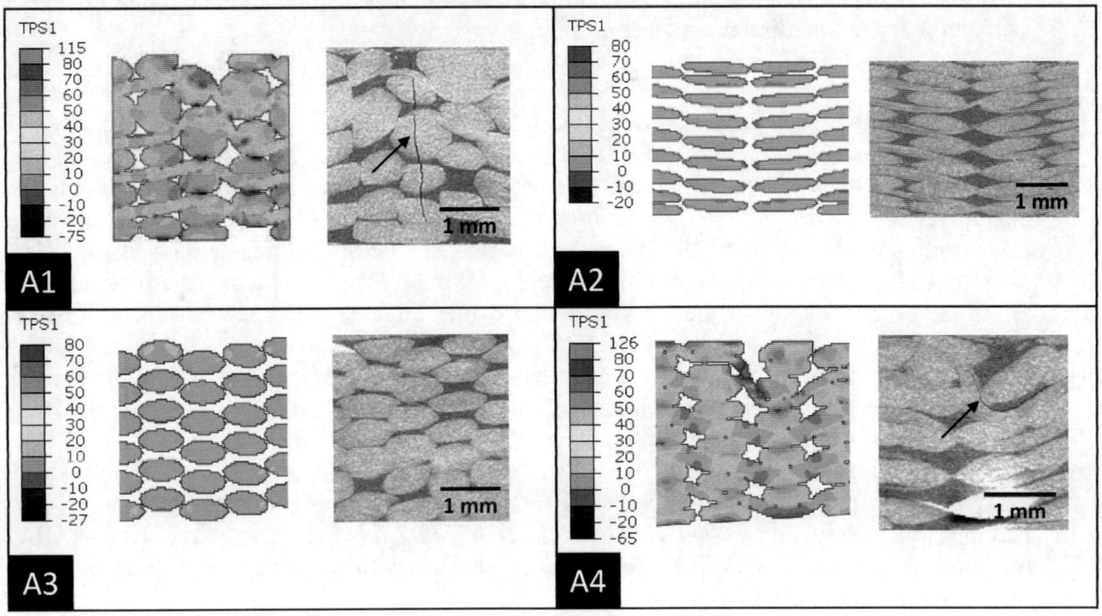

Figure 6. Comparison of areas with high transverse principle stresses and locations of microcracks.

One common feature found in architectures A1 and A4, both of which exhibited microcracking, is an abrupt change in the direction of the top surface yarn relative to other yarns, violating the through-thickness design symmetry and likely contributing to a mismatch in composite properties locally. In case of A1, the symmetry was violated even more by insertion of 12K in place of 6K tows in the second layer from the top surface. These features may not always lead to microcracking in 3D woven composites but can be avoided during the architecture design to reduce the risk of composite failure.

The magnitudes of the second component of the transverse principal stress TPS2 were found to be considerably smaller and, therefore, were not shown. Presumably, the positive (tensile) stresses were largely responsible for creating the microcracks as it appears the tows were pulled apart (TPS1) as opposed to being compressed (TPS2). The strong correlation between the areas of high transverse principal stresses as predicted by meso-scale modeling and the locations of microcracks observed by μCT demonstrates the merits of the approach presented.

4. Conclusions

A mismatch in thermal properties in areas of 3D woven composites, which can vary greatly depending on the architecture selection, can lead to process-induced microcracking in the finished product. A robust finite element simulation approach can add value by providing an ability to evaluate designs for their propensity to microcrack under residual stresses without a need to manufacture physical parts. The meso-scale modeling effort presented in this paper shows that:

- Relying on a parabolic stress failure criterion alone may not fully capture the likelihood of microcracking in the composite part.
- Evaluating transverse principal stresses within tow elements can identify whether intra-tow microcracking is likely to occur, and can be especially useful when comparing alternate architecture designs.

- Incorporating the effects of the macroscopic assembly of 2D and 3D materials can be a critical contributor to evaluating the cure-induced stresses in the composite part, and the coupling of macro- and meso-scale analyses can further separate performance of different architectures.
- Abrupt changes in tow direction and variations in tow size symmetry through the cross section, while perhaps unavoidable in the design of a monolithic preform, can have a non-negligible effect on process induced microcracking.

References

[1] Tong L, Mouritz A P and Bannister M K 2002 *3D Fibre Reinforced Polymer Composites* (Oxford: Elsevier Science)

[2] Dambrine B, Mahieu J-N, Goering J and Ouellette 2008 Development of 3D woven resin transfer molded fanblades *Proc. 9th Int. Conf. Textile Composites TEXCOMP9 Recent Advances in Textile Composites* ed S Advani and J Gillespie Jr (Lancaster, PA: DEStech Publications) pp 19-24

[3] Tsukrov I, Bayraktar H, Giovinazzo M, Goering J, Gross T, Fruscello M and Martinsson L 2011 Finite element modeling to predict cure-induced microcracking in three-dimensional woven composites *Int. J. Fracture* **172** pp 209-16

[4] Ehrlich D, Bayraktar H and Kelley J 2015 Using Different Homogenization Techniques for Different Stages of Product Development for 3D Woven Composites *Proc. 12th Int. Conf. Textile Composites TEXCOMP12* (Raleigh, NC: North Carolina State University)

[5] Cowles B, Backman D, Dutton R 2012 Verification and Validation of ICME Methods and Models for Aerospace Applications *Integrating Materials and Manufacturing Innovation* **1**

[6] Hashin Z 1972 *Theory of Fiber-Reinforced Materials* (Washington DC: NASA). Hashin Z 1979 Analysis of Properties of Fiber Composites with Anisotropic Constituents *J Appl. Mech. Transcations ASME* **46** pp 543-50

[7] Li S 2001 General unit cells for micromechanical analyses of unidirectional composites *Compos. Part A Appl. Sci. Manuf.* **32** pp 815-26

IOP Conf. Series: Materials Science and Engineering **406** (2018) 012054 doi:10.1088/1757-899X/406/1/012054

Effects of misalignments on thermally induced shapes of woven fabric reinforced laminates

L L Warnet[1], **R H Schrijver**[2] **and R Akkerman**[1,2]

[1] University of Twente, Enschede, the Netherlands.
[2] TPRC, ThermoPlastic composite Research Center, Enschede, the Netherlands.

Abstract. Fabric misalignment leads to unsymmetry in fibre reinforced composite laminates. This unsymmetry induces out-of-plane deformation after processing. This paper evaluates the influence of fabric misalignment on the out-of-plane deformation of flat press-consolidated thermoplastic laminates. The results of this analytical and experimentally validated study shows a strong relation between deformations and length/width ratio. In particular narrow laminates show high twisting values. The results obtained are of importance for the prediction of deformations of stamp formed thermoplastic laminates.

1. Introduction

An often demanding requirement when designing and producing a fibre reinforced composite part is the control of the tolerance on the deformation of the part after processing. The source of such deformations is generally well recognised though not always understood in details, being partly originating from the inherently high anisotropic thermo-mechanical behaviour of the material, and from process induced changes in (local) fibre orientation and fraction. These two simple observations are closely linked, as a local change in fibre orientation when for example forming a composite part will lead to local unsymmetry of the laminate, leading not only to a change in thermo-mechanical properties but also to out-of-plane deformation. The control of the relation between the manufacturing process and the change in laminate morphology is therefore necessary to meet tight shape deformation tolerances. This paper is not an attempt to enhance the knowledge on this issue, but addresses the effect of misalignment of woven fabric based composite laminates on their post processing out-of-plane deformations.

Early studies on the room temperature shapes on unsymmetric cross-ply laminates showed that the curvatures measured are a function of the geometry of the laminate [1]. Not only the magnitude of the curvature does vary with the size, but also the type of shape, with even bi-stable shapes. A model was proposed based on a non-linear strain-displacement relation and a minimisation of the total potential energy. Several authors have developed the model further, in order to predict the shapes of other types of laminates [2] as well as rectangular laminates [3].

This paper uses a similar approach to evaluate the shapes of laminates having an originally symmetric lay-up, with a fibre orientation misalignment, making the results applicable for process induced fibre orientation changes. Focus is here on woven fabric based laminates. Both a model and validation are presented in this study. The next part will summarise the modelling which includes an evaluation of the thermo-mechanical properties of misaligned woven fabric laminates and the non-linear model able to predict the relation between degree of misalignment, size and deformation. The third part relates on the

IOP Conf. Series: Materials Science and Engineering **406** (2018) 012054 doi:10.1088/1757-899X/406/1/012054

material, process and measurement equipment used. The fourth part elaborates on both numerical and experimental results, followed by conclusions.

2. Analytical model

This section presents the model used to predict the out-of-plane deformations of rectangular laminates containing misaligned woven fabric layers. The model description is divided in two parts, the first one proposing a procedure to evaluate the properties of laminates having misaligned woven fabric layers, the second using these properties to predict the deformations of these laminates.

2.1. Laminate properties

Laminate basic mechanical properties are often described using the Classical Lamination Theory. The main result of the CLT is the so- called ABD matrix, a laminate stiffness matrix describing relations between deformations, i.e. strains and curvatures, and loading, i.e. forces and moments. This plane stress method is based on summing the stress-strain relation of single plies and is recognised to be well suited for unidirectional ply based laminates subjected to thermo-mechanically induced deformations of limited amplitude. The interlaced fibre structure in a woven fabric layer means that the stress-strain relation is more complex and therefore requires a dedicated approach. An existing multiscale method discretising the woven structures into simple repetitive elements [1] is used in this paper. This method was also developed to analyse the behaviour of non-orthogonal fabric with limited yarn misalignment. A full description of the method is available in [2], and couples the thermo-mechanical properties of the fibre and the matrix and the basic geometrical properties of the impregnated woven fabric to provide an ABD description of the woven fabric ply with possible misalignment considered, as well as a laminate built from these layers. The method consists of a geometrical subdivision of the woven fabric, followed by a rebuilding of the weave with its thermo elastic properties.

2.2. Laminate deformations

The out-of-plane deformations of thin laminates can be predicted using the CLT approach, though it is restricted to small deformations. To this end, a method introduced by Hyer [1] to evaluate the shape of unsymmetric cross-ply UD based laminates is used in this paper. A non-linear displacement-strain definition is used for this method, which is finally solved using a Rayleigh-Ritz approach, minimising the total potential energy. This method successfully described the shapes of square unsymmetric cross-ply laminate as a function of the laminate dimensions, which change from a saddle shape at a low side length / thickness ratio to two stable cylindrical shapes for higher side length / thickness ratio. Several authors have then subsequently extended Hyer's theory in order to apply it to other types of arbitrary layed up unsymmetric laminates [2], as well as rectangular shaped laminates [3].
The current research uses a displacement field description based on the one proposed by Dano & Hyer [2], though adapted for rectangular laminates. The out-of-plane deformations are evaluated by minimising the potential energy, based on a process-induced thermal loading. An expression for the potential energy, adapted in order to use the stiffness characteristics as described in 2.1 is given as

$$\Pi = \int_{\frac{-L_x}{2}}^{\frac{L_x}{2}} \int_{\frac{-L_y}{2}}^{\frac{L_y}{2}} \left(\frac{1}{2} \begin{Bmatrix} \varepsilon^0 \\ \kappa \end{Bmatrix}^T \begin{bmatrix} A & B \\ B & D \end{bmatrix} \begin{Bmatrix} \varepsilon^0 \\ \kappa \end{Bmatrix} - \begin{Bmatrix} \varepsilon^0 \\ \kappa \end{Bmatrix}^T \begin{Bmatrix} N^{Th} \\ M^{Th} \end{Bmatrix} \right) dx\, dy \tag{1}$$

where L_x and L_y respectively denote the length and width of the laminate, $\{\varepsilon^0\}$ and $\{\kappa\}$ the mid-plane strain and curvature vectors which contain all unknowns to be evaluated. $[A]$, $[B]$ and $[D]$ are the extensional stiffness matrix, the coupling matrix and the bending stiffness matrix of the laminate, $\{N^{Th}\}$ and $\{M^{Th}\}$ the process-induced thermal force and moment vectors. Both laminate stiffness matrices and thermal loading vectors are calculated specifically for woven fabrics using the method shortly described in 2.1.

The model uses a non-linear strain-displacement relation for the in-plane strains, also known as the von Karman description, where the approximation for the displacement field contains the unknown to be solved, meant at being applied for arbitrary lay-up and rectangular thin laminates. The in-plane displacements u^0 and v^0 and the out-of-plane displacement w^0 are defined as follows, based on work by Dano and Hyer [2], introducing a set of 14 unknowns.

$$\begin{aligned} u^0(x,y) &= s_1x + s_2y + s_3x^3 + s_4x^2y + s_5xy^2 + s_6y^3 \\ v^0(x,y) &= s_2x + s_7y + s_8x^3 + s_9x^2y + s_{10}xy^2 + s_{11}y^3 \\ w^0(x,y) &= \tfrac{1}{2}(ax^2 + by^2 + cxy) \end{aligned} \tag{2}$$

Strains are derived from these expression according to the von Karman strain relations

$$\begin{aligned} \varepsilon_x &= \varepsilon_x^0 + z\kappa_x^0 = \frac{\partial u^0}{\partial x} + \frac{1}{2}\left(\frac{\partial w^0}{\partial x}\right)^2 - z\left(\frac{\partial^2 w^0}{\partial x^2}\right) \\ &= s_1 + 3s_3x^2 + 2s_4xy + 3s_5y^2 + \frac{(2ax+cy)^2}{8} - za \end{aligned} \tag{3}$$

$$\begin{aligned} \varepsilon_y &= \varepsilon_y^0 + z\kappa_y^0 = \frac{\partial v^0}{\partial y} + \frac{1}{2}\left(\frac{\partial w^0}{\partial y}\right)^2 - z\left(\frac{\partial^2 w^0}{\partial y^2}\right) \\ &= s_7 + 3s_{11}y^2 + 2s_{10}xy + 3s_9x^2 + \frac{(2by+cx)^2}{8} - zb \end{aligned}$$

$$\begin{aligned} \varepsilon_{xy} &= \varepsilon_{xy}^0 + z\kappa_{xy}^0 = \frac{1}{2}\left(\frac{\partial u^0}{\partial y} + \frac{\partial v^0}{\partial x} + \frac{\partial w^0}{\partial x}\frac{\partial w^0}{\partial y}\right) - 2z\left(\frac{\partial^2 w^0}{\partial x \partial y}\right) \\ &= \frac{1}{2}\left(2s_2 + \left(s_4 + 3s_8 + \frac{ac}{2}\right)x^2 + \left(3s_6 + s_{10} + \frac{bc}{2}\right)y^2 \right. \\ &\quad \left. + \left(2s_5 + 2s_9 + ab + \frac{c^2}{4}\right)xy\right) - zc \end{aligned}$$

where ε are the in-plane strains, the superscript ^{0} denotes the mid-plane, κ are the curvatures, s_i and a, b, c are the unknown to be solved, the last three corresponding to the three curvatures.
Substituting equation (3) into (1) gives an algebraic expression of the total potential energy. The deformed shape of a laminate corresponds to a state of minimal potential strain energy. This leads to a set of 14 algebraic nonlinear equations

$$F_i = \frac{\partial \Pi}{\partial c_i} = 0; \quad c_i = s_1, s_2, \dots, s_{11}, a, b, c \tag{4}$$

The 11 first equations can be reorganised and substituted in the remaining equations leading to 3 nonlinear equations to be solved for the 3 curvatures. The equations are solved for each combination of length and width using a Newton-Raphson scheme. All solutions of the system are found by running the scheme using strategically chosen starting values of the unknowns. Stability is checked by examining the second variation of the potential energy with respect to the unknown coefficients.

3. Material and experimental methods

This section presents the experiments performed to validate the model presented in the second chapter. The first part reports on the material used and its relevant properties. The second part describes the production of the laminates with misalignments, while the third part summarises the setup used to measure the curvatures of the laminates.

3.1. Material properties

The laminates considered in this study are based on a generic 280 g/m^2 Toray T300 carbon 5HS woven fabric having a width of 500 mm and used in combination with a Sabic PolyEtherImide Ultem 1000 polymer in 50 μm thick film form. The PEI matrix is amorphous and has a glass transition temperature T_g of 210°C. The thermo-mechanical properties of the constituents and the basic weave description are necessary to evaluate the basic properties of the (misaligned) layer and are given in respectively table 1 and 2. The properties of the fabric concern in particular nominal values as the warp and weft count $C_{x,y}$, the layer thickness h_{lay}, and the layer fibre volume fraction $v_{f,lay}$. Besides, values measured on micrographs of cross section of orthogonal cross-ply laminates are the undulation $U_{x,y}$ and the yarn fibre volume fraction $v_{f,yarn}$. Details for the definition of these parameters can be found in [2].

A sensitivity analysis of the mentioned parameters on the curvature prediction was performed. Though the details of this analysis are omitted in this paper, it is worth adding that the model is particularly sensitive to variation in matrix properties as well as transverse properties of the fibre. The matrix properties are generally well known, in contrast to the transverse fibre properties. The elasticity modulus E_{f22} varies from 10 to 40 GPa in the literature, while the coefficient of thermal expansion α_{f2} varies from 5 to 15 με/°C. These bounds are used when comparing the measured curvatures to the predictions in the fourth part of this paper.

Table 1. Engineering constants of the fibre and matrix material.

Toray T300 carbon fibre [5]		Sabic PEI Ultem 1000	
E_{f11} [GPa]	231	E_m [GPa]	3.3
E_{f22} [GPa]	40		
G_{f12} [GPa]	24	G_m [GPa]	1.2
G_{f23} [GPa]	14.4		
ν_{f12} [-]	0.26	ν_m [-]	0.36
ν_{f23} [-]	0.39		
α_{f1} [με/°C]	-0.4	α_{f12} [με/°C]	56
α_{f2} [με/°C]	5.6		
ρ_f *[kg/m³]*	1760	ρ_m *[kg/m³]*	1270
		T_g *[°C]*	217

Table 2. Nominal (left) and measured (right) fabric specification.

C_x [1/m]	700	U_x [1/m]	0.67
C_y [1/m]	710	U_y [1/m]	0.67
h_{lay} [mm]	0.31	$V_{f,yarn}$ [%]	68
$v_{f,lay}$ [%]	50		
$v_{f,yarn}$ [%]	68		

3.2. Laminate production

The misalignment is performed by shearing the dry woven fabric and laminating the composite plate by a film stacking technique. It is chosen in this work to fix the warp direction of the woven fabric, while varying the weft direction. This results in an anti-symmetric $[(0, 90\text{-}\theta)_N/(90+\theta, 0)_N]$ layup, where 0 corresponds to the warp direction. The misalignment angle θ was given values of 5°, 10° and 15°, while $_N$ was given values of 1, 2 and 3. A total of 9 test laminates was therefore produced for the validation. A 280 g/m² T300 carbon 5HS woven fabric having a width of 500 mm is used in combination with a PolyEtherImide (PEI, Ultem 1000, 50 μm) in film form to layup the laminates. The PEI matrix is amorphous and has a glass transition temperature T_g of 210 °C.

The lay-up of a $[(0,90\text{-}\theta)/(90+\theta,0)]$ laminate is shown in Figure 1. The glass selvage used during the weaving of the fabric is taken as a reference for the warp direction when shearing the fabrics. One side parallel to the warp yarns is fixed to prevent the fabric from shifting. Next, the opposite side is skewed to the desired misalignment angle. The resulting weft misalignment angle is checked using a protractor. A predefined amount of the 50 μm thick PEI film is added to the layup to end up with a nominal fibre volume fraction v_f of 50% in case the fabric is orthogonal. The orientation of the fabric is locked by ultrasonic spot welding the sides of the fabric and matrix layers.

Figure 1. Illustration of the film stacking of a [(0,90-θ)/(90+θ,0)] T300 3k 5HS - PEI laminate. The 0 direction and the x-coordinate corresponds to the warp direction.

The laminate with misalignments are produced using a classical press consolidation process. Release agent (Marbo) coated caul sheets are used as a mould material. The laminates are consolidated at 350°C and 2 MPa pressure on a Fontijne press. The consolidation state of the laminates is controlled with ultrasonic scanning. The resulting fibre volume fraction v_f measured based on the laminate thickness varies from 52 to 54 % depending on the misalignment. A higher value than the nominal 50 % is due to some flow of matrix during the process.

3.3. Set-up and procedure to measure the curvatures

A set-up was built in order to measure the curvatures of the misaligned laminates. The principle was to measure the out-of-place (z) deflection of the laminate while scanning the laminate in the x and y coordinates. The laminate was measured in a vertical position in order to prevent deflection due to the mass of the laminate. As illustrated in Figure 2, the scanning in x and y direction was performed with respectively a manually operated linear slider and a fine screw driven linear motion system. The x and y coordinates were monitored with respectively a Linear Variable Displacement Transducer and a micrometer. The (z) deflection was measured with a 20mm range non-contact deflection transducer based on a laser and a Position Sensing Device (Micro-Epsilon NCDT).

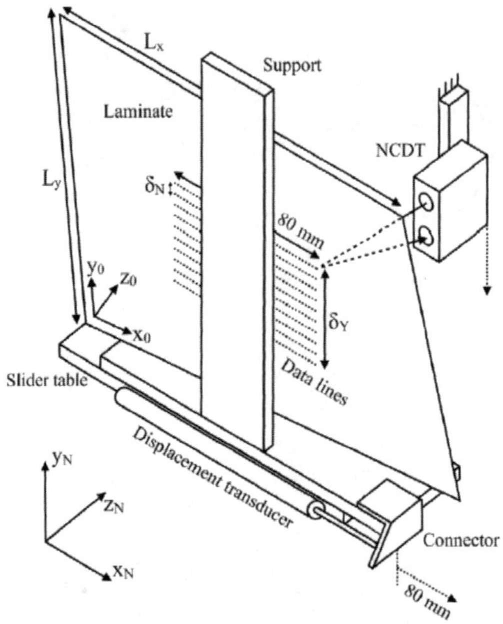

Figure 2. Set-up for the measurement of the laminate deflection.

The surface to be measured was coated with a solvent free carbon spray, which reduced the scattering of the laser by the composite surface. The accuracy of the system was evaluated at 2% by measuring a well- defined cylinder covered with the same coating. The precision of the system was evaluated by measuring a laminate 5 times, leading to a coefficient of variation of 0.1 %. The curvatures were derived by fitting a second order polynomial through the xyz data.
In order to evaluate the effect of the length to width ratio on the deformation of the laminate, it was chosen to fix the length of the laminates (x-direction in Figure 2) and vary the width (y direction). This was performed by cutting the 450 mm x 450 mm laminate in different width, with increments of 50 mm down to a width of 50mm, and with increments of 10mm down to a width of 20 mm.

4. Results

The numerical results are first presented, followed by the experimental results.

4.1. Numerical results

The main result of this research is a relation between the length / width aspect ratio and the curvatures of the misaligned laminates. The loading taken into consideration for the system proposed is restricted to a ΔT from T_g to room temperature. Effects from tool laminate interaction are disregarded in these results. A typical plot of the three relevant curvatures for a 2 layers [(0, 90-θ)/(90+θ, 0)] laminate for misalignments $\theta = 5$, 10, 15° having a fixed length L_x of 500mm is shown in Figure 3. The prediction on these laminate all show bi-stable shapes (B-C and B-D in Figure 3 and in corresponding Figure 4) for width having a value larger than a so-called bifurcation point (B). A third stable shape is found below the bifurcation point, i.e. at low width, as a twisted beam (Shape AB). It is worth adding that the solution at L_y=0 mm corresponds to the solution obtained with the CLT.
The results show a large dependence of the misalignment angle θ on the curvature amplitude. This dependence can be explained by the reduction in bending stiffness in the weft direction with increasing misalignment, leading to an increase in curvature. The same reduction in stiffness also leads to larger curvatures in the κ_y dominant shape (around a factor 2 higher than the curvature in κ_x the dominant shape. The bifurcation point or length, determining a change from stable twisted shape at low width to

bistable shapes at higher width also depends on the misalignment. A low value of the κ_{xy} curvature is predicted for high width/length, while it increases in a non-linear way from the bifurcation point for low width/length ratio, leading to high twisting curvature at low width. Interesting from an engineering point of view is that the width at which the twist becomes significant, i.e. at point 'B' in figure 3, is independent on the amount of misalignment.

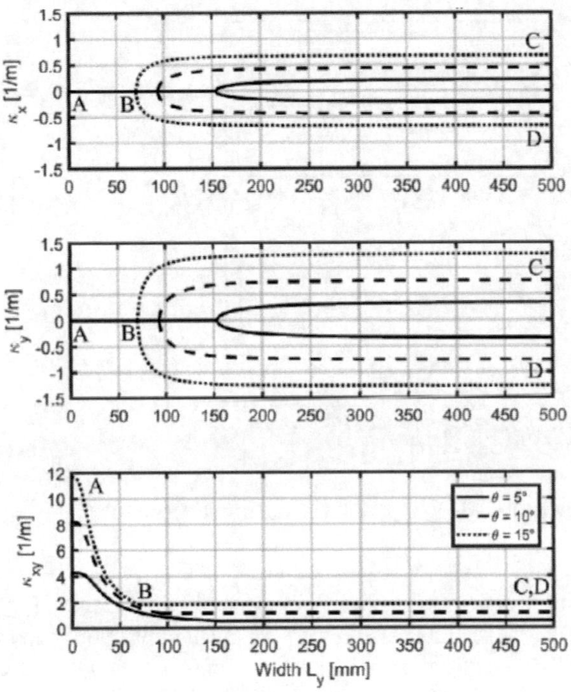

Figure 3. Curvature plots of [(0, 90-θ)/(90+θ, 0)] for three values of the misalignment θ and a fixed L_x=500mm

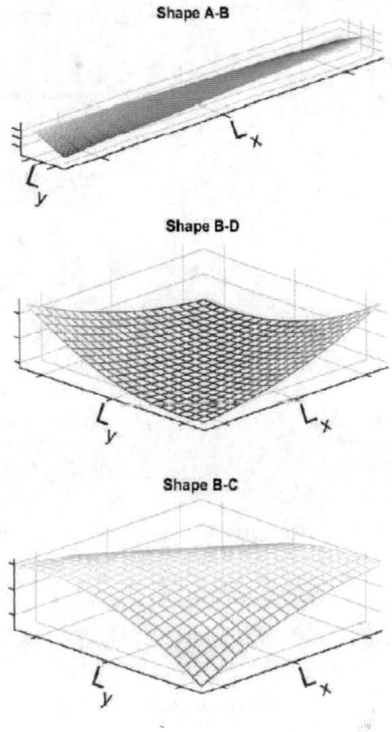

Figure 4. Basic curved shapes corresponding to Figure 3.

The influence of the laminate thickness on the deformation is illustrated with Figure 5, where the curvatures for a 4 layers laminate [(0, 90-θ)$_2$/(90+θ, 0)$_2$] are given. As expected, the behaviour is similar to that of a 2 layers laminate, while the global level of curvature decreases. The bifurcation length changes most significantly. The model also predicts that for small misalignments, a large range of aspect ratios exists where all curvatures stay at a low level, i.e. $\kappa_x = \kappa_y = 0$, $\kappa_{xy} < 0.5$ /m. This trend is extended for thicker laminates, where the bifurcation length exceeds the chosen length L_x of 500 mm for a 6 layer laminate.

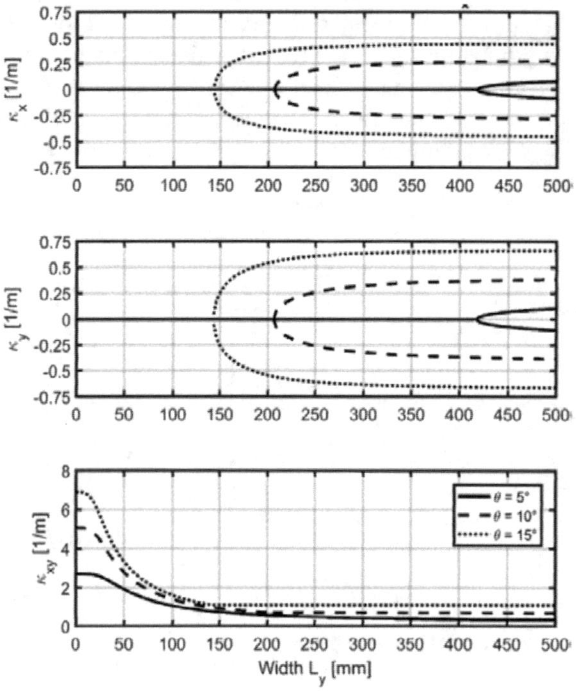

Figure 5. Curvature plots of $[(0, 90\text{-}\theta)_2/(90+\theta, 0)_2]$ for various misalignment θ.

4.2. Experimental results

Experimental results are presented on similar types of figures as used for the numerical results. Figure 6 shows the curvatures of a 2 layers [(0, 75)/(105, 0)] laminate. Both numerical (in continuous lines) and experimental (discrete points) are presented, as well as light-gray continuous lines showing the potential variation on the model following the uncertainty on the fibre transverse properties E_{f2} and α_{f2} values as discussed in the material section. The experimentally measured curvatures are shown to be generally lower than the predicted ones. The prediction of the bifurcation length seems to overestimate the measured bifurcation, though the change from a single torsional curvature to a bistable behaviour as described by the bifurcation length is probably not as discontinuous as the model suggests. Besides, the effect of depicted uncertainty on the fibre transverse properties is such that the measurements are within the modelled boundaries.

Results on the effect of both misalignment and thickness follow the same trends as described on 4.1, and are best described by the lower bound. An example is given in Figure 7 for the thickest laminate considered and a misalignment of 5°. In this case the laminate does only show a twisting curvature with the size considered in this study.

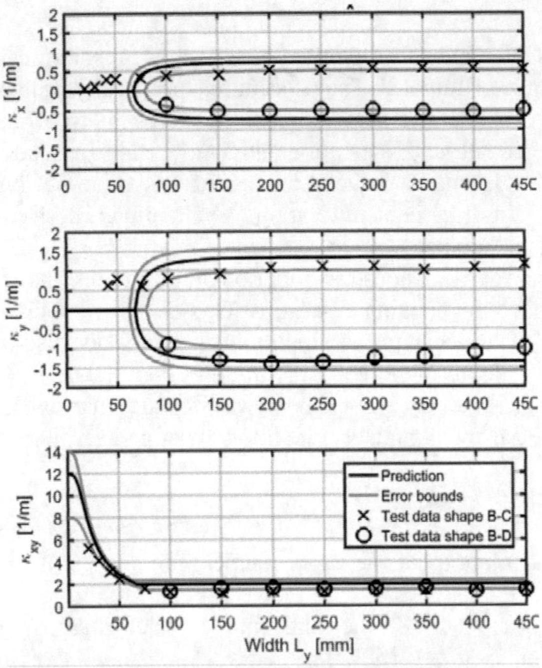

Figure 6. Curvature plots of [(0, 75)/(105, 0)] with numerical and experimental results.

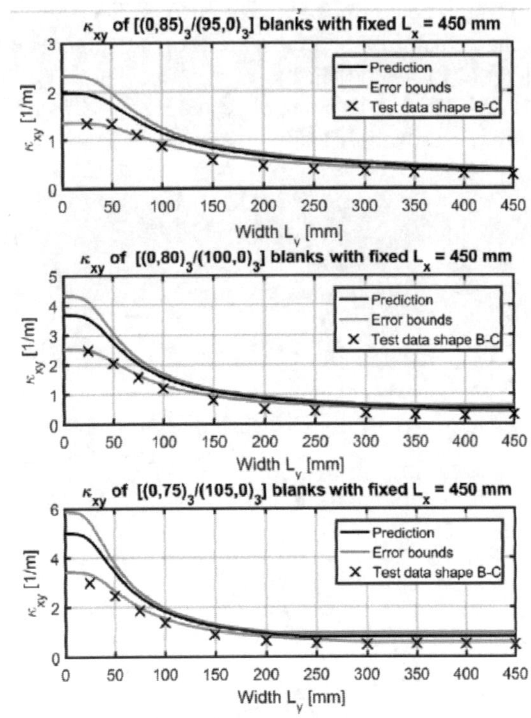

Figure 7. Curvature plots of $[(0, 85)_3/(95, 0)_3]$ with numerical and experimental results.

IOP Conf. Series: Materials Science and Engineering **406** (2018) 012054 doi:10.1088/1757-899X/406/1/012054

5. Conclusions

A numerical procedure is presented to quantify the effect of misalignment on the process induced deformation of fibre reinforced laminates. The procedure, specifically proposed for woven fabric based laminates, is based on two steps. The first provides the basic thermo mechanical properties of the misaligned ply based on a geometrical description and a multiscale approach. The second part concerns a model able to predict the thermally induced shapes of unsymmetric laminates using a non-linear definition of the in-plane strain-displacement relation. A validation is performed on a generic carbon 5H satin reinforced thermoplastic laminate.

The results show out-of-plane deformations as a function of the dimensions of the laminate. Large square laminates tend to show a bi-stable bending curvature. This deformation changes to a twisted curvature for rectangular laminates. The dimensions at which this change takes, also called the bifurcation point, is shown to depend on both the misalignment amplitude as the thickness of the laminate. Last but not least the amount of twist increases in a non-linear way towards narrow laminates. This makes it in general difficult to produce narrow laminates with a low twisting curvature.

References

[1] Hyer M W 1982 Calculations of the room-temperature shapes of unsymmetric laminates *J. of Comp. Mat.* **15** 296

[2] Dano M –L, Hyer M W 1998 Thermally-induced deformation behaviour of unsymmetric laminates *Int. J. of Solids Struct.* **35** 2101

[3] Aimmanee S and Hyer M W 2004 Analysis of the manufactured shape of rectangular THUNDER-type actuators *Smart Mater. Struct.* **13** 1389

[4] Lamers E A D, Wijskamp S and Akkerman R 2000 Modelling the thermo-elastic properties of skewed woven fabric reinforced composites *Proc. ECCM 9 (Brighton)*

[5] Lamers E A D 2004 Shape distortions in fabric reinforced composite products due to processing induced fibre reorientation (Enschede) 134 p

[6] Pandy R and Hahn H T 1992 A micromechanic model for 2D fabrics *Proc. Of the Americam Society for Composites, Lancaster PA,* USA

IOP Conf. Series: Materials Science and Engineering **406** (2018) 012055 doi:10.1088/1757-899X/406/1/012055

Damage characterization during high temperature fatigue of off-axis woven organic matrix composites for aircraft applications

F. Foti[1*], Y. Pannier[1], D. Mellier[1], M. Gigliotti[1]

[1]Institut Pprime, University of Poitiers, ISAE-ENSMA, CNRS 3346 - Department of Physics and Mechanics of Materials - 1, Avénue Clément Ader F-86962 Futuroscope Chasseneuil, FRANCE

* federico.foti@ensma.fr

Abstract. The present work focuses on damage characterization during high temperature fatigue of 8HS woven $[45]_6$ carbon HTS40 carbon fibre/polyimide matrix composite samples. Fatigue tests are performed under controlled environment (air, 2 bar O_2, 2 bar N_2) at 250°C by using a specific experimental setup developed at PPRIME Institute; Digital Image Correlation (DIC) and μ-Computed Tomography (μCT) are used for monitoring the evolution of sample behaviour during fatigue. It is shown that these two techniques are effective and complementary for damage assessment under fatigue at high temperature under aggressive environment: a rich characterisation of damage extent and orientation can be provided which allows damage being formalised in a way that can be naturally integrated within the context of damage-dependent behaviour models. The employment of such description allows performing full assessment of the effects of environmental fatigue.

1. Introduction

In the next future, the employment of woven organic matrix/carbon fibre composites (OMC) is foreseen for the realization of "warm" structures: these parts may be subjected, in service, to mechanical fatigue (e.g. fan/compressor blades turbo-engines), thermal cycling and thermo-mechanical fatigue (e.g. aircraft structural parts). There is a consistent literature concerning the fatigue behaviour of woven composites [1]–[3] with the use of Digital Image Correlation (DIC, see for instance [4]-[6]) and μ-Computed Tomography (μCT, see for instance [7]-[9]) for monitoring of damage and degradation phenomena: for satin woven composites under traction-traction fatigue load these studies show that four degradation stages can be identified (see reference [1] for comprehensive review): onset of matrix cracks in the transverse tows subjected to high local bending-induced tensile stress, crack propagation along resin rich regions or towards the longitudinal tow of the same woven ply bringing to meta-delamination, transverse crack propagation and longitudinal tow failure respectively.

Few works are available concerning high-temperature fatigue of textile composites ([10]), while the interaction between fatigue and environmental degradation of has been poorly explored: previous work on environmental fatigue of composite laminates ([11]) has showed that the effect of thermo-oxidation phenomena occurring at high temperature may severely affect fatigue degradation of such materials.

This work aims at characterizing and modelling the high-temperature (250°C) fatigue behaviour of C/polyimide 8HS textile composites, in particular the development of damage related to cyclic mechanical mechanisms (fatigue) under controlled (temperature and gas) environment.

IOP Conf. Series: Materials Science and Engineering **406** (2018) 012055 doi:10.1088/1757-899X/406/1/012055

An experimental setup is specifically developed in order to perform fatigue tests under controlled environment (2 bar N_2, air, 2 bar O_2). Sample properties evolution during fatigue tests at 250°C are monitored using in-situ Digital Image Correlation (DIC), and damage assessment is provided by ex-situ X-ray microtomography (μCT). Composite behaviour models are employed for rational understanding of the experimental results.

It is shown that these two techniques are effective and complementary for damage assessment under fatigue at high temperature under aggressive environment: a rich characterisation of damage extent and orientation can be provided which allows damage being formalised in a way that can be naturally integrated within the context of damage-dependent behaviour models. The employment of such description allows performing full assessment of the effects of environmental fatigue.

2. Material, samples, experimental setup and methods

The material is an 8-harness satin $[45]_6$ woven carbon/polyimide T650-35/MVK-14 composite material produced by Resin Transfer Moulding. The glass transition temperature of the polymer matrix is around 330°C (DMA results), and the matrix volume fraction in the composite is between 34% and 38%: the tow fibre volume fraction is around 60%. The architecture of a unit cell and of the textile composite ply are schematically reported in Fig. 1.

a) b)

Figure 1. 8 HS woven architecture a) Elementary ply b). Representative Unit Cell. [TexGen®].

All the mechanical tests have been performed using the COMPTINN' test setup illustrated in Figure 2: this is an INSTRON 1251 equipped by a climatic chamber specifically developed to carry out mechanical tests in controlled environment (T_{MAX}: 350°C, p_{MAX}: 5 bar, environment: air, O_2, N_2). The climatic chamber does not allow using gauges for strain measurements, only non-contact techniques can be used for strain measurement.

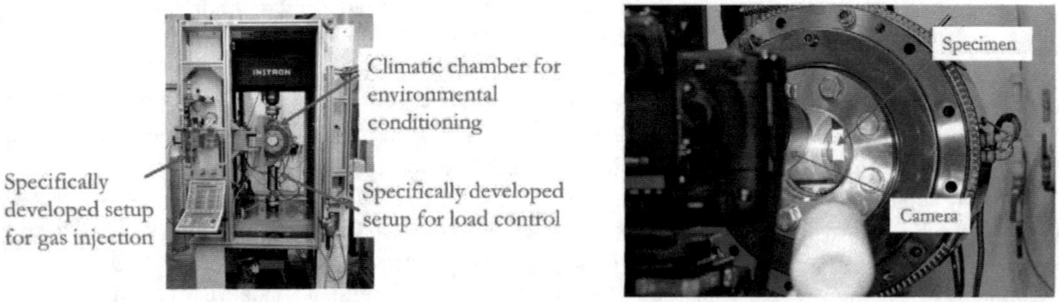

a) b)

Figure 2. a) COMPTINN test setup developed @PPRIME b) detail of the climatic chamber.

IOP Conf. Series: Materials Science and Engineering **406** (2018) 012055 doi:10.1088/1757-899X/406/1/012055

Tension-tension fatigue tests have been performed under three environmental conditions: air, 2 bar N_2 and 2 bar O_2. Fig. 3 illustrates schematically the load-piloted and the fatigue test: Table 1 presents the fatigue test parameters.

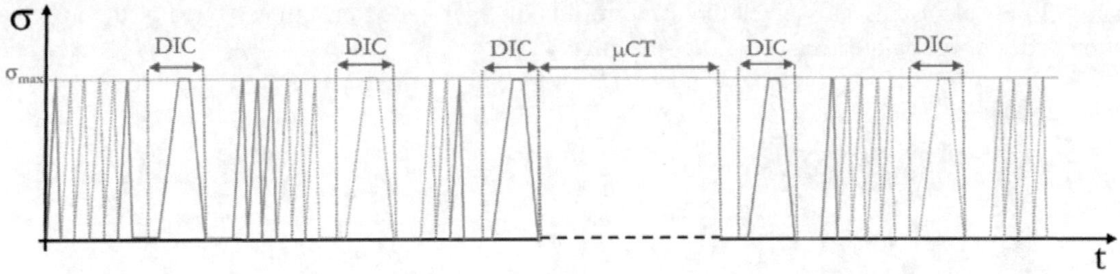

Figure 3. Scheme of the fatigue test with periodic interruption for DIC and μCT scans.

Table 1. Fatigue test parameters.

Stress Ratio R	Max Stress σ_{MAX} [MPa]	Frequency f (Hz)	Maximum number of cycles	Temperature T [°C]
0.1	75 (~ 55% σ_{FAIL})	2	10^6	250

Fatigue tests were periodically interrupted to carry out intermediate quasi-static load/unload cycles and sample periodically removed from the test rig for μCT scans.

2.1. *Digital Image Correlation (DIC)*

For the DIC experimental campaign carried out in this work white particles are directly deposed on the naturally (dark) surface of the composite specimens by using a commercial white spray for use at high temperatures up to 650°C. The average diameter of the particles is 10±7 μm (measured after the speckle pattern application). During tests, the pictures have been taken using a Nikon D3X camera mounting a Sigma 150mm f/2.8 MACRO zoom. For DIC analysis the CORRELA V.2012 software developed at the PPRIME Institute (see for instance [12]) was employed. The maximal specimen area captured by the camera was 2400 × 3700 px^2 (or around 14.5 × 22.2 mm^2) in the central zone of the specimen taking one picture every 6 seconds. A bi-cubic polynomial is selected for the grey levels interpolation inside each subset and the correlations are performed using an iterative algorithm considering also the local displacement gradient. The strain field is calculated from the displacement values using finite difference and the Green-Lagrange tensor [13]. Finally, a low-pass filter is applied on each image to reduce the errors linked to the grey level interpolation. Table 2 summarizes the pixel size, the choice of CORRELA parameters and the errors that could affect the calculated strains. For more details concerning the DIC approach the reader is referred to Reference [13].

Table 2. DIC parameters for fast calculation.

Pixel size	Subset step	Smooth parameter Local gauge length	Correlation error (std)	Environmental error (std)	Max out-of-plane motion error
6-6.2 μm/mm	40 px	1/(2 × 40 px ≈ 0.5 mm)	$\sigma = 2.25 \times 10^{-5}$	$\sigma = 10^{-4}$	5×10^{-5} and 10^{-4} @ $\varepsilon = 1\%$
Max local total error ≈ 2.7 × 10^{-4}					

A typical strain field observed on the tested woven specimens is reported in Fig. 4a. The output strain field from CORRELA is quite heterogeneous. Several inhomogeneities of the strain field are due to several features, in the present woven architecture the tow width is large enough to contain 4 data points. If some subsets fall on a resin rich region or on a tow or even on a crossover tow, the corresponding strain value could be different from the surrounding data points lying on the surface tows. Another source of inhomogeneity of the strain field is represented by cracks: since the employed DIC algorithm allows calculating only continuous strain field, cracks appear as large strain regions. To obtain strain values at the macroscopic scale at a given instant of a traction test an arithmetic mean over the whole stain field is calculated. This gives the same results of a virtual extensometer fixed at the upper and bottom sides of the painted region of the specimen.

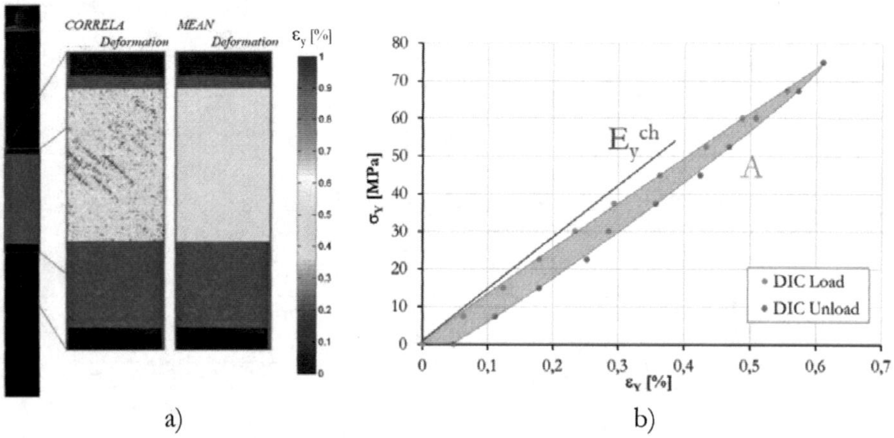

Figure 4. DIC method a) CORRELA output strain field and calculation of mean strain b) stress-strain curve obtained by DIC mean field strain during load-unload traction test.

Fig. 4b. illustrates a typical stress-strain curve obtained by averaging measured DIC strain fields during a traction loading-unloading curve. By this curve it is possible to extract mesoscopic indicators of the material behavior; for instance, the longitudinal chord modulus along the traction direction (y), E_y^{ch}, and the hysteresis area below the whole stress-strain curve, A (Fig. 4b).

DIC can be employed for crack detection by using a separate set of parameters for image correlation, as reported in Table 3.

Table 3. DIC parameters for crack detection.

Pixel size	Subset step	Subset size	Smooth parameter Local gauge length
6-6.2 µm/mm	6 px	6 px	5/(≈ 0.42 mm)

The second part of crack detection is based on image processing. A binary matrix representation of the remaining strain values then analysed by using the Avizo9® software. The binary image is segmented using simple thresholding, a skeletonization is then carried out on the segmentation. By this operation, a line is plotted inside each segmented zone, equidistant to shape boundaries of the element. At the end of this step, the DIC strain field is reduced to a map of segments. Fig. 5 shows the original strain field (Fig. 5a) and the binarization (Fig. 5 b). A MATLAB® script is finally employed to suppress segments oriented along directions different from the expected cracks directions (-45° in Figure 5), hence high strain zones oriented transversally are not considered. The same script is finally employed to measure the length of the remaining segments, reported in blue in Fig. 5c.

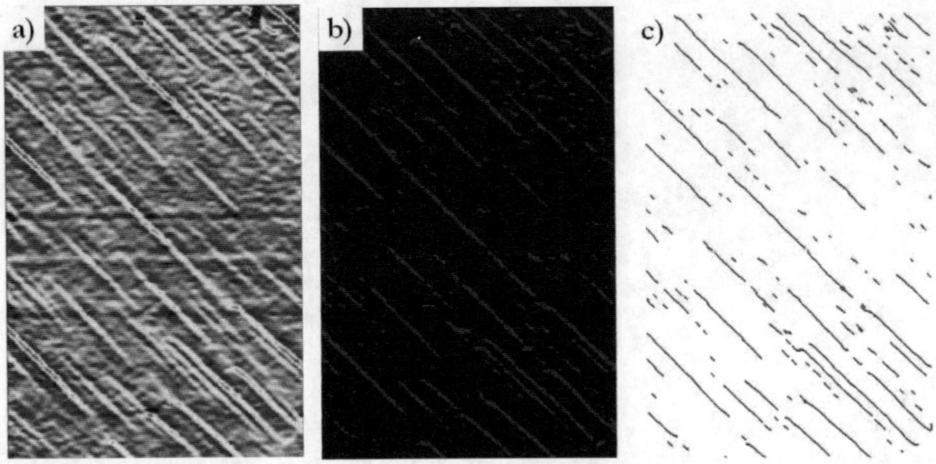

Figure 5. Crack detection by DIC a) original strain field b) binarization c) obtained segments.

2.2. μ-Computed Tomography (μCT)

μCT scans are carried out by employing RX Solution UltraTom XL computed tomography system and the 3D images are reconstructed by the X-Act software. The three X-Ray sources mounted on the μCT system are: a transmission 160 kV nano-focus tube, a reflection 230 kV micro-focus tube and a reflection 150 kV micro-focus tube. For the experimental activity performed in this work, only the last one source has been used. The micro-focus tube Hamamatsu L1216-07 has a maximum power in output of 75 W and an X-ray beam angle of 43°. Between the two available devices for the X-ray detection: a Charge Coupled Device CCD (resolution 4008x2624 px2, pixel size 5.9 μm) and a flat-panel detector Paxscan Varian 2520D (1920x1536 px2, pixel size 127 μm), during the test campaign the Varian detector has been used in agreement with the lower received signal to noise ratio of the Varian detector for the range of used photons energies and flux.

The parameters used for X-Ray generation are reported in Table 4, these parameters are recurrent for this kind of materials. The μCT scans have been performed using the circular stack acquisition and an anti-ring shift option to reduce ring artifact. The duration of each scan is around 11 hours. The 3D image is obtained using the Filter Back-Projection techniques and using some correction tools. One of this allows correcting beam drift and another one allows correcting the offset in the reconstruction of some slices due to relative oscillations of the specimen respect to the source or the detector. During reconstruction, an anti-ring filter was applied and a reduction of the noise by using the filter back projection reconstruction algorithm was obtained.

Table 4. μCT parameters.

Voltage [kV]	Current Intensity [μA]	Power [W]	Averaging	Frame Rate	Spatial Resolution [μm/px]
60	139	8.34	20	4.8	8.79

Fig. 6 illustrates a μCT segmented scan of the external surface ply (exterior ply) of the sample (a non-segmented image is shown in Fig. 6a): Fig. 6b illustrates cracks segmentation; matrix 0° cracks (in red), -45° intra-tow cracks along the -45° tows (in blue), Fig. 6c contains superposition of +45° intra-tow cracks (in green) along the +45° tows. A diffuse pattern of cracks can be noted: 0° matrix cracks are small and concentrated in matrix-rich zone, while intra-tow cracks have the direction of the respective tows. It has been demonstrated by deeper observation that these cracks span the whole tow and have all the same height, which is the height of the tow.

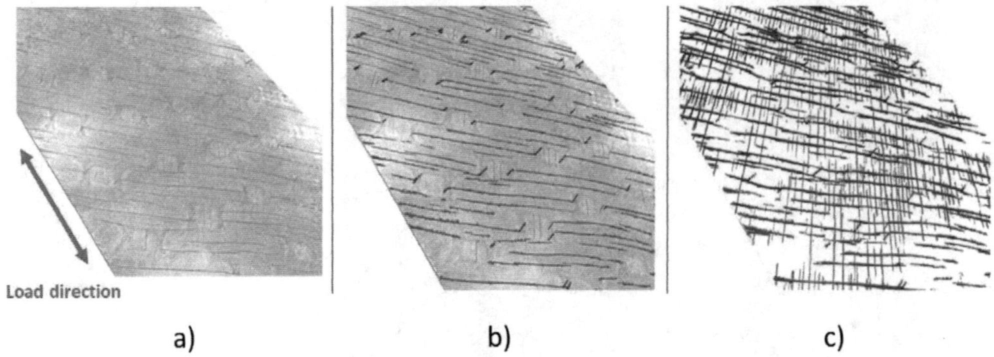

Figure 6. Image resulting from a μCT scan and segmentation a) non-segmented image b) matrix 0° cracks (in red), -45° intra-tow cracks along the -45° tows (in blue) c) +45° intra-tow cracks (in green) along the +45° tows.

The image processing procedure allows describing crack distribution ply-by-ply and for each tow direction, identifying a stacking sequence of semi-plies having approximatively the thickness of a tow, and in which the cracks have the tow orientation. The total crack length (L) can be used to evaluate the damage level in each semi-ply of the sample (Fig. 7). Figs. 7 illustrates the semi-ply-by-semi-ply crack length distribution as a result of the segmentation procedure for three samples tested at 250°C under different environments (2 bar N_2, air, 2 bar O_2) for different cycle numbers. The cracks are mainly narrowed on the external semi-plies (Ply1 -45° and Ply6 +45°), while for the inner plies the crack length distribution is quite homogeneous.

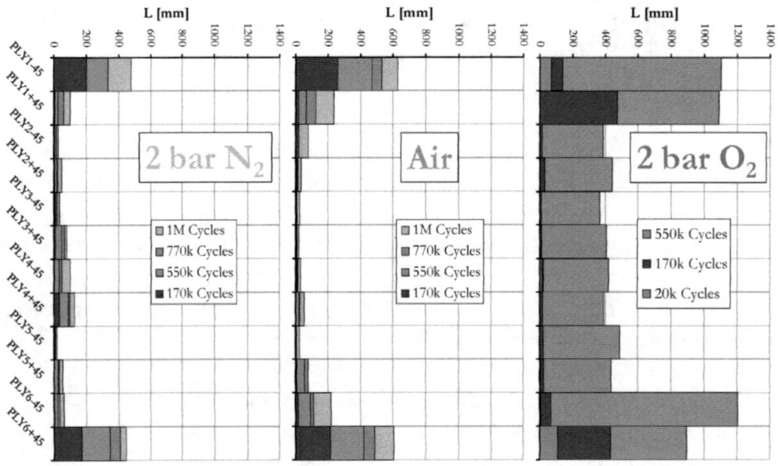

Figure 7. Distribution of crack length along the sample thickness as detected by μCT scans a) 2 bar N_2 b) Air c) 2 bar O_2 samples.

With respect to the inert environment (2 bar N_2), the effect of thermo-oxidation at high temperature (under air or 2 bar O_2 environment), promotes higher crack density. Crack length is higher in the eternal plies directly exposed to the environment.

2.3. DIC vs. μCT crack measurement

Fig. 8a illustrates a superposition of DIC detected cracks (blue) and μCT scan segmentation results (red) for a sample subject to 770k cycles under air environment. The two measures perfectly match for the most important crack lengths.

A quantitative comparison between the capability of the two techniques for crack detection is illustrated in Figure 24 for Ply1 -45° of a sample subject to 770k cycles under air environment. The graph has two vertical axes: on the left side the crack length values obtained by DIC are reported [L_{DIC}], while the values obtained by the image processing method are shown on the vertical axes on the right [$L_{\mu CT}$]. The scaling factor between the two vertical axes is 2.2 for all the printed point, this is obtained by multiplying the ratio between the analysed area between the crack length evaluation procedures (1.8) and the bias scaling factor previously fixed to 1.23. The two-point series in the graph show a similar trend from the beginning of the test to 1M of cycles.

The DIC crack detection carried out on the tested specimens, allows quickly evaluating the crack length on the external surface of the sample that is the most damaged region of the specimens. DIC crack length results are, thus, exploitable for an evaluation of the damage progression for the whole specimen without removing the specimens from the test rig.

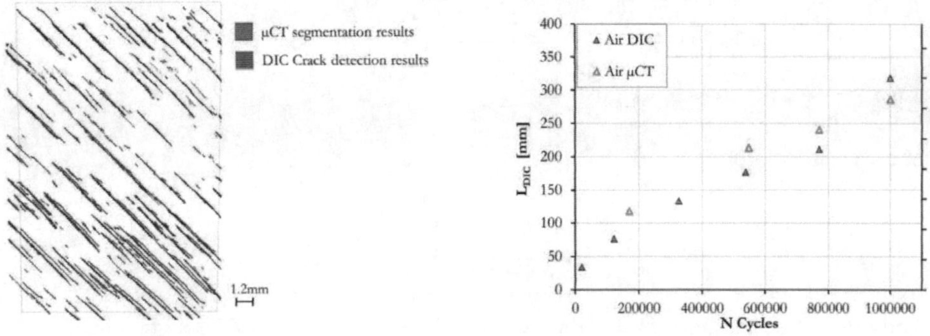

a) b)

Figure 8. Crack detection and measurement: DIC vs. μCT a) surface cracks b) evolution of surface crack length as a function of the cycle number.

2.4. Integration of the measured crack system within a laminated damage-dependent mechanical model

The sample architecture (Fig. 1) allows identifying a ply-by-ply discretization, characterized by elementary woven plies with satin texture (Fig. 1a) arranged in a laminated stacked sequence (Fig. 1b).

On the other hand, the volumetric crack distribution revealed by μCT scans and related segmentation allows identifying a semi-ply-by-semi-ply through-the-thickness crack distribution (Fig. 7), the thickness of a semi-ply being the thickness of a tow which is also the thickness (height) of a crack. Semi-plies are further identified by the orientation or their respective tows which is also the fibre direction and is taken as semi-ply direction 1.

Moreover, μCT scans (and DIC in the external plies) are able to identify an oriented pattern of microcracks, for each semi-ply (Fig. 6c) with density

$$D_{SP}^{\mu CT} = \frac{L_{SP}^{CRACK}}{L_{SP}} \quad (1)$$

where the subscript SP stands for Semi-Ply, where L_{SP}^{CRACK} is the sum of the crack length (along the tow direction) and L_{SP} is the length of the zone observed by μCT (around 42.3 mm). Cracks are aligned along the tow direction (semi-ply direction 1) and open along the direction transverse to the tow (semi-ply direction 2): it is therefore plausible to think that they have an effect essentially on the semi-ply transverse direction (as they directly affect the transverse compliance by crack opening) and the in-plane shear behavior (as cracks may slide with friction in their plane).

It is therefore plausible to construct a semi-ply-based plane stress damage-dependent constitutive law and considering the sample mechanical response as the response of a stacked sequence of damageable semi-plies, having in mind the tows orientation (semi-ply direction = direction 1), the directionality of cracks (along direction 1) and using the lamination theory for constructing the response of the laminate. The damage-dependent constitutive law of the semi-ply is therefore of the form

$$\boldsymbol{\sigma}_{SP} = \mathbf{Q}_{SP}(D)(\boldsymbol{\varepsilon} - \boldsymbol{\varepsilon}_c - \boldsymbol{\varepsilon}_{th}) \tag{2}$$

in which $\boldsymbol{\sigma}_{SP}$ is the semi-ply stress, $\boldsymbol{\varepsilon}$ the total strain, $\boldsymbol{\varepsilon}_{cf}$ and $\boldsymbol{\varepsilon}_{th}$ are, respectively, the inelastic strain related to creep or creep-fatigue interaction and strain of thermal origin (such as those occurring due to material manufacturing at high temperature), $\mathbf{Q}_{SP}(D)$ the semi-ply damage dependent plane stress stiffness matrix given by (in the 1, 2 – tow, transverse-to-tow reference frame)

$$\mathbf{Q}_{SP}(D) = \begin{pmatrix} \frac{E_{11}}{1-\nu_{12}\nu_{21}} & \frac{\nu_{21}E_{11}}{1-\nu_{12}\nu_{21}} & 0 \\ \frac{\nu_{21}E_{11}}{1-\nu_{12}\nu_{21}} & \frac{E_{22}(1-D_{22})}{1-\nu_{12}\nu_{21}} & 0 \\ 0 & 0 & G_{12}(1-D_{12}) \end{pmatrix} \tag{3}$$

in which, for each semi-ply

$$\begin{aligned} D_{22} &= \beta D_{SP}^{\mu CT} \\ D_{12} &= cD_{22} \end{aligned} \tag{4}$$

β and c are interaction parameters. By this way damage acts in a direction perpendicular to that of the tows (D_{22}) and in shear (D_{12}). Following an approach also found in Ref. [14] for satin textile laminates, the stiffness matrix of a ply is given by the algebraic weighted average of the semi-plies: for instance, for the ±45° ply one would write

$$\mathbf{Q}_P^{\pm 45} = n_{+45}\mathbf{Q}_{SP}^{+45} + n_{-45}\mathbf{Q}_{SP}^{-45} \tag{5}$$

To each semi-ply is associated an elastic-damage complementary energy density of the form

$$e_{SP}^* = \frac{1}{2}\left[\frac{\sigma_{11}^2}{E_{11}} + \frac{\sigma_{22}^2}{E_{22}(1-D_{22})} - \frac{2\nu_{12}}{E_{11}}\sigma_{11}\sigma_{22} + \frac{\sigma_{12}^2}{2G_{12}(1-D_{12})}\right] \tag{6}$$

from which it is possible to calculate by derivation the variables (thermodynamic forces) needed to model the kinetic laws of evolution (which can be improperly referred as energy release rates)

$$\begin{aligned} Y_{22} &= \frac{\partial e_{SP}^*}{\partial D_{22}} = \frac{\sigma_{22}^2}{2E_{22}(1-D_{22})^2} \\ Y_{12} &= \frac{\partial e_{SP}^*}{\partial D_{12}} = \frac{\sigma_{12}^2}{2G_{12}(1-D_{12})^2} \end{aligned} \tag{7}$$

3. Mechanical property evolution during high-temperature environmental fatigue

This section focuses on the characterization of material property evolution during high-temperature environmental fatigue and the capability of the damage-dependent modulus of following this evolution, leaving aside other aspects of the material behavior, such as for instance the evolution of hysteresis area or creep inelastic strain during fatigue.

Figure 9 illustrates the evolution of the longitudinal chord modulus, E_y^{ch}, during fatigue, with increasing the number of cycles. A progressive reduction of the moduli is measured for the three environmental conditions: with respect to the inert environment (2 bar N_2) the kinetic of property decrease is more marked under air and under 2 bar O_2, oxidizing conditions promote more rapid degradation related to damage evolution (Fig. 7). The laminated semi-ply-by-semi-ply damage-dependent model (dotted lines in Fig. 9) can follow this evolution for a unique value of the parameters β and c, that is, β = 0.035 and c = 1.

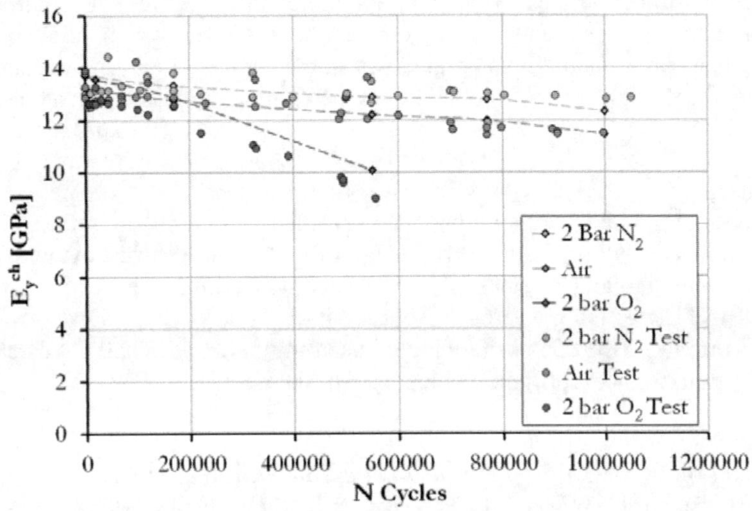

Figure 9. Evolution of the longitudinal chord modulus, E_y^{ch}, during fatigue, with increasing the number of cycles.

Once identified, the model can be employed for the calculation of the thermodynamic forces and to identify the kinetics of damage evolution. For the sake of illustration, Fig. 10 presents the evolution of the thermodynamics forces, $\mathbf{Y_{22}}$ and $\mathbf{Y_{12}}$, in the +45° ply, as a function of number of cycles. The thermodynamic forces increase with increasing number of cycles. Moreover, the highest values of $\mathbf{Y_{22}}$ and $\mathbf{Y_{12}}$ are calculated for the 2 bar O_2 environmental condition.

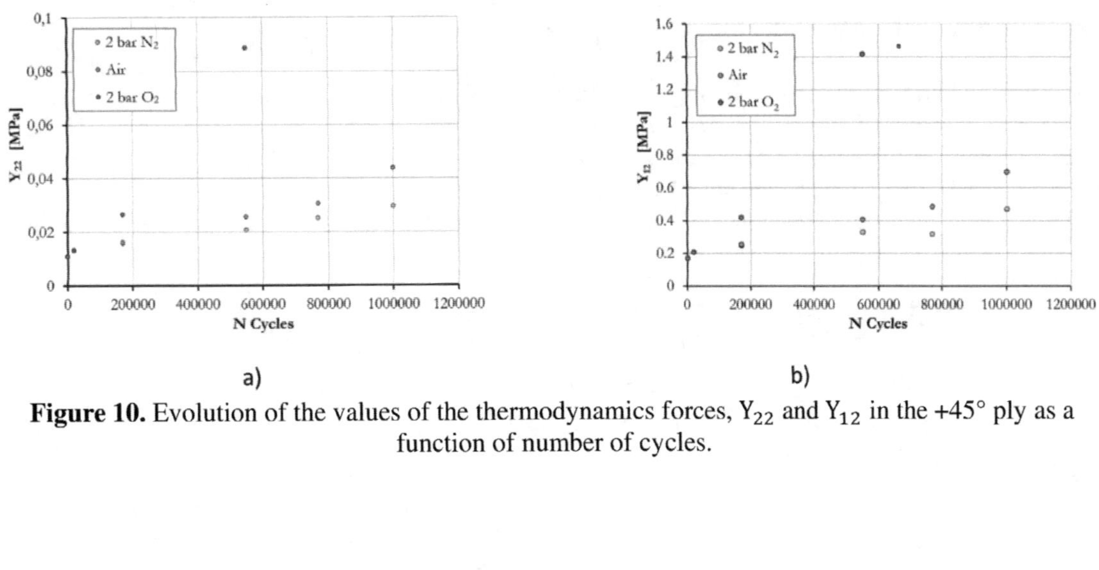

Figure 10. Evolution of the values of the thermodynamics forces, Y_{22} and Y_{12} in the +45° ply as a function of number of cycles.

4. Conclusions and perspectives

This work has been focused on the characterization and modelling of the high-temperature (250°C) fatigue behaviour of C/polyimide 8HS textile composites, the development of damage related to cyclic mechanical mechanisms (fatigue) under controlled (temperature and gas) environment.

An experimental setup is specifically developed in order to perform fatigue tests under controlled environment (2 bar N_2, air, 2 bar O_2). Sample properties evolution during fatigue tests at 250°C are monitored using in-situ Digital Image Correlation (DIC), and damage assessment is provided by ex-situ X-ray microtomography (μCT). It is shown that these two techniques are effective and complementary for damage assessment under fatigue at high temperature under aggressive environment: a rich characterisation of damage extent and orientation can be provided which allows damage being formalised in a way that can be naturally integrated within the context of damage-dependent behaviour models. The employment of such description allows performing full assessment of the effects of environmental fatigue.

Acknowledgements

This work was partially funded by the French Government program "Investissements d'Avenir" (EQUIPEX GAP, reference ANR-11-EQPX-0018). The research subject carried out within this research program falls within the research themes of the French Government program ''Investissements d'Avenir" (LABEX INTERACTIFS, reference ANR-11-LABX- 0017-01). The authors would like to thank SAFRAN for providing the composite material used in this work. Finally, we acknowledge the Poitou-Charentes Region (France) for funding the PhD Thesis of Federico Foti.

References

[1] Carvelli V and Lomov SV 2015 *Fatigue of Textile Composites*. Elsevier
[2] Pandita SD, Huysmans G, Wevers M and Verpoest I 2001 *Composites Part A: Applied Science and Manufacturing* **32** 1533
[3] Yu B, Blanc R, Soutis C and Withers PJ 2016 *Composites Part A: Appied Science and Manufacturing* **82** 279
[4] Lomov SV, Ivanov DS, Verpoest I, Zako M, Kurashiki T, Nakai H, Molimard J and Vautrin A 2008 *Composites Part A: Applied Science and Manufacturing* **39** 1218
[5] Nicoletto G, Anzelotti G and Riva E 2009 *Optics and Lasers in Engineering* **47** 352
[6] Mehdikhani M, Aravand M, Sabuncuoglu B, Callens MG, Lomov SV and Gorbatikh L 2016 *Composites Structures* **140** 192
[7] Pazmino J, Carvelli V and Lomov SV 2014 *Composites Part A: Applied Science and Manufacturing* **61** 76
[8] Barburski M, Straumit I, Zhang X, Wevers X and Lomov SV 2015 *Composites Part A: Applied Science and Manufacturing* **73** 45
[9] Sisodia SM, Garcea SC, George AR, Fullwood DT, Spearing SM and Gamstedt EK 2016 *Composites Science and Technology* **131** 12
[10] Montesano J, Selezneva M, Fawaz Z, Poon C and Behdinan K 2012 *Composites Part A: Applied Science and Manufacturing* **43** 1454
[11] Lafarie-Frenot MC, Rouquié S, Ho NQ and Bellenger V 2006 *Composites Part A: Applied Science and Manufacturing* **37** 662
[12] Dupré JC, Doumalin P, Husseini HA, Germaneau A and Brémand F 2015 *Strain* **51** 391
[13] Foti F 2017 PhD Thesis ISAE-ENSMA, Poitiers, France.
[14] Hochard C and Thollon Y 2010 *International Journal of Fatigue* **32** 1583

An evaluation of damage development for CFRTP by conventional tows and spread tows using acoustic emission

Takayuki Kinoshita[1]*, Tetsusei Kurashiki[1], Kazutaka Mukoyama[1], Koushu Hanaki[1], Masaru Zako[1], Xingsheng Li[1], Stepan V. Lomov[2]

[1]Graduate Student of Eng., Osaka University, Osaka, 565-0871, Japan
[2] Department of Metallurgy and Materials Engineering, KU Leuven, Kasteelpark Arenberg 44, B-3001 Leuven, Belgium

* kinoshita-t@mit.eng.osaka-u.ac.jp

Abstract. Comparison of damage development of textile composites made by combination of two resin (Polyamide and Epoxy) and two kinds of fiber bundles (conventional tows and spread tows) is carried out. The quasi-static tensile tests were carried out with using DIC (digital image correlation) system to estimate the full field strain and AE (acoustic emission) sensors for measurement of the acoustic emission features. The CFRTP specimens were also investigated with optical microscope to clarify the difference of damage. Fig.1 shows characteristic damage of each CFRTP. The fiber debonding and delamination were shown in CF(with spread tows)/PA composites and the damage in weft yarn covered a wide range of test specimens. It is suggested that these features of damage were influenced with effect of fiber opening. This may also be explained by AE events. In weft yarn of CF(with conventional tows)/PA composites, we observed not only transverse crack but also clack which propagated parallel to loading direction. This type of crack was not found in other thermoplastic carbon composites such as CF/PPS described in the literature. From these results, we conclude that the damage development of textile composites is sensitive to the form of fiber bundles and the properties of the matrix.

1. Introduction

Fiber reinforced plastic (FRP) is superior in specific strength, specific rigidity, and high flexibility in material design. Therefore, FRP is used in a wide range of fields, such as aircraft, aerospace and automobile where both light weighting and high strength are required. In particular, it is expected that carbon fiber reinforced thermoplastic (CFRTP) which contains matrix having high formability and high recyclability plays an important role in the composite material market from the environmental point of view. However, it is difficult to impregnate fiber bundles with thermoplastic because of its polymer structure. The focus of this paper is the comparison of damage development of textile composites with two different resin – Polyamide as thermoplastics and Epoxy as thermoset plastics, with two different fiber bundles – conventional tows and spread tows. Spread tows are very thin and wide tows. They are useful for preventing cracks caused by resin shortage in fiber bundles because it is easier to impregnate thin fiber bundles with resin.

In order to compare the damage development, acoustic emission (AE) registration and digital image correlation (DIC) system is used to obtain full field strain mapping during quasi-static tensile tests. After that, the damage pattern is investigated with an optical microscope.

IOP Conf. Series: Materials Science and Engineering **406** (2018) 012056 doi:10.1088/1757-899X/406/1/012056

2. Experimental setup

2.1. Test pieces

Specimens for tensile test were prepared referring to a part of the JIS K 7161: total length 250mm, gage length 150mm, width 30mm. With regard to PA6/CF_Spread specimen, the width is 50mm to take into account the periodicity of fiber bundles because the width of spread tows is approximately 20mm. The 50 × 30mm tabs for specimens with conventional tows and 50 × 50mm tabs for specimens with spread tows are attached with epoxy glue to the specimen ends to avoid failure during test in the grips of the testing machine. Measurements of the number of plies, thickness and fiber volume fraction (Vf) are listed in Table 1(a). The mechanical properties of Epoxy and PA6 are shown in Table 1(b).

Table 1. Material properties

	Epoxy/CF	PA6/CF	PA6/CF_Spread
(a)			
Number of plies	10	10	19
V_f (%)	50.9±	51.1±	50.9±
Thickness (mm)	1.95±0.05	2.08±0.04	1.79±0.03
	Epoxy	PA6	
(b)			
Density (g/cm^3)	1.40	1.13	
Tensile strength (MPa)	80.6	80.0	
Elastic modulus (GPa)	3.18	2.80	

2.2. **Test procedures**

Tensile tests were performed using tensile testing equipment (INSTRON 4505 with a load cell of 100kN), and the crosshead speed was 1mm/min. Five specimens were prepared for each test. In order to obtain full field strain mapping, images of all specimens were acquired using LIMESS system. The images post-processing allowed the measurement of the full field strain on the external surface of the specimen by the DIC. Two AE sensors are attached to a specimen with a distance between their centers 100mm. The amplitude of 100 dB is a saturation limit and the threshold of 40 dB is set for noise reduction.

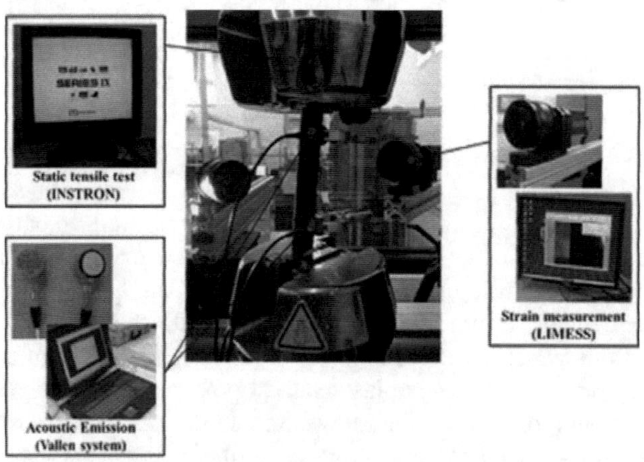

Figure 1. Test set-up

To investigate the damage patterns, samples for optical microscopy observation were prepared. The samples are cut to 20 × 15 mm from the center of the sample using diamond saw and then they are embedded in the plastic sample holder. Finally, they are polished by the automatic polishing machine.

Polishing was done in steps with different gauges of grinding paper, 3 μm diamond suspension and 1 μm diamond suspension.

3. Result of tensile test

3.1. Comparison of PA6/CF and Epoxy/CF

The average values, standard deviations and coefficient of variation (CV) of mechanical properties of the PA6/CF and Epoxy/CF composites in the warp direction were given in table 2. The Young's modulus was calculated by the slope of the stress-strain curves in the 0.1 to 0.3% strain region. Specimens were loaded up to failure.

Table 2. Tensile properties in the warp direction

	PA6/CF			Epoxy/CF		
	Average	Std. dev.	CV	Average	Std. dev.	CV
Ultimate tensile strain (%)	1.52	0.09	0.06	1.85	0.24	0.13
Tensile strength (MPa)	636	43.2	0.07	872	67.0	0.08
Young's modulus (GPa)	54.2	3.68	0.07	73.8	8.21	0.11

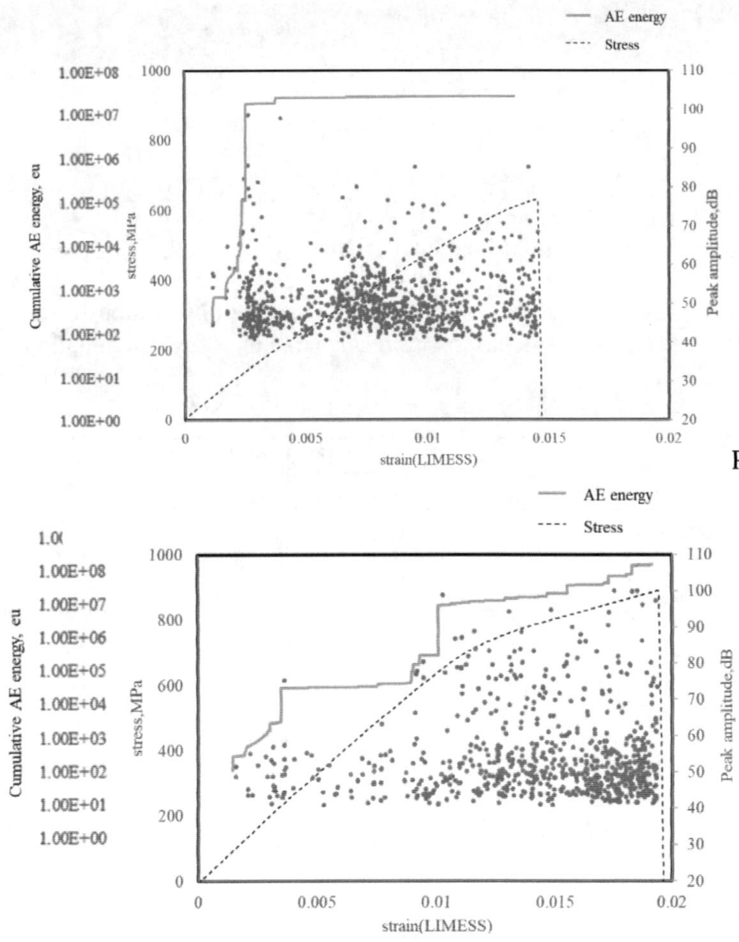

PA6/CF

Epoxy/CF

Figure 2. Representative cumulative AE energy, peak amplitude and stress-strain curve

Cumulative AE energy and peak amplitude are shown in Figure 2 for representative specimen, together with stress-strain curve. From the cumulative AE energy curve, damage initiation and development in PA6/CF composite is earlier than epoxy one. These initial damage may affect ultimate tensile strain and tensile strength. Figure 3 shows the characteristic damage patterns of each specimen. The initial damage in Epoxy/CF composite appears in the form of transverse cracks inside the yarns. It corresponds to the initial AE events with low amplitude. On the other hand, there are cracks parallel to loading direction inside weft yarns in PA6/CF composite. It could be consequence of the resin shortage in fiber bundles or the thermal residual stresses accumulated during manufacturing.

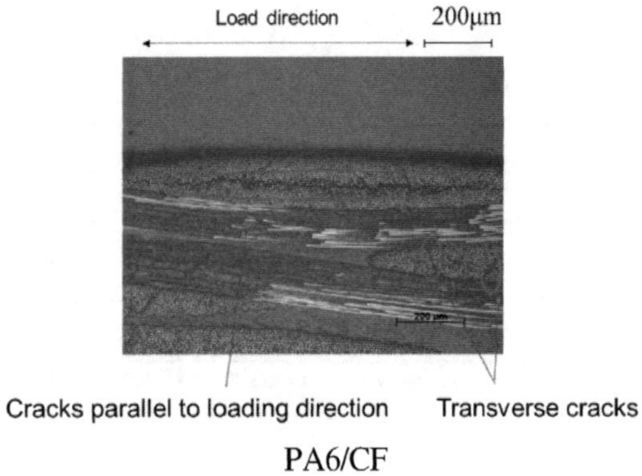

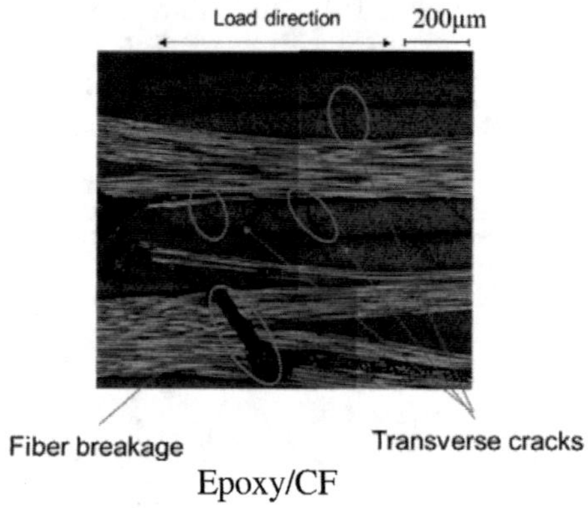

Figure 3. Characteristic damage pattern after tensile tests

3.2. Result of PA6/CF with spread tows

The average values, standard deviations and coefficient of variation (CV) of mechanical properties of the PA6/CF with spread tows in the warp direction are given in Table 3. It shows fiber opening brings remarkable increase in ultimate tensile strain. According to the observation result on specimen after failure, the specimen after failure, the reasons for this is large inter-ply delaminations.

Table 3. Tensile properties in the warp direction

	PA6/CF_Spread		
	Average	Std. dev.	CV
Ultimate tensile strain (%)	3.00	0.66	0.22
Tensile strength (MPa)	768	83.1	0.11
Young modulus (GPa)	51.8	6.16	0.12

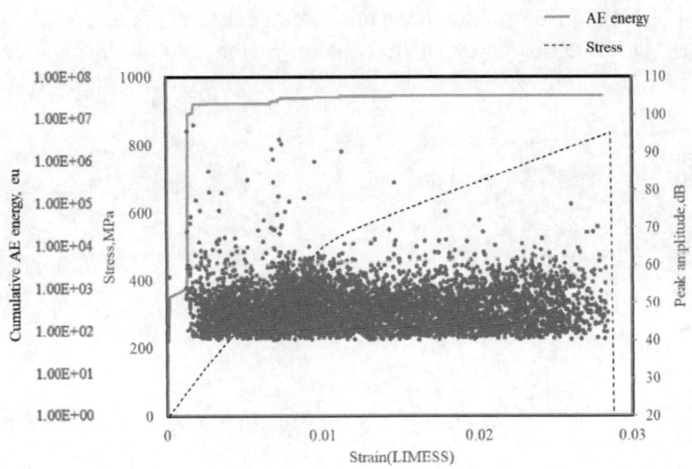

Figure 4. Representative cumulative AE energy, peak amplitude and stress-strain curve

Cumulative AE energy and peak amplitude are shown in Figure 4 for representative specimen with stress-strain curve. The shape of cumulative AE energy curve is similar to the curve of PA6/CF with conventional tows. On the other hand, the cumulative AE energy saturates earlier and the low amplitude enents are corrected continuously. Figure 5 shows the characteristic damage patterns of PA6/ CF with spread tows. Micro-debondings at the fiber-matrix interface and inter-ply delamination were appeared. However, there are not cracks parallel to loading direction inside weft yarns as observed in PA6/ CF with conventional tows specimen. This phenomenon is effected by the spread tows which are superior in impregnation performance.

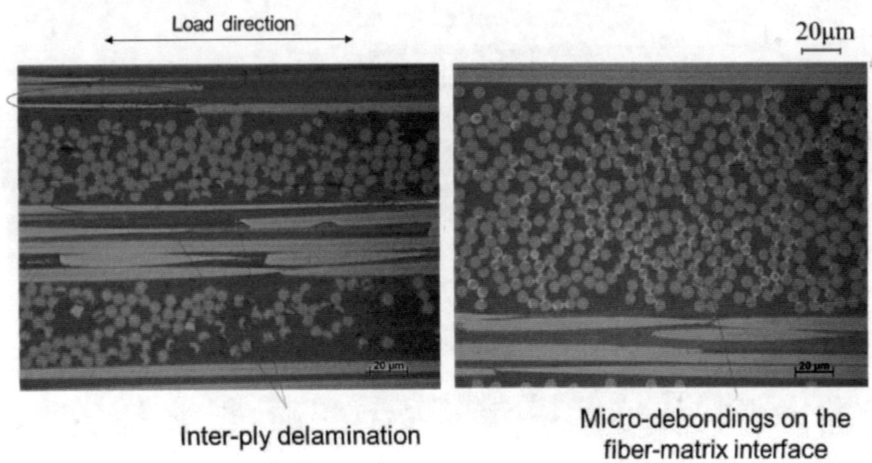

Figure 5. Characteristic damage pattern after tensile tests

4. Damage development analysis

As the damage observations were made on failure samples, FE analysis was carried out to observe damage development of PA6/CF and Epoxy/CF. The FE model is shown in Figure 6 and periodic boundary condition is applied to nodes on all plane of FE model. In this study, Hoffman law is used as failure criterion for elements. It is useful to take anisotropy damage mode into account. The mechanical

IOP Conf. Series: Materials Science and Engineering **406** (2018) 012056 doi:10.1088/1757-899X/406/1/012056

properties of fiber bundles are calculated from reinforcement fiber (TC-33) and each resin, using Chamis law and Rosen law. Damage development analysis imparting a strain of 0.1% step is carried out.

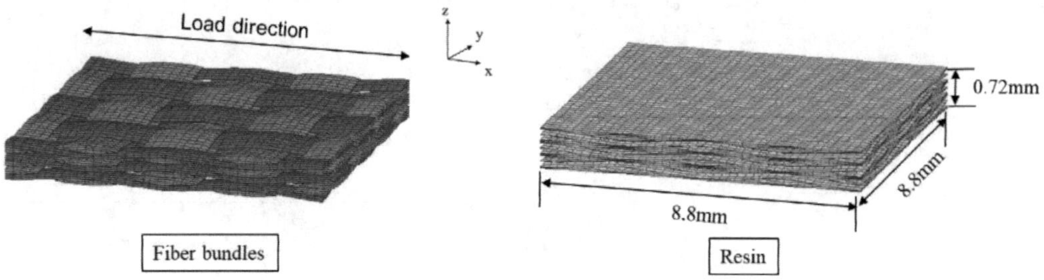

Figure 6. FE model of plain woven fabric

The damage development in fiber bundles of each sample is shown in Figure 7. In this case, damage mode T means transverse crack by normal stress component of transversal direction, mode ZL means delamination by shear stress component of in-plane direction and mode L means fiber breakage by normal stress component of longitudinal direction. Regarding to PA6/CF model, the area of initial damage is much larger than that of Epoxy/CF model. It is considered to be the reason why the difference of AE feature at low strain level appeared. Just before fracture, local delamination and fiber breakage appeared in Epoxy/CF model. As for PA6/CF model, a lot of delaminations in longitudinal yarns by shear stress were observed due to elongation of Polyamide.

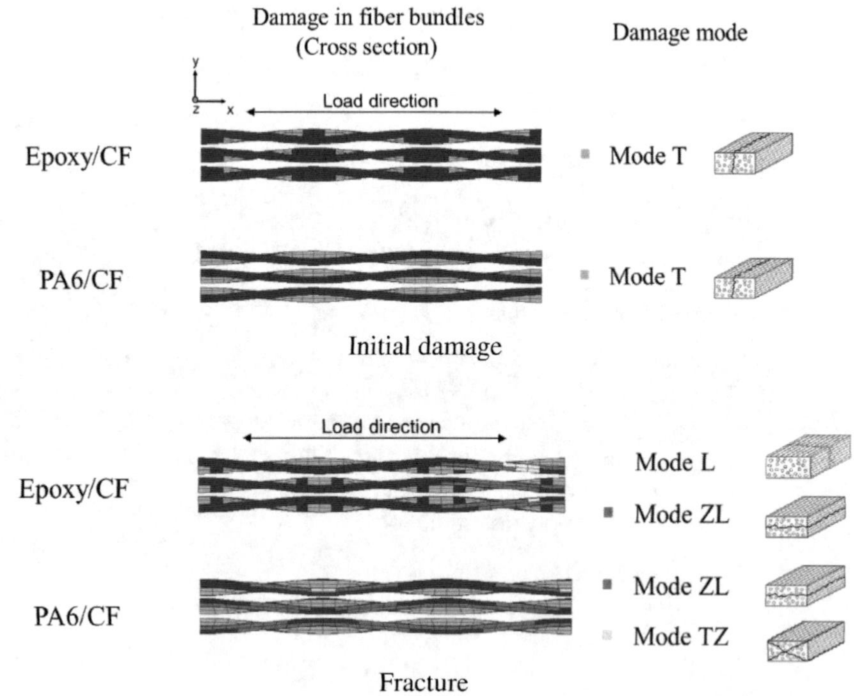

Figure 7. Damage state of each model

IOP Conf. Series: Materials Science and Engineering **406** (2018) 012056 doi:10.1088/1757-899X/406/1/012056

5. Conclusions

Table 4 shows a summary table of damage state for all specimens. The study concludes that the performance of textile composites were highly sensitive to the performance of the matrix such as impregnation characteristics. More specifically, a significant difference in damage development was found in PA6/CF, PA6/CF_Spread and Epoxy/CF composites as indicated by AE data and crack observations. Tensile properties in warp direction were found to be dependent on the matrix behaviour.

Table 4. Characteristics of damage state

	Epoxy/CF	PA6/CF	PA6/CF_Spread
Transverse cracks	○	○	△
Micro-debondings on the fiber-matrix interface	△	×	○
Cracks parallel to warp direction	×	○	×
Inter-ply delamination	○	×	○

Acknowledgments

The authors would like to express our thanks to Maruhachi Co. Ltd., Japan for the cooperation.

References

[1] Sergey G Ivanov, Dries Beyens, et al. Damage development in woven carbon fibre thermoplastic laminates with PPS and PEEK matrices: A comparative study. *Journal of Composite Materials 2017, Vol. 51(5) 637–647.* DOI: 10.1177/0021998316653460.

[2] Valter Carvelli. Acoustic emission and damage mode correlation in textile reinforced PPS composites. *Composite Structures 2017, Vol. 163 399–409.*

[3] M. Zako. Finite element analysis of damaged woven fabric composite materials. *Composite Science and Technology 2003, Vol. 63 507–516*

[4] Yuzo Fujita, Effect of Micro Structure on Mechanical Properties of Woven Fabric Composite, *Safety, Reliability and Risk of Structures, Infrastructures and Engineering Systems 2009, Vol.10 3477-3484*

Multi-instrument multi-scale experimental damage mechanics for fibre reinforced composites

Stepan V Lomov[1]*, Christian Breite[1], Delphine Carrella-Payan[2], Valter Carvelli[3], Nuri Ersoy[4], Marco Gigliotti[5], Raquel Antoranz Gonzalez[5], Sergey Ivanov[1], Malika Kersani[6], Marie-Christine Lafarie-Frenot[5], Mark Mavrogordato[7], Mahoor Mehdikhani[1,8], Arsen Melnikov[1], Francisco Mesquita[1], Fatih Oz[4], Yannick Pannier[5], Lincy Pyl[9], Erich Schöberl[7], Ian Sinclair[7], S Mark Spearing[7], Yentl Swolfs[1], Danny Van Hemelrijk[9], Man Zhu[1,8], Larissa Gorbatikh[1]

[1]Department of Materials Engineering, KU Leuven, Belgium
[2]Siemens PLM Software, Leuven, Belgium
[3]Department of ABC, Politecnico di Milano, Italy
[4]Department of Mechanical Engineering, Bogazici University, Istanbul, Turkey
[5]Institut Pprime, CNRS, ISAE-ENSMA, Université de Poitiers, France
[6]Faculty of Physics, University of Science & Technology Houari Boumediene, Algiers, Algeria
[7]Department of Engineering and Environment, University of Southampton, UK
[8]SIM M3 program, Zwijnaarde, Belgium
[9]Department Mechanics of Materials and Constructions, Vrije Universiteit Brussel, Belgium

*stepan.lomov@kuleuven.be

Abstract. Reliable investigation of damage in fibre reinforced composites requires concurrent *in-* and *ex-situ* application of multiple instruments at different scale: digital image correlation, acoustic emission registration, optical/electron microscopy, C-scan, X-ray imaging and micro-computed tomography. The multi-instrument experimental mechanics allows detailed damage monitoring and inspection.

1. Introduction

The field of experimental mechanics of fibre reinforced composites is confronted with a challenge of damage observation, quantitative characterisation and interpretation not only *post-mortem*, but also during the test. Even at early stages of the specimen loading, with the average specimen deformation of a few thousand micro-strain, the composite starts experiencing cracking at the micro- and meso-scales. This involves multiple interacting damage modes at different scales, such as intra-ply and intra-yarn matrix cracking, fibre-matrix debonding, shear band formation and cracking in the matrix pockets, local and large scale delaminations and stochastic strength-controlled fibre breaking, all leading to specimen failure.

The paper presents several case studies where a multi-instrument methodology is applied to characterisation of the damage initiation and development at different scales (including matrix and fibre cracking dynamics) in composites with various thermoset and thermoplastic matrices, reinforced with glass, carbon and flax fibres. More specifically, the following cases are presented:

— Formation of fibre break clusters observed under synchrotron radiation computed tomography (CT) [1][13];
— Micro-damage on fibre-matrix interface, observed using micro-digital image correlation (DIC) under SEM [2,3];
— Damage thresholding and identification of successive damage phases with concurrent use of acoustic emission, *in-situ* DIC and microscopy, post-mortem microscopy and X-ray inspection, in quasi-static [4-7] tensile tests, including materials with manufacturing defects [8] and fatigue loading [9];
— Delamination onset and propagation in fatigue, identified with DIC [10];
— Thermal cycling and cure-induced cracking via X-ray micro-CT [11].

The paper discusses realisations of the multi-instrument multi-scale experimental mechanics and potential future developments.

2. Formation of fibre break clusters observed under synchrotron radiation CT [1]

Synchrotron radiation CT resolves limitations inherent to standard X-ray micro-CT equipment: long acquisition times and limited space for in-situ testing rigs due to the conical geometry of the beam. Here the beam can be monochromatic and have high brilliance, which allows acquisition times for a complete CT volume of less than a second [12]. The parallel nature of the beam allows to increase the distance between specimen and scintillator, which means more space for loading rigs. Synchrotron radiation computed tomography is therefore a powerful tool for performing in situ tests. A particular case where this can be exploited is in the detection of fibre break development during testing [1, 13].

In [1], a composite laminate was strained and then held at displacement to scan the specimen. Several such steps were done at different load/strain levels to monitor how fibre breaks develop and cluster together. This study revealed new insights into the types of clusters that develop appear (see Figure 1). About 70% of the clusters were found to be co-planar (see Figure 1b) compared to 20-30% in model predictions. This discrepancy between experiments and models has triggered a new body of research trying to identify its potential causes.

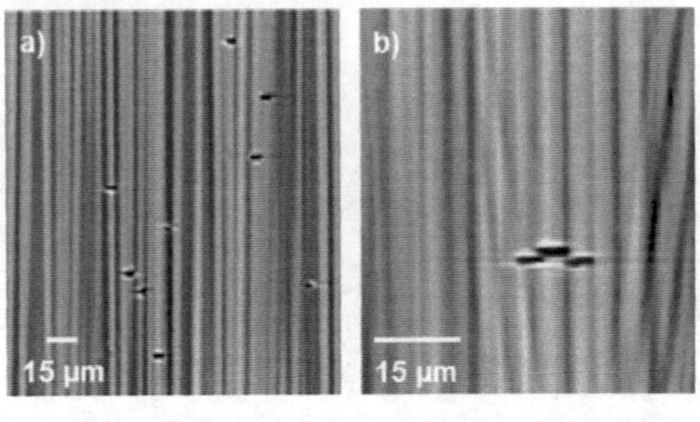

Figure 1. Two types of fibre break cluster found in the synchrotron radiation CT: (a) a diffuse cluster and (b) a co-planar cluster (reprinted from Elsevier with permission).

Similar synchrotron CT measurements were also used in a benchmarking exercise to compare three state-of-the-art models for longitudinal strength of unidirectional composites [13]. Having such data available allowed the models not just to compare fibre break and cluster predictions against each other, but also against experiments.

A second version of this exercise is currently running within FiBreMoD - Marie Skłodowksa-Curie European Training Network. A key improvement targeted within this project and exercise is to avoid the hold at displacement scenarios during scanning. Recent advances at the TOMCAT beamline at the Swiss Light Source synchrotron make it possible to continuously acquire images at reasonably high speeds, therefore enabling realistic strain rates and avoiding the need to hold the sample at load [14]. This allows more scans to be performed on the same specimen, and to scan closer to final failure where large amounts of fibre breaks and clusters can be observed. This methodology will improve the accuracy and reliability of the experimental results and therefore help to advance fibre break models.

Opportunities for more detailed and routine quantitative analysis of CT time series can be identified in the future, exploiting digital volume correlation methods [15], suitably adapted to address the anisotropic and relatively regular microstructure that fibre reinforced composites may present, particularly at spatial resolutions necessary for fibre-by-fibre damage analysis.

3. Micro-damage on fibre-matrix interface, observed using micro-DIC under SEM [2]

The potential of micro-scale DIC to analyse deformation in fibre reinforced composites on the micro scale, including the influence of added nano-reinforcement (carbon nanotubes, CNT) was evaluated in [2, 3].

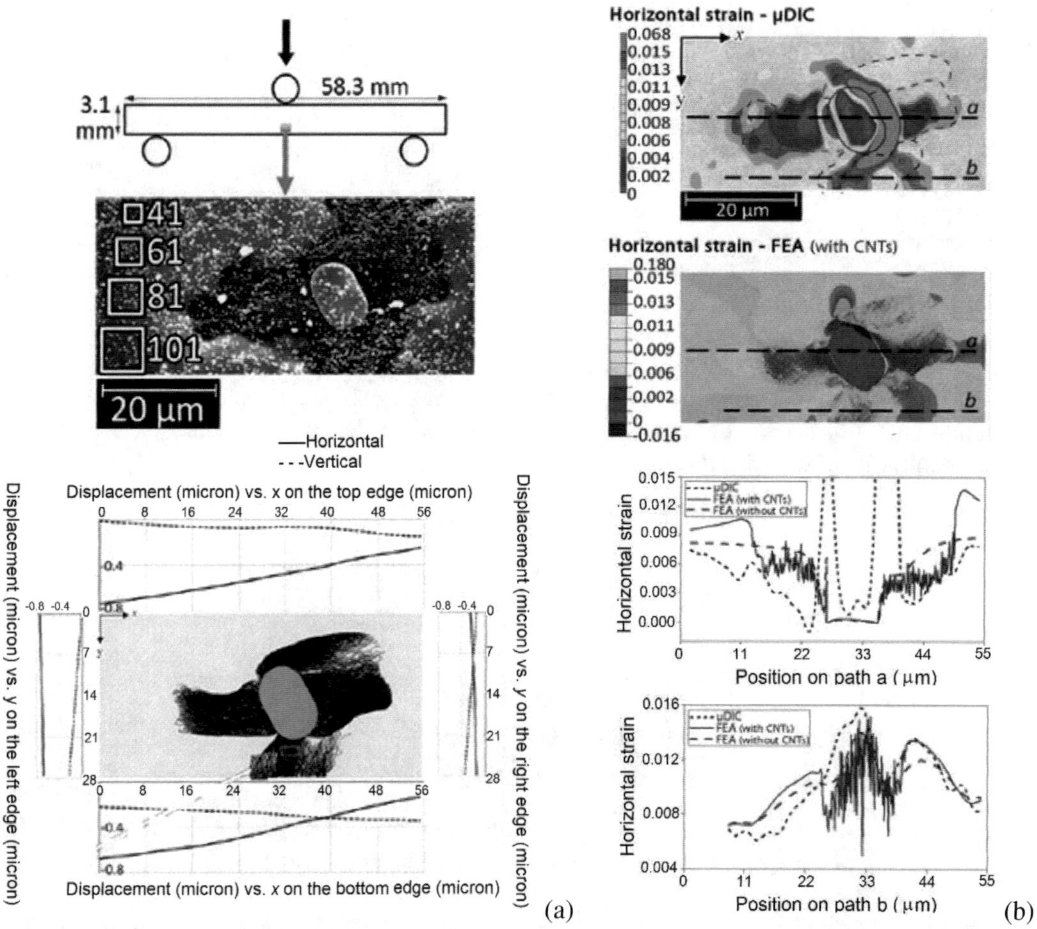

Figure 2. Micro-DIC discovering fibre debonding: (a) schematics on the specimen in three-point bending, region of interest (ROI) and FE model simulating CNTs grown on the surface of the fibre, with the DIC-measured displacements profiles on each edge of the ROI; "horizontal" and "vertical" correspond to the image orientation; (b) DIC-measured and calculated strain fields and profiles [2]

An important pre-requisite for successful DIC measurement is a high-quality random speckle pattern (produced by deposition of alumina nanoparticles), combined with optimized DIC parameters and proper microscopy settings.

Figure 2 illustrates results of the measurements. Features such as fibre, matrix, aligned CNT forest, and debonded regions at the fibre/matrix interface were detected. Openings as small as 35 nm in the debonded regions near the fibres were measured. The image correlation uncertainty in the displacement analysis was found to be below 5 nm, as estimated from the standard deviation confidence of the match. The micro-DIC results were compared with predictions of a two-scale numerical model, and a good agreement between the two was observed.

Micro-DIC measurements revealed propagation of the debond around and/or along the fibre, which manifested itself in a non-linear trend in the opening of the debonded region as function of the applied deformation.

The authors [2] concluded that the micro-DIC was found to be a promising instrument for evaluation of the effects of nano-scale reinforcements on the deformation of hierarchical materials. Further progress in mDIC is needed, particularly in the improvement of correlation methods to distinguish different phases, on which subset averaging is performed. This will allow even higher precisions in capturing deformation gradients at interfaces of materials with high mismatches in properties.

4. Concurrent use of acoustic emission, *in-situ* DIC and microscopy, post-mortem microscopy and X-ray inspection: quasi-static loading

The sequence of damage events in cross-ply and textile laminates [16] (Figure 3) suggests a presence of two thresholds of the applied load. The first, designated as ε_1, corresponds to the onset of the transverse cracking (*t*-cracks), which may not at this stage span the whole width of the specimen, being limited by the yarn crimp and/or presence of stitching in the textile reinforcement structure. The second, designated as ε_2, corresponds to, on one hand, the onset of local delaminations (*l*-cracks) and, on the other hand, to the formation of "strong" transverse cracks, which span the width of the specimen.

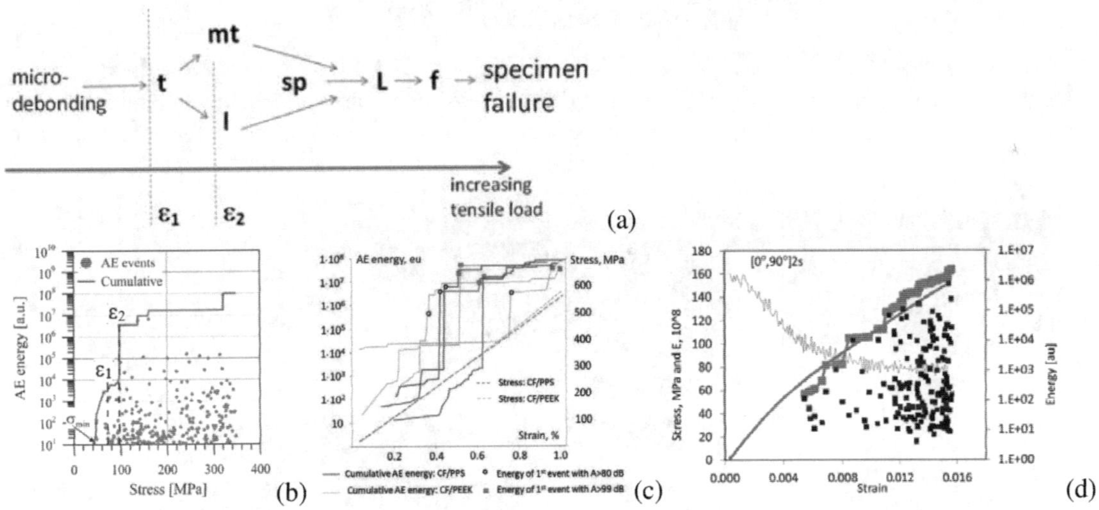

Figure 3. Damage thresholds in cross-ply and textile laminates identified with AE: (a) a sequence of damage modes (*t*: transverse cracks in yarns/plies and the matrix (*mt*); *l*: micro-delaminations; *sp*: longitudinal splitting; *L*: macro-delaminations; *f*: fibre breakage) [16]; (b) typical AE identification of the damage thresholds in a woven glass/epoxy laminate [9]; (c) damage thresholds in carbon/PPS and carbon/PEEK woven laminates [5]; (d) stress-strain, stiffness-strain and AE energy diagrams for tensile loading of flax quasi-UD woven laminates [6]

The damage thresholds can be identified using acoustic emission (AE) registration during the tensile loading, namely, a curve of cumulative energy (E) of the AE events. The damage threshold ε_1 is related to the first increase of the $E(\varepsilon)$ curve, and the damage threshold ε_2 – to the second sharp increase of this curve. This approach was used in 2005-2018 for various textile composites; Figure 3 gives examples for such measurements for composites with carbon, glass and flax fibres with thermoset and thermoplastic matrices. The AE registration allows quantification of the influence of the reinforcement architecture, matrix ductility or fibre properties on the damage initiation behaviour of the composite.

However, use of the cumulative AE energy remains a heuristic approach for the identification of damage thresholds. To be convincing, it is better supported by direct evidence of the damage. In [3] microscope observation of the specimen edge was combined with AE for carbon/PPS woven laminates, in [17] – for carbon/epoxy woven laminates. The crack development observed in a small central portion of the specimen edge was connected to the AE events located in the same zone of the specimen. The number of new intra-yarn matrix cracks generated at a certain load level was compared to the number of acoustic event localized in the zone of observation, with a good agreement found between the number of AE events with low amplitude and low frequency and the number of observed cracks (Figure 4a-c).

The observation of cracks on the specimen edge can be enhanced with the use of DIC. In [8], the cracks were monitored via identification of peaks in the strain profile on the specimen surface in each deformation step, which is carried out semi-automatically. The method was applied to study the influence of voids on the cracking process (Figure 5).

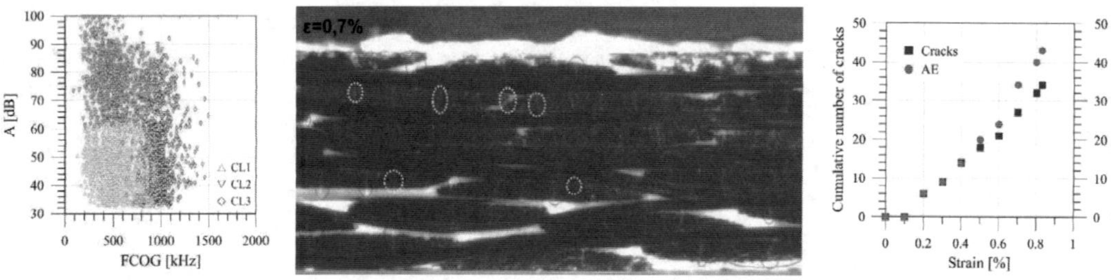

(a) (b) (c)

Figure 4. Concurrent use of AE and edge microscope, tension of carbon/PPS woven laminates: (a) clusters of AE events; (b) cracks observed at the specimen edge; (c) correlation between the number of cracks in an observation zone of 5 mm length and the number of AE events [4].

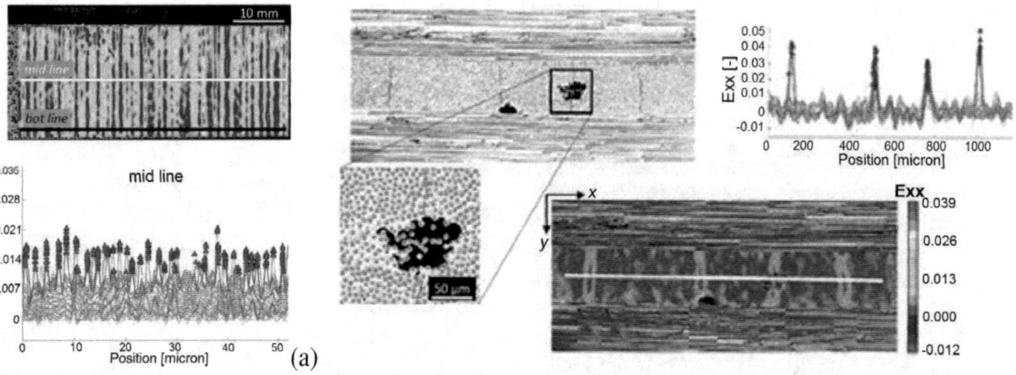

(a) (b)

Figure 5. DIC-enhancement of the edge crack registration: (a) cracks identified as peaks on the strain profile; (b) cracks related to voids in 90° plies [8].

With the laminate structure becoming more complex, the link between the AE events signature and the damage type becomes more complex. The AE registration technique alone during tension tests is not sufficient for damage mode identification in CFRP composites. Even though clustering algorithms can recognise well-separated and dense clusters, in [6] it is found that those clusters cannot be unambiguously used to identify the type of the damage in quasi-isotropic (QI) laminates. This study presents the necessity of multi-instrument optical observations for damage-mode identification and correlation with AE registrations: concurrent use of AE registration, optical edge observations and surface strain mapping by DIC (Figure 6).

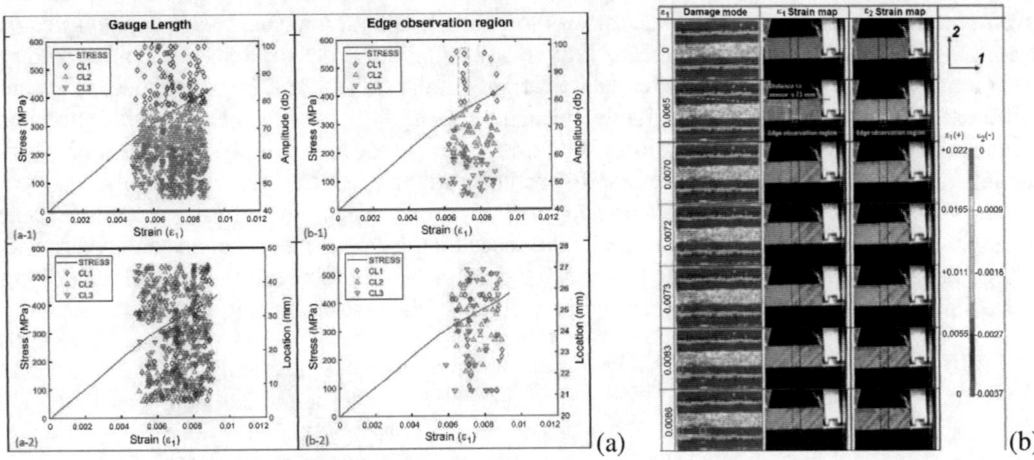

Figure 6. Concurrent use of AE (a), edge microscopy and surface DIC (b) for damage identification during tension of carbon/epoxy QI laminates [7].

5. Delamination onset and propagation in fatigue, identified with DIC [10]

Optical methods are effectively used for following tracking damage development in fatigue loading. In [9] a DIC-based method is proposed to characterize Mode I fatigue delamination onset and propagation in laminated composites. Images of the fatigue delamination on one side of the specimen were automatically recorded at a maximum displacement at specific fatigue cycles. The displacement field around a delamination crack is obtained and further processed to determine the position of the crack tip (Figure 7a). With this method the delamination length can be measured automatically in each cycle with a precision on the order of few hundreds of micrometers. The fatigue delamination onset life is then determined by detecting the increase of the delamination length, and the fatigue delamination propagation rate is calculated. The proposed method produces more conservative fatigue life measurements in comparison with the 5% compliance increase method in ASTM D6115 (Figure 7b).

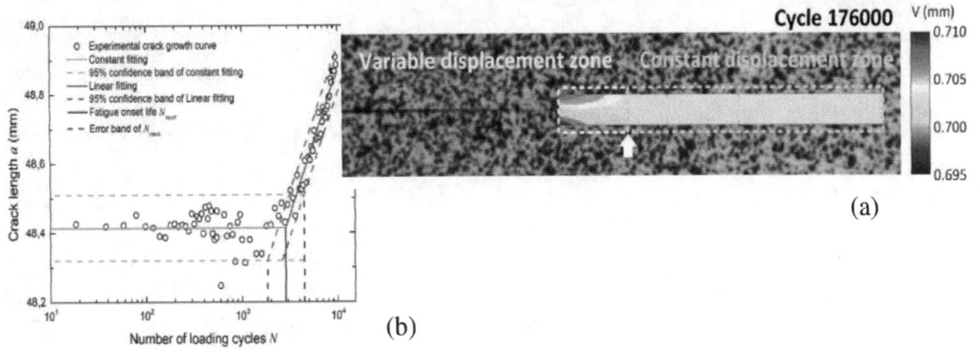

Figure 7. DIC-based method for tracking a fatigue Mode I crack: (a) detection of the crack tip position (white arrow); (b) identification of fatigue onset life N_{onset} [10].

6. Thermal cycling and cure-induced cracking via X-ray micro-CT [11]

The thermal cycling behaviour of fibre reinforced composite can be effectively studied using X-ray micro-CT. The key difficulty is an effective automated recognition of the cracks.

In [11] a carbon fibre/epoxy matrix composite material reinforced with a non-crimp 3D orthogonal woven preform was investigated. The aim was to characterize the damage mechanisms – i.e. matrix cracking – induced by thermal cycling. Qualitative and quantitative descriptions of the morphology and the evolution of cracks with thermal cycling are carried out through the analysis of μCT scans of samples at different cycle numbers. Through suitable image processing, which is of general and wide application, qualitative and quantitative descriptions of the initial porosity, of morphology and evolution of matrix cracking up to 1400 thermal cycles were performed (Figure 8a).

In addition, since the specimens have a certain level of porosity due to the infusion process, a complete description of this defect is carried out, and its influence on the damage mechanisms induced by thermal cycling is analysed. At first sight, the presence of pores seems to facilitate the propagation of cracks, pores acting as "attractors" when they are in the vicinity of the crack path (Figure 8b).

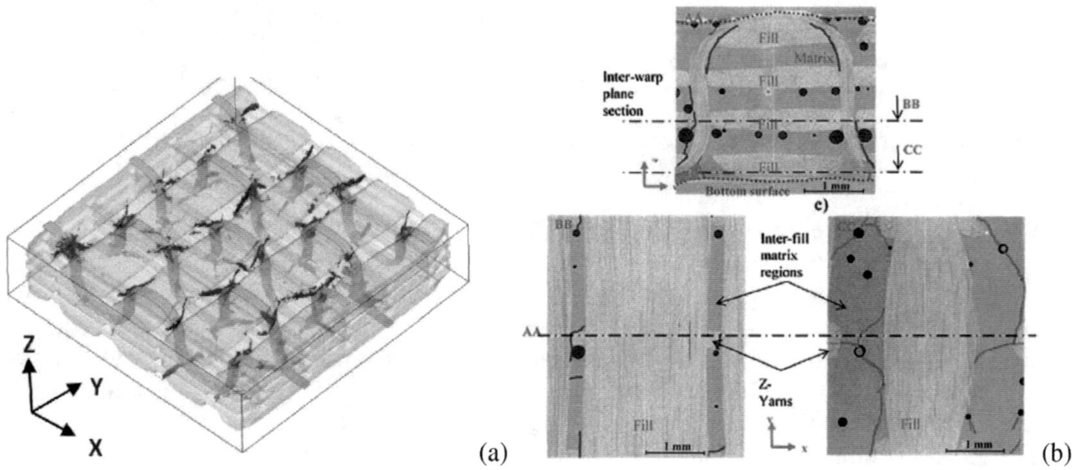

Figure 8. Thermal fatigue of a 3D woven carbon/epoxy composites in the presence of voids: (a) cracks identified using micro-CT; (b) details of the crack morphology [11]

7. Conclusion

Comprehensive methods of the multi-instrument experimental damage mechanics in fibre reinforced composites involve concurrent application of multiple in- and ex-situ instruments at the scale of the specimen and at the micro-scale: digital image correlation, acoustic emission registration, optical/electron microscopy, C-scan and X-ray imaging and micro-computed tomography.

Acknowledgements

The work reported here summarises results of several projects which were supported by various funding sources: FiBreMoD project, EU's Horizon 2020 research and innovation program under the Marie Skłodowska-Curie grant agreement No 722626; Agency for Innovation by Science and Technology in Flanders (IWT), grant funding for Y. Swolfs; Skolkovo Institute of Science and Technology through the No. 335-MRA project linked to the Center for Design, Manufacturing and Materials; Erasmus program of EU and a scholarship of Politecnico di Milano; FP7 NMP program under the M-RECT project (grant no. 246067); Ministry of Higher Education and Research of Algeria (grant number FWOG.035409); Boğaziçi University Research Fund, Istanbul Development Agency (ISTKA), and TUBITAK BIDEB 2214-A under project codes 10020/15A06D3, ISTKA/BIL/2012/58 and 1059B141600673; SIM (Strategic Initiative Materials in Flanders) and VLAIO (Flemish government agency for Innovation and Entrepreneurship), IBO and SBO projects M3Strength, which are part of the research program

MacroModelMat (M3), coordinated by Siemens (Siemens PLM Software, Belgium); French Government program, “Investissements d’Avenir” (LABEX INTERACTIFS, reference ANR-11-LABX-0017-01; EQUIPEX GAP, reference ANR-11-EQPX-0018); S.V. Lomov holds the Toray Chair for composite materials at KU Leuven – the support of all these organisations and agencies is gratefully acknowledged.

References

[1] Swolfs Y, Morton H, Scott AE, Gorbatikh L, Reed PAS, Sinclair I, Spearing SM and Verpoest I 2015 *Composites Part A: Applied Science and Manufacturing*. **77** 106.

[2] Mehdikhani M, Matveeva A, Aravand MA, Wardle BL, Lomov SV and Gorbatikh L 2016 *Composites Science and Technology*. **137** 24.

[3] Mehdikhani, M, Aravand A, Sabuncuoglu B, Callens M.G, Lomov SV and Gorbatikh L 2016, *Composite Structures* **140** 192

[4] Carvelli V, d'Ettore A and Lomov SV 2017 *Composite Structures*. **163** 399.

[5] Ivanov SG, Beyens D, Gorbatikh L and Lomov SV 2017 *Journal of Composite Materials*. **51** 637.

[6] Kersani M, Lomov SV, Van Vuure AW, Bouabdallah A and Verpoest I 2015 *Journal of Composite Materials*. **49** 403.

[7] Oz F, Ersoy N, Mehdikhani M and Lomov SV 2018 *Composite Structures*, **196** 163

[8] Mehdikhani M, Steensels E, Standaert A, Vallons KAM, Gorbatikh L and Lomov SV *Composites Part B*, under review.

[9] Carvelli V, Jain A and S.V. Lomov 2017, *Fatigue of Textile and Short Fiber Reinforced Composites*. Wiley – ISTE, 212 pp

[10] Zhu M, Gorbatikh L, Fonteyn S, Pyl L, Van Hemelrijck D, Carrella-Payan D and Lomov SV 2018 *Proceedings*, **2**(8) 230

[11] Gigliotti M, Pannier Y, Antoranz Gonzalez R, Lafarie-Frenot MC and Lomov SV 2018, *Composites Part A*, **112** 100.

[12] Maire E and Withers PJ 2013 *International Materials Reviews*. **59** 1.

[13] Bunsell A, Gorbatikh L, Morton H, Pimenta S, Sinclair I, Spearing M, Swolfs Y and Thionnet A 2018 Composites Part A: Applied Science and Manufacturing. **111** 138

[14] Mokso R, Marone F, Irvine S, Nyvlt M, Schwyn D, Mader K, Taylor GK, Krapp HG, Skeren M, and Stampanoni 2013 *Journal of Physics D: Applied Physics*. **46** 1.

[15] Borstnar G, Gillard F, Mavrogordato MN, Sinclair I, Spearing SM 2016 *Acta Materialia* **103**, 63

[16] Gorbatikh L and Lomov SV 2016, in *Modeling Damage, Fatigue and Failure of Composite Materials*, R. Talreja and J. Varna, Editors, Elsevier (Woodhead Publishers): Cambridge. 41.

[17] Ivanov SG, Gorbatikh L, Lomov SV, Verpoest I 2013 *Proceedings of the 11th International Conference on Textile Composites (TexComp-11)*, Leuven

IOP Conf. Series: Materials Science and Engineering **406** (2018) 012058 doi:10.1088/1757-899X/406/1/012058

In-situ structural health monitoring of glass fiber reinforced composites by tufted reinforcement

A T Martins*, Z Aboura, A Laksimi, K Khellil

Sorbonne Universités, Université de Technologie de Compiègne, Laboratoire de Mécanique Roberval UMR 7337, Dept GM, CS 60319, 60203 Compiègne Cedex, France

* alan-tulio.martins@utc.fr

Abstract. Through-the-thickness reinforcement (TTR) techniques have been increasingly applied to fiber-reinforced polymer composite laminates to reduce the issues caused by the weak out-of-plane properties which can reduce significantly their load-bearing capacity. The present work investigates the electrical response of carbon tuft threads as reinforcements of GFRP omega stiffened panel. Besides the mechanical contribution of the tuft reinforcements to the composites structures, this study focuses on the evaluation of the electrical response on the tufted yarns as a means of monitor the structural health. The first studies consisted of evaluating GFRP composite laminate plates reinforced with carbon tufts by means of the electrical response through tufts to damage generated by multiple low-velocity impacts. Once the tufts exhibited their ability to respond continuously while increasing damage, pull-off mechanical tests were performed in the omega stiffeners assisted by electrical resistance measurements as well as acoustic emission (AE) and Digital Image Correlation (DIC). The results showed that the measures of electric resistance on the yarns are capable to distinguish important events during the mechanical tests, as the delamination and tuft failure. Furthermore, they are in common agreement with the cumulative energy on AE and strain field response on DIC.

1. Introduction

Conventional composite laminates exhibit moderate fracture toughness especially due to reduced mechanical properties of the polymer matrix and its interaction with fiber reinforcements. Therefore, they are susceptible to interlaminar damages especially by delamination that is one of the major damages in laminate composites and can reduce considerably the load-bearing of a structure. It can be induced when submitted to out-of-plane loading (static or impact loading) and typically combined with stress concentrations (related with structure geometry) or discontinuities such as manufacturing defects, ply drops or free edges [1].

Improving the resistance to crack initiation and propagation throughout fabric plies can improve considerably the interlaminar fracture toughness. Several methods have been developed to counteract the susceptibility of the composite laminates to delamination and can be classified into matrix toughening or 3D fiber architecture. Interesting results can be achieved by performing through-thickness reinforcements (TTR). Among the several methods to reinforce transversally the laminate composites, the main methods are the 3D weaving, stitching, tufting, z-anchoring, and z-pinning.

Tufting method is a one-sided stitching process which consists of inserts needle with the reinforcement yarn is pushed into the fabrics preform and afterward, tufts are held mainly to the friction imposed to the yarn by the fabrics. This method is known by its low-tension under reinforcement which results in a reduction of the penalty effect in the in-plane properties of the stitching process [2]. In general, the tufts firstly contain the crack tip during the structure loading. Subsequently, the fissures are branched by the tufted yarns and finally, they fail as well as the composites. Tufting reinforcement has been shown to enhance the fracture toughness under both modes I [2]–[4] and II [5] loading conditions. Significant improvements to impact resistance and compression after impact strength have also been reported [6]–[8].

The present work reports the manufacture and mechanical characterization of omega stringers made of GFRP and reinforced through-thickness by tufting. The main goal was to monitor the structural health during pull-off tests by means of the electric resistance measurement from the tufted carbon threads. The tests were assisted by Digital Image Correlation (DIC) and Acoustic Emission (AE) in order to validate the electric resistance response. An investigation before to the pull-off tests in omega stiffened panels was performed with GFRP laminate plates so as to evaluate the electric resistance behavior in the carbon tufts when damage is generated intentionally by low-velocity impact tests. Besides the main interest to reinforce transversally a laminate composite with tufts such as increasing the interlaminar fracture toughness, this approach showed capable to acquire in-situ resistance measurements and detect the major damage events and, may avoiding the insertion of sensors into the laminate which can diminish mechanical properties.

2. Materials and methods

2.1. Materials

2.1.1. Laminated composite plates

The plates were manufactured with a stacking sequence of $[0]_6$ in glass fibers 2x2 twill weave fabrics and areal density of 280 g/m^2. The dry preform was stitched following the warp direction of the fabrics with carbon/PBO threads (2K Tenax-J HTA 40 carbon thread wrapped by two PBO yarns). The tufting loops were adjusted to be as smaller as possible avoiding their contact that could change the electrical measurements by means of percolation. Similarly, the tufting density was 10x10mm arranged in a square pattern to avoid this concern. The composite plates were molded by VARTM process with EPOLAM 5015 epoxy resin system. During infusion process, the vacuum pressure was about -1000 mBar at room temperature. The cure cycle was carried out at room temperature for 24 hours and post-cured at 80ºC for 16 hours. Specimens with 100 mm × 150 mm dimensions, with the warp direction aligned with longitudinal side, were prepared for the impact tests.

2.1.2 GFRP omega stiffened panel

Non-crimp fabrics (NCF) made with unidirectional layers of glass fiber tows (0/90) and areal density of 600 g/m^2 were employed to manufacture the omega stringers. The structure layup consisted of $[0]_{10}$ to the stiffener and $[0]_{12}$ to the skin part. The dry preforms were reinforced by tufting with carbon/PBO threads following the weft direction and stitch pitch of 7 mm. VARTM process was also employed with EPOLAM 5015 resin system by maintaining the same parameters of cure cycle as described for the plates. The tufts position and dimensions of the molded specimens are exhibited in Figure 1, with specimens width of 30 mm prepared to pull-off tests.

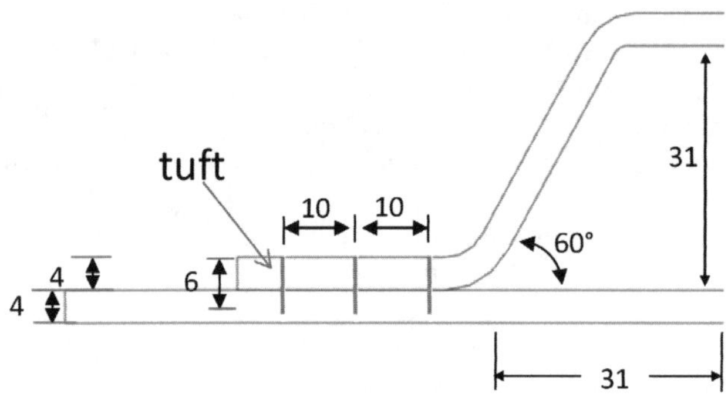

Figure 1. Schematic illustration of the final dimensions of the omega stiffened composites.

2.2. Testing methods

The laminated composite plate was manufactured to evaluate the feasibility of the carbon threads to respond to the damages created by multiple impact tests. Therefore, they have been submitted to various impact loading with energies of 5J, 5J, 10J and 10J so as to create the progressive damage. It was performed by Instron Dynatup 9250HV drop-weight impact machine, using a hemispherical indentor of 50.8 mm diameter and weight impactor adjusted to 14.2 kg. The electrical resistance of the tuft rows (6 tufted lines in the middle of the specimens) was measured before each impact test by using 4-wire sensing to perform more accurate response. The external points (pairs of current-carrying) injected a controlled direct current of 1mA while the internal points (voltage-sensing electrodes) each one placed 10 mm from the external probes, measured the electrical resistance. Moreover, a specific analysis consisted to evaluate the damaged area and heat dissipation before the impact loadings. The heat dissipation was investigated by the infrared camera during the current injection of 400mA between the 6 tuft rows in the middle of the specimen. The damaged area was characterized by C-Scan technique, scanning a zone of 50 mm x60 mm (width x length) centralized in the impact point via ultrasonic. The steps utilized for this approach are shown in Figure 2.

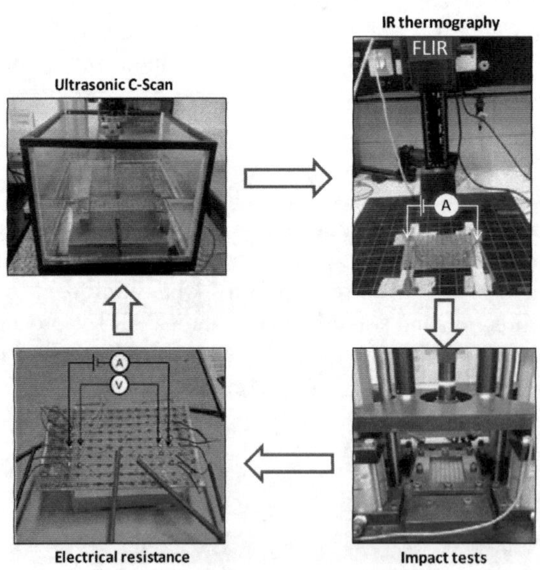

Figure 2. Experimental procedure employed to the investigation of the electrical resistance response before each impact loading

The pull-off tests were carried out at a constant cross-head displacement speed of 1 mm/min. Two fixed rolls (10 mm diameter) were placed 74.5 mm distant from the longitudinal center of the samples and in contact with the skin top surface. The single roll (20 mm diameter) equidistant from the fixed rolls was placed in the bottom surface of the stiffener for loading the structure. The tests were monitored by DIC to evaluate the strain fields in the critical zone of the omega stringer from both sides of the contact flange/skin throughout the loading history and correlated with the software VIC 2D®. The subset and step size employed in the image correlation were 21 and 3 respectively. The AE analysis was also performed for monitoring damage events from the acoustic signals. The signal acquisition was carried out with two wideband sensors (Micro80 - 200-900 kHz) longitudinally positioned on the bottom side of the skin from 80 mm to the center. Furthermore, two-wire measurement was employed to acquire the resistance from both sides of the omega structure by using digital electrometers. Figure 3 exhibits the test apparatus for investigating the omega stringers under pull-off loading and its schematic illustration (symmetrical to the longitudinal center of the sample).

a)

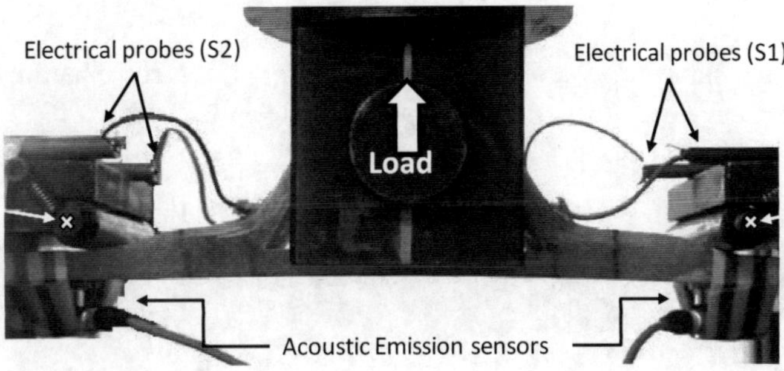

Figure 3. Pull-off test setup showing the electrical probes and AE sensors

3. Results

Figure 4 presents the results of electric resistance obtained before every impact tests to the 6 central tuft rows in the laminated plates. It is remarkable the significant increase of the row 3 and 4 are especially due to their proximity to the impacted zone. The line 4 was exhibited a important increase from the first impact test of 5J energy while line 3 increase gradually its electric resistance. The other rows also responds to impact damages generated during loading even if their response are weaker in comparison to the aforementioned tufts. Furthermore, they present a gradual increase of electric resistance from outside (row 1 and row 6) to inside rows (row 3 and 4), which is related to damage development in impact tests. A comparison between the analysis by C-Scan and infrared thermography (IRT) acquired before and after step III and IV impact tests is shown in Figure 5. The rise on amplitude of C-Scan images is mainly seen from the step III and IV (Figure 5a) and consists of significant damages created in the plate by impact loadings. The events are also reported by spots (rows 3 and 4) in IRT images (Figure 5b) concerning the heat concentration as a consequence of the damages generated during impact in the tufts. These damages difficult the current flow and therefore increase the heating in the specific region. This approach enables monitoring the health of tufts. Once that tufted threads are essential to the out-of-plane properties of laminated composites, monitoring their health may also be important for analyzing the state of health of the tufted structures.

The electrical method as employed in this investigation can also validate the manufacturing process by evaluating the electrical resistance of the tufts. A tuft which presents a significant deviation from the mean value of resistance may have had issues during tufting process, as incomplete insertion of the tuft, unloosening, or significant damage on the thread.

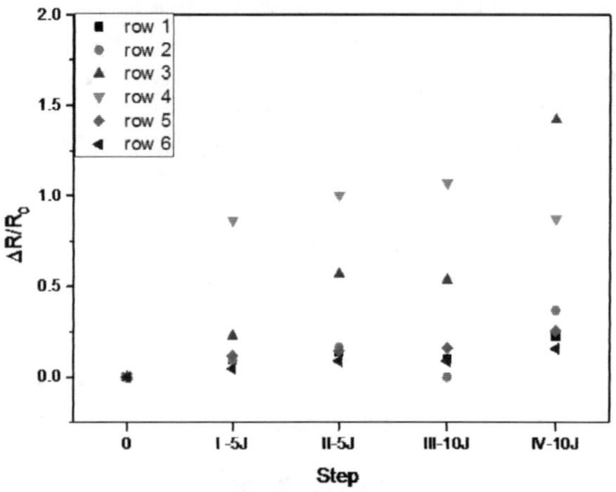

Figure 4. Electrical resistance measurements through different tuft rows before every impact test on the composite plate.

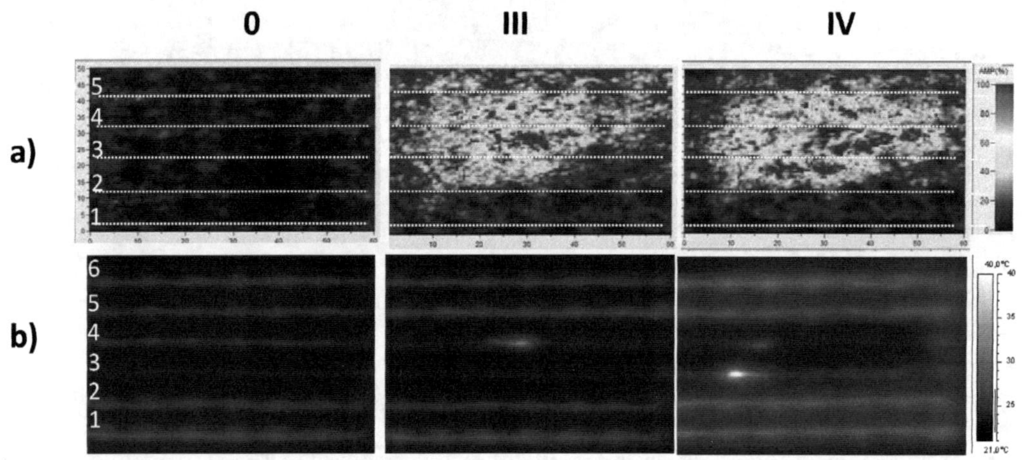

Figure 5. Images acquired before the test and, after step III and IV impact loadings, by a) ultrasonic C-Scan and, b) Infrared thermography.

Figure 6 shows a typical behavior of the electrical resistance from both sides of the omega stiffened composite, named as Resistance S1 and S2, under pull-off test. Acoustic emission activities were also acquired during the test and presented as cumulative energy in the graph. The load drop events exhibited in the curve are also evidenced by AE cumulative energy as a rise in its value, as well as by the electrical resistance on both sides. The mentioned events are significant damages which occur during the tests. Delamination is responsible to the first load drop (steps I-II) and evokes a sharp fall of the electrical resistance, which is better verified in the measurements from the side S1, once the failure occurs on the same side (Figure 7a). This behavior is probably a combination of two simultaneous phenomena: the unloading post damage seen as a sharp drop in the load-time curve that unloads the tuft yarns and therefore enlarges the thread section area, and to the delamination which debonds the resin from the yarns and as consequence diminishes the electric resistance. The reduction of thread cross-section area yarns also leads to an increase in the length (L) when they are submitted to a longitudinal strain. It conducts to the fibers straightening and consequently increases the electrical resistance (R) as described in Eq. (1). Besides, this straightening can induce changes in the contact network by reducing the number of the contact between fibers and also increases the resistance.

$$R = \rho \frac{L}{A} \tag{1}$$

where L is the length of the conductor (m), A is the cross-sectional area of the conductor (m²), and ρ is the electrical resistivity of the material (Ω.m). During delamination, the right side is unloaded, which results in an electrical resistance decrease (Resistance S1), while the left side is slightly loaded and exhibits an increase in its electric resistance (Resistance S2). From the event II, the strain is progressively raised in the left side which amplifies the resistance until the complete rupture of the threads, described as a significant augmentation of the electrical resistance (event III). The events between III and IV characterized by measurements of resistance in the left side (S2) were mainly owing to the delamination generated on the right side (S1) from the analysis performed by DIC. The sharp increase in the electrical resistance seen on the event IV is characterized as delamination on the same side of the S2 (Figure 7b). The increase of the electric resistance was unexpected to a delamination event, which is explained by a continued increase in the opening strain (ε_{yy}) even that delamination was supposed to unload the structure. It is due to a quick sequence of the events, from delamination (event IV) to tuft rupture (event V), and consequently did not let enough time to unload the structure.

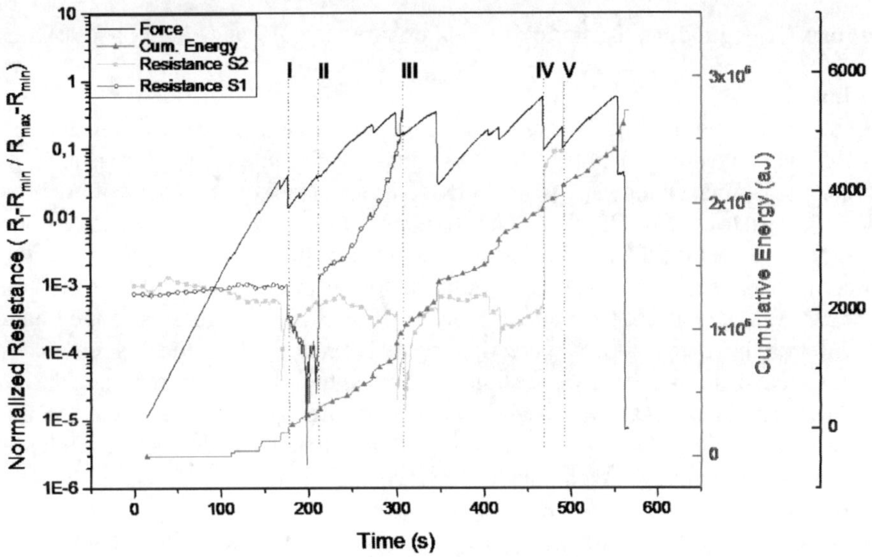

Figure 6. Omega stiffened panel response to pull-off loading assisted by electrical resistance measurements (S2-left side and S1-right side) and AE technique.

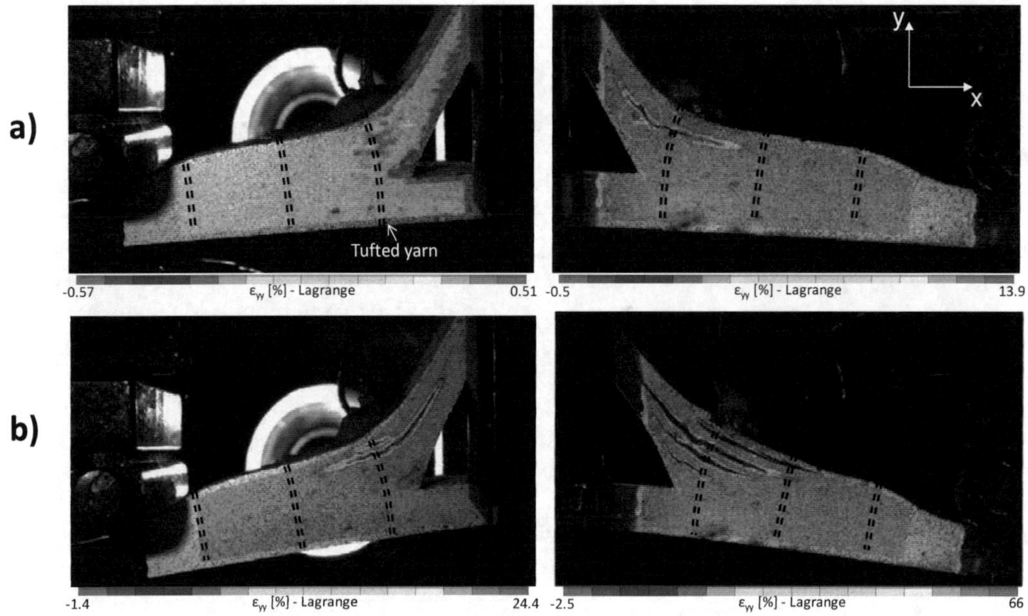

Figure 7. Longitudinal Strain fields (ε_{yy}) measured by DIC at a) Step I and, b) Step IV.

4. Conclusion

The carbon tufted threads employed for reinforcing the GFRP laminate plates were effective to monitor progressive damage by multiple impact tests. An increase in the electrical resistance of the tufts is reported when the laminate composite is subjected to impact loading. This behavior is progressive with the number of impact tests, which consequently increase the damage extent in the composites. It is important to note that despite the monitoring is made through the tuft threads, their structural role in the laminated composites lead to a considerable information of the state of health of the composite from that of the tufts. Future work may explore the electrical resistance response as a method to qualify the state of health of the composite by investigating the relation of residual strength with electrical resistance. It should be employed as a threshold to validate the structure in service.

Moreover, IRT technique was able to localize damage in tufts by imposing an electrical current in the tufted threads and consequently acquiring the temperature gradient by Joule's effect. This approach may be employed as a complementary technique to C-Scan imaging.

Since the investigation on the laminate composite plates reinforced by tufting presented a good response of electrical resistance of the tuft threads to damages generated during impact, the following analysis of this work studied the strain-sensing and damage monitoring of GFRP omega stiffened composites reinforced with carbon tuft threads. The tufts were able to respond to physical phenomena, such as delamination as well as strain-sensing, with electrical resistance variation when loading the specimens under pull-off tests. The results exhibited are promising in order to apply tuft threads as strain sensors of composite structures, however, more studies have to be performed so as to validate this approach.

References

[1] Wisnom, M. R. The role of delamination in failure of fibre-reinforced composites. *Phil. Trans. R. Soc. A* 1850–1870 (2012). doi:10.1098/rsta.2011.0441

[2] Karuppannan, D., Sivaraman, V., Gaddikeri, K. M., Sundaram, R. & Ramesh, A. Effect of Tufting on Mechanical Properties of Laminated Composites. in *ISAMPE National Conference on composite Materials* 47–54 (2012).

[3] Plain, K. P. & Tong, L. The effect of stitch incline angle on mode I fracture toughness –

IOP Conf. Series: Materials Science and Engineering **406** (2018) 012058 doi:10.1088/1757-899X/406/1/012058

Experimental and modelling. *Compos. Struct.* **92,** 1620–1630 (2010).

[4] Lombetti, D. M., Dell'Anno, G., Skordos, A. A. & Partridge, I. K. Delamination performance of tufted carbon / epoxy composites made by automated dry. in *The 19th international conference on composite materials* 1–7

[5] Verdiere, M. C. De, Skordos, A. A., Walton, A. C. & May, M. Influence of loading rate on the delamination response of untufted and tufted carbon epoxy non-crimp fabric composites / Mode II. *Eng. Fract. Mech.* **96,** 1–10 (2012).

[6] Deconinck, P., Capelle, J., Bouchart, V., Chevrier, P. & Ravailler, F. Delamination propagation analysis in tufted carbon fibre-reinforced plastic composites subjected to high-velocity impact. *J. Reinf. Plast. Compos.* **33,** 1353–1363 (2014).

[7] Dell'Anno, G., Cartié, D. D., Partridge, I. K. & Rezai, A. Exploring mechanical property balance in tufted carbon fabric/epoxy composites. *Compos. Part A Appl. Sci. Manuf.* **38,** 2366–2373 (2007).

[8] Scarponi, C., Perillo, A. M., Cutillo, L. & Foglio, C. Advanced TTT composite materials for aeronautical purposes : Compression after impact (CAI) behaviour. *Compos. Part B 38* **38,** 258–264 (2007).

IOP Conf. Series: Materials Science and Engineering **406** (2018) 012059 doi:10.1088/1757-899X/406/1/012059

Laminate fibre wave defect – KDF evaluation

Luciano Brambilla*

Consultant at FACC St. Martin im Innkreis Austria

* l.brambilla@facc.com, brambilla_luciano@libero.it

Abstract. The positive side of using composite materials in Aerospace industry is the weight reduction. The drawback are the defects created on solid laminates, during manufacturing. Structure life and strength reduction, due to the defects, need to be accurately evaluated against the A/C airworthiness requirements.

1. Introduction

The approach, suggested in this paper, consists on using the Conformal Mapping for laminate defects topology modeling. Scope of this approach is to evaluate a KDF for Margin of Safety calculation as a ratio between defect configurations against the intact one. For most of defect scenario, the Conformal Mappings are available in literature. For each type of defect, it is possible to define fiber pattern, angle, derivative, thickness and length.

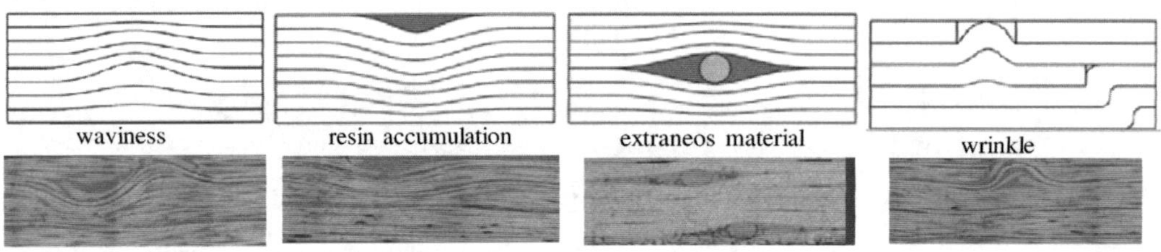

Figure 1. Solid laminate typical defect

This paper reports only a numerical methodology for stress-strain calculation of wavy ply configuration. Fiber and matrix strength check (not included on this paper) is performed in accordance with the applicable criteria and allowable for each type of ply,
The KDF for each *defect topology* is then calculated as follows:

$$KDF = \frac{\left(RF_{\text{intact}-CLT}\right)_{\text{minimum}}}{\left(RF_{\text{wavy}}\right)_{\text{minimum}}}$$

For clarity, the analysis of intra-lamina shear and out-of-plane matrix stress due to wavy ply curvature are not reported on this paper.

2. System approach

2.1. Defect Configuration topology definition

The defect analysed is an embedded resin accumulation shown on Figure 2
The wavy ply profile is defined via complex variable function F(z) (see Ref.[2]).

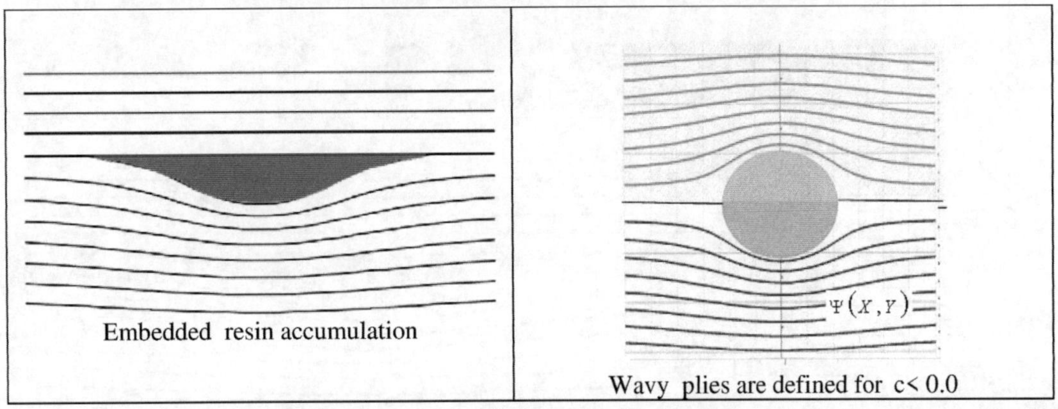

Figure 2. Defect – Resin embedded accumulation

$$F(z) = A\left(z + \frac{1}{z}\right)$$

$$\Psi(X,Y) = \mathbf{Im}\left[X + iY + \frac{1}{X + iY}\right] = Y^3 - cY^2 - Y\left(X^2 - 1\right) - cX^2 = 0.0$$

$$\dot{Y} = \frac{dY}{dX} = -\frac{2XY - 2cX}{3Y^2 - 2cY + X^2 - 1}$$

c: parameters is the variable related to ply thickness.

Note that "streamline" $\Psi(X,Y)$ doesn't have physical meaning rather than representing the profile of the ply, consistently with the defect type:

1. A=1.0
2. Analytical formulation of wavy ply
3. Derivative values in each point of the plies
4. Ply thickness variation
5. Ply curvature
6. Ply length

2.2. Algorithm basic assumption

Defect analysed on this paper is shown on below Figure 3.
Two loading condition are considered separately:

IOP Conf. Series: Materials Science and Engineering **406** (2018) 012059 doi:10.1088/1757-899X/406/1/012059

1. Tension load F_0
2. Shear load S_0

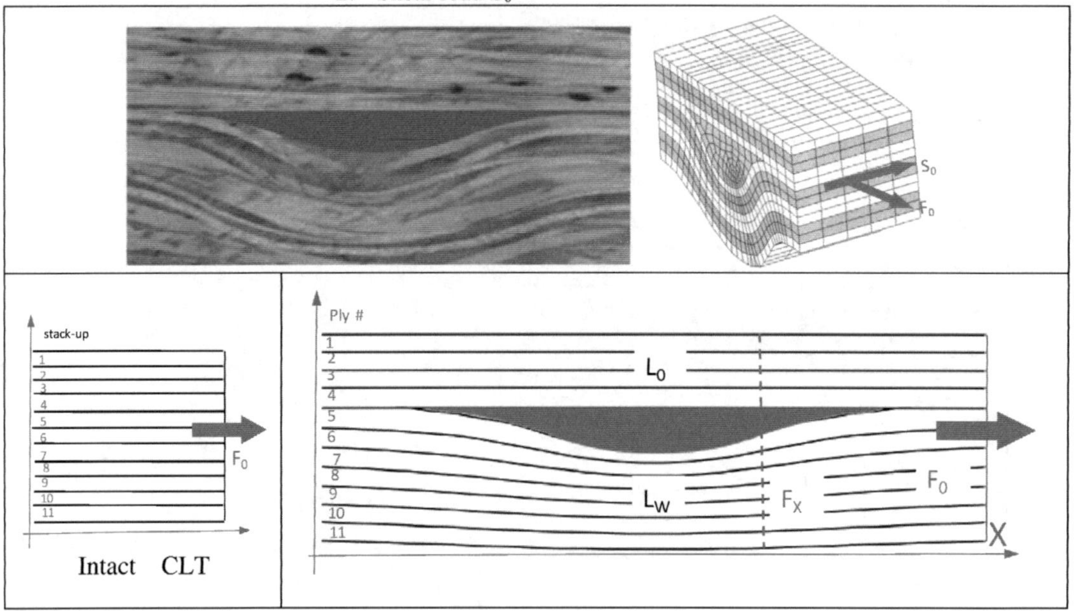

Figure 3. Intact and wavy configuration

The static analysis of wavy configuration, is basically a modification of the ply stress calculated via Classical Laminate Theory (Ref. [1]) for intact configuration.

The following criteria apply (see Figure4).

1. Applied load F_0 is constant in each section along X axis : $F(x) = F_0$
2. Section(X) remain *plane*.
3. Wavy ply mechanical properties, are re-calculated in each section (X)as per fibre-resin mixture criteria due to thickness variation.
4. Ply stress in each section (X) depends on two geometric parameters :
   a. *Derivative Y'*: defines the fibre orientation and the component along X axis.
   b. *Stiffness ratio" λ "*: Wavy longer ply (L_w) have a lower tension stiffness than intact ply (L_0) and consequently "attract" less load. Fibre stress calculated at point a), is therefore scaled down by a factor "λ". (subscript0, refer to CLT).

stiffness $$K = \frac{N}{\delta} = \frac{N}{\varepsilon * l} = \frac{(\varepsilon E * t * 1.)}{\varepsilon * l} = \frac{E * t}{l}$$

Ratio $$\lambda = \frac{K_w}{K_0} = \frac{E_w * t_w}{l_w} * \frac{l_0}{E_0 * t_0} = \frac{t_w * l_0}{t_0 * l_w} \approx \frac{L_0}{L_w}$$

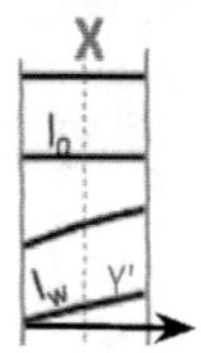

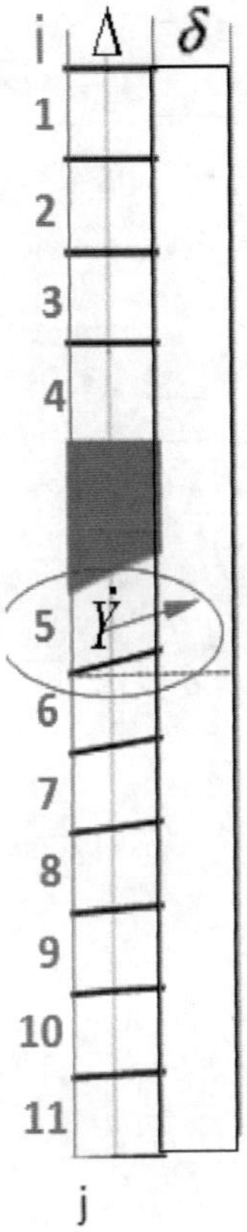

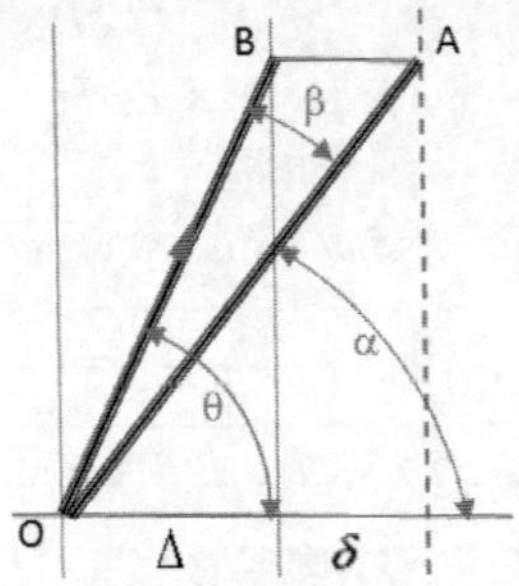

OB: unloaded fibre
OA : loaded fibre

$$\theta = \text{atan}\,\dot{Y}$$

$$\alpha = \text{atan}\,\frac{\dot{Y} * \Delta}{\Delta + \delta}$$

$$\beta = \theta - \alpha$$

$$OA = \frac{\Delta + \delta}{\cos\alpha}$$

$$OB = \frac{\Delta}{\cos\theta}$$

$$\varepsilon_{ij} = \frac{OA - OB}{OB} = \frac{OA}{OB} - 1 = \frac{\cos\theta_{ij}}{\cos\alpha_{ij}}\left(1 + \frac{\delta_j}{\Delta_j}\right) - 1$$

$$\sigma_{ij} = \varepsilon_{ij} * E_{ij} \quad (\textit{see } E_{ij} \text{ mixture criteria})$$

$$f_{ij} = \sigma_{ij} * t_{ij}$$

$$f_{ij-x} = f_{ij} * \cos\alpha_{ij} = \left[\cos\theta_{ij}\left(1 + \frac{\delta_j}{\Delta_j}\right) - \cos\alpha_{ij}\right] * E_{ij} * t_{ij} \quad (X\ component)$$

---

mixture criteria $E_{ij} = \left(E_{xxw}\right)_{ij}$

intact ply $E_{xx} = E_{fo} * V_{fo} + E_m * (1 - V_{fo}) \Rightarrow E_{fo} = \dfrac{E_{xx} - E_m * (1 - V_{fo})}{V_{fo}}$

wavy ply $E_{xxw} = E_{fo} * V_{fw} + E_m * (1 - V_{fw})$ where $V_{fw} = V_{fo} * \left(\dfrac{t_0}{t_w}\right)$

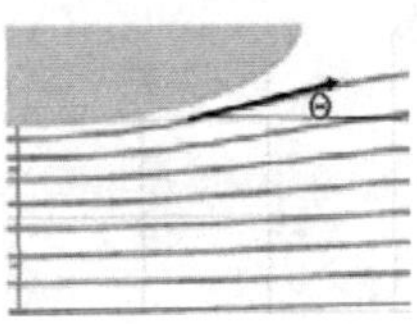

$$F_j = \sum_{i=1}^{n} f_{ij-x} = \sum_{i=1}^{n}\left\{\left[\cos\theta_i\left(1 + \frac{\delta_j}{\Delta_j}\right) - \cos\alpha_i\right]_i * E_i * t_i\right\}$$

Δ_j : constant δ_j : variable ($= \Delta$ *elongation*)

balance $F_j = F_0$ *goalseek on* δ_j

Stiffness : $\delta_{tot} = \sum_{j=1}^{m} \delta_j \qquad K_{wave} = \dfrac{F_0}{\delta_{tot}}$

Figure 4. Plies stress-strain calculation algorithm

2.3. Algorithm: step-by-step definition

In this example, the laminate length is subdivided in 31 section.

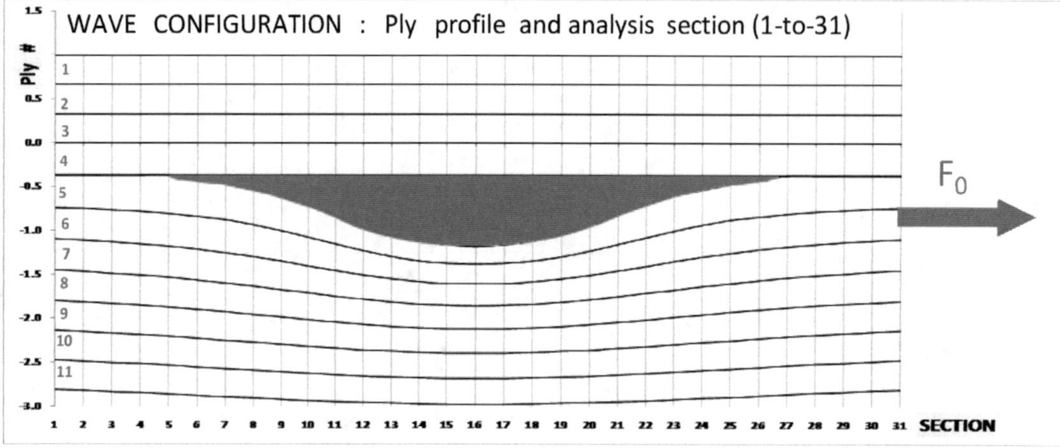

Figure 5. Wavy configuration laminate section

Laminate configuration:

- Ply ID i=1 to 11
- Section ID j=1 to 31
- Ply nominal thickness : .33[mm]
- Length L0: 6. [mm]
- Laminate thickness: 11*.33 = 3.63 [mm]

Stress-strain on each point Pij are calculated according to below step-by-step procedure.

1. Strain ε_{ij} calculation at point $P_{i,j}$
2. Stress calculation $\sigma_{ij} = \varepsilon_{ij}*E(t)_{ij}$ (E: fibre-matrix mixture criteria)
3. Ply force $f_{ij} = \sigma_{ij}*t_{ij}*1$.
4. Resultant force at each section j :

$$F_j = \sum_{i=1}^{n}\left[f(\delta)_i\right]_j$$

5. Balance between applied section force $F_0$and $F_j(F_j=F_0)$ is obtained via trial-and-error process on variable δj.

One single section can be analysed independently from the others, once the following data are available:

1. CLT stress for intact configuration
2. Derivative Y'(i), λ(i) , t(i) , curvature R(i) from conformal mapping .

IOP Conf. Series: Materials Science and Engineering **406** (2018) 012059 doi:10.1088/1757-899X/406/1/012059

3. Example of Numerical Calculation

3.1. Ply mechanical properties

Figure below reports the material properties of 11 plies stack up.

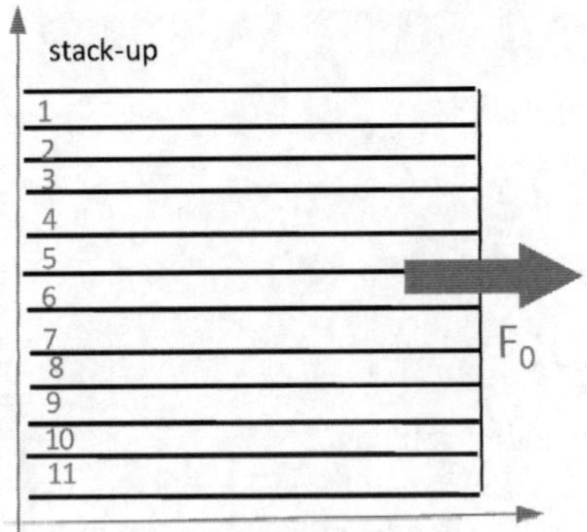

Ply mechanical properties vs angle Θ

$$\begin{Bmatrix} \sigma_x \\ \sigma_y \\ \tau_{xy} \end{Bmatrix} = \begin{bmatrix} E11 & E12 & E13 \\ E21 & E22 & E23 \\ E31 & E32 & E33 \end{bmatrix}_\theta * \begin{Bmatrix} \varepsilon_x \\ \varepsilon_y \\ \gamma_{xy} \end{Bmatrix}$$

	Fabric	
E_{11}	66000	[MPa]
E_{22}	66000	[MPa]
G_{12}	5800	[MPa]
ν_{12}	0.0409	--------
t	0.33	[mm]

$E_{0°}$=

66111	2704	0
2704	66111	0
0	0	5800

$E_{+-45°}$=

40207	28607	0
28607	40207	0
0	0	31703

$E_{90°}$=

66111	2704	0
2704	66111	0
0	0	5800

	UNI	
E_{11}	135000	[MPa]
E_{22}	8550	[MPa]
G_{12}	4200	[MPa]
ν_{12}	0.35	--------
t	0.33	[mm]

$E_{0°}$=

136056	3016	0
3016	8617	0
0	0	4200

$E_{+-45°}$=

41876	33476	31860
33476	41876	31860
31860	31860	34660

$E_{90°}$=

8617	3016	0
3016	136056	0
0	0	4200

Figure 6. Ply mechanical properties

3.2. Section 31 analysis

Figure 7shows the values calculated according to the methodology reported on

Figure4, relevant to section j =1 and j =31. The reason of this section analysis, is to cross check the algorithm against the CLT results on segment where plies axis are in X direction (Y'=0.0) as per CLT.

1. Algorithm is set up on columns under "WAVY CONFIGURATION" label
2. Stack up :UNI [90,0,0,90,90,0, 90,90,0,0,90]
3. Ply properties UNI : see Figure 6
4. Applied force F_0= 1000. [N/mm]
5. "Goal Seek "plug-in, to converge $F_j = F_0$:
   "Set cell" F_j .... "To value" F_0.... "By changing cell" δ .
6. Note that CLT and Algorithm, yield the same stress and force $f_{ij}(=\sigma^*t)$, as expected since for both models the derivative is the same (Y' = 0.0) and length factor ratio in this model isnegligible(λ=1.).

section	J	1 and 31	-------	
	Δ	0.2	mm	
	δ	0.00083	mm	Goal Seek
	ε_{11}	**0.00415**	-------	
	F_0	1000	N/mm	

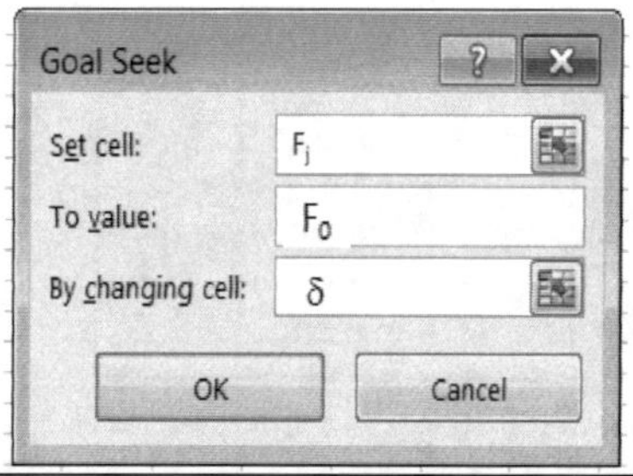

CLT	
UNI laminate	
ε_{11}	0.00415
K	40188.4

WAVE CONFIGURATION											CLT	
Ply #	Y'	θ rad	α rad	β rad	E [MPa]	ε -------	σ [MPa]	$t_{nominal}$ [mm]	$\lambda = L_o/L_w$	$f_{11}=\sigma_{11}$ *t [N/mm]	$f_{11}=\sigma_{11}$ *t [N/mm]	stack-up angle
1	0.000	0.0000	0.0000	0.0000	8616	0.00415	35.8	0.33	1.0000	12	12	90
2	0.000	0.0000	0.0000	0.0000	136055	0.00415	565.0	0.33	1.0000	186	186	0
3	0.000	0.0000	0.0000	0.0000	136055	0.00415	565.0	0.33	1.0000	186	186	0
4	0.000	0.0000	0.0000	0.0000	8616	0.00415	35.8	0.33	1.0000	12	12	90
5	0.000	0.0000	0.0000	0.0000	8616	0.00415	35.8	0.33	**0.9830**	12	12	90
6	0.000	0.0000	0.0000	0.0000	136055	0.00415	565.0	0.33	**0.9897**	185	186	0
7	0.000	0.0000	0.0000	0.0000	8616	0.00415	35.8	0.33	**0.9935**	12	12	90
8	0.000	0.0000	0.0000	0.0000	8616	0.00415	35.8	0.33	**0.9958**	12	12	90
9	0.000	0.0000	0.0000	0.0000	136055	0.00415	565.0	0.33	**0.9973**	186	186	0
10	0.000	0.0000	0.0000	0.0000	136055	0.00415	565.0	0.33	**0.9982**	186	186	0
11	0.000	0.0000	0.0000	0.0000	8616	0.00415	35.8	0.33	**0.9982**	12	12	90
matrix	0	0	0	0	8250	0.00415	0.0	**0.00**	1.0000	0	0	
									F_j	1000	1000.0	

Figure 7. Algorithm

IOP Conf. Series: Materials Science and Engineering **406** (2018) 012059 doi:10.1088/1757-899X/406/1/012059

3.3. Analysis of section from 16 to 31

Because of symmetry, the results are reported, for clarity, only for section from 16 to 31.
The calculation is performed per following parameters:

1. Segment-to-segment calculation is based on constant step ($\Delta = 0.2$ mm).
   Note: the "sharper" is the wave and the "shorter" is the assumed segment step Δ.
2. Stress calculation is performed at each section based on Excel algorithm on Figure 7
3. δ_j is recorded for each section at $F_j = F_0$ (from Goal Seek).
4. Wavy Laminate total elongation:
   $$\delta_{tot} = \sum_{j=1}^{31} \delta_j = 0.0297$$
5. Wavy laminate stiffness:
   $$K_W = \frac{F_0}{\delta_{tot}} = \frac{1000}{.0297} = 33670.$$
6. CLT laminate stiffness(Figure 7):
   $$K_0 = \frac{F_0}{\delta_{tot}} = \frac{F_0}{L_0 * \varepsilon_{11}} = \frac{1000}{6 * 0.00415} = 40188.$$

PLY	STRESS [MPa]															
1	35	36	38	40	42	44	44	44	43	42	41	40	39	38	37	37
2	553	565	596	633	665	689	700	698	684	664	643	625	611	600	592	586
3	553	565	596	633	665	689	700	698	684	664	643	625	611	600	592	586
4	35	36	38	40	42	44	44	44	43	42	41	40	39	38	37	37
5	35	32	25	18	14	12	13	18	23	30	34	37	37	37	37	37
6	553	532	480	419	371	344	345	371	414	461	501	529	547	558	564	567
7	35	35	34	32	31	30	29	30	30	32	33	34	34	35	35	35
8	35	35	36	36	36	35	35	35	34	34	34	34	35	35	35	35
9	553	561	579	597	608	612	610	602	591	578	568	562	559	558	558	559
10	553	563	587	613	632	644	645	638	623	606	590	578	570	566	563	562
11	35	36	37	39	41	42	42	42	41	40	38	37	37	36	36	36
matrix	27	27	27	26	23	19	14	10	7	4	2	0	0	0	0	0
δ	8.12E-04	8.31E-04	8.77E-04	9.30E-04	9.78E-04	1.01E-03	1.03E-03	1.03E-03	1.01E-03	9.76E-04	9.45E-04	9.18E-04	8.98E-04	8.82E-04	8.70E-04	8.61E-04
ε_{11}	4.06E-03	4.16E-03	4.38E-03	4.65E-03	4.89E-03	5.06E-03	5.15E-03	5.13E-03	5.03E-03	4.88E-03	4.73E-03	4.59E-03	4.49E-03	4.41E-03	4.35E-03	4.30E-03
SECTION	16	17	18	19	20	21	22	23	24	25	26	27	28	29	30	31

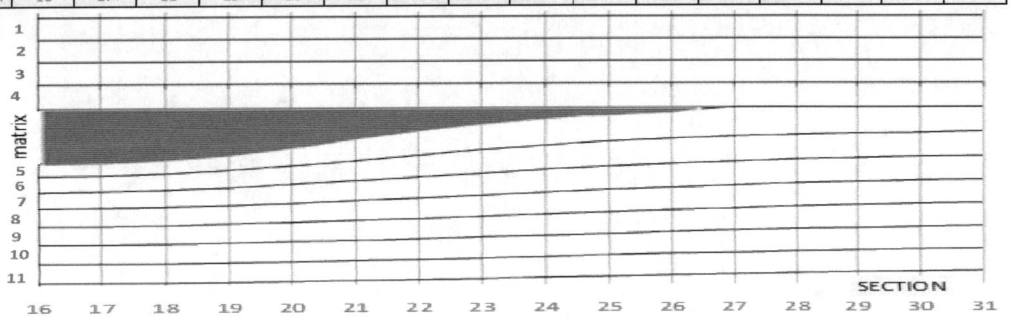

Figure 8. Wavy laminate –stress map-out

IOP Conf. Series: Materials Science and Engineering **406** (2018) 012059 doi:10.1088/1757-899X/406/1/012059

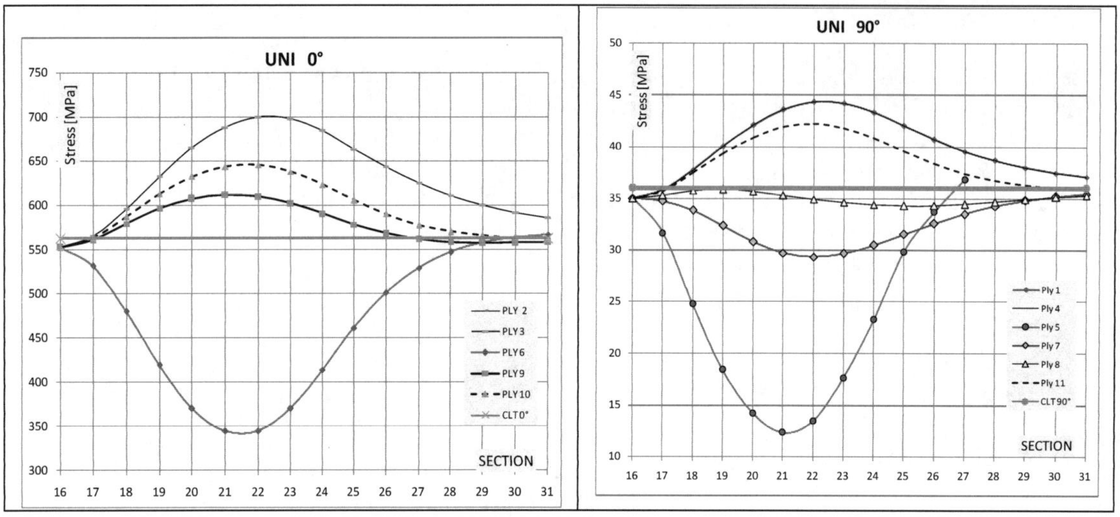

Figure 9. Fibre stress for UNI 0° and 90°

3.4. *3DFEM analysis*

The symmetric 3D FEM is set up with defect scenario "similar" to the one analysed on this paper. Both models show the same "trend" for fibre stress profile, in spite of substantial difference on laminate configuration and mechanical properties: (comment on Par. 5)

1. 12 plies instead of 11
2. stack-up : $[45,0,90,45]_3$
3. ply mechanical properties [Tape, Fabric]
4. $F_0 = 50.$ [N/mm]

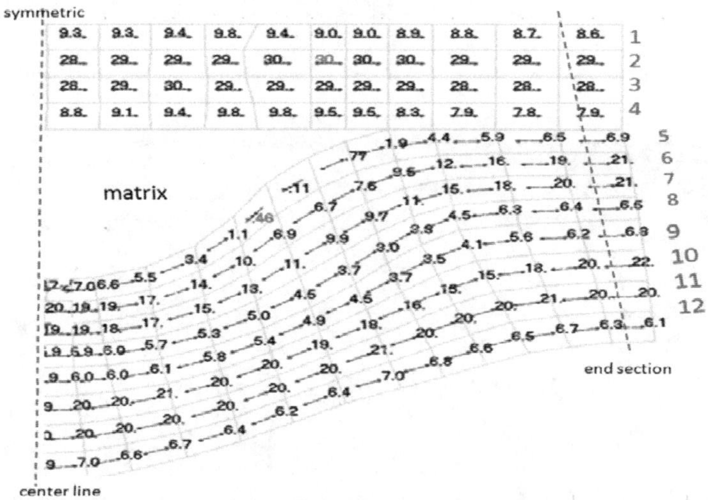

Figure 10. 3DFEM model – plies stress

4. Shear load S_0

This paragraph reports the analysis of laminate defect under shear load. A simplified numerical approach is presented to calculate the shear stress at each ply based on the deformation compatibility. Ply shear load ($f_{xy}=\tau*t$),is proportional to factor [d_0*G] and inversely proportional to the ply length. Example of a 3DFEM model is added for algorithm explanation and evaluation.

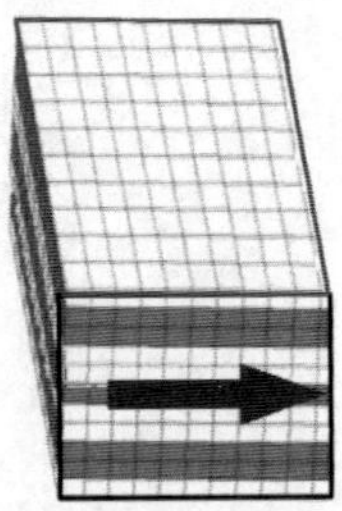

S = 1000. [N/mm] , d_0 = .395[mm]

UNI shear load =1000. [N/mm]								
PLY	angle °	G_{xy}	t	L	G_{xy} / L	t (G_{xy}/L)	τ_{xy}	$f_{xy}=\tau*t$
ID	deg	[MPa]	[mm]	[mm]	[N/mm^3]	[MPa]	[MPa]	[N/mm]
1	90	4200	0.33	6.00	700.0	231.0	276.6	91.3
2	0	4200	0.33	6.00	700.0	231.0	276.6	91.3
3	0	4200	0.33	6.00	700.0	231.0	276.6	91.3
4	90	4200	0.33	6.00	700.0	231.0	276.6	91.3
5	90	4200	0.33	6.10	688.1	227.1	271.9	89.7
6	0	4200	0.33	6.06	692.8	228.6	273.7	90.3
7	90	4200	0.33	6.04	695.5	229.5	274.8	90.7
8	90	4200	0.33	6.03	697.1	230.0	275.4	90.9
9	0	4200	0.33	6.02	698.1	230.4	275.8	91.0
10	0	4200	0.33	6.01	698.7	230.6	276.1	91.1
11	90	4200	0.33	6.01	698.7	230.6	276.1	91.1
					d_0 =	0.3951		1000

SHEAR LOAD ALGORITHM

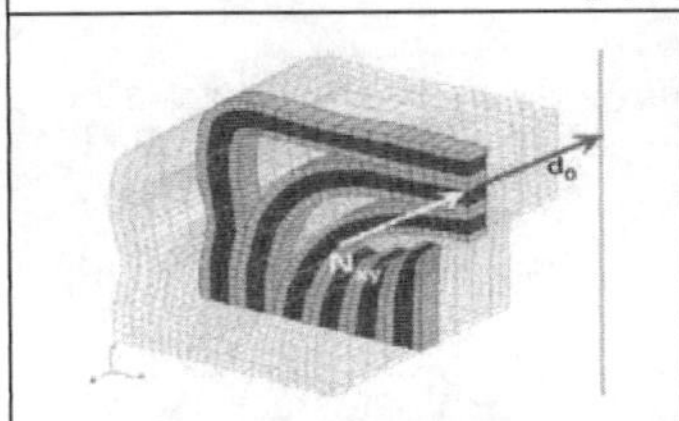

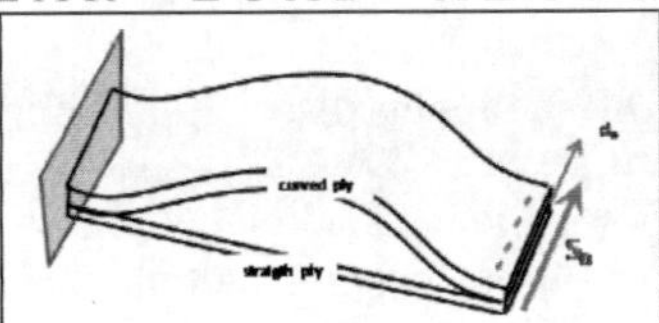

Approach is valid under the validity of deformation compatibility.

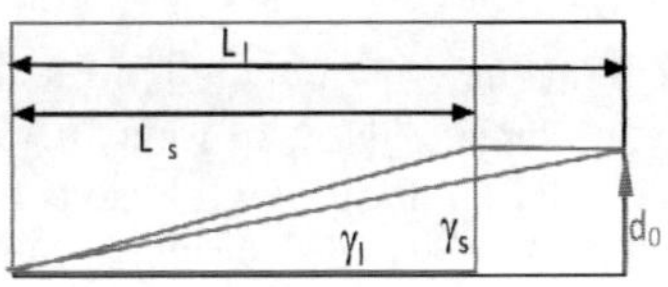

$$\gamma_I = \frac{d_0}{L_I} < \frac{d_0}{L_s} = \gamma_s$$

$$N_{xy} = \sum_{i=1}^{n}(\tau_i * t_i) = \sum_{i=1}^{n}(\gamma_i * G_i * t_i) = \sum_{i=1}^{n}\left(\frac{d_0}{L_i} * G_i * t_i\right) = d_0 * \sum_{i=1}^{n}\left(\frac{G_i}{L_i} * t_i\right) \quad [N/mm]$$

$$d_0 = \frac{N_{xy}}{\sum_{i=1}^{n}\left(\frac{G_i}{L_I} * t_i\right)} \quad [N/mm] \quad \Rightarrow \quad \tau_i = \frac{d_0}{L_i} * G_i \quad [MPa]$$

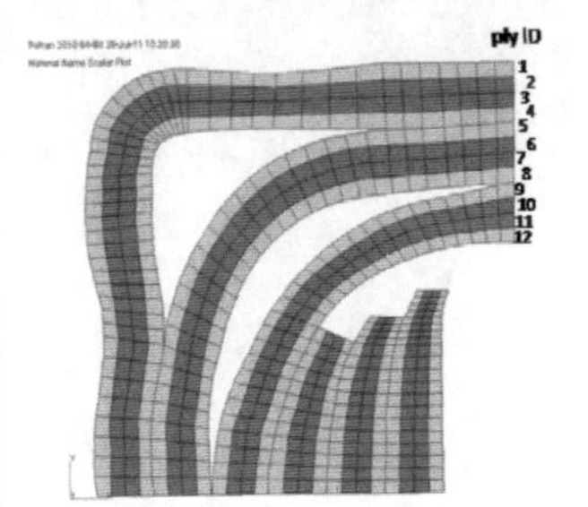

S_0 1000.[N/mm], d_0 = .093 [mm]

PLY	angle	t	G_{xy}	L	G_{xy} /L	t*(G_{xy} /L)	d_0*t*(G_{xy}/L)	3D FEM	ALGO RITHM
	deg	[mm]	[Mpa]	[mm]	[N-mm]	[Mpa]	[N/mm]		
1	45	0.370	31703	11.5	2759.1	1020.9	95.0	255.91	256.8
2	0	0.370	5800	10.9	531.7	196.7	18.3	49.03	49.5
3	0	0.370	5800	10.3	561.6	207.8	19.3	51.46	52.3
4	45	0.370	31703	9.7	3252.7	1203.5	112.0	295.88	302.7
5	45	0.370	31703	9.2	3458.9	1279.8	119.1	312.59	321.9
6	0	0.370	5800	8.6	675.6	250.0	23.3	60.71	62.9
7	0	0.370	5800	8.0	724.7	268.1	25.0	64.68	67.5
8	45	0.370	31703	7.4	4271.5	1580.4	147.1	378.88	397.6
9	45	0.370	31703	6.8	4634.4	1714.7	159.6	407.96	431.3
10	0	0.370	5800	6.3	926.6	342.8	31.9	80.98	86.2
11	0	0.370	5800	5.7	1021.4	377.9	35.2	88.68	95.1
12	45	0.370	31703	5.1	6219.6	2301.3	214.2	536.50	578.9

Figure 11. Ply shear stress_3DFEM vs Numerical model

5. Conclusion

Conclusions are based on comparison between algorithm (Figure 8) and 3DFEM (Figure 10).
The following considerations apply:

1. The defect type is “similar” for both models
2. The algorithm and 3DFEM have a different stack-up and material properties.
3. The analysis highlight the following analogies:
   a. Both model shows on ply (1, 2, 3, 4) “similar trend” for fiber stress on *centerline-end* sections and “similar trend (= stress increase)“on intermediate sections.
   b. Wavy ply 5 shows in both model the same stress on *centerline-end* sections.
   “Similar trend (= stress decrease)“on intermediate section:
   [MPa] $(35_{end}, 12_{middle}, 37_{end})_{ALGORITHM}$, $(7.0_{end}, -0.46_{middle}, 6.9_{end})_{3DFEM}$
   c. The same behavior for ply 10, 11, 12: no stress significant variation on intermediate segment. The reason is that ply slope tends to Y’=0.0 (on both models).
   d. In both model the *centerline-end* sections, show the same stress of ply 1, 2, 3, 4 multiplied by a fiber length ratio λ.
4. For both model, the stress at centerline-end (Y’=0.0) are “close” to the CLT (Figure 7).

Remark:
1. The method can be applied for the strength analysis of single section X (independently from the others), on the basis of following parameters: Y’, λ, thk, $R_{curvature}$.
2. “Similarity” is hinting that defect topology is the leading parameter.
3. Methodology and algorithm, described on this paper, is not uniquely correlated to complex variable mathematical formulation.
4. The method is also effective for defect data (Y’, λ, thk, $R_{curvature}$) taken from laminate micro-section picture.

References

[1] Stephen W. Tsai, Theory of Composites Design Aeronautics & Astronautics Stanford University.
[2] Murray R. Spiegel, Complex Variables SCHAUM’S Outline Series.

Experimental process pressure analysis for model-based manufacturing of composites by resin transfer moulding

E Fauster*, J Thirunavukkarasu, P Hergan and R Schledjewski

Processing of Composites Group, Montanuniversität Leoben,
Otto Glöckel Str. 2/III, 8700 Leoben, Austria

*ewald.fauster@unileoben.ac.at

Abstract. An experimental study of the process pressure characteristics prevailing in an RTM mould during the preform impregnation stage is presented. The pressure data is analysed with respect to a superposition of preform compaction and hydraulic fluid pressure. A methodology is proposed to isolate the hydraulic fluid pressure from the preform compaction pressure involving knowledge about the compressibility behaviour of the reinforcing material. For this purpose, the preform compressibility and relaxation behavior was studied by means of transverse compaction tests on both, unsaturated as well as the saturated sample stacks. The resulting isolated hydraulic fluid pressure is finally compared with numerical predictions derived from flow simulations, revealing good overall agreement. The findings of this work will contribute to realize the concept of model-based processing of fibre-reinforced polymer composites by means of the RTM process.

1. Introduction

Model-based manufacturing of fiber-reinforced polymer composites is based on transferring process-related knowledge from the level of process modelling to the level of active process management. This requires fundamental understanding of the most significant process-related mechanisms in terms of phenomenological models. In the resin transfer moulding (RTM) process, these can be identified as: (i) preform compaction, (ii) preform impregnation and (iii) resin curing [1, 2].

In the preform impregnation stage, when the viscous fluid saturates the preform material, two effects occur at the location of the flow front: a decrease of the preform compaction pressure due to the wetting of the reinforcing material (i.e. lubrication effects) as well as an increase of the hydraulic fluid pressure as the flow front advances. When sensing the in-mould pressure characteristics with pressure sensors during the preform impregnation stage, a superposition of these two effects contribute to the pressure signal. In order to fully exploit the acquired measurement data, the temporally changing interaction of these two contributors needs to be understood. The driving mechanisms are well known to the composites processing community [3–6] and were extensively treated in a recent review paper [7]. However, despite considerable efforts towards modelling of the compaction behavior of reinforcing materials have been made in the past, preform compliance still needs to be characterized by means of experimental techniques [8–13]. On the level of processing composite materials by means of the RTM technique, the process pressure characteristics can be monitored as a part of process documentation. In addition, they can be involved in active process management to support the concept of model-based composite processing. The paper at hand focuses on an experimental process pressure analysis to provide reliable input data for such a process control task.

2. Experimental Work

2.1. Materials

2.1.1. Test Fluid
The experiments described in this paper were realized with plant oil used as test fluid showing a density of about 920 kg/m² and a dynamic viscosity of about 65 mPas at room temperature (23 °C). The actual viscosity-temperature-characteristics of the test fluid were determined with a Couette rheometer system prior to the experimental series.

2.1.2. Reinforcing Materials
The experimental investigations presented in this paper were elaborated on a glass woven fabric (twill 2/2) of type Hexcel HexForce 1202 (see Figure 1) with a nominal areal weight of 0.29 kg/m². Its in-plane permeability as well as out-of-plane compressibility characteristics were determined experimentally as described in the subsequent section.

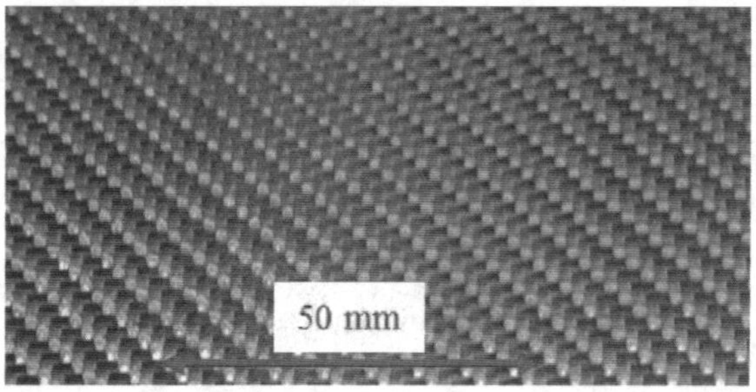

Figure 1. Close-up view of the glass fiber twill woven fabric used for the experiments.

2.2. Material Characterization

2.2.1. In-Plane Permeability
The in-plane permeability characteristics of the glass fiber twill woven fabric were characterized experimentally by means of a testrig following the radial flow method. There, the stack of material layers is placed in a mould and impregnated through a central injection opening. The flow front was tracked by means of eight linear capacitive sensors as described by Grössing et al. [14], which are embedded in the lower part of the mould as shown in Figure 2.

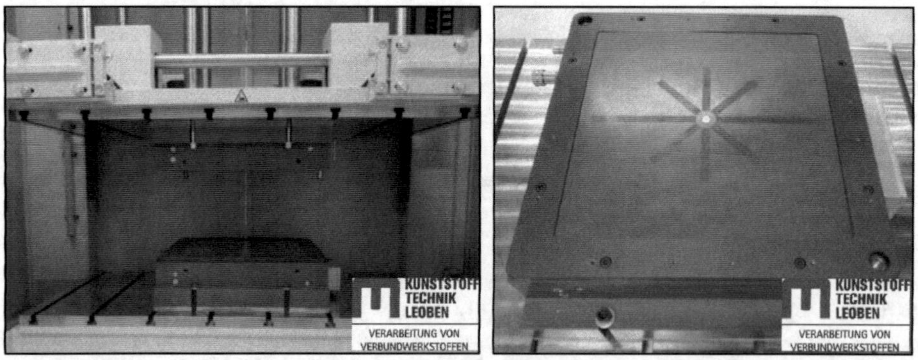

Figure 2. In-plane permeability characterization system mounted to a hydraulic mould carrier (left). Eight linear capacitive sensors are embedded in the lower mould half (right), arranged in a star-like scheme around a pressure point sensor and opposite to the central injection opening.

2.2.2. Out-of-Plane Compressibility

The out-of-plane compressibility behaviour of the glass fiber twill woven fabric was experimentally characterized with the setup depicted in Figure 3, which is mounted to a universal testing machine. Stacks of the material layers, 60 x 60 mm in dimension, are placed on the pot-shaped pressure plate and compressed against the stamp when moving the crosshead of the testing machine. During the experiment, stack height as well as compaction force are continuously measured by means of a Linear Variable Differential Transformer (LVDT) mounted close to the sample stack and the load cell of the testing machine, respectively. Evaluating the stack height and compaction force data, the characteristics of compaction pressure against fiber volume fraction can be derived. The setup allows for investigating the loading/deloading characteristics of the material under test as well as the relaxation behaviour which is of particular interest for the RTM process. In addition, the tests can be run with unsaturated as well as saturated sample stacks in order to study the impact of wetting on the out-of-plane compressibility behaviour.

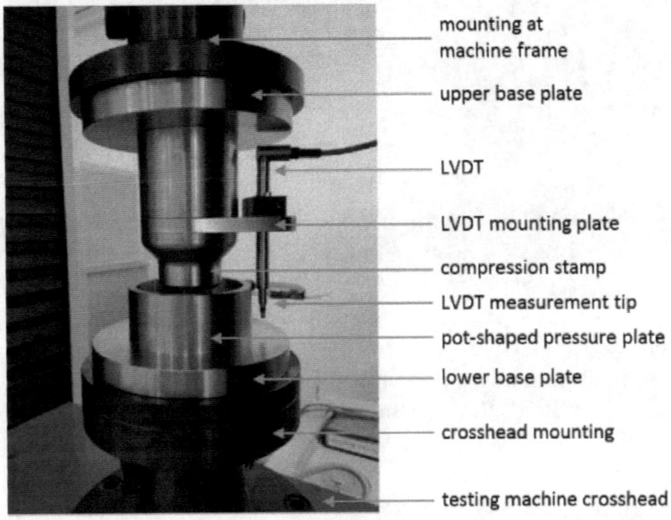

Figure 3. Test equipment for out-of-plane compressibility characterization.

2.3. Manufacturing of Flat Composite Plates by means of Resin Transfer Moulding

2.3.1. Experimental RTM Processing

For the work at hand, RTM process pressure characteristics were studied by means of an RTM mould with dimensions of 270 x 270 x 2 mm. It provides seven openings in the upper part of the mould allowing for a flexible mounting of sensors as shown in Figure 4. The test fluid or chemically reactive resin system can be injected either from one of the side edges of the plate (resulting in linear flow front advancement) or through the central opening in the upper part of the mould (resulting in radial flow front advancement). For the experiments discussed in this paper, linear flow tests were realized, whereas the test fluid was injected by means of a pressure pot with an injection pressure set value of 0.4 MPa.

A fiber volume fraction of about 51 % was realized by placing nine layers of the glass fiber twill woven fabric described in Section 2.1. inside the mould. The preform was then impregnated with the test fluid and the process pressure characteristics were captured by means of three pressure point sensors mounted along the desired flow path (see Figure 4). These pressure characteristics were finally studied in terms of partitioning the contributions of preform compaction pressure and hydraulic fluid pressure, both varying with location and experimental time.

IOP Conf. Series: Materials Science and Engineering **406** (2018) 012060 doi:10.1088/1757-899X/406/1/012060

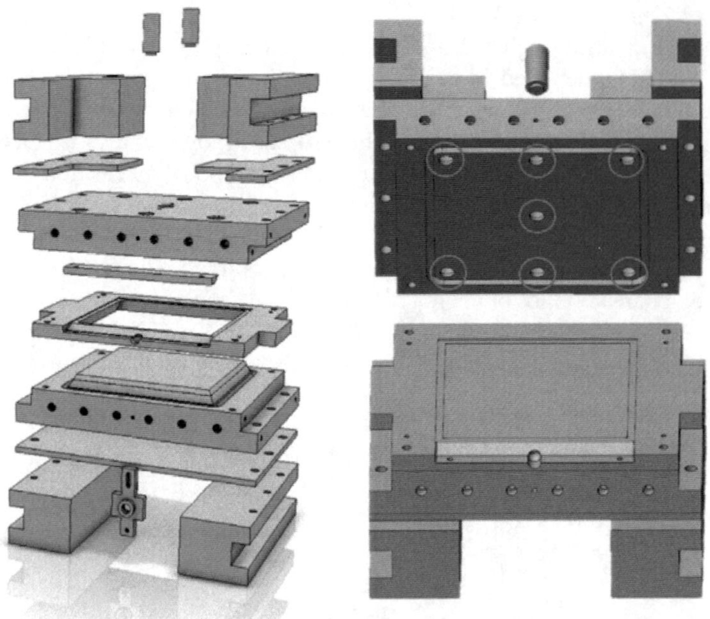

Figure 4. CAD design of the RTM mould for manufacturing composite plates (left) showing openings for integration of sensors in the upper part of the mould (right).

2.3.2. Flow Simulation

In the composites processing community, it is commonly accepted to model the fibrous preform material as porous media and the flow of fluids through the preform according to Darcy's law [2, 15, 16], a simple mathematical model relating the flow velocity to the driving pressure gradient involving the dynamic viscosity of the fluid as well as the permeability of the reinforcing material [17, 18].

For the work presented in this paper, numerical flow simulations were run on a model of the RTM mould cavity shown in Figure 4. The preform was modelled by means of nine layers representing the materials layers. Both, permeability of the reinforcing material and dynamic viscosity of the test fluid used for the mould filling tests were characterized experimentally as described in the results section of this paper. Further details about the numerical flow simulations are omitted here due to space limitations for this paper.

Supplementing information about the mould filling trend and the overall mould filling time, the fluid pressure characteristics at the location of the pressure point sensors in the RTM mould were extracted as major results of the numerical flow simulations runs.

3. Results

3.1. Test Fluid Viscosity

As shown in Figure 5, the temperature dependent viscosity characteristics of the test fluid were characterized between 10 °C and 35 °C, a range relevant for experiments at conventional laboratory conditions. In this temperature range, the viscosity characteristics decrease from about 0.10 Pas to about 0.04 Pas following a slightly nonlinear trend.

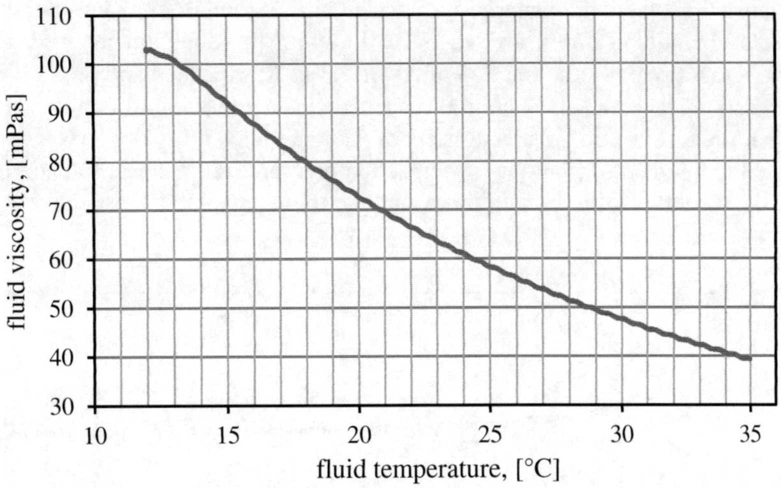

Figure 5. Temperature-dependent viscosity characteristics of the test fluid.

3.2. In-Plane Permeability

The glass fiber twill woven fabric was characterized at three levels of nominal fiber volume fraction in the range of 0.45 to 0.53 with five repetitions at each level. Plotted at logarithmic scale against fiber volume fraction, the major and minor in-plane permeability data k1 and k2, respectively, follow linear trends as shown in Figure **6**.

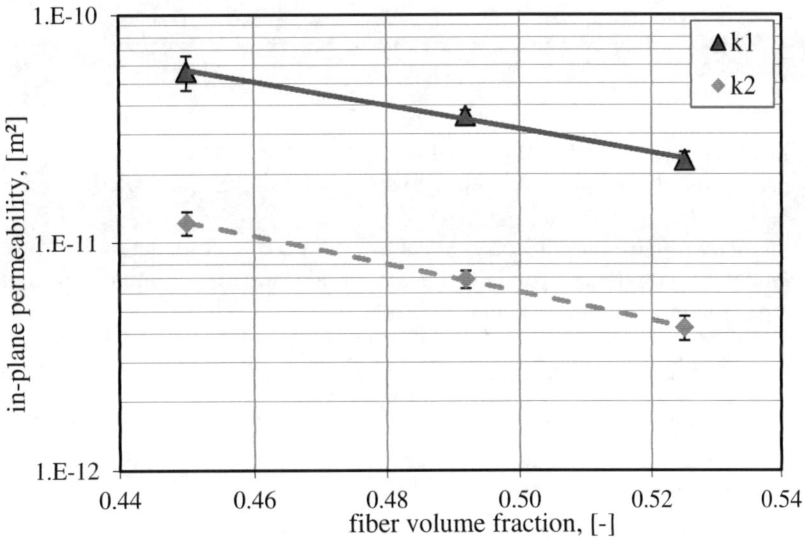

Figure 6. In-plane permeability characteristics of the glass fiber twill woven fabric.

3.3. Out-of-Plane Compressibility

The glass fiber twill woven fabric was experimentally characterized with respect to its out-of-plane compressibility in the dry, i.e. unsaturated, and saturated state. For both of the two configurations, ten material stacks, each comprising nine layers of the material, were tested in order to add statistical significance to the resulting data. For the experiments in the saturated state, the sample stacks were wetted in a pot of test fluid for three minutes, followed by a dripping period of five minutes. All of the material relaxation tests were run with the following test parameters:

IOP Conf. Series: Materials Science and Engineering **406** (2018) 012060 doi:10.1088/1757-899X/406/1/012060

- in the compaction stage, the material was loaded with a compression speed of 1 mm/min until a final stack height of 2 mm (which is in line with the situation in the RTM mould and corresponds to a fiber volume fraction of about 51 %) was reached and
- in the relaxation stage, the stack height was frozen for a period of 10 minutes in order to study the internal stress relaxation characteristics of the material.

Figure 7 shows the trends of dry and wet compaction pressure with respect to experimental time in terms of average data together with the minimum and maximum envelopes, respectively, from each set of ten experiments.

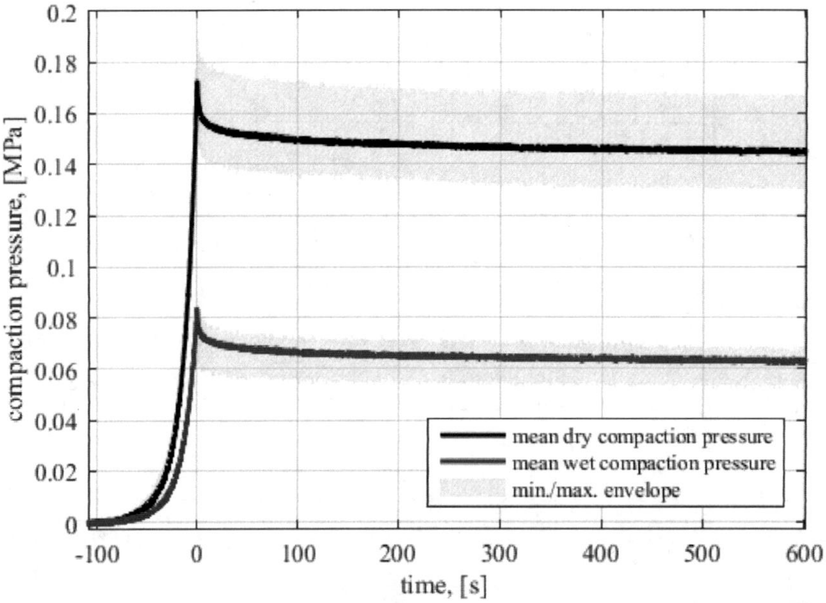

Figure 7. Relaxation characteristics of the glass fiber twill woven fabric stacks as plots of average unsaturated and saturated compaction pressure together with the associated confidence envelopes.

The result plots shown in Figure 7 clearly reveal the decrease of compaction pressure as a result of wetting the woven fabric with the test fluid. The average and standard deviation values of maximum as well as final compaction pressure are collected in Table 1.

Table 1. Overview of dry and wet compaction pressure data.

material state	max. compaction pressure, [MPa]		final compaction pressure, [MPa]		compaction pressure reduction, [MPa]	
	mean	std. dev.	mean	std. dev.	mean	std. dev.
dry	0.171	0.016	0.143	0.013	0.028	0.004
wet	0.085	0.008	0.063	0.005	0.022	0.003
drop	0.085 (50.1 %)		0.080 (55.7 %)		0.006 (21.4 %)	

Comparison of the average maximum and final compaction pressure in the dry and wet material state reveals a drop of about 50 % and 56 %, respectively, caused by the wetting. This level of relative compaction pressure drop was used for analyzing the process pressure characteristics as will be described in the subsequent section.

3.4. Analysis of RTM Process Pressure Characteristics

In Figure 8, process pressure characteristics are shown as acquired by means of three pressure point sensors mounted to the RTM mould described in Section 2.3.1. From these characteristics, the stages of preform compaction (due to closing of the RTM mould) and preform impregnation can be separated.

During the preform compaction stage, an increase of compaction pressure can be observed which reaches a steady level for all of the three sensors. However, the distinct compaction pressure levels vary among the sensors as a result of local preform inhomogeneity: locations of fiber bundle crossing points obviously show significantly higher levels of preform compaction pressure as regions in between such crossing points.

In the preform impregnation stage however, the three sensors show strongly varying pressure characteristics. The first pressure sensor, which is saturated by the test fluid immediately after starting fluid injection, shows a rapid increase of the pressure data which results from the fluid injection pressure level dominating over the preform compaction pressure. The other two pressure sensors show different characteristics: At first, a slight and steady decrease of the signal can be observed, which is attributed to capillary effects causing partial wetting of the preform at the location of the sensors. This is followed by a rapid decrease of the pressure signal as soon as the macroscopic flow front reaches the location of the sensor. Following this short period of preform wetting, the pressure signal increases as the flow front advances further towards the location of the vents. As soon as the fluid reaches the vents, another rapid increase of the pressure signal can be observed. Here, backpressure is building as a result of increased flow resistance inside the venting tubes. As a result, the amount of pressure increase is highest at the sensor mounted closest to the vents while it is smallest at the sensor closest to the point of fluid injection.

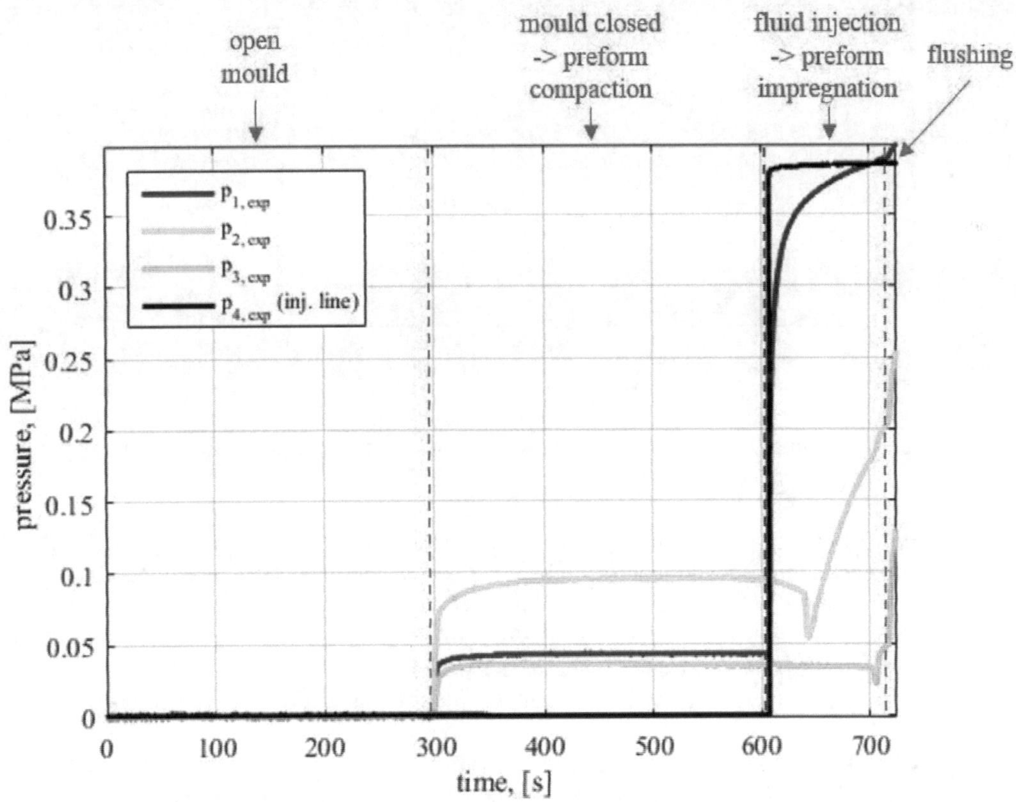

Figure 8. Process pressure characteristics from pressure point sensors mounted to the RTM mould.

IOP Conf. Series: Materials Science and Engineering **406** (2018) 012060 doi:10.1088/1757-899X/406/1/012060

As discussed in Section 3.3, preform compaction pressure drops by a factor of $s_{Drop} \approx 0.56$ as a result of wetting of the glass fibre twill woven fabric. This finding can be used to isolate the hydraulic fluid pressure $p_{F,k}$ (index k denoting the number of the pressure sensor, i.e.: $k = \{1..3\}$) from the preform compaction pressure $p_{C,k}$ in the characteristics p_k during the fluid impregnation stage:

- at first, the level of local dry preform compaction pressure $p_{C,dry,k}$ is extracted for each of the three pressure point sensors by calculating the average pressure value from the latest period of data in the preform compaction stage,
- from this, the local pressure drop caused by preform wetting can be compensated from the pressure data acquired during the preform impregnation stage in order to isolate the portion of hydraulic fluid pressure: $p_{F,k} = p_k - (1 - s_{Drop})p_{C,dry,k}$.

Figure 9 shows the isolated fluid pressure data (dashed lines) as a result of compensating the wetting effect from the process pressure characteristics (thin solid lines).

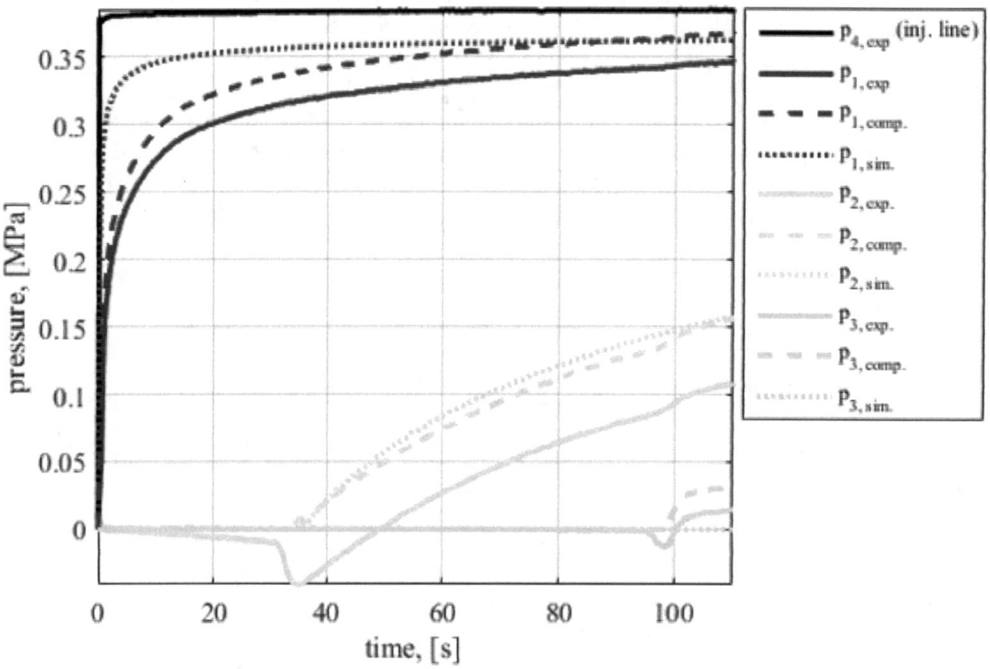

Figure 9. Process pressure characteristics during the preform impregnation stage with a compensation of the wetting effect on the preform compaction pressure.

In addition, the resulting hydraulic fluid pressure characteristics can now be compared with fluid pressure data obtained from flow simulation runs as shown in Figure 9. There, the data obtained from the simulation runs (thick solid lines) are in good agreement with the hydraulic fluid pressure data derived from the process pressure characteristics. Although a temporal offset is apparent in the direct comparison of the data, the overall trend as well as the final level of fluid pressure is well comparable. The temporal offset can be attributed to experimental imperfections such as unwanted handling effects on the rather flexible glass fiber woven fabric or race tracking effects along the side edges of the preform inside the mould. However, this is not affecting the validity of the approach presented in this work: RTM process pressure data can be partitioned into their contributions from preform compaction and hydraulic fluid pressure, respectively.

4. Summary and Conclusions

This paper focusses on an experimental analysis of process pressure characteristics acquired in RTM experiments involving a two-sided mould with integrated pressure sensors. The process pressure data acquired with the sensors shows two contributions: the preform compaction and fluid pressure, respectively, both varying with time and location inside the mould. The work reveals an approach for partitioning the fluid pressure from the preform compaction pressure involving a simple model for the preform compaction pressure drop caused by the wetting of the preform material. The amount of compaction pressure drop depends on the fibrous reinforcing material and needs to be experimentally determined by means of compaction tests. The resulting hydraulic fluid pressure characteristics are compared to fluid pressure data predicted by flow simulation, revealing a reasonable level of agreement.

The approach presented in this paper allows for subsequent usage of the isolated fluid pressure characteristics for online computation of fluid flow velocity and its active control following the concept of model-based processing of composites. Future work will concentrate on validating the methodology presented in this paper on other types of preform materials, a study of process parameters as well as tests with chemically reactive resin systems.

Acknowledgments

The work presented in this paper has been elaborated within the context of project "MoVeTech", financed through the Austrian Ministry for Transport, Innovation and Technology in the frame of the program "Take Off", which is greatly acknowledged.

References

[1] Neitzel M and Mitschang P (eds) 2004 *Handbuch Verbundwerkstoffe: Werkstoffe, Verarbeitung, Anwendung* (München: Hanser)
[2] Advani S G and Sozer E M 2011 *Process modeling in composites manufacturing* 2nd edn (Boca Raton, FL: CRC Press)
[3] Robitaille F and Gauvin R 1998 *Polymer Composites* **19** 198–216
[4] Robitaille F and Gauvin R 1998 *Polymer Composites* **19** 543–57
[5] Robitaille F and Gauvin R 1999 *Polymer Composites* **20** 48–61
[6] Yenilmez B, Senan M and Murat Sozer E 2009 *Composites Science and Technology* **69** 1710–9
[7] Michaud V 2016 *Transp Porous Med* **115** 581–601
[8] Potluri P and Sagar T V 2008 *Composite Structures* **86** 177–85
[9] Chen B and Chou T-W 1999 *Composites Science and Technology* **59** 1519–26
[10] Chen B and Chou T-W 2000 *Composites Science and Technology* **60** 2223–31
[11] Chen B, Lang E J and Chou T-W 2001 *Materials Science and Engineering: A* **317** 188–96
[12] Chen B, Cheng A-D and Chou T-W 2001 *Composites Part A: Applied Science and Manufacturing* **32** 701–7
[13] Lekakou C, Johari, M. A. K. and Bader M G 1996 *Polymer Composites* **17** 666–72
[14] Grössing H, Becker D, Kaufmann S, Schledjewski R and Mitschang P 2015 *Polymer Express Letters* **9** 129–42
[15] Henry Darcy 1856 *Les fontaines publiques de la ville de Dijon: Exposition at application des principes a suivre et des formules a employer dans les questions de: Distribution d'eau ouvrage terminé par un appendice relatif aux fournitures d'eau des plusieurs villes au filtrage des aux et a la fabrication des tuyeaux de fonte, de plomb, de tole et de bitume* (Paris: Victor Dalmont, Libraire des Corps imperiaux des ponts et chaussées et des mines)
[16] Bear J 1988 *Dynamics of fluids in porous media* (New York: Dover)
[17] Fauster E, Berg D C, May D, Blößl Y and Schledjewski R 2018 *Advanced Manufacturing: Polymer & Composites Science* 24–40
[18] Konstantopoulos S, Hueber C, Mühlbachler E and Schledjewski R 2016 *Advanced Manufacturing: Polymer & Composites Science* **2** 57–66

IOP Conf. Series: Materials Science and Engineering **406** (2018) 012061 doi:10.1088/1757-899X/406/1/012061

Mechanical and permeability properties of NCF with different knitting condition

Akio Ohtani[1]*, Shintaro Agata[2], Masayuki Kitamura[3] Asami Nakai[2]

[1] Faculty of Fiber Science and Engineering, Kyoto Institute of Technology, Japan
[2] Mechanical and Engineering, Faculty of Engineering, Gifu University, Japan
[3] Hokuriku Fiber Glass Co., Ltd., Japan

*a-ohtani@kit.ac.jp

Abstract. A Non-Crimp Fabric (NCF) is composed of some layers which have large number of fiber bundles aligning in a certain direction and are stacked and fixed with a thin thread compared with the reinforcing fiber bundle by warp knitting technology. In the reinforced form of the composite material, there is no crimp of the fiber bundle compared with other fabric such as woven, knitted fabric or etc., and composite materials reinforced by NCF shows better mechanical properties. Since it is easy to control the orientation angle, it is also possible to control the mechanical properties as required. Therefore, in recent years, it has been widely used as a reinforcement of large-sized composite material structural members such as watercraft, automobile, wind turbine blades, and etc. Many studies on the mechanical properties of composite materials using NCF have been made so far, however, little attention has been paid to the difference in the tension and structure of the warp knitting yarn binding the reinforcing fiber bundles in NCF.

In this study, the effect of difference of tension and structure of warp knitting yarn on the structure of reinforcing fibers and the mechanical properties of composite materials were investigated. In addition, the effects of those differences in NCF on permeability properties were also investigated.

1. Introduction

A Non-Crimp Fabric (NCF) is one of a textile in which some layers of aligned large number of fiber bundles in same direction were stacked in different direction according to the requirement for the reinforcement of composite materials. These stacked layers are fixed together with a thin thread by using warp knitting technique. In the reinforcement of the composite material, there is no crimp of the fiber bundle compared to the woven fabric, and it shows better mechanical properties. Since the orientation angle can be controlled, it is also possible to control the mechanical properties according to the requirement. Therefore, in recent years, it has been widely used as a reinforcement of large-sized composite material structural members such as watercraft, automobile, blades of windmill, and etc. Many studies on the mechanical properties of composite materials using NCF have been made so far, however, little attention has been paid to the difference in the tension of the knitting yarn and the warp knit structure during the production of NCF [1].

IOP Conf. Series: Materials Science and Engineering **406** (2018) 012061 doi:10.1088/1757-899X/406/1/012061

In this study, the effect of the difference of the tension of warp knitted thread and its structure on the structure of reinforcing fiber bundles and the mechanical properties of the composite was investigated. In addition, permeability of the resin during the RTM process was also investigated.

2. Experimental method

2.1 Reinforced substrate

Table 1 shows details and appearance photographs of four types of NCFs with different warp knitting structure and knitting tension used in this study. In both cases, glass fibers were used as reinforcing fibers and reinforcing fibers were oriented in directions of 0° and 90°. The weight per unit area was 480 g/m^2 for the warp yarns and 480 g/m^2 for the weft yarns. When manufacturing 0°/90° NCF, 90° fiber bundle is crossed with a large number of 0° fiber bundles that are aligned and both are fixed by knitting yarn. At that time, the knitting yarn is crossed obliquely with respect to the 0° fiber bundle to fix the 90° fiber bundle. The warp knitting structure at this time is called tricot. On the other hand, it is also possible to fix the 90° fiber bundle by a warp knit structure called a chain, in which case the knitting yarns do not intersect the 0° fiber. Among the four types in Table 1, the upper two are NCFs that alternately apply tricot and chain, and this NCF is called Combination and represented by “C”. On the other hand, NCF in which 90° fiber bundles are fixed only with tricot structure is called Tricot and represented by “T”. Therefore, the amount of knitting yarn crossing over the 0° fiber bundle (referred to as the knitting yarn density for convenience) is twice that of the combination of tricot. The knitting yarn tension during NCF preparation was 1.8 cN and 7.0 cN. Due to the difference between these structures and tension, those with warp knitting structures in combination of 1.8 cN and 7.0 cN are called C_W and C_S, respectively, and tricot is likewise called T_W and T_S.

Table 1. Detail of each NCF.

	Construction	Tension [cN]	Picture
Cw	Combination	1.8	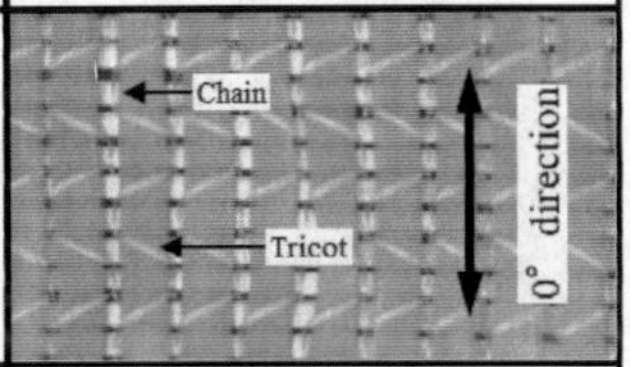
Cs		7	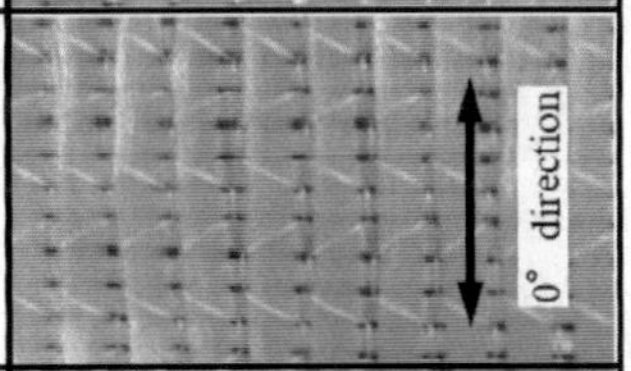
Tw	Tricot	1.8	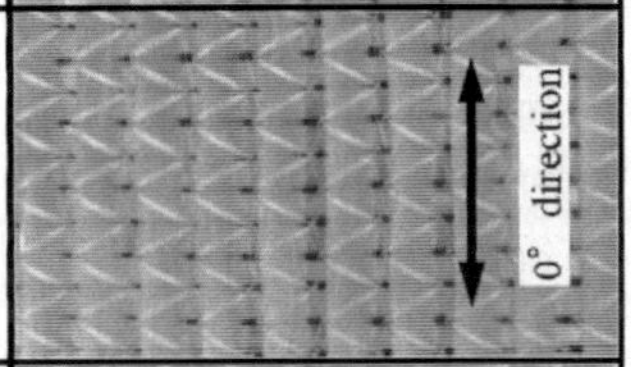
Ts		7	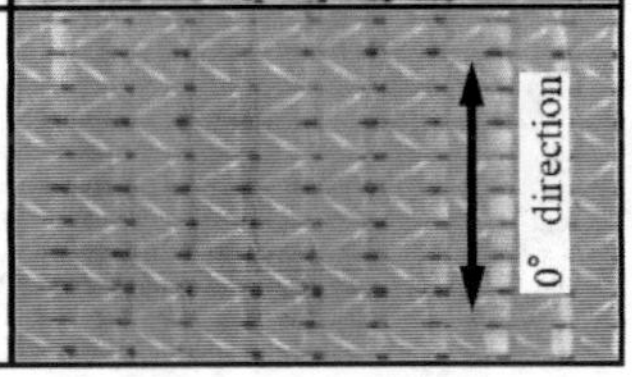

IOP Conf. Series: Materials Science and Engineering **406** (2018) 012061 doi:10.1088/1757-899X/406/1/012061

2.2 Test specimen preparation method and experiment method

The specimen used in this study was molded by Va-RTM using unsaturated polyester resin (Ligolac 150 HR BQNTNW manufactured by Showa Denko K.K.) as the matrix resin (Figure 1). Two layers of NCF shown in Table 1 were stacked symmetrically and the spiral tubes for supplying resin were arranged at both ends in the 0° direction of the reinforcement and covered with a vacuum bag and evacuated, and then impregnated with a resin and cured at room temperature. The permeability K_x was calculated by the following formula based on the Darcy's rule.

$$K_x = \frac{k \cdot \mu \cdot \phi}{2P}$$

Here, k [m^2/s] is the slope of the graph of the square of the flow front position and time, μ [Pa · s] is the viscosity of the resin, φ [-] is the porosity, and P [Pa] is the injection pressure is there. Flow front position was acquired from moving images shot at resin injection. Both the case where the curing agent was added and the case where it was not added were examined.

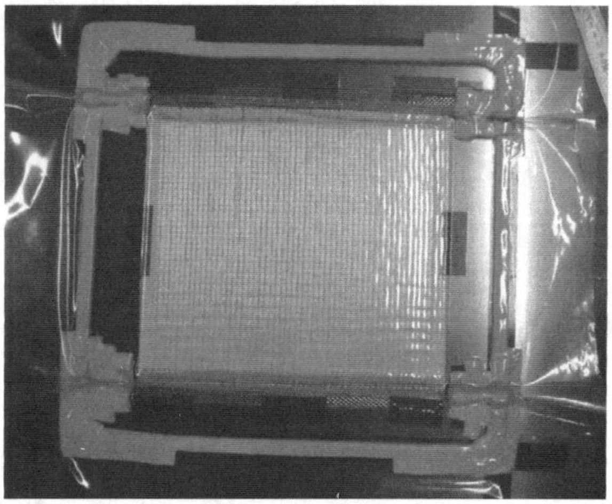

Figure 1. Va-RTM.

3. Experiment Result

An example of the relationship between the square of flow front position and time is shown in Figure 2. The plot was approximated to the least squares and the slope of the obtained straight line was substituted into the above equation to calculate the permeability. The obtained results are shown in Figure 3.

It was found that the permeability was higher in any of the specimens when the curing agent was added. It can be said that this is because the reaction heat was generated by adding the curing agent and the resin became easy to flow because the resin viscosity decreased. In both C and T, it was found that the higher the knitting yarn tension, the higher the permeability and the higher the permeability of T than C.

Next, the relationship between the permeability and the test piece thickness is shown in Figure 4. As the thickness increased, the permeability tended to increase. From the above results, it was found that when the thickness of the test piece is small, it is difficult for the resin to flow, and when the thickness of the test piece is large, the resin flows easily. This is because the higher the knitting yarn density and the higher the knitting yarn tension is, the more the fiber bundle converges and the fiber bundle sectional shape becomes closer to a circle, so the thickness of the test piece increases and the area of the resin flowing between the fiber bundles As a result, the permeability became higher.

IOP Conf. Series: Materials Science and Engineering **406** (2018) 012061 doi:10.1088/1757-899X/406/1/012061

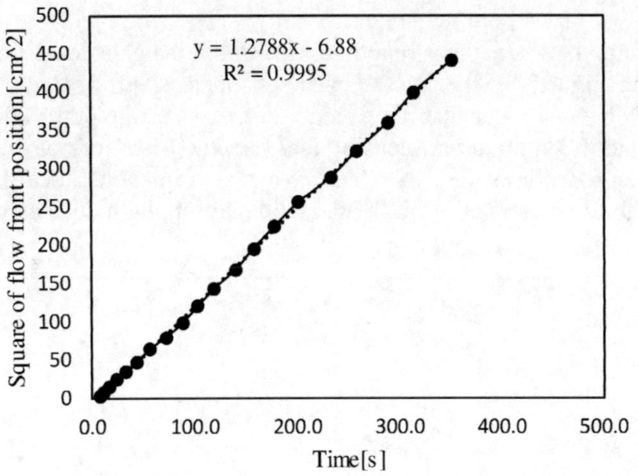

Figure 2. Relationship between square flow front position and time (C_W).

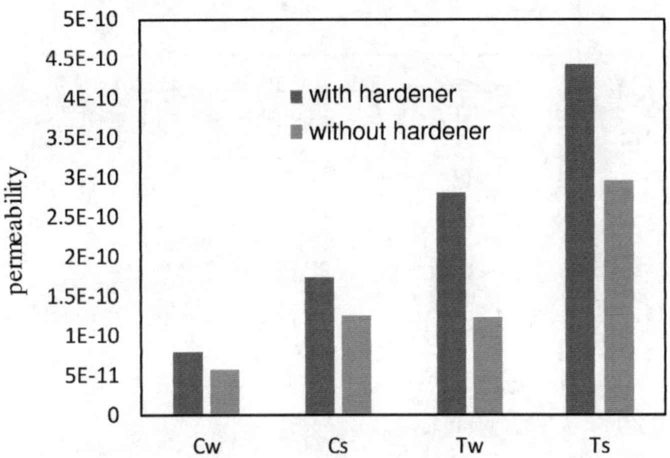

Figure 3. Permeability of each specimen

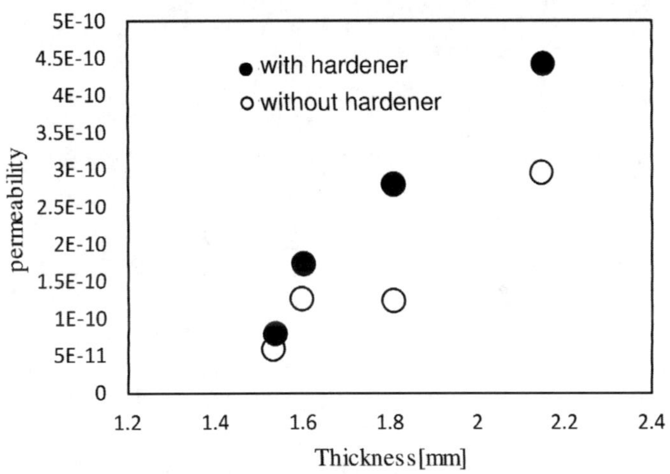

Figure 4. Relationship between permeability and thickness.

IOP Conf. Series: Materials Science and Engineering **406** (2018) 012061 doi:10.1088/1757-899X/406/1/012061

Next, the relationship between the strength obtained by the tensile test and the permeability coefficient is shown in Figure 5. Since Vf of each specimen is different, this intensity is a value normalized using Vf. It became clear that the strength decreases as the permeability increases. As the knitting yarn density and the knitting yarn tension increase, the force for converging the fiber bundle increases, so that nonuniformity in the test piece becomes conspicuous, local fibrous meandering is promoted , The permeability increased but at the same time the strength decreased.

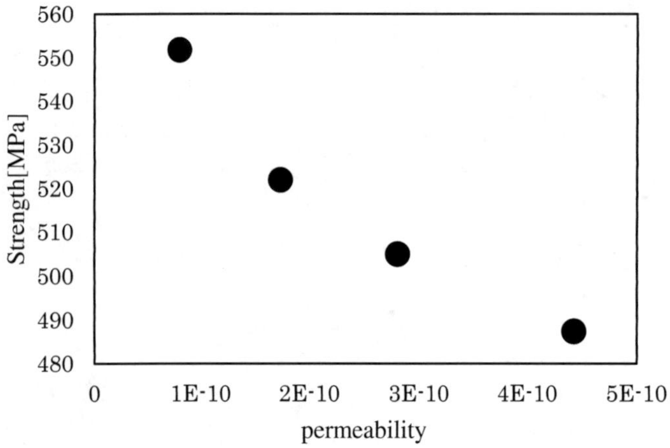

Figure 5. Relationship between strength and permeability.

References

[1] Lomov S V, 2011, *Non-crimp fabric composites: manufacturing, properties and applications*, p. 42-83.

Characterization and modeling of composite Vacuum Infusion process: influence of fabric type, resin viscosity and strain rate

Claire-Isabelle Zénone[1], Philippe Le Grognec[2] and Chung Hae Park[1]*

[1] IMT Lille Douai, Department of Polymers and Composites Technology and Mechanical Engineering, 941 rue Charles Bourseul, CS 10838, F-59508 Douai Cedex, France

[2] ENSTA Bretagne, Institut de Recherche Dupuy de Lôme, UMR CNRS 6027, 2 rue François Verny, F-29806 Brest Cedex 9, France

*chung-hae.park@imt-lille-douai.fr

Abstract. In Vacuum Infusion (VI) process, used to manufacture large composite structures, the fibrous preform undergoes large deformations during the impregnation due to the use of a flexible vacuum bag. Moreover, the resin pressure difference between the upstream and downstream sides of the preform leads to a non-uniform preform thickness during the mold filling process, which may degrade the part quality. Thus, the prediction of resin flow and preform thickness variations is a very important issue. However, fluid-structure interaction modeling is a difficult task, mainly due to the complex mechanical behavior of the preform. Indeed, the through-thickness deformation of fabrics depends on many factors such as loading type, fabric saturation and fabric type, and also on strain rate effects, as already observed in the literature. In this study, one focuses on the preform deformation during decompaction. Experimental measurements were performed using a universal testing machine with different strain rates, for two different types of fiberglass fabrics, i.e. a random mat and a twill weave fabric, in a dry or wet state saturated by liquids with different viscosities. From these fabric characterization tests, a constitutive model was deduced, which was subsequently implemented in a numerical software to simulate preform decompaction during VI process. Some VI process experiments were also conducted in order to measure the liquid pressure gradient and the preform thickness profile during the mold filling stage and validate the numerical results. It was found from fabric characterization tests that strain rate and viscosity have no significant impact on the preform response in decompaction, in the tested ranges of strain rate and viscosity, for both the random mat and twill weave fabric. However, the analysis of VI process experimental results highlights that, in the case of the twill weave fabric, a time lag appeared between the resin pressure rise and the preform thickness increase, which suggests that a viscoelastic behavior was nevertheless present; whereas in the case of the random mat, it seems thus sufficient to model the preform deformation occurring during the impregnation of VI process with a simple time-independent constitutive law.

1. Introduction

Composite manufacturing processes by Liquid Composite Molding (LCM) are closed-mold techniques where a liquid resin is impregnated into a dry fibrous reinforcement pre-placed in a mold. LCM processes offer some technical advantages such as applicability to complex part geometries, high structural performance and low manufacturing cost. In addition, health risks to operators due to styrene vapors are eliminated contrary to open-mold techniques. Vacuum Infusion (VI) process, also known as

Vacuum Assisted Resin Infusion (VARI) or Resin Infusion under Flexible Tooling (RIFT), is one of LCM processes particularly suitable for the production of large composite structures. Tooling cost is reduced by the use of a single side mold on which the preform is laid up. A thin plastic film, referred to as vacuum bag, replaces the upper rigid half mold [1].

Due to the flexibility of the vacuum bag, the fibrous preform undergoes large deformations during the process. Firstly, while air is extracted from the closed cavity by a vacuum pump before the resin infusion, the preform is compacted by the atmospheric pressure applied on the flexible vacuum bag; then, once the resin valve is opened, the liquid resin impregnates the compacted dry reinforcement and inlet resin pressure leads to a local decompaction state of the preform during the mold filling process. Compaction pressure on the fibrous reinforcement gradually reduces due to fluid pressure which increases as the resin flows, inducing variations in fiber volume fraction and therefore in permeability [2]. The differential pressure between the upstream and downstream sides of the preform may degrade the final part quality as it gives rise to non-uniform final thickness and fiber volume fraction distribution [3]; whereas permeability changes make it difficult to predict process cycle times [4]. Thus, this strong coupling between the resin flow and the fabric deformation has to be properly considered in the numerical simulation of the process to improve final part quality and optimize process cycle time. It is however a difficult task mainly due to the complex mechanical behavior of the preform. Indeed, the through-thickness deformation of fabrics is affected by many factors such as fiber architecture, loading type and saturation state. Furthermore, available results in the literature do not present a consistent agreement on the presence of significant viscoelastic effects during the mechanical response of preforms submitted to compaction pressure. Modeling of the mold filling stage of VI process is therefore challenging, due to the complex characterization of the mechanical behavior of preform throughout the impregnation. As a matter of fact, the preform is decompacted during the impregnation by the resin, i.e. in wet conditions; whereas it is initially dry when compacted at the pre-filling stage of VI process.

This study focuses on the factors which may influence the through-thickness behavior of the fabric during the VI process filling stage. Different effects on the mechanical response of the preform were examined, in particular during decompaction, such as the influence of strain rate, viscosity of the liquid resin and fabric architecture. The objective is to dispose of a more reliable characterization of fabrics behavior using a universal testing machine. The testing conditions may be chosen so as to reproduce at best the real conditions occurring during an impregnation experiment. The final purpose is to improve the modeling of VI process through the implementation of the most appropriate constitutive law for the mechanical behavior of the fabric into the numerical model.

To this end, experimental measurements of preform compaction and decompaction were carried out on a universal testing machine, with different strain rates relevant to the impregnation experiments of VI process, for two different types of fiberglass fabrics, i.e. a random mat and a twill weave fabric, in a dry or wet state saturated by liquids with different viscosities. Some VI process experiments were performed in parallel in order to obtain the liquid pressure gradient within the fabric and the preform thickness profile during the mold filling stage. Fabric characterization experiments were performed in the same conditions as VI experiments, i.e. with the same fabric types, numbers of plies, compaction pressure and impregnating liquid. The liquid pressure gradient and the preform thickness profile were measured respectively by pressure transducers and by vertical displacement sensors, integrated into the mold set-up. Finally, these two data were compared to numerical results obtained with the previously characterized constitutive law.

2. Literature review

The mechanical behavior of fabrics has already been studied by many researchers with the objective to better understand the physics of VI process and to improve its modeling. Several authors performed compaction experiments on fabrics so as to highlight the relevant factors which may influence their mechanical response. These main factors are listed below.

2.1. Fabric architecture
Many parameters affect the reinforcement behavior such as the fabric architecture, as seen in Chen et al. [5]. These authors conducted compaction experiments on three types of dry fiberglass preforms with different architectures, i.e. continuous strand mats, plain woven materials and unidirectional knitted fabrics. They pointed out the major factors contributing to the compaction behavior according to the reinforcement architecture, such as nesting, which was particularly dominant for woven fabrics but negligible for continuous strand mats.

2.2. Influence of loading type and saturation
The mechanical behavior of the preform is also dependent on the loading type. Correia et al. [6] identified different behaviors between compaction and decompaction of fiber reinforcements. Furthermore, they highlighted differences in fabric properties, according to whether experiments were carried out in dry or wet conditions. As an example, fiber friction coefficient is reduced by the lubrication effect in wet conditions, leading to a higher preform compaction degree at the same applied pressure as in dry conditions. Yenilmez and Sozer [7] reproduced and distinguished, in the same characterization experiment using a special set-up, fabric dry loading and fabric unloading once impregnated, to investigate fabric behavior during both the pre-filling and filling stages of VI process. Saunders et al. [8] analyzed the effect of saturation on the preform behavior, by testing three different resin viscosities on several woven fabrics. No significant differences were detected between the compaction data associated with the different resins, which indicated that the resin viscosity had probably no effect on the mechanical behavior of the preform in compaction. As experiments were not conducted for unloading cases, the conclusion cannot apply to the reinforcement mechanical behavior in decompaction.

2.3. Viscoelastic effects
In the literature, most of researchers assumed a time-independent relationship between compaction pressure and local fiber volume fraction, considering an instantaneous response of the preform when the applied pressure varies. Some authors, though, proposed a time-dependent model displaying a viscoelastic constitutive behavior for the preform. Saunders et al. [9] suggested that a time-independent relationship was only appropriate to describe the constitutive response of dry compaction (which does not exhibit dependence on the strain rate), and that it was not suitable in the case of wet hysteresis experiments which revealed viscoelasticity effects. The same authors [8-9] highlighted the influence of strain rate, which was proved to affect the wet plain weave compaction behavior. As the strain rate increased, a greater compaction pressure was needed to reach the same fiber volume fraction, suggesting that a part of the applied pressure was used to remove the resin. They interpreted that viscoelastic characteristics of wet compaction were mainly due to the resin flow and they developed a viscoelastic model which considered elastic and viscoelastic effects from the fiber preform, and also viscous effects from resin. Bickerton et al. [10] also investigated the apparent viscoelasticity of three dry common LCM reinforcements submitted to strain-rate dependent and hysteresis experiments. They observed that the stress could be generally expressed as a function of strain, strain rate and time, as formulated in Yenilmez et al. [11], where the authors described the behavior of a woven fabric during all stages of VI process by testing several viscoelastic models.

3. Materials and experimental procedures
Two sets of experiments were carried out: fabric decompaction characterization experiments and fabric impregnation experiments by VI process. The decompaction constitutive law of the impregnated preform derived from the characterization experiments will serve as an input in the subsequent numerical model.

3.1. Materials

3.1.1. Fabrics. Two types of fiberglass fabrics with different architectures were tested: a random mat and a twill weave fabric, whose main characteristics are presented in Table 1. The same number of plies was retained for both experiments, whatever the fabric considered. Each fabric was selected in such a way that both preforms display the same thickness when compacted with the same pressure of 0.09 MPa. Therefore, their behavior in decompaction should be compared easily.

Table 1. Fabric properties.

	Random mat	Twill weave
Fabric reference	Unifilo® 450	1500 S3
Supplier	Vetrotex	Chomarat
Fabric pattern	Random	3/3 oblique pattern
Fiber type	E-glass	E-glass
Areal weight (g.m^{-2})	450	1500
Average ply thickness (mm)	2.40 ± 0.08	1.80 ± 0.07
Width and length plies for compaction test (mm x mm)	80 x 80	80 x 80
Width and length plies for infusion test (mm x mm)	150 x 300	150 x 300
Number of plies for both experiments	4	3
Average thickness of dry preform compacted at 0.09 MPa (mm)	2.81 ± 0.01	2.80 ± 0.08
Average fiber volume fraction of dry preform compacted at 0.09 MPa	0.24 ± 0.01	0.62 ± 0.02

3.1.2. Liquids used to impregnation of fabrics. Instead of synthetic resins classically used for VI process, two different mineral oils were chosen here so as to better evaluate the influence of viscosity on fiber lubrication and avoid resin polymerization during characterization and impregnation experiments: a low viscosity oil and a medium viscosity oil, whose viscosities are respectively in the range of 0.060-0.070 Pa.s and 0.250-0.350 Pa.s at room temperature (~ 20°C).

3.2. Fabric deformation characterization experiments

Fabric deformation characterization tests were performed on a universal testing machine (Zwick 1474). The experimental set-up in dry conditions consisted in two compression platens between which the dry preform was laid. In the case of wet experiments, the set-up was completed with a perforated compression platen, on which the preform was laid, allowing for the outflow of excess oil which was collected into a receptacle placed below. Two plastic films were put above and below the preform in order to ensure oil flow only from the side (Figure 1). The tested preforms were initially impregnated ply by ply using a syringe. In all compaction test cases listed in Table 2, the same loading pattern was used, namely a compressive loading at the chosen strain rate until the compaction state under 0.09 MPa, then a waiting period of 10 min in the compacted state and finally an unloading step until the fully decompacted state at the same strain rate as before. Each characterization experiment was performed three times so as to check repeatability.

Figure 1. Set-up for characterization experiments on fabrics in wet conditions.

Table 2. Compaction experiment conditions.

	Fabric type	Saturation state	Viscosity (Pa.s)	Strain rate (mm.min^{-1})
Compaction 1	Random mat	Dry	-	0.2
Compaction 2	Random mat	Dry	-	1.0
Compaction 3	Random mat	Wet	0.300	0.2
Compaction 4	Random mat	Wet	0.300	1.0
Compaction 5	Random mat	Wet	0.070	0.2
Compaction 6	Random mat	Wet	0.070	1
Compaction 7	Twill weave	Dry	-	0.2
Compaction 8	Twill weave	Dry	-	1
Compaction 9	Twill weave	Wet	0.300	0.2
Compaction 10	Twill weave	Wet	0.300	1
Compaction 11	Twill weave	Wet	0.070	0.2
Compaction 12	Twill weave	Wet	0.070	1

3.3. VI process experiments

The experimental set-up for the infusion tests is shown in Figure 2. The test bench consisted in a special set-up, whose bottom was a PMMA platen with five aligned holes to insert pressure transducers and two other holes for an oil inlet and an air outlet. The preform, composed of several plies (Table 1), was laid on the bottom platen. A thick distribution medium was placed between the oil inlet and preform, and also between the air outlet and the preform, in order to ensure a straight oil flow front and avoid pressure losses. The oil inlet was connected by pipes to the oil pot and the air outlet to a vacuum pump. Valves on the inlet and outlet pipes allowed to control the start time of the preform compaction and oil impregnation. A vacuum bag was laid on the preform and the mold was closed by a sealing tape positioned between the bottom platen and the vacuum bag. One pressure transducer (sensor 0) measured the oil pressure just before the tip of the preform and the other ones were placed respectively at 20 mm (sensor 1), 60 mm (sensor 2), 95 mm (sensor 3) and 135 mm (sensor 4) from the tip of the preform. Four vertical displacement sensors, supported by a metallic structure, were put in contact with the top surface of the mold, i.e. the vacuum bag, and were positioned at the same planar locations as the pressure transducers. Three VI experiments were carried out, whose characteristics (oil viscosity and filling time) are listed in Table 3. Both types of fabric were tested with the two different oils, but only the data listed in Table 3 were recorded, because the impregnation of random mat was too fast to obtain meaningful

IOP Conf. Series: Materials Science and Engineering **406** (2018) 012062 doi:10.1088/1757-899X/406/1/012062

measurement data with the low viscosity oil. Thus, the twill weave fabric has been tested using both low and medium viscosity oils, whereas the random mat was tested only with the medium viscosity oil. For all experiments, vacuum pressure was -0.09 MPa, which was equivalent to a compaction pressure of 0.09 MPa on the top surface of the preform. Vacuum was applied during 10 min so as to obtain a stable and homogenous compaction of the preform.

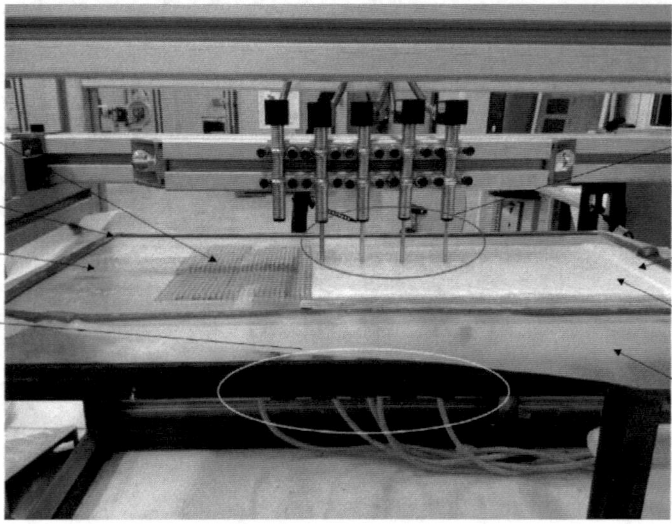

Figure 2. Test bench of vacuum infusion experiments.

Table 3. Vacuum infusion experiment conditions.

	Fabric type	Oil viscosity (Pa.s)	Filling time (s)
Infusion 1	Random mat	0.264	63
Infusion 2	Twill weave	0.249	853
Infusion 3	Twill weave	0.070	229

4. Results and discussion

Owing to the non-uniformity in fiber structure of fabrics, each set of characterization experiment was conducted at least three times. In each case, the average value of $h/h_{compacted}$ is depicted vs. the compaction pressure, where h is the current thickness measured instantaneously in response to the compaction pressure and $h_{compacted}$ is the thickness of the specimen measured just before decompaction. Only the decompaction stage of fabrics will be studied in the sequel.

Experimental results for random mat characterization are shown in Figure 3, for all values of saturation and strain rate. All the curves present a very similar tendency. At strain rates of 1 mm.min^{-1} and 0.2 mm.min^{-1}, the random mat shows the same decompaction behavior, regardless of saturation, which means that viscosity, in the considered range, has no significant impact on the decompaction behavior.

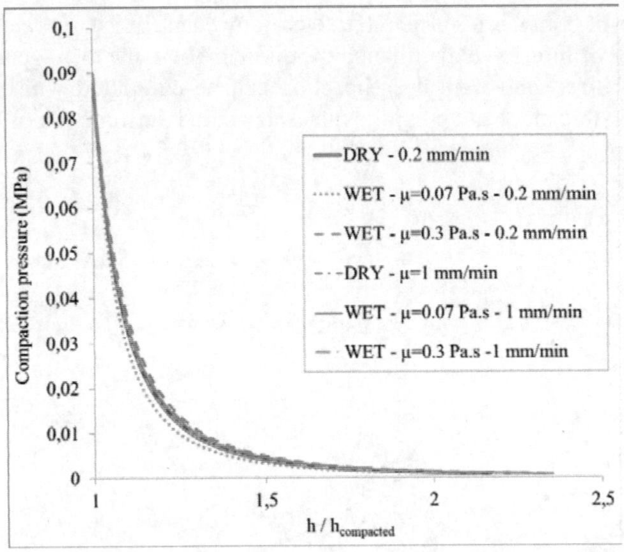

Figure 3. Decompaction test results for random mat with strain rates of 0.2 mm.min^{-1} and 1 mm.min^{-1} for all the saturation cases.

The same conclusion can be drawn for experimental results concerning twill weave fabric characterization presented in Figure 4. No significant effect of strain rate or viscosity appears to affect the mechanical behavior during the decompaction of the twill weave fabric in the chosen ranges.

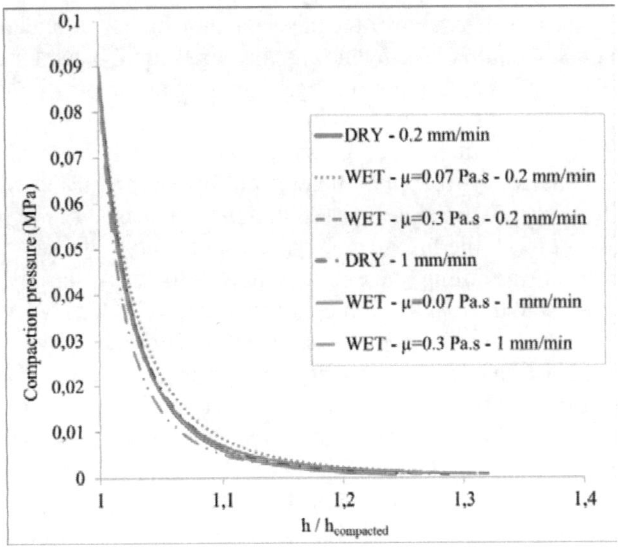

Figure 4. Decompaction test results for twill weave fabric with strain rates of 0.2 mm.min^{-1} and 1 mm.min^{-1} for all the saturation cases.

Comparison between Figure 3 and Figure 4 reveals however that the two fabrics (with different architectures) exhibit a different behavior in decompaction even if both fiberglass fabrics were unusually compacted to the same thickness (see Table 1). The final thickness of the specimens were 6.45 mm for random mat and 3.65 mm for twill weave fabric at the end of the tests, which shows that fabric type architecture strongly influences the mechanical response in decompaction.

The characterization of both fabrics is not affected by any change in strain rate and saturation, at least in the considered ranges of infusion experimental conditions. Then, the average mechanical behavior of random mat and twill weave fabric in decompaction can be compared with the results of infusion experiments, namely the thickness and pressure values measured during infusion experiments by sensor 1 (Figures 5 and 6).

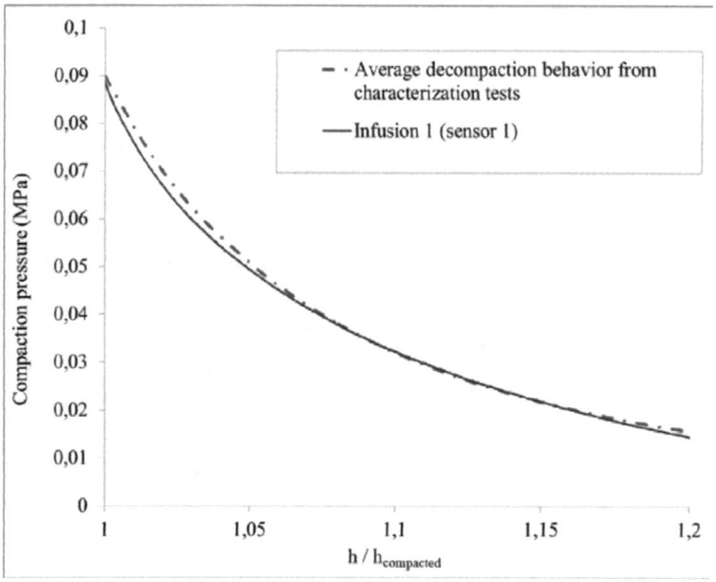

Figure 5. Comparison between the average mechanical behavior of random mat specimens obtained by characterization experiments and the mechanical behavior of random mat measured during impregnation tests.

For random mat, infusion experimental conditions correspond to the ranges of viscosities and strain rates employed during its characterization, i.e. the preform was impregnated with a medium viscosity oil (μ=0.264 Pa.s) and the average strain rate was 0.6 mm.min^{-1}at sensor 1. The behavior of the preform measured at sensor 1 did not reach the fully decompacted state, due to the rapidity of oil impregnation into the random mat preform (the filling stage is only 63 s). However, the resin pressure reached the value of 0.07 MPa, equivalent to the compaction pressure of 0.2 MPa. Characterization results depicted in Figure 5 are consequently focused until this compaction pressure. These results reveal a good agreement with the decompaction behavior observed during infusion tests. Thus, a time-independent constitutive law seems to be appropriate for random mat and may be characterized by simple decompaction tests in the average conditions of viscosity and strain rate observed during the mold filling stage of VI process.

On the contrary, comparison between characterization tests and infusion experiments on twill weave fabric (Figure 6) shows strong discrepancies between both infusion tests (Infusion 2 and Infusion 3) and the average mechanical behavior provided by the compression testing machine. A constitutive law derived from characterization tests does not seem to be representative of the real behavior occurring during the impregnation of the twill weave fabric, even in experimental conditions close to infusion conditions. In addition, a higher viscosity appears to affect even more the mechanical behavior during infusion. As it has been shown (see Figure 4) that neither the viscosity nor the strain rate in the considered ranges have any effect on the mechanical behavior, the differences in fabric behavior are probably induced by another phenomenon. For example, the long filling time, which could result in some relaxation during the filling stage, may be responsible for this disagreement. Thus, it is not finally relevant to consider only a time-independent model to represent the mechanical behavior of twill weave fabric during infusion.

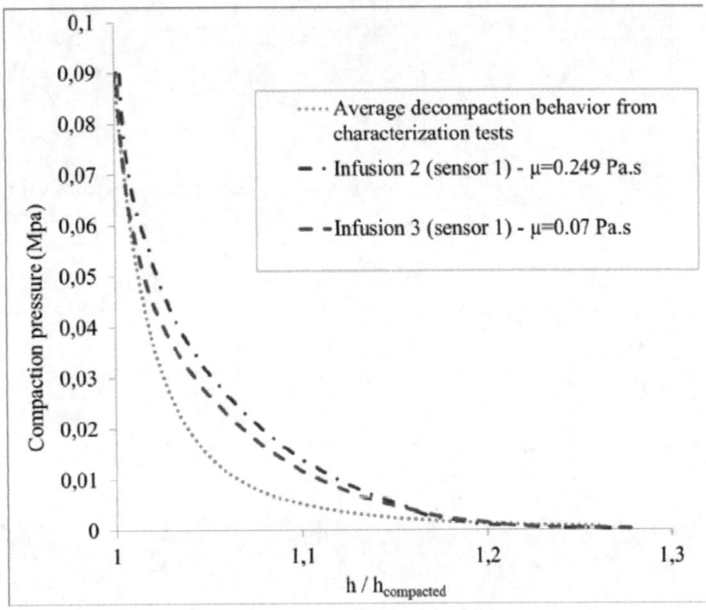

Figure 6. Comparison between the average mechanical behavior of twill weave fabric specimens obtained by characterization experiments and the mechanical behavior of twill weave fabric measured during impregnation tests.

5. Conclusion

The comparison between two fiberglass fabric types, a random mat and a twill weave fabric, whose thickness is the same in dry conditions and compacted at 0.09 MPa, demonstrated that the mechanical behavior in decompaction is highly influenced by the fabric architecture. Characterization of the decompaction behavior of these two fabrics using a universal testing machine, with two strain rates and two different viscosities, has highlighted that strain rate and viscosity had no particular effect on the response of the preform in decompaction, at least in the tested ranges of strain rate and viscosity. To identify the constitutive law of preforms undergoing transverse deformations in VI process, it seems therefore not necessary to reproduce strictly the same conditions of saturation and strain rate.

However, the comparison between these constitutive behaviors identified through characterization experiments and the fabric response occurring during infusion experiments shows that only the experiments performed on random mat appear to be representative of the real fabric behavior during impregnation, whereas the twill weave infusion experiments reveal some differences in terms of thickness, when compared to characterization experiments, which are not due to lubrication or strain rate effects.

These observations were confirmed by numerical simulations, where the constitutive law of the average fabric behavior was implemented into a numerical model so as to simulate infusion experiments. Oil pressure gradient and preform thickness profile (Δh) were calculated numerically and compared to experimental values. These results are presented with respect to t*, which is equivalent to the ratio between current time and final filling time for an accurate comparison of thicknesses and pressures, regardless of time lag. The first case study was the impregnation of random mat (Figure 7), whose experimental filling time was 63 s and numerical filling time was 67 s; whereas the second case study was the impregnation of twill weave fabric (Figure 8), which presented an experimental filling time of 853 s and a numerical one of 766 s.

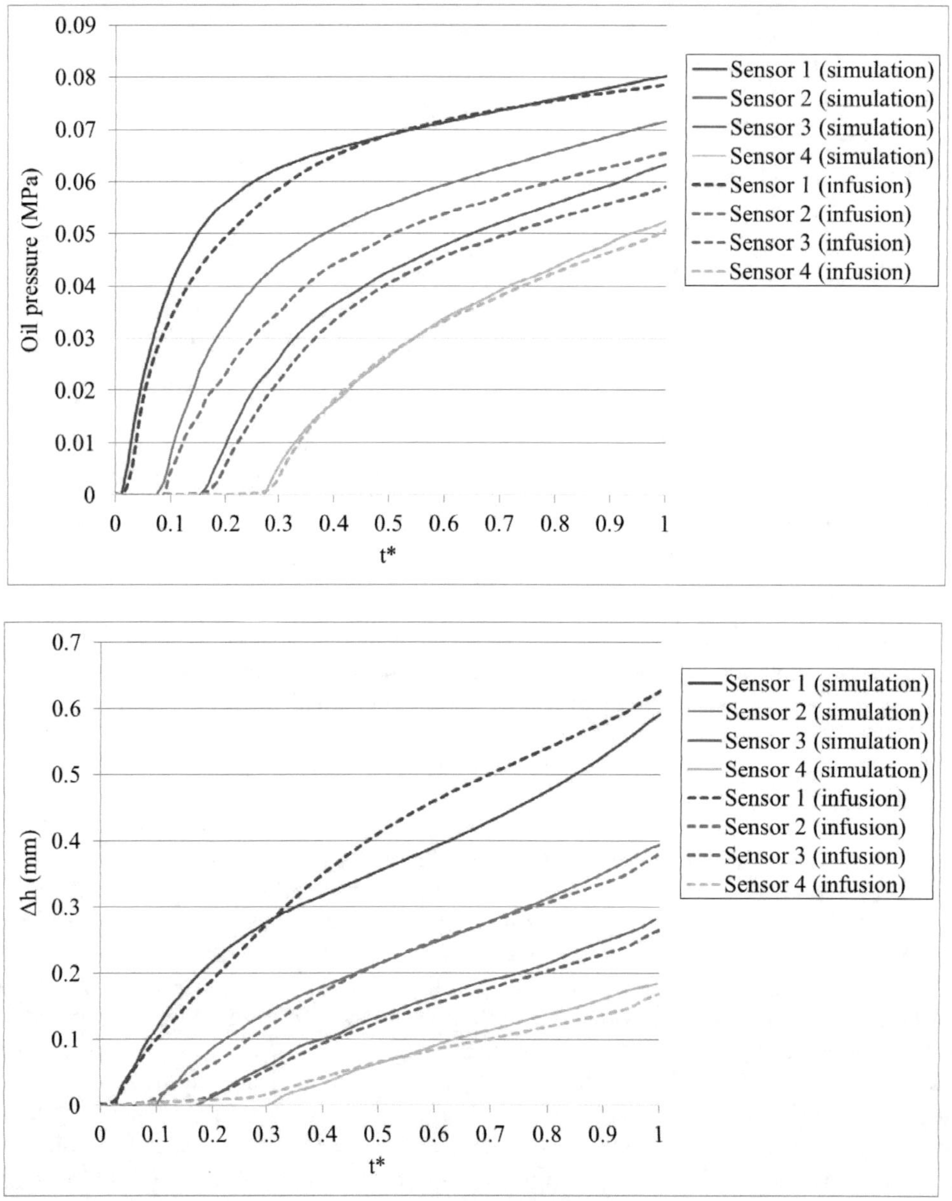

Figure 7. Comparison between experimental and numerical results of thickness and pressure for random mat impregnation (μ=0.264 Pa.s)

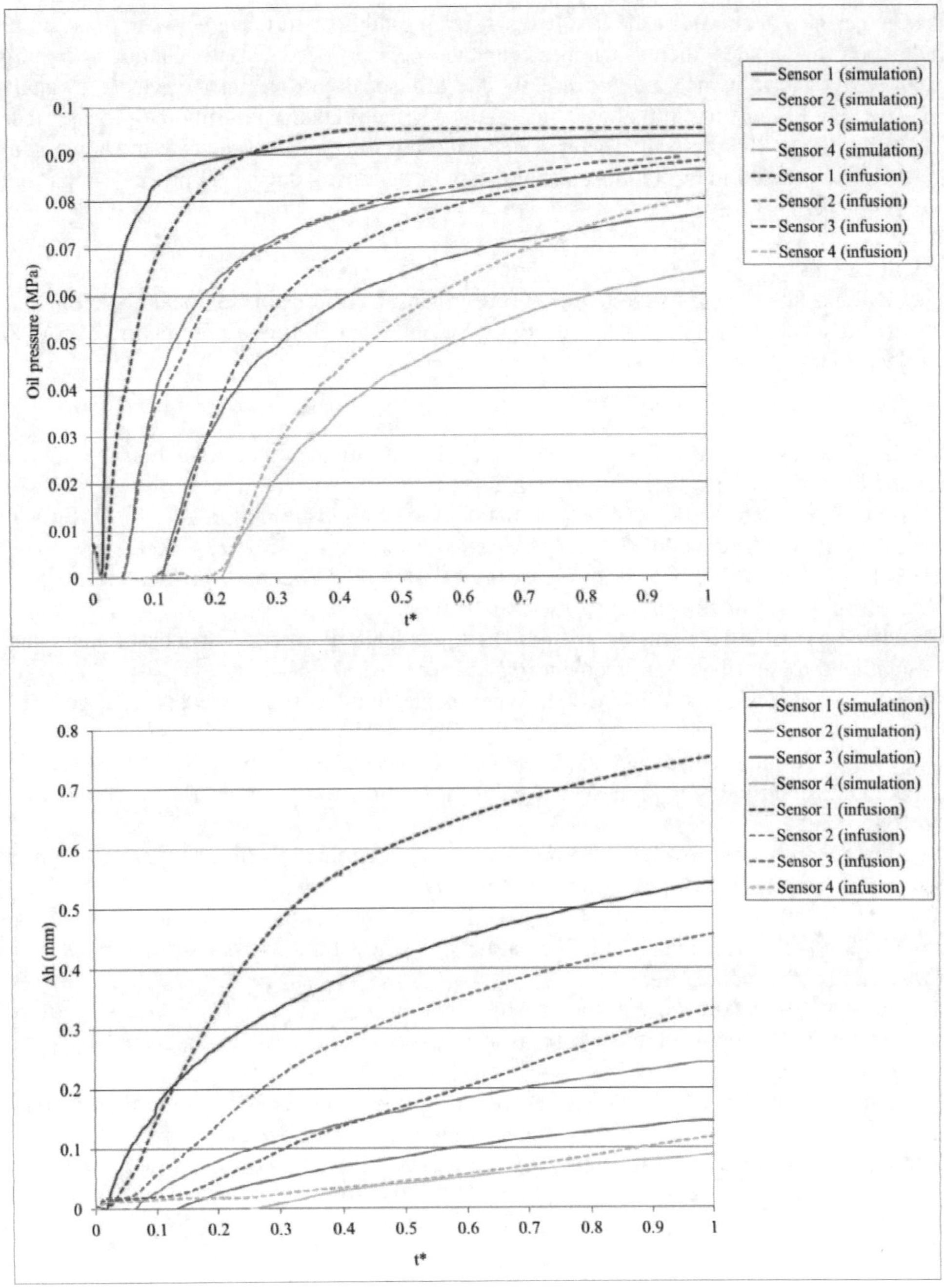

Figure 8. Comparison between experimental and numerical results of thickness and pressure for twill weave fabric impregnation (μ=0.249 Pa.s)

As expected, numerical results obtained for random mat are close to corresponding experimental results, whereas the discrepancies observed for the twill weave fabric confirm that the constitutive law implemented into the numerical model is not sufficient to reproduce the real behavior of such a fabric during infusion. The calculated pressures of twill weave fabric are slightly lower than the experimental ones in particular for sensors 3 and 4, whereas sensors 1 and 2 lead to a good agreement. However, the analysis of experimental curves of the first two sensors highlights a time lag between the resin pressure rise and the preform thickness increase. It means that viscoelastic effects may occur during impregnation of the twill weave fabric, which are not directly due to viscosity or strain rate effects, as analyzed in Figure 4. The long filling time in that case may induce some relaxation during the process, increasing thus the deformation of the preform. Therefore, considering viscoelastic effects in the numerical model could improve the prediction of the mechanical behavior occurring during VI process, in particular for fabrics with low permeability.

Acknowledgements
The authors would like to acknowledge the European Union (European Regional Development Fund, FEDER), the French State and the Hauts-de-France Region Council for co-funding the ELSAT2020 by CISIT project (POPCOM action).

References

[1] Williams C, Summerscales J and Grove S 1996 Resin Infusion under Flexible Tooling (RIFT): a review *Compos. Part* A *Appl. Sci. Manuf.* **27** 517–24
[2] Park CH and Souab A 2009 Analytical modeling of composite molding by Resin Infusion with Flexible Tooling: VARI and RFI processes *J. Compos. Mater.* **43** 1877–1900
[3] Tackitt KD and Walsh SM 2005 Experimental study of thickness gradient formation in the VARTM process *Mater. Manuf. Processes* **20** 607–27
[4] Gauvin R, Trochu F, Lemenn Y and Diallo L 1996 Permeability measurement and flow simulation through fiber reinforcement *Polym. Compos.* **17** 34–42
[5] Chen B, Lang E.J and Chou T-W. 2001. Experimental and theoretical studies of fabric compaction behavior in resin transfer molding *Mater. Sci. Eng.,* A **317** 188-196
[6] Correia NC, Robitaille F, Long AC, Rudd CD, Simacek P and Advani SG 2005 Analysis of the vacuum infusion moulding process: I. Analytical formulation *Compos. Part* A *Appl. Sci. Manuf.* **36** 1645–56
[7] Yenilmez B and Sozer EM 2009 Compaction of e-glass fabric preforms in the Vacuum Infusion Process, A: Characterization experiments *Compos. Part* A *Appl. Sci. Manuf.* **40** 499–510
[8] Saunders RA, Lekakou C and Bader MG 1999 Compression in the processing of polymer composites 1. A mechanical and microstructural study for different glass fabrics and resins *Compos. Sci. Technol.* **59** 983–93
[9] Saunders RA, Lekakou C and Bader MG 1999 Compression in the processing of polymer composites 2. Modeling of the viscoelastic compression of resin-impregnated fibre networks *Compos. Science. Technol.* **59** 1483–94
[10] Bickerton S, Buntain MJ and Somashekar AA 2003 The viscoelastic compression behavior of liquid composite molding preforms *Compos. Part* A *Appl. Sci. Manuf.* **34** 431–44
[11] Yenilmez B, Caglar B and Sozer EM 2017 Viscoelastic modeling of fiber preform compaction in vacuum infusion process. *J. Compos. Mater.* **51** 4189–4203

IOP Conf. Series: Materials Science and Engineering **406** (2018) 012063 doi:10.1088/1757-899X/406/1/012063

Adaptive fiber-reinforced plastics based on open reed weaving functionalization

M Ashir*, DMP Vo, A Nocke and C Cherif

Technische Universität Dresden, Faculty of Mechanical Science and Engineering, Institute of Textile Machinery and High Performance Material Technology, 01069 Dresden, Germany

* Moniruddoza.ashir@tu-dresden.de

Abstract. It is aimed at realizing adaptive, and thus mechanically adjustable fiber-reinforced plastic (FRP) components with complex 2D layouts so that complex 3D deformation of adaptive FRPs can be achieved. Therefore, the textile-technical integration of shape memory alloys into reinforcing fabrics is required. Hence, this paper presents the development of adaptive FRPs, where nickel-titanium-based shape memory alloy was structurally integrated into textile-reinforced semi-finished products by means of open reed weaving technology to generate adaptive FRP components. After manufacturing of the functionalized preform, it was infused by both a thermosetting resin system and a thermosetting system modified by means of plasticizer. Results of thermo-mechanical characterization revealed that the deformation of adaptive FRPs increased fourfold by adding the plasticizer to the reference thermosetting resin system.

1. Introduction

The innovation potential of fiber-reinforced plastics (FRPs) is based on the design of load-bearing lightweight constructions in accordance with the end product. These load-bearing structures for FRPs can be implemented by the specific arrangement of high-performance fibers in the load direction. The enormous potential of FRPs can furthermore be increased by means of functionalization. By developing FRPs with adaptive characteristics, smart lightweight structures with a high application potential can be realized. These structures possess highly tailored energy absorption capabilities and their mechanical characteristics, such as geometry and stiffness, can be adjusted to external influences.

The adaptive characteristics of FRP can be generated integrating shape memory alloys (SMAs) into reinforcing fabrics during the production process [1-3]. After plastic deformation at low temperature, SMAs return to their original shape during the thermal induced activation. The textile-technical integration of SMAs into reinforcing fabrics is mainly executed by means of the weaving technology because weaving has the advantage of high manufacturing flexibility and productivity compared to other textile techniques such as knitting, braiding or stitching. Furthermore, woven structures have a notable lightweight strength and long-term durability. By utilizing appropriate weave patterns and parameters, a wide variety of structures can be achieved using the weaving technique without requiring complex constructive modifications.

In previous works, the authors presented the integration of SMAs into reinforcing fabrics by weaving technology, whereby the SMAs were parallel to warp or weft yarns [4-6]. However, it is aimed at realizing adaptive, and thus mechanically adjustable fiber-reinforced plastic (FRP) components with

complex 2D layouts so that complex 3D deformation of adaptive FRPs can be achieved. Therefore, the textile-technical integration of shape memory alloys into reinforcing fabrics by means of open reed weaving technology (ORW) is required.
Technically, the ORW technology is developed on the basis of a conventional weaving machine. Besides essential working elements, such as warp let-off device, conventional shafts or heald frames, weft feeder and fabric take-off system, an ORW weaving machine is equipped with an open reed and a so-called special shaft that is mounted between the open reed and other conventional shafts (Figure 1). In the special shaft, flat needles are used as thread guides and coupled with a linear motor to form a sideward movable system. In combination with the open reed, offset of warp yarns that are guided in the special shaft during weaving is possible. In this way, woven fabrics with diagonally integrated yarns can be achieved in a one-stage process [7].

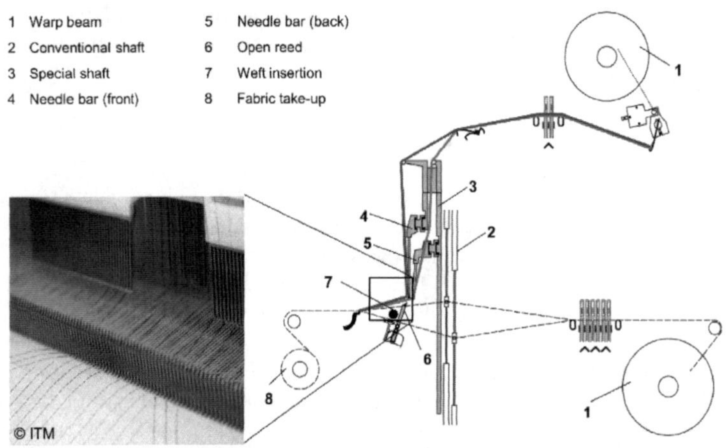

Figure 1. Schematic and actual representation of an ORW weaving machine [8].

There is a diversity of potential applications for ORW products. One example is decorative fabrics with various embroidery patterns [9]. In the field of composites, multiaxial fabrics, locally reinforced fabrics, grid-like non-crimp woven reinforcing fabrics and 2D net shape fabrics for different end uses can be manufactured [9-11]. Another very promising approach is to employ ORW technology for the fabrication of functionalized textiles. Known products are woven structures with integrated CF-strain sensors for on-line controlling of composite components [12]. However, the use of ORW technology for the development of functionalized preforms with SMA for the fabrication of adaptive FRP has not yet been reported.
Hence, in the framework of this research project, adaptive FRPs were developed, where SMA was structurally integrated into textile-reinforced semi-finished products by means of ORW technology. Next, the functionalized preforms were infiltrated by two different resin systems, and finally, the thermomechanical characterization of both types of adaptive FRP was executed.

2. Materials and methods

2.1. Materials

In this research project, a SMA (Alloy H ox. sa., Memry GmbH, Germany) in the form of a wire was selected for the development of adaptive FRP. The transition temperature, diameter, tensile strength and the elongation at break of the selected SMA were 95-110°C, 0.305 mm, 1152.7 MPa and 11.1%, respectively. In order to ensure free and even mobility of SMA within FRPs, which is necessary to fully exploit the deformation capability of SMA and therefore the adaption potential of the entire FRP

structure, the wire-shaped SMA was coated by a separating agent – Dexcoat 8 (Tag chemicals, Germany). This separating agent is suitable for all thermoset resins.
Subsequently, the coated SMA was integrated into the reinforcing fabric that consisted of glass rovings (Glas EC17-1200-350, Glasseiden GmbH Oschatz, Germany). The parameter of glass rovings for the realization of adaptive FRPs, which were tested by the authors at the research institute, is stated in Table 1. These glass rovings were chosen due to their price-performance of 1-2€/Kg, high applicability in FRP applications of around 98%, electrical insulation and high elongation at break of up to 5.4%.

Table 1. Mechanical parameters of the glass rovings (SD…Standard deviation).

	Average	SD	Norm applied
Fineness (tex)	1200	2.5	DIN EN ISO 2060
Tensile strength (N)	540	34	ISO 3341
Elongation at break (%)	1.96	0.08	ISO 3341

In this research project, a cold-curing thermosetting matrix system was used. This system can withstand temperatures of up to 130°C or even 200°C, which is significantly higher compared to the transition temperature of SMA. In order to prevent the SME during the infusion process which is an exothermic reaction, a cold-curing system was used. In this case, the commercially available resin system MGS® RIMR 135 in combination with curing agent RIMH® 137 (Hexion, USA) were selected. The combination of materials displays an interconnection temperature of 70°C, which is significantly below the transition temperature of the selected SMA. In order to increase the flexibility of the matrix material, a plasticizer - Heloxy (Hexion, USA) was mixed with the resin system.

2.2. Development of functionalized preform by means of ORW technology

The rapier weaving machine PTS 4/SOD with ORW function from Lindauer Dornier GmbH, Lindau, Germany was employed for the development of functionalized preforms. To manufacture a functionalized reinforcing fabric with integrated SMAs, conventional shafts were used to form the base woven fabric, whereas needles of the special shaft interweave SMAs on its surface. Two sideward movable systems with independent horizontal movements are available in mentioned machine, hence woven structures with two sets of SMAs, each having a separate pathway, can be achieved. It is necessary to note that SMAs were preferably considered as weft yarns in the resulting fabric due to their offset movement. During the production of the functionalized preform, the offset of the SMA was 20 mm. Therefore, different fabric thicknesses had to be realized along the fabric width to enable the plastic deformation of the final FRP component. Technologically, varied fabric thickness can be achieved by alternately implementing single and multi-layer structures in different fabric areas. Based on the desired thickness ratio, an appropriate layer count with the corresponding warp and weft ratio in every single fabric area was determined. Subsequently, suitable weave patterns together with corresponding drafting and lifting plans were developed. For the weaving process, it was important to synchronize the needle movement with the shedding motions of other working elements, so that the desired structure could be attained. The machine used for the development of functionalized preforms is shown in Figure 2.

Figure 2. Rapier weaving machine PTS 4/SOD with ORW function.

One challenge to the fabrication of reinforcing fabric with integrated SMAs was the specific material stiffness of the selected SMAs. This stiffness limits the material flexibility during the weaving process and may lead to yarn damage in case of extreme movements. To overcome this issue, technological solutions were implemented to enable the supply of SMAs at different rates and to compensate the difference in yarn length when the working elements move relatively to each other. Furthermore, experiments have been carried out to determine the maximal offset movement and yarn tension, at which the particular SMAs can be processed. The results were carefully considered when setting fabric and machine parameters to attain functionalized reinforcing fabrics with minimized damage. The functionalized preform with integrated SMA is shown in Figure 3.

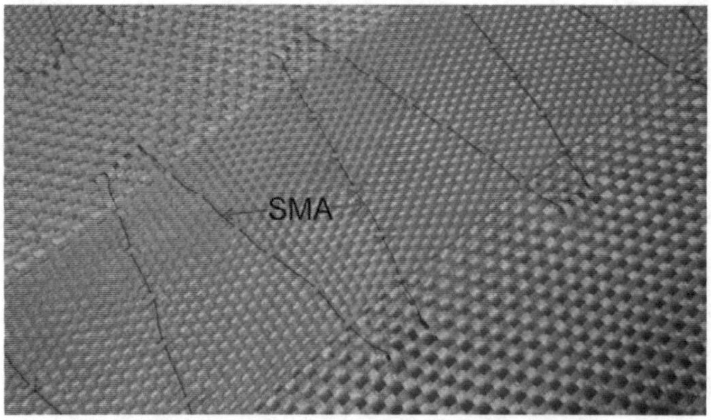

Figure 3. Functionalized preform with integrated SMA.

2.3. *Infusion*

Considering its cost-effectiveness and suitability for the production of small composite parts, in this research project, the Seeman Composites Resin Infusion Molding Process (SCRIMP) was used for the infusion of the functionalized reform. An additional flow aid was used in order to ensure the proper distribution of the resin throughout the reinforced fabrics and to avoid dry areas on the structure surface. Prior to the infusion process, resin and curing agent were mixed at a ratio of 10:3 for the formation of reference adaptive FRP. The plasticizer was mixed at a ratio of 7:3 with the matrix system for the fabrication of adaptive FRP with plasticizer. In this case, resin and curing agent were mixed at a ratio of 10:2.4.

After preparing the matrix material, the functionalized preform was prepared by cutting the floating weft yarns. The free ends of SMA were secured against slippage and to generate a local force transmission area during the thermal induced activation of SMA. This was executed by fixing the free ends of SMA with a screw, as shown in Figure 4 (Fixing of the free ends of SMA). Subsequently, the functionalized preform was laid onto the metal plate, and the free ends of SMA were isolated by means of tacky tape for their contacting during electrical characterization. Later, the perforated foil, distribution channel and non-permeable film were placed on top, and the resin was flown by means of the help of negative pressure. The whole infusion system is shown in Figure 4. In a next step, the whole sealed system was cured for 12 hours at room temperature and then tempered in a hot gas chamber at 50°C for 3 hours in order to reduce the residual stress of adaptive FRP. After the infusion process, the adaptive FRP was tailored by means of a laboratory wet say to a size of 210 mm x 110 mm.

Figure 4. Infusion process for the fabrication of adaptive FRP.

2.4. Electromechanical characterization of adaptive FRP

The electromechanical characterization of both types of adaptive FRP was executed by means of the deformation measurement. The maximum deformation was measured by a laser triangulator of type LJ7200 (Kyence, America), featuring a contact-free and therefore mechanically reactionless measuring principle. The adaptive FRP and the laser triangulator were mounted on the test stand. The deformation was measured at a distance of 180 mm from the clamping distance.

The thermal induced activation of SMA was executed by a laboratory power supply unit (BT, Germany). Based on preliminary results considering the generated temperature of SMA, in this research, the SMA was activated by 14.5 V/ 1.6 A current flow. The periodic heating cycle was set to 60 s, and the periodic cooling cycle was set to 90 s. The periodic cooling was executed at room temperature. Both periodic heating and cooling were controlled by a LilyPad programmed by an Arduino environment unit due to the ease of use and flexibility. The adaptive FRP during the heating and the cooling cycle is shown in Figure 5.

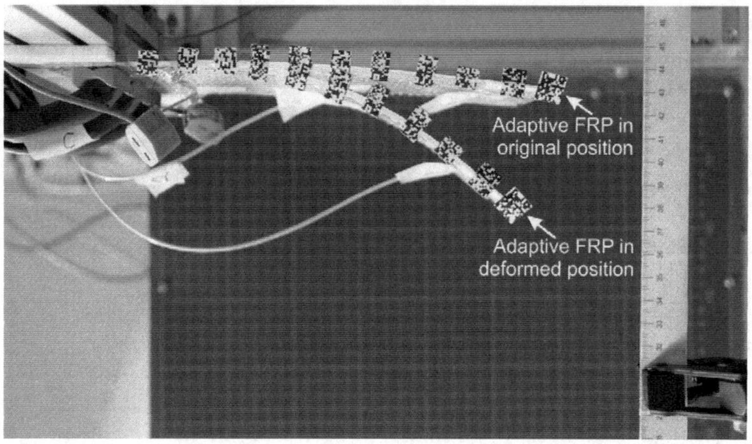

Figure 5. Adaptive FRP (here, in two setting positions – deformed and original) during the heating and the cooling cycle.

3. Results and discussions

The deformation behavior of the adaptive FRP over the entire measurement period is demonstrated in Figure 6. This figure shows that during the thermal induced activation, the adaptive FRP deforms due to the phase change of integrated SMA from martensite to austenite state.

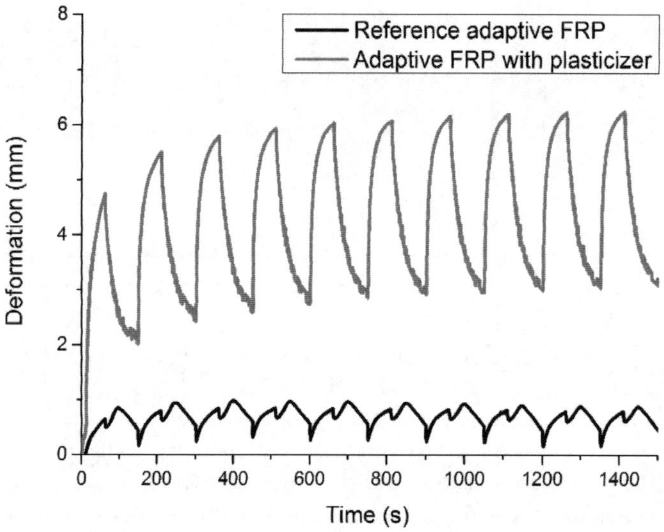

Figure 6. Deformation behavior of adaptive FRPs over the entire measurement period.

Figure 6 reveals as well that, by adding the plasticizer to the reference resin, the deformation of adaptive FRPs was increased fourfold. In order to evaluate this relationship, a three point bending testing of reference resin and resin with plasticizer was executed. Result reveal that bending modulus of reference resin with plasticizer was reduced significantly compared to reference resin as shown in Figure 7. The deformation potential of adaptive FRP was increased by creating locally soft bending areas, which was enabled by the mixing of plasticizer into the reference resin system.

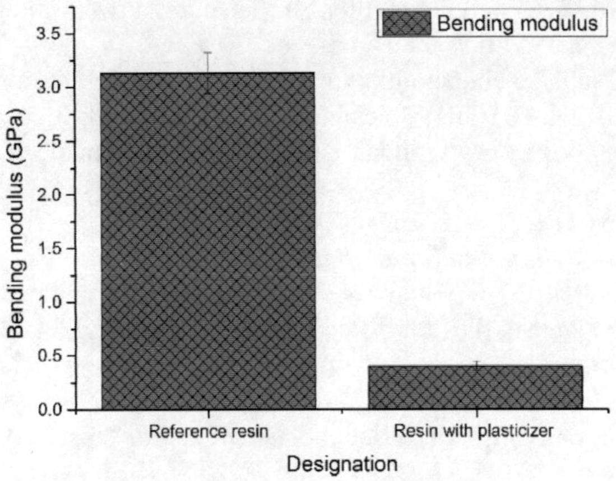

Figure 7. Bending modulus of both reference resin and reference resin with plasticizer.

4. Conclusions

The deformation characterization of adaptive FRP by a thermosetting resin system and another thermosetting resin system modified by means of plasticizer was executed in this work. For the optimum use of SMA in adaptive FRP, it was coated with a coating agent. Furthermore, to increase the deformation behavior, a variable thickness of the functionalized preform was realized. After manufacturing of the functionalized preform, it was infused by the thermosetting resin system as well as the thermosetting system modified with plasticizer and reference resin system at a mixing ratio of 7:3. The deformation of adaptive FRPs increased fourfold by adding the plasticizer to the reference resin because the plasticizer reduces the bending stiffness of the reference resin system.

Acknowledgements

The IGF research projects 18808 BR and 19832 BR of the Forschungsvereinigung "Forschungskuratorium Textil e. V." are funded through the AiF within the program for supporting the „Industrielle Gemeinschaftsforschung (IGF)" from funds of the Federal Ministry of Economics and Energy (BMWi) by a resolution of the German Bundestag. Financial support is gratefully acknowledged. We would like to thank all the participating companies for their technical support and the supply of test material as well as all further partners supporting our research work within this application area.

Supported by:
Federal Ministry for Economic Affairs and Energy
on the basis of a decision by the German Bundestag

References

[1] Ashir M, Hahn L et al. 2016 Development of innovative adaptive 3D Fiber Reinforced Plastics based on Shape Memory Alloys, *Composites Science and Technology* **126**, 43-51

[2] Ashir M, Hahn L et al. 2016 Electro-bending characterization of adaptive 3D Fiber Reinforced Plastics based on Shape Memory Alloys, *Smart Materials and Structures* **25**, 035 – 041

[3] Ashir M, Nocke A et al. 2017 Development of adaptive hinged fiber reinforced plastics based on shape memory alloys, *Composite Structures* **170**, 243-249

[4] Ashir M, Hindahl J et al. 2017 Development of adaptive pleated fiber reinforced plastic composites. *Composites Science and Technology* **148**, 27–34

[5] Ashir M, Nocke A et al. 2018 Development and mechanical properties of adaptive fiber-reinforced plastics. *Journal of Industrial Textiles*. Epub ahead of print. DOI: 10.1177/1528083718757523.

[6] Ashir M, Hindahl J et al. 2017 Time dependent deformation behavior of adaptive fiber reinforced plastics. In: 21st International Conference on Composite Materials. Xi'an, China 20-25th August 2017

[7] Wahhoud A 2011 Neuartige Herstellung textiler Flächen durch Prozessintegration im Webvorgang. *Melliand Textilberichte* **4**, 195–197
[8] Lindauer DORNIER GmbH: Benutzeranleitung ORW-Greiferwebmaschine P1
[9] DORNIER. Open Reed Weave (ORW) Technology - Weave and Stich in one Process. Retrieved 08.11.2017. Website: https://www.lindauerdornier.com/en/weaving-machine/open-reed-weave-orw-technology
[10] Weise D, Schneider B et al. 2015 ORW Technology For Non-Crimped Heavy Tow Woven Reinforcement Fabrics. *Melliand International* **1**, 32–33
[11] Vo DMP, Hoffmann G et al. 2018 Novel weaving technology for the manufacture of 2D net shape fabrics for cost effective textile reinforced composites. *AUTEX Research Journal.* DOI: 10.1515/aut-2018-0005
[12] Häntzsche E, Schneider B et al. 2014. Integrally manufactured sensor systems for structural health monitoring using ORW weaving technology. *Technical Textiles* **5**, 175-177

IOP Conf. Series: Materials Science and Engineering **406** (2018) 012064 doi:10.1088/1757-899X/406/1/012064

Intrinsic fibre heating: a novel approach for automated dry fibre placement

F Helber[1*], A Amann[1], S Carosella[1], P Middendorf[1]

[1] Institute of Aircraft Design, University of Stuttgart, Stuttgart, 70569, Germany

* helber@ifb.uni-stuttgart.de

Abstract. The present study deals with a novel binder activation approach for automated dry fibre placement. Via application of an electrical current within carbon fibre tows, ohmic heating is used for activation of binder systems. The working principle of intrinsic fibre heating is demonstrated within two prototype devices: A handheld tape laying device for the manual application of pre-bindered carbon fibre tows is introduced. Furthermore the automation potential of intrinsic fibre heating is demonstrated within a unique fibre laying approach: the Advanced Ply Placement (APP).

1. Introduction

Modern production technologies for carbon fibre reinforced plastic components, such as Automated Tape Laying (ATL) and Automated Fibre Placement (AFP) have been established as standard manufacturing processes in commercial aircraft production and the automotive industry. These state-of-the-art large-scale industrial fibre placement processes mainly utilize pre-impregnated unidirectional fibre materials for the manufacturing of preforms with low scrap-rates. Main goal is the production of load-path optimized lightweight components with minimal utilization of reinforcement fibers. A near net shape geometry is the target to reduce cutoff waste and therefore minimize material and manufacturing costs. Compared to a prepreg process chain, preforms made from dry carbon fibres offer substantial potential to reduce lead time and costs. The almost unrestricted processing time of dry fibres allows the production of large parts without restricting layup rates.

The dry carbon fibres utilized in this paper are prepared with a binder system that is necessary for the tack between fibre and the mold or in-between the different fibre layers. The automated placement methods for dry fibres commonly utilize infrared (IR) radiators or diode lasers for binder activation during the layup process. Thereby, heat is generated on top of the carbon fibre tow in order to activate the binder systems. However, diode lasers involve high costs and IR radiators generally heat up the surrounding environment, resulting in a loss of efficiency.

For that reason, a new approach for binder activation during fibre layup is investigated in the scope of this paper. The electrical properties of carbon fibres are being utilized and due to the application of an electric current ohmic heat is generated and subsequently binder systems can be activated.

2. State of the Art and Fundamentals

The following chapter describes the necessary fundamentals for understanding the principle of the intrinsic fibre heating approach described in this paper.

IOP Conf. Series: Materials Science and Engineering **406** (2018) 012064 doi:10.1088/1757-899X/406/1/012064

2.1. Binder Systems

Binder systems are auxiliaries in a fibre-reinforced preform to achieve structural and / or geometrical fixation [1]. The binding effect depends on the adhesion of the binder system with the fabrics' and fibres' surface and its cohesion. For the stabilization of rovings, fabrics as well as semi-finished products and subsequent use and fixation in textile processes, different binder systems are commercially available and can be distinguished in powder, veil and hot-melt binder systems. The melting range of binder systems lies between 75°C and 200°C, the majority however is found within 100°C and 130°C.

Powder binder systems consisting of thermoplastic or thermoset powders vary in different particle sizes between 32 µm and 250 µm. Different particle sizes can result in varying preform permeability, due to blocking of flow channels by small particles. Large particles however can increase preform permeability by increasing distance between layers. After binder application onto the semi-finished product, the binder needs to be activated via heat input. The activation process melts the binder and after cool-down and solidification subsequent layers are bonded together.

Binder veils consist of thermoplastic threads or mats, with or without any preferred orientation. Various binder veils are commercially available and vary in polymer composition and areal weight. Typical areal weights can be found in the area between 4 g/m^2 and 50 g/m^2. They are generally composed of co-polymers. The binder veil usually is placed between the reinforcement layers. Subsequently the preform is activated via heat input and consolidation (e.g. thermoforming, calender processes, etc.).

Binder activation can be distinguished in four different operating principles [2]: Heat transfer, radiation, mechanical activation and electrical activation, as it can be seen in Figure 1.

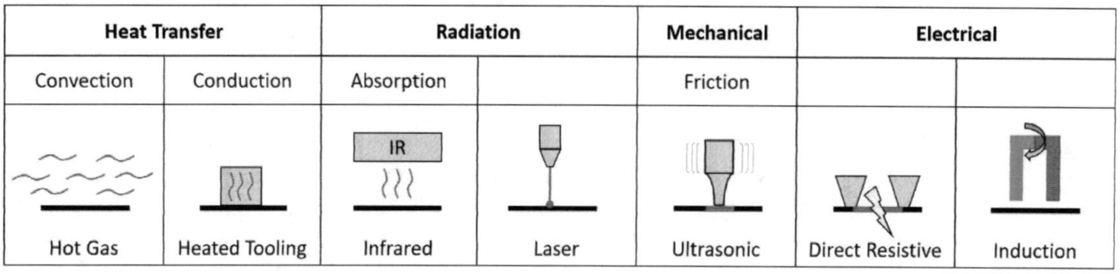

Figure 1. Binder activation methods

Among the different principles, some binder activation approaches are more commonly used in fibre placement machinery due to better controllability, integrability and lower complexity. Binder activation via infrared radiation (IR) is mostly used. State of the art AFP and ATL machinery utilize this approach due to long time experience, good controllability and efficient IR radiators. However, additional periphery at the lay-up head is necessary and efficiency is lost due to heat convection into the surrounding environment. Another commonly used process is the heat transfer via hot gas. A hot gas torch is pointed towards the nip point of the lay-up head and hot compressed air activates the binder system just before compaction. The disadvantages of this approach are the additional gas pipes and the additional periphery attached to the lay-up head, resulting in lesser flexibility during the lay-up process. Furthermore introduces the gas flow kinetic energy that causes fibre misalignment, especially before initial touchdown of the fibre tape. Ultrasonic (US) vibrations and the subsequent mechanical friction can also be used for binder activation. However, cost-intensive generators and sonotrodes, special end-effectors and the additional periphery prevent US binder activation from extensive use. Furthermore the ultrasonic vibrations might lead to fibre breakage and tool damage, if welding parameters are not set accurately. However, the local binder activation opens new possibilities for fast preform manufacturing with adjustable draping properties.

In the scope of this paper, special focus is paid on the electrical operating principle. Intrinsic fibre heating is based on ohmic heating of electrical conductive carbon fibres. This approach allows fast, efficient and local binder activation.

2.2. Joule Heating

Resistive heating, also known as ohmic heating or joule heating, describes the heat generation if an electric current passes through a conductor. Joule's first law states that the power P of generated heat is proportional to the product of its resistance R and the square of the current I ($P \sim R * I^2$). Joule heating can be described as follows: A voltage difference between two points of a conductor (in this case two electrodes connected to the conductive carbon fibres) creates an electric field that accelerates charge carriers (electrons) in the direction of the electric field, giving them kinetic energy. If electrons are colliding with ions in the conductive material, the kinetic energy is converted into thermal energy.

Generally known, carbon fibres have a good electrical conductivity. The specific electrical resistance lies between $\rho = 8\ \Omega mm^2/m$ (HM-fibres) and $20\ \Omega mm^2/m$ (HT-fibres) [3]. This means, resistive heating of laminates due to electric currents is possible and is widely used in heated composite tools. Selected layers of carbon fibres are electrically contacted before lamination. After part consolidation an electric current can be applied and the tool can be used for part curing.

2.3. Advanced Ply Placement

State of the art AFP processes depend on small tape widths, leading to various abutting tape edges. Therefore, requirements concerning tape width are demanding for avoiding effects like gaps or overlaps [4]. Double curved surfaces require varying roving lengths within one tape to equalize length deviations between adjacent fibres. For that reason wide tapes in combination with complex part geometries usually lead to wrinkling and buckling defects. Therefore ATL processes are mainly used for large geometries with little complexity.

The Advanced Ply Placement (APP) process avoids these problems due to a unique forming approach. By allowing longitudinal tape shearing it allows draping of wide unidirectional tapes over double curved and complex surfaces.

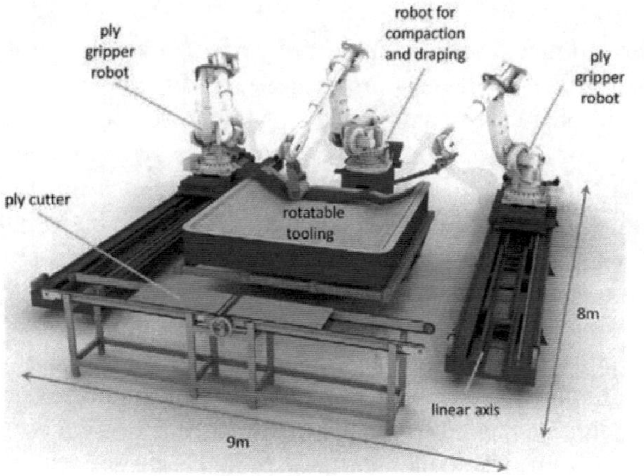

Figure 2. APP process scheme [5]

The APP production cell consists of two ply gripper robots on linear axes, a stationary draping / roller robot, a heated mould on a rotary axis and a ply cutting and pick-up station that can be adapted to different ply lengths. The APP process scheme can be seen in Figure 2. During part production, the two ply gripper robots move to the ply cutter and pick up the tape. The two coordinated robots move parallel to the designated placement position. At the same time the mould rotates to the defined angle for the required fibre orientation of the current ply. Having reached the placement position, the draping robot pushes the ply towards the mould surface applying a pretension that evens out present wrinkles. After initial touchdown of the tape, one gripping robot relaxes the tape, while the roller robot moves along the mould surface towards the gripping robot. Subsequently, the roller robot moves back to the initial point of touchdown and repeats the same procedure towards the second ply gripper robot. This process is repeated until the designated preform is completed.

One key element of this unique draping approach are segmented gripping units at the two ply gripper robots. Various clamping fingers can move independently and allow tape shearing at predefined tape tension, allowing compensation of tape length differences.

The second key element is the segmented compaction roller that applies a defined compaction pressure onto the tape during fibre layup. The self-adjusting roller geometry is able to adapt to the mould surface ensuring constant and equal pressure over the whole contact surface between compaction roller and mould surface. This ensures tape adhesion and preform consolidation.

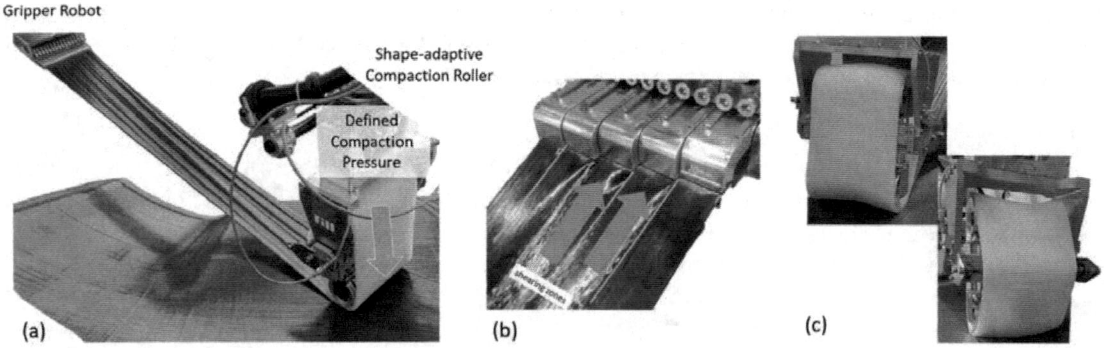

Figure 3. Working principle APP. (a) Wrinkle compensation (b) Tape shearing (c) Shape-adaptive compaction roller [4]

3. Intrinsic Fibre Heating

Intrinsic fibre heating is no new development, since joule heating is a widely used principle for heat generation. In the scope of the CFRP part production it is already used in heated composite tooling. However, there is no widespread use in automated tape laying devices. Cost intensive diode lasers or IR radiators are still state of the art for ATL and AFP processes. Most binder activation systems require additional periphery that limits the operating space of the layup head (radiators, spotlights, hot gas torches, etc.). Furthermore additional piping and generators need to be installed.

The intrinsic fibre heating approach grants several advantages compared to other binder activation systems, allowing local binder activation. Since the heat is generated directly within the carbon fibres, there is no significant efficiency loss due to convection. The overall process is well controllable due to proven equipment like adjustable power supply, etc. By disconnecting the power supply, fibre heating is instantly turned off without any delay. The electrodes can be integrated into the robotic end-effector. And at last, no special equipment needs to be provided, since potent power supply is already available (presumed the layup-head is integrated into a robotic production cell).

3.1. Relation between Power Input and Heat Generation

Carbon fibres can be described as NTC (negative temperature coefficient thermistors). This means, with increasing fibre temperature the electrical resistance decreases [6]. If carbon fibres are heated and a constant current is applied, the heating power P will increase as the resistance decreases. If the fibres are attached to a constant voltage source U, the thermal output will increase successively due to the thermistor behaviour.

Figure 4 shows the relation between fabric temperature T and the applied power input P. A unidirectional (UD) fabric with TENAX-E-HTS40 F13 12 k fibres has been used. Different tape widths are investigated (16 mm, 27,5 mm and 40 mm). A controllable laboratory power supply (VOLTCRAFT PS 2403 Pro) has been attached to the fabric and the temperature is measured with a thermal imaging camera (FLIR Systems Thermovision A 40). The graph shows the increasing temperature with increasing input power. Furthermore, it can be seen that wider tapes require increased power input.

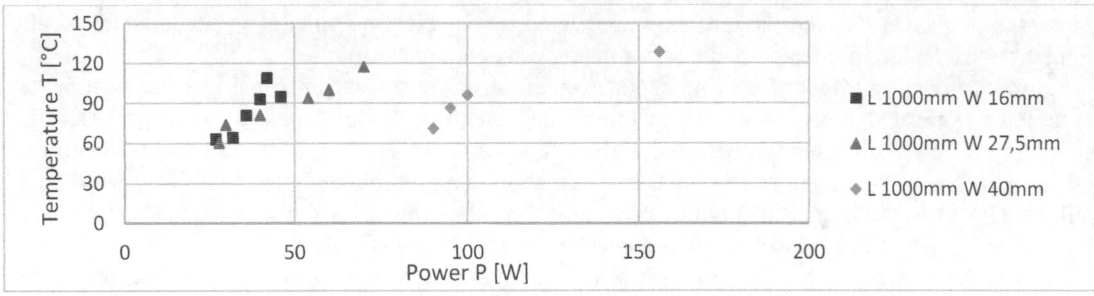

Figure 4. Relation temperature T and power input P

3.2. Electrical Contacting

In terms of electrical conductivity, silver and copper are the superior conductive metallic materials (silver: 61 * 10^6 S/m; copper: 58 * 10^6 S/m). Considering material costs, copper is the preferred electrode material. To use carbon fibres for resistive / ohmic heating with a homogeneous heat distribution, the transition resistance at the contact points should be as low as possible. For joule heating of carbon fibres, preferably low voltages and high currents (> 50 A) are used [6]. A higher transition resistance will subsequently lead to excessive heating at the contact points. State of the art carbon fibre contacting in tool making is mostly realized via metallic sleeves or integration of metallic knitting threads. Adhesion of the thermoplastic fraction of the semi-finished product and power dissipation at the contact points needs to be minimized. Lowest contact resistance can be achieved via large contacting areas and high electrical conductivity of the electrodes' material. The goal therefore needs to be the implementation of an optimal power introduction, as well as detachable and cost efficient contacting approaches.

Generally speaking, contact partners do not touch each other at the maximal possible area. Waviness, roughness and impurities especially in fabrics made from carbon fibres caused by knitting yarns, weave style and textile behaviour of fibres results in several contact areas. Forceless contact will only result in contact of outer peaks. With increasing contact pressure, elastic and plastic changes of waviness will occur, resulting in increased bearing surface. This leads to power transmission via a variety of small areas. The number, distribution and its size depends on several factors: contact force, contact shape, temperature, material properties and surface structure. The resulting effective contact surface results from the actual bearing surface minus the surface fractions covered with impurity layers (thermoplastic binder, knitting threads, etc.) [6] At these many, but small, contact areas power bottlenecks are created at which a higher current density has to pass, leading to excessive heat generation and hot spots. This so called constriction resistance can be reduced with higher compaction forces resulting in higher contact areas, as well as increasing electrical conductivity.

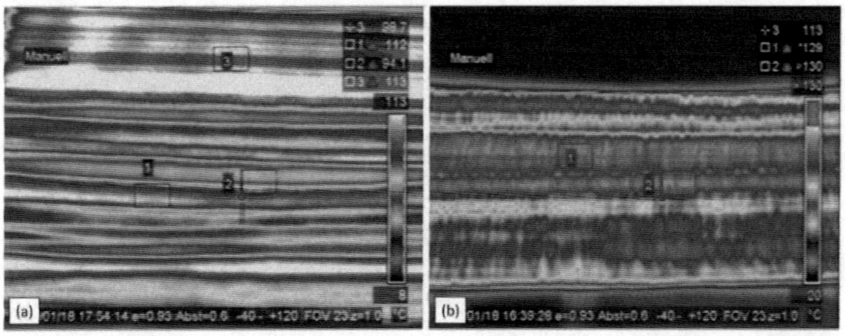

Figure 5. Effects of varying contact pressure during intrinsic fibre heating. (a) Insufficient pressure – Uneven heat distribution (b) Sufficient pressure – More even heat distribution

Furthermore, filaments that are not perfectly aligned in tape length will also cause misleading of currents and the creation of hot spots, since electrical currents always go along the path with the least resistance. These cross-currents have to be prevented. One approach is the segmentation of contact surfaces, in a manner that only few rovings are heated by one electrode-pair at a time. Several parallel aligned electrodes along the tape width result in an even heat generation.

Figure 5 shows the effect of varying contact forces during intrinsic fibre heating. In the left picture (a) various overheated areas can be seen, while at the same time colder fibres are occurring. The right picture (b) shows a sufficiently contacted carbon fibre semi-finished product with a more even heat distribution. The fabric had a width of 300 mm and was manufactured with TENAX-E-HTS40 F13 12 k fibres. However, the increasing contact pressure, in combination with rigid copper contact elements and the created heat can lead to undesired deformations in the semi-finished product.

Most of the fibre contacting is done via mechanical clamping with copper elements (sleeves, plates, profiles, etc.). This is state of the art and well experienced. However, the rigid copper contact element cannot adapt to varying mould surfaces and the waviness of textile semi-finished products.

One approach for better contacting is the achievement of electrical conductive silicones that can be manufactured as blocks, membranes, etc. With good electrical conductivity and the resilience and stretchability of silicones, adequate contacting due to sufficient bearing surfaces can be achieved at lower contact forces and local hot spots can be avoided.

There are few electrical conductive silicones available, eg. ELASTOSIL LR 3162 [7]. To achieve electrical conductive silicones, it is possible to add conductive additives to silicones. For achieving contact between the added particles and therefore formation of a current path, additives must be added in high quantities. However, high concentrations will affect the composition of the silicone. The percolation curve describes the relation between electric conductivity and additive concentration. Increasing the weight-fraction of conductive particles will result in touching of the particles. Above a critical point K_c, further increasing of additive concentration will only achieve little improvement of conductivity, while the consistency of the silicone will change [8]. If the fraction of added particles is only little compared to the overall mass, electron transport happens via tunnel-effect [9]. If the distance between conductive particles is large compared to the atomic dimensions, the silicone's resistance is dominant. Decreasing the distance (< 10 nm), the electrons are able to penetrate the silicone.

In the scope of this paper various additives are investigated. The commercial silicone ELASTOSIL VARIO 15 / CAT PT is used as base material. Subsequently, different additives have been added with increasing weight fractions, increasing in steps of 5 %w. One approach was the addition of a graphite powder (GraphCOND 15/98). Another approach was the adding of silver coated additives, like glass flakes (ECKART eConduct Glass 35200), copper (ECKART eConduct Copper 12200) and aluminium balls (ECKART eConduct Aluminium 202000). Different industrial carbon black pigments have also been investigated (Co. PyroPowder; Printex XE-2B), as well as copper and iron powder.

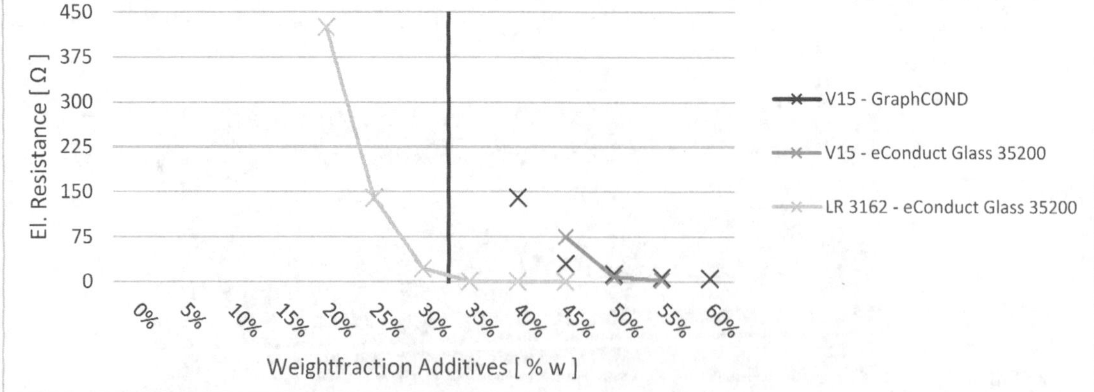

Figure 6. Electrical resistance of silicones

Successful establishment of conductivity has only been realized with two additives: GraphCOND 15/98 and ECKART eConduct Glass 35200). Figure 6 shows the development of resistance with increasing weight fractions. As it can be seen, the electric resistance decreases with increasing additive fraction. Especially the ELASTOSIL LR3162 in combination with silver-coated glass flakes looks promising, since already low weight fractions of the glass flakes results in acceptable conductivity. Different series of experiments have been showing, that increasing the additive weight fraction beyond 30 %w results in severe impacts of the silicone texture. Up to 30 %w the VARI 15 has its silicone-like consistency. By further adding of additives the consistency became more porous and fragile. Going beyond 50 %w the silicone did not cure properly and maintained a dough-like consistency. Similar results have been observed by Schüler [9].

3.3. Effects of Intrinsic Fibre Heating on Carbon Fibres

As described earlier, insufficient contacting can lead to power bottlenecks, at which higher current density has to pass, which subsequently will lead to increased heat generation (hot spots). Figure 7 shows a hot spot in the area of electrical contacting with burned knitting yarns, binder system and possibly damaged fibres. Insufficient contact pressure during power application has been applied and electric arcs have been observed. Increasing pressure or flexible electrode materials can resolve this problem.

Figure 7. Hotspot; local melting of knitting yarns

4. Handheld Tape Laying Device

In the scope of this work a prototype-device for manual layup of dry carbon fibres has been developed, see Figure 8 a. Main focus was put on using ohmic heating for fast and efficient binder activation. A lightweight and small construction is necessary for one-handed use and sufficient flexibility.

The fibres are stored on a coil on the upper end of the device. Fibre separation is realized via a cutting mechanism. The mechanism can be released either via compressed air or CO_2 cartridges. Cartridges offer the benefit of greater mobility, since pneumatic hoses are not necessary. However, the maximum number of releases is limited to approx. 60 cuts. A quick reload mechanism allows fast replacement of empty cartridges for minimization of downtime.

Electrical power can be supplied by a power supply unit or lithium polymer batteries. For greater mobility and flexibility the lithium polymer accumulators are preferred since additional wiring can be avoided. A commercial model making battery (SLS XTRON 3200 mAh 3S1P 11,1V 20C+/40C) has been installed, due to its light weight and efficient characteristics.

The handheld tape laying device allows one handed tape layup of varying tape widths and different binder systems [10]. In this case a spread tape of 20 mm, equipped with the EPIKOTE TRAC 06720 powder binder has been use for tape layup, see Figure 8 b. The activation temperature of 120°C has been achieved with the lithium polymer accumulator. An overall power input of about 40 W has been measured for successful binder activation and tape layup onto a glass fabric.

The tape is led from the fibre storage coil via a feeding mechanism into the cutting section. For fibre cutting only the trigger at the handle needs to be released and the compressed air will then release the cutting mechanism. After the cutting section, the first electrical contact is installed before the tape reaches a silicone compaction roller. The second electrode is placed shortly after the nip point, the point at which the compaction roller applies pressure onto the tape and attaches it to the substrate. With this constellation, the tape is being contacted and joule heat generated during pressure application, ensuring sufficient binder activation and optimal adhesion.

Figure 8. Handheld tape laying device prototype (a); Manual tape layup (b)

5. Automation Potential

To demonstrate the automation potential of the intrinsic fibre heating approach, a unique end-effector for the APP production cell has been designed and constructed (Figure 9 a). Goal is the manufacturing of omega-shaped stiffener elements in various lengths. The two gripping robots maintain their original functionality. The compaction roller however is replaced with the new end-effector.

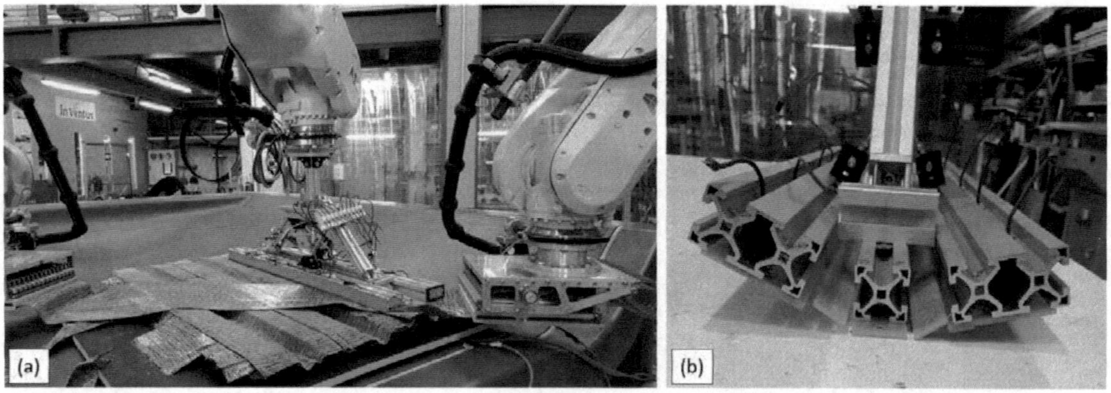

Figure 9. Automated omega stringer production using APP periphery and intrinsic fibre heating approach (a); Side view of end-effector

The end-effector consists of five segments (Figure 9 b) and imitates the manual draping process of omega stiffener elements. Meaning that initially the fabric will be consolidated at the lowest point of the omega. Subsequently, the lateral surfaces and upper flanges are pressed against the mould. For manipulation of the draping mechanism, pneumatic cylinders are installed. The automated end-effector can reach two positions. In the first position, the flanges are lifted up, preventing fibre friction and damage during initial contact and layup. In the second position, the end-effector already pushes the fibres at the bottom segment against the surface of the mould. The pneumatic cylinders are then activated and pressure is build up, so that the lateral surfaces and flanges are pressed against the mould and consolidation is achieved.

The electrical contacting is realized via copper plates, integrated into the surfaces of the five segments. The applied pressure of the pneumatic cylinders also ensure proper contacting pressure. As soon as sufficient consolidation is achieved, the power supply is connected and the binder can be activated via intrinsic fibre heating. Depending on the fibre angle of the various layers, different contacts located in tooling and end-effector have to be connected to the power supply unit for binder activation.

6. Discussion and Conclusion

As shown in the scope of this paper, intrinsic fibre heating can be used for automated placement of dry carbon fibres. The application of a power supply in combination with the electrical properties of carbon fibres is used for ohmic heat generation within the fibres. The direct heating of carbon fibres allows fast, efficient and local binder activation during fibre layup.

In two prototype end-effectors the fundamental functionality of intrinsic fibre heating in the operational environment have been demonstrated. Intrinsic fibre heating can be used for AFP and ATL processes for dry carbon fibres. Furthermore integration in the APP process has been shown.

Fibre contacting is an ongoing field of research. Insufficient contacting due to missing compaction force will lead to electric arcs and local overheating. This leads to possible fibre damaging and melting of knitting yarns. Effects concerning mechanical properties on the final part have to be validated. Further work will include optimization and automation of the handheld tape laying device and possible benchmarking against state of the art AFP processes. Moreover tests with conductive silicones will be performed, since this is a crucial factor in optimization of the intrinsic fibre heating process.

Acknowledgments

The work addressed by this paper has been performed within the project ecoTECH (www.cleansky2ecotech.eu), an international EU-project of the aeronautical research programme Clean Sky 2 (www.cleansky.eu). The project was funded directly by the European Commission as well as the participating industry ITD leaders (public private partnership). The author wants to acknowledge the funding and support by all the partners involved in this project.

References

[1] S. Schmidt, T. Mahrholz, A. Kühn, P. Wierach, Powder binders used for the manufacturing of wind turbine blades. Part 1. Characterization of resin-binder interaction and preform properties, Polymer Composites 2016

[2] M. Bulat, F. Heieck, Binder application methods for textile pr, Stuttgart: LTH Faserverbund-Leichtbau, 2014

[3] H. Schürmann, Konstruieren mit Faser-Kunststoff-Verbunden, TU Darmstadt, 2. Bearbeitete und erweiterte Auflage, Springer Verlag Berlin Heidelberg, S. 41, (2007), ISBN: 978-3-540-72189-5

[4] K. Heudorfer, M. Engelfried, J. Fial, S. Carosella, P. Middendorf, Investigation of a new out of autoclave prepreg placement process, mould concepts / release strategies and resuable heated bagging material, 7th EASN International Conference, Warsaw, 2017

[5] M. Szcesny, F. Heieck, S. Carosella, P. Middendorf, H. Sehrschön, M. Schneiderbauer, The advanced ply placement process – an innovative direct 3D placement technology for plies and tapes, Advanced Manufacturing: Polymer & Composites Science, https://doi.org/10.1080/20550340.2017.1291398

[6] A. Hemmen, Direktbestromung von Kohlenstofffasern zur Minimierung von Zykluszeiten und Energieaufwand bei der Herstellung von Karbonbauteilen, Dissertation, University of Augsburg, S. 39-44, (2016)

[7] WACKER, ELASTOSIL® LR 3162 A/B, TDS, WACKER, 2018

[8] U. Leute, Elektrisch leitfähige Polymerwerkstoffe, Springer Fachmedien Wiesbaden, ISBN: 978-

3-658-10538-9, 2015

[9] R. Schüler, Entwicklung polymerer Verbundwerkstoffe mit elektrischer Leitfähigkeit, Dissertation, University of Hamburg-Harburg, 1994

[10] F. Helber, B. Schrick, S. Carosella, P. Middendorf, Tailored preform production with customizable UD materials for hybrid material design, SAMPE Europe, Stuttgart, 2017

IOP Conf. Series: Materials Science and Engineering **406** (2018) 012065 doi:10.1088/1757-899X/406/1/012065

Investigating the unwinding behavior of technical yarns and development of a new sensor system for the braiding process

S Maidl[1,*], M Sabieraj[1], A Mierzwa[1], C Ebel[1] and K Drechsler[1]

[1] Chair of Carbon Composites, TUM Department of Mechanical Engineering, Technical University of Munich, Boltzmannstr. 15, 85748 Garching, Germany

* maidl@lcc.mw.tum.de

Abstract. A new cost-efficient sensor module for the detection of thread tension anomalies in braiding machines was developed. The sensor module is mainly attached to the body of the braiding machine and works by contactlessly detecting the positions of the levers of the yarn tensioning units of the bobbin carriers through magnets and stationary Hall effect sensors as the bobbins pass by. This way, time-discrete estimations of the tension of the moving braiding yarns can be calculated. The sensor module was validated by investigating the unwinding behavior of several kinds of technical yarns from bobbins on a stationary test stand which simulates the unwinding process during braiding. Flawless reference measurements revealed that the signals from the Hall probe are in good agreement with precise yarn tension measurements obtained simultaneously from a deflection roller based yarn tension measurement device. Further measurements with purposefully provoked unwinding-related irregularities showed that braiding defects are foreshadowed by prominent variations in yarn tension which the Hall effect sensors are able to detect. Finally, experiments with the sensors installed into a running braiding machine were conducted. In this near-production environment, the sensor module was capable of identifying irregularities soon enough before major braiding defects evolved.

1. Introduction

The quality of braided textiles from reinforcement fibers and the stability of the process are negatively affected by irregularities that occur during braiding. Since irregularities can lead to braiding defects which cause material waste and machine downtime, machine productivity is reduced and additional costs for error cause analysis and error correction time arise. Previous investigations conducted by Ebel et al. [1] have shown that braiding defects are often induced by a small cause and evolve through various stages to a major failure event. The more advanced a defect is when it is detected, the higher the aforementioned additional error costs are. Thus, the development of an online monitoring system for the braiding process which is able to detect the evolution of braiding defects already in early stages is highly desirable.

As typical braiding defects, Ebel et al. [1] mention a generally fuzzy braid due to fiber damage, loops in the fabricated preform due to a loss of yarn tension, gaps in the preform as well as yarn breakages. Whereas the fuzzy braid and loops in the preform can be seen as an optical damage, gaps in the preform and yarn breakages have a significant impact on the mechanical performance of the finished part and on machine uptime. They elaborate that these gaps and yarn breakages are commonly caused by a specific effect named fibrous ring. This effect is described as a ring-shaped accumulation of carbon filaments which impedes the yarn from unwinding from the bobbin and consequently increases the yarn tension. It origins primarily from yarn predamage and partly from unsuitable rewinding parameters.

IOP Conf. Series: Materials Science and Engineering **406** (2018) 012065 doi:10.1088/1757-899X/406/1/012065

Furthermore, they point out that especially the occurrence of yarn breakages as a result of the formation of fibrous rings significantly reduces the productivity of a braiding machine. In their endurance braiding tests, which were performed on an axial braiding machine with 60 carriers, a horn gear diameter of 120 mm and a horn gear speed of 120 rpm, they manufactured triaxial hoses with glass fiber as braiding yarns (braiding angle: 45°) and carbon fiber as UD yarns. They observed a machine downtime of up to 26 % of the total production time due to the necessity of manually repairing or rethreading yarns which had broken due to fibrous rings. However, Ebel explains in his work [2] that in a production environment with trained workers, this portion of machine downtime may be lower compared to the investigated research environment. Mierzwa et al. [3] delineate the extent of the deterioration in mechanical properties due to local yarn gaps in braided preforms (braiding angle: 45°) fabricated from T700SC 24k 60E carbon fiber yarns from Toray Industries, Inc. In their test program, they purposefully introduced 4 mm wide gaps into the preforms and infiltrated them using the VAP (Vacuum Assisted Process) and the RTM (Resin Transfer Molding) process. In their subsequent coupon tests of the specimens manufactured by the VAP method, they observed a 36 % reduction in tensile and a 33 % reduction in compressive strength when the gap was oriented perpendicularly to the loading direction. However, they did not observe any significant reduction in strength of the specimens produced by the RTM process. They concluded that – in contrast to the rigid RTM tooling on both sides of the preform – the flexible vacuum bag promoted fiber undulations in the vicinity of the gap which lead to the loss of strength.

2. Existing systems for process irregularities detection during braiding

In order to avoid the effects of braiding defects described above, some sensor systems for the monitoring of the braiding process already exist. On the one hand, there are systems which make use of tactile sensors that are stationary attached to the body of the braiding machine (bobbin carrier independent systems). On the other hand, there are approaches which include the installation of sensors onto the bobbin carriers (bobbin carrier dependent systems).

One of the commercially available bobbin carrier independent systems comprises rudimentary switches which jut into the tracks of the bobbin carriers. These switches are activated by extensions of the levers or sliders which are part of the yarn tensioning mechanism of the bobbin carriers when a total loss of yarn tension arises (cf. Figure 1). Such a system may serve as a trigger to stop the braiding machine on occurrence of a yarn breakage to avert the production of a braid with loose or missing yarns. An advantage of such a system is its simplicity and cost-efficiency. However, this kind of system causes a considerable amount of machine downtime because it only responds when a yarn has already broken. Due to the above-mentioned necessity of manually repairing or rethreading yarns after a breakage has occurred, resolving a braiding process defect in this final stage is much more labor intensive than in earlier stages like gap formation due to a moderate increase in yarn tension. The latter defect can in most cases simply be resolved by removing a fibrous ring on the respective bobbin.

Figure 1. Rudimentary switch jutting into track of bobbin carriers.

Another bobbin carrier independent system invented by Lenkeit [4] makes use of a force sensor with a skid attached to it. The sensor and its skid are arranged between the plane spanned by the uppermost thread guiding elements of the bobbin carriers and the braiding point in a way that the yarns periodically touch and slide along the skid of the force sensor as the bobbins travel through the braiding machine along their closed tracks. In doing so, the skid deflects the yarns by a defined angle. Hence, the tension of each thread can be calculated from the force measured by the sensor at discrete time intervals. Such an arrangement can detect process irregularities that result in a variation in yarn tension before a thread has already broken. Nonetheless, a major drawback is the yarn damage that may be caused when the yarns touch the skid at high speed. This point is particularly relevant when processing carbon fiber yarns which consist of thin, brittle filaments.

An example for a bobbin carrier dependent sensor system is provided by Braeuner [5]. He designed a whole new bobbin carrier with a slide that is displaced by the yarn tension against a resilient element. The yarn tension is determined by sensing the position of the slide along its track on the bobbin carrier. Furthermore, Braeuner's invention also comprises a communication module and an actively driven material buffer so that the yarn tension can be controlled wirelessly while the braiding machine is running.

A similar bobbin carrier dependent sensor system was developed by von Reden [6]. He mounted a linear potentiometer as a force sensor, an actively driven bobbin, a control unit, a data transfer unit and an energy supply unit onto a specially constructed bobbin carrier (cf. Figure 2). Moreover, the spindle and its motor which move the yarn deflection element up and down parallelly to the central axis of the bobbin is characteristic for his bobbin carrier. This way, the yarn is always unwound perpendicularly from the bobbin and a defined angle of attack of the yarn tension on the yarn deflection element is ensured. This is needed to obtain precise measurements of the yarn tension from the linear potentiometer.

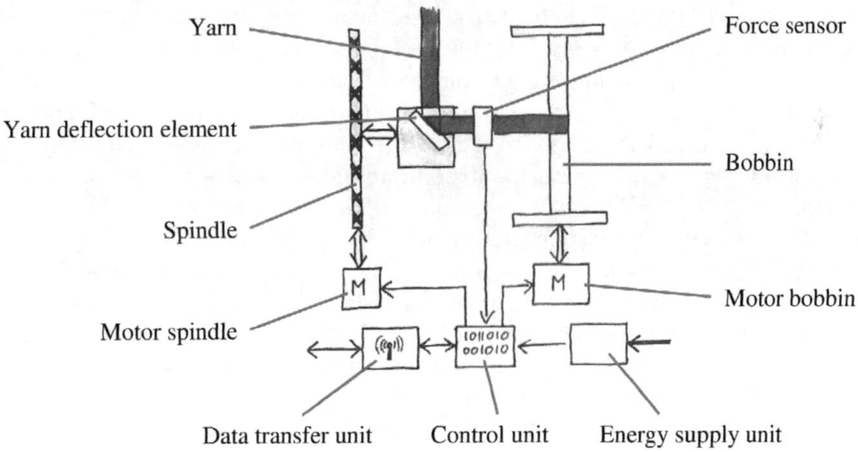

Figure 2. Schematic of the components and their linkages of the electronically controlled bobbin carrier designed by von Reden; image adapted from [6].

Both of these bobbin carrier dependent systems are able to measure the yarn tension during the braiding process very accurately and are therefore capable of detecting irregularities at short response times. Major drawbacks of this kind of systems are however the comparatively high costs for energy supply units, sensor and communication hardware as well as the effort it takes to install the hardware on all bobbins (up to several hundreds) of a braiding machine.

Conscious of the existing sensor systems and their inherent strengths and weaknesses, we are currently working on an online monitoring system for the braiding process which is cost-efficient on the one hand while being sufficiently precise to determine unusual variations in yarn tension on the other hand. Furthermore, the system shall be on a modular basis so that it can be used in the production of a broad spectrum of products ranging from mass goods such as shoelaces (trimmed-down version of the monitoring system) to high performance carbon fiber reinforced composite parts (full version of the monitoring system). The aim is to be able to predictively stop the braiding machine before process irregularities lead to yarn breakages and consequently to labor intensive error correction. The first sensor module of the system in development with its measurement principle and validation experiments are depicted in the paper at hand.

3. New concept for thread tension anomalies detection during braiding

During operation of the braiding machine, the bobbin carriers are moved through the machine by rotating horn gears along closed, intersecting tracks. At the same time, yarn is tangentially unwound from the bobbins. If the thread tension drops below a desired value, a thread tensioning unit at each bobbin carrier prevents the bobbin from rotating around its central axis by means of a locking pin (cf. Figure 1) which engages with lateral notches in the bobbins. As more yarn is pulled by the braiding machine, the yarn tension increases. Two times the unwinding yarn tension plus a frictional component is applied to the lever of the yarn tensioning unit by a 180° deflection pulley (cf. Figure 1 and Figure 3). This way, the increasing yarn tension lifts the lever against a spring incorporated inside the bobbin carrier until the release force is reached. At this point, the lever retracts the locking pin. The given yarn tension then causes the bobbin to rotate around its central axis and yarn is unwound from the bobbin. This in turn leads to another drop in yarn tension which causes the lever to move downwards and reengage the locking pin. During braiding, cycles of engaging and disengaging the locking pin according to the given yarn tension succeed each other.

The newly developed sensor module for the detection of thread tension anomalies makes secondary use of the above-depicted and already existing yarn tensioning unit of each bobbin carrier as a kind of spring balance. Thereby, the number of required additional components is reduced. To measure the thread tension as the bobbins travel through the machine, the position of the lever of the yarn tensioning unit is detected by the new sensor module. For this, the module comprises magnets which are mounted onto the levers of the yarn tensioning units of the bobbin carriers, Hall effect sensors which are stationary attached to the body of the braiding machine and an Arduino microcontroller as a computing device. Additionally, LEDs as a visual indicator to mark the position of an anomalous bobbin carrier to maintenance personnel were arranged near the braiding machine. As magnets, permanent, cylindrical neodymium magnets with a diameter of 8 mm, a height of 3 mm, an energy density of approximately 342-366 kJm^{-3} and a maximum service temperature of 80 °C (quality class N45) were used. In order to reduce their susceptibility to corrosion, the magnets were coated with an epoxy resin film. Firstly, the magnets need to be attached to the yarn tensioning units of the bobbin carriers. For this, a single magnet is pressed into a recess in a 3D printed housing (cf. Figure 3). The housing in turn features a slot so that it can tightly be pushed onto the extension of the lever of the thread tensioning unit of the bobbin carriers. For a more elaborate version than the prototype described herein, a lever of the thread tensioning mechanism with an integrated magnet is conceivable. Secondly, the Hall probe (an Iduino SE022 analog Hall sensor module) needs to be held in place by a 3D printed, height-adjustable fixture in such a way that it is able to detect the magnetic flux density of the field created by the magnet that is attached to the lever of the yarn tensioning unit. Due to the fact that this lever is rotatably placed to the bobbin carrier, the distance between the magnet and the hall probe as well as the orientation of the magnet to the probe alter with varying yarn tension. Hence, the analog signal from the hall probe is a non-linear function of the yarn tension. The corresponding mapping function can be obtained from experiments.

Every time a bobbin carrier passes by the stationary sensor, the corresponding yarn tension can now contactlessly be estimated. Since the sensor is arranged next to the closed track along which the bobbin carriers are travelling during operation of the braiding machine, a single sensor may serve to determine thread tension anomalies of all bobbins on one track. Due to the fact that there are two of these closed tracks with opposite directions of bobbin movement in a rotational braiding machine, at least two Hall sensors are required to monitor all braiding yarns of a machine. However, the system can only measure the corresponding yarn tension of a bobbin in discrete time intervals. If multiple sensors are arranged along the track, "blind spots" can be reduced and along with that the overall response time of the system can be improved. Certainly, all sensor fixtures need to be adjusted to exactly the same height in this case to generate comparable sensor signals.

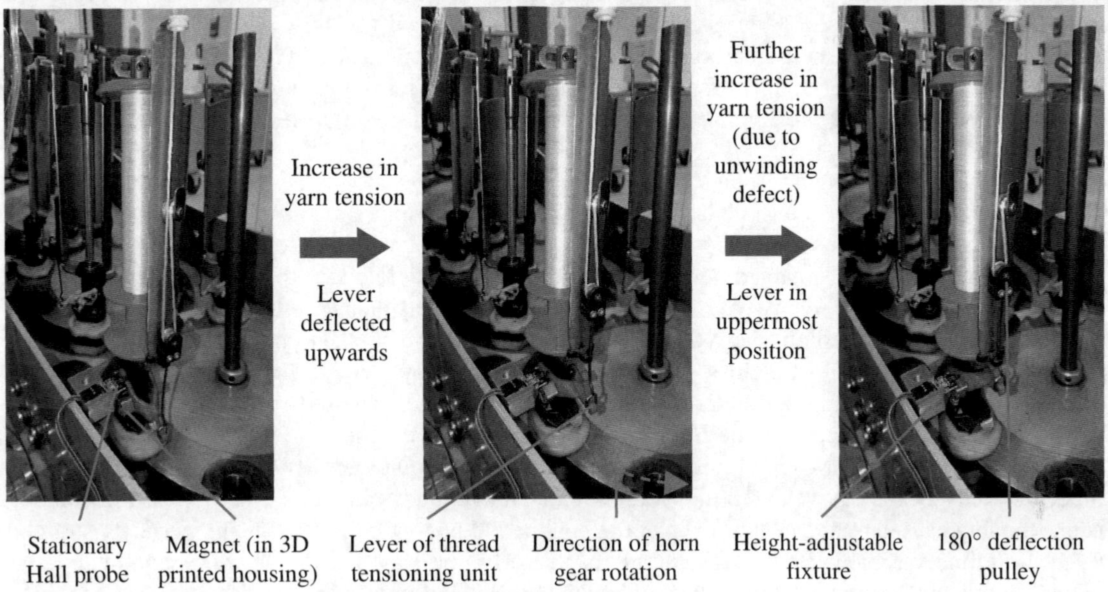

Figure 3. Arrangement of movable bobbin carrier, magnet as well as stationary Hall effect sensor in an RF 1/128-100 braiding machine from Herzog GmbH (all images) and cinematic of lever of yarn tensioning unit with increasing yarn tension (from left to right).

To conclude this chapter, it is to note that the measurement principle described above can be applied to radial as well as axial braiding machines as long as their bobbin carriers comprise a spring based thread tensioning unit. Other sensor types such as optical, acoustic or inductive sensors to determine the position of the lever of the thread tensioning unit were also considered. The given requirements regarding precision and especially costs were, however, best met by the chosen approach via magnets and Hall effect sensors.

4. Validation experiments of the sensor concept

To simulate and to closely study the unwinding process during the braiding process, an unwinding test stand is available at the Chair of Carbon Composites at TU Munich. In order to validate the new sensor module, the stationary bobbin carrier of this unwinding test stand was equipped with the components of the senor module mentioned above (housing for magnet, neodymium magnet, height-adjustable fixture and Hall probe). Furthermore, a self-constructed rotary position transducer was added to keep track of the length of the yarn that is being unwound as well as to precisely adjust the unwinding speed. The test stand works as follows: A NEMA-23 bipolar precision stepper motor with a 15:1 gear box from Phidgets Inc. winds the yarn onto a reel, thereby unwinding it from a bobbin which is located on a stationary bobbin carrier. The unwinding process is recorded by an SLR camera, the position of the lever of the thread tensioning unit is determined by the Hall sensor and the yarn tension is measured by a load cell mounted onto a 90° deflection roller (M1391 from Tensometric Messtechnik GmbH). The data is acquired using a USB-6009 data acquisition device from National Instruments and MATLAB R2015b.

Three different yarn materials were investigated, as there are: a double folded polyester (PES) monofilament yarn with a diameter of 0.25 mm, a double folded PES multifilament yarn with a titer of 300 tex of the individual yarns and a carbon fiber yarn of the type Tenax®-E HTS40 F13 12K with a titer of 800 tex. All of the three yarns were tested at unwinding speeds of 40 mms^{-1} and 80 mms^{-1}, respectively. To keep the duration of a single test at about 20 minutes, 50 m of yarn were wound onto the bobbins for the configurations with the lower and 100 m of yarn were wound onto the bobbins for the configurations with the higher speed. Moreover, all configurations of yarn material and unwinding speed were investigated with a compression spring of the yarn tensioning mechanism with a release force equivalent to a mass of 350 g. Additionally, the PES monofil was tested with a 130 g-spring and the PES multifil as well as the carbon fiber yarn were analyzed with a 700 g-spring. Finally, three reference measurements of each configuration to study the behavior of the yarn materials when they are unwound flawlessly as well as five measurements with provoked irregularities were conducted.

4.1. Flawless reference measurements on a stationary test stand

The first question that needed to be answered by the validation experiments was if the measurements from the cost-efficient sensor module described in the previous chapter were in reasonable agreement with the measurements obtained simultaneously from the deflection roller based yarn tension measurement device. Exemplary measurement results of yarn tension and voltage of the Hall sensor of a flawless test with the carbon fiber yarn are shown in Figure 4. The diagram in Figure 4a reveals that the periodic oscillations in yarn tension with the higher frequency – created by cycles of engaging and disengaging the locking pin and determined by the deflection roller based yarn tension measurement device – are well represented by the Hall sensor module. In the diagram in Figure 4b, a superimposed fluctuation in yarn tension with a lower frequency, which was coincidentally discovered, is noticeable. Since the frequency is equivalent to the rotation frequency of the bobbin, this superimposed fluctuation is presumably caused by the central axes of the bobbin and bobbin carrier not being perfectly straight. Even these fine variations in yarn tension are captured by the Hall sensor. Measurements with other materials and unwinding parameters were in line with these findings.

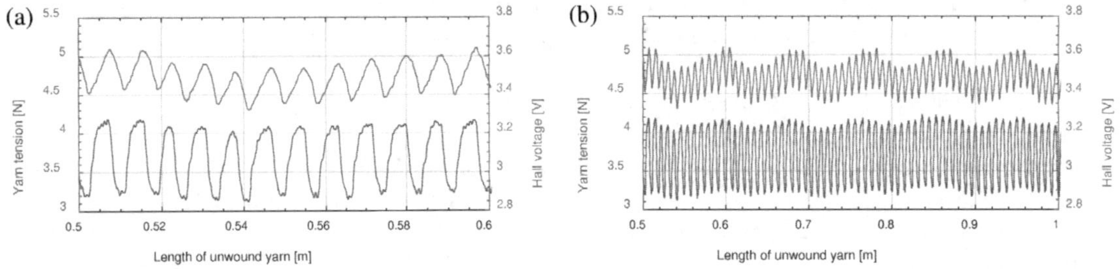

Figure 4. Yarn tension (blue) and Hall voltage (red) obtained from unwinding experiments with a pristine carbon fiber yarn at 40 mms^{-1} unwinding rate and 350 g-spring at different levels of detail; higher level of detail (a), lower level of detail (b).

Apart from this slight superimposed variation in yarn tension, a second, stronger superimposed fluctuation in yarn tension was observed when conducting experiments with the 700 g-spring. To illustrate this, a comparison between measurements obtained from test runs with a 350 g-spring and a 700 g-spring is shown in Figure 5. Whereas the yarn tension remained in a corridor with constant lower and upper bounds of about 3 N and 4.5 N for the 350 g-spring (cf. Figure 5a), especially the upper bound of the yarn tension varied between about 9 N and 16 N for the 700 g-spring (cf. Figure 5b). Matching the measurements with the videos obtained from the SLR camera revealed that the values of the upper bound of the yarn tension depend on the position in longitudinal direction of the bobbin where the yarn is being unwound. Unwinding the yarn from the lower (upper) part of the bobbin lead to a lower (higher) upper bound of the oscillating yarn tension. This is due to the fact that the first yarn deflection element is – in contrast to the one displayed in Figure 2 – fixed on the bobbin carrier that was used. Since there is always a bit of play of the bobbin on the bobbin carrier, the varying angle of attack of the yarn tension on the bobbin causes the bobbin to be lifted from (pushed down towards) the locking pin. This means that the locking pin needs to be retracted less (more) to disengage with the lateral notches in the bobbin. Such behavior was particularly pronounced when the testing the PES multifilament as well as the carbon fiber yarn with the 700 g-spring. When testing all of the materials with the 350 g-spring, this behavior was partly observable and it was not observable when testing the PES monofilament with the 130 g-spring. Since the measurements obtained from the Hall effect sensor showed similar patterns, the authors concluded that this measurement concept was in principle suitable for detecting thread tension anomalies during braiding.

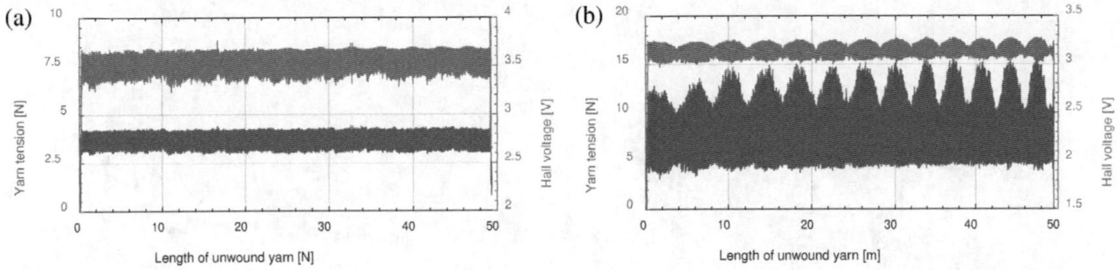

Figure 5. Overviews of measurements of the yarn tension (blue) and Hall voltage (red) of a pristine carbon fiber yarn at 40 mms^{-1} unwinding rate and 350 g-spring (a) as well as 700 g-spring (b).

4.2. Measurements with provoked irregularities on a stationary test stand

The second question that had to be addressed was if the sensor module, when integrated into a braiding machine, was capable of detecting unwinding related braiding process irregularities soon enough before defects have reached their final stage (yarn breakage). In order to clarify this issue, flaws were purposefully introduced into the yarns. The carbon fiber yarn was manipulated by predamaging it during the rewinding step with sand paper with a particle size of 800 Mesh. This reinforces the tendency of the yarn to form fibrous rings during unwinding. The double-folded PES multifil was manipulated by rewinding it onto the bobbins with diverging yarn tensions of 3.8 N and 8.0 N (determined by a portable yarn tension measurement device of the type DTMX-500-U from Hans Schmidt & Co. GmbH). Shortly before the unwinding test, the yarn with the lower tension during rewinding was unwound one revolution from the bobbin while the other yarn remained unaffected – a flaw that can be introduced by maintenance personnel when replacing an empty bobbin in a machine, for example. The double-folded PES monofil was manipulated by rewinding it at diverging yarn tensions only, namely 4.6 N and 12.1 N. In a production environment, only slight yarn tension differences occur. However, with several kilometers of yarn being wound onto a bobbin, even small differences in yarn strain can accumulate to large differences in yarn length. The extreme manipulation procedures cause both of the PES yarns to reliably develop loops at the bobbin during unwinding at the test stand. Eventually, the loops become knotted and impede the unwinding process.

A simple, hypothetical trigger criterion for the sensor module was then formulated: As soon as a Hall sensor detects a lever of the thread tensioning unit of a bobbin carrier in a running braiding machine which is in its uppermost position, the corresponding bobbin is considered to show a process irregularity. The idea behind this criterion is as follows: During braiding, the machine constantly pulls the yarn. The yarn tension ultimately reaches the release force of the spring of the bobbin carrier, the locking pin is retracted, the bobbin is then free to rotate and yarn can be unwound. Consequently, the yarn tension must drop and the lever of the yarn tensioning unit also moves to a lower position. If an unwinding-related irregularity occurs, the yarn cannot be unwound properly from the bobbin, although the locking pin is retracted. Since the braiding process goes on, the machine keeps pulling the yarn. Therefore, the yarn tension increases even more, causing the lever of the yarn tensioning unit to move further upwards than the point of the release force. This position of the lever beyond the point of the release force (lever deflection) occurs when an irregularity is present. In the diagram in Figure 6a, an exemplary measurement of an unwinding test conducted with a carbon fiber yarn with a provoked fibrous ring is shown. Prominent rises in yarn tension throughout the test run precede the final yarn breakage. The rises in yarn tension are accompanied by comparatively long lever deflections that are detected by the Hall effect sensor. The diagram in Figure 6b shows a representative measurement curve of an unwinding experiment with a manipulated PES multifilament. There are also rises in yarn tension during the experiment observable that precede the final yarn breakage. However, the lever deflections are shorter than the deflections during the experiments with the carbon fiber yarn. In the diagram in Figure 6c, an exemplary measurement of a PES monofilament is depicted. In general, few lever deflections shortly before the yarn broke were observable with this kind of material.

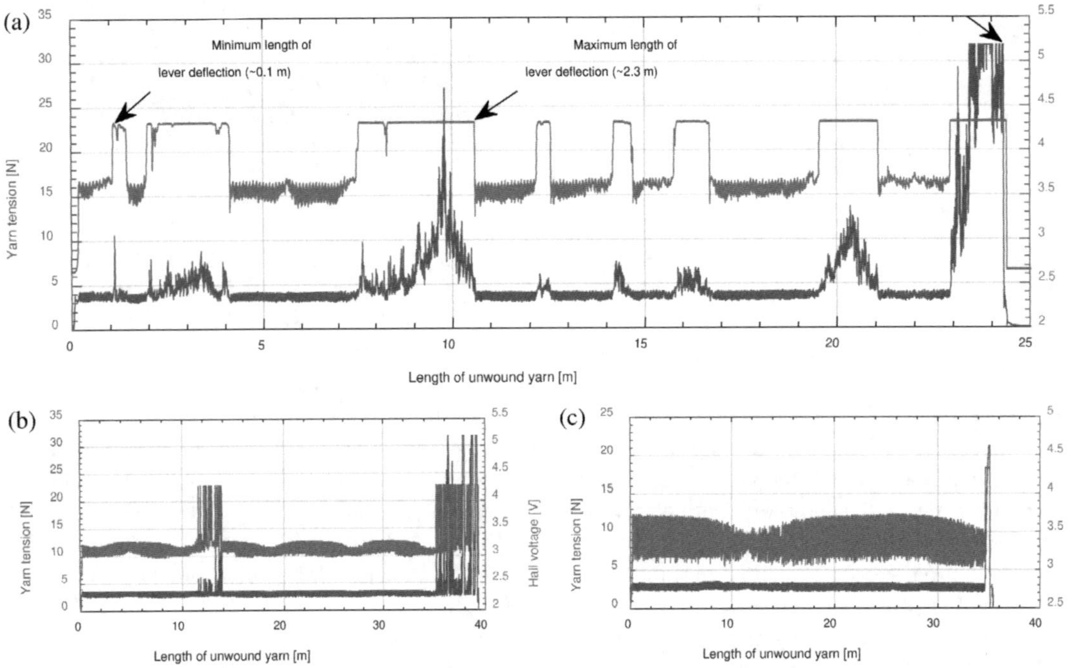

Figure 6. Measurements of the yarn tension (blue) and the Hall voltage (red) during unwinding a manipulated carbon fiber yarn (a), a manipulated PES multifilament yarn (b) and a manipulated PES monofilament (c) at an unwinding rate of 40 mms^{-1} and a 350 g-spring.

The unwound lengths of yarn from the bobbins when the lever was detected in its uppermost position were analyzed for all tests which showed unwinding irregularities. The condensed results of this analysis are depicted in Table 1. The threshold above which the lever was assumed to be in its uppermost position was set to 4.2 V for the data shown in the table. Different thresholds may lead to slightly different results. However, the general findings remain the same. For reasons of space, the detailed results concerning the influences of unwinding speeds and release forces of the compression springs are not discussed herein. The table reveals that there is a considerable number of lever deflections which foreshadow a yarn breakage or an overload of the stepper motor. Since the sensor module only acquires time-discrete estimations of the yarn tension, it is crucial to know how long in terms of unwound yarn length the lever deflections last. The table shows that there are in fact very short, and therefore almost undetectable, minimum lever deflections (1-2 mm) for all of the three yarn materials. However, the significantly higher mean values of the unwound yarn lengths during single lever deflections suggest that most of the deflections are detectable by the sensor system. This statement is underpinned when regarding the mean cumulated length of unwound yarn – a measure for the likelihood that any lever deflection of a moving bobbin carrier is detected by a stationary sensor. It says that there is on average at least 1.1 m of yarn unwound from a manipulated bobbin when the lever of the yarn tensioning mechanism is deflected upwards. Nevertheless, the overall minimum of the longest lever deflection per test indicates that the stationary sensors may not always predictively detect all major braiding defects. With this key figure being in the range of about 0.5 m for the carbon fiber yarn and considering typical circumferences and fiber angles of braided carbon composite parts, a single sensor per track is expected to work very well for the detection of fibrous rings. This is because the length of the yarn unwound from a bobbin during a full circulation through the machine is in most cases less than the determined lead time. The lead time in terms of unwound yarn length is already shorter in the case of the PES multifil (237 mm). Due to the fact that typical products made from this material (e.g. shoelaces) also show significantly smaller circumferences, the detection probability of unwinding irregularities may still be characterized as sufficient. However, the detection of loops which entangle during unwinding of the double-folded PES monofil is not fully guaranteed since the lead time was – in the worst case measurement – only 47 mm.

Table 1. Analysis of lever deflections during unwinding tests.

Material		Carbon	PES multifil	PES monofil
Number of specimens [-]		20	20	20
Number of tests with yarn breakages / stepper motor overloads [-]		20	16	20
Mean number of lever deflections per test with unwinding irregularities [-]		5.5	23.9	3.5
Unwound yarn length during single lever deflection [mm]	Min	1	1	2
	Mean	1038	129	522
	Max	4362	1351	6655
Cumulated unwound yarn length during all lever deflections per test [mm]	Min	510	614	47
	Mean	4753	2049	1101
	Max	14058	4363	6691
Overall minimum of unwound yarn length during longest lever deflection per test [mm]		510	237	47

4.3. Test of the sensor module in an operating braiding machine

On the stationary test stand, the measurement principle has shown to theoretically be capable of detecting fibrous rings when processing the carbon fiber yarn. The sensor module was then tested in a running braiding machine to validate its functionality in a near-production environment. For this, 64 bobbin carriers of an RF 1/128-100 braiding machine from Herzog GmbH were equipped with bobbins onto which 100 m of the carbon yarn were wound. For the first test, no bobbin was manipulated. For the second test, one bobbin was manipulated in a way that it already featured a fibrous ring before braiding. Four bobbin carriers from the same track, including the one carrying the manipulated bobbin, were equipped with magnets. One stationary Hall effect sensor was installed and the height of its fixture adjusted in a way that the sensor was able to detect the positions of the levers of the yarn tensioning units of the bobbin carriers when they pass by. A cylindrical mandrel with a diameter of 65 mm and a length of approximately 2 m was then overbraided. The speed of horn gear rotation was set to 60 rpm and the haul-off speed of the mandrel was set to 12 mms^{-1}. This resulted in a braiding angle of approximately 47°. The measurement curve from the first, flawless test is depicted in Figure 7a. In the beginning of the experiment, the braiding yarns are not fully tensed, yet. Due to the fact that there are 32 horn gears in the machine, it takes 16 s for a bobbin carrier to complete one circulation through the braiding machine at the adjusted speed. Since four bobbin carriers from the same track were equipped with magnets, there are peaks in the signal from the Hall effect sensor every 4 s. Also, every fourth peak is induced by the same bobbin carrier. The maximum values of the peaks are around 3 V. These values indicate that the yarn is unwound regularly from the bobbins. By contrast, in the diagram which is shown in Figure 7b, there are two prominent peaks in the Hall voltage with maximum values of more than 4 V

IOP Conf. Series: Materials Science and Engineering **406** (2018) 012065 doi:10.1088/1757-899X/406/1/012065

at about 32 s and 48 s. These high peak values are caused by the increased yarn tension which is induced by the fibrous ring that was purposefully introduced before braiding. Already observable at second 64 and clearly visible from the regular sequence of small peaks starting at second 80, the yarn that is excessively tensioned by the fibrous ring eventually breaks. This means that the sensor module was capable of detecting an unusual rise in yarn tension well before a yarn breakage occurred.

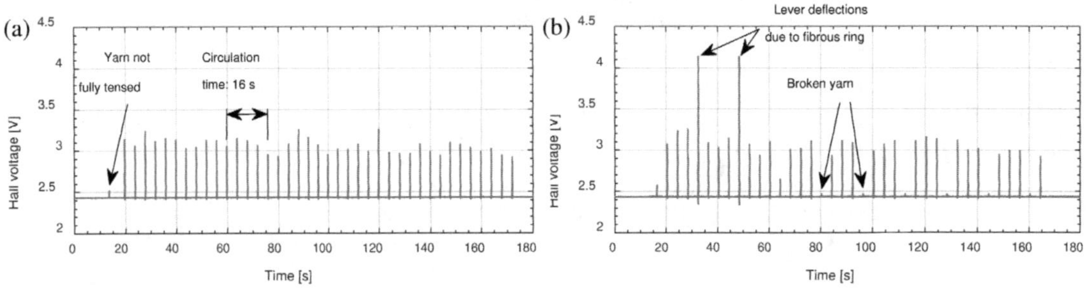

Figure 7. Comparison of measurement curves obtained from a hall sensor implemented into a braiding machine during regular operation (a) and during operation with a provoked fibrous ring (b).

5. Conclusion and outlook

The development of a first sensor module as part of a larger, comprehensive sensor system for the braiding process was presented. The sensor module is bobbin carrier independent and works by contactlessly detecting the position of the lever of the yarn tensioning mechanism of the bobbin carriers through magnets and Hall effect sensors as the bobbins pass by the sensors. This way, the module enables estimating the yarn tension of braiding yarns in discrete time intervals. Validation experiments on a stationary test stand with carbon fiber yarns, a double-folded PES multifil and a double-folded PES monofil were conducted. The experiments with several spring release forces and unwinding speeds as variation parameters proved that the detection of the position of the said lever by means of the sensor module provides a good estimation of the yarn tension. Furthermore, braiding irregularities were purposefully provoked by incorporating flaws into the bobbins that cause irregularities which hamper the unwinding process. The results on the mean cumulated unwound length of the yarn during a lever deflection obtained from the experiments with provoked irregularities revealed that the sensor module is in general very well capable of detecting process irregularities before they have led to major braiding defects like yarn breakage. However, if the worst case test runs are taken as the assessment basis, it becomes apparent that the lead time, in terms of braidable yarn length until final yarn failure, is significantly lower for the PES monofil (47 mm) than for PES multifil and the carbon fiber yarn (237 mm and 510 mm, respectively). If a high error detection reliability is required, the number of stationary Hall sensors along the tracks of the bobbin carriers, which acquire the data in a time-discrete manner, has to be increased according to the circumference of the braid and the braiding angle. The sensor system was also tested during operation of an RF 1/128-100 braiding machine from Herzog GmbH. In this a near-production environment, the sensor module was able to detect an anomalous rise in yarn tension caused by a purposefully introduced fibrous ring soon enough before the yarn broke.

Future work will involve the development of additional sensor modules which will be part of the striven, integrated sensor system. Ideas for these modules include the measurement of reaction forces at the braiding ring, the optical observation of the braid formation zone as well as the development of a tension control unit for the rewinding step. Subsequently, real-time capable algorithms which process the data gathered by all sensor modules need to be implemented. Finally, the cost-effectiveness and the economic benefit of the whole system has to be evaluated under near-production conditions.

Acknowledgements
The authors thankfully acknowledge the funding provided by the German Federal Ministry of Education and Research under the scheme "KMU-innovativ: Produktionsforschung" on an integrated sensor system for the monitoring of the braiding process ("Integriertes Sensorsystem zur Überwachung des Flechtprozesses – InSensoFlecht", funding code: 02P16K553).

References

[1] Ebel C, Mierzwa A and Kind K 2016 13 - Yarn damage during braiding of reinforcement fibers for composites *Advances in Braiding Technology* ed Y Kyosev (Woodhead Publishing) pp 319–54

[2] Ebel C S 2017 Hochgeschwindigkeitsumflechten für die Fertigung von Faserverbundbauteilen *Dissertation* Chair of Carbon Composites, Technical University of Munich, Urn: http://nbn-resolving.de/urn/resolver.pl?urn:nbn:de:bvb:91-diss-20170424-1335638-1-6

[3] Mierzwa A, Ebel C and Drechsler K 2018 Auswirkungen lokaler Garnlücken auf die mechanischen Eigenschaften geflochtener Kohlenstofffaser-Kunststoff-Verbunde *Zeitschrift Kunststofftechnik / Journal of Plastics Technology* **14** 146–73

[4] Lenkeit J 1997 Verfahren zur Zugkraftmessung an laufenden Fäden einer Fadengruppe sowie Vorrichtung zur Ausübung des Verfahrens *German Patent DE000019730965C1*

[5] Braeuner M O 2014 Klöppel für eine Flechtmaschine, Flechtmaschine sowie Verfahren zum Steuern und Regeln einer auf ein Flechtmaterial aufgebrachten Spannung *German Patent DE102014016381B4*

[6] von Reden T 2015 Erweiterung der Systemgrenzen der Flechttechnik durch elektronisch gesteuerte Klöppel *Dissertation* Chair of Carbon Composites, Technical University of Munich, Urn: http://nbn-resolving.de/urn/resolver.pl?urn:nbn:de:bvb:91-diss-20150929-1239240-0-9

IOP Conf. Series: Materials Science and Engineering **406** (2018) 012066 doi:10.1088/1757-899X/406/1/012066

Effect of configuration in intermediate materials on textile processability by TFP technology

Asami Nakai[1*], Akio Ohtani[2], Kenya Morie[1]

[1] Mechanical and Engineering, Faculty of Engineering, Gifu University, Japan
[2] Faculty of Fiber Science and Engineering, Kyoto Institute of Technology, Japan

* nakai@gifu-u.ac.jp

Abstract. Continuous fiber reinforced thermoplastic (c-FRTP) has been focused on because of not only the excellent specific rigidity and strength but also the recyclability and the secondary workability due to a feature of the thermoplastic resin as the matrix. However, unlike the thermosetting resin, it is very difficult to impregnate thermoplastic resin into continuous fiber bundles because of its high melt viscosity. In order to solve this problem, various intermediate materials have been developed. On the other hand, Tailored Fiber Placement (TFP) is a textile manufacturing technology for continuous fiber preforming. Compared to other textile manufacturing processes, the continuous fiber material can be placed in any direction according to the requirement. From this characteristic, it is possible to change the local fiber orientation in the fiber reinforced composite material, so that mechanical properties of reinforcing fibers can be utilized to maximum. Applying fibrous intermediate material for c-FRTP to TFP technology, the high cycle c-FRTP molding with optimized fiber orientation can be achieved. However, in order to prepare a preform of c-FRTP, textile processability (flexibility) is needed for fibrous intermediate materials as dry fiber. Therefore, to develop fibrous intermediate material with both better impregnation property and flexibility for TFP technology is required. In this study, the effect of fabrication condition of newly developed fibrous intermediate material called PCY (Partially Impregnated Commingled Yarn) on preforming by TFP technology was investigated. The textile processability was evaluated by measuring the deformation resistance of various PCYs. The unidirectional (UD) preforms were prepared using PCY with better textile processability and impregnation property. In order to investigate the effect of stitch yarn on impregnation property during molding, the impregnation state was quantified by cross-sectional observation of various specimens. In addition, a tensile test using UD specimen was carried out to investigate the effect of stitch yarn on mechanical properties.

1. Introduction

Continuous fiber reinforced thermoplastic (cFRTP) has been focused on because of not only the excellent specific rigidity and strength but also the recyclability and the secondary workability due to a feature of the thermoplastic resin as the matrix. However, unlike the thermosetting resin, it is very difficult to impregnate thermoplastic resin into continuous fiber bundles because of its high melt viscosity. In order to solve this problem, various intermediate materials have been developed.

On the other hand, Tailored Fiber Placement (TFP) is a textile manufacturing technology for continuous fiber preforming. Compared to other textile manufacturing processes, the continuous fiber material can be placed in any direction according to the requirement. From this characteristic, it is possible to change the local fiber orientation in the fiber reinforced composite material, so that mechanical properties of reinforcing fibers can be utilized to maximum. Applying fibrous intermediate material for c-FRTP to TFP technology, the high cycle c-FRTP molding with optimized fiber orientation can be achieved [1]. However, in order to prepare a preform of c-FRTP, textile processability (flexibility) is needed for fibrous intermediate materials as dry fiber. Therefore, to develop fibrous intermediate material with both better impregnation property and flexibility for TFP technology is required.

In this study, the effect of fabrication condition of newly developed fibrous intermediate material called PCY (Partially Impregnated Commingled Yarn) on preforming by TFP technology was investigated. The textile processability was evaluated by measuring the deformation resistance of various PCYs. The unidirectional (UD) preforms were prepared using PCY with better textile processability and impregnation property. In order to investigate the effect of stitch yarn on impregnation property during molding, the impregnation state was quantified by cross-sectional observation of various specimens. In addition, a tensile test using UD specimen was carried out to investigate the effect of stitch yarn on mechanical properties.

2. Materials and Molding Method

2.1. Fabrication Method of PCY and Investigation of Fabrication Condition

The carbon fiber (TR50S-6L, Mitsubishi Chemical Corporation) was used as reinforcing fiber, and the polyamide resin (LEXTER8500, MITSUBISHI GAS CHEMICAL COMPANY, Inc.) was used as resin fiber. After commingled these fibers at the same rate (Vf =50%), surface of the fiber bundle was heated and melted by a heat roller to prepare PCY. The contact time between the fiber bundle and the heat roller was controlled by changing the production speed, and the impregnation state in PCY was changed.

The experiments measuring the resistance load when PCY passed through a thin hole to quantify the deformability were conducted for various PCYs with different initial impregnation states. Here, deformability is better as the load value is lower. The relationship between the resistance load and the heat roller velocity is shown in Figure 1. As shown in this figure, the resistance load decreased from 5 to 10 m/min in the heat roller speed, and showed almost constant value in the load over 10 m/min of roller velocity. This result means the textile processability greatly increases when the heat roller speed change from 5 to 10 m/min.

The relationship between Un-impregnation ratio obtained from cross-sectional observation of UD specimens and roller velocity is shown in Figure 2. As shown in this figure, the un-impregnation ratio increased with increasing roller velocity, and the rate of increase became larger after 15 m/min. Based on these results, PCY produced with a heat roller speed of 10 m/min was chosen for preform fabrication by TFP technology.

2.2. Stitching Conditions and Molding Method

Unidirectional preform having a length of 200 mm and width of 20 mm was produced and pressed to prepare a specimen. A preform with PCY was produced using the embroidery machine (TCWM101 Tajima Industries Ltd.). The effect of existence of stitch yarn on impregnation property after molding was examined by using polyamide fiber that be melted during the molding process and aramid fiber as stitch yarn. UD specimens were molded by using following molding conditions; the molding temperature was 260 ℃, molding pressure was 3MPa and molding time was 3 min. Furthermore, the effect of using the TFP technology on mechanical properties was examined by comparing the un-impregnation ratio and the mechanical properties with non-stitched specimens.

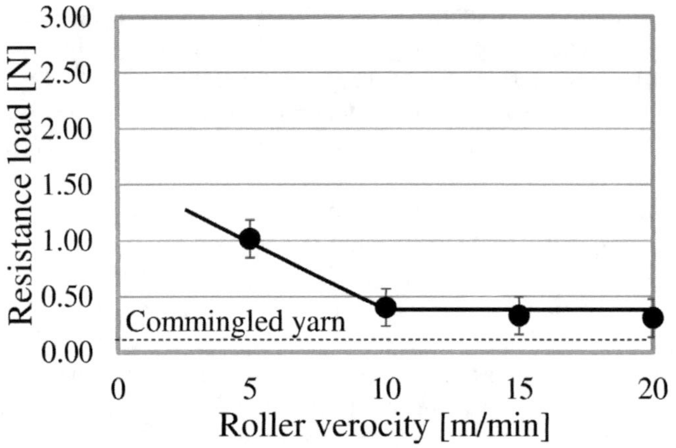

Figure 1. Relationship between resistance load of each PCY and roller velocity.

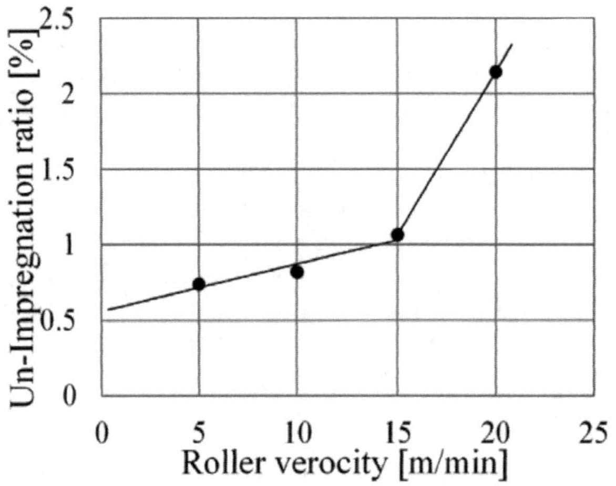

Figure 2. Relationship between Un-Impregnation ratio and roller velocity.

3. Results and Discussion

Un-impregnation ratio from the cross-sectional observation of each specimen is shown in Figure 3. The un-impregnation ratio in all specimens was less than 0.6%, but in the specimen with aramid stitch yarn, the un-impregnation ratio was slightly higher than the other specimens. From this result, unlike polyamide, the aramid stitch yarn does not melt by the heat during molding and remains in the specimen. Therefore, the impregnation state of the specimen would be affected by stitch yarn remaining in the specimen. As a result, it seemed that the waviness of PCY was generated, PCY was tied up and Vf in PCY was increased by the stitch yarn, as compared with the non-stitched one. Therefore, more molding time or molding pressure or molding temperature will be needed for better impregnation state for the molding with TFP preform due to the increase of Vf in the fiber bundle by stitch yarn.

In order to investigate the effect of stitch yarn on mechanical properties, a tensile test was conducted for various specimens. Achievement ratio of mechanical properties to the theoretical value of each specimen is shown in Figure 4. The achievement ratio is the ratio of measured value to the theoretical elastic modulus and strength in the 0 degree direction obtained from the rule of mixture with Vf set at 50%. As shown in Figure 4, achievement ratio of elastic modulus is more than 90% in all specimens.

The achievement ratio of tensile strength for all specimens was also about 70% and the specimen showed a few percent difference. In previous studies, the unimpregnation ratio of less than 1% does not affect the strength of composite material. Consequently, the difference in structure of reinforcing fiber due to the stitch yarn does not affect the strength.

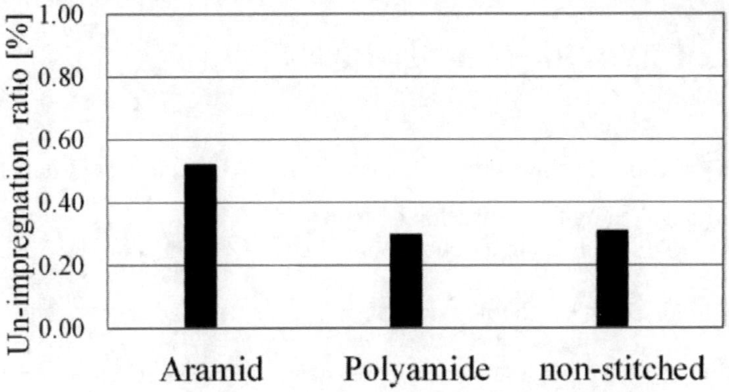

Figure 3. Un-impregnation ratio of various specimens.

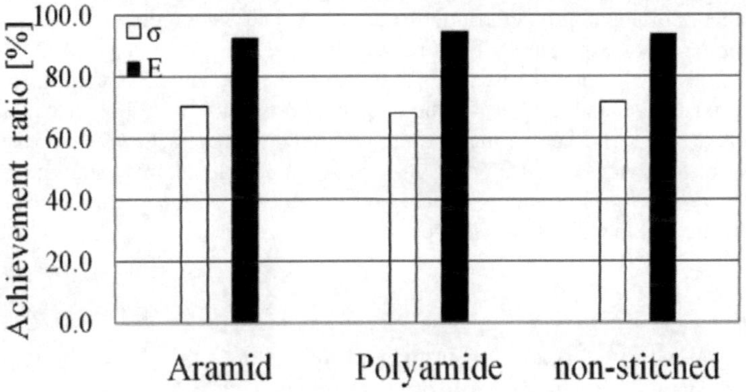

Figure 4. Achievement ratio of mechanical properties to the theoretical value.

References

[1] Mattheji P, Gliesche K, Feltin D, 1998, Tailored Fiber Placement-Mechanical Properties and Applications, *J. Reinforced Plastics and Composites*, **17**, 774-786.

Fabrication of hybrid thin ply tapes

J. Quintana Casanovas [1,2]*, J. Costa [2], J.A. Mayugo [2], A. Galan Llongueras [1]

[1] Frimal Trading S.L, Barcelona, Spain
[2] AMADE - Polytechnic School, University of Girona, Girona, Spain

* jquintana@twistechnology.com

Abstract. Thin ply hybrids are composed of at least three different materials: two or more yarns and one binder to fix the final thin ply tape. This study is divided in the steps required to convert the raw materials (carbon, glass, aramid and polyester) into the hybrid thin ply tapes: unwinding of the yarn; spreading of each individual yarn; hybridization, blending and spreading of all the materials, and finally binding and winding the final thin ply hybrid. All these processes are studied individually to determine the effect of each one on different yarns. The more important steps on this hybrid spreading line are the spreading methods and the blending of different materials. Different processes such as ultrasounds, mechanical vibration, convex rollers, air flow, etc, exist to spread the yarns. The choice of this study is to use mechanical vibrations in the individual yarns whereas the second step is performed with vacuum, paying attention to the control of the tension during the entire process. The developed technique allows for production rates from 5 to 20 m/min of hybrid fiber depending on the characteristics of the raw material, showing that the fabrication process of these materials is suitable for industry.

1. Introduction

In the search for composite materials with improved mechanical properties, thin ply based laminates have raised a great interest due to its ability to delay matrix cracking, to improve the free edge delamination or to extend the fatigue lifetime, among other benefits. A further step in this field is to mix different fibers in the thin ply. This can be done to reduce the cost of the material by reducing the use of expensive fibers, or to bring specific properties to the composite material (mechanical or functional properties) adapted to the specific needs of each application. In particular, there is the prospect to be able to enhance the fracture toughness of the material by tailoring the mixture of different fibers. There are many theoretical studies on the damage progression of thin ply hybrids. However, despite of this relevant theoretical knowledge, there is a lack of experimental evidences. Basically, because these materials have not yet been implemented in industry, showing how difficult is to produce these new type of materials.

This communication presents a developed method to obtain this type of materials. In addition to obtaining these materials, it has been possible to obtain them with applicability to the industry, both technically and economically.

2. Spreading hybridization methodology

In an analogous methodology to obtain tow spread fibers of a single material, it is possible to use the same techniques to obtain thin ply hybrid tows. The three forms of hybridization that we have studied to achieve the hybrids fibers are vacuum, mechanical vibrations and the convex rollers.

The method used in this study allows to achieve and study the process to manufacture hybrid tapes of surface density between 40 and 100 g/m^2 and a width between 10 and 50 mm by combining two equal or different fibers, dimensions of filaments and density count.

2.1. Parts of the machine

The machine is divided mainly into 3 sections. The first corresponding to the unwinding of the raw materials, the second in the opening, hybridization and fixation of the tape and the last in the winding of the final product. Despite the great importance of the three sections, this study focuses on the second part that is the essence of the final product and where there is more uncertainty.

The machine used to obtain the first samples of hybrid industrial fiber materials follows the operating scheme that can be seen below.

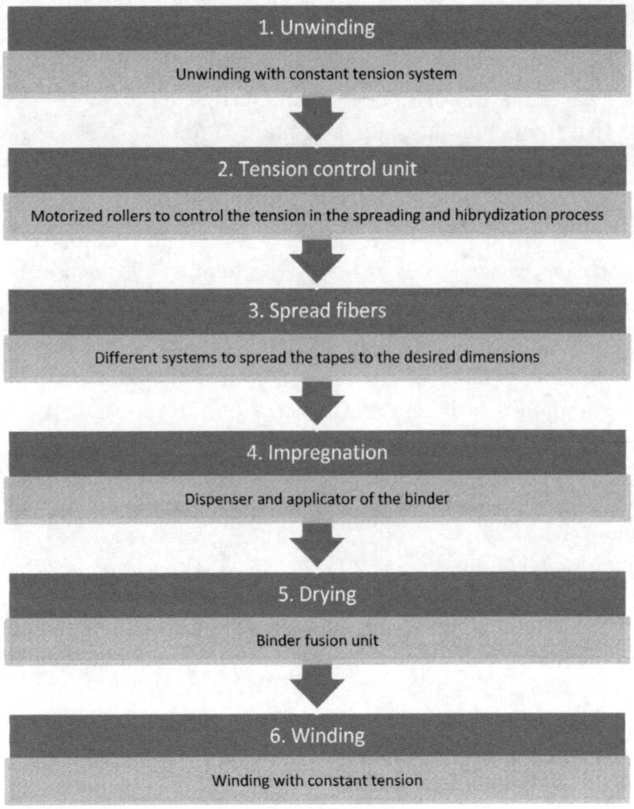

The unwinding process is made individual for each one of the fibers that are used. They are performed at a constant tension from the start to the end of the product. For the correct operation of the system is important to control the unwinding tension. Experimentally it has been verified that with low tension defects from the unwinding process were less than 1% of the defects detected.

During the process of hybridization and opening of the tapes, the tension is controlled at all times, it is one of the most important variables and it effects severally to the final result.

The opening of the tapes consists of two steps, the first one where the different tows are open at the same width and the next step this tows are joined and spread to the desired width. The different methods used to achieve the hybridization are mechanical vibrations, vacuum and convex rollers.

Impregnation is necessary to ensure the dimensions of the tapes once manufactured, in addition to ensuring these dimensions for subsequent processes. It is used a Copolyester because is compatible with epoxy matrices. In this study the amount of binder used in the samples is 6% + - 1% with respect to the total mass of the fibers and the binder.

Finally, the finished hybrid tape is winded with the desired format. The winding tension is very low to not affect the width reached.

3. Experimental results

Experimental tests have been performed using the fabrication line exposed in point 2 to hybridize thin ply tapes. What I wanted to achieve in this study is to hybridize two different carbon fiber tows and on

IOP Conf. Series: Materials Science and Engineering **406** (2018) 012067 doi:10.1088/1757-899X/406/1/012067

the other hand, carbon fiber and aramid tows. Both focused on improving the mechanical properties of the final product and reducing the cost. The experiments were carried out to obtain 50 g/m^2 specimens.

3.1. Different types of fiber

The fibers for this feasibility study of the hybrid thin ply tapes are carbon fibers and aramid fibers. In concrete, the models used are summarized in the following table.

Table 1. Types of fibbers used as raw material

Denomination	Manufacturer	*Type of material*	Density (g/cm^3)	*Yield tex (g/1000 m)*	Tensile strenght (Mpa)	*Tensile modulous (GPa)*	*Strain (%)*
A-42 12k	AKSACA	Carbon fiber	1,78	800	4200	240	1,8
Sigrafil CT24-5.0	SGL	Carbon fiber	1,81	1600	5000	270	1,9
Twaron	Teijin	Aramid	1,42-1,46	336	1650-2200	60-80	3-4,4

The experiments will be done mixing first the two carbons together and then AKSACA carbon and Twaron. Also each of these materials has some mechanical properties as well as a specific format as a raw material. This fact forces to open each combination of fibers to certain dimensions to obtain the desired surface weight of 50 g/m^2 with the two fibers that make up the final hybrid.

Table 2. Wide calculation A-42 12k and Sigrafil CT24-5.0

Denomination	Manufacturer	*Type of material*	Density (g/cm^3)	*Yield tex (g/1000 m)*	Areal weight (g/m^2)	*Spread width (mm)*
A-42 12k	AKSACA	Carbon fiber	1,78	800	16,65	48
Sigrafil CT24-5.0	SGL	Carbon fiber	1,81	1600	33,35	48

Table 3. Wide calculation A-42 12k and Twaron

Denomination	Manufacturer	*Type of material*	Density (g/cm^3)	*Yield tex (g/1000 m)*	Areal weight (g/m^2)	*Spread width (mm)*
A-42 12k	AKSACA	Carbon fiber	1,78	800	14,8	22,7
Twaron	Teijin	Aramid	1,42-1,46	336	35,2	22,7

According to the previous tables, the test with two carbons should get a 48 mm wide tape and the next two we should get a hybrid of 22.7 mm wide.

3.2. Baseline data for testing (INPUTS)

The data considered to be relevant to study that have a direct effect on the final result are the following:

- Linear mass density (tex): It depends of fiber type
- Initial width (mm): It depends of fiber type
- Production speed (m/min): It is fixed to 10 m/min for all the test
- Tension between spread and hybridization (cN)

3.3. Data obtained (OUTPUTS)

The data registered to determine the quality of the tape and information to quantify or valuate the tape are the following:

- Thickness (μm)
- Width (mm)

IOP Conf. Series: Materials Science and Engineering **406** (2018) 012067 doi:10.1088/1757-899X/406/1/012067

- Width regularity
- Empty area (mm^2)
- Parallelism of fibers

3.4. Test results

The first result that indicates the quality of the tape is the average width reached and the deviation over it. These parameters are directly related to the dimensions of the tape both thickness and the width. The result in the two cases of the hybrids manufactured are the following:

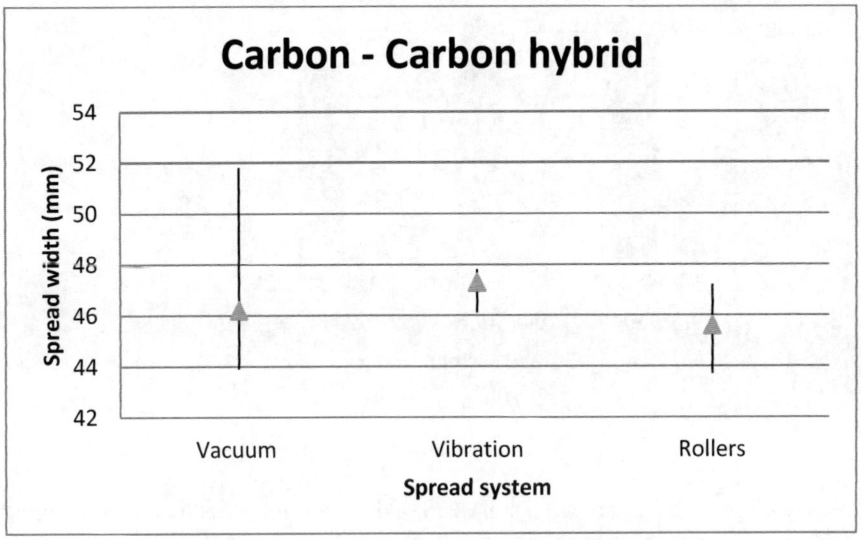

Figure 1. Carbon – carbon hybrid width results

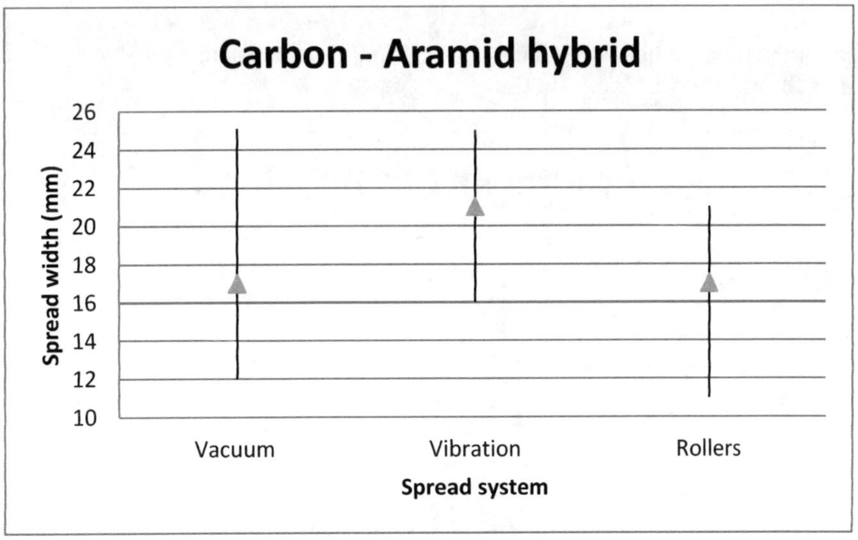

Figure 2. Carbon – aramid hybrid width results

It can be seen about the variability of the tape is higher in the case of the use of vacuum. On the other hand mechanical vibration with rollers has the best results in the range of the final tape and also in the reach the medium dimension desired. Finally, although the regularity of the rollers is good enough, most of the tape does not reach the desired width.

In the case of carbon – aramid hybridization with these methods is more complicated. The deviation of the width and, consequently, the thickness is quite variable throughout the production. And although the average case is close enough to the objective value, the profitability of the tape is very bad.

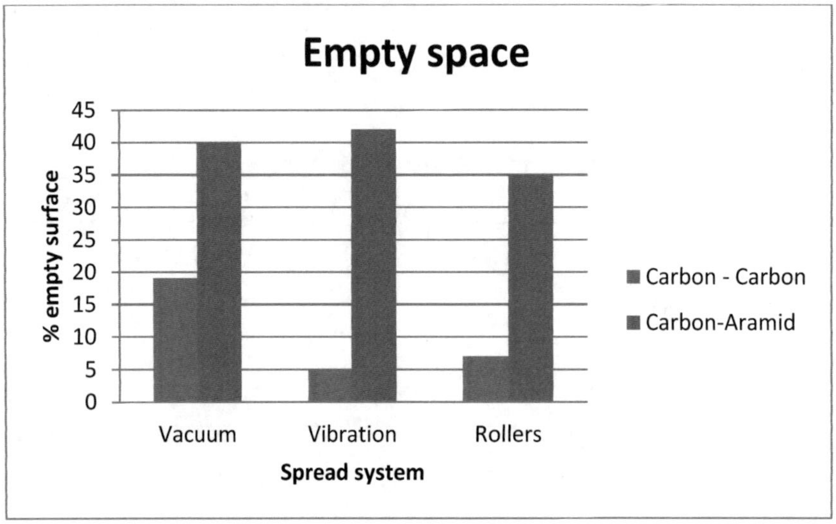

Figure 3. Empty space of the specimens

The graph shows how to combine two materials of a different nature is much more complicated than trying to hybridize two materials much closer in terms of length and diameter of fibers. The spaces without material that we find throughout the final product are smaller in fibers equal than in different ones, and a priori the method that brings us closer to the "perfect" hybrid is that use mechanical vibrations.

Parallelism of the fibers is quite constant and it does not provide enough information to compare the results, since in all cases they are very similar.

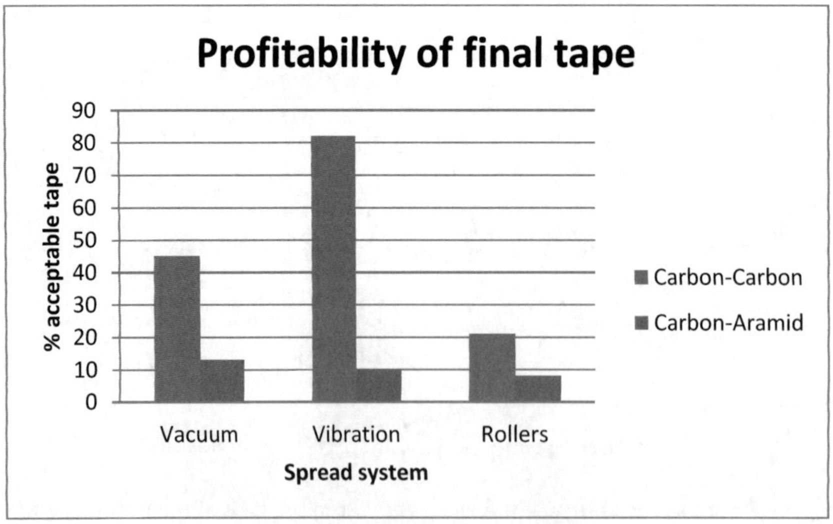

Figure 4. Profitability of final tape

4. Summary

The results obtained indicate that in order to industrialize the hybrid tape the best system is the mechanical vibration, it is the one with the least empty areas of material that could be defects in the final application of the product, as well as having the best regularity of the methods studied.

As shown in the graph following the profitability of the produced hybrid tapes it is much better in hybrids with a single type of material, as well as with mechanical vibration with respect to the others methods.

5. Concluding remarks

The hybrid thin ply tapes have gained a lot of interest for the industry in recent years. Knowing all the advantages that could be obtained causes the need to manufacture these fibers in an industrial environment.

The first relevant result is that it is possible to manufacture a 2400 tex carbon-carbon hybrid with a width of 48 mm at a speed of 10 m/s and also this system is viable technically and economically.

If mechanical vibration is used as the main method to open and hybridize the fibers, the regularity of the final tape is more stable (variation less than 10%) and the presence of errors is more than 37% less respect to other methods. This implies that it is the most appropriate method for industrializing the process. The maintenance of the equipment is cheap and the cost of manufacture is estimated around 0.18 €/m.

The fibers of different types are more difficult to hybridize, when the fibers do not have the same origin, the efficiency of the system used decreases considerably.

Acknowledgements

This applied research was possible with the collaboration of Frimal Trading SL and University of Girona. I encourage both to continue promoting this type of collaboration. This Project wasn't possible with the PhD in industry environment projects created of the Generalitat de Catalunya.

References

[1] Daniel, I. M.; Ishai, O. Engineering mechanics of composite materials. USA: Oxford University Press, 2005. ISBN 9780195150971.
[2] Gay, Daniel. Composite materials: design and applications. 3rd ed. Boca Raton, FL: Taylor, 2015. ISBN 9781466584877.
[3] Hyer, M. W. Stress analysis of fiber-reinforced composite materials. Boston, Massachusets: McGraw-Hill, 1998. ISBN 9789339205317.
[4] Jones, Robert M. Mechanics of composite materials. 2nd ed. New York: Taylor & Francis, 1999. ISBN 9781560327127.
[5] Tsai, Stephen W. Strength & life of composites. Stanford: Composites Design Group, Composites Design Group. ISBN 9780981914305.
[6] ADHERENT TECHNOLOGIES INC. *Method and apparatus for spreading fiber bundles.* LIKKE Joseph Lee, BUSSELLE Lincoln Drake, FINLEY Dana Jesse, GORDON Brent William. United States, Classe A, US6049956. 18 d'Abril de 2000.
[7] UNIV MICHIGAN STATE. *Method and system for spreading a tow of fibers.* YER Shridhar, DRZAL Lawrewnce. United States, Classe A, US5042111. 27 d'Agost de 1991.
[8] GARLOCK INC. *Fiber impregnation process.* CALKINS Richard. United States, Classe A, US4994303. 19 de Febrer de 1991.
[9] UBE INDUSTRIES. *Fiber separator for producing fiber reinforced metallic orresin body.* NAKAGAWA Narihito, OHSORA Yasumasa. Japan, Classe A, US5101542. 07 d'Abril de 1992.
[10] HARMON INDUSTRY CO. *Fiber opening apparatus for mass fibers.* SHINKADO Hiroaki. Japan, Classe A1, US7596834. 08 de Juny de 2006.
[11] MAYER MALIMO TEXTILMASCHF. *Device and method for spreading a carbon fiber hank.* NESTLER Juergen, VETTERMANN Frank, REUCHSEL Dietrmar. Germany, Classe A1,

US7536761. 10 de Maig de 2007.

[12] Sangwook Sihn, Ran Y. Kim, Kazumasa Kawabe i Stephen W. Tsai. (2006). Experimental studies of thin-ply laminated composites. *Composites Science and Technology* 67 (2007) 996–1008.

[13] R. Amacher, J. Cugnoni, J. Botsis, L. Sorensen, W. Smith i C. Dransfel. (2014). Thin ply composites: Experimental characterization and modeling of size-effects. *Composites Science and Technology* 101 (2014) 121–132.

[14] Yentl Swolfs, Larissa Gorbatikh i Ignaas Verpoest. (2014). Fibre hybridisation in polymer composites: A review. *Composites* Part A 67 (2014) 181–200.